Bacterial Invasion into Eukaryotic Cells

Subcellular Biochemistry

Volume 33

SUBCELLULAR BIOCHEMISTRY

Recent Volumes in This Series

Volume 24 **Proteins: Structure, Function, and Engineering**
Edited by B. B. Biswas and Siddhartha Roy

Volume 25 **Ascorbic Acid: Biochemistry and Biomedical Cell Biology**
Edited by J. Robin Harris

Volume 26 ***myo*-Inositol Phosphates, Phosphoinositides, and Signal Transduction**
Edited by B. B. Biswas and Susweta Biswas

Volume 27 **Biology of the Lysosome**
Edited by John B. Lloyd and Robert W. Mason

Volume 28 **Cholesterol: Its Functions and Metabolism in Biology and Medicine**
Edited by Robert Bittman

Volume 29 **Plant–Microbe Interactions**
Edited by B. B. Biswas and H. K. Das

Volume 30 **Fat-Soluble Vitamins**
Edited by Peter J. Quinn and Valerian E. Kagan

Volume 31 **Intermediate Filaments**
Edited by Harald Herrmann and J. Robin Harris

Volume 32 **α-Gal and Anti-Gal: α1,3-Galactosyltransferase, α-Gal Epitopes, and the Natural Anti-Gal Antibody**
Edited by Uri Galili and José Luis Avila

Volume 33 **Bacterial Invasion into Eukaryotic Cells**
Edited by Tobias A. Oelschlaeger and Jörg Hacker

Bacterial Invasion into Eukaryotic Cells

Subcellular Biochemistry
Volume 33

Edited by

Tobias A. Oelschlaeger

and

Jörg Hacker
University of Würzburg
Würzburg, Germany

Kluwer Academic / Plenum Publishers
New York, Boston, Dordrecht, London, Moscow

The Library of Congress cataloged the first volume of this title as follows:

Sub-cellular biochemistry.
London, New York, Plenum Press.
v. illus. 23 cm. quarterly.
Began with Sept. 1971 issue. Cf. New serial titles.
1. Cytochemistry—Periodicals. 2. Cell organelles—Periodicals
QH611.S84 574.8'76 73-643479

ISSN 0306-0225

ISBN 0-306-46290-7

This series is a continuation of the journal *Sub-Cellular Biochemistry*, Volumes 1 to 4 of which were published quarterly from 1972 to 1975

233 Spring Street, New York, New York 10013

http://www.wkap.nl

10 9 8 7 6 5 4 3 2 1

A C.I.P. record for this book is available from the Library of Congress

Printed in the United States of America

Contributors

Rudolf I. Amann Max-Planck-Institut für Marine Mikrobiologie, 28359 Bremen, Germany

Burt E. Anderson Department of Medical Microbiology and Immunology, College of Medicine, University of South Florida, Tampa, Florida 33612-4799

Andreas J. Bäumler Department of Medical Microbiology and Immunology, College of Medicine, Texas A&M University, College Station, Texas 77843-1114

Raul Barletta Department of Veterinary and Biomedical Sciences, Institute of Agriculture and National Resources, University of Nebraska, Lincoln, Nebraska 68583-0905

Luiz E. Bermudez Kuzell Institute for Arthritis and Infectious Diseases, California Pacific Medical Center Research Institute, San Francisco, California 94115

Blaine L. Beaman Department of Medical Microbiology and Immunology, University of California School of Medicine, Davis, California 95616

LoVelle Beaman Department of Medical Microbiology and Immunology, University of California School of Medicine, Davis, California 95616

Anne Boland Microbial Pathogenesis Unit, Christian de Duve Institute of Cellular Pathology, Université Catholique de Louvain, Faculté de Medecine, B1200 Brussels, Belgium

Bettina Brand Institut für Molekulare Infektionsbiologie, Universität Würzburg, 97070 Würzburg, Germany

William J. Broughton LBMPS, Université de Genève, 1292 Chambery/Geneva, Switzerland

Ofelia Chacón Department of Veterinary and Biomedical Sciences, University of Nebraska, Lincoln, Nebraska 68583 and Department of Veterinary Pathobiology, Texas A&M University, College Station, Texas 77843

Jeffrey D. Cirillo Department of Veterinary and Biomedical Sciences, University of Nebraska, Lincoln, Nebraska 68583

P. Patrick Cleary Department of Microbiology, University of Minnesota, Minneapolis, Minnesota 55455

Guy R. Cornelis Microbial Pathogenesis Unit, Christian de Duve Institute of Cellular Pathology, Université Catholique de Louvain, Faculté de Medecine, B1200 Brussels, Belgium

David Cue Department of Microbiology, University of Minnesota, Minneapolis, Minnesota 55455

Gregory A. Dasch Naval Medical Research Institute, Bethesda, Maryland 20813

Cristoph Dehio Department Infektionsbiologie, Max-Plank-Institut für Biologie, D-72076 Tübingen, Germany

Guido Dietrich Lehrstuhl für Mikrobiologie, Theodor-Boveri-Institut für Biowissenschaften, Universitat Würzburg, 97074 Würzburg, Germany and Chiron Behring GmbH & Co, Preclinical Vaccine Research, 35006 Marburg, Germany

Ruben O. Donis Department of Veterinary and Biomedical Sciences, University of Nebraska, Lincoln, Nebraska 68583

Gordon Dougan Department of Biochemistry, Imperial College of Science, Technology and Medicine, University of London, Londón SW7 2AZ, United Kingdom

Marina E. Eremeeva Department of Microbiology and Immunology, University of Maryland School of Medicine, Baltimore, Maryland 21201

Gad Frankel Department of Biochemistry, Imperial College of Science, Technology and Medicine, University of London, London SW7 2AZ, United Kingdom

Werner Goebel Lehrstuhl für Mikrobiologie, Universität Würzburg, 97074 Würzburg, Germany

Scott D. Gray-Owen Department Infektionsbiologie, Max-Planck-Institut für Biologie, D-72076 Tübingen, Germany; *present address*: Department of Medical Genetics and Microbiology, University of Toronto, Toronto, Ontario, M5S 1A8, Canada

Dorothee Grimm Institut für Molekulare Infektionsbiologie, Universität Würzburg, 97070 Würzburg, Germany

Jörg Hacker Institut für Molekulare Infektionsbiologie, Universität Würzburg, 97070 Würzburg, Germany

Ted Hackstadt Host-Parasite Interactions Section, Laboratory of Intracellular Parasites, National Institute of Allergy and Infectious Diseases, National Institutes of Health Rocky Mountain Laboratories, Hamilton, Montana 59840

Kwang Sik Kim Division of Infectious Diseases, Childrens Hospital Los Angeles, Los Angeles, California 90027

Robert A. Kingsley Department of Medical Microbiology and Immunology, College of Medicine, Texas A&M University, College Station, Texas 77843-1114

Stuart Knutton Institute of Child Health, University of Birmingham, Birmingham B4 6NH, United Kingdom

Dennis J. Kopecko Laboratory of Enteric and Sexually Transmitted Diseases, Food and Drug Administration, Bethesda, MD 20892

Michael Kuhn Lehrstuhl für Mikrobiologie, Universität Würzburg, 97074 Würzburg, Germany

Yousef Abu Kwaik Department of Microbiology and Immunology, University of Kentucky, Chandler Medical Center, Lexington, Kentucky 40536-0084

Catherine A. Lee Department of Microbiology & Molecular Genetics, Harvard Medical School, Boston, Massachusetts 02115

Reinhard Marre Department of Medical Microbiology and Hygiene, University of Ulm, 89070 Ulm, Germany

Thomas F. Meyer Department Infektionsbiologie, Max-Planck-Institut für Biologie, D-72076 Tübingen, Germany and Molekulare Biologie, Max-Planck-Institut für Infektionsbiologie, D-10117 Berlin, Germany

Michael F. Minnick Division of Biological Sciences, University of Montana, Missoula, Montana 59812-1002

Kirsten Niebuhr Unité de Pathogénie Microbienne Moléculaire, Institut Pasteur, 75724 Paris Cedex 15, France

Tobias A. Oelschlaeger Institut für Molekulare Infektionsbiologie, Universität Würzburg, 97070 Würzburg, Germany

X. Perret LBMPS, Université de Genève, 1292 Chambery/Geneva, Switzerland

Yasuko Rikihisa Department of Veterinary Biosciences, College of Veterinary Medicine, Ohio State University, Columbus, Ohio 43210-1093

Axel Ring Department of Infectious Diseases, St. Jude Children's Research Hospital, Memphis, Tennessee 38105

Ilan Rosenshine Department of Molecular Genetics and Biotechnology, Faculty of Medicine, The Hebrew University, Jerusalem 91120, Israel

Shlomo Rottem Department of Membrane and Ultrastructure Research, The Hebrew University-Hadassah Medical School, Jerusalem 91120, Israel

Felix J. Sangari Kuzell Institute for Arthritis and Infectious Diseases, California Pacific Medical Center Research Institute, San Francisco, California 94115

Philippe Sansonetti Unité de Pathogénie Microbienne Moléculaire, Institut Pasteur, 75724 Paris Cedex 15, France

Lisa M. Schechter Department of Microbiology and Molecular Genetics, Harvard Medical School, Boston, Massachusetts 02115

M. Scidmore-Carlson Host-Parasite Interactions Section, Laboratory of Intracellular Parasites, National Institute of Allergy and Infectious Diseases, National Institutes of Health, Rocky Mountain Laboratories, Hamilton, Montana 59840

Homayoun Shams Department of Veterinary and Biomedical Sciences, University of Nebraska, Lincoln, Nebraska 68583

David J. Silverman Department of Microbiology and Immunology, University of Maryland School of Medicine, Baltimore, Maryland 21201

Michael Steinert Institut für Molekulare Infektionsbiologie, Universität Würzburg, 97070 Würzburg, Germany

Elaine Tuomanen Department of Infectious Diseases, St. Jude Children's Research Hospital, Memphis, Tennessee 38105

V. Viprey LBMPS, Université de Genève, 1292 Chambery/Geneva, Switzerland

David Yogev Department of Membrane and Ultrastructure Research, The Hebrew University-Hadassah Medical School, Jerusalem 91120, Israel

Contents

Chapter 2

Interaction of Enteropathogenic *Escherichia coli* with Host Cells
Ilan Rosenshine, Stuart Knutton, and Gad Frankel

Chapter 3

***E. coli* Invasion of Brain Microvascular Endothelial Cells as a Pathogenetic Basis of Meningitis**
Kwang Sik Kim

Chapter 4

Host Cell Invasion by Pathogenic *Neisseriae*
Christoph Dehio, Scott D. Gray-Owen, and Thomas F. Meyer

Chapter 5

Bartonella **Interactions with Host Cells**

Michael F. Minnick and Burt E. Anderson

Chapter 6

Host Cell Invasion by *Streptococcus pneumoniae*

Axel Ring and Elaine Tuomanen

Chapter 7

High Frequency Invasion of Mammalian Cells by β Hemolytic Streptococci
P. Patrick Cleary and David Cue

Chapter 8

***Nocardia asteroides* as an Invasive, Intracellular Pathogen of the Brain and Lungs**
Blaine L. Beaman and LoVelle Beaman

Chapter 9

Mycoplasma Interaction with Eukaryotic Cells
Shlomo Rottem and David Yogev

Part II: Professional Facultative Intracellular Bacteria

Chapter 10

Mycobacterial Invasion of Epithelial Cells
Luiz E. Bermudez and Felix J. Sangari

Chapter 11

Invasion of Epithelial Cells by Bacterial Pathogens: The Paradigm of Shigella
Kirsten Niebuhr and Philippe J. Sansonetti

Chapter 12

Salmonella Invasion of Non-Phagocytic Cells
Lisa M. Schechter and Catherine A. Lee

Chapter 13

Salmonella Interactions with Professional Phagocytes
Robert A. Kingsley and Andreas J. Bäumler

Chapter 14

Interaction of Yersinia with Host Cells
Anne Boland and Guy R. Cornelis

Chapter 15 383–90

Invasion of Mammalian and Protozoan Cells by *Legionella pneumophila*
Yousef Abu Kwaik

Chapter 16

Internalization of *Listeria monocytogenes* by Nonprofessional and Professional Phagocytes
Michael Kuhn and Werner Goebel

Chapter 17

Host-Plant Invasion by Rhizobia
V. Viprey, X. Perret, and W.J. Broughton

Part III: Obligate Intracellular Bacteria

Chapter 18

Chlamydia: Internalization and Intracellular Fate
M. Scidmore-Carlson and T. Hackstadt

Chapter 19 479–90

Interaction of Rickettsiae with Eukaryotic Cells: Adhesion, Entry, Intracellular Growth, and Host Cell Responses
Marina E. Eremeeva, Gregory A. Dasch, and David J. Silverman

Chapter 24

New Approaches for Diagnosis of Infections by Intracellular Bacteria
Reinhard Marre

Introduction

Strategies of Bacterial Interaction with Eukaryotic Cells

*Tobias A. Oelschlaeger and Jörg Hacker

1. BENEFICIAL BACTERIAL-HOST INTERACTIONS

Already during birth and soon thereafter mammals are colonized by bacteria belonging to the resident microbial flora. Cutaneous and mucosal surfaces and the gastrointestinal tract are the areas which become colonized. These indigenous or autochthonous bacteria have a variety of beneficial effects on their hosts. They play a protective role by bacterial antagonism in fighting infections (Hoszowski and Truszczynski, 1997; Hentges, 1979). Production of vitamin K is another essential contribution of the resident microbial flora to the health of the host (Hill, 1997). Even more important, studies with germ-free animals demonstrated the involvement of the microbial flora on the development of the immune system. Such animals have underdeveloped and relatively undifferentiated lymphoid tissues and low concentrations of serum immune globulins (Cebra *et al.*, 1998). They

TOBIAS A. OELSCHLAEGER and JÖRG HACKER Institut für Molekulare Infektionsbiologie, Universität Würzburg, 97070 Würzburg, Germany. *Corresponding author; Phone: (0)931-312150; FAX: (0)931-312578; E-mail: t.oelschlaeger@mail.uni-wuerzburg.de also show defects in specific immune responsiveness and in nonspecific resistance induced by endotoxin, which may account for their lowered resistance. A more typical example of symbiotic interaction of bacteria with a host are bacteria like Ruminococcus in the gut of ruminants, essential for degradation of cellulose (Hobson, 1988). The closest benefical bacterial-host interactions are those of intracellular symbiotic bacteria and their host cells. The best studied examples are rhizobia and legume plant cells, Buchnera and aphids, and Symbiodinium and corals and other related animals. In each of these symbioses with intracellular bacteria the microorganisms provide certain metabolic products to their respective host (cell). Rhizobia, able to fix atmospheric nitrogen export NH_3 into the cytosol of the plant cell and only there, not in the bacterial cell, ammonia is assimilated into glutamine (Streeter, J.G., 1989; see also the Rhizobium chapter). Bacteria of the genus *Buchnera* overproduce essential amino acids and export these into their host cells. The Buchnera hosts, aphids, depend in this respect on their symbionts, because they are not able to synthezise essential amino acids by themselves (Mittler, 1971). Finally, the dinoflagellate alga Symbiodinium supports the corals and other related animals with glycerol (Sutton and Hoegh-Guldberg, 1990). Interestingly, after separation of the symbiotic microorganisms from their host cells, they no longer secrete the compounds their respective host relies on.

2. DETRIMENTAL BACTERIAL-HOST INTERACTIONS

2.1. Classical Adhesins

For most types of bacterial-host interactions (e.g. colonization) physical contact between the two partners is necessary. In order to initiate physical contact, bacteria have developed surface structures capable of binding to host receptors and thereby enable the baceria to adhere to host cells. These bacterial surface structures are called adhesins. They are either filamentous surface appendices termed pili or fimbriae or so-called nonfimbrial adhesins (Oelschlaeger *et al.*, 1998; Hacker, 1992). Pili are composed of a major subunit protein, constituting more than 90% of the pilus structure and one or more minor subunit proteins. One minor subunit is usually the adhesin proper (Hacker and Morschhäuser, 1994). However in certain pili the major subunit can function as the adhesin alone (Jacobs *et al.*, 1987a, b) or together with a minor subunit (Khan *et al.*, 1996). The adhesin binds to carbohydrate residues on the host cell thereby anchoring the bacteria to the host's surface. Because a single bacterium can encode different types of pili, the alternative expression of pili with different receptor specificities might be an important factor for host tropism and, after infection, also for tissue tropism in a given host. Additionally, switching from one immunogenic type of pili to another helps the pathogen to evade the host's immune response.

Furthermore, adherence is also a prerequisite for a variety of bacterial effects on host cells. These effects include the action of LT and ST enterotoxins, destruction of microvilli, reorganization of the cytoskeleton and inhibition of phagocytosis by professional phagocytes. Some of these effects are the result of bacterial signalling. Adherence of *Helicobacter pylori* to the gastric mucosa results in signal transduction to the host cell leading to activation of NF-κB and IL-8 induction requiring tyrosine kinase activity (Beales and Calam, 1997). On the other hand, the expression of adhesins is also regulated by host signals. For example, receptor binding specificity of *Helicobacter pylori* changes under acidic pH, typical for the gastric environment, and this change is prevented by inhibitors of protein synthesis (Huesca *et al.*, 1996).

2.2. Adhesins as Invasins

The molecular crosstalk between bacteria and host cells based on certain adhesins was shown to lead even to internalization of the bacteria. Important afimbrial adhesins functioning as invasins are the AfaD protein of the adhesive sheath of certain pathogenic *E. coli* (Jouve *et al.*, 1997; see also chapter 1), Dr-II of pyelonephritis associated *E. coli* (Pham *et al.*, 1997; see also chapter 1), the invasin, an outermembrane protein of Yersinia (Isberg and Leong, 1988; see also the Yersinia chapter), or internalin A of *Listeria monocytogenes* (Gaillard *et al.*, 1991; see also the Listeria chapter). Pili known to mediate invasion are the Dr fimbriae of uropathogenic *E. coli* (Goluszko *et al.*, 1997; see also chapter 1) and pili of the periodontal pathogen *Porphyromonas gingivalis* (Weinberg *et al.*, 1997; see also chapter 1). Even certain variants of the most prominent fimbriae of *Enterobacteriaceae*, type 1 pili, are able to mediate invasion (Mulvey *et al.*, 1998; Baorto *et al.*, 1997; see also chapter 1). Adhesins functioning as invasins might help bacteria to cross the epithelial cell layer of the mucosa (i.e. transcytosis). At the basal site of the epithelial cells, bacteria face another host barrier, the basal membrane. However, most pili are additionally able to activate plasminogen via binding tissue plasminogen activator. The activity of the bacterium-associated plasmin might help these pathogens to cross even the basal membrane and reach deeper tissue (Parkkinen *et al.*, 1991).

2.3. Bacterial Effectors

Besides adhesins, there is a variety of factors and functions specific for pathogenic bacteria. These abilities and structures typically not found in nonpathogenic bacteria are called virulence factors. Rather often, several virulence factors are encoded together on defined units of the chromosome (i.e. pathogenicity islands). Futhermore, these entities might be deleted or transfered to other bacteria of the same or different species by horizontal gene transfer (reviewed in Hacker *et al.*, 1997). In contrast to adhesins, which stay connected with the bacterium, most other virulence factors important for interaction with eukaryotic host cells are secreted by the pathogenic microorganisms. Many of these secreted compounds function as effectors based on certain enzymatic activities they exert on the host cell. Such effectors, e.g. toxins, might stay active for prolonged periods of time after they have been secreted. Therefore, effects of certain toxins can be observed even after the bacterial producer is no longer alive or present, in contrast to effects depending on the presence of (live) bacteria.

2.3.1. Secreted Effectors

Probably the most well known bacterial virulence factors are toxins. Most of these protein toxins belong to the family of secreted toxins, the so-called exotoxins. They can be divided into three groups according to their architecture (A-B toxins) and their mode of action (membrane active toxins and superantigens).

In contrast to the other two groups of toxins, A-B toxins must enter the host cell in order to exert their toxic effects. The B subunit of these toxins is responsible for binding to the appropriate host cell receptor and to induce internalization. It also determines the host cell specificity. The A portion of most A-B toxins catalyzes the removal of the ADP-ribosyl group from NAD and attaches it covalently to some host cell protein. This ADP-ribosylation either inactivates or causes an abnormal behavior of the modified host protein. The type of ADP-ribosylated protein determines the effect on the host cell. Well characterized examples of A-B toxins are the diphtheria toxin of *Corynebacterium diphtheriae* (ADP-ribosylation of elongation factor 2), the cholera toxin of *Vibrio cholerae* (ADP-ribosylation of regulatory protein), the tetanus toxin of *Clostridium tetani* (Proteolytic activity) and the shiga toxin of *Shigella dysenteriae* (Cleaves host cell rRNA).

2.3.2. Translocated Effectors

A rather recent and suprising finding is that some of the effectors as those responsible for cytoskeletal reorganization and inhibition of phagocytosis are injected into host cells by type III secretion systems (TTSS) after adherence has been established. TTSS form a channel through the cytoplasmic and the outer membrane of gram-negative bacteria and assemble into bacterial surface structures resembling the hollow needle of a syringe. Through this "needle", the effectors are translocated into the cytoplasm of the host cell where they exert their detrimental activity. Even more surprisingly, this delivery strategy of effectors is wide spread among pathogenic bacteria. TTSS were not only discovered in bacteria pathogenic for animals but also in the four major genera of plant-pathogenic bacteria Erwinia, Pseudomonas, Ralstonia and Xanthomonas (Bonas 1994; Collmer and Bauer, 1994). However, the kind of effector molecules delivered by TTSS varies considerably among the various bacterial species reflecting the differences in target species/target cell types and the different effects they cause in host cells (TTSS are reviewed in: Hueck, 1998).

2.3.2.1. Inhibitors of Phagocytosis One kind of effector translocated into host cells is of particular importance. These are inhibitors of phagocytosis. The bacterial ability to inhibit phagocytosis impairs a very efficient antibacterial component of the host's immune system. The presence of such inhibitors in the arsenal of virulence factors of pathogenic bacteria provides the respective baceria with a big advantage in the battle with the host. The best sudied examples of antiphagocytic effectors are certain Yops (**Y**ersinia **o**uter **p**roteins). For some of them the mode of action as been elucidated. For example, *Yersinia* spp. translocate YopE, a cytotoxin which disrupts microfilaments, and YopH, which dephosphorylates paxilin, FAK, and $p130^{cas}$, both of which are involved in inhibition of phagocytosis (Persson *et al.*, 1997; Forsberg and Wolf-Watz, 1990; see also chapter 14). The exoenzyme S of *Pseudomonas aeruginosa* not only shares sequence homology with YopE of Yersinia but obviously has also an antiphagocytic effect on host cells. In order to induce this antiphagocytic effect it also has to be translocated into host cells via a TTSS (Frithz-Lindsten *et al.*, 1997).

2.3.2.2. Inducers of Apoptosis Several pathogenic bacterial species are able to trigger a suicide program in host cells termed apoptosis (apoptosis is review by several articles in Science 267; 1995). This ability might allow bacteria to fight off the attack of phagocytic cells or to achieve release from phagocytic or nonphagocytic cells. Yersiniae are able to induce apoptosis by injecting YopI/P via TTSS into the cytosol of macrophages (Mills *et al.*, 1997; see also the Yersinia chapter). Similarly, IpaB of *Shigella flexneri* is also injected into host cells and induces there apoptosis by directly binding to ICE (i.e. **I**nterleukin-1β **C**onverting **E**nzyme) (Chen *et al.*, 1996). ICE is an already rather downstream located link of the signalling cascade leading to apoptosis. Interestingly, certain members of the ICE family are also activated to cause cell death by the cytotoxic T cell product granzym B (Chinnaiyan *et al.*, 1996; Duan *et al.*, 1996). Recently, it was also shown that *Salmonella* spp. induce apoptosis in macrophages. The bacterial effector was demonstrated to be SipB (**S**almonella **i**nvasion **p**rotein) interacting with the proapoptotic protease caspase-1 (Hersh *et al.*, 1999). Subsequently, the activated caspase-1 processes its substrate interleukin-1β finally leading to apoptosis. However, nonbacterial pathogens, i.e. certain viruses like lymphotropic viruses interfere with apoptosis of their host cells. This leads to enhanced viral replication and evasion of cytotoxic T-cell effects (Meinl *et al.*, 1998).

2.3.2.3. Invasins The translocated effectors IpaB and SipB of Shigella and Salmonella, respectively, were first characterized as invasion proteins of these facultative intracellular bacteria and only recently also identified as inducers of apoptosis. In order to induce internalization of Shigella, IpaB has to act in concert with the other invasion proteins IpaA, C, and D. All these invasion proteins are secreted and translocated into host cells via the Shigella plasmid encoded TTSS (Tran Van Nhieu *et al.*, 1997; Menard *et al.*, 1993; see also the Shigella chapter). For *Salmonella typhimurium* invasion several invasins have to be translocated by a TTSS encoded on pathogenicity island 1 of the chromosome (Collazo and Galan, 1997; see also the Salmonella chapters). Some of these proteins are Sip/SspB, Sip/SspC, and Sip/SspD which show sequence homology to the Shigella invasins IpaB, C, and D, respectively (Kaniga *et al.*, 1995a, b).

3. CONCLUSIONS

Bacterial interaction with eukaryotic cells can be beneficial not only for the bacteria but also for the host cell, if the bacterial partner is nonpathogenic. This is most clearly demonstrated by the interaction of endosymbionts like rhizobia, Buchnera and Symbiodinium with their respective host. Unfortunatly, there is not as much known about extracellularly located probably symbiotic bacteria, especially of the interaction of the human resident microbial flora and host cells. Much more insight was already gained in the interaction of pathogenic bacteria with eukaryotic cells. Even without ever colonizing a host, bacteria are able to cause harm by preformed toxins contained in contaminated food or water. In general, however, bacteria must colonize a host before they are able to cause any detrimental effect. A prerequisite for colonization are bacterial adhesins. As for many other virulence factors, adhesins can have more than the predominant function. Obviously, bacteria rather often have developed dual- or even multi-use tools, termed virulence factors, to achieve infection. An important general strategy in this kind of interaction is that the bacteria try to evade the host's immune response. This is achieved by toxins, inducers of apoptosis, inhibitors of phagocytosis or by invasion. With respect to duration or persistence of a bacterial infection, those species are most successful, which are able to invade host cells. The intracellular location is obviously a privileged niche, well protected not only from the immune system but also the action of many antibiotics. Furthermore, invasion ability is employed by several bacterial species to cross host barriers like the gut mucosa or even the blood-brain barrier. Some of the bacterial effects on eukaryotic cells are accomplished by the bacteria by abusing host cell structures and/or processes intended for very different purposes by the host cell. For an increasing number of invasive bacteria, it became clear that invasion and intracellular survival in cells is not restricted to cells of higher eukaryotes. Rather, invasion ability was acquired early in evolution and is used by bacteria to survive in the environment under unfavorable conditions inside protists and their cysts (Steinert *et al.*, 1998; Kilvington and Price, 1990). Therefore, the environment might be a reservoir for pathogenic or potentially pathogenic bacteria.

The investigation of the different kinds of interaction between bacteria and eukaryotic cells will not only result in a much deeper understanding of bacterial pathogenicity and ecology but will also create the basis for the development of bacterial tools to disect and elucidate eukaryotic processes by what is now emerging as "cellular microbiology". Furthermore, the identification of virulence factors and the characterization of their mode of action will be crucial for the development of new antimicrobial drugs directed to new bacterial targets, of new vaccines with engineered species and tissue tropism as well as directed immune responses.

In particular, one strategy of bacterial interaction with eukaryotic cells is the focus of this volume: bacterial invasion into host cells. The reader will find (almost?) all important and well studied invasive bacteria. However, bacterial invasion is here not viewed as a separate event but set in the context of pathogenicity or, in the case of rhizobia, of symbiosis.

4. REFERENCES

Baorto, D.M., Gao, Z., Malaviya, R., Dustin, M.L., van der Merwe, A., Lublin, D.M., and Abraham, S.N., 1997, Survival of FimH-expressing enterobacteria in macrophages relies on glycolipid traffic, *Nature* **389**:636–639.

Beales, I.L., and Calam, J., 1997, Stiumlation of IL-8 production in human gastric epithelial cells by *Helicobacter pylori*, IL-1β and TNFα requires tyrosine kinae activity, but not protein kinase C, *Cytokine* **9**:514–520.

Bonas, 1994, *hrp* genes of phytopathogenic bacteria, *Curr. Top. Microbiol. Immunol.* **192**: 79–98.

Cebra, J.J., Periwal., S.B., Lee, G., Lee, F., and Shroff, K.E., 1998, Development and maintenance of the gut-associated lymphoid tissue (GALT): the role of enteric bacteria and viruses, *Dev. Immunol.* **6**:13–18.

Chen, Y., Smith, M.R., Thirumalai, K., and Zychlinsky, A., 1996, A bacterial invasin induces macrophage apoptosis by binding directly to ICE, *EMBO J.* **15**:3853–3860.

Chinnaiyan, A.M., Hanna, W.L., Orth, K., Duan, H., Poirier, G.G., Froelich, C.J., and Dixit, V.M., Cytotoxic T-cell-derived granzym B activates the apoptotic protease ICE-LAP3, *Curr. Biol.* **6**:897–899.

Collazo, C.M., and Galan, J.E., 1997, The invasion-associated type III system of *Salmonella typhimurium* directs translocation of Sip proteins into the host cell, *Mol. Microbiol.* **24**:747–756.

Collmer, A., and Bauer, D.W., 1994, *Erwinia crysanthemi* and *Pseudomonas syringae*: plant pathogens trafficking in extracellular virulence proteins, *Curr. Top. Microbiol. Immunol.* **192**:43–78.

Duan, H., Orth, K., Chinnaiyan, A.M., Poirier, G.G., Froelich, C.J., He, W.W., and Dixit, V.M., 1996, ICE-LAP6, a novel member of the ICE/Ced-3 gene family, is activated by the cytotoxic T cell protease granzym B, *J. Biol. Chem.* **271**:16720–16724.

Frithz-Lindsten, E., Du, Y., Rosqvist, R., and Forsberg, A., 1997, Intracellular targeting of exoenzyme S of *Pseudomonas aeroginosa* via type III-dependent translocation induces phagocytosis resistance, cytotoxicity and disruption of actin microfilaments, *Mol. Microbiol.* **25**:1125–1139.

Forsberg, A., and Wolf-Watz, H., 1990, Genetic analysis of the *yopE* region of *Yersinia* spp.: identification of a novel conserved locus, *yerA*, regulating *yopE* expression, *J. Bacteriol.* **172**:1547–1555.

Gaillard, J.-L., Berche, P., Freihel, C., Gouin, E., and Cossart, P., 1991, Entry of *L. monocytogenes* into cells is mediated by internalin, a repeat protein reminiscent of surface antigens from gram-positive cocci, *Cell* **65**:1127–1141.

Goluszko, P., Popov, V., Selvarangan, R., Novicki, S., Pham, T., and Nowicki, B.J., 1997, Dr fimbriae operon of uropathogenic *Escherichia coli* mediate microtubule-dependent invasion to the HeLa epithelial cell line, *J. Infect. Dis.* **176**:158–167.

Hacker, J., 1992, Role of fimbrial adhesins in the pathogenesis of *Escherichia coli* infections, *Can. J. Microbiol.* **38**:720–727.

Hacker, J., and Morschhäuser, J., 1994, S and F1C fimbriae, in: *Fimbriae: adhesion, genetics, biogenesis and vaccines* (P. Klemm, ed.), CRC Press, Boca Raton, FL, pp. 27–36.

Hacker, J., Blum-Oehler, G., Mühldorfer, I., and Tschäppe, H., 1997, Pathogenicity islands of virulent bacteria: structure, function and impact on microbial evolution, *Mol. Microbiol.* **23**:1089–1097.

Hentges, D.J., 1979, The intestinal flora and infant botulismus, *Rev. Infect. Dis.* **1**:668–673.

Hersh, D., Monack, D.M., Smith, M.R., Ghori, N., Falkow, S., and Zychlinsky, A., 1999, The Salmonella invasin SipB induces macrophage apoptosis by binding to caspase-1, *Proc. Natl. Acad. Sci. USA* **96**:2396–2401.

Hill, M.J., 1997, Intestinal flora and endogenous vitamin synthesis, *Eur. J. Cancer Prev.* **6(Suppl. 1)**:S43–S45.

Hobson, P.N., 1988, *The rumen microbial ecosystem*, Elsevier Applied Science, London.

Hoszowski, A., and Truszczynski, M., 1997, Prevention of *Salmonella typhimurium* caecal colonization by different preparations for competitive exclusion, *Comp. Immunol. Microbiol. Infect. Dis.* **20**:111–117.

Hueck, C.J., 1998, Type III secretion systems in bacterial pathogens of animals and plants, *Microbiol. Mol. Biol. Rev.* **62**:379–433.

Huesca, M., Borgia, S., Hoffman, P., and Lingwood, C.A., 1996, Acidic pH changes receptor binding specificity of *Helicobacter pylori*: a binary adhesion model in which surface heat shock (stress) proteins mediate sulfatide recognition in gastric colonization, *Infect. Immun.* **64**:2643–2648.

Isberg, R.R., and Leong, J.M., 1988, Cultured mammalian cells attach to the invasin protein of *Yersinia pseudotuberculosis*, *Proc. Natl., Acad. Sci. USA* **85**:6682–6686.

Jacobs, A.A.C., Roosendaal, B., van Breemen, H.F.L., and deGraf, F.K., 1987a, Role of phenylalanine 150 in the receptor-binding domain of the K88 fimbrial subunit. *J. Bacteriol.* **169**:4907–4911.

Jacobs, A.A.C., Simons, L.H., and de Graaf, F.K., 1987b, The role of lysine 132 and arginine 136 in the receptor-binding domain of the K99 fibrillar subunit, *EMBO J.* **6**:1805–1808.

Jouve, M., Garcia, M.-I., Courcoux, P., Labigne, A., Gounon, P., and Bouguenec, C.L., 1997, Adhesion to and invasion of HeLa cells by pathogenic *Escherichia coli* carrying the *afa-3* gene cluster are mediated by the AfaE and AfaD proteins, respectively, *Infect. Immun.* **65**:4082–4089.

Kaniga, K., Trollinger, D., and Galan, J.E., 1995a, Identification of two targets of the type III protein secretion system encoded by the *inv* and *spa* loci of *Salmonella typhimurium* that have homology to the Shigella IpaD and IpaA proteins, *J. Bacteriol.* **177**:7078–7085.

Kaniga, K., Tucker, S., Trollinger, D., and Galan, J.E., 1995b, Homologs of the Shigella IpaB and IpaC invasins are required for *Salmonella typhimurium* entry into cultured epithelial cells, *J. Bacteriol.* **177**:3965–3971.

Khan, A.S., Johnston, N.C., Goldfine, H., and Schifferli, D.M., 1996, Porcine 987P glycolipid receptors on intestinal brush borders and their cognate bacterial ligands, *Infect. Immun.* **64**:3688–3693.

Kilvington, S., and Price, J., 1990, Survival of *L. pneumophila* within cysts of *Acanthamoeba polyphaga* following chlorine exposure, *J. Appl. Bacteriol.* **68**:519–525.

Meinl., E., Fickenscher, H., Thome, M., Tschopp, J., and Fleckenstein, B., 1998, Anti-apoptotic strategies of lymphotropic viruses, *Immunol. Today* **19**:474–479.

Menard, R., Sansonetti, P.J., and Parsot, C., 1993, Nonpolar mutagenesis of the *ipa* genes defines IpaB, IpaC, and IpaD as effectors of *Shigella flexneri* entry into epithelial cells, *J. Bacteriol.* **175**:5899–5906.

Mills, S.D., Boland, A., Sory, M.P., van der Smissen, Kerbouch, C., Finlay, B.B., and Cornelis, G.R., 1997, *Yersinia enterocolitica* induces apoptosis in macrophages by a process requiring functional type III secretion and translocation mechanisms and involving YopP, presumably acting as an effector protein, *Proc. Natl. Acad Sci. USA* **94**:12638–12643.

Mittler, T.E., 1971, Dietary amino acid requirements of the aphid *Myzus persicae* affected by antibiotic uptake, *J. Insect Physiol.* **101**:1023–1028.

Mulvey, M.A., Lopez-Boado, Y.S., Wilso, C.L., Roth, R., Parks, W.C., Heuser, J., and Hultgren, S.J., 1998, Induction and evasion of host defenses by type1-piliated uropathogenic *Escherichia coli*, *Science*, **282**:1494–1497.

Oelschlaeger, T.A., Khan, A.S., Meier, C., and Hacker, J., 1997, Receptors and ligands in adhesion and invasion of *Escherichia coli*, *Nova Acta Leopoldina NF75*, **301**:195–205.

Parkkinen, J., Hacker, J., and Korhonen, T.K., 1991, Enhancement of tissue plasminogen activator-catalyzed plasminogen activation by *Escherichia coli* S fimbriae associated with neonatal septicemia and meningitis, *Thromb. Haemost.* **65**:483–486.

Pham, T.Q., Goluszko, P., Popov, V., Nowicki, S., and Nowicki, B.J., 1997, Molecular cloning and characterization of Dr-II, a nonfimbrial adhesin-I-like adhesin isolated from gestational pyelonephritis-associated *Escherichia coli* that binds to decay-accelerating factor, *Infect. Immun.* **65**:4309–4318.

Persson, C., Carballeira, N., Wolf-Watz, H., and Fallman, M., 1997, The PTPase YopH inhibits uptake of Yersinia, tyrosine phosphorylation of p130Cas and FAK, and the associated accumulation of these proteins in peripheral focal adhesions, *EMBO J.* **16**:1207–2318.

Steinert, M., Birkness, K., White, E., Fields, B., and Quinn, F., 1998, *Mycobacterium avium* bacilli grow saprozoically in coculture with *Acanthamoeba polyphaga* and survive within cyst walls, *Appl. Environ. Microbiol.* **64**:2256–2261.

Streeter, J.G., 1989, Estimation of ammonium concentration in the cytosol of soybean nodules, *Plant Physiol.* **90**:779–782.

Sutton, D.C., and Hoegh-Guldberg, 1990, Host-zooxanthella interactions in four temperate marine invertebrate symbioses: assessment of effect of host extracts on symbionts, *Biol. Bull.* **175**:178–186.

Tran Van Nhieu, G., Ben-Ze'ev, A., and Sansonetti, P.J., 1997, Modulation of bacterial entry into epithelial cells by association between vinculin and the Shigella IpaA invasin, *EMBO J.* **16**:2717–2729.

Weinberg, A., Belton, C.M., Park, Y., and Lamont, R.J., 1997, Role of fimbriae in *Porphyromonas gingivalis* invasion of gingival epithelial cells, *Infect. Immun.* **65**:313–316.

Part I

Nonprofessional Facultative Intracellular Bacteria

Chapter 1

Microtubule Dependent Invasion Pathways of Bacteria

Tobias A. Oelschlaeger and Dennis J. Kopecko

1. INTRODUCTION

Bacterial pathogens have evolved mechanisms to resist standard host defenses and to subvert normal host machinery in order to initiate disease. Invasive bacterial pathogens are now known to interact, via biochemical crosstalk, with the host, stimulating a signal transduction cascade(s) that results in host cytoskeletal rearrangements that lead to internalization of the pathogen. Until about five years ago, virtually all bacterial uptake pathways involved the exclusive requirement for host microfilaments (MFs). Since then, internalization into host cells for several bacterial genera has been reported to require microtubules (MTs) alone or together with MFs. Despite striking differences in the host cell cytoskeletal requirements for invasion, all bacterial uptake pathways have followed a common mechanistic scheme. In this general scheme, a bacterial "invasion

TOBIAS A. OELSCHLAEGER Institut für Molekulare Infektionsbiologie, Universität Wuerzburg, 97070 Wuerzburg, Germany. **DENNIS J. KOPECKO** Laboratory of Enteric and Sexually Transmitted Diseases, Food and Drug Administration, Bethesda, MD 20892, USA.

Subcellular Biochemistry, Volume 33: Bacterial Invasion into Eukaryotic Cells, edited by Oelschlaeger and Hacker. Kluwer Academic / Plenum Publishers, New York, 2000.

effector ligand(s)" interacts with an eukaryotic receptor to induce a signal transduction cascade leading ultimately to the cytoskeletal rearrangements necessary for internalization. Of special interest are the receptors and intermediary host molecules which transduce the initial bacterial signal through the host cell thereby activating cytoskeletal changes and uptake of the bacterium, intracellular survival and movement of the bacterium, and sometimes exocytic release of the pathogen. In contrast to the many plasma membrane receptors with well-documented connection to cytoskeletal microfilaments, there are only a few examples of known membrane components which are connected with the microtubular cytoskeleton. It should not be surprising that our understanding of these more recently described MT-dependent invasion mechanisms are limited. Furthermore, it is unclear if these MT-mediated uptake pathways simply offer the pathogen an alternative (i.e. other than MF-dependent entry) cell invasion mechanism or if they are of unique pathogenic advantage.

In this chapter, we will present an overview of the host cell structures and processes, as well as corresponding bacterial components, known to be involved in MT-dependent bacterial invasion of host cells.

2. ROLE OF THE HOST CELL

2.1. Role of the Cytoskeleton

Absent the entry of some *Chlamydia* serovars by receptor-mediated endocytosis which does not require MTs or MFs, all bacterial internalization pathways require a component(s) of the host cytoskeleton. The cytoskeleton of eukaryotic cells consists of MFs, MTs, and intermediate filaments. However, due to lack of experimental tools nothing is known about the involvement of intermediate filaments in bacterial uptake. The role of the other two cytoskeletal components in internalization can be tested biochemically by the employment of specific inhibitors in a quantitative assay for bacterial invasion (i.e. the gentamicin protection assay; Elsinghorst, 1994). If invasion into cells is inhibited after depolymerization of MFs by cytochalasin D (i.e. at concentrations which do not affect host cell viability), then invasion by the bacteria under investigation can be considered to be MF-dependent. The requirement for MTs in bacterial internalization can be tested by depolymerization of MTs with substances such as colchicine, nocodazole, vincristine, or vinblastine. Another tool to probe the involvement of microtubules in uptake pathways is taxol. Taxol "freezes" the microtubule network by crosslinking tubulin and thereby blocking the ability of the host cell to restructure the microtubular framework.

The involvement of MTs in early events of host cell infection by *Chlamydia trachomatis* was recently demonstrated by indirect immunofluorescence microscopy (Clausen *et al.*, 1997). Employment of the above mentioned inhibitors in the gentamicin protection invasion assay demonstrated that internalization of *Campylobacter jejuni* and *Citrobacter freundii* into certain cell lines was only inhibited by depolymerization of MTs (Oelschlaeger *et al.*, 1993). In contrast, entry of *Klebsiella pneumoniae* and the newborn meningitis-causing *Escherichia coli* K1 strain IHE3034 is additionally inhibited by "freezing" MTs with taxol (Oelschlaeger and Tall, 1997; Meier *et al.*, 1996). These results indicate the existence of, at least, two different overall MT-dependent uptake mechanisms (see Fig. 1). However, these uptake pathways are likely to be even more complex. The involvement in bacterial internalization of both MFs and MTs has been demonstrated for *Campylobacter jejuni* (Oelschlaeger *et al.*, 1993), enterohemorrhagic *E. coli* (EHEC; Oelschlaeger *et al.*, 1994), enteropathogenic *E. coli* (EPEC; Donnenberg *et al.*, 1990), newborn meningitis-causing *E. coli* (MENEC; Meier *et al.*, 1996), *K. pneumoniae* (Oelschlaeger and Tall, 1997), *Mycobacterium bovis* BCG (Buchwalow *et al.*, 1997), *Mycobacterium tuberculosis* (Bermudez and Goodman, 1996), *Mycoplasma penetrans* (Borovsky *et al.*, 1998), *Neisseria gonorrhoeae* (Grassme *et al.*, 1996), and *Porphyromonas gingivalis* (Lamont *et al.*, 1995). Also, *Legionella pneumophila* internalization into Vero cells was shown to involve both MTs and MFs (Maruta *et al.*, 1998). Even *Listeria monocytogenes*, thought to represent a classical example of MF-dependent, MT-independent uptake into host epithelial cells, is internalized into mouse dendritic cells in a strongly MT-dependent and weakly MF-dependent manner (Guzman *et al.*, 1995). Similarly, engulfment of *L. monocytogenes* by P338D1 macrophages is sensitive to treatment with cytochalasin D as well as colchicine and nocodazole, but not to taxol (Kuhn, 1998). The specific requirements for the host cell cytoskeleton depend not only upon the bacterial strain, but also upon the type of host cell utilized. Therefore, not surprisingly, *Citrobacter freundii* uptake into the human T24 urinary bladder epithelial cells is solely MT-dependent, but internalization of the same bacteria into INT407 human embryonic intestinal epithelial cells depends upon both MTs and MFs (Oelschlaeger *et al.*, 1993). Currently, it is not evident if these bacteria exhibiting dual MT- and MF-dependent invasion require simultaneous interaction of both elements of the cytoskeleton during internalization and/or represent bacteria encoding multiple, separate invasion pathways (i.e. MT-dependent alone, or MF-dependent uptake) that recognize different host receptors which are not expressed on all host cell types.

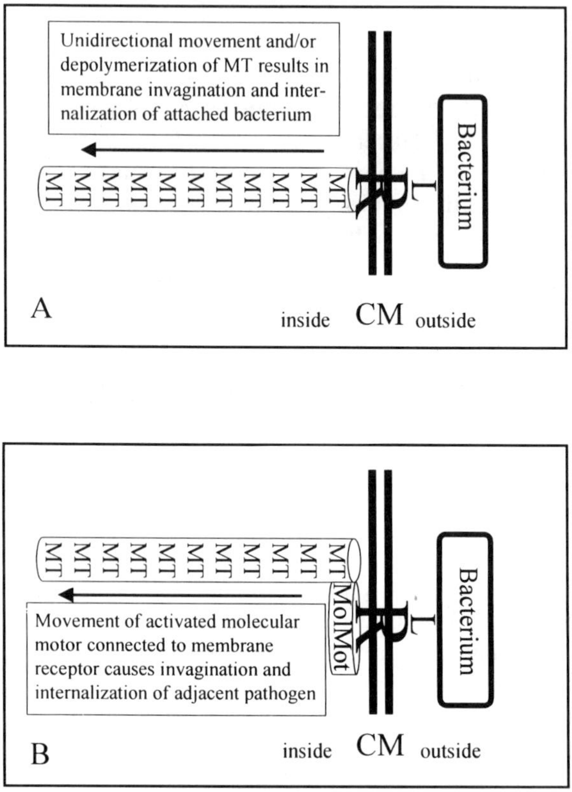

FIGURE 1. Models for microtubule-dependent internalization of bacteria. Bacteria interact via a bacterial ligand (L) with an eukaryotic receptor (R). This receptor is connected to microtubules (MTs) directly, only if it crosses the host cell cytoplasmic membrane (CM), or via linker proteins. If such MTs are shortened, they will engulf bacteria into the host cell via membrane invagination (A). Such an uptake pathway is not only inhibited by MT depolymerization (e.g. with colchicine, nocodazole) but also by freezing MTs with taxol. Such an inhibitor pattern was observed for *Klebsiella pneumoniae* and MENEC. Alternatively, the eukaryotic receptor (R) is connected to a molecular motor (MolMot) which in turn can move along the MTs, thereby tracking the external bacteria into the host cell (B). Such molecular motors (MolMot) can be members of the dynein or the kinesin family. This internalization mechanism would be inhibited only by MT depolymerization (by e.g. colchicine, nocodazole) but not by freezing the MTs with taxol. Such an inhibitor pattern was reported for *Citrobacter freundii*, *Campylobacter jejuni* and other bacerial pathogens (Ölschläger *et al.*, 1993; see also text). For *Chlamydia trachomatis* inhibition of dynein had a pronounced effect on infectivity (Clausen *et al.*, 1997), indicating an internalization mechanism for *Chlamydia trachomatis* as illustrated in part B.

2.2. Signal Transduction

The involvement of the host cell cytoskeleton in internalization is considered to be the result of a host cell signal transduction cascade induced by the invasive bacterium. As observed in many signal transduction processes initiated by bacteria, kinases and/or phosphatases are typically involved. Additionally, in some cases the transient alteration of the intracellular Ca^{++} concentration results from such signalling events. Specific biochemical inhibitors can be employed to test the potential involvement of kinases and/or phosphatases in signal transduction leading to internalization. In addition, the quantitative appearance or disappearance of corresponding phosphorylated/dephosphorylated host cell proteins during the internalization process can be detected by gel electrophoretic and immunological techniques.

Specific protein kinase activation has been observed during the internalization of most of the bacteria which are taken up by MT-dependent mechanisms. Protein tyrosine kinase activation seems to be necessary for MT-dependent internalization of *Bordetella bronchiseptica* (Guzman *et al.*, 1994), *C. jejuni* (Woolridge *et al.*, 1996), *Chlamydia trachomatis* (Clausen *et al.*, 1997; Birklund *et al.*, 1994), *Ehrlichia risticii* (Zhang and Rikihisa, 1997), EPEC (Kenny and Finlay, 1997; Foubister *et al.*, 1994), *L. monocytogenes* (Tang *et al.*, 1998), *Mycoplasma penetrans* (Andreev *et al.*, 1995), and *P. gingivalis* (Guzman *et al.*, 1995). Additionally, protein kinase C activation was reported to be important for the MT-dependent internalization of *B. bronchiseptica* (Guzman *et al.*, 1994), EPEC (Kenny and Finlay, 1997), *M. bovis* BCG (Buchwalow *et al.*, 1997), and *N. gonorrhoeae* (Grassme *et al.*, 1997). Calcium flux and/or calmodium activation participate in signal transduction stimulated by *E. risticii* (Rikihisa *et al.*, 1995), EPEC (Kenny and Finlay, 1997), and *P. gingivalis* (Itzutsu *et al.*, 1996). How any of these different signal transduction events lead to specific microtubule activity resulting in bacterial internalization is unknown. However, there is strong evidence that heterotrimeric G-proteins affect MT formation and, therefore, these G-proteins are candidates for transmitting intracellular signals to MTs (Ravindra, 1997).

2.3. Internalization Receptors

From the previous section, it should be evident that only isolated portions of the signalling pathways have been elucidated and the precise link to MTs is still unclear. Ironically, there is seemingly more information available about the host receptor end of the signal transduction pathway, despite the paucity of data on how MTs are connected to the host membrane.

Apparently, the interaction of a specific bacterial "invasion ligand(s)" with its host receptor(s) initiates the signaling cascade that results in bacterial uptake. Numerous host receptors have been identified as being important for the internalization of bacteria simultaneously utilizing MTs and MFs. However, it is not known which, if any, of these receptors are responsible for activating MTs versus MFs. The integrin $\alpha_M\beta_2$ (i.e. complement III receptor) on macrophages is the trigger for the internalization of *B. bronchiseptica* (Leininger *et al.*, 1992). For *K. pneumoniae* the receptor structures are proximal GlcNAc residues of an N-glycosylated protein (s) (Fumagalli *et al.*, 1997). Similarly, *E. coli* causing newborn meningitis interact with GlcNAc epitopes of brain microvascular endothelial cell glycoproteins (Prasadarao *et al.*, 1996). In contrast, Dr fimbriae-expressing uropathogenic *E. coli* (UPEC) bind to the SCR-3 domain of the decay-accelerating factor receptor for internalization (Nowicki *et al.*, 1993). The attenuated *M. bovis* BCG subverts the host $\alpha_5\beta_1$ integrin as its internalization receptor (Kuroda *et al.*, 1993). Wildtype *M. tuberculosis* makes use of CD51 and CD29 (i.e. β_1 integrin) receptors to induce its internalization into alveolar epithelial cells (Bermudez and Goodman, 1996). It is not clear which, if any, of the different identified receptors for *Neisseria* (e.g. syndecan-like proteoglycan, CD66a, lutotropin receptor, $\alpha_5\beta_3$ or $\alpha_5\beta_1$ integrin) is the relevant one for MT-dependent internalization. A 48 kDa protein of normal human gingival epithelial cells seems to represent the internalization receptor for *P. gingivalis* (Weinberg *et al.*, 1997). A remarkable internalization receptor for EPEC is Tir (Translocated intimin receptor). Tir is actually not an eukaryotic protein, but a protein of EPEC which is translocated into the host cell cytoplasmic membrane (Kenny *et al.*, 1997). Besides Tir, there might well be other receptors involved in internalization of EPEC, e.g. βGal (1-4 or 1-3)βGlcNAc lactosamines (Vanmaele *et al.*, 1995) or β_1 integrins (Frankel *et al.*, 1996). This fact is further strenghtened by the observation of homology between a 55 amino acid region of Tir and a β_1 integrin. In Tir, this 55 amino acid stretch is the binding region on Tir for intimin (Kenny *et al.*, 1999).

For any of the above-mentioned receptors, it is not understood whether the receptor is specifically associated with the microtubular skeleton and, if so, whether the connection is direct or indirect. Nevertheless, there are host membrane proteins with documented connections to microtubules. Transferrin uptake via the transferrin receptor in hemopoetic, but not adherent, cell lines is inhibitable by MT depolymerization (Subtil and Dautry-Varsat, 1997). Furthermore, the glycine receptor as well as the γ-aminobutyric acid receptor in neuronal cells seem to be connected to MTs. This connection is not direct, but achieved in both cases via receptor-associated proteins which bind to tubulin (Wang *et al.*, 1999; Kirsch *et al.*, 1991). Another interesting example of a plasma membrane protein that is associated with MTs is NRAMP (natural resistance-associated macrophage protein) (Kishi *et al.*, 1996). It is the product of the single dominant gene locus for natural resistance to infection with intracellular parasites such as *Leishmania*, *Salmonella*, and *Mycobacterium*, and is present in all macrophage/monocytes and B-and T-lymphocytes of mice. Finally, the adenomatous polyposis coli (APC) membrane protein is thought to be connected to MTs and is involved in normal intestinal cell migration from the crypts to the gut lumen (Nathile *et al.*, 1996). Unfortunately, no role in MT-dependent bacterial invasion has yet been defined for this latter group of MT-connected proteins.

3. ROLE OF THE BACTERIUM

Invasive bacteria apparently synthesize and export effector molecules which are surface-bound and can interact with host receptors, and/or which are secreted directly into the host cell in order to trigger internalization. The invasion systems of bacteria internalized via strict MF-dependent processes have been studied most intensively. Arguably the simplest invasion system, the outer membrane protein invasin of Yersinia, induces internalization by binding to the β_1 subunit of certain integrins (Isberg and Leong, 1990; also see the *Yersinia* chapter). Invasin is necessary and sufficient to induce the internalization even of dead Yersinia. It is exported across the bacterial cytoplasmic membrane via a Sec-dependent process and inserted into the outer membrane via its N-terminal domain. Internalization via invasin mimics the "membrane-zippering" process seen in phagocytosis, except that sequential reactions occur between numerous invasin molecules, distributed over the bacterial surface, and the host integrin receptors, each subsequent interaction zippering the host membrane tightly around the pathogen. Other strict MF-dependent bacterial invasion pathways are apparently more complex. For example, Salmonella and Shigella invasion systems consist of several bacterial effectors with corresponding chaperones (see the corresponding chapters). Upon appropriate host cell recognition, the bacterial effectors and chaperones are secreted and translocated into host cells, via type III secretion systems comprised of >20 proteins (Hueck, 1998). These effectors initiate a signal transduction cascade(s) that results in MF-rearrangement and bacterial engulfment. The bacterial systems requiring MTs for invasion are not yet as well characterized.

3.1. Adhesins as Invasins

Bacterial effectors for MT-dependent invasion are often surface structures, which also act as or resemble adhesins. For *B. bronchiseptica* filamentous hemagglutinin (FHA) and pertactin not only act as adhesins, but both types of proteins can also mediate invasion. FHA seems to be responsible for invasion of macrophages via C3 receptor, whereas pertactin might be responsible for invasion of epithelial cells via integrins (Leininger *et al.*, 1992). For *C. trachomatis* a glycan of the major outer membrane protein (MOMP) and/or terminal mannose structures on the bacterial surface have been reported to be involved in the invasion process by separate research groups (Amin *et al.*, 1995; Swanson and Kuo, 1994). It is unclear, however, if the mannose structures identified by one group are part of the glycan of the MOMP described by the other group. By molecular cloning of chromosomal DNA of *C. trachomatis* into *E. coli* K-12 and screening for an adherence phenotype, a third research group has identified a protein, related to the Hsp70 family of proteins that mediates adhesion and might be necessary for invasion (Raulston *et al.*, 1993). Binding to and invasion of host cells by EPEC requires the outer membrane protein intimin, an *eae* gene product. Intimin binds to Tir in the host cell cytoplasmic membrane, but also to β_1 integrins. Additionally, there is a report of conferring invasiveness to a noninvasive *E. coli* K-12 strain by cloning the gene for a 32 kDa protein of EPEC serotype O111 (Scaletsky *et al.*, 1995). Furthermore, certain other adhesin complexes of pathogenic *E. coli* strains are able to mediate invasion. One such adhesin complex is the *afa*-3 gene cluster, which encodes the AfaD protein as a component of an adhesive sheath. However, AfaD is not only involved in adhesion but also responsible for the invasion ability of *afa*-3 carrying *E. coli* strains causing either intestinal or urinary tract infections (Jouve *et al.*, 1997). Another nonfimbrial adhesin of pyelonephritis-associated *E. coli* is Dr-II which has been shown to mediate internalization into HeLa cells (Pham *et al.*, 1997). Related fimbrial adhesins mediating invasion are the Dr fimbriae of uropathogenic *E. coli* stains. Although the Dr fimbriae and Afa adhesin families share the same adhesin receptor, only Dr fimbriae induce internalization via DAF (decay accelerating factor) receptor (Goluszko *et al.*, 1997). Another example in which fimbriae mediate MT-dependent invasion are the pili of *P. gingivalis*, a periodontal pathogen. Non-piliated *P. gingivalis* mutants are impared by 8-fold in invasion ability, but only 2-fold in adherence to primary cultures of normal human gingival epithelial cells (Weinberg *et al.*, 1997). A protein with adhesin properties in *Haemophilus influenzae* also seems to be able to mediate invasion. This Hap protein shares homology with IgA proteases and is a member of the autotransporter family of proteins exported by a type IV secretion pathway (St. Geme *et al.*, 1994). A tip-like surface structure of *Mycoplasma penetrans*, not yet further characterized, functions also as an adhesin and simultaneously as an invasion complex (Borovsky *et al.*, 1998). Surprisingly, even type 1 fimbriae can function as an invasin. Baorto *et al.* (1997) reported that type 1 fimbriae mediate internalization of *E. coli* into macrophages in the absence of antibodies by an uptake pathway which advantageously leads to intracellular survival of *E. coli*, in contrast to normal phagocytosis of *E. coli* following opsonization. However, it is important to note that the ability to mediate invasion is not a general property of adhesins, and is found only in certain variants of adhesins.

3.2. Soluble Extracellular Matrix Proteins—a Linker Strategy for Bacterial Invasion

Another bacterial strategy utilized to induce internalization is to bridge a surface-attached bacterial ligand with a host soluble protein which can then bind to an eukaryotic internalization receptor. This linker strategy is followed by *M. bovis* BCG in binding fibronectin to its own surface via a 55 kDa protein. Fibronectin bound-BCG can, in turn, bind to an eukaryotic receptor ($\alpha_5\beta_1$ integrin) to induce internalization into T24 human bladder epithelial cells (Kuroda *et al.*, 1993). BCG employs a different host component for its internalization into macrophages by this linker strategy. This host component has been identified as C type lectin SP-A (surfactant-associated protein A) of the human lung which can bind to an unidentified bacterial surface component which then interacts with its eukaryotic receptor to stimulate internalization of BCG into macrophages (Weikert *et al.*, 1997). "Linker strategies" for invasion are also found in *Neisseria* species. For *N. meningitidis* the bacterial ligand is the outer membrane protein Opc, which binds vitronectin, thereby enabling *Neisseria* to induce its uptake via interaction with the $\alpha_5\beta_3$ or $\alpha_5\beta_1$ intergrins of human umbilical vein endothelial cells (Virji *et al.*, 1994). *N. gonorrhoeae* are internalized into CHO cells only in the presence of vitronectin, which is bound to the bacteria via OpaA. The corresponding host receptor has, however, not been identified (Duensing and van Putten, 1997). In addition to utilizing "linker strategy" invasion pathways, *Neisseriae* utilize other outer membrane proteins (e.g. OpaC, OpaI, and Opa52) which can function as invasins by binding directly to eukaryotic receptors without any bridging molecules (Virji *et al.*, 1996a, b; also see the *Neisseria* chapter). In contrast, *Salmonella typhimurium* and *Staphylococcus aureus* are also able to bind to fibronectin, but this binding is not sufficient to enable uptake into several tested phagocytic cell types (Van de Water *et al.*, 1983).

Table I
Examples of Bacterial Species Inducing Microtubule-Dependent Internalization Mechanisms in Host Cells. Only those Species are Listed for which the Bacterial Invasion Proteins and Their Corresponding Eukaryotic Internalization Receptors have been Identified

Species	Uptake Pathway	Invasion Protein	Internalization Rezeptor
Citrobacter freundii	MFs and MTs	Type 1 pili	GlcNAc residues of N-glycosylated proteins
EPEC	MFs and MTs	Intimin	Tir β1 Integrins
MENEC	MFs and MTs	OmpA	GlcNAcβ1-4GlcNAc of BMEC glycoprotein(s)
UPEC	MTs	Dr Fimbriae	SCR-3 domain of DAF
Listeria monocytogenes	MFs and MTs	Internalin	E-Cadherin
Mycobacterium bovis BCG	MFs and MTs	Fibronectin bound to 55 kDa Omp	$\alpha_5\beta_1$ Integrins
Neisseria spp.	MFs and MTs	Opa30/Opa50 Opa52/Opa60 Vitronectin bound to Opc	Proteoglycan CD66 $\alpha_v\beta_3$ or $\alpha_5\beta_1$ Integrins
Porphyromonas gingivalis	MFs and MTs	Pilus	48 kDa Protein of gingival epithelial cells

Thus, some bacteria can trigger their internalization by binding certain soluble host components of the extracellular matrix, which subsequently bind to their corresponding eukaryotic cell surface receptors. However, other microorganisms may bind extracellular host factors, but do not invade the host. One potential explanation is that binding of such host factors is a necessary initial step for bacterial adherence to the host, but other specific interactions between bacteria and host cells are essential for inducing the internalization process. Alternatively, the relative affinity (K_m) of binding may be important, with high affinity required for subsequent internalization (Tran Van Nhieu and Isberg, 1993). The presence of more than one invasion system in a single bacterial species has already been described for several bacterial pathogens. It seems plausible that such systems are not just redundant, but each system may be suited for invasion of a certain cell type (i.e. tissue tropism). The results from studies with *Neisseriae*, which express several invasion systems, support this hypothesis. *N. gonorrhoeae* expressing Opa50 preferentially interact with proteoglycan molecules of and are taken up by epithelial cells, whereas Opa52 or Opa60-expressing *N. gonorrhoeae* are preferentially internalized by granulocytes (Gray-Owen *et al.*, 1997; also see chapter 4).

4. INTRACELLULAR FATE

The intracellular fate of engulfed bacteria can vary greatly. Following the uptake of low infectious doses of many pathogens, the bacteria are often typically degraded intracellularly via phagosome-lysosome fusion. However, many invasive bacteria appear able to survive intracellularly, at least if they have reached a suitable host cell type. For some bacteria, such as *M. tuberculosis*, intracellular survival might entail either multiplication or conversion to a chronic dormant state. The mechanisms utilized by pathogens for intracellular survival (e.g. inhibition of endosomal acidification or fusion with lysosomes), are just beginning to be understood. In contrast to *M. tuberculosis*, other bacteria multiply extensively in the appropriate host cell, as exemplified by Salmonella in hepatocytes or Shigella in colonic epithelial cells. For those invasive bacteria able to cross host barriers such as the epithelial lining of the gut or the blood-brain barrier, internalization is only the first step in the process of transcytosis. Certain bacteria undergo transcytosis by entering a particular cell type, passing through the cell barrier without damaging it, and exocytosing basolaterally (e.g. *Salmonella typhi* and *C. jejuni* undergo transcytosis *in vitro* which is thought to mimic the silent passage of these organisms across the gut epithelium during disease pathogenesis). Intracellular multiplication is, in general, not necessary for efficient transcytosis. To date there is little information available about the intracellular fate of bacteria following MT-dependent entry. However, *C. freundii* and *K. pneumoniae* have been isolated intracellularly from host cells one week after invasion, at the same titer as at the beginning of the experiment (Oelschlaeger, unpublished). Similarly, *Haemophilus influenzae* are able to persist in human umbilical vein endothelial cells (Virji *et al.*, 1992). During either MT- or MF-dependent mechanisms of internalization of bacteria, invagination of the host membrane results in the bacterium being surrounded by inverted host plasma membrane. Therefore, intracellular survival and multiplication abilities are likely to be inherent bacterial properties and not typically affected by MT- versus MF-dependent entry per se. Finally, even if bacteria are not multiplying intracellularly, they often are able to survive in host cells for prolonged periods of time.

Some bacteria typically remain within endosomes after invasion and are able to avoid fusion with lysosomes. This was demonstrated for *C. trachomatis* (Ooji *et al.*, 1997), *Ehrlichia risticii* (Wells and Rikihisa, 1988), *Legionella pneumophila* in human monocytes as well as for *Acanthamoeba castellanii* (Bozue and Johnson, 1996; Horwitz, 1983) and *M. tuberculosis* (Xu *et al.*, 1994; Hart *et al.*, 1987). Additionally, *M. tuberculosis* might inhibit phagosome acidification by exclusion of the vesicular proton-ATPase, as was demonstrated for *M. avium* (Strugill-Koszycki *et al.*, 1994). In contrast, other pathogens exhibit the unique ability to lyse the endosomal vacuole and multiply/survive free in the host cell cytoplasm, as has been well documented for shigellae, rickettsiae, and *L. monocytogenes*. For example, *B. bronchiseptica* internalized into Caco-2 and A549 human cells were observed free in the cytoplasm shortly after invasion (Schipper *et al.*, 1994). Similarly, *C. jejuni* in monkey colonic cells (Russel *et al.*, 1993) and *P. gingivalis* in human oral epithelial cells have been observed occasionally to be free in the cytoplasm (Madiano *et al.*, 1996; Lamont *et al.*, 1995; Sandros *et al.*, 1993). The abilities of pathogenic bacteria to invade specific cell types, to survive intracellularly and mature, to enter a chronic dormant state, to remain within an endosome and be released via exocytosis, and/or to lyse the endosome which may allow for intracellular multiplication and eventually death of the host cell, are now well documented with numerous bacteria. However, the relative importance of each of these bacterial abilities in the pathogenesis of specific disease is, at present, unclear.

5. THE *in Vivo* SITUATION

Early molecular studies of bacteria that cause histopathologically confirmed invasive diseases *in vivo* (e.g. Shigella, Salmonella, Yersinia) revealed the presence of bacterial invasion mechanisms, which could be analyzed *in vitro* in cultured cells and which have now been documented as important pathogenic determinants. It is important to note that for many bacteria, invasion ability has only been demonstrated *in vitro* (i.e. involvement of invasion in typical disease pathogenesis is unclear). Nevertheless, several bacteria exhibiting MT-dependent entry *in vitro* have been observed, by histopathologic techniques, within infected animal cells. For example, *C. jejuni* have been found within gut epithelial cells of infected monkeys (Russel *et al.*, 1993) and colonic biopsies of infected patients (Van Spreeuwel *et al.*, 1985). The obligate intracellular *Chlamydia* species were demonstrated by fluorescence, as well as electron microscopy to reside inside host cells of patients (see the chapter 18). Similarly, the *in vivo* intracellular localization of different *Rickettsia* spp. is well documented (see the chapter 19). Numerous studies have demonstrated limited cellular invasion by enteropathogenic *E. coli* in animal models (Tzipori *et al.*, 1985; Moon *et al.*, 1983; Staley *et al.*, 1969) and in human infant intestinal biopsies (Uhlsen and Rollo, 1980). The intracellular localization of *M. tuberculosis in vivo* has also been well documented (see the chapter 10). Thus, it would appear that MT-dependent mechanisms of bacterial entry into the host are relevant to disease, but the data, at present, are still limited.

6. CONCLUDING REMARKS

A large number of diverse bacterial genera have now been observed to trigger internalization, into various human cell types, that is dependent upon MTs alone or upon both MTs and MFs. Furthermore, invasion by *C. jejuni* is considered to be an essential component of disease pathogenesis, providing initial support to suggest that MT-dependent entry mechanisms are important pathogenicity determinants. Both bacterial and eukaryotic cell components which are involved in these internalization mechanisms have been identified for many of these organisms, but the linked series of biochemical events that lead to bacterial uptake remain to be determined. Does simultaneous requirement for MTs and MFs signal a complex single mechanism of entry or indicate the presence of separate MF- or MT-dependent uptake pathways within a single bacterium? How do bacteria initiate contact with host receptors (i.e. via bacterial surface bound ligands or via secreted proteins)? Which host receptors are important for MT- or MF-dependent uptake and how do they differ on host cells of different origin? How are the host receptors connected to the microtubular network and how does the host signal transduction pathway lead to cytoskeletal reorganization and bacterial entry? Are MTs involved in exocytosis?

Further molecular understanding of how these bacteria subvert normal host cell processes to trigger disease will undoubtedly be aided by current intensive investigations into the molecular cell biology of the involvement of MTs in host cell movement and intracellular vesicle trafficking.

7. REFERENCES

Amin, K., Beillevaire, D., Mahmoud, E., Hammar, L., Mardh, P.H., and Froman, G., 1995, Binding of *Galanthus nivalis* lectin to *Chlamydia trachomatis* and inhibition of *in vitro* infection, APMIS **103**:714–720.

Andreev, J., Borovsky, Z., Rosenshine, I., and Rottem, S., 1995, Invasion of HeLa cells by *Mycoplasma penetrans* and the induction of tyrosine phosphorylation of a 145kDa host cell protein, *FEMS Microbiol. Lett.* **132**:189–194.

Baorto, D.M., Gao, Z., Malaviya, R., Dustin, M.L., van der Merwe, A., Lublin, D.M., and Abraham, S.N., 1997, Survival of FimH-expressing enterobacteria in macrophages relies on glycolipid traffic, *Nature* **389**:636–639.

Bermudez, L.E., and Goodman, J., 1996, *Mycobacterium tuberculosis* invades and replicates within type II alveolar cells, *Infect. Immun.* **64**:1400–1406.

Birkelund, S., Johnsen, H., and Christiansen, G., 1994, *Chlamydia trachomatis* serovar L2 induces protein tyrosine phosphorylation during uptake by HeLa cells, *Infect. Immun.* **62**:4900–4908.

Borovsky, Z., Tarshis, M., Zhang, P., and Rottem, S., 1998, Protein kinase C activation and vacuolation in HeLa cells invaded by *Mycoplasma penetrans*, *J. Med. Microbiol.* **47**:915–922.

Bozue, J.A., and Johnson, W., 1996, Interaction of *Legionella pneumophila* with *Acanthamoeba castellanii*: uptake by coiling phagocytosis and inhibition of phagosome-lysosome fusion, *Infect. Immun.* **64**:668–673.

Buchwalow, I.B., Brich, M., and Kaufmann, S.H., 1997, Signal transduction and phagosome biogenesis in human macrophages during phagocytosis of *Mycobacterium bovis* BCG, *Acta. Histochem.* **99**:63–70.

Clausen, J.D., Christiansen, G., Holst, H.U., and Birkelund, S., 1997, *Chlamydia trachomatis* utilizes the host cell microtubule network during early events of infection, *Mol. Microbiol.* **25**:441–449.

Donnenberg, M.S., Donohue-Rolfe, A., and Keusch, G.T., 1990, A comparison of Hep-2 cell invasion by enteropathogenic and enteroinvasive *Escherichia coli*, *FEMS Microbiol. Lett.* **69**:83–86.

Duensing, T.D., and van Putten, J.P.M., 1997, Vitronectin mediates internalization of *Neisseria gonorrhoeae* by chinese hamster ovary cells, *Infect. Immun.* **65**:964–970.

Elsinghorst, E.A., 1994, Measurement of invasion by gentamicin resistance, *Methods. Enzymol.* **236**:405–420.

Foubister, V., Rosenshine, I., and Finaly, B.B., 1994, A diarrheal pathogen, enteropathogenic *Escherichia coli* (EPEC), triggers a flux of inositol phosphates in infected epithelial cells, *J. Exp. Med.* **179**:993–998.

Fumagalli, O., Tall, B.D., Schipper, C., and Oelschlaeger, T.A., 1997, N-glycosylated proteins are involved in efficient internalization of *Klebsiella pneumoniae* by cultured human epithelial cells, *Infect. Immun.* **65**:4445–4451.

Grassme, H.U., Gulbins, E., Brenner, B., Ferlinz, K., Sandhoff, K., Harzer, K., Lang, F., and Meyer, T.F., 1997, Acidic sphingomyelinase mediates entry of *N. gonorrhoeae* into non-phagocytic cells, *Cell* **91**:605–615.

Grassme, H.U., Ireland, R.M., and van Putten, J.P., Gonococcal opacity protein promotes bacterial entry-associated rearrangement of the epithelial cell actin cytoskeleton, *Infect. Immun.* **64**:1621–1630.

Gray-Owen, S.D., Dehio, C., Haude, A., Grunert, F., and Meyer, T.F., 1997, CD66 carcinoembryonic antigens mediate interaction between Opa-expressing *Neisseria gonorrhoeae* and human polymorphonuclear phagocytes, *EMBO J.* **16**:3435–3445.

Goluszko, P., Popov, V., Selvarangan, R., Novicki, S., Pham, T., and Nowicki, B.J., 1997, Dr fimbriae operon of uropathogenic *Escherichia coli* mediate microtubule-dependent invasion to the HeLa epithelial cell line, *J. Infect. Dis.* **176**:158–167.

Guzman, C.A., Rhode, M., Chakraborty, T., Domann, E., Hudel, M., Wehland, J., and Timmis, K.N., 1995, Interaction of *Listeria monocytogenes* with mouse dendritic cells, *Infect. Immun.* **63**:3665–3673.

Guzman, C.A., Rhode, M., and Timmis, K.N., 1994, Mechanisms involved in uptake of *Bordetella brònchiseptica* by mouse dentritic cells, *Infect. Immun.* **62**:5538–5544.

Hart, P.D., Young, M.R., Gordon, A.H., and Sullivan, K.H., 1987, Inhibition of phagosome-lysosome fusion in macrophages by certain mycobacteria can be explained by inhibition of lysosomal movements observed after phagocytosis, *J. Exp. Med.* **166**:933–946.

Horwitz, M.A., 1983, The legionaires' disease bacterium (*Legionella phneumophila*) inhibits phagosome-lysosome fusion in human monocytes, *J. Exp. Med.* **158**:2108–2126.

Hueck, C.J., 1998, Type III secretion systems in bacterial pathogens of animals and plants, *Microbiol. Mol. Biol. Rev.* **62**:379–433.

Isberg, R.R., and Leong, J.M., 1990, Multiple b1 chain integrins are receptors for invasin, a protein that promotes bacterial penetration into mammalian cells, *Cell* **60**:861–871.

Itzutsu, K.T., Belton, C.M., Chan, A., Fatherazi, S., Kanter, J.P., Park, Y., and Lamont, R.J., 1996, Involvement of calcium in interactions between gingival epithelial cells and *Porphyromonas gingivalis*, *FEMS Microbiol. Lett.* **144**:145–150.

Jouve, M., Garcia, M.-I., Courcoux, P., Labigne, A., Gounon, P., and Bouguenec, C.L., 1997, Adhesion to and invasion of HeLa cells by pathogenic *Escherichia coli* carrying the *afa-3* gene cluster are mediated by the AfaE and AfaD proteins, respectively, *Infect. Immun.* **65**:4082–4089.

Kenny, B., 1999, Phosphorylation of tyrosine 474 of the enteropathogenic *Escherichia coli* (EPEC) Tir receptor molecule is essential for actin nucleation activity and is preceded by additional host modifications, *Mol. Microbiol.* **31**:1229–1241.

Kenny, B., and Finlay, B.B., 1997a, Intimin-dependent binding of enteropathogenic *Escherichia coli* to host cells triggers novel signaling events, including tyrosine phosphorylation of phospholipase C-γ1, *Infect. Immun.* **65**:2528–2536.

Kenny, B., DeVinney, R., Stein, M., Reinscheid, D.J., Frey, E.A., and Finlay, B.B., 1997b, Enteropathogenic *E. coli* (EPEC) transfer its receptor for intimate adherence into mammalian cells, *Cell* **91**:511–520.

Kirsch, J., Langosch, D., Prior, P., Littauer, U.Z., Schmitt, B., and Betz, H., 1991, The 93 kDa glycine receptor-associated protein binds to tubulin, *J. Biol. Chem.* **266**:22242–22245.

Kishi, F., Yoshida, T., and Aiso, S., 1996, Location of NRAMP1 molecule on the plasma membrane and its association with microtubules, *Mol. Immunol.* **33**:1241–1246.

Kuhn, M., 1998, The microtubule depolymerizing drugs nocodazole and colchicine inhibit the uptake of *Listeria monocytogenes* by P388D1 macrophages, *FEMS Microbiol. Lett.* **160**:87–90.

Kuroda, K., Brown, E.J., Telle, W.B., Russell, D.G., and Ratliff, T.L., 1993, Characterization of the internalization of bacillus Calmette-Guerin by human bladder tumor cells, *J. Clin. Invest.* **91**:69–76.

Lamont, R.J., Chan, A., Belton, C.M., Izutzu, K.T., Vasel, D., and Weinberg, A., 1995, *Porphyromonas gingivalis* invasion of gingival epithelial cells, *Infect. Immun.* **63**:3878–3885.

Leininger, E., Ewanowich, C.A., Bhargava, A., Peppler, M.S., Kenimer, J.G., and Brennan, M.J., 1992, Comparative roles of the Arg-Gly-Asp sequence present in the *Bordetella pertussis* adhesins pertactin and filamentous hemagglutinin, *Infect. Immun.* **60**:2380–2385.

Madianos, P.N., Papapanou, P.N., Nannmark, U., Dahlen, G., and Sandros, J., 1996, *Porphyromonas gingivalis* FDC381 multiplies and persists within human oral epithelial cells *in vitro*, *Infect. Immun.* **64**:660–664.

Maruta, K., Ogawa, M., Miyamoto, H., Izu, K., and Yoshida, S.I., 1998, Entry and intracellular localization of *Legionela dumoffii* in Vero cells, *Mircob. Pathog.* **24**:65–73.

Meier, C., Oelschlaeger, T.A., Merkert, H., Korhonen, T.K., and Hacker, J., 1996, Ability of the newborn meningitis isolate *Escherichia coli* IHE3034 (O18:K1:H7) to invade epithelial and endothelial cells, *Infect. Immun.* **64**:2391–2399.

Moon, H.W., Whipp, S.C., Argenzio, R.A., Levine, M.M., and Gianella, R.A., 1983, Attaching and effacing activities of rabbit and human enteropathogenic *Escherichia coli* in pig and rabbit intestines, *Infect. Immun.* **41**:1340–1351.

Nathke, I.S., Adams, C.L., Polakis, P., Sellin, J.H., and Nelson, W.J., 1996, The adenomatous polyposis coli tumor suppressor protein localizes to plasma membrane sites involved in active cell migration, *J. Cell. Biol.* **134**:165–179.

Nowicki, B., Hart, A., Coyne, K.E., Lublin, D.M., and Nowicki, S., 1993, Short consensus repeat-3 domain of recombinant decay-accelerating factor is recognized by *Escherichia coli* recombinant Dr adhesin in a model of cell-cell interaction, *J. Exp. Med.* **178**:2115–2121.

Oelschlaeger, T.A., and Tall, B.D., 1997, Invasion of cultured human epithelial cells by *Klebsiella pneumoniae* isolated from the urinary tract, *Infect. Immun.* **65**:2950–2958.

Oelschlaeger, T.A., Barrett, T.J., and Kopecko, D.J., 1994, Some structures and processes of human epithelial cells involved in uptake of enterohemorrhagic *Escherichia coli* O157:H7 strains, *Infect. Immun.* **62**:5142–5150.

Oelschlaeger, T. A., Guerry, P., and Kopecko, D.J., 1993, Unusual microtubule-dependent endocytosis mechanisms triggered by *Campylobacter jejuni* and *Citrobacter freundii*, *Proc. Natl. Acad. Sci. USA* **90**:6884–6888.

Ooij, C., Apodaca, G., and Engel, J., 1997, Characterization of the *Chlamydia trachomatis* vacuole and its interaction with the host endocytic pathway in HeLa cells, *Infect. Immun.* **65**:758–766.

Pham, T.Q., Goluszko, P., Popov, V., Nowicki, S., and Nowicki, B.J., 1997, Molecular cloning and characterization of Dr-II, a nonfimbrial adhesin-I-like adhesin isolated from gestational pyelonephritis-associated *Escherichia coli* that binds to decay-accelerating factor, *Infect. Immun.* **65**:4309–4318.

Prasadarao, N.V., Wass, C.A., and Kim, K.S., 1996, Endothelial cell GlcNAcβ1–4GlcNAc epitopes for outer membrane protein A enhances traversal of *Escherichia coli* actross the blood brain barrier, *Infect. Immun.* **64**:154–160.

Raulston, J.E., Davis, C.H., Schmiel, D.H., Morgan, M.W., and Wyrick, P.B., 1993, Molecular characterizationand outermembrane association of a *Chlamydia trachomatis* protein related to the hsp70 family of proteins, *J. Biol. Chem.* **268**:23139–23147.

Ravindra, R., 1997, Is signal transduction modulated by an interaction between heterotrimeric G-proteins and tubulin?, *Endocrine* **7**:127–143.

Russel, R.G., O'Donnoghue, M., Blake, D.C., Zulty, J., and DeTolla, L.J., 1993, Early colonic damage and invasion of *Campylobacter jejuni* in experimentally challenged infant *Macaca mulatta*, *J. Infect. Dis.* **168**:210–215.

Rikihisa, Y., Zhang, Y., and Park, J., 1995, Role of Ca^{2+} and calmodulin in ehrlichial infection in macrophages, *Infect. Immun.* **63**:2310–2316.

Sandros, J., Papapanou, P., and Dahlen, G., 1993, *Porphyromonas gingivalis* invades oral epithelial cells *in vitro*, *J. Periodontal Res.* **28**:219–226.

Scalettsky, I.C.A., Gatti, M.S.V., da Silveira, J.F., DeLuca, I.M.S., Freymuller, E., and Travassos, L.R., 1995, Plasmid coding for drug resistance and invasion of epithelial cells in enteropathogenic *Escherichia coli* O111:H^-, *Microb. Pathog.* **18**:387–399.

Schipper, H., Krohne, G.F., and Gross, R., 1994, Epithelial cell invasion and survival of *Bordetella bronchiseptica*, *Infect. Immun.* **62**:3008–3011.

Staley, T.E., Jones, E.W., and Corley, L.D., 1969, Attachment and penetration of *Escherichia coli* into intestinal epithelium of the ileum in newborn pigs. *Am. J. Pathol.* **56**:371–392.

St Geme, III, J.W., de la Morena, M.L., and Falkow, S., 1994, A *Haemophilus influenzae* IgA protease-like protein promotes intimate interaction with human epithelial cells, *Mol. Microbiol.* **14**:217–233.

Sturgill-Koszycki, S., Schlesinger, P.H., Chakraborty, P., Haddix, P.L., Collins, H.L., Fok, A.K., Allen, R.D., Gluck, S.L., Heuser, J., and Russel, D.G., 1994, Lack of acidification in Mycobacterium phagosomes produced by exclusion of the vesicular proton-ATPase, *Science* **263**:678–681.

Subtil, A., and Dautry-Varsat, A., 1997, Microtubule depolymerization inhibits clathrin coated-pit internalization in non-adherent cell lines while interleukin 2 endocytosis is not affected; *J. Cell Sci.* **110**:2441–2447.

Swanson, A.F., and Kuo, C.-C., 1994, Binding of the glycan of the major outer membrane protein of *Chlamydia trachomatis* to HeLa cells, *Infect. Immun.* **62**:24–28.

Tang, P., Sutherland, C.L., Gold, M.R., and Finlay, B.B., 1998, *Listeria monocytogenes* invasion of epithelial cells requires the MEK-1/ERK-2 mitogen-activated protein kinase pathway, *Infect. Immun.* **66**:1106–1112.

Tran Van Nhieu, G., and Isberg, R.R., 1993, Bacterial internalization mediated by β1 chain integrins is determinend by ligand affinity and receptor density, *EMBO J.* **12**:1887–1895.

Tzipori, S., Robins-Browne, R.M., Gonis, G., Hayes, J., Withers, M., and McCartney E., 1985, Enteropathogenic *Escherichia coli* enteritis: evaluation of the gnotobiotic piglet as a model of human infection, *Gut* **26**:570–578.

Uhlsen, M.H., and Rollo, J.L., 1980, Pathogenesis of *Escherichia coli* gastroenteritis in man—another mechanism, *N. Engl. J. Med.* **302**:99–101.

Van de Water, L., Destree, A.T., and Hynes, R.O., 1983, Fibronectin binds to some bacteria but does not promote their uptake by phagocytic cells, *Science* **220**:201–204.

Van Spreeuwel, J.P., Duursma, G.C., Meijer, C.J., Bax, R., Rosekrans, P.C., and Lindeman, J., 1985, Campylobacter colitis: histological immunohistochemical and ultrastructural findings, *Gut* **26**:945–951.

Virji, M., Käythy, H., Ferguson, D.J.P., Alexandrescu, C., and Moxon, E.R., 1992, Interactions of *Hämophilus influenzae* with human endothelial cells *in vitro*, *J. Infect. Dis.* **165**(suppl 1):S115–S116.

Virji, M., Makepeace, K., and Moxon, E.R., 1994, Distinct mechanisms of interactions of Opc-expressing meningococci at apical and basolateral surfaces of human endothelial cells; the role of integrins in apical interactions, *Mol. Microbiol.* **14**:173–184.

Virji, M., Makepeace, K., Ferguson, D.J., and Watt, S.M., 1996a, Carcinoembryonic antigens (CD66) on epithelial cells and neutrophils are receptors for Opa proteins of pathogenic neisseriae, *Mol. Microbiol.* **22**:941–950.

Virji, M., Makepeace, K., Ferguson, D.J., and Watt, S.M., 1996b, The N-domain of the human CD66a adhesion molecule is a target for Opa proteins of *Neisseria meningitidis* and *Neisseria gonorrhoeae*, *Mol. Microbiol.* **22**:929–939.

Wang, H., Bedford, F.K., Brandon, N.J., Moss, S.J., and Olsen, R.W., 1999, $GABA_A$-receptor-associated protein links $GABA_A$ receptors and the cytoskeleton, *Nature* **397**:69–72.

Weinberg, A., Belton, C.M., Park, Y., and Lamont, R.J., 1997, Role of fimbriae in *Porphyromonas gingivalis* invasion of gingival epithelial cells, *Infect. Immun.* **65**:313–316.

Weikert, L.F., Edwards, K., Chroneos, Z.C., Hager, C., Hoffman, L., and Shepherd, V.L., 1997, SP-A enhances uptake of bacillus Calmette-Guerin by macrophages through a specific SP-A receptor, *Am. J. Physiol.* **272**(5Pt1):L989–L995.

Wells, M.Y., and Rikihisa, Y., 1988, Lack of lysosomal fusion with phagosomes containing *Ehrlichia risticii* in P388D1 cells: abrogation of inhibition with oxytertacycline. *Infect. Immun.* **56**:3209–3215.

Woolridge, K.G., Williams, P.H., and Ketley, J.M., 1996, Host signal transduction and endocytosis of *Campylobacter jejuni*, *Microb. Pathog.* **21**:299–305.

Xu, S., Cooper, A., Sturgill-Koszycki, S., van Heyningen, T., Chatterjee, D., Orme, I., Allen, P., and Russell, D.G., 1994, Intracellular trafficking in *Mycobacterium tuberculosis* and *Mycobacterium avium* infected macrophages, *J. Immunol.* **153**:2568–2578.

Zhang, Y., and Rikihisa, Y., 1997, Tyrosine phosphorylation is required for ehrlichial internalization and replication in P388D1 cells, *Infect. Immun.* **65**:2959–2964.

Chapter 2

Interaction of Enteropathogenic *Escherichia coli* with Host Cells

Ilan Rosenshine*, Stuart Knutton, and Gad Frankel

1. INTRODUCTION

Most *E. coli* strains are non-pathogenic and constitute part of the normal gut flora. However, other *E. coli* strains may be pathogenic and can cause either bladder infection, meningitis, or diarrhea. At least five different classes of diarrheagenic *E. coli* have thus far been identified, including, enterotoxigenic *E. coli* (ETEC), enterohemorragic *E. coli* (EHEC), enteroaggregative *E. coli* (EAEC), enteroinvasive *E. coli* (EIEC), and enteropathogenic *E. coli* (EPEC). These *E. coli* strains, causing symptoms ranging from cholera-like to extreme colitis, possess distinct sets of virulence factors, including specific adhesins, invasins and/or toxins, which are responsible for the characteristic symptoms and diarrhea associated with each class.

EPEC refers to certain serotypes of *E. coli* that were first incriminated in epidemiological studies in the 1940s and 1950s as causes of epidemic and

---

ILAN ROSENSHINE Department of Molecular Genetics and Biotechnology, The Hebrew University, Faculty of Medicine, The Hebrew University, Jerusalem 9112, Israel. **STUART KNUTTON** Institute of Child Health, University of Birmingham, Birmingham B4 6NH, United Kingdom. **GAD FRANKEL** Department of Biochemistry, Imperial College of Science, Technology and Medicine, University of London, London SW7 2AZ, United Kingdom. * Corresponding author.

Subcellular Biochemistry, Volume 33: Bacterial Invasion into Eukaryotic Cells, edited by Oelschlaeger and Hacker. Kluwer Academic / Plenum Publishers, New York, 2000.

sporadic infantile diarrhea. On the other hand, EHEC which are closely related to EPEC, were first identified as a human pathogen in 1982 in the USA. Whilst EHEC is regarded as an emerging zoonotic pathogen which can cause acute gastro-enteritis and hemorrhagic colitis, and produce severe/fatal renal and neurological complications as a result of the translocation of Shiga-like toxins (Stx 1 and Stx 2) across the gut, EPEC is an established etiological agent of human diarrhea, and remains an important cause of infant mortality in developing countries (Nataro and Kaper, 1998).

Traditionally, EPEC are considered to be comprised of strains belonging to twelve different O serogroups: O26, O55, O86, O111, O114, O119, O125, O126, O127, O128, O142 and O158 whilst the most prominent serotype among EHEC is O157:H7. However, epidemiological studies have recently shown that isolation of other EHEC serogroups (e.g. O26, O103, O111) is on the increase, and may be more prevalent than O157:H7 in some countries. Population genetic surveys have shown that EPEC strains can be divided into two major groups of related clones, designated EPEC clone 1 and EPEC clone 2. Within each group, a variety of O antigens are present while the somatic flagellar (H) antigens are conserved; EPEC clone 1 belong to flagellar antigen H6 whereas EPEC clone 2 belong to flagellar antigen H2. EHEC strains are similarly divided into two divergent clonal groups. EHEC clone 1 includes the O157:H7 serotype (and the atypical EPEC O55:H7) whilst EHEC clone 2 comprises other Stx-producing serogroups (including O26 and O111). (Whittam and McGraw, 1996).

Infections with EPEC and EHEC induce an "attaching and effacing" (A/E) histopathology on gut enterocytes characterized by localized destruction of brush border microvilli, intimate bacterial adhesion and gross cytoskeletal reorganization (Donnenberg *et al.*, 1997). A/E lesion formation is essential for EPEC pathogenicity and a similar histopathology has been associated with several other bacterial mucosal pathogens, including the human pathogen *Hafnia alvei*, and several animal pathogens including RDEC-1 (a rabbit specific EPEC), and *Citrobacter rodentium* a specific murine pathogen. Importantly, all the A/E lesion forming pathogens harbor a unique pathogenicity island, the locus of enterocyte effacement (LEE) which is necessary and sufficient for A/E lesion formation on epithelial cells *in vitro* (McDaniel and Kaper, 1997; McDaniel *et al.*, 1995).

2. VIRULENCE GENES OF EPEC

Several putative and confirmed virulence factors have been identified in EPEC. Most of these genes are grouped either in the LEE pathogenicity island, or in two gene clusters located on the EAF virulence plasmid; the bundle forming pilus (BFP) locus and the *perABC/bfpTVW* gene cluster that encodes for a transcriptional activator. Whilst the LEE region appears to be common to all the A/E lesions forming pathogens, the EAF plasmid encoded virulence factors are found only amongst EPEC strains.

The LEE of EPEC E2348/69, and of other EPEC and EHEC strains belonging to their respective clone 1, is inserted at minute 82 of the *E. coli* K-12 chromosome, adjacent to the *selC* locus (McDaniel *et al.*, 1995) which is also the site of insertion of a pathogenicity island in uropathogenic *E. coli*. The insertion site of the LEE in strains belonging to EPEC and EHEC clone 2 was recently shown to be at minute 94 of the K-12 chromosome at the *pheU* locus, and there is at least a third, as yet unidentified, insertion site for the LEE (Kaper *et al.*, 1998). These results indicate that the LEE has inserted at multiple times during the evolution of the EPEC/EHEC family of pathogens and subsequently acquired additional virulence factors encoded on bacteriophage and on large ca. 90 kb plasmids found in EPEC and EHEC.

2.1. The LEE Pathogenicity Island

The entire 35.6 Kb LEE region of EPEC E2348/69 (O127:H6) has been cloned on a plasmid which, when present in *E. coli* K-12, is sufficient to produce the A/E phenotype (McDaniel and Kaper, 1997). The G + C content of the LEE (38.3%) is significantly lower than that of the *E. coli* chromosome (50.8%) suggesting that the LEE arose by horizontal gene transfer of this pathogenicity island from another species. The LEE from EPEC E2348/69 contains 41 open reading frames of greater than 50 amino acids (Elliott *et al.*, 1998). These genes are organized into three major regions with known functions. The middle region contains the *eae* and *tir* genes, the products of which are involved in intimate adherence, as discussed below. Upstream of *eae* and *tir* are genes encoding components of a type III secretion system (Jarvis *et al.*, 1995). The third major region of the LEE, located downstream of *eae* and *tir*, encodes proteins that are secreted via the type III secretion system and that are needed for protein translocation. The most prominent of these proteins are EspA, EspB, EspD (Lai *et al.*, 1997; Kenny *et al.*, 1996).

2.2. The LEE Encoded Type III Secretion/Translocation System

Several animal and plant pathogen employ a type III protein secretion system to secrete key virulence factors, some of which are injected (translocated) directly from the pathogens cytoplasm into the host cell cytosol (Hueck, 1998). The type III secretion pathway is independent of the *sec* system and secretion requires the involvement of a large number of accessory proteins (Hueck, 1998). So far, type III secretion systems have been identified in a number of pathogenic bacteria including Salmonella, Shigella, Yersinia, EHEC/EPEC, Chlamydia, *Pseudomonas aeruginosa* and more recently in *Bordetella* species (Hueck, 1998; Yuk *et al.*, 1998; Hsia *et al.*, 1997). Type III secretion systems are also essential for the virulence of several plant pathogens including *Pseudomonas solanacearum* and *P. syringae*, and *Xanthomonas capestris* (Alfano and Collmer, 1997).

EPEC use a type III secretion system to secrete several LEE-encoded proteins including Tir, EspA, EspB, and EspD (Deibel *et al.*, 1998; Lai *et al.*, 1997; Kenny *et al.*, 1996; Kenny and Finlay, 1995). Secretion of these proteins is essential for EPEC virulence. Based on experimental data and homology with other type III secretion systems 12 LEE genes including *escR,S,T,U,C,J,V,N,D,F,* and *sepQ,Z* appear to encode for components of a type III secretion system (Elliott *et al.*, 1998). The *esc* genes are homologous to *ysc* genes encoding type III secretion system components in Yersinia and carry the same suffix as in the homologous gene in Yersinia (Elliott *et al.*, 1998). Other genes involved in the EPEC type III secretion system that are not *ysc* homologues were termed *sep* genes. Six of the *esc/sep* genes including *escU,C,VN* and *sepQ,Z* were confirmed experimentally to be required for protein secretion by EPEC (Knutton *et al.*, 1998; Wolff *et al.*, 1998; Lai *et al.*, 1997; Kenny *et al.*, 1996; Jarvis *et al.*, 1995; Kenny and Finlay, 1995). Another protein termed CesD is specifically required for secretion of EspD and EspD (Wainwright and Kaper, 1998).

Some of the genes involved in forming the type III secretion apparatus are homologues to proteins involved in assembly of the flagella basal body (Hueck, 1998). Electron microscopical studies of the Salmonella type III secretion system indicate that it has a basal-body-like morphology with two outer membrane rings, two inner membrane rings and a connecting cylindrical structure (Kubori *et al.*, 1998). EscC may form one of the outer membrane rings in EPEC's type III secretion system. EscC is homologous to YscC and PulD, that forms a ring-shaped oligomeric complex in the outer bacterial membrane with an ~20 nm diameter central pore (Koster *et al.*, 1997). Other members of PulD-like proteins are involved in morphogenesis of filamentous bacteriophages, biogenesis of type IV pili, and type II and type III protein secretion (Koster *et al.*, 1997; Pugsley *et al.*, 1997). Thus EscC may also be a channel-forming protein in the EPEC outer membrane. EscN appears to be an ATPase and the ATPase activity of its homologue in the Yersinia and Salmonella type III secretion systems is needed for secretion (Hueck, 1998). Not much is known yet about the function of the other components of type III secretion apparatuses including the EPEC's type III system.

2.3. The Translocation Apparatus

In addition to secretion, the *esc* and *sep* genes appear to be also required for delivery of proteins by EPEC into the host cytosol and cell membrane. Two proteins have thus far been shown to be translocated during EPEC infection of epithelial cells in culture: EspB and Tir (Wolff *et al.*, 1998; Kenny *et al.*, 1997). The *escV,N,* and *U* genes have been shown to be required for translocation of Tir and EspB (Wolff *et al.*, 1998; Kenny *et al.*, 1997). However, protein secretion by the *esc/sep* system is not sufficient for protein translocation. Mutants in *espA*, *espB* and *espD* although capable of secretion are defective in protein translocation (Wolff *et al.*, 1998).

Our current model proposes that at least three components are needed to bring about protein translocation: (i) an energized basal-body-like structure made out of the Esc/Sep protein (ii) a channel in the host cell membrane made out of EspB and perhaps other proteins, and (iii) a hollow filamentous structure connecting the bacterial basal-body-like structure and the host cell membrane. EspA and EspD appear to be involved in forming this filament.

The EspA polypeptide is a major component of a filamentous organelle which interacts with the host cell during the early stage of A/E lesion formation (Knutton *et al.*, 1998). In addition to being required for EspB and Tir translocation the EspA-filaments, because they interact with host cells, may also contribute to bacterial adhesion. Each ~50 nm diameter and up to 2 μm long filament is composed of a number of smaller ~7 nm diameter fimbrial-like structures (Knutton *et al.*, 1998). Thus EspA filaments appear to have a hollow cylindrical structure that could form a channel through which proteins are delivered into the host cell. Moreover it is possible that the EspA-filaments are connected to the bacterial surface via the putative EscC outer membrane ring. EspD, whose function is currently unknown, may be another component of the filament because an EspD mutant was found to secrete only low levels of EspA and produced barely detectable filaments (Knutton *et al.*, 1998). EspA, in common with many other proteins secreted by Type III systems, is predicted to contain a coiled-coil domain (Pallen *et al.*, 1997). This suggests that assembly of the EspA filament might involve coiled-coil interaction between helices of different sub-units. EspA homologues exist in all A/E bacteria examined and EspA amino acid sequences from different clinical EPEC isolates appear highly conserved within the two EPEC clones (Neves *et al.*, submitted).

Another protein that is needed for translocation but not for secretion is EspB. EspB is unlikely to be a component of the filamentous structure because an antiserum made to this EspB did not stain EspA-filaments. Moreover EspB mutants still produce normal EspA-filaments (Knutton *et al.*, 1998). Immediately following bacterial attachment, EspB is translocated into the host cell where it is distributed in both membrane and cytosol fractions; translocation of EspB was not dependent on but was strongly enhanced by intimate bacterial attachment (Wolff *et al.*, 1998). An EspB mutant was able to secrete EspA, EspD, Tir and an EspB-adenylate cyclase (CyaA) fusion protein but was unable to translocate Tir and EspB-CyaA suggesting that a functional EspB is required both for its own translocation and for the translocation of Tir and perhaps other yet to be identified effector proteins (Wolff *et al.*, 1998). Since EspB is translocated to the host cell membrane and has weak structural similarity to YopB of Yersinia (a protein involved in forming the translocation-channel in the host cell membrane) (Hakansson *et al.*, 1996) it is possible that EspB serves a similar function in EPEC and that it is involved in directing the translocated proteins from the EspA filament and through/into the host cell membrane. The cytoplasmic localization of a sub population of EspB may indicate additional functions.

2.4. The LEE-Encoded Intimin and the Translocated Intimin Receptor (Tir)

The product of the LEE *eae* gene, intimin, is an outer membrane protein adhesin homologous to invasins, proteins which promote eukaryotic cell invasion by Yersinia (Jerse *et al.*, 1990). Like in invasin, the cell binding activity of the intimin family of proteins is localized to the C-terminal 280 amino acids (Int280) (Frankel *et al.*, 1995). Within this domain lies a 76-amino acid loop formed by a disulphide bridge between two cysteines at positions 862 and 937, that is required for intimin-mediated intimate attachment and invasion into cultured mammalian cells (Kelly *et al.*, in press). Introducing small in-frame mutations at the C-terminus of EPEC intimin could dramatically reduce intimin-mediated cell invasion without detectably effecting A/E lesion formation. In particular, deletion of the last amino acid (Lys 939) from the intimin C-terminus segregated intimin-mediated A/E lesion formation from intimin-mediated HEp-2 cell invasion (Frankel *et al.*, 1998).

Studies using serological and molecular approaches found the presence of at least five distinct intimin subtypes: intimin α, intimin β, intimin γ, intimin δ and an nontypeable intimin expressed by EPEC O127:H40 (Adu-Bobie *et al.*, 1998; Agin and Wolf, 1997). Importantly, intimin α was specifically expressed by strains which belong to the EPEC clone 1, and intimin β was mainly associated with EPEC and EHEC strains belonging to their respective clone 2. Intimin γ was associated with EHEC O157:H7, EPEC O55:H7 and O55:H-, while intimin δ was expressed only by EPEC O86:H34.

The 78–80 kDa Tir polypeptide was shown to be secreted by the type III secretion system and translocated into the host cell where they are localized to the plasma membrane (Deibel *et al.*, 1998; Kenny *et al.*, 1997). It is believed that Tir consists of at least three functional regions, an extracellular domain that interacts with intimin, a transmembrane domain and a cytoplasmic domain that can interact with the host cell cytoskeleton (Kenny *et al.*, 1997). From this new data the essence of EPEC infection now appears to be interaction between the carboxy-terminal region of intimin and translocated Tir which in turn results in intimate EPEC attachment, actin accretion and A/E lesion formation.

Tir becomes tyrosine phosphorylated immediately upon translocation, causing the translocated polypeptide to appear as a 90 kDa protein on SDS-PAGE (hence Tir was initially referred to as Hp90, and thought to be a host cell-derived associated intimin receptor) (Rosenshine *et al.*, 1992). The physiological significance of Tir phosphorylation is at present not clear for several reasons. Intimin can bind the unphosphorylated form of Tir and Tir from O157:H7 EHEC does not appear to become tyrosine phosphorylated in the host cell (Deibel *et al.*, 1998; Kenny *et al.*, 1997). Experiments with tyrosine protein kinase inhibitors did not inhibit A/E lesion formation (Rabinowitz *et al.*, 1996). Furthermore, several reports have shown that purified intimin can bind mammalian cells in the absence of Tir which suggests there may be an additional intimin receptor of cellular origin; *in vitro* studies have shown that intimin, like Yersinia invasin, has the ability to bind to β1 integrins (Frankel *et al.*, 1996). However, the role of intimin-integrin interactions in colonization and A/E lesion formation has yet to be determined.

2.5. The BFP Gene Cluster

BFP is a very flexible type IV pilus that forms bundles and mediates bacterial aggregation. It is encoded by a cluster of 14 genes that encompasses 11.5 kb of the EAF virulence EPEC plasmid (Sohel *et al.*, 1996; Stone *et al.*, 1996). The function of some of these genes has been determined experimentally. The major pilus structural subunit, bundlin is encoded by *bfpA* (Donnenberg *et al.*, 1992). The BfpA preprotein is processed by a prepilin peptidase encoded by the *bfpP* gene (Zhang *et al.*, 1994). Several genes including *bfpD* a nucleotide binding protein and *bfpL* are needed for the biogenesis of BFP (Bieber *et al.*, 1998; Donnenberg *et al.*, 1997). Not much is known about the biogenesis of type IV pili including BFP but it appears to be a complex process and in some aspects resembles type II protein secretion and DNA transfer systems (Donnenberg *et al.*, 1997). In addition to these plasmid-encoded genes, formation of a functional BFP structure requires the presence of the chromosomal *dsbA* gene which encodes a periplasmic oxido-reductase necessary for disulphide bond formation (Donnenberg *et al.*, 1997).

BFP expression occurs in tissue culture media at the mid-logarithmic growth phase at 37°C and expression is associated with bacterial auto-aggregation. The bacterial aggregates dissociate back to single cells after entering the stationary growth phase or in response to a drop in temperature (Puente *et al.*, 1996). The product of *bfpF*, a nucleotide binding protein, is needed for the dispersal of the auto-aggregates and for full virulence in volunteers (Bieber *et al.*, 1998). Interestingly, *bfpF* is homologous to *pilT* that has been implicated in switching motility of *P. aeruginosa* in infected tissues.

2.6. The *perABC/bfpTVW* Gene Cluster

In addition to the BFP gene cluster the EAF plasmid contains the *perABC* operon (also referred to as *bfpTVW*) (Tobe *et al.*, 1996; Gomez-Duarte and Kaper, 1995). These genes act as a transcriptional activator of the BFP genes and several other genes located within the LEE region including *espB* and *eae* (Tobe *et al.*, 1996; Gomez-Duarte and Kaper, 1995). The PerA protein belongs to the AraC family of regulators but all three ORFs are required for efficient transcriptional activation (Gomez-Duarte and Kaper, 1995). Per may serves as a global regulator affecting transcription of both chromosomal and plasmid-encoded genes, thereby allowing EPEC to respond to different environmental conditions and different phases of growth.

2.7. Other Putative Virulence Genes

Several other putative EPEC virulence genes have been described including *espC* and *bipA/typA*.

BipA/TypA: BipA (termed also TypA) is a GTP binding protein found in all *E. coli* strains including non-pathogenic strains (Farris *et al.*, 1998; Freestone *et al.*, 1998). In EPEC it appears to be tyrosine phosphorylated, which enhances its GTPase activity (Farris *et al.*, 1998). BipA appears to interact with a global regulatory network. Accordingly, inactivation of *bipA* alters the pattern of protein synthesis (Freestone *et al.*, 1998), and causes pleothrophic effects including inability to form A/E lesions, increased sensitivity to bactericidic peptides and enhanced motility (Farris *et al.*, 1998).

EspC: EspC is a chromosomally encoded secreted protein. EspC is not part of the LEE region and its secretion is not dependent upon the type III secretion system (Stein *et al.*, 1996). The primary sequence of EspC is highly related to an autotransporter protein family found in different gram-negative pathogens, which includes the Pet enterotoxin of EAEC (Eslava *et al.*, 1998), the Tsh protein of an avian-pathogenic *E. coli* strain (Provence and Curtiss, 1994), the SepA protein from *Shigella flexneri* (Benjelloun-Touimi *et al.*, 1995), the plasmid-encoded EspP/PssA protein from EHEC (Brunder *et al.*, 1997; Djafari *et al.*, 1997), and IgA1 proteases from *Neisseria* spp. and *Haemophilus influenzae* (Lomholt *et al.*, 1995; Poulsen *et al.*, 1992). EspE/PssA has been shown to be a serine protease and PssA to be cytotoxic for Vero cells (Brunder *et al.*, 1997; Djafari *et al.*, 1997). Moreover, a specific immune response against EspP was detected in sera from patients suffering from EHEC infections (Brunder *et al.*, 1997). On this basis it seems that the activity of EspC may be of functional importance during infection of the mucosal cell layer by the bacterial pathogen.

3. THE COLONIZATION PROCESS

3.1. Regulation of Expression of EPEC Virulence Genes

Expression of LEE and EAF genes is a prerequisite for successful colonization by EPEC. The expression of these genes is regulated by several environmental cues including temperature, growth media and growth phase.

Thermoregulation: Temperature is a typical environmental signature of the host and several pathogens, including EPEC, use a temperature shift to 37°C as the signal to activate virulence mechanisms. EPEC induced A/E lesion formation is dependent on growth at 37°C (Rosenshine *et al.*, 1996). In accordance with this, *espB* and *eae* and other LEE genes are expressed at 37°C but not at 27°C (Kenny and Finlay, 1995; Knutton *et al.*, 1997). Expression of the BFP is also thermoregulated (Puente *et al.*, 1996).

Growth phase regulation: EPEC induces rapid formation of A/E lesions in mid-logarithmic growth phase, but becomes inactive at late-logarithmic growth phase and upon entering the early stationary phase (Rosenshine *et al.*, 1996). BFP expression is also growth phase regulated being expressed best at mid-logarithmic growth phase (Puente *et al.*, 1996).

Medium composition: In addition to dependency on temperature and growth phase, optimal expression of LEE and BFP genes is dependent upon the composition of the culturing medium, including calcium, ammonium ions, and pH (Kenny *et al.*, 1997; Ebel *et al.*, 1996; Puente *et al.*, 1996; Rosenshine *et al.*, 1996).

Thus, it appears that environmental cues dictate concerted expression of LEE and BFP genes. The *perABC* genes may play a role in responding to these environmental cues although the molecular mechanism(s) behind the growth phase regulation of BFP and LEE genes is still not known.

3.2. Initial Attachment and Activation of the Type III Secretion System

EPEC grown in Dulbecco's modified Eagle's medium (DMEM) at 37°C to mid-logarithmic growth phase were termed "activated". Activated EPEC express mature BFP, type III secretion systems and EspA-filaments (Knutton *et al.*, 1998; Knutton *et al.*, 1997; Rosenshine *et al.*, 1996; Vuopio-Varkila and Schoolnik, 1991). However the secretion system exhibits only basal levels of secretory activity. Contact of activated EPEC with the host cell activates the pre-formed secretion system to secrete and translocate proteins into the host cell (Wolff *et al.*, 1998). So far the translocation of EspB and Tir has been demonstrated but it is expected that these will not be the only translocated proteins. The contact activation of the type III secretion system, EspB translocation, and formation of A/E lesions is inhibited by chloramphenicol (Rosenshine *et al.*, 1996; and unpublished results), indicating that in EPEC activation of the type III secretion system is dependent on *de novo* protein synthesis upon its contact with the host cell. It is not yet clear whether this dependency on *de novo* protein synthesis involves transcriptional activation of specific gene(s) upon contact with the host cell, or whether it involves the replacement of a very labile protein. Contact activation of type III secretion systems has also been reported in Salmonella, Shigella, and Yersinia (Hueck, 1998).

In the tissue culture cell model of EPEC infection, BFP appears to play an important role in activating the type III system by mediating rapid initial bacterial contact with the host cells. However, recent evidence using *in vitro* organ culture of paediatric small intestine, a model which more closely resembles the *in vivo* situation, indicates that adhesin(s) other than BFP initiate colonization of the mucosal surface and that BFP serves later to allow formation of three dimensional bacterial micro-colonies via inter-bacterial interactions (Hicks *et al.*, 1998). It is possible that the initial bacteria-host cell contact is mediated by EspA-filaments that transverse the mucous layer covering the enterocytes thereby allowing transfer of Tir to the host cell membrane. This may be followed by intimin-Tir interaction and intimate attachment. Alternatively initial attachment may be mediated by an as yet to be identified adhesive factor.

3.3. Amplification of the Colonization Process

Once a few individual EPEC have formed intimate attachment the colonization process is amplified by two synergistic mechanisms. The first mechanism involves strong enhancement of the translocation efficiency upon establishment of intimate attachment. This results in an excess of translocated Tir in the vicinity of the intimately attached individual bacteria providing receptors for recruitment of additional intimin expressing bacteria and for the formation of two dimensional micro-colonies. A second mechanism for colonization amplification involves using BFP to recruit more EPEC to the site of colonization via intra-bacterial interaction and formation of three dimensional micro-colonies (Hicks *et al.*, 1998). It is likely that colonization will be inefficient in the absence of intimin and BFP and accordingly the virulence of EPEC mutants deficient in *eae* or *bfpA* is highly attenuated (Bieber *et al.*, 1998; Donnenberg *et al.*, 1993).

3.4. Invasion of the Host Epithelial Cells

Invasion of enterocytes by colonizing EPEC was first described by Staley *et al.* (1969), and this observation was later confirmed in *in vitro* EPEC infections using several different culture cell lines. Only about 1% of attached EPEC invade infected epithelial cells (Rosenshine *et al.*, 1992). However, since EPEC attachment is very efficient their invasion rate is high and comparable to the level of host cell invasion by Salmonella or invasin-mediated invasion of Yersinia strains. Intracellular EPEC are usually seen in a network of large vacuoles whilst retaining intimate contact with the surrounding membrane (Andrade, 1996). In the intracellular site, EPEC do not grow or grow only at a very slow rate. EPEC can penetrate monolayers of polarized epithelial cell lines by invading through the apical surface and exiting at the basolateral surface (Canil *et al.*, 1993). Penetration of the epithelial monolayer was also described in *in vivo* experimental infections (Staley *et al.*, 1969). The invasion process can be inhibited by agents that disrupt several processes in host cells including inhibitors of protein tyrosine kinases, inhibitors of small GTP binding proteins, and by agents that disrupt the host cell actin cytoskeleton (Ben-Ami *et al.*, 1998; Donnenberg *et al.*, 1997; Rosenshine *et al.*, 1992).

EPEC invasion is dependent on concerted activity of BFP, intimin and the type III secretion system. Mutants that do not produce BFP or intimin are 10-fold less invasive than the wild type strain and mutants defective in the type III secretion system are more than 100-fold less invasive than the wild type strain. Most of the genes that are needed for the formation of A/E lesions are also required for invasiveness. Formation of A/E lesions without invasion was observed in host cells treated with inhibitors of small GTP binding proteins (Ben-Ami *et al.*, 1998), in cells infected with EPEC *sepZ* mutants (Rabinowitz *et al.*, 1996) and with EPEC that express a mutated intimin (Frankel *et al.*, 1998). However, invasion without formation of A/E lesions has not been observed. This may indicate that invasion and formation of A/E lesions represent different outcomes of EPEC-induced rearrangements of the host cell cytoskeleton. The contribution of invasion to EPEC pathogenicity is not yet clear although it is possible that intracellular bacteria maintain a reservoir that enables chronic EPEC infections.

3.5. Disaggregation and Spreading

Mutants in *bfpF* produce normal BFP but aggregate better than the wild type EPEC (Anantha *et al.*, 1998; Bieber *et al.*, 1998). In addition they are deficient in disaggregation upon entering the stationery phase (Bieber *et al.*, 1998). Thus spreading of previously aggregated bacteria can not take place in these mutants. Interestingly, a *bfpF* mutant exhibited attenuated virulence in an experimental infection of human volunteers (Bieber *et al.*, 1998). These results may indicate that virulence is also dependent on disaggregation of the microcolonies and expansion of the colonized area after initial colonization, a process that may involve lateral spreading from the initial disaggregated infection foci. An additional mechanism that may aid in lateral spreading is the actin-based cell surface motility of EPEC. As discussed later in this chapter EPEC induce formation of actin pedestals that propel attached bacteria along the host cell surface (Sanger *et al.*, 1996). It is possible that EPEC use this mechanism to move from cell to cell along the brush border thereby minimizing the need for detachment when moving from one cell to another. In conclusion, it appears that initial attachment that enables initiation of protein translocation is the rate limiting step of the colonization process. Once bacteria have achieved a foothold they use several mechanisms to amplify the colonization process and to expand the colonized area.

3.6. Down Regulation of Virulance Factors

Intimin and EspA are expressed by all EPEC at early stages of infection of tissue cultured cells. However, at late stages of infection, following A/E lesion formation, intimin and EspA expression was greatly reduced (Knutton *et al.*, 1997; 1998). Genetic analysis indicate that PerABC play a role in this down regulation process (Knutton *et al.*, 1997). Intimin and EspA have been shown to be highly immunogenic and both serum and colostral antibodies to these proteins have been detected (Loureiro *et al.*, 1998; Jarvis *et al.*, 1995). Accordingly, if the observed down regulation of intimin expression following A/E lesion formation occurs *in vivo*, it could be an important EPEC regulatory mechanism for overcoming host immune responses.

4. CELL BIOLOGY OF EPEC INFECTED CELL

4.1. The Actin Pedestal

One of the most striking effects of EPEC on the host cell is the induced formation of actin pedestals just beneath intimately attached bacteria (Knutton *et al.*, 1989). These structures are formed immediately upon establishment of intimate attachment and are composed of actin filaments and several actin-binding proteins. The typical morphology of these actin structures evolves within a few hours from flat, cup-like structures into a short actin pedestal and then into elongated pedestals on which the EPEC rest (Rosenshine *et al.*, 1996; Sanger *et al.*, 1996). These pedestals can grow longer or shorter whilst remaining tethered in place on the cell surface (Sanger *et al.*, 1996). Alternatively, these actin stalks propel attached extracellular EPEC along the cell surface, reaching a speed of up to 0,07 μm/sec (Sanger *et al.*, 1996). The base of the pedestals is rich in myosin and tropomyosin (Sanger *et al.*, 1996; Manjarrez-Hernandez *et al.*, 1992), while actin filaments, villin, and α-actinin are distributed uniformly along it (Sanger *et al.*, 1996; Finlay *et al.*, 1992). The tip of this structure is rich in tyrosine phosphorylated Tir (Rosenshine *et al.*, 1996). High concentrations of ezrin, plastin, and talin within the pedestals have also been reported (Adam *et al.*, 1995; Finlay *et al.*, 1992), but their distribution along the stalk was not examined.

4.2. Formation of the Actin Pedestal

EPEC must induce several processes in order to form the actin pedestals, including localized actin polymerization, bundling of the newly polymerized actin filaments, and organization of the polymerized bundles beneath attached bacteria. Genetic analysis indicates that actin polymerization and bundling is dependent upon the type III secretion system and the EspA/EspB translocation system (Donnenberg *et al.*, 1997). Thus it is expected that actin polymerization is mediated by a translocated effector protein. The signal(s) and the mechanisms responsible for the actin rearrangement are currently unknown although Rac, Rho and Cdc42-

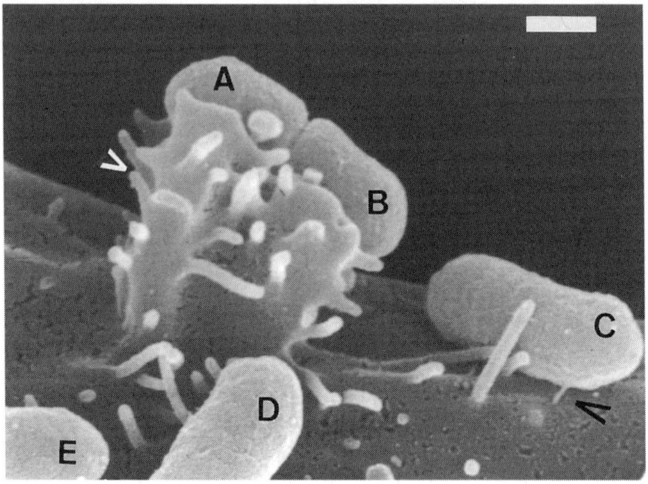

FIGURE 1. Scanning electron microscopy of HeLa cells infected with EPEC. Bacteria A and B are in advanced stages of infection and rested on top of pedestals (white arrowhead). C, D and E are in initial stages of infection and C is connected to the host cell by thin filament which may represent the EspA-filaments that mediate protein translocation into the host cells (black arrowhead). The bar represent 0.4 μm.

dependent pathways do not appear to be involved (Ben-Ami *et al.*, 1998; Ebel *et al.*, 1998).

Mutants in *eae* or *tir* still induce a disorganized "shadow" of polymerized actin indicating that these proteins are not directly involved in induction of actin polymerization. However, Tir and intimin are needed to organize the pedestal beneath attached bacteria since mutants in either *eae* or *tir* fail to focus the actin structure beneath the bacterium (Rosenshine *et al.*, 1996; Rosenshine *et al.*, 1992; Jerse *et al.*, 1990). The intracellular domain of Tir may be associated with the actin bundles and therefore the association of the extracellular domain of Tir with intimin provides a physical link between the extracellular EPEC and the actin bundles. However direct or indirect interaction of actin with Tir has yet to be shown.

4.3. Signal Transduction

Stimulation of several signal transduction pathways in the host cell appear to be associated with EPEC infection including phosphorylation and dephosphorylation of several proteins, release of second messengers, activation of cytokine expression and secretion, and changes in the epithelial permeability.

4.3.1. Protein Phosphorylation and Dephosphorylation

Activation of protein kinase C (PKC): Transient PKC activation was demonstrated by several methods including translocation to the plasma membrane and enzymatic activity (Crane and Oh, 1997). However, activation was weak and transient, peaking at 30 min post infection. Both intimin and the EAF plasmid were required for PKC activation.

Phosphorylation of myosin light chain (MLC): Following infection, MLC is phosphorylated at two sites by protein kinase C (PKC) and possibly by the calmodulin dependent MLC kinase (MLCK) (Yuhan *et al.*, 1997; Manjarrez-Hernandez *et al.*, 1996). MLC phosphorylated by MLCK becomes associated with the cell cytoskeleton (Manjarrez-Hernandez *et al.*, 1996). Moreover by indirect fluorescence microscopy myosin co-localized with actin in the EPEC induced pedestals (Sanger *et al.*, 1996; Manjarrez-Hernandez *et al.*, 1992).

Other phosphrylation events: Whilst assaying for PKC activity Crane and Oh (1997) detected increases in other kinase activities in infected cells although the kinase(s) involved were not identified. Other groups have also reported an increase in the phosphorylation state of several small and medium size proteins upon EPEC infection (Yuhan *et al.*, 1997; Manjarrez-Hernandez *et al.*, 1991).

Tyrosine phosphorylation of phospholipase C-γ1 (PLC-γ1): Tyrosine phosphorylation of PLC-γ1 was detected in cells infected with wild type EPEC but not in cells infected with mutants in *espA*, *espB*, *eae*, or in *esc/sep* genes (Kenny and Finlay, 1997). Tyrosine phosphorylation is known to activate PLC-γ1 to generate two second messengers: phosphatidylinositol (IP) metabolites, and diacylglycerol (DAG).

Tyrosine phosphorylation of Tir: Immediately following translocation to the host cell Tir becomes tyrosine phosphorylated (Kenny and Finlay, 1997; Rosenshine *et al.*, 1996). The significance of this event is not clear yet since phosphorylation does not appear to be required for interaction with intimin (Kenny and Finlay, 1997).

Other tyrosine phosphorylation and dephosphorylation events: Tyrosine phosphorylation of a group of proteins about 150 kDa in size and tyrosine dephosphorylation of a protein about 200 kDa in size has been detected in EPEC infected cells (Kenny and Finlay, 1997). In all of these cases phosphorylation or dephosphorylation was dependent on the type III secretion system and the EspA/EspB translocon.

4.3.2. Release of Second Messengers

Increases in the level of intracellular Ca^{2+} and IPs has been reported in EPEC infected cells (Baldwin *et al.*, 1991). IP release appeared to be dependent on protein translocation since IP fluxes were not induced by an *espB* mutant (Foubister *et al.*, 1994). In a more recent study increased Ca^{2+} levels were not detected at early stages of infection nor during the formation of A/E lesions (Bain *et al.*, 1998), suggesting that increased Ca^{2+} levels and possibly also the IP flux, represent late events perhaps related to general intoxication of the host cells.

4.3.3. Activation of NFκB

Activation of the transcriptional factor NFκB in the host cell was associated with infection with wild type EPEC but not with an *espB* mutant (Savkovic *et al.*, 1997). Thus this event is probably dependent on protein translocation by EPEC. The activation of NFκB is associated with increased IL-8 production and transmigration of polymorphonuclear cells (PMN) (Savkovic *et al.*, 1996).

4.3.4. Changes in the Epithelia Permeability

Two classes of changes in the permeability of EPEC infected epithelia have been reported; opening of membrane channels and an increase in parcellular permeability.

Opening of membrane channels: Two types of experiments indicate opening of channels in the plasma membrane of EPEC infected cells. EPEC induce a rapid transient increase in short circuit current (Isc) across monolyers of polarized Caco2 intestinal cells (Collington *et al.*, 1998). The Isc increase appears to involve a flow of Cl^-, Na^+ and other electrolytes through EPEC induced channels in the host cell plasma membrane. Moreover, rapid decrease in the resting membrane potential (RMP) of infected cells detected by direct patch clamping of individual infected cells also indicates that EPEC can alter the relative ion distribution across the membrane of infected cells via opening of channels in the plasma membrane of the host cells (Stein *et al.*, 1996). Genetic analysis indicates that the increases in Isc and decrease in RMP were dependent on EspB and/or an intact type III secretion system (Collington *et al.*, 1998; Stein *et al.*, 1996).

Increase in the parcellular permeability: A gradual but steady decrease in the electric resistance across monolayers of polarized epithelial cells (trans epithelial electric resistance: TEER) has been detected upon EPEC infection (Collington *et al.*, 1998; Spitz *et al.*, 1995; Canil *et al.*, 1993). Dual $^{22}Na^+$ and [^{3}H]mannitol flux studies indicate that this event involves an increase in paracellular permeability due to disruption of the tight junctions (Spitz *et al.*, 1995). However, EPEC infected monolyers still maintain partial TEER indicating only subtle disruption of tight junction structure. Moreover, no gross alterations in the distribution of actin and ZO1 protein that are localized along the tight junction to form a honey-comb shape were observed (Canil *et al.*, 1993). Genetic analysis indicates that EspB and/or the type III secretion system are needed to induce a decrease in TEER (Collington *et al.*, 1998; Canil *et al.*, 1993).

4.3.5. Signaling Pathways

Work in many laboratory using different approaches including inhibitor studies indicate that one of the EPEC-induced signaling pathways involves tyrosine phosphorylation and thus activation of PLC-γ1. It is not yet clear how EPEC induce the PLC-γ1 phosphorylation. However, activated PLC-γ1 generates fluxes of IP_3 and DAG. IP_3 may induce release of Ca^{2+} from intracellular stores that together with DAG activate PKC. PKC in turn may phosphorylate several proteins. The increase in Ca^{2+} concentration may lead also to activation of the calmodulin dependent MLCK and phophorylation of MLC. Phophorylated MLC becomes associated with actin and may induce actin bundling and contraction. Contraction of the actin perijunctional belt was implicated in causing disruption of the tight junction and the decrease of TEER upon EPEC infection (Yuhan *et al.*, 1997). However, this pathway seems not to be directly involved in the assembly of the actin pedestal but a late event that develops slowly. On the other hand, activation of MLC my contribute to the bundling of actin fibers in the pedestals. Clearly more work and careful kinetic studies are needed to establish the relationships between the above different molecules and signalling events.

A second pathway may involve activation of NFκB to induce IL-8 production which in turn activates PMN migration across the epithelial monolyer and secretion of products that may have a very complex secondary effect on infected epithelial cells. Clearly little is yet known about the complex interaction between different host cells following EPEC infection.

A common theme running through all of the host cell responses to EPEC infection is their dependence on the type III secretion system and on the EspA/EspB translocon. We predict that in all cases the signalling pathways are triggered by translocated effector(s) proteins or by the translocation process. For example, like YopB in the Yersinia translocon, EspB may be a pore-forming protein and play a direct role in disturbing the Isc or RMP. EspB can also be found in the cytoplasm of infected cells (Wolff *et al.*, 1998), and therefore may also play a direct role in triggering signalling events. Alternatively the only function of EspB may be restricted to aid translocation of yet to be identified effector protein(s) that are the direct or indirect cause of disruption in epithelial permeability, actin polymerization, and other signalling events.

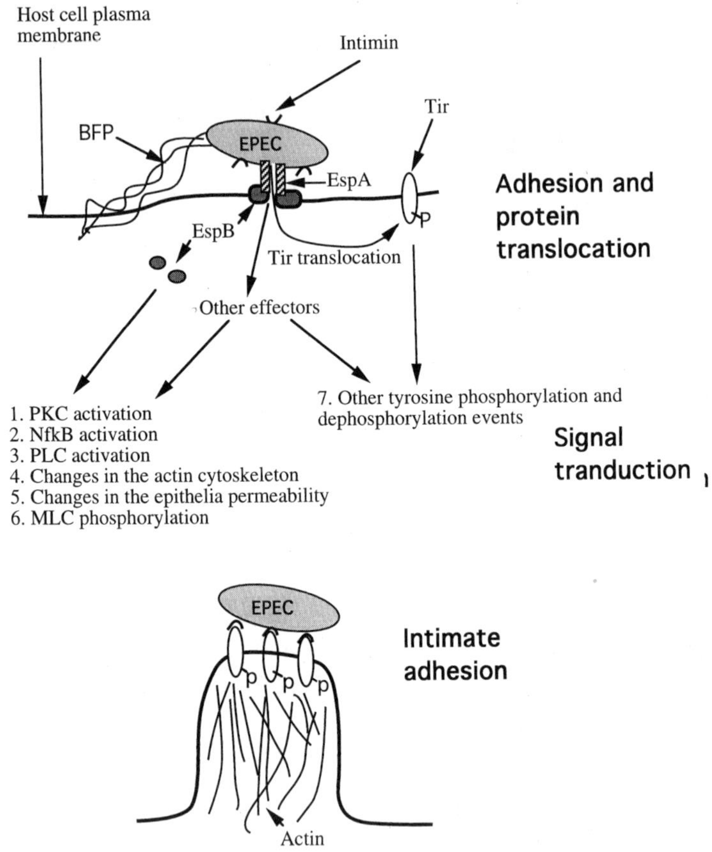

FIGURE 2. Model of EPEC/EHEC protein translocation and A/E lesion formation. The proposed translocation apparatus (translocon) consists of pores in the bacterial envelope (EscC generated pore) and in the host cell membrane (EspB generated pore) with the pores connected by a hollow EspA filament thereby providing a continuous channel from the bacterial to the host cell cytosol. Energy for protein translocation is thought to be provided by the EscN protein. The translocon is used to translocate Tir(EspE) into the host cell where it becomes inserted into the host cell membrane; EPEC Tir, but not O157:H7 Tir, becomes phosphorylated on tyrosine residues following translocation. Translocated Tir and/or other yet to be identified effector proteins transduce signals which induce breakdown of the brush border microvillous actin cytoskeleton with consequent vesiculation of the microvillous membrane. Although the mechanisms responsible are not known, localized translocation of effector proteins result in localized cytoskeletal changes. Intimate adhesion and pedestal formation results from the interaction of intimin and Tir and the accumulation of actin (and other cytoskeletal proteins) beneath intimately attached bacteria following microvillous effacement.

5. CONCLUDING REMARKS

An emerging theme in the pathogenesis of bacterial infections is subversion by bacterial pathogens of host cell functions including signal-transduction pathways and cytoskeletal organization. A particularly good example of this is shown by EPEC and EHEC which have the ability to induce a characteristic A/E lesion on gut enterocytes after translocating effector proteins across the host cell plasma membrane. Horizontal acquirement of the LEE, of virulence plasmids and of the temperate phage that carries the *stx* genes, played a crucial role in the evolution of EPEC and EHEC. Interestingly, a high proportion (~2%) of the EHEC O157 clinical isolates are genetically unstable due to deficiencies in either *mutS, mutH* or *mutL* (LeClerc *et al.*, 1996). This may aid in horizontal gene transfer and generate a high mutation rate (Modrich and Lahue, 1996), that may represent the two driving forces in EPEC/EHEC evolution.

Like several other gram-negative pathogens of animals and plants, EPEC relies for its exploitation of host cell machinery on a complex specialized type III secretion system. Recently, there has been a burst of new data that has revolutionized some basic concepts of the molecular basis of bacterial pathogenesis in general (including identification of Tir and the EspA filaments) and EPEC pathogenesis in particular. Major breakthroughs and developments in the genetic basis of A/E lesion formation, signal transduction, protein translocation, host cell receptors, and colonization have been highlighted in this review. However, there is still much work to be performed before we have a complete understanding of EPEC/EHEC pathogenesis or have identified all the gene products that contribute to colonization of the host and virulence and we can expect more breakthroughs in this important area in the near future.

6. REFERENCES

Adam, T., Arpin, M., Prevost, M.C., Gounon, P., and Sansonetti, P.J., 1995, Cytoskeletal rearrangements and the functional role of T-plastin during entry of *Shigella flexneri* into HeLa cells, *J. Cell Biol.* **129**:367–381.

Adu-Bobie, J., Frankel, G., Bain, C., Goncalves, A.G., Trabulsi, L.R., Douce, G., Knutton, S., and Dougan, G., 1998, Detection of intimins alpha, beta, gamma, and delta, four intimin derivatives expressed by attaching and effacing microbial pathogens, *J. Clin. Microbiol.* **36**:662–668.

Agin, T.S., and Wolf, M.K., 1997, Identification of a family of intimins common to *Escherichia coli* causing attaching-effacing lesions in rabbits, humans, and swine, *Infect. Immun.* **65**:320–326.

Alfano, J.R., and Collmer, A., 1997, The type III (Hrp) secretion pathway of plant pathogenic bacteria: trafficking harpins, Avr proteins, and death, *J. Bacteriol.* **179**:5655–5662.

Anantha, R.P., Stone, K.D., and Donnenberg, M.S., 1998, Role of BfpF, a member of the PilT family of putative nucleotide-binding proteins, in type IV pilus biogenesis and in interactions between enteropathogenic *Escherichia coli* and host cells, *Infect. Immun.* **66**:122–131.

Andrade J.R., 1996, Invasion by EPEC. *Rev. Micobiol.* **27**:63–66

Bain, C., Keller, R., Collington, G.K., Trabulsi, L.R., and Knutton, S. 1998, Increased levels of intracellular calcium are not required for the formation of attaching and effacing lesions by enteropathogenic and enterohemorrhagic *Escherichia coli*, *Infect. Immun.* **66**:3900–3908.

Baldwin, T.J., Ward, W., Aitken, A., Knutton, S., and Williams, P.H., 1991, Elevation of intracellular free calcium levels in HEp-2 cells infected with enteropathogenic *Escherichia coli*, *Infect. Immun.* **59**:1599–1604.

Benjelloun-Touimi, Z., Sansonetti, P.J., and Parsot, C., 1995, SepA, the major extracellular protein of *Shigella flexneri*: autonomous secretion and involvement in tissue invasion, *Mol. Microbiol.* **17**:123–135.

Ben-Ami, G., Ozeri, V., Hanski, E., Hofmann, F., Aktories, K., Hahn, K.M., Bokoch, G.M., and Rosenshine, I., 1998, Agents that inhibit Rho, Rac, and Cdc42 do not block formation of actin pedestals in HeLa cells infected with enteropathogenic *Escherichia coli*, *Infect. Immun.* **66**:1755–1758.

Bieber, D., Ramer, S.W., Wu, C.Y., Murray, W.J., Tobe, T., Fernandez, R., and Schoolnik, G.K., 1998, Type IV pili, transient bacterial aggregates, and virulence ofenteropathogenic *Escherichia coli*, *Science* **280**:2114–2118.

Brunder, W., Schmidt, H., and Karch, H., 1997, EspP, a novel extracellular serine protease of enterohaemorrhagic *Escherichia coli* O157:H7 cleaves human coagulation factor V, *Mol. Microbiol.* **24**:767–778.

Canil, C., Rosenshine, I., Ruschkowski, S., Donnenberg, M.S., Kaper, J.B., and Finlay, B.B., 1993, Enteropathogenic *Escherichia coli* decreases the transepithelial electrical resistance of polarized epithelial monolayers, *Infect. Immun.* **61**:2755–2762.

Collington, G.K., Booth, I.W., and Knutton, S., 1998, Rapid modulation of electrolyte transport in Caco-2 cell monolayers by enteropathogenic *Escherichia coli* (EPEC) infection, *Gut* **42**:200–207.

Crane, J.K., and Oh, J.S., 1997, Activation of host cell protein kinase C by enteropathogenic *Escherichia coli*, *Infect. Immun.* **65**:3277–3285.

Deibel, C., Kramer, S., Chakraborty, T., and Ebel, F., 1998, EspE, a novel secreted protein of attaching and effacing bacteria, is directly translocated into infected host cells, where it appears as a tyrosine-phosphorylated 90 kDa protein, *Mol. Microbiol.* **28**:463–474.

Djafari, S., Ebel, F., Deibel, C., Kramer, S., Hudel, M., and Chakraborty, T., 1997, Characterization of an exported protease from Shiga toxin-producing *Escherichia coli*, *Mol. Microbiol.* **25**:771–784.

Donnenberg, M.S., Giron, J.A., Nataro, J.P., and Kaper, J.B., 1992, A plasmid-encoded type IV fimbrial gene of enteropathogenic *Escherichia coli* associated with localized adherence, *Mol. Microbiol.* **6**:3427–3437.

Donnenberg, M.S., Kaper, J.B., and Finlay, B.B., 1997, Interactions between enteropathogenic *Escherichia coli* and host epithelial cells, *Trends Microbiol.* **5**:109–114.

Donnenberg, M.S., Tacket, C.O., James, S.P., Losonsky, G., Nataro, J.P., Wasserman, S.S., Kaper, J.B., and Levine, M.M., 1993, Role of the *eaeA* gene in experimental enteropathogenic *Escherichia coli* infection [see comments], *J. Clin. Invest.* **92**:1412–1417.

Donnenberg, M.S., Zhang, H.Z., and Stone, K.D., 1997, Biogenesis of the bundle-forming pilus of enteropathogenic *Escherichia coli*: reconstitution of fimbriae in recombinant *E. coli* and role of DsbA in pilin stability — a review, *Gene* **192**:33–38.

Ebel, F., Deibel, C., Kresse, A.U., Guzman, C.A., and Chakraborty, T., 1996, Temperature- and medium-dependent secretion of proteins by Shiga toxin-producing *Escherichia coli*, *Infect. Immun.* **64**:4472–4479.

Ebel, F., von Eichel-Streiber, C., Rohde, M., and Chakraborty, T., 1998, Small GTP-binding proteins of the Rho- and Ras-subfamilies are not involved in the actin rearrangements induced by attaching and effacing *Escherichia coli*, *FEMS Microbiol. Lett.* **163**:107–112.

Elliott, S.J., Wainwright, L.A., McDaniel, T.K., Jarvis, K.G., Deng, Y.K., Lai, L.C., McNamara, B.P., Donnenberg, M.S., and Kaper, J.B., 1998, The complete sequence of the locus of enterocyte effacement (LEE) from enteropathogenic *Escherichia coli*, *Mol. Microbiol.* **28**: 1–4.

Eslava, C., Navarro-Garcia, F., Czeczulin, J.R., Henderson, I.R., Cravioto, A., and Nataro, J.P., 1998, Pet, an autotransporter enterotoxin from enteroaggregative *Escherichia coli*, *Infect. Immun.* **66**:3155–3163.

Farris, M., Grant, A., Richardson, T.B., and O'Connor, C.D., 1998, BipA: a tyrosine-phosphorylated GTPase that mediates interactions between enteropathogenic *Escherichia coli* (EPEC) and epithelial cells, *Mol. Microbiol.* **28**:265–279.

Finlay, B.B., Rosenshine, I., Donnenberg, M.S., and Kaper, J.B., 1992, Cytoskeletal composition of attaching and effacing lesions associated with enteropathogenic *Escherichia coli* adherence to HeLa cells, *Infect. Immun.* **60**:2541–2543.

Foubister, V., Rosenshine, I., and Finlay, B.B., 1994, A diarrheal pathogen, enteropathogenic *Escherichia coli* (EPEC), triggers a flux of inositol phosphates in infected epithelial cells, *J. Exp. Med.* **179**:993–998.

Frankel, G., Candy, D.C., Fabiani, E., Adu-Bobie, J., Gil, S., Novakova, M., Phillips, A.D., and Dougan, G., 1995, Molecular characterization of a carboxy-terminal eukaryotic-cell-binding domain of intimin from enteropathogenic *Escherichia coli*, *Infect. Immun.* **63**:4323–4328.

Frankel, G., Lider, O., Hershkoviz, R., Mould, A.P., Kachalsky, S.G., Candy, D., Cahalon, L., Humphries, M.J., and Dougan, G., 1996, The cell-binding domain of intimin from enteropathogenic *Escherichia coli* binds to beta1 integrins, *J. Biol. Chem.* **271**:20359–20364.

Frankel, G., Philips, A.D., Novakova, M., Batchelor, M., Hicks, S., and Dougan G., 1998, Generation of *Escherichia coli* intimin-derivatives with differing biological activities using site-directed mutagenesis of the intimin C-terminus domain, *Mol. Microbiol.* **29**:559–570.

Freestone, P., Trinei, M., Clarke, S.C., Nystrom, T., and Norris, V., 1998, Tyrosine phosphorylation in *Escherichia coli*, *J. Mol. Biol.* **279**:1045–1051.

Gomez-Duarte, O.G., and Kaper, J.B., 1995, A plasmid-encoded regulatory region activates chromosomal *eaeA* expression in enteropathogenic *Escherichia coli*, *Infect. Immun.* **63**:1767–1776.

Hakansson, S., Schesser, K., Persson, C., Galyov, E.E., Rosqvist, R., Homble, F., and Wolf-Watz, H., 1996, The YopB protein of *Yersinia pseudotuberculosis* is essential for the translocation of Yop effector proteins across the target cell plasma membrane and displays a contact-dependent membrane disrupting activity, *EMBO J.* **15**:5812–5823.

Hicks, S., Frankel, G., Kaper, J.B., Dougan, G., and Phillips, A.D., 1998, Role of intimin and bundle-forming pili in enteropathogenic *Escherichia coli* adhesion to pediatric intestinal tissue *in vitro*, *Infect. Immun.* **66**:1570–1578.

Hsia, R.C., Pannekoek, Y., Ingerowski, E., and Bavoil, P.M., 1997, Type III secretion genes identify a putative virulence locus of Chlamydia, *Mol. Microbiol.* **25**:351–359.

Hueck, C.J., 1998, Type III protein secretion systems in bacterial pathogens of animals and plants, *Microbiol. Mol. Biol. Rev.* **62**:379–433.

Jarvis, K.G., Giron, J.A., Jerse, A.E., McDaniel, T.K., Donnenberg, M.S., and Kaper, J.B., 1995, Enteropathogenic *Escherichia coli* contains a putative type III secretion system necessary for the export of proteins involved in attaching and effacing lesion formation, *Proc. Natl. Acad. Sci. USA* **92**:7996–8000.

Jerse, A.E., Yu, J., Tall, B.D., and Kaper, J.B., 1990, A genetic locus of enteropathogenic *Escherichia coli* necessary for the production of attaching and effacing lesions on tissue culture cells, *Proc. Natl. Acad. Sci. USA* **87**:7839–7843.

Kaper, J.B., Elliot, S., V., S., Perna, T.P., F., M.G., and Blatner, F.R., 1998, Attaching-and-effacing intestinal histopathology and locus of enterocyte effacement, in: *Escherichia coli and other Shiga toxin-producing E. coli stains*, (J.B. Kaper and A.D. O'Brien, eds.), ASM press Washington DC, pp. 163–182.

Kelly, G., Prasannan, S., Daniel, S., Frankel, G., Dougan, G., Connerton, I., and Mathews, S., 1998, Sequential assignment of the triple labelled 30.1 kDa cell adhesion domain of intimin from enteropathogenic *E. coli*, *J. Biomolecular NMR*. In press.

Kenny, B., Abe, A., Stein, M., and Finlay, B.B., 1997, Enteropathogenic *Escherichia coli* protein secretion is induced in response to conditions similar to those in the gastrointestinal tract, *Infect. Immun.* **65**:2606–2612.

Kenny, B., DeVinney, R., Stein, M., Reinscheid, D.J., Frey, E.A., and Finlay, B.B., 1997, Enteropathogenic *E. coli* (EPEC) transfers its receptor for intimate adherence into mammalian cells, *Cell* **91**:511–520.

Kenny, B., and Finlay, B.B., 1995, Protein secretion by enteropathogenic *Escherichia coli* is essential for transducing signals to epithelial cells, *Proc. Natl. Acad. Sci. USA* **92**:7991–7995.

Kenny, B., and Finlay, B.B., 1997, Intimin-dependent binding of enteropathogenic *Escherichia coli* to host cells triggers novel signaling events, including tyrosine phosphorylation of phospholipase C-gamma1, *Infect. Immun.* **65**:2528–2536.

Kenny, B., Lai, L.C., Finlay, B.B., and Donnenberg, M.S., 1996, EspA, a protein secreted by enteropathogenic *Escherichia coli*, is required to induce signals in epithelial cells, *Mol. Microbiol.* **20**:313–323.

Knutton, S., Adu-Bobie, J., Bain, C., Phillips, A.D., Dougan, G., and Frankel, G., 1997, Down regulation of intimin expression during attaching and effacing enteropathogenic *Escherichia coli* adhesion, *Infect. Immun.* **65**:1644–1652.

Knutton, S., Baldwin, T., Williams, P.H., and McNeish, A.S., 1989, Actin accumulation at sites of bacterial adhesion to tissue culture cells: basis of a new diagnostic test for enteropathogenic and enterohemorrhagic *Escherichia coli*, *Infect. Immun.* **57**:1290–1298.

Knutton, S., Rosenshine, I., Pallen, M.J., Nisan, I., Neves, B.C., Bain, C., Wolff, C., Dougan, G., and Frankel, G., 1998, A novel EspA-associated surface organelle of enteropathogenic *Escherichia coli* involved in protein translocation into epithelial cells, *EMBO J.* **17**:2166–2176.

Koster, M., Bitter, W., de Cock, H., Allaoui, A., Cornelis, G.R., and Tommassen, J., 1997, The outer membrane component, YscC, of the Yop secretion machinery of *Yersinia enterocolitica* forms a ring-shaped multimeric complex, *Mol. Microbiol.* **26**:789–797.

Kubori, T., Matsushima, Y., Nakamura, D., Uralil, J., Lara-Tejero, M., Sukhan, A., Galan, J.E., and Aizawa, S.I., 1998, Supramolecular structure of the *Salmonella typhimurium* type III protein secretion system, *Science* **280**:602–605.

Lai, L.C., Wainwright, L.A., Stone, K.D., and Donnenberg, M.S., 1997, A third secreted protein that is encoded by the enteropathogenic *Escherichia coli* pathogenicity island is required for transduction of signals and for attaching and effacing activities in host cells, *Infect. Immun.* **65**:2211–2217.

LeClerc, J.E., Li, B., Payne, W.L., and Cebula, T.A., 1996, High mutation frequencies among *Escherichia coli* and Salmonella pathogens [see comments], *Science* **274**:1208–1211.

Lomholt, H., Poulsen, K., and Kilian, M., 1995, Comparative characterization of the *iga* gene encoding IgA1 protease in *Neisseria meningitidis, Neisseria gonorrhoeae* and *Haemophilus influenzae*, *Mol. Microbiol.* **15**:495–506.

Loureiro, I., Frankel, G., Adu-Bobie, J., Dougan, G., Trabulsi, L.R., and Carneiro-Sampaio, M.M.S., 1998, Human colostrum contains IgA antibodies reactive to enteropathogenic *Escherichia coli*-virulence-associated proteins: intimin, BfpA; EspA and EspB. *J. Pediatr. Gastorenterol. Nutr.* **27**:166–171.

Manjarrez-Hernandez, H.A., Amess, B., Sellers, L., Baldwin, T.J., Knutton, S., Williams, P.H., and Aitken, A., 1991, Purification of a 20 kDa phosphoprotein from epithelial cells and identification as a myosin light chain. Phosphorylation induced by enteropathogenic *Escherichia coli* and phorbol ester, *FEBS Lett.* **292**:121–127.

Manjarrez-Hernandez, H.A., Baldwin, T.J., Aitken, A., Knutton, S., and Williams, P.H., 1992, Intestinal epithelial cell protein phosphorylation in enteropathogenic *Escherichia coli* diarrhoea, *Lancet* **339**:521–523.

Manjarrez-Hernandez, H.A., Baldwin, T.J., Williams, P.H., Haigh, R., Knutton, S., and Aitken, A., 1996, Phosphorylation of myosin light chain at distinct sites and its association with the cytoskeleton during enteropathogenic *Escherichia coli* infection, *Infect. Immun.* **64**:2368–2370.

McDaniel, K.T., and Kaper, J.B., 1997, A cloned pathogenicity island from enteropathogenic *Escherichia coli* confers the attaching and effacing phenotype on *E. coli* K-12, *Mol. Microbiol.* **23**:399–407.

McDaniel, T.K., Jarvis, K.G., Donnenberg, M.S., and Kaper, J.B., 1995, A genetic locus of enterocyte effacement conserved among diverse enterobacterial pathogens, *Proc. Natl. Acad. Sci. USA* **92**:1664–1668.

Modrich, P., and Lahue, R., 1996, Mismatch repair in replication fidelity, genetic recombination, and cancer biology, *Annu. Rev. Biochem.* **65**:101–133.

Nataro, J., and Kaper, J.B., 1998, Diarrheagenic *Escherichia coli*, *Clin. Microbiol. Rev.* **11**:142–210.

Neves, B.C., Knutton, S., Trabulsi, L.R., Sperandio, V., Kaper, J.K., Dougan, G., and Frankel, G., Molecular and Ultrastructural Characterisation of EspA From Different Enteropathogenic *Escherichia coli* Serotypes. *Infect. Immun.* Submitted.

Pallen, M.J., Dougan, G., and Frankel, G., 1997, Coiled-coil domains in proteins secreted by type III secretion systems [letter], *Mol. Microbiol.* **25**:423–425.

Poulsen, K., Reinholdt, J., and Kilian, M., 1992, A comparative genetic study of serologically distinct *Haemophilus influenzae* type 1 immunoglobulin A1 proteases, *J Bacteriol.* **174**:2913–2921.

Provence, D.L., and Curtiss, R., 1994, Isolation and characterization of a gene involved in hemagglutination by an avian pathogenic *Escherichia coli* strain, *Infect. Immun.* **62**:1369–1380.

Puente, J.L., Bieber, D., Ramer, S.W., Murray, W., and Schoolnik, G.K., 1996, The bundle-forming pili of enteropathogenic *Escherichia coli*: transcriptional regulation by environmental signals, *Mol. Microbiol.* **20**:87–100.

Pugsley, A.P., Francetic, O., Hardie, K., Possot, O.M., Sauvonnet, N., and Seydel, A., 1997, Pullulanase: model protein substrate for the general secretory pathway of gram-negative bacteria, *Folia Microbiol.* **42**:184–192.

Rabinowitz, R.P., Lai, L.C., Jarvis, K., McDaniel, T.K., Kaper, J.B., Stone, K.D., and Donnenberg, M.S., 1996, Attaching and effacing of host cells by enteropathogenic *Escherichia coli* in the absence of detectable tyrosine kinase mediated signal transduction, *Microb. Pathog.* **21**:157–171.

Rosenshine, I., Donnenberg, M.S., Kaper, J.B., and Finlay, B.B., 1992, Signal transduction between enteropathogenic *Escherichia coli* (EPEC) and epithelial cells: EPEC induces tyrosine phosphorylation of host cell proteins to initiate cytoskeletal rearrangement and bacterial uptake, *EMBO J.* **11**:3551–3560.

Rosenshine, I., Ruschkowski, S., and Finlay, B.B., 1996, Expression of attaching/effacing activity by enteropathogenic *Escherichia coli* depends on growth phase, temperature, and protein synthesis upon contact with epithelial cells, *Infect. Immun.* **64**:966–973.

Rosenshine, I., Ruschkowski, S., Stein, M., Reinscheid, D.J., Mills, S.D., and Finlay, B.B., 1996, A pathogenic bacterium triggers epithelial signals to form a functional bacterial receptor that mediates actin pseudopod formation, *EMBO J.* **15**:2613–2624.

Sanger, J.M., Chang, R., Ashton, F., Kaper, J.B., and Sanger, J.W., 1996, Novel form of actin-based motility transports bacteria on the surfaces of infected cells, *Cell Motil. Cytoskeleton* **34**:279–287.

Savkovic, S.D., Koutsouris, A., and Hecht, G., 1996, Attachment of a noninvasive enteric pathogen, enteropathogenic *Escherichia coli*, to cultured human intestinal epithelial monolayers induces transmigration of neutrophils, *Infect. Immun.* **64**:4480–4487.

Savkovic, S.D., Koutsouris, A., and Hecht, G., 1997, Activation of NF-kappaB in intestinal epithelial cells by enteropathogenic *Escherichia coli*, *Am. J. Physiol.* **273**:C1160–C1167.

Sohel, I., Puente, J.L., Ramer, S.W., Bieber, D., Wu, C.Y., and Schoolnik, G.K., 1996, Enteropathogenic *Escherichia coli*: identification of a gene cluster coding for bundle-forming pilus morphogenesis, *J. Bacteriol.* **178**:2613–2628.

Spitz, J., Yuhan, R., Koutsouris, A., Blatt, C., Alverdy, J., and Hecht, G., 1995, Enteropathogenic *Escherichia coli* adherence to intestinal epithelial monolayers diminishes barrier function, *Am. J. Physiol.* **268**:G374–G379.

Staley, T.E., Jones, E.W., and Corley, L.D., 1969, Attachment and penetration of *Escherichia coli* into intestinal epithelium of the ileum in newborn pigs. *Am. J. Pathol.* **56**:371–392.

Stein, M., Kenny, B., Stein, M.A., and Finlay, B.B., 1996, Characterization of EspC, a 110-kilodalton protein secreted by enteropathogenic *Escherichia coli* which is homologous to members of the immunoglobulin A protease-like family of secreted proteins, *J. Bacteriol.* **178**:6546–6554.

Stein, M.A., Mathers, D.A., Yan, H., Baimbridge, K.G., and Finlay, B.B., 1996, Enteropathogenic *Escherichia coli* markedly decreases the resting membrane potential of Caco-2 and HeLa human epithelial cells, *Infect. Immun.* **64**:4820–4825.

Stone, K.D., Zhang, H.Z., Carlson, L.K., and Donnenberg, M.S., 1996, A cluster of fourteen genes from enteropathogenic *Escherichia coli* is sufficient for the biogenesis of a type IV pilus, *Mol. Microbiol.* **20**:325–337.

Tobe, T., Schoolnik, G.K., Sohel, I., Bustamante, V.H., and Puente, J.L., 1996, Cloning and characterization of *bfpTVW*, genes required for the transcriptional activation of *bfpA* in enteropathogenic *Escherichia coli* [see comments], *Mol. Microbiol.* **21**:963–975.

Vuopio-Varkila, J., and Schoolnik, G.K., 1991, Localized adherence by enteropathogenic *Escherichia coli* is an inducible phenotype associated with the expression of new outer membrane proteins, *J. Exp. Med.* **174**:1167–1177.

Wainwright, L.A., and Kaper, J.B., 1998, EspB and EspD require a specific chaperone for proper secretion from enteropathogenic *Escherichia coli*, *Mol. Microbiol.* **27**: 1247–1260.

Whittam, T.S., and McGraw, E.A., 1996, Clonal analysis of EPEC serogroups. *Rev. Microbiol. Sao Paulo* **27**(Suppl l):7–16.

Wolff, C., Nisan, I., Hanski, E., Frankel, G., and Rosenshine, I., 1998, Protein translocation into host epithelial cells by infecting enteropathogenic *Escherichia coli*, *Mol. Microbiol.* **28**:143–155.

Yuhan, R., Koutsouris, A., Savkovic, S.D., and Hecht, G., 1997, Enteropathogenic *Escherichia coli*-induced myosin light chain phosphorylation alters intestinal epithelial permeability, *Gastroenterology* **113**:1873–1882.

Yuk, M.H., Harvill, E.T., and Miller, J.F., 1998, The BvgAS virulence control system regulates type III secretion in *Bordetella bronchiseptica*, *Mol. Microbiol.* **28**:945–959.

Zhang, H.Z., Lory, S., and Donnenberg, M.S., 1994, A plasmid-encoded prepilin peptidase gene from enteropathogenic *Escherichia coli*, *J. Bacteriol.* **176**:6885–6891.

Chapter 3

E. coli Invasion of Brain Microvascular Endothelial Cells as a Pathogenetic Basis of Meningitis

Kwang Sik Kim

1. INTRODUCTION

Bacterial meningitis still results in a high mortality and morbidity despite advances in antimicrobial chemotherapy and supportive care (Durand *et al.*, 1993; Unhanand *et al.*, 1993). Both clinical and experimental data indicate limited efficacy with antimicrobial chemotherapy alone (Kim, 1985; McCracken *et al.*, 1984). A major contributing factor is the incomplete understanding of the pathogenesis and pathophysiology associated with the bacterial meningitis. For example, most cases of bacterial meningitis develop as a result of hematogenous spread, but it is not clear how circulating bacteria cross the blood-brain barrier. We have utilized *E.coli* as a paradigm to examine how circulating bacteria traverse the blood-brain barrier. In addition, *E. coli* is the most common gram-negative bacterium that causes meningitis, particularly during the neonatal period. Our inves-

KWANG SIK KIM Professor of Pediatrics/Molecular Microbiology and Immunology, USC School of Medicine and Head, Division of Infectious Diseases, Childrens Hospital Los Angeles, Los Angeles, CA 90027

Subcellular Biochemistry, Volume 33: Bacterial Invasion into Eukaryotic Cells, edited by Oelschlaeger and Hacker. Kluwer Academic / Plenum Publishers, New York, 2000.

tigations have become feasible with the availability of both *in vitro* and *in vivo* models of the blood-brain barrier.

2. MODELS OF BLOOD-BRAIN BARRIER

We have developed *in vitro* models of the blood-brain barrier by isolation and cultivation of brain microvascular endothelial cells (BMEC) from cows, rats and humans (Stins *et al.*, 1997, 1994). Bovine BMEC was initially isolated and cultivated because of easy availability of bovine brains. Since our *in vivo* model of the blood-brain barrier utilizes rats, we also isolated and cultivated rat BMEC. Our *in vitro* studies using BMEC would be more relevant to human pathogenesis if the *in vitro* model used human tissues. We, therefore, obtained small fragments of human cerebral cortex from surgical resections and human BMEC was isolated as described previously (Stins *et al.*, 1997). Briefly, brain specimens devoid of large blood vessels were homogenized in DMEM containing 2% bovine calf serum (DMEM-S) and centrifuged in 15% dextran in DMEM-S for 10 min at 1000 × g. The pellets containing crude microvessels were further digested in a solution containing collagenase/dispase (1 mg/ml) for 1 hr at 37°C. Microvascular capillaries were isolated by absorption to a column of glass bead and washing off the beads, and recovered in growth medium. Cell viability was greater than 95% as judged by trypan blue exclusion test. The human brain microvessels were plated on collagen coated dishes or glass coverslips and cultured in RPMI 1640 based growth medium at 37°C in a humid atmosphere of 5% CO_2. The resulting human BMEC were positive for factor VIII, carbonic anhydrase IV, *Ulex europaeus* Agglutinin I, and took up acetylated low density lipoprotein (AcLDL), demonstrating their endothelial origin. Human BMEC also expressed gamma-glutamyl transpeptidase, indicating their brain origin. Human BMEC were purified by FACS using fluorescently labeled DiL-AcLDL and found to be > 99% pure after studying non-endothelial cell types using markers for astrocyte (glial fibrillary acidic protein), oligoglia (galactocerebroside-C), pericyte (smooth muscle actin), epithelial cell (cytokeratin) and microglia (macrophage antigens) as well as by studying the morphology for fibroblast contamination. Upon cultivation on collagen-coated Transwell inserts these human BMEC formed continuous lining of endothelial cells (largely a single monolayer with occasional 2 or 3 endothelial cells being overlaying one to another) and exhibited transendothelial electrical resistance of 100 to 600 Ω-cm^2 (7), a unique property of the brain microvascular endothelial monolayer compared to systemic vascular endothelium.

The *in vivo* model of the blood-brain barrier was established by induction of hematogenous meningitis in 5-day-old rats (Prasadarao *et al.*, 1996a; Huang *et al.*, 1995; Kim *et al.*, 1992). In this experimental meningitis model, bacteria are injected via intracardiac or subcutaneous injection, resulting in bacteremia and subsequent entry into the central nervous system, which most likely occurs at the sites of the blood-brain barrier. The development of techniques for atraumatic collection of blood and CSF specimens enables us to use this *in vivo* animal model to examine the pathogenetic mechanisms involved in crossing of the blood-brain barrier by circulating bacteria.

3. *E. coli* STRUCTURE CONTRIBUTING TO INVASION OF BRAIN MICROVASCULAR ENDOTHELIAL CELLS (BMEC)

We presently do not know how circulating *E. coli* cross the blood-brain barrier. Using the above-mentioned *in vitro* (BMEC in culture) and *in vivo* (hematogenous meningitis model in newborn rats) models of the blood-brain barrier, we have shown that successful traversal of *E. coli* across the blood-brain barrier requires multiple steps of *E. coli*-BMEC interactions, e.g. binding to BMEC, invasion of BMEC, and intracellular survival and exit as live bacteria. In this chapter, I'll describe *E. coli*-BMEC characteristics pertaining to invasion process (Figure 1).

3.1. OmpA Protein

Our first attempt to identify potential *E. coli* structures contributing to invasion of BMEC was by comparison of *E. coli* with the known proteins involved in invasion of eukaryotic cells by meningitis causing bacteria. Neiserria have been shown to invade eukaryotic cells via Opa and *E. coli* OmpA has been shown to have a sequence homology with Neiserria Opa

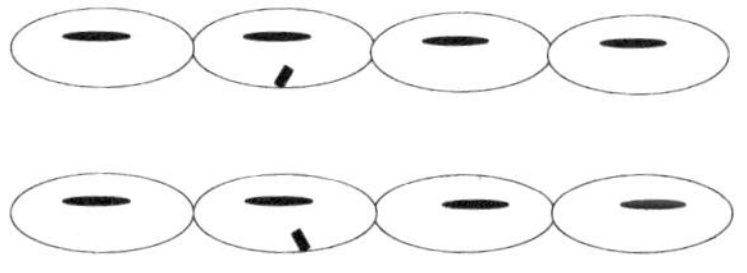

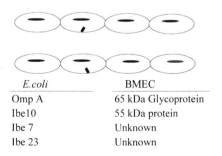

E.coli	BMEC
Omp A	65 kDa Glycoprotein
Ibe10	55 kDa protein
Ibe 7	Unknown
Ibe 23	Unknown

FIGURE 1. *E. coli* invasion of BMEC.

proteins. We have shown that OmpA contribute to *E. coli* invasion of BMEC. This was shown by our demonstration that the ability of OmpA+ *E. coli* to invade BMEC is 25 to 50 fold greater than that of $OmpA^-$ *E. coli* and that the invasive capability of $OmpA^-$ *E. coli* is restored by complementation with the OmpA gene (Prasadarao *et al.*, 1996b). These studies were done by using $OmpA^-$ *E. coli* generated from the CSF isolate of *E. coli* K1 strain RS218 (018:K1:H7) by P1 transduction (i.e., by inserting tetracycline resistance marker into the *E. coli* K-12 strain BRE51 (ΔsulA-OmpA) at near the OmpA locus and then a P1 lysate was used to transduce strain RS218, selecting for OmpA-phenotype) and OmpA complementation by transformation of OmpA-mutant with pRD 87 which contains the OmpA gene on pUC 9 as described previously (Prasadarao *et al.*, 1996b). Recently, we constructed an isogenic OmpA-mutant from *E. coli* K1 strain RS218 by an allelic exchange, in which the total OmpA gene was deleted without affecting other parts of the genome. The isogenic OmpA-mutant was also shown to be approximately 5-fold less invasive in BMEC compared to the parent strain RS218, clearly establishing a role of OmpA in *E. coli* invasion of BMEC both *in vitro* and *in vivo*. Moreover, OmpA+ *E. coli* invasion of BMEC was inhibited by purified OmpA proteins and anti-OmpA antibodies. OmpA is one of the major proteins in the outer membrane of *E. coli*. This heat-modifiable 35 kDa protein is highly conserved among gram-negative bacteria, and its N-terminal domain encompassing aminoacids residues 1 to 177 is thought to cross the membrane eight times in antiparallel β-strands with four relatively large and hydrophilic surface-exposed loops and three short periplasmic turns. We showed that N-terminal portion of OmpA, not the C-terminal potion, contributes to *E. coli* invasion of BMEC and that synthetic peptides representing a part of the first and second extracellular loops of OmpA inhibited *E. coli* invasion of BMEC, suggesting that they represent the essential domains of OmpA structures enhancing *E. coli* invasion of BMEC (Prasadarao *et al.*, 1996b). However, as shown below, $OmpA^+$ *E. coli* invasion was unique to BMEC and no such invasion was observed in endothelial cells of systemic vessels (e.g., HUVEC). We have subsequently identified a novel 65 kDa receptor for OmpA, which is present on BMEC, not on systemic vascular endothelial cells such as HUVEC. This specific OmpA-BMEC interaction contributes to *E. coli* invasion of BMEC and traversal of *E. coli* across the blood-brain barrier.

3.2. Ibe Proteins

Our next approach to identify potential *E. coli* structures contributing to invasion of BMEC was by the use of the transposon Tn*phoA*. Tn*phoA* is a modified transposon which is engineered by inserting the *phoA* gene into one end of Tn*5*. The gene fusion can be randomly generated by Tn*phoA* insertion into the target gene in the chromosome or plasmid. To facilitate the identification of the genes contributing to *E. coli* invasion of BMEC, Tn*phoA* was used to generate a collection of non-invasive mutants from strain RS218 as described previously (Huang *et al.*, 1995). Our *in vitro* invasion assay utilizes approximately 10^7 bacteria and confluent BMEC grown in each well of a 24-well plate with a multiplicity of infection of 50 to 100. The plates were incubated for 1.5 hours at 37°C to allow invasion to occur. The wells were washed and incubated with medium containing gentamicin (100 μg/ml) at 37°C for 1 hour to kill extracellular bacteria. The BMEC were subsequently washed, lysed and plated for determination of intracellular CFU. Several non-invasive mutants with single Tn*phoA* insertions were examined for their phenotypic characteristics compared with the parent strain. Because of possible polar effect associated with transposon mutagenesis, we also examined whether non-invasive mutants do not have insertions in the operons encoding known virulence factors by Southern hybridization. Those non-invasive mutants with single Tn*phoA* insertions which retained the same characteristics as the parent strain were examined in our newborn rat model of hematogenous meningitis for their ability to enter the CNS. From these *in vitro* and *in vivo* assays, we have identified three non-invasive mutants (10A-23, 7A-23 and 23A-20) (Table 1), suggesting that the DNA flanking the transposon insertions in these mutants may encode gene(s) necessary if not sufficient for invasion. These putative genes have been designated as *ibe10*, *ibe7* and *ibe23*, respectively (termed after **i**nvasion of **b**rain **e**ndothelial cells derived from **10**A-23, **7**A-23, **23**A-20).

Of note, at the time of collection of blood and CSF specimens for quantitative cultures all animals infected with the parent strain E44

Table I
Comparison of *E. coli* Mutants and Their Parent Strain for Their Ability to Invade BMEC *in Vitro* and *in Vivo*

E. coli strains	*In vitro*	*In vivo*	P value
Parent (E44)	0.1%	12/21 (57%)	<0.05
E44 ΔOmpA	<0.001%	4/20 (20%)	
Parent (E44)	0.1%	15/18 (83%)	<0.01
E44 ibe10::TnphoA	<0.001%	7/21 (33%)	
Parent (E44)	0.1%	15/22 (68%)	<0.01
E44 *ibe7*::Tn*phoA*	<0.001%	4/25 (29%)	
Parent (E44)	0.1%	18/24 (75%)	<0.05
E44 *ibe23*::Tn*phoA*	<0.001%	10/24 (42%)	

(spontaneous rifampin resistant mutant of strain RS218) or its mutants developed bacteremia of 10^5 to 10^8 CFU/ml of blood. This level of bacteremia was shown to be sufficient for allowing circulating *E. coli* to enter the CNS (Kim *et al.*, 1992). In addition, the magnitudes of bactermia were similar between the two groups (parent vs. its mutant), indicating that the inability of mutant strains to enter the CSF in the animals did not reflect decreased bacterial counts in the bloodstream and that OmpA and other genetic loci in the mutants, 10A-23, 7A-23, 23A-20 which were affected by Tn*phoA* mutagenesis are most likely to confer the ability to invade BMEC both *in vitro* and *in vivo*. We subsequently identified novel open reading frames (23 kDa, 50 kDa and 67 kDa, respectively) for *ibe10*, *ibe7*, and *ibe23*. In addition, genomic mapping of *ompA* and *ibe* loci revealed that these *E. coli* genes are located at different regions (Figure 2).

It is important to note that the invasion frequency of BMEC by the invasive parent strain RS218 or E44 (spontaneous rifampin resistant mutant of RS218) (approximately 0.1%) is considerably less than the reported epithelial invasion frequencies by other gram-negative bacteria such as *Shigella* and *Salmonella* species (usually 1 to 10%). However, as shown in Table 1, the *in vitro* invasion frequency of 0.1% corresponds with the enhanced bacterial penetration through the blood-brain barrier in the infant rat model of experimental hematogenous meningitis. Thus, the 0.1% invasion frequency observed for invasive *E. coli* into BMEC is biologically relevant.

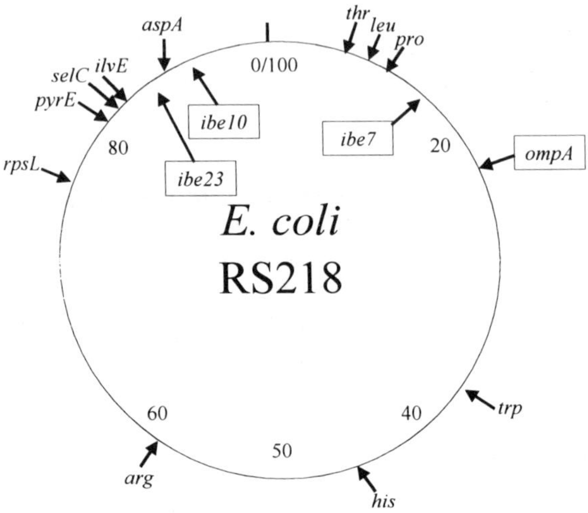

FIGURE 2. Genomic mapping of *ompA* and *ibe* loci.

4. BMEC STRUCTURES INVOLVED IN *E. coli* INVASION

One of the most intriguing findings of our studies has been that the above-mentioned characteristics of *E. coli* invasion of endothelial cells are specific to BMEC, and no such invasion characteristics were observed for endothelial cells of non-brain origin, e.g. human umbilical vein endothelial cells (HUVEC) and human aortic arterial endothelial cells (HAAEC) (Table 2). These findings suggest that the presence or differential expression of specific endothelial receptors contribute to tissue tropism associated with meningitis.

4.1. OmpA Interaction with the GlcNAcβ1–4 GlcNAc Epitope of the 65 kDa BMEC Glycoprotein

We have previously shown that the invasion of *E. coli* strain E44 (spontaneous rifampin resistant mutant of strain RS218) into BMEC is completely abolished by treating the BMEC with periodate (Prasadarao *et al.*, 1996b), suggesting the involvement of a carbohydrate epitope on BMEC. Various lectins were used to identify the BMEC carbohydrate structures involved in invasion. Only wheat germ agglutinin (WGA), specific for GlcNAcβ1–4 GlcNAc, blocked the invasion by >95%, whereas other lectins showed no blocking activities (Prasadarao *et al.*, 1996a). The role of sialic acid resides in WGA interactions was excluded by showing no blocking effect on *E. coli* invasion of BMEC with neuraminidase treatment of BMEC and no effect of *Maackia amurensis* lectin, specific for NeuAcα2,3-Gal.

In contrast, pretreatment of BMEC monolayers with PNGase F, an ezyme that cleaves asparagine-linked (N-linked) high-mannose, hybrid, and complex oligosaccharides from glycoproteins, resulted in a marked decrease in invasion of *E. coli* (Prasadarao *et al.*, 1996a). These results suggest that invasive *E. coli* recognizes the GlcNAcβ1–4GlcNAc epitopes linked to asparagine on the glycoproteins. GlcNAcβ1–4GlcNAc epitopes often contain α-fucose linked to asparigine-bound GlcNAc via α1, 6

Table II
K1 *E. coli* Invasion of Human Endothelial Cells Derived from Brain and Systemic Origin

E. coli strains	Phenotypic Characteristics	HBMEC	HUVEC	HAAEC
RS218, E44, E69	018:K1 ($OmpA^{+}$, Ibe^{+})	0.08 ± 0.02	<0.0001	<0.0001
E91, E98	018:K1 ($OmpA^{-}$, Ibe^{+})	<0.0001	<0.0001	<0.0001
10A-23, 7A-23, 23A-20	018:K1 ($OmpA^{+}$, Ibe^{-})	<0.0001	<0.0001	<0.0001

linkage, but neither *Pisum sativum* agglutinin or UAE lectin, which interacts with L-fucose, nor free fucose was able to block the invasion (Prasadarao *et al.*, 1996a). In fact, the *E. coli* invasion of BMEC was not affected by treating the brain endothelial cells with α-fucosidase with or without neuraminidase. These findings indicate that the GlcNAcα1–4 GlcNAc epitopes of complex oligosaccharides on BMEC glycoproteins interact with *E. coli* for invasion of BMEC. This concept was further supported by our demonstration that chitin hydrolysate containing GlcNAc polymers of various lengths or chitotriose completely blocked *E. coli* K1 invasion of BMEC in a dose-dependent manner, whereas GlcNAc and mannose monosaccharides were ineffective. More importantly, chitin hydrolysate blocked the invasion of *E. coli* into the central nervous system in the newborn rat model of experimental hematogenous meningitis (Prasadarao *et al.*, 1996a). Thus GlcNAcβa1–4 GlcNAc epitopes of BMEC indeed appears to mediate the traversal of *E. coli* across the blood-brain barrier *in vitro* and *in vivo.*

We next showed that *E. coli* invasion of BMEC is due to the interaction of OmpA with the GlcNAcβ1–4 GlcNAc epitope of the 65 kDa BMEC glycoprotein. This was shown by our demonstration (Prasadarao *et al.*, 1996a) that (a) the chitotriose-bound membrane proteins from $OmpA^+$ and $OmpA^-$ *E. coli* strains revealed a 35 kDa protein reactive with the anti-OmpA antibody only in chitotriose-bound proteins from $OmpA^+$ *E. coli*, which inhibited *E. coli* invasion of BMEC and (b) the WGA-bound BMEC membrane proteins inhibited invasion of $OmpA^+$ *E. coli* into BMEC. These proteins were separated by using a size exclusion column on a HPLC system. We subsequently identified a 65 kDa glycoprotein present on BMEC, but absent (or undetectable) in systemic vascular endothelial cells. The N-terminal amino acid sequences (XLMSVAIVQLVFQAVFDLXE, X is an unidentified amino acid) revealed that this 65 kDa protein represents a novel BMEC glycoprotein (Prasadarao *et al.*, 1997), contributing to *E. coli* OmpA interaction with BMEC, resulting in invasion.

4.2. Ibe 10 Interaction with the 55 kDa BMEC Protein

We also identified the surface BMEC protein interacting with the *E. coli* invasin, Ibe 10 by subjecting BMEC membrane proteins to Ibe10-Ni-Sepharose affinity chromatography (the recombinant Ibe10 protein generated by pQE contains histidine tag and Ibe10 protein was coupled to Ni-sepharose). This 55 kDa protein exhibited a dose-dependent inhibition of *E. coli* invasion into BMEC (50% inhibition was achieved with 10 μg/ml protein). Both N-terminal and endopeptidase cleaved amino acid sequences of this 55 kDa protein revealed that it is a novel albumin-like protein present on BMEC interactive with the *E. coli* invasin protein, Ibe10. The surface location of this 55 kDa protein was demonstrated by immunocytochemistry with anti-55 kDa antibody and also by the use of the surface biotinylated BMEC membrane proteins in affinity column chromatography (Prasadarao *et al.*, 1999). Thus, *E. coli* invasion of BMEC involves specific receptors present on BMEC, not on systemic vascular endothelial cells.

5. MECHANISMS OF *E. coli* INVASION OF BMEC

5.1. Role of BMEC Cytoskeletons

Previous studies have revealed a number of different strategies evolved by many pathogenic bacteria to exploit host cell function and cause diseases. We showed by transmission electron microscopy that *E. coli* transmigrates through BMEC in an enclosed vacuole without intracellular multiplication. Double immunofluorescence staining with TRITC-labeled phalloidin for F-actin and anti-K1 monoclonal antibody for *E. coli* revealed invasive *E. coli*-associated F-actin condensation, while non-invasive *E. coli* failed to exhibit the recruitment of F-actin. The microfilament disrupting agent, cytochalasin D, completely blocked the *E. coli* invasion of BMEC. Of interest, the microtubule inhibitors, nocodazole and colchicine, exhibited approximately 50% inhibition of *E. coli* invasion into BMEC (Figure 3). These findings suggest that cytoskeleton rearrangement of BMEC is involved in *E. coli* entry into the BMEC, which appears to be dependent upon both microfilaments and microtubules. We next showed that invasive *E. coli* rapidly stimulated tyrosine phosphorylation of focal adhesion kinase (FAK), Src kinase and Paxcillin in BMEC. Of interest, *E. coli* failed to invade BMEC stably overexpressing mutant FAK proteins defective in binding to Src kinases. Work is in progress to understand how *E. coli* invasion of BMEC is associated with activation of BMEC signaling and cytoskeleton rearrangement.

5.2. Environmental Regulation of *E. coli* Invasion of BMEC

As stated before, most cases of bacterial meningitis develop as a result of hematogenous spread. We examined if environmental growth conditions similar to those that the bacteria might be exposed to in the blood could influence the ability of *E. coli* K1 to invade BMEC *in vitro* and to cross the

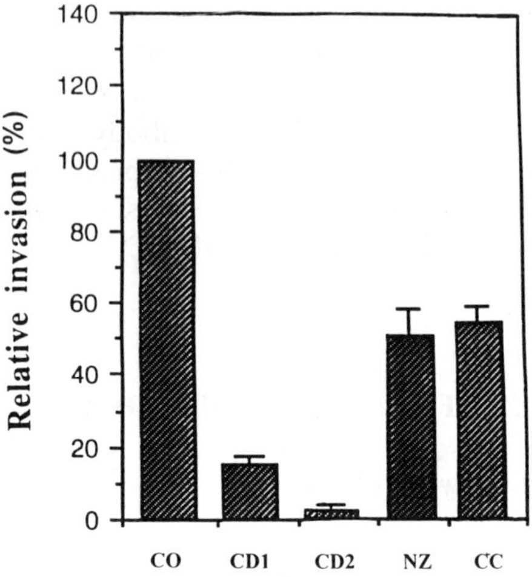

FIGURE 3. Effect of cytochalasin D, 0.5 μM (CD1) and 1.0 μM (CD2), nocodazole 20 μM (NZ) and colchicine 5 μM (CC) on *E. coli* invasion of BMEC.

blood-brain barrier *in vivo*. We found the following bacterial growth conditions enhanced *E. coli* K1 invasion 3–10 fold in BMEC; microaerophilic growth, media buffered at pH 6.5, or media supplemented with 50% newborn bovine serum (NBS), magnesium, or iron. Conditions that significantly repressed invasion (i.e., 2–250 fold) included iron chelation, pH8.5, and high osmolarity. More importantly, *E. coli* K1 traversal of the blood brain barrier was significantly greater for the growth condition enhancing BMEC invasion (50% NBS) than for the condition repressing invasion (osmolarity) in newborn rats with experimental hematogenous meningitis (Table 3).

Table III
Comparison of Bacterial Counts in Blood (mean ± SD) and Number of Animals with Positive CSF Cultures between the Two Groups Receiving *E. coli* K1 Grown in Media with 50% NBS or 0.2 M NaCl

Groups	N	Magnitude of Bacteremia (log CFU/ml of blood)	No. of animals with positive CSF culture (%)
50% NBS	40	6.86 ± 0.47	26 (65)†
0.2 M NaCl	33	7.17 ± 0.60	12 (33)

† $p < 0.05$.

This is the first demonstration that bacterial ability to enter the central nervous system can be affected by environmental growth conditions (Badger and Kim 1998). A complete understanding of the basis for this environmental regulation should enhance our knowledge about the pathogenesis of *E. coli* meningitis.

6. AGE-DEPENDENCY OF *E. coli* MENINGITIS

E. coli meningitis is common during the neonatal period but the basis of this age-dependency is not clear. Previous studies have shown that most *E. coli* meningitis occurs as a result of hematogenous spread. We have shown that a high degree of bacteremia is a primary determinant of meningeal invasion by *E. coli* K1 (Kim *et al.*, 1992), suggesting that one possible explanation for *E. coli* being predominant in the neonatal period to neonates relative susceptibility to high-level bacteremia compared to adults. Another possibility is that *E. coli* may be able to interact (e.g., invasion) with neonatal BMEC compared with adults BMEC. We have shown that the abilities of *E. coli* to invade BMEC were similar between BMEC derived from fetus, children, adults, and geriatric population, suggesting that the age-dependency of *E. coli* meningitis is not due to differences in host tissues to interact with *E. coli*. We have previously shown that the induction of a high degree of bacteremia for the development of meningitis requires different bacterial inocula for newborn vs. adult animals, i.e. approximately 10^6-fold greater inocula of *E. coli* K1/gram of body weight are required in adult rats compared to newborn rats to induce a similar level of high-degree of bacteremia (e.g., $>10^4$ CFU/ml of blood) (8). These findings indicate that one of the reasons for the close association of newborns with *E. coli* meningitis is the relative easiness of *E. coli* K1 to escape from neonatal host defenses and to achieve a threshold level of bacteremia necessary for meningeal invasion.

7. SUMMARY

A major limitation to advances in prevention and therapy of bacterial meningitis is our incomplete understanding of the pathogenesis of this disease. Successful isolation and cultivation of BMEC, which constitute the blood brain barrier, and the development of experimental hematogenous meningitis animal model, which mimics closely the pathogenesis of human meningitis, enabled us to dissect the pathogenetic mechanisms of bacterial meningitis. We have shown for the first time using *E. coli* as a paradigm the mechanisms of bacterial crossing of the blood-brain barrier into the central nervous system. We have shown that invasion of BMEC is a requirement for *E. coli* K1 crossing of the blood-brain barrier *in vivo* (Prasadarao *et al.*, 1996b; Huang *et al.*, 1995). We have identified several novel *E. coli* proteins (i.e., Ibe10, Ibe7, and Ibe23) contributing to invasion of BMEC. We have also established a novel phenotype, i.e., invasion of BMEC, of a well known major *E. coli* protein, OmpA. In addition, we have shown that some of these *E. coli* proteins (i.e., OmpA, Ibe10) interact with novel endothelial receptors present on BMEC, not on systemic vascular endothelial cells. Further understanding and characterization of these *E. coli*-BMEC interactions should allow us to develop novel strategies to prevent this serious infection. In addition, the *in vitro* and *in vivo* models of the blood-brain barrier and the information derived from our study should be beneficial to investigating the pathogenesis of meningitis due to other organisms such as group B streptococci, *Listeria monocytogenes*, *Streptococcus pneumoniae* and Citrobacter.

ACKNOWLEDGMENTS. I thank all members of our laboratory (Drs. Badger, Das, Hoffman, Huang, Linsangan, Nemani, Reddy, Stins, Wang and Wass) who have contributed to generation of data contained in this chapter. This work was supported by USPHS grants RO1-NS26310 and RO1-HL61951.

8. REFERENCES

Badger, J.L., and Kim, K.S., 1998, Environmental growth conditions influence the ability of *E. coli* K1 to invade brain microvascular endothelial cells and confer serum resistance, *Infect. Immun.* **66**:5692–5697.

Durand, M.L., Calderwood, S.B., Weber, D.J., Miller, S.I., Southwick, F.S., Caviness, V.S. Jr, and Swartz, M.N., 1993, Acute bacterial meningitis in adults. A review of 493 episodes, *N. Engl. J. Med.* **328**:21–28.

Huang, S.H., Wass, C.A., Fu, Q., Prasadarao, N.V., Stins, M., and Kim, K.S, 1995, *E. coli* invasion of brain microvascular endothelial cells *in vitro* and *in vivo*: Molecular cloning and characterization of *E. coli* invasion gene *ibe10*, *Infect. Immun.* **63**:4470–4475.

Kim, K.S., 1985, Comparison of cefotaxime, imipenem-cilastatin, ampicillin-gentamicin and ampicillin-chloramphenicol in the treatement of experimental *Escherichia coli* bacteremia and meningitis, *Antimicrob. Agents Chemother.* **28**:433–436.

Kim, K.S., Itabashi, H., Gemski, P., Sadoff, J., Warren, R.L., and Cross, A.S., 1992, The K1 capsule is the critical determinant in the development of *Escherichia coli* meningitis in the rat, *J. Clin. Invest.* **90**:897–905.

McCracken, G.H. Jr., Threlkeld, N., Mize, S., Baker, C.J., Kapal, S.L., Fraingezicht, I., Feldman, W.F., and Schad, U., The Neonatal Meningitis Cooperative Study Group, 1984, Moxalactam therapy for neonatal meningitis due to gram-negative sepsis enteric bacilli, *JAMA* **252**:1427–1437.

Nizet, V., Kim, K.S., Stins, M., Jonas, M., Nguyen, D., and Rubens, C.E., 1997, Invasion of brain microvascular endothelial cells by group B stretococci, *Infect. Immun.* **65**:5074–5081.

Prasadarao, N.V., Wass, C.A., and Kim, K.S., 1996a, Endothelial cell GlcNAcβ1–4 GlcNAc epitopes for outer membrane protein A traversal of *E. coli* across the blood-brain barrier, *Infect. Immun.* **64**:154–160.

Prasadarao, N.V., Wass, C., Stins, M.F., Weiser, J., Huang, S.H., and Kim, K.S., 1996b, Outer membrane protein A of *E. coli* contributes to invasion of brain microvascular endothelial cells, *Infect. Immun.* **64**:146–153.

Prasadarao, N.V., Huang, S.H., Wass, C.A., and Kim, K.S., 1999, Identification and characterization of a novel Ibe10 binding protein on brain microvascular endothelial cells contributing to *E. coli* invasion, *Infect. Immun.* (In press).

Prasadarao, N.V., Wass, C.A., and Kim, K.S., 1997, Identification and characterization of S-fimbriae binding sialoglycoproteins on brain microvascular endothelial cells, *Infect. Immun.* **65**:2852–2860.

Stins, M.F., Gilles, F., and Kim, K.S., 1997, Selective expression of adhesion molecules on human brain microvascular endothelial cells, *J. Neuroimmunol.* **76**:81–90.

Stins, M.F., Prasadarao, N.V., Ibric, L., Wass, C.A., Luckett, P., and Kim, K.S., 1994, Binding characteristics of S-fimbriated *Escherichia coli* to isolated brain microvascular endothelial cells, *Amer. J. Pathol.* **145**:1228–1236.

Unhanand, M., Musatafa, M.M., McCracken, G.H., and Nelso, J.D., 1993, Gram-negative enteric bacillary meningitis: A twenty-one year experience, *J. Pediatr.* **122**:15–21.

Chapter 4

Host Cell Invasion by Pathogenic *Neisseriae*

Christoph Dehio, Scott D. Gray-Owen*, and
Thomas F. Meyer

1. INTRODUCTION

Neisseria gonorrhoeae and *Neisseria meningitidis* have become exquisitely adapted to life within humans, their only natural host. Both species possess the ability to colonize human mucosal tissues without generating any detectable clinical manifestations, and this carrier state likely contributes to their maintenance within the population. They are, however, also capable of persisting during the massive inflammatory responses, representing a hallmark of neisserial diseases. Despite their close evolutionary relationship (Tinsley and Nassif, 1996), *N. gonorrhoeae* primarily infects the uro- or anorectal mucosa following intimate sexual contact, while *N. meningitidis* colonizes the nasopharynx after the inhalation of infected

CHRISTOPH DEHIO and SCOTT D. GRAY-OWEN Dept. Infektionsbiologie, Max-Planck-Institut für Biologie, D-72076 Tübingen, Germany. **THOMAS F. MEYER** Dept. Infektionsbiologie, Max-Planck-Institut für Biologie, D-72076 Tübingen, Germany and Molekulare Biologie, Max-Planck-Institut für Infektionsbiologie, D-10117 Berlin, Germany. *Present Address: Dept. Medical Genetics & Microbiology, University of Toronto, Toronto, Ontario M5S 1A8, Canada.

Subcellular Biochemistry, Volume 33: Bacterial Invasion into Eukaryotic Cells, edited by Oelschlaeger and Hacker. Kluwer Academic / Plenum Publishers, New York, 2000.

respiratory droplets. This association is at least partially the result of their respective modes of transmission rather than a tropism restricted to these loci, since gonococcal pharyngitis and meningococcal anogenital infections have both also been described (Janda *et al.*, 1980; Givan *et al.*, 1977). Nasopharyngeal and fallopian tube organ cultures, and ureteral tissue models of infection indicate that both species penetrate to submucosal layers *in vitro* (Mosleh *et al.*, 1997; McGee *et al.*, 1983), and this is consistent with the detection of sloughed epithelial cells containing intracellular gonococci in urethral exudates obtained from men with symptomatic gonorrhea (Apicella *et al.*, 1996). Complications arising from the spread of gonococci from the primary loci of infection can result in significant morbidity and, potentially, sterility. Similarly, although localized meningococcal infection of the mucosa is typically asymptomatic, dissemination from this site can lead to the rapidly advancing and often fatal meningococcal meningitis and/or meningococcemia.

The ability of pathogenic *Neisseriae* to persist in the human host relies to a large extent on their extraordinary ability to vary their surface structures. The pathogens may use this property to alter their antigenic makeup, and thereby avoid attack by the immune system. As important, this variation can change the functional characteristics of several factors which influence bacterial interactions within different microenvironments encountered in the host. Such changes may affect the pathogen's ability to adhere to different host cell types (epithelial cells, endothelial cells or phagocytes), invade into cells, transcytose across the mucosal barrier, or persist extracellularly by the expression of polysaccharide structures (e.g. capsule or sialylated lipopolysaccharide) which protect it against the bactericidal activity of serum. In recent years, a concentrated effort has been made to dissect these neisserial interactions with host cells, tissues and compartments at the cellular and molecular level. This has lead to the establishment of a model of mucosal colonization (Figure 1) and the description of a number of bacterial surface components which represent putative bacterial virulence determinants that mediate or influence bacterial adherence and cellular invasion (depicted in Figure 2). In some cases the corresponding host cell receptors have also been identified. Moreover, these studies have revealed distinct genetic mechanisms which are responsible for the impressive antigenic variation and phase-variable (on- and off-) expression which is observed in many neisserial surface components (Robertson and Meyer, 1992). In this review, we will outline the mechanisms by which each of these virulence determinants contributes to the colonization, persistence and penetration into host tissues.

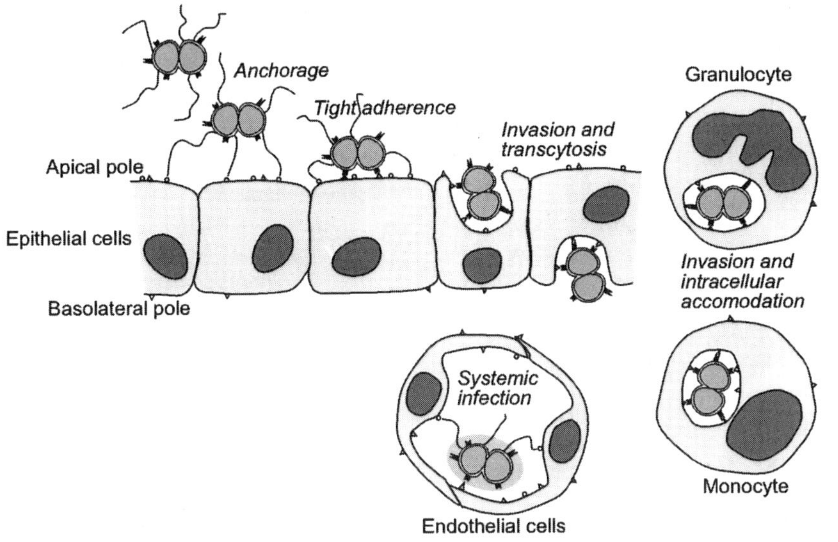

FIGURE 1. Model of the sequential interactions between pathogenic *Neisseriae* and host cells during the colonization of a mucosal tissue. Primary adherence to epithelial cells probably occurs via hair-like surface appendages, i.e. pili. The bacteria can then establish an intimate contact with the host cells via additional adhesins, an interaction that may allow their engulfment and transcytotic passage to subepithelial tissues. Interactions between bacteria and professional phagocytes (e.g. granulocytes and/or monocytes) also lead to the opsonin-independent uptake of bacteria by these cells. Neisserial interactions with the endothelia may result in its entry into the bloodstream and subsequent systemic dissemination. Expression of a polysaccharide capsule by meningococci or sialylation of meningococcal or gonococcal lipopolysaccharide renders the bacteria resistant to killing by serum and blocks opsonin-dependent uptake by host phagocytes, likely because these structures can mask the integral outer membrane adhesins.

2. PILUS-MEDIATED INTERACTIONS

Neisserial type-4 pili are generally thought to be necessary for the primary colonization of host mucosal epithelia (Ward *et al.*, 1974; Swanson, 1973; Kellogg *et al.*, 1968). *N. meningitidis* isolated from the nasopharynx of both asymptomatic carriers and meningococcal disease patients are piliated (3–34 pili/diplococci) (Stephens and McGee, 1981; McGee *et al.*, 1979), and meningococcal pili mediate primary interactions with the microvilli of nonciliated epithelial cells in primary nasopharyngeal organ cultures *in vitro* (Rayner *et al.*, 1995; Stephens *et al.*, 1983). Pili also mediate primary adherence of *N. gonorrhoeae* to surface cells in fallopian tube organ culture

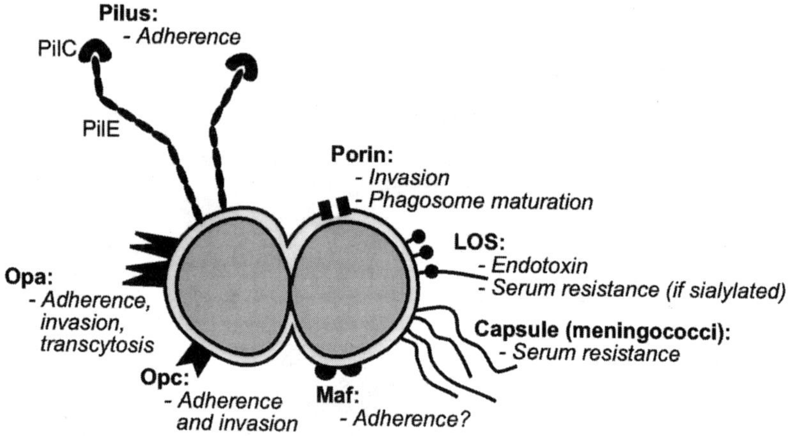

FIGURE 2. Putative virulence factors of pathogenic *Neisseriae* and their roles in host-pathogen interactions. The neisserial type-4 pilus is composed of the variable pilus subunit PilE forming the pilus fiber, and the pilus-associated/tip-located adhesin PilC. PilC mediates pilus attachment to epithelial and phagocytic human cell types, and also plays a role in pilus biogenesis. The phase-variable opacity-associated Opa proteins are outer membrane proteins mediating intimate binding to and invasion into various host cells (e.g. epithelial, endothelial and phagocytic cells), which may also result in transcytotic traversal of mucosal epithelia. Opc is an outer membrane protein structurally distinct from Opa, which also mediates host cell binding and invasion. Glycolipid adhesins such as members of the multiple adhesin family (Maf) may further contribute to neisserial interactions with host cells, however their role is less well characterized. The polysaccharide capsule of meningococci renders bacteria resistant to killing by serum, but may mask integral outer membrane adhesins such as Opa from binding to their host cell receptors. The variable lipooligosaccharide (LOS) acts as an endotoxin, and the sialylation of a terminal galactose residue present in some variants of neisserial LOS renders these bacteria serum resistant. The porin PorB translocates into the host cell membrane and forms an ATP-regulated ion channel which plays a role in epithelial cell invasion and the intracellular accommodation of *Neisseriae*.

(McGee *et al.*, 1981), ureteral tissue (Mosleh *et al.*, 1997) and human corneal (Tjia *et al.*, 1988) models of infection. Consistently, challenge studies with human volunteers indicate that pili are important bacterial determinants for establishing infection *in vivo* (Cohen *et al.*, 1994; Hook and Holmes, 1985).

2.1. Pilus Biogenesis and Antigenic Variation

At least 22 gene products are typically involved in type-4 pilus assembly and function (Mattick *et al.*, 1996), and many homologues of the well characterized *Pseudomonas aeruginosa* type-4 machinery have been identified in the pathogenic *Neisseria* (Tonjum and Koomey, 1997). Each pilus fiber is composed of a single small subunit pilin protein (PilE) (Meyer *et al.*, 1984) which is assembled into an α-helical structure to form the extended pilus structure (Parge *et al.*, 1995). This arrangement buries the highly conserved N-terminal residues of pilin, but exposes the O-linked carbohydrate and hypervariable sequences (Forest and Tainer, 1997; Forest *et al.*, 1996; Parge *et al.*, 1995) which are responsible for the remarkable inter- and intrastrain size (13–20 kDa) and antigenic variability of this protein (Hagblom *et al.*, 1985; Olafson *et al.*, 1985; Zak *et al.*, 1984; Robertson *et al.*, 1977; Novotny and Turner, 1975). Such variation can occur both *in vitro* and *in vivo*, and results from a typical non-reciprocal exchange of variant "minicassettes" from one of the 17 silent gene copies to *pilE* (Seifert *et al.*, 1994; Haas *et al.*, 1992; Gibbs *et al.*, 1989; Haas and Meyer, 1986; Swanson *et al.*, 1986). Only in certain cases reciprocal changes between *pilS* and *pilE* have been observed (Gibbs *et al.*, 1989). The non-expressing *pil* copies are tandemly arranged within several silent loci (*pilS*) and contain the structural information to encode PilE but lack any promoter sequences. The Sma/Cla repeats found at the 3′ end of all gonococcal and type I meningococcal *pilE* and *pilS* genes are homologous to several recombinase-binding sites, and have been shown to facilitate the recombination between pilin genes (Wainwright *et al.*, 1994). The resulting recombination can generate pilus subunits which are antigenically and/or functionally distinct from the parental form, and often also affects the binding efficacy seen *in vitro* (Marceau *et al.*, 1995; Rayner *et al.*, 1995; Jonsson *et al.*, 1994; Virji *et al.*, 1993*b*; Nassif *et al.*, 1993; Rudel *et al.*, 1992; Virji *et al.*, 1992a; Pinner *et al.*, 1991). In some strains (e.g. *N. gonorrhoeae* MS11) there are two separate expression loci encoding PilE (denoted *pilE1* and *pilE2;* Meyer *et al.*, 1984 and 1982), and possibly more than one pilin type may be co-expressed by a single bacterium. Importantly, recombinational events involving *pilE* can also result in the phenotypic conversion of a bacterium from piliated to non-piliated, and vice versa. This can be the result of a missense mutation (Bergström *et al.*, 1986), or of the generation of an extra-long L-pilin (Manning *et al.*, 1991; Hill *et al.*, 1990) or an alternatively processed, soluble, secreted S-pilin (Haas *et al.*, 1987; Perry *et al.*, 1987). The non-reciprocal recombination into *pilE* can be the result of transformation of living cells by DNA released upon lysis of other cells within the population (Gibbs *et al.*, 1989; Seifert *et al.*, 1988). However, transformation-mediated pilin variation may not only be a driving force of pilin diversity, but also result in allelic homogenisation (Hill, 1996) which may be important in order to maintain antigenic synchrony during infection. It appears that pilin variation can readily occur also in transformation-deficient mutants (Facius *et al.*, 1993; Zhang *et al.*, 1992), indicating that the extracellular and intracellular recombination pathways are of similar efficiency *in vitro*.

PilC, a minor 110 kDa pilus-associated protein that co-purifies with PilE in pili preparations, has been shown to function in pilus biogenesis since mutagenesis of the two (*pilC1* and *pilC2*) loci may result in a non-piliated phenotype (Rudel *et al.*, 1995*b*; Nassif *et al.*, 1994; Jonsson *et al.*, 1991). However, $PilC^-$ mutants are capable of assembling pili (Rudel *et al.*, 1995a and 1992). PilC expression is also phase variable, with translation from both alleles being turned on or off by frame shift mutations in a poly-guanine stretch within the signal peptide coding sequence (Jonsson *et al.*, 1991).

2.2. Cellular Interactions Mediated by Pilus

Pili can extend up to 6 µm from the bacterial surface (Stephens *et al.*, 1985). This structure should facilitate its proposed function in overcoming the electrostatic barrier which occurs because the surface of bacteria and cells are both negatively charged (Heckels *et al.*, 1976). Although neisserial pili are known to bind erythrocytes (Salit, 1981; Koransky *et al.*, 1975), this hemagglutinating activity is distinct from its binding to other cell types and is independent of PilC (Rudel *et al.*, 1992). In *N. gonorrhoeae*, the expression of PilC from either expression locus generates adherent pili (Jonsson *et al.*, 1994; Rudel *et al.*, 1992), whereas in one meningococcal strain only the PilC1 variant was found to mediate meningococcal adherence to epithelial and endothelial cells (Nassif *et al.*, 1994). Ultrastructural analysis resulted in the location of PilC at the tip of pili (Rudel *et al.*, 1995*c*), however, PilC has also been microscopically and biochemically localised on the bacterial cell surface of both piliated and non-piliated *Neisseria* strains (Rahman *et al.*, 1997; Rudel *et al.*, 1995*b*) and was purified from the outer membrane of a non-piliated over-producing strain (Rudel *et al.*, 1995*c*). PilC protein purified from either recombinant non-piliated *N. gonorrhoeae* or *E. coli* was capable of binding to human epithelial and endothelial cells, exhibiting the same (human) species-specificity as piliated gonococci. Moreover, binding of gonococcal PilC to human epithelial cells prevented pilus-mediated, but not Opa protein-mediated, binding of both *N. gonorrhoeae* and *N. meningitidis*. This indicates that PilC is likely to account for the species-specificity of both organisms and, that pili of both species recognize identical receptors (Ryll *et al.*, 1997; Rudel *et al.*, 1995*c*).

The membrane co-factor protein (MCP or CD46) has been proposed as a cellular receptor for pilus of both pathogenic *Neisseria* (Kallstrom *et al.*, 1997). MCP is a transmembrane C3b/C4b-binding glycoprotein which functions to control the activation state and deposition of complement, and thereby protects host cells from damage caused by the human complement system (Liszewski *et al.*, 1991). It is expressed on almost every human cell and tissue type with the exception of erythrocytes, and a soluble form of unknown function is also found in serum (Seya *et al.*, 1995). Whether MCP binds to the pilus-associated PilC adhesin or to the pilus fiber (i.e. pilin) has not yet been investigated. Two splice variants of MCP, designated as BC1 and BC2, are both recognized by pili, while two others, C1 and C2 are not (Kallstrom *et al.*, 1997). Interestingly, BC1 and BC2 express different C-terminal cytoplasmic tails (Liszewski *et al.*, 1991), suggesting that adherence to them may result in distinct intracellular signaling and cellular responses. This is consistent with the fact that a major rearrangement of the actin cytoskeleton follows pilus-mediated neisserial binding, resulting in a foot-print-like appearance below adherent bacteria (Pujol *et al.*, 1997), although other neisserial adhesins may also contribute to this remodeling of the cell surface.

After adherence to the apical side of polarized T84 epithelial cell monolayers, meningococci show a clear reduction in piliation which results in the bacterial and cellular surfaces becoming closely juxtaposed (Pujol *et al.*, 1997). Subsequent to this, gonococci and meningococci have both been shown to be engulfed and transcytosed across the monolayers, eventually being released from the basolateral surface of these cells. This process does not lead to the disruption of tight intercellular junctions, and intracellular bacteria are evident within a phagosome, suggesting that it occurs via a transcellular rather than a paracellular route (Pujol *et al.*, 1997; Merz *et al.*, 1996). This result is surprising considering that previous studies have shown a correlation between LOS-induced cytotoxicity and pilus-mediated adherence to human epithelial and endothelial cells grown in non-polarized cultures (Dunn *et al.*, 1995; Virji and Everson, 1981). It may, therefore, suggest that the cellular mediator of this effect is not expressed on the apical surface of polarized epithelial cells to which the pathogenic *Neisseriae* adhere. Furthermore, the mechanisms of LOS-mediated cytotoxicity requires clarification since epithelial cells are considered to lack the CD14 receptor for LPS, but soluble CD14 present in serum may possibly replace the CD14 receptor function. Very likely, other neisserial factors besides LPS and pili (e.g. porins—see Section 8), also contribute to the observed cytotoxic effects.

3. Opa-MEDIATED INTERACTIONS

3.1. Diversity and Phase-Variable Expression of Neisserial Opa Proteins

Neisserial Opa proteins were originally identified because their expression changes the color and opacity of gonococcal colonies (Swanson, 1978;

James and Swanson, 1978). This effect has since been found to be due to an increased inter-bacterial aggregation which results from the lectin-like ability of Opa proteins to bind to LOS on adjacent bacteria (Blake *et al.*, 1995). The Opa proteins were, however, subsequently shown to constitute a family of closely related but size variable (Heckels, 1981) integral outer membrane proteins which are predicted to span the membrane 8 times to expose 4 surface loops (Malorny *et al.*, 1998; Bhat *et al.*, 1991). An essential role for Opa proteins in neisserial pathogenesis is suggested by the finding that gonococci recovered after urogenital, cervical or rectal infections typically express at least one Opa protein, as do bacteria recovered after the inoculation of human volunteers with transparent (Opa^-) bacteria (Jerse *et al.*, 1994; Swanson *et al.*, 1988). The exception to this *in vivo* selection is that Opa^- bacteria predominate in the cervix early in the menstrual cycle (James and Swanson, 1978), thus implying that their expression is detrimental under these conditions.

A single strain can possess as many as 3–4 (in meningococci) (Stern and Meyer, 1987) or 11 (in gonococci, Table 1) (Kupsch *et al.*, 1993; Bhat *et al.*, 1991; Stern *et al.*, 1986) unlinked chromosomal alleles that encode distinct Opa variants. Although these sequences are approximately 70% identical, their well described antigenic variability (Wang *et al.*, 1993; Diaz and Heckels, 1982) results from the exposure of one semi-variable and two hypervariable domains at the cell surface (Malorny *et al.*, 1998; Kupsch *et al.*, 1993; Bhat *et al.*, 1991; Stern *et al.*, 1986). Expression from each *opa* allele is phase variable due to the generation of RecA-independent changes which alter the number of pentanucleotide coding repeat units in the leader sequence, which thereby influences the reading frame of these constitutively transcribed genes (Murphy *et al.*, 1989; Stern *et al.*, 1986). This phenomenon maintains a heterogenous population of bacteria which express none, one or multiple Opa proteins. The complexity of this situation is further increased because both intra- and inter-strain recombination can occur, although at a low rate compared to pilin variation. A vast array of alleles has thereby been generated which may be differentially distributed between strains (Hobbs *et al.*, 1998; Morelli *et al.*, 1997; Connell *et al.*, 1990). This has led Achtman and co-workers (Malorny *et al.*, 1998; http://novell-ti.rz-berlin.mpg.de) to systematize all alleles described to date using a nomenclature scheme based on that of Kupsch *et al.* (1993).

3.2. Opa-Mediated Interactions with Cellular Heparan Sulfate Proteoglycan Receptors

A subset of Opa protein variants expressed by the pathogenic *Neisseria* bind to the host cell surface-associated heparan sulfate proteoglycans

(HSPGs) (de Vries *et al.*, 1998; van Putten and Paul, 1995; Chen *et al.*, 1995). HSPGs consist of long repeating sulfated glycoconjugates known as the heparan sulfate glycosaminoglycans (HS-GAGs), which can be linked to various core proteins (Carey, 1997). Among the 11 variant Opa proteins encoded by *N. gonorrhoeae* strain MS11, several appear to bind cellular HS-GAGs of HSPGs (Bos *et al.*, 1997; van Putten and Paul, 1995; Chen *et al.*, 1995; Kupsch *et al.*, 1993), while only one, Opa_{50}, mediates a strong adherence sufficient to trigger uptake into most cultured epithelial cell lines (Kupsch *et al.*, 1993; Weel *et al.*, 1991; Weel and van Putten, 1991). This process of binding and internalization, which is mediated by HSPG receptors, has been studied in most detail for this neisserial Opa variant.

3.2.1. HSPG-Dependent Uptake of Opa_{50}-Expressing *Neisseria* into Epithelial Cells

HSPG-dependent internalization of Opa_{50}-expressing gonococci into cultured epithelial cells has been shown to occur by at least three alternative mechanisms. Invasion into some epithelial cell lines (e.g. Chang human conjunctiva or Me-180 human cervical carcinoma cells), depends on the activation of phosphatidylcholine-dependent phospholipase C (PC-PLC) and acidic sphingomyelinase (ASM), which results in the generation of the second messengers diacylglycerol and ceramide, respectively (Grassme *et al.*, 1997) (Figure 3a). In many other epithelial cell lines (e.g. HeLa human cervical carcinoma or Chinese hamster ovary (CHO) cells), this signaling pathway appears to be less prominent and bacterial entry is poor. In these cell lines, however, an alternative pathway of HSPG-dependent invasion is triggered in the presence of serum. One serum-derived factor that stimulates invasion has been identified as the extracellular matrix protein vitronectin (VN) (Duensing and van Putten, 1997; Gomez-Duarte *et al.*, 1997). VN binds specifically to Opa_{50}-expressing gonococci and stimulates bacterial uptake into HeLa cells in an $\alpha_v\beta_3$ and $\alpha_v\beta_5$ integrin-dependent manner. This signaling process also appears to depend on the activity of protein kinase C (PKC) (Dehio *et al.*, 1998*b*) (Figure 3b). A specific role for HSPG ligation in the two different Opa_{50}-dependent bacterial uptake mechanisms represented by Chang cells and HeLa cells has been confirmed using latex beads coated with antibodies directed against the HS-GAG side chains of HSPGs, thereby mimicking the binding activity of Opa_{50}. Consistent with bacterial uptake, these beads were efficiently internalized by Chang cells in the absence of any additional factor, while serum or purified VN (which non-specifically associates with the beads) was necessary to stimulate efficient uptake into HeLa cells (Dehio *et al.*, 1998*a*). Besides VN, the functionally related extracellular matrix protein fibronectin (FN) is also bound

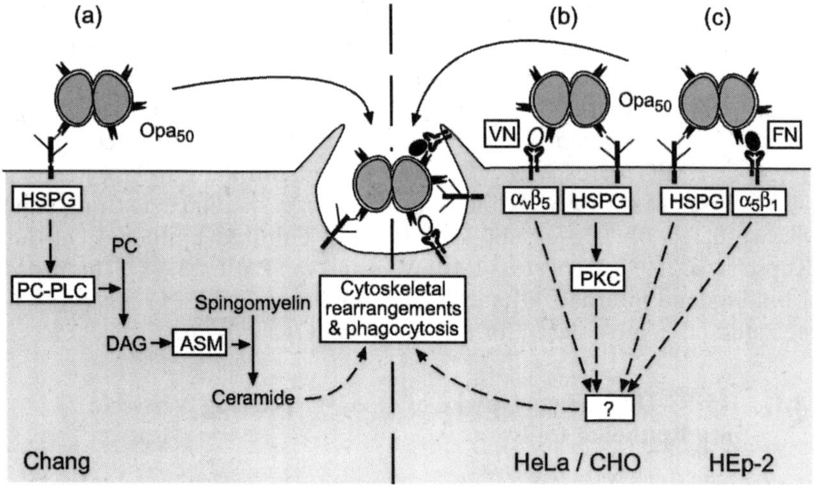

FIGURE 3. Opa_{50}-mediated internalization into cultured epithelial cells lines may occur by at least three distinct mechanisms. (a) In Chang conjunctiva epithelial cells, heparan sulfate proteoglycan (HSPG)-dependent internalization involves the activation of phosphatidylcholine-dependent phospholipase C (PC-PLC) which results in the generation of the second messenger diacylglycerol (DAG) from phosphatidylcholine (PC). DAG activates the acidic sphingomyelinase (ASM), which then generates ceramide from sphingomyelin. By an unknown process, ceramide is implicated in mediating cytoskeletal reorganization and bacterial uptake by a mechanism that resembles conventional phagocytosis. (b) Efficient bacterial uptake into HeLa cervical carcinoma cells and Chinese hamster ovary (CHO) cells also relies on the ability of Opa_{50} to mediate binding to the extracellular matrix protein vitronectin (VN) and to thereby co-ligate HSPGs and α_v integrin-containing VN receptors, including $\alpha_v\beta_5$. This internalization process appears to be dependent on the activity of protein kinase C (PKC). (c) In HEp-2 larynx carcinoma cells, efficient bacterial uptake of Opa_{50}-expressing gonococci requires binding of the extracellular matrix protein fibronectin (FN), which results in a coligation of HSPGs and the FN receptor $\alpha_5\beta_1$ integrin, however, the mechanism of entry is still poorly understood.

by Opa_{50}-expressing gonococci. This interaction triggers bacterial uptake into HEp-2 cells by co-ligating HSPGs and the FN-integrin receptor $\alpha_5\beta_1$ (van Putten *et al.*, 1998*b*) (Figure 3c).

The core protein of different HSPGs may either possess a transmembrane and intracellular domain, composing the syndecan receptor family, or may be glycosylphosphatidylinositol (GPI)-anchored to the membrane, composing the glypican family of HSPGs (Carey, 1997). Syndecan-4 is widely expressed by many epithelial cell lines. Interestingly, the overexpression of this receptor by HeLa cells increases both the VN-triggered uptake of Opa_{50}-expressing gonococci and the basal level of uptake in the absence of VN. In contrast, over-expression of a mutant form of syndecan-4 which carries a deletion of the cytoplasmic domain instead reduces bacterial uptake (E. Freissler, C.D. and T.F.M., unpublished). Hence, syndecan-4 seems to play a major role in mediating bacterial uptake, and the cytoplasmic domain appears to be critical for this process. Interestingly, Oh *et al.* (1997) have demonstrated that PKC, which appears to be necessary for bacterial uptake into HeLa cells (Dehio *et al.*, 1998*b*), binds directly to this cytoplasmic domain, resulting in its increased kinase activity. The roles of other syndecans (e.g. syndecan-1) and the glypicans in mediating gonococcal interaction with epithelial cells still remain to be investigated.

3.2.2. HSPG-Mediated Uptake of Opa_{50}-Expressing *Neisseria* into other Cell Types

Owing to the ubiquitous expression of HSPGs on eukaryotic cells, Opa-mediated interactions with HSPGs are not likely to be limited to epithelial cells. Opa_{50}-expressing bacteria bind strongly to endothelial cells, but uptake is inefficient in the absence of additional factors. Recruitment of either VN or FN to the surface of bacteria triggers efficient bacterial internalization in an integrin-dependent manner (M. Dehio, K.T. Preissner, T.F.M., and C.D., unpublished). Such a process could contribute to neisserial entry into the bloodstream and/or extravascularization of the bacteria leading to the colonization of non-mucosal tissues during disseminated disease. Fibroblasts, being potential targets in the submucosal tissue, may represent a further target for Opa_{50}-mediated infection (Grassmé *et al.*, 1997). Opa_{50} also mediates efficient uptake of gonococci into monocytes but has little effect on the interaction with polymorphonuclear leukocytes. Moreover, measurements of luminol-enhanced chemiluminescence demonstrated that phagocytosis of Opa_{50}-expressing gonococci was accompanied by a release of oxygen-reactive metabolites (Knepper *et al.*, 1997).

3.3. Opa-Mediated Interactions with the Cellular CD66 Receptors

Recently, several groups have clearly shown that most neisserial Opa variants specifically bind to the CD66 receptors which are differentially expressed on multiple tissues within the human host. CD66a (biliary glycoprotein, BGP), CD66c (non-specific cross-reacting antigen, NCA), CD66d (CEA gene family member 1, CGM1) and CD66e (carcinoembryonic antigen, CEA) can all function as receptors for *N. gonorrhoeae* and *N. meningitidis*, while CD66b (CGM6) is not recognized by any Opa protein tested to date. As shown in Table 1 using the well characterized Opa

Table I
Cellular Receptors for Opa Proteins Expressed by *N. gonorrhoeae* MS11

Opa allele[a]	Opa protein[a]	Cellular receptor	References[b]
C30 / C50[c]	30 (50)[c]	HSPG	1, 2
B51	51	CD66e	3, 4
G52	52	CD66a/c/d/e	3, 4
A53	53	CD66a	3, 4
I54	54	CD66a/e	3, 4
E55	55	CD66e	3, 4
F56	56	CD66e	3, 4
K57	57	CD66a/c/d/e	3, 4
J58	58	CD66a/c/d/e	3, 4
D59	59	CD66a/e	3, 4
H60	60	CD66a/c/d/e	3, 4

[a] Nomenclature used is as described in Malorny *et al.* (1998) and was provided by M. Achtman of the Max-Planck-Institut für Molekulare Genetik, Berlin, Germany. A list of nomenclature for all currently described *opa* alleles can be found at http://novell-ti.rz-berlin.mpg.de.
[b] (1) van Putten and Paul, 1995; (2) Chen *et al.*, 1995; (3) Bos *et al.*, 1997; (4) Gray-Owen *et al.*, 1997*b*.
[c] $opaC_{30}$/Opa_{30} and $opaC_{50}$/Opa_{50} refer to chromosomal and recombinant forms of the same allele, respectively.

variants of *N. gonorrhoeae* strain MS11 as an example, individual gonococcal and meningococcal Opa proteins may display various patterns of reactivity with one or more CD66 receptors (Gray-Owen *et al.*, 1997*a* and b; Bos *et al.*, 1997; Chen *et al.*, 1997; Virji *et al.*, 1996*a*; Chen and Gotschlich, 1996). Although most Opa proteins bind to either CD66 or HSPG receptors, a few appear to interact with both types of cellular receptors (de Vries *et al.*, 1998). The gonococcal Opa variants which do bind to both receptors are, however, able to mediate cellular invasion only via CD66 (Bos *et al.*, 1997; Gray-Owen *et al.*, 1997*b;* Kupsch *et al.*, 1993; Weel *et al.*, 1991).

CD66 proteins represent a subset of the CEA receptor family, which itself belongs to the immunoglobulin superfamily. Each receptor consists of a single, highly-conserved amino-terminal immunoglobulin variable region (Ig_V)-like domain, followed by a variable number of immunoglobulin Ig_{C2}-like constant domains exposed at the cell surface. The carboxyl-terminal domains of CD66a and CD66d contain transmembrane and cytoplasmic domains, while anchorage of CD66b, CD66c and CD66e occurs via a glycosylphosphatidylinositol (GPI)-anchor attached to the protein's carboxyl-terminus (Thompson *et al.*, 1991). Although the primary role of CD66 receptors *in vivo* is still unclear, they appear to mediate intercellular adhesion via both homotypic (CD66c and CD66e) and heterotypic (CD66b–CD66c and CD66c–CD66e) interactions (Oikawa *et al.*, 1991; Benchimol *et al.*, 1989). This function may or may not be related to their apparent influence on cell cycle control and cellular differentiation (Luo *et al.*, 1997; Screaton *et al.*, 1997). CD66a and CD66c have also been shown to present the sialyl Lewis^x (sLe^x) blood group antigen to E-selectin, and CD66c can also stimulate the upregulation of CD18-integrins on Huvecs. Both of these processes should facilitate the adherence of PMNs to inflammatory cytokine-stimulated endothelial cells *in vivo* (Kuijpers *et al.*, 1992), perhaps targetting the phagocyte to loci of infection. Interestingly, carbohydrate structures expressed by CD66a, CD66c and CD66e have also been shown to function as a cellular receptor for the type 1 fimbriae of *E. coli* and Salmonella strains, an interaction which may contribute to the commensal colonization of the colon by these species (Leusch *et al.*, 1990, 1991).

Although the individual proteins are highly glycosylated, these sugar structures do not function in binding to Opa proteins (Bos *et al.*, 1998). Chimeras of CD66b and CD66c indicate that Opa proteins bind to a surface composed of four β-strands (CC'FG) in the amino-terminal domain of the predicted CD66 protein structure (A. Popp, C.D., F. Grunert, T.F.M. and S.D.G., unpublished), a surface which is void of any glycosylation sites (Bates *et al.*, 1992). The differential specificity of Opa variants to each CD66 receptor is also determined by a divergent tripeptide sequence within this region (A. Popp, C.D., F. Grunert, T.F.M. and S.D.G., unpublished). Importantly, the CC'FG β-sheet has also been proposed to function as the ligand binding site of other members of the immunoglobulin superfamily, including the closely related membrane-bound CD2 (Bodian *et al.*, 1994) and CD4 (Wang *et al.*, 1990) receptor proteins, suggesting that the natural ligand for CD66 receptors may also bind here. Over 95% of Opa-expressing mucosal and disease isolates of *N. gonorrhoeae* and *N. meningitidis* have been shown to bind CD66a (Virji *et al.*, 1996*b*), thus implying the importance of these interactions for neisserial infection. It is also significant that meningococcal strains expressing both capsule and sialylated lipopolysaccharide are still able to bind CD66a (Virji *et al.*, 1996*b*), since these structures had previously been thought likely to sterically hinder any potential Opa-mediated interactions with a cellular receptor.

3.3.1. Opa/CD66-Dependent Interaction with Epithelial and Endothelial Cells

Individual CD66 receptors are differentially distributed on human tissues (Prall *et al.*, 1996; Thompson *et al.*, 1991; Berling *et al.*, 1990), suggesting that the distinct specificities of individual Opa variants may influence both tissue tropism and the cellular response to bacterial binding. Characterization of the Opa binding specificites has been achieved using stably transfected cell lines which express each CD66 receptor in isolation. In each case, adherence leads to a subsequent engulfment of the bound bacteria, thus demonstrating each family member is itself capable of facilitating cellular invasion (Bos *et al.*, 1997; Chen *et al.*, 1997; Gray-Owen *et al.*, 1997*b*). The singular importance of Opa proteins in this process is indicated by the fact that *E. coli* strains expressing recombinant Opa proteins are also efficiently internalized (Chen *et al.*, 1997; Gray-Owen *et al.*, 1997*b*).

Several cell lines which naturally express CD66 receptor proteins have also been identified, and some of these have been employed in infection assays to determine whether the Opa-expressing bacteria are also capable of invading into cells in a somewhat less artificial system. Consistent with the transfected cell lines, CD66 receptors expressed by human colonic (HT29) and lung (A549) epithelial cell lines can mediate meningococcal binding and engulfment (Virji *et al.*, 1996*a*), while Opa-expressing gonococci have been shown to be taken up by the CD66e-expressing LS174T colonic adenocarcinoma cells (Chen *et al.*, 1997). Recently, we have shown that bacteria expressing CD66-specific Opa proteins are capable of passing from the apical to basolateral surface of polarized T84 epithelial cell line monolayers without disrupting its transepithelial barrier function (Wang *et al.*, 1998). Electron microscopic analysis of infected monolayers clearly shows that this transmigration occurs via the transcellular route, since gonococci were evident within a tightly adherent phagocytic vacuole within the cell cytoplasm. This process depends on one or more of the CD66a, CD66c and CD66e receptors which were seen to be expressed exclusively on the apical surface of the monolayers, since CD66-specific antisera can block both bacterial binding and cellular invasion (Wang *et al.*, 1998). This remarkable process suggests that the presence of CD66 on epithelia of the cervix and uterus (Prall *et al.*, 1996) may mediate an analogous penetration of the epithelia to allow gonococcal dissemination from these mucosal surfaces.

Although primary endothelial cells (Huvecs) grown in culture typically express very low amounts of CD66a, there is a substantial upregulation of its expression following stimulation of these cells with the proinflammatory cytokine tumor necrosis factor alpha (TNFα; Gray-Owen *et al.*, 1997*b*). Although there is a report suggesting that treatment of Huvecs with TNFα may also upregulate CD66e (Majuri *et al.*, 1994), we have been unable to find the expression of CD66c, CD66d or CD66e receptors by various techniques, including flow cytometry, Western blotting and semi-quantitative reverse transcriptase-polymerase chain reaction (RT-PCR) analyses (P. Muenzner, S.D.G., C.D. and T.F.M., unpublished). The upregulated expression of CD66a results in a marked increase in binding by recombinant *N. gonorrhoeae* strains which express CD66a-specific Opa proteins (Gray-Owen *et al.*, 1997*b*), and our preliminary results indicate that this adherence may lead to bacterial uptake by the Huvecs.

3.3.2. CD66-Dependent Interactions with Phagocytic Cells

Clinical specimens from patients with gonorrhea typically display human polymorphonuclear neutrophils (PMNs) containing intracellular gonococci. *In vitro*, neisserial binding to PMNs results in the opsonin-independent phagocytosis of these bacteria (Fischer and Rest, 1988; Virji and Heckels, 1986; King and Swanson, 1978). The CD66 receptors are thought to be constitutively expressed at low levels on the surface of PMNs, however degranulation of these cells results in the release of large amounts of CD66 proteins from primary and secondary granules (Kuroki *et al.*, 1995; Ducker and Skubitz, 1992). Consistent with this, neisserial binding to PMNs increases significantly following their stimulation either by adherence to glass or treatment with phorbol myristate acetate (Farrell and Rest, 1990; Densen and Mandell, 1978).

Opa-mediated binding to CD66 receptors on PMNs correlates with the generation of an enhanced respiratory burst in comparison to that which is stimulated by non-opaque or piliated gonococci, however one that is reduced when compared to that seen with commensal neisserial species (Hauck *et al.*, 1997; Belland *et al.*, 1992; Virji and Heckels, 1986). The presence of Opa-specific $F(ab')_2$ antibody fragments prevents bacterial binding and the subsequent oxidative response, confirming a role for these adhesins in both of these events (Virji and Heckels, 1986). The fact that purified Opa protein can also abrogate this response also suggests that the cellular CD66 receptors must be ligated by multiple Opa proteins expressed on the bacterial surface in order to induce an oxidative burst (Naids *et al.*, 1991). Whether, and in which way, Opa-directed phagocytosis via CD66 receptors could provide a survival advantage to the bacterial population is currently still a matter of intriguing speculations.

3.3.3. CD66-Dependent Intracellular Signaling

The cytoplasmic domains of CD66a and CD66d contain sequences which have homology to the immunoreceptor tyrosine-based inhibitory (ITIM) and activation (ITAM) motifs, respectively (Beauchemin *et al.*, 1997; Nagel *et al.*, 1993). The presence of these structures on other receptors has been shown to modulate the response of various immune cells. For example, the clustering of ITAM-containing T cell receptor complexes results in the Src kinase-mediated activation of the T cells, whereas the coligation of an ITIM-containing receptor abrogates this response (Olcese *et al.*, 1996; Kolanus *et al.*, 1993). Similarly, the stimulation of B-cells by immunoglobulin binding to Fcγ RIIa receptors can be blocked by the inhibitory action of Fcγ RIIb receptors (Daeron *et al.*, 1995). Consistent with this (Binstadt *et al.*, 1996), CD66a has been shown to associate with the Src-family kinases Lyn and Hck (Skubitz *et al.*, 1995), and the Opa-mediated binding of *N. gonorrhoeae* to neutrophils does result in the activation of the Hck and Fgr in a process which is essential for bacterial phagocytosis (Hauck *et al.*, 1998). The Src homology 2-containing tyrosine phosphatase 1 (SHP-1) also interacts with CD66a (Beauchemin *et al.*, 1997), and appears to be inactivated by receptor cross-linking (C.R. Hauck, E. Gulbins and T.F.M., unpublished observations). Subsequent to these events, the GTPase Rac1, the p21-activated protein kinase (PAK) and Jun-N-terminal kinase (JNK) have also been shown to be stimulated. Bacterial uptake can be reduced either by the presence of Src kinase inhibitors or by the pre-incubation of cells with anti-sense oligonucleotides which down-regulate the expression of Rac1, indicating that both activities are essential for this process (Hauck *et al.*, 1998). This is consistent with the role of the small G-protein Rac1 in regulating cytoskeletal rearrangements (Hall, 1994), such as those which might be required for neisserial phagocytosis. The stress-activated protein kinase JNK has previously been implicated in the induction of activator protein 1 (AP-1)-regulated transcription (Koj, 1996), suggesting that a long-term adaptive response may also be triggered by neisserial CD66 binding. In epithelial cells, the JNK/AP-1 pathway may also be activated independent of Opa-mediated interactions via the Rho family of GTPases and the cellular kinases PAK, MKKK and MKK4 (Naumann *et al.*, 1998). Figure 4 summarizes the signaling cascade mediated by the interaction of Opa and cellular CD66 receptors in phagocytic cells.

3.4. Opa-Mediated Binding to Intracellular Pyruvate Kinase

Although intracellular gonococci are generally considered to remain inside a phagosomal compartment, occasional reports indicate that they may have the capacity to escape into the cytoplasm (Shaw and Falkow, 1988). Recently, gonococcal Opa proteins were reported to bind human pyruvate kinase (PK) subtype M2 *in vitro*, and this cytoplasmic enzyme appears to associate with intracellular Opa-expressing gonococci (Williams *et al.*, 1998). PK is expressed by many cell types including epithelial cells,

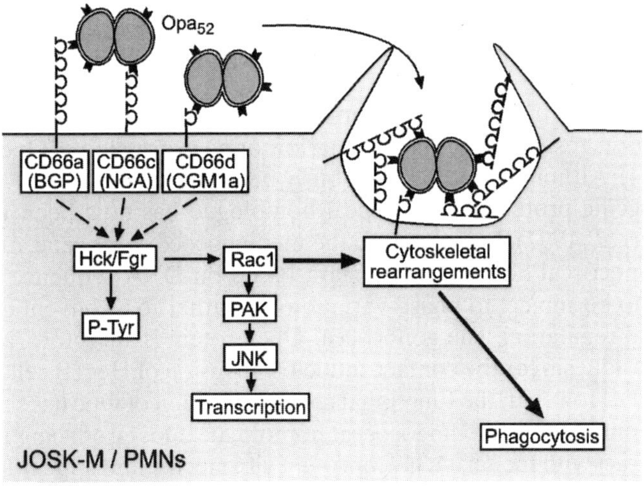

FIGURE 4. Schematic representation of Opa-mediated signaling by CD66 receptors in phagocytic cells. Opa_{52} is shown as a representative of the CD66-binding Opa protein family. In the myelomonocytic cell line JOSK-M and in polymorphonuclear neutrophils (PMNs), Opa-mediated binding to surface-expressed CD66 receptors [either CD66a (biliary glycoprotein, BGPa), CD66c (non-specific cross-reacting antigen, NCA) or CD66d (CEA gene family member 1, CGM1a)] results in the activation of the Src-family non-receptor protein tyrosine kinases Hck and Fgr. This results in an increased cellular protein tyrosine phoshorylation and the activation of the small G-protein Rac1, which is implicated in the cytoskeletal rearrangements ultimately leading to the phagocytic uptake of bound bacteria. Rac1 also activates the p21-activated protein kinase (PAK) and Jun-N-terminal kinase (JNK), probably leading to a subsequent activation of nuclear transcription.

and this enzyme catalyzes the irreversible conversion of phosphoenol pyruvate to pyruvate with the resulting generation of ATP. Interestingly, a *N. gonorrhoeae* mutant that is unable to use pyruvate or lactate is unaffected in its uptake into epithelial cells, but does not survive intracellularly (Williams *et al.*, 1998). It is thus possible that intracellular gonococci bind PK to gain an ample source of pyruvate, one of only three carbon sources known to be used by *N. gonorrhoeae*. Whether pyruvate may also play a role in the intracellular survival of *N. meningitidis* or other intracellular pathogens is still unknown.

4. Opc-MEDIATED INTERACTIONS

Opc is an outer membrane adhesin of the pathogenic *Neisseriae* which is similar in size to the Opa proteins, but is structurally and antigenically distinct from them (Merker *et al.*, 1997; Achtman *et al.*, 1988). The phase variable expression of the meningococcal Opc has been shown to be mediated by the generation of slip-strand errors in a homopolymeric cytidine tract residing within the promoter of the *opc* gene. The resulting changes in spacing between the –10 and –35 promoter elements affects the strength of transcription, resulting in either strong, low or no expression of Opc (Sarkari *et al.*, 1994). Although Opc has previously been considered to be a meningococcal-specific protein, a gonococcal homologue has now been identified (Merker *et al.*, 1997). In contrast to the meningococcal *opc* gene, expression of the gonococcal homologue appears, based on DNA sequence information, not to be subject to phase-variation but rather to a conventional transcriptional regulation. The gonococcal Opc protein is, therefore, a possible candidate adhesin for the contact-inducible invasion of Hec1B cells (Spence *et al.*, 1997). However, the functional characterization of Opc has so far been reported only for the meningococcal homologue. Opc-producing meningococci interact with the serum glycoprotein vitronectin, and appear to use this molecule to attach to $\alpha_v\beta_3$ integrins that are present on the apical pole of endothelial cells (Virji *et al.*, 1994). This molecular bridging may result in bacterial uptake into endothelial cells (Virji *et al.*, 1995*b*). Opc-expression also promotes meningococcal binding to and entry into certain epithelial cell lines (i.e. Chang conjunctiva cells) in the absence of additional factors (de Vries *et al.*, 1996; Virji *et al.*, 1992*b*). The host cell receptor for this process has recently been identified as being syndecan-like heparan sulfate proteoglycans (HSPGs) (de Vries *et al.*, 1998). The apparently analogous functions of the HSPG/VN-binding Opa proteins (e.g. Opa_{50}; see Section 3.2) and Opc seen *in vitro* makes the reason for neisserial maintenance of these two otherwise unrelated adhesins an intriguing question.

5. INTERACTIONS MEDIATED BY A NOVEL MULTIPLE ADHESIN FAMILY

Screening of known host cell surface components for their capacity to bind gonococci has lead to the description of several lacto- and ganglio-series glycolipids as being putative adhesion receptors for *N. gonorrhoeae* (Deal and Krivan, 1990; Strömberg *et al.*, 1988). For one of these putative receptors (asialo-$_{Gm1}$:Gal(β1-3)GalNAc(β1-4)Gal(β1-4)Glc(β1-1)Cer), a corresponding bacterial ligand has been identified. The heterologous expression of a plasmid encoding a 36 kDa neisserial protein in *E. coli* was suggested to confer bacterial adhesion to the immobilized glycolipid (Purachuri *et al.*, 1990). More recent experiments carried out in this laboratory indicate that a 36 kDa lipoprotein exhibits a yet undefined binding function and that another closely linked adhesin comprises the above cited glycolipid specificity (S. Eickernjäger, E. Fischer, J. Maier, T. Schwan and T.F.M., unpublished). Both genetically linked adhesins are encoded by multiple genes in the neisserial genomes, and as members of a multiple adhesin family have therefore been termed MafA and MafB, respectively. P36/MafA is clearly different from the sialic acid-specific 27 kDa adhesin, Sia-1, which is expressed by the commensal species *Neisseria flava* and recognizes the structure NeuAc(α2-3)Gal(β1-4)Glc on erythrocytes (Nyberg *et al.*, 1990). Although the ubiquitous expression of such glycolipids could potentially contribute to neisserial interactions with many host tissues, the current lack of described function for MafA and MafB makes their role in neisserial infection uncertain.

6. THE INFLUENCE OF LIPOOLIGOSACCHARIDE ON HOST CELL INTERACTIONS

Neisseria produce a short type of lipopolysaccharide, known as lipooligosaccharide (LOS), which lacks any repetitive O-side chains. Nonetheless, its structural heterogeneity is evident by the multiple size classes and antibody reactivity patterns seen in bacteria cultured *in vitro* (Schneider *et al.*, 1988). LOS variation relies on the phase-variable expression of the glycosyl transferases which are involved in the biosynthesis of the variable α-chain of LOS. This phenomenon occurs due to a slipped-strand mispairing within a homo-polymeric tract of guanines within the coding sequence, which thereby influences the reading frame of these genes (Burch *et al.*, 1997; Yang and Gotschlich, 1996; Jennings *et al.*, 1995; Danaher *et al.*, 1995). *In vitro*, the phase variation of LOS occurs spontaneously at a frequency of 0.02–0.2% (Schneider *et al.*, 1988). Several lines of evidence suggest that this variation also occurs *in vivo*. In meningococcal carriers, more than 70% of bacteria isolated from the nasopharynx preferentially express a short LOS species (Broome, 1986), whereas 97% of clinical isolates from the blood and CSF instead display the long form of LOS (Jones *et al.*, 1998). Likewise, experimental gonococcal infection of human volunteers demonstrates that bacteria isolated early in the infection have short LOS, whereas after the development of inflammatory response a long LOS species predominates (Schneider *et al.*, 1991).

6.1. The Defensive Role of LOS

A major difference between the variant LOS molecules is the presence of additional carbohydrates in the longer LOS forms, including a terminal galactose residue. This galactose can be externally modified by the membrane-associated bacterial sialyltransferase using host-derived or endogenous cytidine 5′-monophospho-*N*-acetylneuraminic acid (CMP-NANA) as a sialyl donor (Mandrell and Apicella, 1993; van Putten, 1993). LOS sialylation does occur during natural infection, and greatly affects the biological properties of bacteria seen *in vitro*, including both their ability to enter epithelial cells and to resist the host immune defenses. Opa_{50}-mediated entry into host cells is less efficient if LOS is sialylated, whereas invasion levels can be enhanced if sialylation is prevented due either to the absence of a terminal galactose residue or the unavailability of CMP-NANA substrate (de Vries *et al.*, 1996; van Putten, 1993). This effect may be due to the steric masking of outer membrane proteins such as Opa or Opc (Virji *et al.*, 1995*b*). Unsialylated bacteria are, however, more susceptible to killing by antibodies and complement (Moran *et al.*, 1994), likely due to its antigenic similarities with host cell structures. Indeed, the Gal(β1-4)Glc-NAc(β1-3)Gal(β1-4)Glc (lacto-*N*-neotetraose) carbohydrate structure present at the non-reducing terminus of LOS variants is identical to the non-reducing terminus of oligosaccharides found on many human glycolipids and glycoproteins, and its sialylation on the terminal galactose mimics human I and i antigens (Mandrell *et al.*, 1990; Nairn *et al.*, 1988). On the basis of these findings, it has been postulated that LOS variation may serve as a mechanism that enables bacterial switching between invasive and immunoresistant phenotypes (van Putten, 1993). This hypothesis is supported by clinical observations showing poor virulence of sialylated gonococci in experimentally infected human volunteers (Schneider *et al.*, 1996), despite the fact that such phenotypes do predominate during inflammatory disease (Schneider *et al.*, 1995).

Purified LOS has been shown to activate and/or be cytotoxic to a variety of host cell types. In primary endothelial cells, meningococcal LOS-mediated toxicity was found to be modulated by the pilus-dependent but Opc-independent adherence, suggesting that pili have a synergistic effect contributing to the overall damage caused by LOS (Dunn *et al.*, 1995). Interestingly, a viable meningococcal mutant producing no LOS or endotoxin was recently reported (Steeghs *et al.*, 1998). This mutant should allow to dissect between LOS-mediated cytotoxicity and cytotoxicity caused by other mechanisms, such as porin-induced apoptosis (see below). It should also allow the study of the interactions occurring between neisserial surface adhesins (e.g. pili, PilC, Opa, Opc and Maf) and host cell receptors in the absence of any interference by toxicity or the steric hindrance of the LOS structure.

6.2. Neisserial LOS as an "Adhesin"

In addition to its influence on cellular interactions mediated by Opa, Opc and pilus, the unsialylated form of lacto-*N*-neotetraose-containing

LOS may also play a more direct role in host cell binding in the absence of these proteins. Near the end of their functional life, human glycoproteins and glycolipids are converted from the sialo-[NANA(a2-3)Gal(β1-4)Glc-NAc(β1-3)Gal(β1-4)Glc] to the asialo-[Gal(β1-4)Glc-NAc(β1-3)Gal(β1-4)Glc] form, by the loss of their terminal NANA moieties. When this occurs, these compounds are bound by and removed from the circulation via asialo-glycoprotein receptors. Similarly, the asialo-lacto-*N*-neotetraose-containing LOS has been demonstrated to interact with both the asialoglycoprotein receptor and with an additional 70 kDa receptor protein on the surface of hepatic HepG2 cells. Furthermore, this variant LOS type does mediate gonococcal adherence and invasion into HepG2 cells by an Opa-independent mechanism (Porat *et al.*, 1995*a* and *b*). Whether a similar LOS-dependent uptake mechanism can also operate at the level of the mucosa has not yet been investigated.

7. THE MENINGOCOCCAL CAPSULE

In contrast to the gonococci, *N. meningitidis* and many commensal *Neisseria* spp. may express a polysaccharide capsule. The capsular polysaccharides of the *N. meningitidis* serogroups B and C, which predominate in the Northern hemisphere, are homopolymers of sialic acids with α-2,8 and α-2,9 linkages, respectively (Jennings *et al.*, 1977). In contrast, *N. meningitidis* serogroup A is responsible for most meningococcal epidemics and expresses a capsule which lacks sialic acid (Achtman, 1995). Polysialic capsules mediate resistance to both phagocytosis and complement-mediated killing via the alternative pathway of complement activation (Jarvis, 1995; Hammerschmidt *et al.*, 1994). Although CD66-specific Opa proteins have been shown to bind their receptors even in the presence of capsule, this interaction was enhanced in unencapsulated variants (Virji *et al.*, 1996*a*). HSPG-specific Opa- and Opc-mediated invasion into host cells is also blocked by the expression of a capsule. Consistent with this, selection of meningococcal variants which efficiently invade into primary mucosal cells results in the recovery of variants which are non-piliated, unencapsulated, and express a short LOS form which is not sialylated and an Opa protein which can bind to both HSPG and CD66 host cell receptors (de Vries *et al.*, 1998 and 1996). Together these findings suggest that capsule expression must be down-regulated following primary contact with the mucosa (Stephens *et al.*, 1993; Virji *et al.*, 1993*a*), and then be re-expressed following transmigration across the epithelial barrier in order to provide protection against the host's immune defenses.

Certain capsule types appear to be selected for by various environmental stimuli (Brener *et al.*, 1981), however, spontaneous phase variation of capsule expression is also observed under standard growth conditions (Hammerschmidt *et al.*, 1996*b*). Recently, a novel mechanisms of genetic variation that allows a reversible concurrent phase (on- and off-) switching of both polysialic capsule synthesis and endogenous LOS sialylation has been identified in meningococci (Hammerschmidt *et al.*, 1996*a*). This fluctuation between a sialylated and a non-sialylated bacterial phenotype operates by the reversible insertion/excision of a naturally occurring insertion sequence element into the essential sialic acid biosynthesis gene *siaA*. The *in vivo* relevance of this particular adaptive mechanism for the regulation of meningococcal penetration into the mucosal barrier does, however, remain to be demonstrated.

8. ROLE OF THE PORIN Porb IN CELLULAR INTERACTIONS

Porins are the most abundant proteins in the neisserial outer membrane. While *N. meningitidis* is capable of producing two porin species simultaneously, i.e. the phase variable PorA (previously knowns as class 1 outer membrane protein) and one of the two PorB allels (class 2 and class 3) (Tsai *et al.*, 1981), in *N. gonorrhoeae* only the PorB homologue with its two alleles P.1A and P.1B was known (Sandstrom *et al.*, 1984). Recently, however, the gene for a PorA homologue has been found in *N. gonorrhoeae* (D. Günther, A. Kahrs and T.F.M., unpublished), which led us to also adopt the PorA/PorB nomenclature (Hitchcock, 1989) for *N. gonorrhoeae*.

Several lines of *in vivo* evidence suggested a role of PorB in gonococcal virulence and, in particular, an association of the $PorB_{1A}$ allele with the invasive potential of these pathogens (Cannon *et al.*, 1983). In contrast to the neisserial Opa and pili proteins, PorB is constitutively expressed and undergoes neither antigenic nor phase variation. Despite the fact that a relatively large number of sub-alleles have been described, the relative stability of each allele within individual clones has allowed the antigenic classification of PorB to be used as the basis for serotyping of gonococcal strains (Knapp *et al.*, 1984). The porins are predicted to form a homotrimer in the bacterial outer membrane, with each monomer possessing 16 membrane-spanning sequences arranged into an amphipathic β-barrel which exposes eight loops at the bacterial surface (Ward *et al.*, 1992; van der Ley *et al.*, 1991; Jeanteur *et al.*, 1991). In the bacterial membrane, the porins serve as a classical ion and nutrient transport channel (Tommassen *et al.*, 1990; Benz, 1988), however, PorB also possesses the remarkable ability to translocate into artificial lipid bilayers (Song *et al.*, 1998; Mauro *et al.*, 1988) and eukaryotic cell membranes (Rudel *et al.*, 1996; Weel and van Putten, 1991; Blake and Gotschlich, 1987). Its vectorial transport requires a close juxtaposition of bacterial and cellular membranes, and leads to the portion of PorB which is normally exposed at the bacterial surface to become cytoplasmically-facing in the target cell membrane (Blake and Gotschlich, 1987). Once in the cellular membrane, the channel function of this porin is tightly regulated by the cytosolic nucleoside triphosphates GTP and ATP, either of which can interact with the portion of PorB exposed in the mammalian cell cytoplasm and thereby modulate its pore size, voltage-dependent gating and ion selectivity (Rudel *et al.*, 1996).

The transfer of PorB into the membrane of PMNs causes a transient change in membrane potential which ultimately leads to an inhibition of host cell degranulation without influencing the oxidative response of the phagocyte (Bjerknes *et al.*, 1995; Haines *et al.*, 1991; Haines *et al.*, 1988). In addition, purified PorB has been seen to block actin polymerization and the subsequent phagocytosis of bound meningococci by phagocytic cells (Bjerknes *et al.*, 1995). Recent studies in our laboratory have also shown that purified PorB can arrest the maturation of latex bead-containing phagosomes in human macrophage, as evidenced by a reduction in the association of late endocytic markers with the phagosome (I. Mosleh, L.A. Huber, P. Steinlein, C. Pasquali, D. Günther and T.F.M, unpublished). It is conceivable that PorB functions in concert with the CD66-specific Opa proteins in order to direct gonococci to a priviledged phagosomal niche. We have also seen that the translocated neisserial porin induces apoptosis of epithelial and phagocytic cell lines *in vitro*, an activity that is mediated by its generation of a rapid Ca^{2+} influx and its subsequent induction of the calcium-dependent cysteine protease calpain and members of the caspase family (A. Müller, D. Günther, F. Düx, M. Naumann, T.F.M and T. Rudel, unpublished). We assume a prime importance of this mechanism for several processes taking place during infection of the mucosal lining, such as cytotxic effects (Dunn *et al.*, 1995) and sloughing of epithelial cells (Mosleh *et al.*, 1997; Appicella *et al.*, 1996; McGee *et al.*, 1981). The induction of apoptosis and the concommitant tissue destruction may, however, be only one mechanism by which porin promotes the infection process. The potential of porin to trigger a Ca^{2+} influx may enhance the potential to invade epithelial cells by directly activating PKC (compare Figure 3). This notion is supported by our recent finding that a site-specific deletion in one of the surface loops of PorB leads to a dramatic decrease of Opa/HSPG-dependent epithelial cell invasion (F.J. Bauer, T. Rudel, M. Stein and T.F.M., unpublished). In line with, van Putten *et al.* (1998*a*) reported that the P.1A allele of PorB can, in the absence of phosphate, trigger gonococcal uptake into Chang cells even independently of Opa_{50}, supporting the idea that PorB provides at least a co-stimulatory function for epithelial cell invasion.

Together these data suggest that tight adherence mediated by the neisserial adhesins and the subsequent insertion of PorB into the target cell membrane may simultaneously affect bacterial and clonal viability by altering bacterial invasion, phagosomal maturation, degranulation and cellular viability. How these functions contribute to neisserial pathogenesis within the human host remains an exciting topic for future study.

9. SUMMARY

As outlined in this review, various experimental techniques have been employed in an attempt to understand neisserial pathogenesis. *In vitro* genetic analysis has been used to study the genetic basis for the structural variability of cell surface components. Transformed or primary epithelial cell cultures have provided the simplest model to analyze bacterial adherence and invasion, while the infection of polarized epithelial monolayers, fallopian tube and nasopharyngeal organ cultures, and ureteral tissue have each been used to more closely represent the events which occur *in vivo*. Finally, the *in vivo* infection of human volunteers with *N. gonorrhoeae* has provided a powerful means to confirm and expand the results obtained *in vitro*. By these various approaches, a number of neisserial adhesins (i.e. pili, Opa, Opc and P36) and additional putative virulence determinants which affect bacterial adherence and invasion into host cells (i.e. LOS, capsule, PorB) have been identified. Clearly, neisserial surface variation serves as an adaptive mechanism which can modulate tissue tropism, immune evasion and survival in the changing host environment. Important progress has been made in recent years with respect to the host cellular receptors and subsequent signal transduction processes which are involved in neisserial adherence, invasion and transcytosis. This has led to the identification of (i) CD46 as a receptor for pilus which allows adherence to epithelial and endothelial cells, (ii) HSPGs, in cooperation with vitronectin and fibronectin, as receptors for a particular subset of Opa proteins and Opc, which may both mediate invasion into most epithelial and endothelial cells, and (iii) CD66 as the receptors for most Opa variants, potentially being involved in cellular interactions including adherence, invasion and transcytosis with epithelial, endothelial and phagocytic cells. As most of these data have been obtained using transformed cell lines growing *in vitro*, attempts must be made to translate these basic observations into a more natural situation. It can be expected that the successful ongoing integration of laboratory findings from the various infection models with human volunteer studies will further increase our understanding of the biology of neisserial infection. Perhaps the most difficult but also most rewarding challenge for the future will be to use volunteer studies to identify and understand the role of host factors which are important for the infectious process. Hopefully, insights gained from each of these studies will reveal new and useful strategies for the preventative and/or therapeutic intervention into infection and disease by these fascinating microbes.

10. REFERENCES

Achtman, M., 1995, Epidemic spread and antigenic variability of *Neisseria meningitidis*, Trends Microbiol. **3**:186–192.

Achtman, M., Wall, R.A., Bopp, M., Kusecek, B., Morelli, G., Saken, E., and Hassan-King, M., 1991, Variation in class 5 protein expression by serogroup A meningococci during a meningitis epidemic, *J. Infect. Dis.* **164**:375–382.

Apicella, M.A., Ketterer, M., Lee, F.K., Zhou, D., Rice, P.A., and Blake, M.S., 1996, The pathogenesis of gonococcal urethritis in men: confocal and immunoelectron microscopic analysis of urethral exudates from men infected with *Neisseria gonorrhoeae*, *J. Infect. Dis.* **173**:636–646.

Bates, P.A., Luo, J., and Sternberg, M.J.E., 1992, A predicted three-dimensional structure for the carcinoembryonic antigen (CEA), *FEBS Lett.* **301**:207–214.

Beauchemin, N., Kunath, T., Robitaille, J., Chow, B., Turbide, C., Daniels, E., and Veillette, A., 1997, Association of biliary glycoprotein with protein tyrosine phosphatase SHP-1 in malignant colon epithelial cells, *Oncogene* **14**:783–790.

Belland, R.J., Chen, T., Swanson, J., and Fischer, S.H., 1992, Human neutrophil response to recombinant neisserial Opa proteins, *Mol. Microbiol.* **6**:1729–1737.

Benchimol, S., Fuks, A., Jothy, S., Beauchemin, N., Shirota, K., and Stanners, C.P., 1989, Carcinoembryonic antigen, a human tumor marker, functions as an intercellular adhesion molecule, *Cell* **57**:327–334.

Benz, R., 1988, Structure and function of porins from gram-negative bacteria. *Ann. Rev. Microbiol.* **42**:359–393.

Bergstrom, S., Robbins, K., Koomey, J.M., and Swanson, J., 1986, Piliation control mechanisms in *Neisseria gonorrhoeae*, *Proc. Natl. Acad. Sci. USA* **83**:3890–3894.

Berling, B., Kolbinger, F., Grunert, F., Thompson, J.A., Brombacher, F., Buchegger, F., von Kleist, S., and Zimmerman, W., 1990, Cloning of a carcinoembryonic antigen gene family member expressed in leukocyte of chronic myeloid leukemia patients and bone marrow, *Cancer Res.* **50**:6534–6539.

Bhat, K.S., Gibbs, C.P., Barrera, O., Morrison, S.G., Jahnig, F., Stern, A., Kupsch, E.M., Meyer, T.F., and Swanson, J., 1991, The opacity proteins of *Neisseria gonorrhoeae* strain MS11 are encoded by a family of 11 complete genes, *Mol. Microbiol.* **5**:1889–1901. Published erratum, 1992, in *Mol. Microbiol.* **6**:1073–1076.

Binstadt, B.A., Brumbaugh, K.M., Dick, C.J., Scharenberg, A.M., Williams, B.L., Colonna, M., Lanier, L.L., Kinet, J.P., Abraham, R.T., and Leibson, P.J., 1996, Sequential involvement of Lck and SHP-1 with MHC-recognizing receptors on NK cells inhibits FcR-initiated tyrosine kinase activation, *Immunity* **5**:629–638.

Bjerknes, R., Guttormsen, H.K., Solberg, C.O., and Wetzler, L.M., 1995, Neisserial porins inhibit human neutrophil actin polymerization, degranulation, opsonin receptor expression, and phagocytosis but prime the neutrophils to increase their oxidative burst, *Infect. Immun.* **63**:160–167.

Blake, M.S., and Gotschlich, E.C., 1987, Functional and immunological properties of pathogenic *Neisseria* surface proteins, in: *Bacterial Outer Membranes as Model Systems* (M. Inouye, ed.), John Wiley and Sons, New York, pp. 377–400.

Blake, M.S., Blake, C.M., Apicella, M.A., and Mandrell, R.E., 1995, Gonococcal opacity: lectin-like interactions between Opa proteins and lipooligosaccharide, *Infect. Immun.* **63**:1434–1439.

Bodian, D.L., Jones, E.Y., Harlos, K., Stuart, D.I., and Davis, S.J., 1994, Crystal structure of the extracellular region of the human cell adhesion molecule CD2 at 2.5 A resolution, *Structure* **2**:755–766.

Bos, M.P., Grunert, F., and Belland, R.J., 1997, Differential recognition of members of the carcinoembryonic antigen family by Opa variants of *Neisseria gonorrhoeae*, *Infect. Immun.* **65**:2353–2361.

Bos, M.P., Kuroki, M., Krop-Watorek, A., Hogan, D., and Belland, R.J., 1998, CD66 receptor specificity exhibited by neisserial Opa variants is controlled by protein determinants in CD66 N-domains, *Proc. Natl. Acad. Sci. USA* **95**:9584–9589.

Brener, D., DeVoe, I.W., and Holbein, B.E., 1981, Increased virulence of *Neisseria meningitidis* after *in vitro* iron limited growth at low pH, *Infect. Immun.* **33**:59–66.

Broome, C.V., 1986, The carrier state: *Neisseria meningitidis, J. Antimicrob. Chemother.* **18A**:25–34.

Burch, C.L., Danaher, R.J., and Stein, D.C., 1997, Antigenic variation in *Neisseria gonorrhoeae*: production of multiple lipooligosaccharides, *J. Bacteriol.* **179**:982–986.

Cannon, J.G., Buchanan, T.M., and Sparling, P.F., 1983, Confirmation of association of protein I serotype of *Neisseria gonorrhoeae* with ability to cause disseminated infection, *Infect. Immun.* **40**:816–819.

Carey, D.J., 1997, Syndecans: multifunctional cell-surface co-receptors, *Biochem. J.* **327**:1–16.

Chen, T., and Gotschlich, E.C., 1996, CGM1a antigen of neutrophils, a receptor of gonococcal opacity proteins, *Proc. Natl. Acad. Sci. USA* **93**:14851–14856.

Chen, T., Belland, R., Wilson, J., and Swanson, J., 1995, Adherence of pilus-Opa$^+$ gonococci to epithelial cells *in vitro* involves heparan sulfate, *J. Exp. Med.* **182**:511–517.

Chen, T., Grunert, F., Medina-Marino, A., and Gotschlich, E.C., 1997, Several carcinoembryonic antigens (CD66) serve as receptors for gonococcal opacity proteins, *J. Exp. Med.* **185**:1557–1564.

Cohen, M.S., Cannon, J.G., Jerse, A.E., Charniga, L.M., Isbey, S.F., and Whicker, L.G., 1994, Human experimentation with *Neisseria gonorrhoeae*: rationale, methods, and implications for the biology of infection and vaccine development, *Infect. Dis.* **169**:532–537.

Connell, T.D., Shaffer, D., and Cannon, J.G., 1990, Characterization of the repertoire of hypervariable regions in the Protein II (opa) gene family of *Neisseria gonorrhoeae*, *Mol. Microbiol.* **4**:439–449.

Daeron, M., Latour, S., Malbec, O., Espinosa, E., Pina, P., Pasmans, S., and Fridman, W.H., 1995, The same tyrosine-based inhibition motif, in the intracytoplasmic domain of Fc gamma RIIB, regulates negatively BCR-, TCR-, and FcR-dependent cell activation, *Immunity* **3**:635–646.

Danaher, R.J., Levin, J.C., Arking, D., Burch, C.L., Sandlin, R., and Stein, D.C., 1995, Genetic basis of *Neisseria gonorrhoeae* lipooligosaccharide antigenic variation. *J. Bacteriol.* **177**:7275–7279.

de Vries, F.P., van Der, E., van Putten, J.P., and Dankert, J., 1996, Invasion of primary nasopharyngeal epithelial cells by *Neisseria meningitidis* is controlled by phase variation of multiple surface antigens, *Infect. Immun.* **64**:2998–3006.

de Vries, F.P., Cole, J., Dankert, J., Frosch, M., and van Putten, J.P.M., 1998, *Neisseria meningitidis* producing the Opc adhesin binds epithelial cell proteogylcan receptors, *Mol. Microbiol.* **27**:1203–1212.

Deal, C.D., and Krivan, H.C., 1990, Lacto- and ganglio-series glycolipids are adhesion receptors for *Neisseria gonorrhoeae*, *J. Biol. Chem.* **265**:12774–12777.

Dehio, C., Freissler, E., Lanz, C., Gomez-Duarte, O.G., David, G., and Meyer, T.F., 1998*a*, Ligation of cell surface heparan sulfate proteoglycans by antibody-coated beads stimulates phagocytic uptake into epithelial cells: a model for cellular invasion by *Neisseria gonorrhoeae*, *Exp. Cell Res.* **242**:528–539.

Dehio, M., Gomez-Duarte, O.G., Dehio, C., and Meyer, T.F., 1998*b*, Vitronectin-dependent invasion of epithelial cells by *Neisseria gonorrhoeae* involves α_v integrin receptors, *FEBS Lett.* **424**:84–88.

Densen, P., and Mandell, G.L., 1978, Gonococcal interactions with polymorphonuclear neutrophils: importance of the phagosome for bactericidal activity, *J. Clin. Invest.* **62**: 1161–1171.

Diaz, J.L., and Heckels, J.E., 1982, Antigenic variation of outer membrane protein II in colonial variants of *Neisseria gonorrhoeae* P9, *J. Gen. Microbiol.* **128**:585–591.

Ducker, T.P., and Skubitz, K.M., 1992, Subcellular localization of CD66, CD67, and NCA in human neutrophils, *J. Leuko. Biol.* **52**:11–16.

Duensing, T.D., and van Putten, J.P., 1997, Vitronectin mediates internalization of *Neisseria gonorrhoeae* by Chinese hamster ovary cells, *Infect. Immun.* **65**:964–970.

Dunn, K.L., Virji, M., and Moxon, E.R., 1995, Investigations into the molecular basis of meningococcal toxicity for human endothelial and epithelial cells: the synergistic effect of LPS and pili, *Microb. Pathog.* **18**:81–96.

Facius, D., and Meyer, T.F., 1993, A novel determinant (comA) essential for natural transformation competence in *Neisseria gonorrhoeae* and the effect of a comA defect on pilin variation, *Mol. Microbiol.* **10**:699–712.

Farrell, C.F., and Rest, R.F., 1990, Up-regulation of human neutrophil receptors for *Neisseria gonorrhoeae* expressing PII outer membrane proteins, *Infect. Immun.* **58**:2777–2784.

Fischer, S.H., and Rest, R.F., 1988, Gonococci possessing only certain PII outer membrane proteins interact with human neutrophils, *Infect. Immun.* **56**:1574–1579.

Forest, K.T., and Tainer, J.A., 1997, Type-4 pilus-structure: outside to inside and top to bottom—a minireview, *Gene* **192**:165–169.

Forest, K.T., Bernstein, S.L., Getzoff, E.D., So, M., Tribbick, G., Geysen, H.M., Deal, C.D., and Tainer, J.A., 1996, Assembly and antigenicity of the *Neisseria gonorrhoeae* pilus mapped with antibodies, *Infect. Immun.* **64**:644–652.

Gibbs, C.P., Reimann, B.Y., Schultz, E., Kaufmann, A., Haas, R., and Meyer, T.F., 1989, Reassortment of pilin genes in *Neisseria gonorrhoeae* occurs by two distinct mechanisms, *Nature* **338**:651–652.

Givan, K.F., Thomas, B.W., and Johnston, A.G., 1977, Isolation of *Neisseria meningitidis* from the urethra, cervix, and anal canal: further observations, *Brit. J. Ven. Dis.* **53**: 109–112.

Gomez-Duarte, O.G., Dehio, M., Guzman, C.A., Chhatwal, G.S., Dehio, C., and Meyer, T.F., 1997, Binding of vitronectin to Opa-expressing *Neisseria gonorrhoeae* mediates invasion of HeLa cells, *Infect. Immun.* **65**:3857–3866.

Grassme, H., Gulbins, E., Brenner, B., Ferlinz, K., Sandhoff, K., Harzer, K., Lang, F., and Meyer, T.F., 1997, Acidic sphingomyelinase mediates entry *of N. gonorrhoeae* into nonphagocytic cells, *Cell* **91**:605–615.

Gray-Owen, S.D., Dehio, C., Haude, A., Grunert, F., and Meyer, T.F., 1997*a*, CD66 carcinoembryonic antigens mediate interactions between Opa-expressing *Neisseria gonorrhoeae* and human polymorphonuclear phagocytes, *EMBO J.* **16**:3435–3445.

Gray-Owen, S.D., Lorenzen, D.R., Haude, A., Meyer, T.F., and Dehio, C. 1997*b*, Differential Opa specificities for CD66 receptors influence tissue interactions and cellular response to *Neisseria gonorrhoeae*, *Mol. Microbiol.* **26**:971–980.

Haas, R., and Meyer, T.F., 1986, The repertoire of silent pilus genes in *Neisseria gonorrhoeae*: evidence for gene conversion, *Cell* **44**:107–115.

Haas, R., Schwarz, H., and Meyer, T.F., 1987, Release of soluble pilin antigen coupled with gene conversion in *Neisseria gonorrhoeae*, *Proc. Natl. Acad. Sci. USA* **84**:9079–9083.

Haas, R., Veit, S., and Meyer, T.F., 1992, Silent pilin genes of *Neisseria gonorrhoeae* MS11 and the occurrence of related hypervariant sequences among other gonococcal isolates, *Mol. Microbiol.* **6**:197–208.

Hagblom, P., Segal, E., Billyard, E., and So, M., 1985, Intragenic recombination leads to pilus antigenic variation in *Neisseria gonorrhoeae*, *Nature* **315**:156–158.

Haines, K.A., Yeh, L., Blake, M.S., Cristello, P., Korchak, H., and Weissmann, G., 1988, Protein I, a translocatable ion channel from *Neisseria gonorrhoeae*, selectively inhibits exocytosis from human neutrophils without inhibiting O_2-generation, *J. Biol. Chem.* **263**:945–951.

Haines, K.A., Reibman, J., Tang, X.Y., Blake, M., and Weissmann, G., 1991, Effects of protein I of *Neisseria gonorrhoeae* on neutrophil activation: generation of diacylglycerol from phosphatidylcholine via a specific phospholipase C is associated with exocytosis, *J. Cell Biol.* **114**:433–442.

Hall, A., 1994, Small GTP-binding proteins and the regulation of the actin cytoskeleton, *Ann. Rev. Cell Biol.* **10**:31–54.

Hammerschmidt, S., Birkholz, C., Zahringer, U., Robertson, B.D., van Putten, J., Ebeling, O., and Frosch, M., 1994, Contribution of genes from the capsule gene complex (cps) to lipooligosaccharide biosynthesis and serum resistance in *Neisseria meningitidis*, *Mol. Microbiol.* **11**:885–896.

Hammerschmidt, S., Hilse, R., van Putten, J.P., Gerardy-Schahn, R., Unkmeir, A., and Frosch, M., 1996*a*, Modulation of cell surface sialic acid expression in *Neisseria meningitidis* via a transposable genetic element, *EMBO J.* **15**:192–198.

Hammerschmidt, S., Muller, A., Sillmann, H., Muhlenhoff, M., Borrow, R., Fox, A., van Putten, J., Zollinger, W.D., Gerardy-Schahn, R., and Frosch, M., 1996*b*, Capsule phase variation in Neisseria meningitidis serogroup B by slipped-strand mispairing in the polysialyltransferase gene (siaD): correlation with bacterial invasion and the outbreak of meningococcal disease, *Mol. Microbiol.* **20**:1211–1220.

Hauck, C.R., Lorenzen, D., Saas, J., and Meyer, T.F., 1997, An *in vitro*-differentiated human cell line as a model system to study the interaction of *Neisseria gonorrhoeae* with phagocytic cells, *Infect. Immun.* **65**:1863–1869.

Hauck, C.R., Meyer, T.F., Lang, F., and Gulbins, E., 1998, CD66-mediated phagocytosis of Opa_{52} *Neisseria gonorrhoeae* requires a Src-like tyrosine kinase- and Rac1-dependent signalling pathway, *EMBO J.* **17**:443–454.

Heckels, J.E., 1981, Structural comparison of *Neisseria gonorrhoeae* outer membrane proteins. *J. Bacteriol.* **145**:736–742.

Heckels, J.E., Blackett, B., Everson, J.S., and Ward, M.E., 1976, The influence of surface charge on the attachment of *Neisseria gonorrhoeae* to human cells, *J. Gen. Microbiol.* **96**:359–364.

Hill, S.A., 1996, Limited variation and maintenance of tight genetic linkage characterize heteroallelic *pilE* recombination following DNA transformation of *Neisseria gonorrhoeae*, *Mol. Microbiol.* **20**:507–518.

Hill, S.A., Morrison, S.G., and Swanson, J., 1990, The role of direct oligonucleotide repeats in gonococcal pilin gene variation, *Mol. Microbiol.* **4**:1341–1352.

Hitchcock, P.J., 1989, Unified nomenclature for pathogenic *Neisseria* species, *Clin. Microbiol. Rev.* **2**:64–65.

Hobbs, M.M., Malorny, B., Prasad, P., Morelli, G., Kusecek, B., Heckels, J.E., Cannon, J.G., and Achtman, M., 1998, Recombinational reassortment among *opa* genes from ET-37 complex *Neisseria meningitidis* isolates of diverse geographical origins, *Microbiol.* **144**:157–166.

Hook III, E.W., and Holmes, K.K., 1985, Gonococcal infections, *Ann. Intern. Med.* **102**:229–243.
James, J.F., and Swanson, J., 1978, Studies on gonococcus infection. XIII. Occurrence of color/opacity colonial variants in clinical cultures, *Infect. Immun.* **19**:332–340.
Janda, W.M., Bohnoff, M., Morello, J.A., and Lerner, S.A., 1980, Prevalence and site-pathogen studies of *Neisseria meningitidis* and *N. gonorrhoeae* in homosexual men, *JAMA* **244**:2060–2064.
Jarvis, G.A., 1995, Recognition and control of neisserial infection by antibody and complement, *Trends Microbiol.* **3**:198–201.
Jeanteur, D., Lakey, J.H., and Pattus, F., 1991, The bacterial porin superfamily: sequence alignment and structure prediction, *Mol. Microbiol.* **5**:2153–2164.
Jennings, H.J., Battacharjee, A.K., Kenne, L., Kenny, C.P., and Calver, G., 1977, Structures of the capsular polysaccharides of *Neisseria meningitidis* as determined by ^{13}C-nuclear magnetic resonance spectroscopy, *J. Infect. Dis.* 136, S78-S83.
Jennings, M.P., Hood, D., Peak, I.R.A., Virji, M., and Moxon, E.R., 1995, Molecular analysis of a locus for the biosynthesis and phase variable expression of the lacto-*N*-neotetraose terminal LPS structure in *Neisseria meningitidis*, *Mol. Microbiol.* **18**:729–740.
Jerse, A.E., Cohen, M.S., Drown, P.M., Whicker, L.G., Isbey, S.F., Seifert, H.S., and Cannon, J.G., 1994, Multiple gonococcal opacity proteins are expressed during experimental urethral infection in the male, *J. Exp. Med.* **179**:911–920.
Jones, D.M., Borrow, R., Fox, A.J., Gray, S., Cartwright, K.A., and Pollman, J.T. The lipooligosaccharide immunotype as a virulence determinant in *Neisseria meningitidis*, *Microb. Pathog.* **13**:219–224.
Jonsson, A.B., Nyberg, G., and Normark, S., 1991, Phase variation of gonococcal pili by frameshift mutation in pilC, a novel gene for pilus assembly, *EMBO J.* **10**:477–488.
Jonsson, A.B., Ilver, D., Falk, P., Pepose, J., and Normark, S., 1994, Sequence changes in the pilus subunit lead to tropism variation of *Neisseria gonorrhoeae* to human tissue, *Mol. Microbiol.* **13**:403–416.
Kallstrom, H., Liszewski, M.K., Atkinson, J.P., and Jonsson, A.B., 1997, Membrane cofactor protein (MCP or CD46) is a cellular pilus receptor for pathogenic *Neisseria*, *Mol. Microbiol.* **25**:639–647.
Kellogg, D.S.J., Cohen, I.R., Norins, L.C., Schroeter, A.L., and Reising, G., 1968, *Neisseria gonorrhoeae*. II. Colonial variation and pathogenicity during 35 months *in vitro*, *J. Bacteriol.* **96**:596–605.
King, G.J., and Swanson, J., 1978, Studies on gonococcus infection. XV. Identification of surface proteins of *Neisseria gonorrhoeae* correlated with leukocyte association, *Infect. Immun.* **21**:575–583.
Knapp, J.S., Tam, M.R., Nowinski, R.C., Holmes, K.K., and Sandstrom, E.G., 1984, Serological classification of *Neisseria gonorrhoeae* with use of monoclonal antibodies to gonococcal outer membrane protein I, *J. Infect. Dis.* **150**:44–48.
Knepper, B., Heuer, I., Meyer, T.F., and van Putten, J.P., 1997, Differential response of human monocytes to *Neisseria gonorrhoeae* variants expressing pili and opacity proteins, *Infect. Immun.* **65**:4122–4129.
Koj, A., 1996, Initiation of acute phase response and synthesis of cytokines, *Biochim. Biophy. Acta* **1317**:84–94.
Kolanus, W., Romeo, C., and Seed, B., 1993, T cell activation by clustered tyrosine kinases, *Cell* **74**:171–170.
Koransky, J.R., Scales, R.W., and Kraus, S.J., 1975, Bacterial hemagglutination by *Neisseria gonorrhoeae*, *Infect. Immun.* **12**:495–498.
Kuijpers, T.W., Hoogerwerf, M., van der Laan, L.J.W., Nagel, G., van der Schoot, C.E., Grunert,

F., and Roos, D., 1992, CD66 Nonspecific cross-reacting antigens are involved in neutrophil adherence to cytokine-activated endothelial cells, *J. Cell Biol.* **118**:457–466.

Kupsch, E.-M., Knepper, B., Kuroki, T., Heuer, I., and Meyer, T.F., 1993, Variable opacity (Opa) outer membrane proteins account for the cell tropisms displayed by *Neisseria gonorrhoeae* for human leukocytes and epithelial cells, *EMBO J.* **12**:641–650.

Kuroki, M., Yamanaka, T., Matsuo, Y., Oikawa, S., Nakazato, H., and Matsuoka, Y., 1995, Immunochemical analysis of carcinoembryonic antigen (CEA)-related antigens differentially localized in intracellular granules of human neutrophils, *Immunol. Invest.* **24**:829–843.

Leusch, H.G., Hefta, S.A., Drzeniek, Z., Hummel, K., Markos-Pusztai, Z., and Wagener, C., 1990, *Escherichia coli* of human origin binds to carcinoembryonic antigen (CEA) and nonspecific crossreacting antigen (NCA), *FEBS Lett.* **261**:405–409.

Leusch, H.G., Drzeniek, Z., Markos-Pusztai, Z., and Wagener, C., 1991, Binding of *Escherichia coli* and *Salmonella* strains to members of the carcinoembryonic antigen family: differential binding inhibition by aromatic alpha-glycosides of mannose, *Infect. Immun.* **59**:2051–2057.

Liszewski, M.K., Post, T.W., and Atkinson, J.P., 1991, Membrane cofactor protein (MCP or CD46): newest member of the regulators of complement activation gene cluster, *Ann. Rev. Immunol.* **9**:431–455.

Luo, W., Wood, C.G., Earley, K., Hung, M.C., and Lin, S.H., 1997, Suppression of tumorigenicity of breast cancer cells by an epithelial cell adhesion molecule (C-CAM1): the adhesion and growth suppression are mediated by different domains, *Oncogene* **14**:1697–1704.

Majuri, M.-L., Hakkarainen, M., Paavonen, T., and Renkonen, R., 1994, Carcinoembryonic antigen is expressed on endothelial cells. A putative mediator of tumor cell extravasation and metastasis, *APMIS* **102**:432–438.

Malorny, B., Morelli, G., Kusecek, B., Kolberg, J., and Achtman, M., 1998, Sequence diversity, predicted two-dimensional protein structure, and epitope mapping of neisserial Opa proteins, *J. Bacteriol.* **180**:1323–1330.

Mandrell, R.E., Lesse, A.J., Sugai, J.V., Shero, M., Griffiss, J.M., Cole, J.A., Parsons, N.J., Smith, H., Morse, S.A., and Apicella, M.A., 1990, *In vitro* and *in vivo* modification of *Neisseria gonorrhoeae* lipooligosaccharide epitope structure by sialylation, *J. Exp. Med.* **171**:1649–1664.

Mandrell, R.E., and Apicella, M.A., 1993, Lipo-oligosaccharides (LOS) of mucosal pathogens: molecular mimicry and host-modification of LOS, *Immunobiol.* **187**:382–402.

Manning, P.A., Kaufmann, A., Roll, U., Pohlner, J., Meyer, T.F., and Haas, R., 1991, L-pilin variants of *Neisseria gonorrhoeae* MS11, *Mol. Microbiol.* **5**:917–926.

Marceau, M., Beretti, J.L., and Nassif, X., 1995, High adhesiveness of encapsulated *Neisseria meningitidis* to epithelial cells is associated with the formation of bundles of pili, *Mol. Microbiol.* **17**:855–863.

Mattick, J.S., Whitchurch, C.B., and Alm, R.A., 1996, The molecular genetics of type-4 fimbriae in *Pseudomonas aeruginosa*—a review, *Gene* **179**:147–155.

Mauro, A., Blake, M., and Labarca, P., 1988, Voltage gating of conductance in lipid bilayers induced by porin from outer membrane of *Neisseria gonorrhoeae*, *Proc. Natl. Acad. Sci. USA* **85**:1071–1075.

McGee, Z.A., Street, C.H., Chappell, C.L., Cousar, E.S., Morris, F., and Horn, R.G., 1979, Pili of *Neisseria meningitidis*: effect of media on maintenance of piliation, characteristics of Pili, and colonial morphology, *Infect. Immun.* **24**:194–201.

McGee, Z.A., Johnson, A.P., and Taylor-Robinson, D., 1981, Pathogenic mechanisms of *Neisseria gonorrhoeae*: observations on damage to human fallopian tubes in organ culture by gonococci of colony type 1 or type 4, *J. Infect. Dis.* **143**:413–422.

McGee, Z.A., Stephens, D.S., Hoffman, L.H., Schlech, W.F., and Horn, R.G., 1983, Mechanisms of mucosal invasion by pathogenic *Neisseria*, *Rev. Infect. Dis.* **5**:Supp. l–14.

Merker, P., Tommassen, J., Kusecek, B., Virji, M., Sesardic, D., and Achtman, M., 1997, Two-dimensional structure of the Opc invasin from *Neisseria meningitidis*, *Mol. Microbiol.* **23**:281–293.

Merz, A.J., Rifenbery, D.B., Arvidson, C.G., and So, M., 1996, Traversal of a polarized epithelium by pathogenic *Neisseriae*: facilitation by type IV pili and maintenance of epithelial barrier function, *Mol. Medicine* **2**:745–754.

Meyer, T.F., Mlawer, N., and So, M., 1982, Pilus expression in *Neisseria gonorrhoeae* involves chromosomal rearrangement, *Cell* **30**:45–52.

Meyer, T.F., Billyard, E., Haas, R., Storzbach, S., and So, M., 1984, Pilus genes of *Neisseria gonorrhoeae*: chromosomal organization and DNA sequence, *Proc. Natl. Acad. Sci. USA* **81**:6110–6114.

Moran, E.E., Brandt, B.L., and Zollinger, W.D., 1994, Expression of the L8 lipopolysaccharide determinant increases the sensitivity of *Neisseria meningitidis* to serum bactericidal activity, *Infect. Immun.* **62**:5290–5295.

Morelli, G., Malorny, B., Muller, K., Seiler, A., Wang, J.F., del Valle, J., and Achtman, M., 1997, Clonal descent and microevolution of *Neisseria meningitidis* during 30 years of epidemic spread, *Mol. Microbiol.* **25**:1047–1064.

Mosleh, I.M., Boxberger, H.J., Sessler, M.J., and Meyer, T.F., 1997, Experimental infection of native human ureteral tissuc with *Neisseria gonorrhoeae*: adhesion, invasion, intracellular fate, exocytosis, and passage through a stratified epithelium, *Infect. Immun.* **65**:3391–3398.

Murphy, G.L., Connell, T.D., Barritt, D.S., Koomey, M., and Cannon, J.G., 1989, Phase variation of gonococcal protein II: regulation of gene expression by slipped-strand mispairing of a repetitive DNA sequence, *Cell* **56**:539–547.

Nagel, G., Grunert, F., Kuijpers, T.W., Watt, S.M., Thompson, J., and Zimmerman, W., 1993, Genomic organization, splice variants and expression of CGM1, a CD66-related member of the carcinoembryonic antigen gene family, *Eur. J. Biochem.* **214**:27–35.

Naids, F.L., Belisle, B., Lee, N., and Rest, R.F., 1991, Interactions of *Neisseria gonorrhoeae* with human neutrophils: studies with purified PII (Opa) outer membrane proteins and synthetic Opa peptides, *Infect. Immun.* **59**:4628–4635.

Nairn, C.A., Cole, J.A., Patel, P.V., Parsons, N.J., Fox, J.E., and Smith, H., 1988, Cytidine 5′-monophospho-*N*-acetylneuraminic acid or a related compound is the low Mr factor from human red blood cells which induces gonococcal resistance to killing by human serum, *J. Gen. Microbiol.* **134**:3295–3306.

Nassif, X., Lowy, J., Stenberg, P., O'Gaora, P., Ganji, A., and So, M., 1993, Antigenic variation of pilin regulates adhesion of *Neisseria meningitidis* to human epithelial cells, *Mol. Microbiol.* **8**:719–725.

Nassif, X., Beretti, J.L., Lowy, J., Stenberg, P., O'Gaora, P., Pfeifer, J., Normark, S., and So, M., 1994, Roles of pilin and PilC in adhesion of *Neisseria meningitidis* to human epithelial and endothelial cells, *Proc. Natl. Acad. Sci. USA* **91**:3769–3773.

Novotny, P., and Turner, W.H., 1975, Immunological heterogeneity of pili of *Neisseria gonorrhoeae*, *J. Gen. Microbiol.* **89**:87–92.

Nyberg, G, Strömberg, N., Jonsson, A., Karsson, K.A., and Normark S., 1990, Erythrocyte gangliosides act as receptors for *Neisseria flava*: identification of the Sia-1 adhesin, *Infect. Immun.* **58**:2555–2563.

Oh, E.S., Woods, A., and Couchman, J.R., 1997, Multimerization of the cytoplasmic domain of syndecan-4 is required for its ability to activate protein kinase C, *J. Biol. Chem.* **272**:11805–11811.

Oikawa, S., Inuzuka, C., Kuroki, M., Arakawa, F., Matsuoka, Y., Kosaki, G., and Nakazato, H., 1991, A specific heterotypic cell adhesion activity between members of carcinoembryonic antigen family, W272 and NCA, is mediated by N-domains, *J. Biol. Chem.* **266**:7995–8001.

Olafson, R.W., McCarthy, P.J., Bhatti, A.R., Dooley, J.S., Heckels, J.E., and Trust, T.J., 1985, Structural and antigenic analysis of meningococcal piliation, *Infect. Immun.* **48**:336–342.

Olcese, L., Lang, P., Vely, F., Cambiaggi, A., Marguet, D., Blery, M., Hippen, K.L., Biassoni, R., Moretta, A., Moretta, L., Cambier, J.C., and Vivier, E., 1996, Human and mouse killer-cell inhibitory receptors recruit PTP1C and PTP1D protein tyrosine phosphatases, *J. Immunol.* **156**:4531–4534.

Parge, H.E., Forest, K.T., Hickey, M.J., Christensen, D.A., Getzoff, E.D., and Tainer, J.A., 1995, Structure of the fibre-forming protein pilin at 2.6 A resolution, *Nature* **378**:32–38.

Perry, A.C., Hart, C.A., Nicolson, I.J., Heckels, J.E., and Saunders, J.R., 1987, Inter-strain homology of pilin gene sequences in *Neisseria meningitidis* isolates that express markedly different antigenic pilus types, *J. Gen. Microbiol.* **133**:1409–1418.

Pinner, R.W., Spellman, P.A., and Stephens, D.S., 1991, Evidence for functionally distinct pili expressed by *Neisseria meningitidis*, *Infect. Immun.* **59**:3169–3175.

Porat, N., Apicella, M.A., and Blake, M.S., 1995*a*, A lipooligosaccharide-binding site on HepG2 cells similar to the gonococcal opacity-associated surface protein Opa, *Infect. Immun.* **63**:2164–2172.

Porat, N., Apicella, M.A., and Blake, M.S., 1995*b*, *Neisseria gonorrhoeae* utilizes and enhances the biosynthesis of the asialoglycoprotein receptor expressed on the surface of the hepatic HepG2 cell line, *Infect. Immun.* **63**:1498–1506.

Prall, F., Nollau, P., Neumaier, M., Haubeck, H.-D., Drzeniek, Z., Helmchen, U., Loning, T., and Wagener, C., 1996, CD66a (BGP), an adhesion molecule of the carcinoembryonic antigen family, is expressed in epithelium, endothelium, and myeloid cells in a wide range of normal human tissues, *J. Histochem. Cytochem.* **44**:35–41.

Pujol, C., Eugene, E., de Saint, M., and Nassif, X., 1997, Interaction of *Neisseria meningitidis* with a polarized monolayer of epithelial cells, *Infect. Immun.* **65**:4836–4842.

Purachuri, D.K., Seifert, H.S., Ajioka, R.S., Karlsson, K.A., and So, M. 1990, Identification and characterization of a *Neisseria gonorrhoeae* gene encoding a glycolipid-binding adhesin, *Proc. Natl. Acad. Sci. USA* **87**:333–337.

Rahman M., Källström H., Normark S., and Jonsson A., 1997, PilC of pathogenic *Neisseria* is associated with the bacterial surface, *Mol. Microbiol.* **25**:11–25.

Rayner, C.F., Dewar, A., Moxon, E.R., Virji, M., and Wilson, R., 1995, The effect of variations in the expression of pili on the interaction of *Neisseria meningitidis* with human nasopharyngeal epithelium, *J. Infect Dis.* **171**:113–121.

Robertson, B.D., and Meyer, T.F., 1992, Genetic variation in pathogenic bacteria, *Trends Genet.* **8**:422–427.

Robertson, J.N., Vincent, P., and Ward, M.E., 1977, The preparation and properties of gonococcal pili. *J. Gen. Microbiol.* **102**:169–177.

Rudel, T., van Putten, J.P., Gibbs, C.P., Haas, R., and Meyer, T.F., 1992, Interaction of two variable proteins (PilE and PilC) required for pilus-mediated adherence of *Neisseria gonorrhoeae* to human epithelial cells, *Mol. Microbiol.* **6**:3439–3450.

Rudel, T., Boxberger, H.J., and Meyer, T.F., 1995*a*, Pilus biogenesis and epithelial cell adherence of *Neisseria gonorrhoeae pilC* double knock-out mutants, *Mol. Microbiol.* **17**:1057–1071.

Rudel, T., Facius, D., Barten, R., Scheuerflug, I., Nonnenmacher, E., and Meyer, T.F., 1995*b*, Role of pili and the phase-variable PilC protein in natural competence for transformation of *Neisseria gonorrhoeae, Proc. Natl. Acad. Sci.* **92**:7986–7990.

Rudel, T., Scheuerpflug, I., and Meyer, T.F., 1995*c*, Neisseria PilC protein identified as type-4 pilus tip-located adhesin, *Nature* **373**:357–359.

Rudel, T., Schmid, A., Benz, R., Kolb, H.A., Lang, F., and Meyer, T.F., 1996, Modulation of *Neisseria* porin (PorB) by cytosolic ATP/GTP of target cells: parallels between pathogen accommodation and mitochondrial endosymbiosis, *Cell* **85**:391–402.

Ryll, R.R., Rudel, T., Scheuerpflug, I., Barten, R., and Meyer, T.F., 1997, PilC of *Neisseria meningitidis* is involved in class II pilus formation and restores pilus assembly, natural transformation competence and adherence to epithelial cells in PilC-deficient gonococci, *Mol. Microbiol.* **23**:879–892.

Salit, I.E., 1981, Hemagglutination by *Neisseria meningitidis*, *Can. J. Microbiol.* **27**:586–593.

Sandstrom, E.G., Knapp, J.S., Reller, L.B., Thompson, S.E., Hook, E.W., and Holmes, K.K., 1984, Serogrouping of *Neisseria gonorrhoeae*: correlation of serogroup with disseminated gonococcal infection, *Sex. Trans. Dis.* **11**:77–80.

Sarkari, J., Pandit, N., Moxon, E.R., and Achtman, M., 1994, Variable expression of the Opc outer membrane protein in *Neisseria meningitidis* is caused by size variation of a promoter containing poly-cytidine, *Mol. Microbiol.* **13**:207–217.

Schneider, H., Hammack, C.A., Apicella, M.A., and Griffiss, J.M., 1988, Instability of expression of lipooligosaccharides and their epitopes in *Neisseria gonorrhoeae*, *Infect. Immun.* **56**:942–946.

Schneider, H., Griffiss, J.M., Boslego, J.W., Hitchcock, P.J., Zahos, K.M., and Apicella, M.A., 1991, Expression of paragloboside-like lipooligosaccharides may be a necessary component of gonococcal pathogenesis in men, *J. Exp. Med.* **174**:1601–1606.

Schneider, H., Cross, A.S., Kuschner, R.A., Taylor, D.N., Sadoff, J.C., Boslego, J.W., and Deal, C.D., 1995, Experimental human gonococcal urethritis: 250 *Neisseria gonorrhoeae* MS11mkC are infective, *J. Infect. Dis.* **172**:180–185.

Schneider, H., Schmidt, K.A., Skillman, D.R., Van De Verg, L., Warren, R.L., Wylie, H.J., Sadoff, J.C., Deal, C.D., and Cross, A.S., 1996, Sialylation lessens infectivity of *Neisseria gonorrhoeae* MS11mkC, *J. Infect. Dis.* **173**:1422–1427.

Screaton, R.A., Penn, L.Z., and Stanners, C.P., 1997, Carcinoembryonic antigen, a human tumor marker, cooperates with Myc and Bcl-2 in cellular transformation, *J. Cell Biol.* **137**:939–952.

Seifert, H.S., Ajioka, R.S., Marchal, C., Sparling, P.F., and So, M., 1988, DNA transformation leads to pilin antigenic variation in *Neisseria gonorrhoeae*, *Nature* **336**:392–395.

Seifert, H.S., Wright, C.J., Jerse, A.E., Cohen, M.S., and Cannon, J.G., 1994, Multiple gonococcal pilin antigenic variants are produced during experimental human infections, *J. Clin. Invest.* **93**:2744–2749.

Seya, T., Hara, T., Iwata, K., Kuriyama, S., Hasegawa, T., Nagase, Y., Miyagawa, S., Matsumoto, M., Hatanaka, M., and Atkinson, J. P., 1995, Purification and functional properties of soluble forms of membrane cofactor protein (CD46) of complement: identification of forms increased in cancer patients' sera, *Internat. Immunol.* **7**:727–736.

Shaw, J.H., and Falkow, S., 1988, Model for invasion of human tissue culture cells by *Neisseria gonorrhoeae*, *Infect. Immun.* **56**:1625–1632.

Skubitz, K.M., Campbell, K.D., Ahmed, K., and Skubitz, A.P.N., 1995, CD66 family members are associated with tyrosine kinase activity in human neutrophils, *J. Immunol.* **155**:5382–5390.

Song, J., Minetti, C.A., Blake, M.S., and Colombini, M., 1998, Successful recovery of the normal electrophysiological properties of PorB (class 3) porin from *Neisseria meningitidis* after expression in *Escherichia coli* and renaturation, *Biochim. Biophys. Acta* **1370**:289–298.

Spence, J.M., Chen, C.-R., and Clark, V., 1997, A proprosed role for the lutropin receptor in contact-inducible gonococcal invasion of Hec1B cells, *Infect. Immun.* **65**:3736–3742.

Steeghs, L., den Hartog, R., den Boer, A., Zomer, B., Roholl, P., and van der Ley, P., 1998, Meningitis bacterium is viable without endotoxin, *Nature* **392**:449–450.

Stephens, D.S., and McGee, Z.A., 1981, Attachment of *Neisseria meningitidis* to human mucosal surfaces: influence of pili and type of receptor cell, *J. Infect. Dis.* **143**:525–532.

Stephens, D.S., Hoffman, L.H., and McGee, Z.A., 1983, Interaction of *Neisseria meningitidis* with human nasopharyngeal mucosa: attachment and entry into columnar epithelial cells, *J. Infect. Dis.* **148**:369–376.

Stephens, D.S., Whitney, A.M., Rothbard, J., and Schoolnik, G.K., 1985, Pili of *Neisseria meningitidis*. Analysis of structure and investigation of structural and antigenic relationships to gonococcal pili, *J. Exp. Med.* **161**:1539–1553.

Stephens, D.S., Spellman, P.A., and Swartley, J.S., 1993, Effect of the (alpha2–8)-linked polysialic acid capsule on adherence of *Neisseria meningitidis* to human mucosal cells, *J. Infect. Dis.* **167**:475–479.

Stern, A., and Meyer, T.F., 1987, Common mechanism controlling phase and antigenic variation in pathogenic neisseriae, *Mol. Microbiol.* **1**:5–12.

Stern, A., Brown, M., Nickel, P., and Meyer, T.F., 1986, Opacity genes in *Neisseria gonorrhoeae*: control of phase and antigenic variation, *Cell* **47**:61–71.

Stromberg, N., Deal, C., Nyberg, G., Normark, S., So, M., and Karlsson, K., 1988, Identification of carbohydrate structures that are possible receptors for *Neisseria gonorrhoeae*, *Proc. Natl. Acad. Sci. USA* **85**:4902–4906.

Swanson, J., 1973, Studies on gonococcus infection. IV. Pili: their role in attachment of gonococci to tissue culture cells, *J. Exp. Med.* **137**:571–589.

Swanson, J., 1978, Studies on gonococcus infection. XIV. Cell wall protein differences among color/opacity colony variants of *Neisseria gonorrhoeae*, *Infect. Immun.* **21**:292–302.

Swanson, J., Bergstrom, S., Robbins, K., Barrera, O., Corwin, D., and Koomey, J.M., 1986, Gene conversion involving the pilin structural gene correlates with pilus$^+$ in equilibrium with pilus$^-$ changes in *Neisseria gonorrhoeae*, *Cell* **47**:267–276.

Swanson, J., Barrera, O., Sola, J., and Boslego, J., 1988, Expression of outer membrane protein II by gonococci in experimental gonorrhoea, *J. Exp. Med.* **168**:2121–2129.

Thompson, J.A., Grunert, F., and Zimmerman, W., 1991, Carcinoembryonic antigen gene family: molecular biology and clinical perspectives, *J. Clin. Lab. Anal.* **5**:344–366.

Tinsley, C.R., and Nassif, X., 1996, Analysis of the genetic differences between *Neisseria meningitidis* and *Neisseria gonorrhoeae*: two closely related bacteria expressing two different pathogenicities, *Proc. Natl. Acad. Sci. USA* **93**:11109–11114.

Tjia, K.F., van Putten, J.P., Pels, E., and Zanen, H.C., 1988, The interaction between *Neisseria gonorrhoeae* and the human cornea in organ culture. An electron microscopic study, *Graefes Arch. Clin. Exp. Ophthalmol.* **226**:341–345.

Tommassen, J., Vermeij, P., Struyve, M., Benz, R., and Poolman, J. T., 1990, Isolation of *Neisseria meningitidis* mutants deficient in class 1 (porA) and class 3 (porB) outer membrane proteins, *Infect. Immun.* **58**:1355–1359.

Tonjum, T., and Koomey, M., 1997, The pilus colonization factor of pathogenic neisserial species: organelle biogenesis and structure/function relationships–a review, *Gene* **192**:155–163.

Tsai, C.M., Frasch, C.E., and Mocca, L.F., 1981, Five structural classes of major outer membrane proteins in *Neisseria meningitidis*, *J. Bacteriol.* **146**:69–78.

van der Ley, P., Heckels, J.E., Virji, M., Hoogerhout, P., and Poolman, J.T., 1991, Topology of outer membrane porins in pathogenic *Neisseria* spp, *Infect. Immun.* **59**:2963–2971.

van Putten, J.P., 1993, Phase variation of lipopolysaccharide directs interconversion of invasive and immuno-resistant phenotypes of *Neisseria gonorrhoeae*, *EMBO J.* **12**:4043–4051.

van Putten, J.P., and Paul, S.M., 1995, Binding of syndecan-like cell surface proteoglycan receptors is required for *Neisseria gonorrhoeae* entry into human mucosal cells, *EMBO J.* **14**:2144–2154.

van Putten, J.P.M., Duensing, T.D., and Carlson, J., 1998*a*, Gonococcal invasion of epithelial cells driven by P.IA, a bacterial ion channel with GTP binding properties, *J. Exp. Med.* **188**:941–952.

van Putten, J., Duensing, T.D., and Cole, R.L., 1998*b*, Entry of Opa+ gonococci into HEp-2 cells requires concerted action of glycosaminoglycans, fibronectin and integrin receptors, *Mol. Microbiol.* **29**:369–379.

Virji, M. and Everson, J.S., 1981, Comparative virulence of opacity variants of *Neisseria gonorrhoeae* strain P9, *Infect. Immun.* **31**:965–970.

Virji, M. and Heckels, J.E., 1986, The effect of protein II and pili on the interaction of *Neisseria gonorrhoeae* with human polymorphonuclear leucocytes, *J. Gen. Microbiol.* **132**:503–512.

Virji, M., Alexandrescu, C., Ferguson, D.J., Saunders, J.R., and Moxon, E.R., 1992*a*, Variations in the expression of pili: the effect on adherence of *Neisseria meningitidis* to human epithelial and endothelial cells, *Mol. Microbiol.* **6**:1271–1279.

Virji, M., Makepeace, K., Ferguson, D.J.P., Achtman, M., Sarkari, J., and Moxon, E.R., 1992*b*, Expression of the Opc protein correlates with invasion of epithelial and endothelial cells by *Neisseria meningitidis*, *Mol. Microbiol.* **6**:2785–2795.

Virji, M., Makepeace, K., Ferguson, D.J., Achtman, M., and Moxon, E.R., 1993*a*, Meningococcal Opa and Opc proteins: their role in colonization and invasion of human epithelial and endothelial cells, *Mol. Microbiol.* **10**:499–510.

Virji, M., Saunders, J.R., Sims, G., Makepeace, K., Maskell, D., and Ferguson, D.J., 1993*b*, Pilus-facilitated adherence of *Neisseria meningitidis* to human epithelial and endothelial cells: modulation of adherence phenotype occurs concurrently with changes in primary amino acid sequence and the glycosylation status of pilin, *Mol. Microbiol.* **10**:1013–1028.

Virji, M., Makepeace, K., and Moxon, E.R., 1994, Distinct mechanisms of interactions of Opc-expressing meningococci at apical and basolateral surfaces of human endothelial cells; the role of integrins in apical interactions, *Mol. Microbiol.* **14**:173–174.

Virji, M., Makepeace, K., Peak, I., Payne, G., Saunders, J.R., Ferguson, D.J., and Moxon, E.R., 1995*a*, Functional implications of the expression of PilC proteins in meningococci, *Mol. Microbiol.* **16**:1087–1097.

Virji, M., Makepeace, K., Peak, I.R.A., Ferguson, D.J.P., Jennings, M.P., and Moxon, E.R., 1995*b*, Opc- and pilus-dependent interactions of meningococci with human endothelial cells: molecular mechanisms and modulation by surface polysaccharides, *Mol. Microbiol.* **18**:741–754.

Virji, M., Makepeace, K., Ferguson, D.J.P., and Watt, S., 1996*a*, Carcinoembryonic antigens (CD66) on epithelial cells and neutrophils are receptors for Opa proteins of pathogenic neisseriae, *Mol. Microbiol.* **22**:941–950.

Virji, M., Watt, S.M., Barker, S., Makepeace, K., and Doyonnas, R., 1996*b*, The N-domain of the human CD66a adhesion molecule is a target for Opa proteins of *Neisseria meningitidis* and *Neisseria gonorrhoeae*, *Mol. Microbiol.* **22**:929–939.

Wainwright, L.A., Pritchard, K.H., and Seifert, H.S., 1994, A conserved DNA sequence is required for efficient gonococcal pilin antigenic variation, *Mol. Microbiol.* **13**:75–87.

Wang, J., Gray-Owen, S.D., Knorre, A., Meyer, T.F., and Dehio, C., 1998, Opa binding to cellular CD66 receptors mediates the transcellular traversal of *Neisseria gonorrhoeae* across polarised T84 epithelial cell monolayers, *Mol. Microbiol.*, **in press**.

Wang, J.F., Caugant, D.A., Morelli, G., Koumare, B., and Achtman, M., 1993, Antigenic and epidemiologic properties of the ET-37 complex of *Neisseria meningitidis*, *J. Infect. Dis.* **167**:1320–1329.

Wang, J.H., Yan, Y.W., Garrett, T.P., Liu, J.H., Rodgers, D.W., Garlick, R.L., Tarr, G.E., Husain, Y., Reinherz, E.L., and Harrison, S.C., 1990, Atomic structure of a fragment of human CD4 containing two immunoglobulin-like domains, *Nature* **348**:411–418.

Ward, M.E., Watt, P.J., and Robertson, J.N., 1974, The human fallopian tube: a laboratory model for gonococcal infection, *J. Infect. Dis.* **129**:650–659.

Ward, M.J., Lambden, P.R., and Heckels, J.E., 1992, Sequence analysis and relationships between meningococcal class 3 serotype proteins and other porins from pathogenic and non-pathogenic *Neisseria* species, *FEMS Microbiol. Lett.* **73**:283–289.

Weel, J.F., and van Putten, J.P., 1991, Fate of the major outer membrane protein P.IA in early and late events of gonococcal infection of epithelial cells, *Res. Microbiol.* **142**:985–993.

Weel, J.F.L., Hopman, C.T.P., and van Putten, J.P.M., 1991, *In situ* expression and localization of *Neisseria gonorrhoeae* opacity proteins in infected epithelial cells: apparent role of Opa proteins in cellular invasion, *J. Exp. Med.* **173**:1395–1405.

Williams, J.M., Chen, G.C., Zhu, L., and Rest, R.F., 1998, Using the yeast two-hybrid system to identify human epithelial cell proteins that bind gonococcal Opa proteins: intracellular gonococci bind pyruvate kinase via their Opa proteins and require host pyruvate for growth, *Mol. Microbiol.* **27**:171–186.

Yang, Q.L., and Gotschlich, E.C., 1996, Variation of gonococcal lipooligosaccharide structure is due to alterations in poly-G tracts in *lgt* genes encoding glycosyl transferases, *J. Exp. Med.* **183**:323–327.

Zak, K., Diaz, J.L., Jackson, D., and Heckels, J.E., 1984, Antigenic variation during infection with *Neisseria gonorrhoeae*: detection of antibodies to surface proteins in sera of patients with gonorrhea, *J. Infect. Dis.* **149**:166–174.

Zhang, Q.Y., DeRyckere, D., Lauer, P., and Koomey, M., 1992, Gene conversion in *Neisseria gonorrhoeae*: evidence for its role in pilus antigenic variation, *Proc. Natl. Acad. Sci. USA* **89**:5366–5370.

Chapter 5

Bartonella Interactions with Host Cells

Michael F. Minnick and Burt E. Anderson

1. INTRODUCTION

Bartonella species are versatile bacterial pathogens that can parasitize a variety of cells within the circulatory system of mammals. Following transmission by the bite of an arthropod, or through wounds inflicted by an infected mammal, bartonellae colonize the blood vasculature and produce an acute episode characterized by bacteremia and hemotrophy, low-grade fever and malaise, vascular lesions (hemangiomas, papules or peliosis) and lymphadenopathy. The course of disease can be life-threatening and frequently culminates in a carrier state within the host. Chronic *Bartonella* infections have been reported in humans, dogs, cats, rodents, and insect vectors such as lice and fleas. The wide variety of mammalian reservoirs and arthropod vectors (Table 1) suggests that infected animals and their ectoparasites will continue to be a perennial public health risk for humans.

Bartonella species are responsible for several emerging, infectious cardiovascular diseases of humans. Maladies caused by these bacteria are more

---

MICHAEL F. MINNICK Division of Biological Sciences, University of Montana, Missoula, Montana 59812-1002. **BURT E. ANDERSON** Department of Medical Microbiology and Immunology, College of Medicine, University of South Florida, Tampa, Florida 33612.

Subcellular Biochemistry, Volume 33: Bacterial Invasion into Eukaryotic Cells, edited by Oelschlaeger and Hacker. Kluwer Academic / Plenum Publishers, New York, 2000.

Table I
A Synopsis of *Bartonella* Species, Their Diseases and Epidemiology

Species	Symptomology	Vector(s)	Reservoir(s)
HUMAN PATHOGENS-			
bacilliformis	Oroya fever, verruga peruana	sandflies	humans
elizabethae[a]	endocarditis	unknown	unknown
henselae[a]	cat-scratch disease, endocarditis, bacillary angiomatosis, bacillary peliosis, bacteremic syndrome in humans and cats	cats, fleas	cats
quintana[a]	trench fever, endocarditis, bacillary angiomatosis, bacteremic syndrome	body louse	body louse
clarridgeiae	bacteremic syndrome in cats, cat-scratch disease?	fleas ?	cats
NON-HUMAN PATHOGENS-			
doshiae[b]	bacteremia	fleas	rodents, insectivores
grahamii[b]	bacteremia	fleas	rodents, insectivores
peromysci[b]	bacteremia	fleas	rodents, insectivores
talpae[b]	bacteremia	fleas	rodents, insectivores
taylorii[b]	bacteremia	fleas	rodents, insectivores
vinsonii[a]	bacteremia, endocarditis in dogs	ticks ? mites	dogs voles

[a] previously classified as a *Rochalimaea* species.
[b] previously classified (or at one time would have been classified) as *Grahamella* species.

common than most people realize, and include cat-scratch disease (CSD), bacillary angiomatosis (BA), urban trench fever and Oroya fever. Bartonellae also cause a chronic bacteremic syndrome and are frequently responsible for "unculturable" cases of endocarditis (Raoult *et al.*, 1996). Immunodeficient patients are particularly at risk of contracting opportunistic bartonellosis, but immunocompetent individuals can also be infected. The recent increased incidence of human bartonellosis probably reflects both an augmented frequency of infection plus a heightened awareness of the group's pathogenic potential. Three of the five *Bartonella* species known to cause human disease (Table 1), are spreading into habitats that have not been historically exploited. *B. quintana*, the agent of trench fever, was a significant cause of morbidity during World War's I and II (Kostrzewski, 1950; McNee and Renshaw, 1916). Poor sanitation, overcrowding and louse infestation all contributed to the high incidence trench fever among troops in the field. Following nearly a fifty-year hiatus, trench fever is now re-emerging as an agent of BA, endocarditis and "urban trench fever" among homeless, inner-city, alcoholic males (Jackson and

Spach, 1996). The vector and reservoir contributing to *B. quintana*'s current re-emergence are unknown, but exposure to lice and cats is suspected (Brouqi *et al.*, 1996). *B. bacilliformis*, the agent of Oroya fever, has been historically endemic to the high-altitude regions of the South American Andes, presumably due to the geographical restriction of its phlebotamine sandfly vector, *Lutzomyia verrucarum* (Kreier and Ristic, 1981). However, recent cases of Oroya fever reported in coastal areas of South America, suggest that the pathogen is actively spreading into new territories (Alexander, 1995). The emergence of Oroya fever is thought to be due to additional sandfly species serving as vectors and novel strains of the pathogen (Amano *et al.*, 1997; Alexander, 1995). Finally, *B. henselae* is emerging as an opportunistic pathogen of mainly immunocompromised patients and can cause endocarditis, peliosis and BA (Drancourt *et al.*, 1996; Koehler and Tappero, 1993). Undoubtedly, the increase in AIDS has fostered the coincidental emergence of *B. henselae* infection in humans.

In spite of *Bartonella*'s ability to infect several host cells and organisms, and the group's rising importance as etiologic agents of human infectious disease, very little is known regarding the molecular basis for virulence and emergence. This chapter details the current state of knowledge on the molecular biology of host cell adhesion and invasion by bartonellae and its contribution to disease.

2. HOST CELL TYPES

Bartonellae are highly adaptable pathogens that employ a generalist approach to parasitism, thereby allowing them to exploit a variety of hosts and cell types for replication. *Bartonella*'s parasitic versatility likely reflects its phylogenetic relationship to other α-*Proteobacteria*, including *Rhizobium*, *Agrobacterium*, and *Brucella* (Brenner *et al.*, 1993). Species within these genera are very adept at interacting with eukaryotic cells in mutualistic or parasitic associations.

2.1. Parasitism of Erythrocytes (Hemotrophy)

Although several bacterial genera are known to parasitize mammalian erythrocytes (e.g. *Anaplasma* and *Haemobartonella* species), only *Bartonella* demonstrates hemotrophy in humans. The most stark example of erythrocyte parasitism occurs during Oroya fever, the disease caused by *B. bacilliformis*. Two disparate and sequential stages occur during the course of this disease, however only the primary phase involves erythrocytes. This stage is characterized by an acute bacteremia, wherein nearly all the circulating erythrocytes are parasitized and the hematocrit is significantly reduced. Although not as dramatic as infections with *B. bacilliformis*, other bartonellae parasitize (invade) their respective host's erythrocytes. *B. henselae* has been shown to enter cat erythrocytes (Mehock *et al.*, 1998; Kordick and Breitschwerdt, 1995), but apparently only a small fraction of the bacteria are intracellularly located within red cells and many bacteria are located epicellularly or in the serum (Mehock *et al.*, 1998). The five *Bartonella* species previously classified as *Grahamella* (Table 1) can invade rodent erythrocytes with one to ten percent of the circulating erythrocytes infected with one to twenty bartonellae (Fay and Rausch, 1969; Tyzzer, 1942; Graham-Smith, 1905). Finally, *B. quintana* is known to epicellularly associate with human erythrocytes (Merrell *et al.*, 1978).

Previous studies using *B. quintana* showed that an unusually high concentration of heme (20–40 ug/ml), but not protoporphyrin or serum, was required for growth (Myers *et al.*, 1969). Erythrocyte parasitism undoubtedly fulfills the blood or hemin requirement necessary for the growth of all bartonellae. In fact, heme uptake is a strategy employed by several pathogenic bacteria to acquire iron and/or porphyrin. Iron acquisition is probably the underlying impetus for hemotrophy in bartonellae, however more studies are needed.

2.2. Parasitism of the Vascular Endothelium

A common feature of human bartonellosis is parasitism of vascular endothelial cells within the microvasculature. The resulting vascular lesions can affect many different organs including the skin, liver, brain, bone, lymph nodes, and eyes (Golnik *et al.*, 1994; Tappero *et al.*, 1993; Perkocha, Koehler *et al.*, 1992; Spach *et al.*, 1992).

In Oroya fever, the endothelial (tissue) phase of disease occurs two to four weeks following the hematic stage and results from *Bartonella*'s invasion of the endothelial cells lining the capillary beds. The resulting bacteria-filled vacuoles, termed rocha lima inclusions, are the predecessors to a localized cellular proliferation leading to formation of nodules or papules, termed verruga peruana (Arias-Stella *et al.*, 1986). Veruga lesions are essentially hemangiomas or soft, blood-filled tumors that can persist for several weeks to months and may cause bleeding and scarring (Weinman, 1965). Anemia is no longer present during this stage of disease (Ricketts, 1949), but viable bacteria can be recovered from the blood, marrow, and hemangioma tissue of patients. Continual shedding of the organism from the vascular endothelium into the circulatory system contributes to the chronic carrier state in humans.

Bacillary angiomatosis (BA), resulting from infection with *B. henselae* or *B. quintana*, is histologically similar to verruga peruana and superficially resembles the vascular neoplasms of Kaposi's sarcoma (Webster *et al.*, 1992; Leboit *et al.*, 1988). BA lesions contain extensive vascular channels containing protuberant, cuboidal endothelium. The lesions also contain a multicellular inflammatory infiltrate consisiting of neutrophils with leukocytoclastic characteristics (Cockerell *et al.*, 1990; Leboit *et al.*, 1989). BA may last for several months (Koehler *et al.*, 1992), and the disease can be life-threatening in immunocompromised patients.

Bacillary peliosis (BP) lesions are cystic blood-filled cavities within multiple or single organs including the liver, spleen and lymph nodes. BP may be accompanied by gastrointestinal distress, fever and chills, and the syndrome occurs alone or in combination with BA or bacteremic syndrome. Hepatomegaly and elevated levels of liver enzymes in the serum (e.g., γ-glutamyl transferase and alkaline phosphatase) are common and BP-induced liver failure can be fatal (Perkocha *et al.*, 1990).

Unlike veruga or BA, cat-scratch disease (CSD) is characterized by granulomatous skin lesions that develop in 7–10 days at the site of a bite or scratch from an infected cat. The lesions appear as papules or pustules, and contain necrotic areas bordered by histiocytes, lymphocytes and giant cells (Johnson and Helwig, 1969). Lymphadenopathy involving the proximal draining lymph nodes is characteristic, and occurs within two to three weeks of infection. Like the skin papules, lymphatic lesions are granulomatous microabcesses that containing infiltrates of lymphocytes, giant cells, and follicular hyperplasia (Carithers, 1985). CSD patients often present with a mild fever, malaise, and gastrointestinal distress. Although most cases are self-resolving, complications involving the central nervous system, bone, lung, liver, spleen, and eyes have been reported (Caniza *et al.*, 1995; McCrary, 1994; Milam *et al.*, 1990; Carithers, 1985).

2.3. Parasitism of Insect Cells

Details regarding the relationship between *Bartonella* and its insect vectors are largely unknown. Early reports with *B. bacilliformis* showed that wild-caught phlebotamine sandflies that did not appear to have taken a recent blood meal, were capable of transmitting the bacterium to experimental monkeys, suggesting that they were actively infected. However, *B. bacilliformis* could only rarely be cultured from the probosci of wild-caught sandflies (Hertig, 1942). Later work showed that sandflies that fed on patients with acute Oroya fever acquired an infection of their midgut lumen characterized by bacteria adhering to the midgut epithelium and heavily-contaminated feces (Hertig, 1942). With rare exception, bartonellae within these insects did not multiply, invade the midgut epithelium, or spread to other organs. The ability of sandflies to serve as an alternate host for *B. bacilliformis* is still in need of clarification. A recent report suggests that many more species of these insects may be facilitating the spread of Oroya fever into low-altitude regions of South America, than was previously thought (Alexander, 1995).

Fleas can serve as vectors for *B. henselae* (Chomel *et al.*, 1996). Given the coincidental infection of cats by *B. henselae* and *B. clarridgeiae* (Gurfield *et al.*, 1997), fleas probably serve as vectors for *B. clarridgeiae*, however this remains to be verified. A recent study by Higgins *et al.*, showed that artificial feeding of cat fleas with blood meals containing *B. henselae* can generate active infections in the gut of the cat flea. Viable *B. henselae* were found in the gut within three hours and up to nine days following infection. In addition, bartonellae were present in the feces. It is important to note that bacterial counts within the flea increased with time, suggesting that fleas are transient host for *B. henselae*; not only maintaining the bacteria within their gut, but fostering bacterial replication for several days (Higgins *et al.*, 1996). Undoubtedly this transient parasitism of the flea gut and contamination of the fecal material fosters transmission from the flea to mammalian hosts. Although *B. henselae* adheres to midgut epithelial cells, invasion of insect cells has not been demonstrated.

Bartonellae that are endemic to wild rodent and insectivore populations (previously termed *Grahamella* spp.; see Table 1) are also transmitted by fleas (Fay and Rausch, 1969; Tyzzer, 1942; Krampitz, 1962). Infection of small mammals peaks during periods of heavy flea infestation and can reach prevalence levels as high as 62% (Birtles *et al.*, 1994). These bartonellae are able to infect not only small mammals, but also the gut epithelium of the flea vector. Like *B. henselae*, these bartonellae are shed into the feces which, together with flea bites, can promote transmission to mammals (Kreier and Ristic, 1981).

Although it is still unclear if human body lice are responsible for *B. quintana*'s current re-emergence, this arthropod was clearly the vector of transmission for classical cases of trench fever (Byam, 1919). Humans become infected with *B. quintana* through louse bites or by contact with contaminated louse feces through skin abrasions, and incidence of trench fever is correlated with overcrowding and poor hygiene. In a classical experiment, Vinson and Fuller (1961) showed that *B. quintana* was capable of infecting and multiplying in the louse gut. Both Vinson and Fuller's work and that of Wolbach *et al.* (1922) suggest that *B. quintana*'s replication in the insect is strictly extracellular, with masses of bacteria in the lumen and adhering to the gut cells. Infected lice do not suffer any apparent pathology or death from the *Bartonella* infection (Wolbach *et al.*, 1922). Because of the carrier state of the louse, these insects can transmit the trench fever agent for extended periods of time.

3. ADHERENCE TO HOST CELLS

3.1. *Bartonella* Adhesins

Bartonella's ability to parasitize several cell types suggests that these bacterial pathogens possess a variety of surface molecules that facilitate adhesion to host cells. To date, most of the data regarding adhesins from bartonellae have been obtained using *in vitro* adherence assays together with monospecific antibodies or spontaneous mutants of the bacterium. Recent successes using random, Tn5-mediated mutagenesis (Dehio *et al.*, 1997a) and site-directed allelic exchange mutagenesis of the flagellin gene (Battisti and Minnick, 1997) are providing new avenues to study the role of putative virulence determinants in *in vitro* and *in vivo* models of *Bartonella* pathogenesis.

3.1.1. Flagella

B. bacilliformis and *B. clarridgeiae* possess a polar tuft of flagella that confers a high degree of motility upon the bacterium (Figure 1A). Early observations noting that motile *B. bacilliformis* bound human erythrocytes to a much greater extent than natural, non-motile mutants (Benson *et al.*, 1986), and that a polar tuft of fibers mediated erythrocyte adhesion (Walker and Winkler, 1981) prompted our lab to investigate whether the appendage was serving as an adhesin for host cells. In these studies, monospecific antibodies generated against the purified flagellin were used to treat the bacterium prior to red cell association assays. Data showed that antiflagellin antibodies significantly reduced *B. bacilliformis*' erythrocyte-binding capability by approximately 40% and 50% relative to pre-immune serum or saline controls, respectively (Scherer *et al.*, 1993). To more fully understand the contribution of this structure to adherence, we recently generated flagellin-minus mutants of *B. bacilliformis* by allelic exchange mutagenesis using a suicide vector construct that contained an internal portion of the flagellin gene (Battisti and Minnick, 1997). The resulting mutants do not synthesize flagellin protein, lack flagella (Figure 1B), and are completely non-motile. Flagellin$^-$ mutants of *B. bacilliformis* showed a 75% reduction in their ability to adhere to human red cells when compared to wild-type strains of the bacterium. This significant reduction in adhesiveness suggests

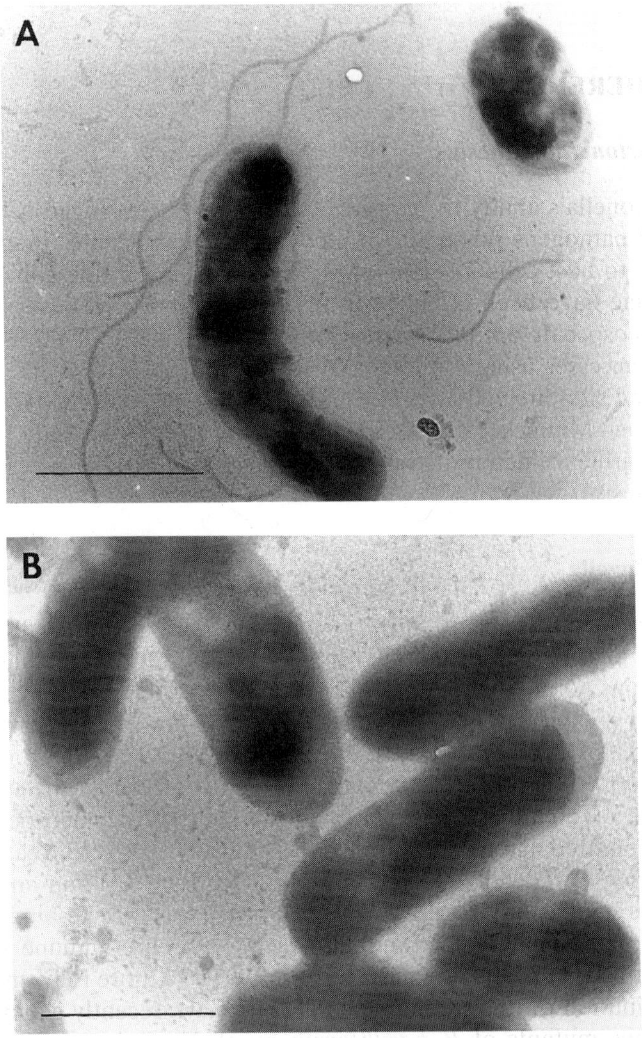

FIGURE 1. Transmission electron micrographs depicting: (A) a flagellated, wild-type *B. bacilliformis* strain (KC584), and (B) a "bald" strain (JB585) constructed by allelic exchange mutagenesis of the flagellin gene using a suicide vector (Battisti and Minnick, 1997). Bars = 0.5 μm.

that the flagellin plays an active role in adherence to erythrocytes, or at very least increases the likelihood of host cell-bacterium interactions, thereby increasing intercellular association. To fulfill Koch's molecular postulates, we also complemented these strains in *trans* using a flagellin-containing shuttle vector. Complementation restores expression of flagella. The resulting transcomplemented strains display an intermediate adhesive phenotype, perhaps due to expression and/or assembly anomalies (Battisti and Minnick, 1997). The role of flagella in adherence to other host cells is currently under investigation by our group.

3.1.2. Pili

Twitching motility displayed by *B. henselae*, together with the propensity of *Bartonella* species to autoagglutinate clearly implicated a type IV-like pilin in the virulence determinants of the group. In addition, early observations with *B. bacilliformis* suggested that a polar tuft of fibrous material (flagella? pili?) was serving as an erythrocyte adhesin (Walker and Winkler, 1981). With these characteristics in mind, Peek *et al.* (1995) discovered that low-passaged *B. henselae* expressed bundle-forming pili (BFP) on the surface of their cells (Figure 2), and that these structures would laterally autoaggregate to form bundles of pili. Subsequent work by our group showed that low-passaged *B. bacilliformis* also contained BFP with similar characteristics (McCallister *et al.*, 1995). BFP expression in both *B. bacilliformis* and *B. henselae* exhibits passage-dependent phase variation, where synthesis of the appendage is lost after repeated passage of the bacterium. In *B. henselae*, there is a concomitant change in colony morphology and autoagglutination phenotype during repeated passage; low-passage isolates autoagglutinate and form dry embedded colonies, whereas high-passage strains do not agglutinate and form unembedded, mucoid colonies. In *B. bacilliformis*, colonies freely interconvert between a small, translucent round morphology (T1) to a larger colony with an irregular edge (T2) (Walker and Winkler, 1981). Erythrocyte adherence rates for colony type T2 were nearly twice that of bacteria derived from colony type T1 (Walker and Winkler, 1981), but whether BFP expression is involved in the T1–T2 interconversion is unknown. The correlation between expression of BFP and *B. henselae*'s ability to adhere to, and invade HEp-2 cells is noteworthy (Batterman *et al.*, 1995). In these experiments, *B. henselae* (strain 87-66; piliated) adhered five times more frequently to Hep-2 cells than a high-passaged isogenic counterpart that was sparsely expressing pili. Recent studies with low-passaged (piliated) *B. quintana* (Oklahoma strain) show that the bacteria begin adhering to endothelial cells within the first minute of incubation (Brouqui and Raoul, 1996). In contrast, a non-piliated strain

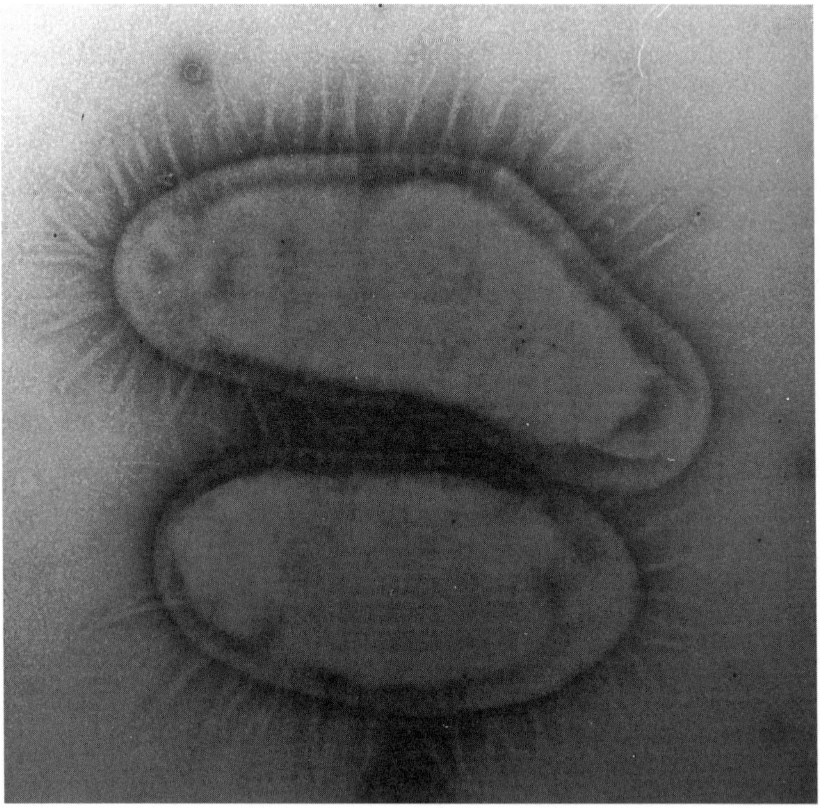

FIGURE 2. Transmission electron micrograph showing low-passaged, piliated cells *of B. henselae* (Houston-1 strain). Note the laterally aggregating bundles of pili on the cell surface. Gold particles (small black dots) = 10 nm. (Courtesy of Joel Peek and Stan Falkow, Stanford University).

of *B. quintana* (strain VR-358) lacks the capacity to adhere to endothelial cells (Batterman *et al.*, 1995). Whether the BFP from bartonellae are actually type IV will require biochemical and genetic analyses; prospects that have proven difficult, given the resistance of the BFP bundles to solubilization with detergents (Peek *et al.*, 1995; McCallister *et al.*, 1995). Clearly results to date are encouraging, but a more thorough investigation of the BFP's role in virulence is needed.

3.1.3. Other Surface-Exposed Proteins

Several biochemical methods have been used to determine the number of surface-exposed proteins on the surface of bartonellae. Estimates range from nine in *B. henselae* (Burgess and Anderson, 1998) to fourteen proteins in *B. bacilliformis* (Minnick, 1994). The role of these polypeptides in adhesion is largely unknown. However, recent work with *B. henselae* has identified five biotinylated proteins ranging from 28 to 58 kDa that are capable of binding to intact human umbilical vein endothelial cells (HUVECs). Of these, a 43 kDa polypeptide was identified as the major adhesin of the pathogen. It is significant to note that the 43 kDa protein was also recognized by reciprocal probing with biotinylated HUVEC surface proteins (Burgess and Anderson, 1998). The exact nature of the 43 kDa adhesin is currently under investigation.

3.2. Host Cell Association and Receptors

Very little is known regarding *Bartonella*'s host cell receptors. *B. bacilliformis* can associate with human umbilical vein endothelial cells (HUVECs) or epithelial cells (HEp-2) with equal binding efficiency, and most of the adhesion occurs within the first 60 minutes of incubation (McGinnis-Hill *et al.*, 1992). By contrast, erythrocyte adherence is most rapid within the first thirty minutes of incubation (Walker and Winkler, 1981) but requires about six hours for maximal complexing (Benson *et al.*, 1986). These distinct binding characteristics suggest that unique receptor—ligand interactions are taking place as *Bartonella* associates with different cell types. Early work by Walker and Winkler (1981) showed that adherence to red cells could be inhibited if *B. bacilliformis* is treated with UV light, or reagents that inactivate proton-motive force (*N*-ethyl maleimide) or respiration (KCN). However, adherence is unaffected by pre-treating the red cell with inhibitors of erythrocyte glycolysis (NaF) or proton-motive force (*N*-ethyl maleimide). In short, these observations suggest that adherence to erythrocytes requires a viable bacterium undergoing respiration and that the red cell's role in the process is passive.

The identity of the erythrocyte receptor for *Bartonella* is not known, but two studies imply that it is a glycolipid. The receptor is probably not proteinaceous, as erythrocyte association by *B. bacilliformis* or *B. henselae* is augmented if red cells are pre-treated with pronase or subtilisin, and trypsin treatment has no effect on binding (Mehock *et al.*, 1998; Walker and Winkler, 1981). By contrast, α- or β-glucosidase treatment of erythrocytes significantly decreases bacterial adhesion. Sequential treatment with protease and glucosidase suggests that proteolysis exposes a glycolipid receptor on the red cell that can be subsequently destroyed by glucosidase. The sialic acid of red cells is apparently not involved, as neuraminidase does not affect *B. henselae* or *B. Bacilliformis* adhesion (Mehock *et al.*, 1998; Walker and Winkler, 1981). It is also known that *B. bacilliformis* preferentially binds human erythrocytes relative to red cells from rabbits or sheep, suggesting that human cells possess a more appropriate receptor or possess greater receptor density/accessibility for *B. bacilliformis* (Walker and Winkler, 1981). Given *B. henselae*'s ability to infect cats and humans, it would be interesting to compare its binding efficiency with cat and human erythrocytes. More recently, studies show that *B. bacilliformis* and *B. henselae* recognize five or six proteins, respectively, from human erythrocyte membranes (Iwaki-Egawa and Ihler, 1997).

The endothelial cell receptor for bartonellae has not been characterized. The observation that *B. bacilliformis* binding to epithelial cells and endothelial cells displays a similar degree of efficiency, suggests that the pathogen's apparent predilection for endothelial cells may actually be due to tissue site (e.g., circulatory system), rather than receptor-mediated, constraints (McGinnis Hill *et al.*, 1992). In addition, binding data also suggest that both cell types contain a suitable, if not the same, receptor(s).

4. ENTRY INTO HOST CELLS

All *Bartonella* species can invade a variety of eukaryotic cells, but the reasons for the invasive phenotype are unclear. Undoubtedly, the cellular and humoral arms of the host's immune system plus nutritional demands were strong selective pressures favoring evolution of this phenotype. Invasion of erythrocytes has been documented for *B. bacilliformis* (Benson *et al.*, 1986), *B. henselae* (Mehock *et al.*, 1998; Kordick and Breitschwerdt, 1995), and those species endemic to rodents (Table 1) (Birtles *et al.*, 1995). Red cell invasion by *B. quintana* and *B. elizabethae* has not been demonstrated, although both require blood or hemin for growth. In addition to red blood cells, invasion of other host cell types (epithelial and endothelial cells) has been demonstrated for *B. henselae* (Dehio *et al.*, 1997b; Batterman *et al.*, 1995) and *B. bacilliformis* (Garcia *et al.*, 1992; McGinnis-Hill *et al.*, 1992) and *B. quintana* (Brouqui and Raoult, 1996).

4.1. Erythrocyte Invasion

Erythrocytes are uncommon host cells for bacteria. The task of invading relatively fragile, non-phagocytic cells while somehow preventing their lysis, is not trivial. *Bartonella*'s erythrocyte invasiveness ranges from a strictly epicellular association between *B. quintana* and human red cells (Merrell *et al.*, 1978) to *B. henselae*'s mild invasion of about five percent of circulating cat erythrocytes (Mehock *et al.*, 1998; Kordick and Breitschwerdt, 1995), to the marked invasion of nearly all circulating human erythrocytes by *B. bacilliformis* (Cuadra and Takano, 1969). The genetic basis for this species-specific differential invasiveness is unknown, but studies implicate a variety of virulence determinants that contribute to *Bartonella*'s invasion of erythrocytes.

4.1.1. Flagella

Early work by Benson *et al.* (1986) suggested that *B. bacilliformis* uses its flagella to force its way into the human red blood cell. These data prompted our lab to generate monospecific antibodies against the flagellin subunit to determine if they could inhibit the invasion process. The resulting antibody was used to treat *B. bacilliformis* prior to *in vitro* invasion assays, and nearly abolished invasion of human erythrocytes (Scherer *et al.*, 1993). These data suggest that the motive force provided by flagella are likely involved in the invasion process. We are currently conducting erythrocyte invasion assays using flagellin-minus allelic exchange mutants of the pathogen to more firmly establish the contribution of flagella to invasion.

4.1.2. Deformin

One of the remarkable features of *B. bacilliformis* hemotrophy is the marked production of pits and trenches in the red cell membrane (Figure 3) (Benson *et al.*, 1986). This phenomenon is the direct result of a bacterial protein termed deformation factor or deformin (Mernaugh and Ihler, 1992). Deformin generates red cell lesions that are indistinguishable from those observed in infected erythrocytes, and acts independently of the bacterium. Recent work indicates that *B. henselae* also produces the deformin protein (Iwaki-Egawa and Ihler, 1997), however, red cell invagination has never been reported during infection by this agent. Deformin is actively secreted during growth of *Bartonella* and is a 130 kDa homodimer in its native state. Deformin is sensitive to heat (70–80°C) and protease treatment and its activity is augmented if erythrocytes are pre-treated with trypsin or neuraminidase. Activity is destroyed if red cells are pre-treated with phospholipase D. The deformin-induced invaginations can be reversed by treatment with vanadate, DLPC, or by increasing intracellular Ca^{2+} levels with ionophores (Xu *et al.*, 1995; Mernaugh and Ihler, 1992). Whether deformin activity is restricted to erythrocytes is not known. Deformin-induced invaginations probably results in entry portals for colonizing the erythrocyte, but *Bartonella* cannot enter the red cell cytosol unless they are motile (Mernaugh and Ihler, 1992). The gene for deformin has not been characterized.

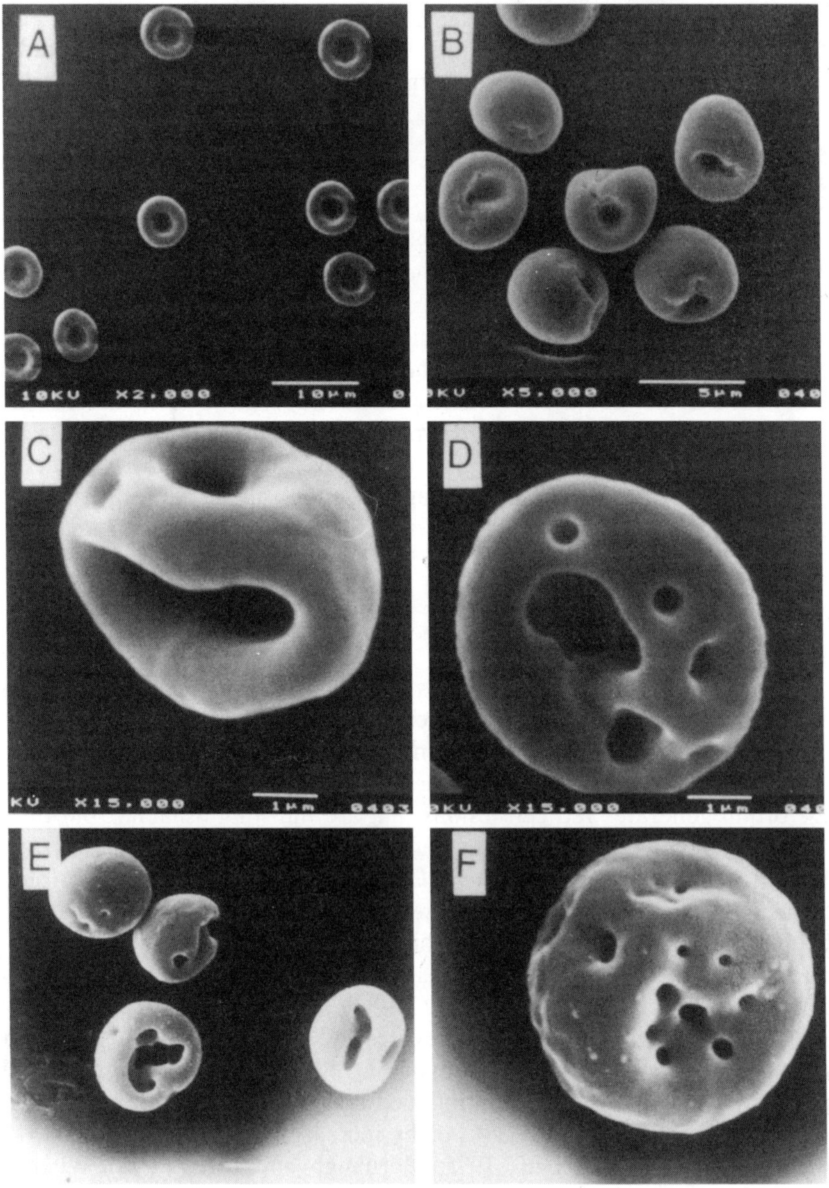

FIGURE 3. Scanning electron micrographs showing the effect of deformin from *B. bacilliformis* on human erythroctyes. (A) trypsinized normal red cells, (B–D) trypsinized cells incubated two hours with filtrate containing deformin, and (E–F) trypsinized erythrocytes incubated with purified deformin for two hours. (Reprinted from Xu, Y.-H., Lu, Z.-Y., and Ihler, G.M., 1995, *Biochim. Biophys. Acta*, **1234**:173–183. With permission.)

4.1.3. The Invasion-Associated Locus Genes

Recent work shows that *B. bacilliformis* possesses an invasion-associated locus (*ial*) (Mitchell and Minnick, 1995). The locus is approximately 1500 bp in length and contains two ORFs, termed *ialA* and *ialB*. Plasmids containing both genes can render an erythrocyte-invasive phenotype on minimally invasive strains of *E. coli* (strains HB101 and DH5.a), and both genes are necessary to confer invasiveness. The *ialA* gene codes for a 21 kDa protein with predicted NTPase activity, based upon homology to a consensus domain found in a variety of NTPases (Koonin, 1993). The *ialB* gene codes for a protein with approximately 60% amino acid sequence similarity to the adhesion and invasion locus (Ail) protein of *Yersinia enterocolitica* (Miller *et al.*, 1990) and the resistance to complement killing (Rck) protein of *Salmonella typhimurium* (Heffernan *et al.*, 1994). Mature IalB, Ail and Rck proteins are all approximately 18 kDa in mass. Both Ail and Rck are implicated in host cell attachment, invasion, and serum resistance. *B. henselae*, *B. quintana*, *B. vinsonii* and *B. elizabethae* all possess homologues of *ialA* and *ialB* when analyzed by low stringency DNA hybridizations (approx. 30% DNA mismatch) (Mitchell and Minnick, 1997a). A subsequent report by Murakawa *et al.* determined that the *ialAB* locus from *B. henselae* shares 70–85% nucleotide identity to the *B. bacilliformis* locus and that these genes can also confer a 100-fold increase in the invasive phenotype of *E. coli* in assays using human erythrocytes and epithelial cells (Murakawa, 1997; Murakawa *et al.*, 1996). These are exciting data suggesting that a conserved invasion locus is present in several *Bartonella* species and that it may facilitate entry into more than one human cell type. To further analyze the *ialAB* locus, we recently determined that the IalB protein is surface exposed on *B. bacilliformis*; in keeping with its Ail-like qualities (Coleman *et al.*, 1998), and generated *ialB*-minus mutants of *B. bacilliformis* for testing their virulence potential *in vitro*.

Characterization of genes that flank the *ialAB* locus indicates that *ialA* and *ialB* are components of a larger pathogenicity gene cluster. A gene encoding a carboxy-terminal protease, *ctpA*, lies immediately upstream of *ialA* and *ialB* (Figure 4). Results show that the CtpA protease is autocatalytic and that the protein is secreted and post- translationally modified (Mitchell and Minnick, 1997a). Like other C-terminal proteases, CtpA probably degrades abnormally-folded stress-response proteins (Barber and Andersson, 1992; Hara *et al.*, 1991). It is important to note that Prc, a C-terminal protease from *Salmonella typhimurium*, enhances intracellular survival in macrophages (Bäumler *et al.*, 1994). The conserved directionality of *ctpA* and close linkage to the *ialAB* genes from two *Bartonella* species (Mitchell and Minnick, 1997a; Murakawa *et al.*, 1996), and precedence for

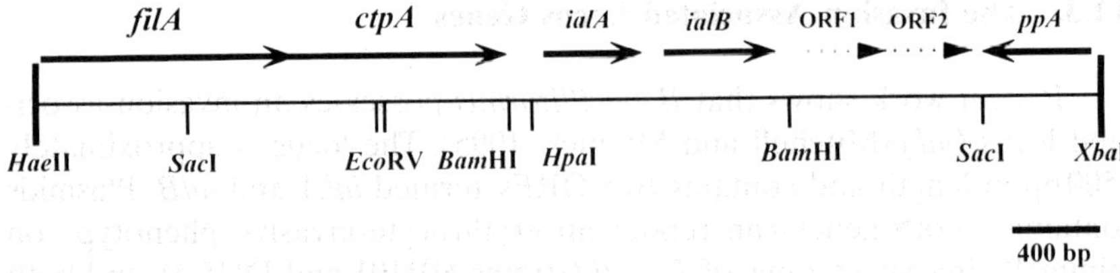

FIGURE 4. Linkage and partial restriction map of the invasion-associated locus of *B. bacilliformis* plus flanking sequences characterized to date. Relative positions of the genes are indicated by arrows. Gene abbreviations-*filA*, filament A gene; *ctpA*, carboxy-terminal processing protease; *ialA* and *B*, invasion-associated locus genes A and B, respectively; ORFs 1 and 2, open reading frames 1 and 2, respectively; and *ppa*, inorganic pyrophosphatase. [GenBank accession nos. U73652, L37094, L25276, and L46591, respectively]. Identical linkage and approximately 85% nucleotide sequence identity is observed in homologous gene clusters from *B. quintana* and *B. henselae*. (Reprinted from Minnick *et al.*, 1996, *Trends Microbiol.* **4**:343–347. With permission.)

virulence function by homologues from other bacteria, suggests that CtpA is involved in virulence.

Immediately upstream of *ctpA* is a 1200-bp gene termed *filA* (Figure 4) (Minnick, unpublished data). Characteristics of the encoded FilA polypeptide include: 1) a typical secretory signal sequence that follows the (–3,–1) rule, 2) a remarkable 60% alpha-helical secondary structure, 3) a C-terminal hydrophobic domain that could anchor the protein in a membrane, 4) leucine-rich composition (12%) with numerous leucine repeats; a characteristic found in proteins that engage in protein-protein interactions (Kobe and Deisenhofer, 1995), and 5) amino acid similarity to a variety of filamentous proteins including smooth muscle myosin and the M1 protein of *Streptococcus pyogenes*; a known virulence determinant involved in adhesion and invasion (Kehoe, 1994; Fischetti *et al.*, 1990). The M1 protein, like FilA, contains a C-terminal hydrophobic sequence that serves as a transmembrane anchor (Fischetti *et al.*, 1990). FilA's unidirectionality and close linkage to *ialAB*, the sequence and hydropathic similarities to M1 protein, its potential surface location and filamentous nature, plus the possibility that FilA provides for protein-protein interaction, argues that the protein is a possible virulence determinant.

Two ORFs of approximately 500 bp lie downstream of *ialAB* (Figure 4). Neither ORF, nor the predicted proteins encoded, has homologues with known function in various databases. In addition, neither ORF produced visible translation products in an *in vitro* transcription/translation system, nor did they have a significant effect on *in vitro* virulence assays conducted with *E. coli* transformants and human erythrocytes (Mitchell and Minnick, 1997a). Their potential role in virulence is unknown.

The 3' end of the putative pathogenicity gene cluster is presumably demarcated by a gene, lying in opposite orientation to the other six genes in the cluster, which encodes the inorganic pyrophosphatase (PPase) enzyme (Figure 4) (Mitchell and Minnick, 1997b). PPase is not likely involved in virulence, since this enzyme plays a metabolic housekeeping role in all organisms. To date, *B. quintana* (Minnick, unpublished data), *B. henselae* (Murakawa *et al.*, 1996) and *B. bacilliformis* have been shown to possess the gene cluster shown in Figure 4, and there is conservation of linkage and sequence (>85% nucleotide sequence identity).

4.2. Invasion of Other Host Cells

Early virulence studies using *B. bacilliformis* and cultured epithelial or endothelial cell monolayers demonstrated that host cells could be induced by *Bartonella* to reconfigure the cytoskeleton, thereby enhancing bacterial internalization. Uptake of *Bartonella* was significantly reduced (~30% of controls) if actin filament formation was inhibited by using cytochalasin D, or if the bacteria were pre-treated with anti-*bartonella* antiserum (McGinnis-Hill *et al.*, 1992). This partial inhibition of invasion with cytochalasin or antiserum suggests that *Bartonella* is playing an active role in the process, and that surface-borne molecule(s) which are accessible to antibody are involved. More recently, actin was shown to recognize and bind to five outer membrane proteins from *B. henselae*, and four of these proteins could be isolated by actin-based affinity chromatography (Iwaki-Egawa and Ihler, 1997). The nature of the actin-associating proteins remains to be determined.

Recent work by Dehio *et al.* (1997b) shows that *B. henselae* enters endothelial cells by a novel structure termed an invasome (Figure 5). In *in vitro* models of infection, *B. henselae cells* are contacted and moved rearward to form an aggregate on the leading lamella of the endothelial cell. The "clumps" of bacteria are subsequently engulfed by membrane protrusions that are rich in cortical F-actin, ICAM1 and phosphotyrosine. Based upon chemical inhibition studies, invasome activity is actin-dependent and microtubule-independent. Although most clinical strains of *B. henselae* were internalized via formation of the invasome, a natural mutant of *B. henselae* was internalized via an alternate process and was ultimately located in a perinuclear phagosome. The authors hypothesized that invasome-mediated internalization may somehow interfere with the perinuclear

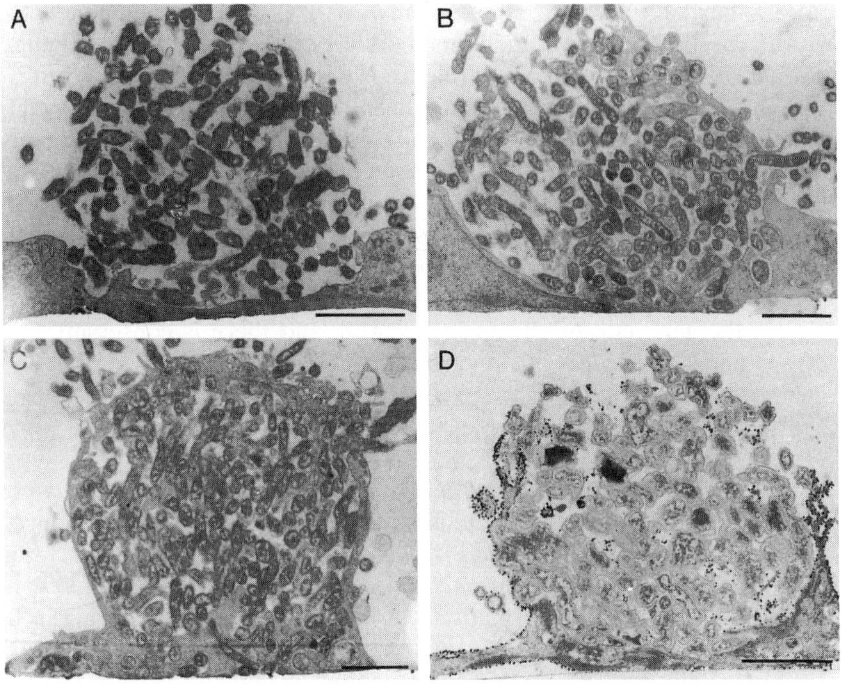

FIGURE 5. The invasome; a structure formed by endothelial cells during internalization of *B. henselae*. Sequential transmission electron microscopy analysis of: (A) initial invasome formation, (B,D) engulfment of *B. henselae*, and (C) internalization of the bacteria. The electron-dense granules seen in panel D reflect immunogold-labeling with monoclonal antibodies to ICAM-1, prior to embedding. (Reprinted from Dehio, C., Meyer, M., Schwarz, H., and Lanz, C., 1997, *J. Cell Sci.* **110**:2141–2154. With permission.)

phagosome formation. It is also possible that the mutant lacked the appropriate surface ligand for triggering formation of the structure.

B. quintana was shown to invade epithelial cells in classical *in vitro* experiments using Hep-2 cells (Vinson and Fuller, 1961). More recent work has shown that the pathogen is also invasive for human endothelial cells *in vitro* and, more importantly, in cardiac valve tissue from endocarditis patients (Brouqui and Raoult, 1996). *In vitro* studies with *B. quintana* and human endothelial cells show that bacteria can become internalized within only one minute of co-incubation. Concurrently, host cells exhibit ruffling and the bacterial cell wall becomes modified to produce surface appendages (20–40 nm wide by up to 500 nm length) (Figure 6). The appendages are apparently lost after adherence or internalization of the pathogen, and are similar to those observed in *Salmonella typhimurium*.

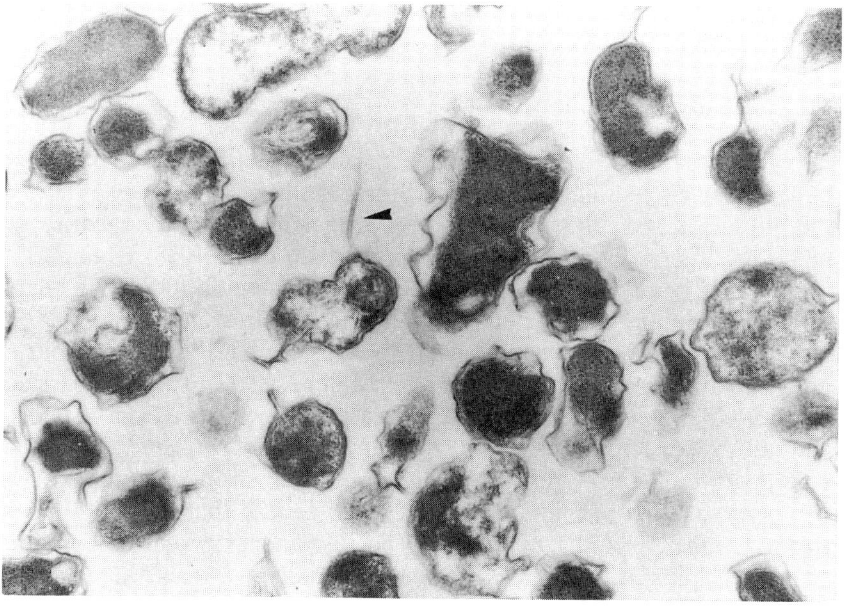

FIGURE 6. *B. quintana* appendages formed during association with vascular endothelium. (Reprinted from Brouqui P., and Raoult, D., 1996, *Res. Microbiol.* **147**:719–731. With permission.)

The authors hypothesize that the appendage is mediating or directing host cell adherence and endocytosis. Following host cell uptake, *B. quintana* multiplies in a vacuole, culminating in formation of a morulae resembling those seen during infection by ehrlichiae or chlamydiae. Older *B. quintana* and endothelial cell co-cultures showed that morulae contain both bacteria and vesicle-like blebs presumably derived from the bacterial membrane. However, membrane blebs were not observed in cardiac tissue samples. It is interesting to note that *B. bacilliformis* (Minnick, unpublished data), *B. henselae* and *B. quintana* (Brouqui and Raoult, 1996) all produce blebs during *in vitro* growth, but their potential role in pathogenesis remains a mystery.

B. bacilliformis invasion of endothelial cells can occur within one hour of co-culture, producing small membrane-bound inclusions. By three hours, numerous endosomes are observed in the host cell. By 12 hours, large cytoplasmic vacuoles, termed rocha lima inclusions, are formed that contain multiple bartonellae. Rocha lima inclusions can occupy the majority of the cytosol (Garcia *et al.*, 1992) and resemble the morulae produced by *B. quintana* (Brouqui and Raoult, 1996).

5. INTRACELLULAR REPLICATION AND HOST CELL DEATH; CONTRIBUTION TO BARTONELLOSIS

5.1. Hemolytic Anemia

The acute phase of Oroya fever is characterized by bacterial parasitism of nearly all circulating erythrocytes. More significantly, as many as 80% of these infected red cells ultimately lyse; producing one of the most severe hemolytic anemias known in humans (Hurtado *et al.*, 1938). The marked reduction in hematocrit that occurs during Oroya fever is thought to be due to splenic culling of infected red cells (Reynafarje and Ramos, 1961), however, a recently discovered β-hemolysin released by the pathogen during growth may be involved in the process (Minnick, 1997). The primary stage lasts for typically two to four weeks, and case fatalities range from 40 to 88% without proper antibiotic therapy (Weinman, 1965; Gray *et al.*, 1990). The high death rate from Oroya fever probably results from a combination of hemolytic anemia and immunosuppression; a state that probably results from the anemia and leukocytosis. The immunocompromised condition renders the patient susceptible to a host of secondary or reactivated illnesses including toxoplasmosis, tuberculosis, shigellosis, and salmonellosis; all illnesses that could compromise patient recovery (Gray *et al.*, 1990; Cuadra, 1956; Urteaga and Payne, 1955). The acute phase can be accompanied by anorexia, headache, and coma in fatal cases (Roberts, 1995).

5.2. Angiogenesis in the Vascular Endothelium

A number of studies using bartonellae and cultured human umbilical vein endothelial cells (HUVECs) have shown that these bacteria markedly change the morphology of the host cell during infection and enhance endothelial cell proliferation and migration. *In vivo*, these characteristics contribute to production of proliferative neovascular lesions including bacillary angiomatosis, verruga peruana, and peliosis. Endothelial cells infected with bartonellae are typically enlarged, elongated and fusiform; all characteristics resulting from reorganization of the cytoskeletal F actin (Palmari *et al.*, 1996). A modified topographical organization of the infected monolayer accompanies individual cellular modifications; an outcome analogous to the disrupted growth that accompanies oncogenesis and which undoubtedly contributes to the pseudoneoplastic characteristics of the vascular lesions. It is possible that these actin-based reorganizations of the cytoskeleton are initially triggered by formation of the actin-based invasome structure described above.

Bartonella species synthesize an angiogenic protein, termed the *Bartonella* angiogenic protein (BAP), that is specifically mitogenic for vascular endothelial cells and can induce neovascularization *in vivo*. Early studies using *B. bacilliformis* and human umbilical vein endothelial cells (HUVECs) showed that HUVECs treated with fractions containing BAP grew at rates nearly three times faster than uninduced control cells. BAP fractions were also found to stimulate production of tissue type plasminogen from HUVECs in a dose-dependent fashion and stimulated neovascularization in rat models of angiogenesis (Garcia *et al.*, 1990). Further work showed that live bartonellae including *B. bacilliformis*, *B. henselae*, or *B. quintana* were able to stimulate endothelial cell proliferation when co-cultured with HUVECs (Garcia *et al.*, 1992). Endothelial cells have also been shown to migrate towards bartonellae in co-cultures with HUVECs (Conley *et al.*, 1994). BAP's specificity for endothelial cells was shown using mitogenicity assays, wherein mitogenicity was observed with endothelial cells, but not fibroblasts, smooth muscle, or mesenchyme cells. The proteinaceous nature of BAP was established by its sensitivity to heat (56°C for 30 min or 100°C for 5 min) or trypsin treatment, and its ability to precipitate with $(NH_4)_2SO_4$ (Conley *et al.*, 1994; Garcia *et al.*, 1990). The protein does not have affinity for, nor is its activity enhanced by, heparin (Garcia *et al.*, 1990). Attempts to localize BAP in the *Bartonella* cell indicate that the protein is present in the cell wall fraction of *B. henselae* (Conley *et al.*, 1994) and in the cytosol of *B. bacilliformis* (Garcia *et al.*, 1990). Clearly, much remains to be determined regarding BAP's molecular nature and mechanism of angiogenic and tumorigenic activities.

5.3. Endocarditis

Several *Bartonella* species can invade the endothelial cells lining valve tissue and produce infectious endocarditis (Table 1). Endocarditis has been reported in both immunocompromised and immunocompetent humans (Drancourt *et al.*, 1996; Raoult *et al.*, 1996) and in domestic dogs (Breitschwerdt *et al.*, 1995). Recovery and culturing bartonellae from tissue biopsies can be difficult, and bacterial identification usually requires molecular or serological protocols. Endocarditis can produce vegetative lesions on the valve, and can spread to produce a generalized myocardia. A high titer antibody response often results that can be cross reactive with chlamydial antigens; making diagnosis troublesome (Maurin *et al.*, 1997). Valve replacement is often required in acute cases of endocarditis, and the disease can be fatal if left untreated.

6. CONCLUSIONS AND FUTURE DIRECTIONS

The emerging nature of bartonellosis as an infectious disease, together with the recent taxonomic reorganizations of *Bartonella*, *Rochalimaea* and *Grahamella* genera into a single genus, has spawned vigorous interest among microbiologists. Nevertheless, many questions remain unanswered. Comparative studies between clinical manifestations of bartonellosis in humans and other mammals are still in their infancy. Animal models are still being developed. Little is known regarding the relationship between bartonellae and their arthropod vectors. The molecular basis for pathogenesis is also an area of keen interest, but one with little data. How do these bartonellae generate such diverse and disparate symptomologies? How do they induce angiogenesis? The questions seem endless. To be sure, recent advances in our ability to manipulate the *Bartonella* genome by random and site-directed methods of mutagenesis will play a pivotal role in unraveling the molecular biology of these enigmatic pathogens.

ACKNOWLEDGMENTS. The authors gratefully acknowledge their past and present lab personnel and the financial support of the National Institutes of Health (Public Health Service grants AI34050 and RR10169 to MFM and AI38178 to BEA).

7. REFERENCES

Alexander, B., 1995, A review of bartonellosis in Ecuador and Colombia, *Am. J. Trop. Med. Hyg.* **52**:354–359.

Amano, Y., Rumbea, J., Knobloch, J., Olson, J., and Kron, M., 1997, Bartonellosis in Ecuador: serosurvey and current status of cutaneous verrucous disease, *Am. J. Trop. Med. Hyg.* **57**:174–179.

Arias-Stella, J., Lieberman, P.H., Erlandson, R.A., and Arias-Stella, J.J., 1986, Histology, immunohistochemistry, and ultrastructure of the verruga in Carrion's disease, *Am. J. Surg. Pathol.* **10**:595–610.

Barber, J., and Andersson, B., 1992, Too much of a good thing: light can be bad for photosynthesis, *Trends Biochem. Sci.* **17**:61–66.

Batterman, H.J., Peek, J.A., Loutit, J.S., Falkow, S., and Tompkins, L.S., 1995, *Bartonella henselae* and *Bartonella quintana* adherence to and entry into cultured human epithelial cells, *Infect. Immun.* **63**:4553–4556.

Battisti, J., and Minnick, M.F., 1997, Site-directed mutagenesis of the *Bartonella bacilliformis* flagellin gene, in: *Abstracts of the 13th National Meeting of the American Society for Rickettsiology*, Abstract no. 13.

Bäumler, A.J., Kusters, J.G., Stojiljkovic, I., and Heffron, F., 1994, *Salmonella typhimurium* loci involved in survival within macrophages, *Infect. Immun.* **62**:1623–1630.

Benson, L.A., Kar, S., McLaughlin, G., and Ihler, G.M., 1986, Entry of *Bartonella bacilliformis* into erythrocytes, *Infect. Immun.* **54**:347–353.

Birtles, R.J., Harrison, T.G., and Molyneux, D.H., 1994, *Grahamella* in small woodland mammals in the U.K.: isolation, prevalence and host specificity, *Ann. Trop. Med. Parasit.* **88**:317–327.

Birtles, R.J., Harrison, T.G., Saunders, N.A., and Molyneux, D.H., 1995, Proposals to unify the genera *Grahamella* and *Bartonella*, with descriptions of *Bartonella talpae* comb. nov., *Bartonella peromysci* comb. nov., and three new species, *Bartonella grahamii* sp. nov., *Bartonella taylorii* sp. nov., and *Bartronella doshiae* sp. nov., *Int. J. Syst. Bacteriol.* **45**:1–8.

Breitschwerdt, E.B., Kordick, D.L., Malarkey, D.E., Keene, B., Hadfield, T.L., and Wilson, K., 1995, Endocarditis in a dog due to infection with a novel *Bartonella* subspecies, *J. Clin. Microbiol.* **33**:154–160.

Brenner, D.J., O'Connor, S.P., Winkler, H.H., and Steigerwalt, A.G., 1993, Proposals to unify the genera *Bartonella* and *Rochalimaea*, with descriptions of *Bartonella quintana* comb. nov., *Bartonella vinsonii* comb. nov., *Bartonella henselae* comb. nov., and *Bartonella elizabethae* comb. nov., and to remove the family *Bartonellaceae* from the order *Rickettsiales*, *Int. J. Syst. Bacteriol.* **43**:777–786.

Brouqui, P., Houpikian, P., Dupont, H.T., Toubiana, P., Obadia, Y., Lafay, V., and Raoult, D., 1996, Survey of the seroprevalence of *Bartonella quintana* in homeless people, *Clin. Infect. Dis.* **23**:756–759.

Brouqui, P., and Raoult, D., 1996, *Bartonella quintana* invades and multiplies within endothelial cells *in vitro* and *in vivo* and forms intracellular blebs, *Res. Microbiol.* **147**:719–731.

Burgess, A.W.O., and Anderson, B.E., 1998, Outer membrane proteins of *Bartonella henselae* and their interaction with human endothelial cells, *Microb. Pathogen.* (in press).

Byam, W., 1919, Trench fever. In *Lice and their menace to man.* pp. 120–130. (LL Loyd, ed.) Oxford University Press, Oxford.

Caniza, M.A., Granger, D.L., Wilson, K.H., Washington, M.K., Kordick, D.L., and Frush, D.P., 1995, *Bartonella* (*Rochalimaea*) *henselae*: etiology of pulmonary nodules in a patient with depressed cell-mediated immunity, *Clin. Infect. Dis.* **20**:1505–1511.

Carithers, H.A., 1985, Cat scratch disease: an overview based on a study of 1200 patients, *Am. J. Dis. Child.* **139**:1124–1133.

Chomel, B.B., Kasten, R.W., Floyd-Hawkins, K., Chi, B., Yamamoto, K., Roberts-Wilson, J., Gurfield, A.N., Abbott, R.C., Pedersen, N.C., and Koehler, J.E., 1996, Experimental transmission of *Bartonella henselae* by the cat flea, *J. Clin. Microbiol.* **34**:1952–1956.

Cockerell, C.J., Bergstresser, P.R., Myrie-Williams, C., and Tierno, P.M., 1990, Bacillary epithelioid angiomatosis occurring in an immunocompetent individual, *Arch. Dermatol.* **126**:787–790.

Coleman, S., Mitchell, S.J., and Minnick, M.F., 1998, Surface localization of the invasion associated locus B protein in *Bartonella bacilliformis*, in: *Abstracts of the 12th General Meeting of the American Society for Microbiology*, Abstract no. B76.

Conley, T., Slater, L., and Hamilton, K., 1994, *Rochalimaea* species stimulate human endothelial cell proliferation and migration *in vitro*, *J. Lab. Clin. Med.* **124**:521–528.

Cuadra, M.S., 1956, Salmonellosis complication in human bartonellosis, *Tex. Rep. Biol. Med.* **14**:97–113.

Cuadra, M., and Takano, J., 1969, The relationship of *Bartonella bacilliformis* to the red blood cell as revealed by electron microscopy, *Blood.* **33**:708–716.

Dehio, C., and Meyer, M., 1997a, Maintenance of broad-host-range incompatibility group P and group Q plasmids and transposition of Tn*5* in *Bartonella henselae* following conjugal plasmid transfer from *Escherichia coli*, *J. Bacteriol.* **179**:538–540.

Dehio, C., Meyer, M., Schwarz, H., and Lanz, C., 1997b, Interaction of *Bartonella henselae* with endothelial cells results in bacterial aggregation on the cell surface and the subsequent engulgment and internalisation of the bacterial aggregate by a unique structure, the invasome, *J. Cell Sci.* **110**:2141–2154.

Drancourt, M., Birtles, R., Chaumentin, G., Vandenesch, F., Etienne, J., and Raoult, D., 1996, New serotype of *Bartonella henselae* in endocarditis and cat-scratch disease, *Lancet* **347**:441–443.

Fay, F.H., and Rausch, R.L., 1969, Parasitic organisms in the blood of arvicoline rodents in Alaska. *J. Parasit*, **55**:1258–1265.

Fischetti, V.A., Pancholi, V., and Schneewind, O., 1990, Conservation of a hexapeptide sequence in the anchor region of surface proteins from gram-positive cocci, *Mol. Microbiol.* **4**:1603–1605.

Garcia, F.U., Wojta, J., Broadley, K.N., Davidson, J.M., and Hoover, R.L., 1990, *Bartonella bacilliformis* stimulates endothelial cells *in vitro* and is angiogenic *in vivo*, *Am. J. Pathol.* **136**:1125–1135.

Garcia, F.U., Wojta, J., and Hoover, R.L., 1992, Interactions between live *Bartonella bacilliformis* and endothelial cells, *J. Infect. Dis.* **165**:1138–1141.

Golnik, K.C., Marotto, M.E., Fanous, M.M., Heitter, D., King, L.P., Halpern, J.I., and Holly, P.H., 1994, Opthalmic manifestations of *Rochalimaea* species, *Am. J. Opthalmol.* **118**:145–151.

Graham-Smith, G.S., 1905, A new form of parasite found in the red blood corpuscles of moles, *J. Hyg.* **5**:453–459.

Gray, G.C., Johnson, A.A., Thornton, S.A., Smith, W.A., Knobloch, J., Kelley, P.W., Escudero, L.O., Huayda, M.A., and Wignall, F.S., 1990, An epidemic of Oroya fever in the Peruvian Andes, *Am. J. Trop. Med. Hyg.* **42**:215–221.

Gurfield, A.N., Boulouis, H.J., Chomel, B.B., Heller, R., Kasten, R.W., Yamamoto, K., and Piemont, Y., 1997, Coinfection with *Bartonella clarridgeiae* and *Bartonella henselae* and with different *Bartonella henselae* strains in domestic cats, *J Clin Microbiol.* **35**:2120–2123.

Hara, H., Yamamoto, Y., Higashitani, A., Suzuki, H., and Nishimura, Y., 1991, Cloning, mapping and characterization of the *Escherichia coli prc* gene, which is involved in the C-terminal processing of penicillin-binding protein 3, *J. Bacteriol.* **173**:4799–4813.

Heffernan, E.J., Wu, L., Louie, J., Okamoto, S., Fierer, J., and Guiney, D.G., 1994, Specificity of the complement resistance and cell association phenotypes encoded by the outer membrane protein genes *rck* from *Salmonella typhimurium* and *ail* from *Yersinia enterocolitica*, *Infect. Immun.* **62**:5183–5186.

Herrer, A., 1953, Carrion's disease. II. Presence of *Bartonella bacilliformis* in the peripheral blood of patients with the benign tumor form, *Am. J. Trop. Med.* **2**:645–649.

Hertig, M., 1942, Phlebotomus and Carrion's disease, *Am. J. Trop. Med.* **22**:1–81.

Higgins, J.A., Radulovic, S., Jaworski, D.C., and Azad, A.F., 1996, Acquisition of the cat scratch disease agent *Bartonella henselae* by cat fleas (*Siponaptera: Pulicidae*), *J. Med. Entomol.* **33**:490–495.

Hurtado, A., Musso, J.P., and Merino, C., 1938, La anemia en la enfermedad de Carrion (verruga peruana), *Ann. Fac. Med. Lima* **28**:154–168.

Iwaki-Egawa, S., and Ihler, G.M., 1997, Comparison of the abilities of proteins from *Bartonella bacilliformis* and *Bartonella henselae* to deform red cell membranes and to bind to red cell ghost proteins, *FEMS Microbiol. Lett.* **157**:207–217.

Jackson, L.A., and Spach, D.H., 1996, Emergence of *Bartonella quintana* infection among homeless persons, *Emerg. Infect. Dis.* **2**:141–144.

Johnson, W.T., and Helwig, E.B., 1969, Cat-scratch disease (histopathologic changes in the skin), *Arch. Dermatol.* **100**:148–154.

Kehoe, M.A., 1994, Cell wall-associated proteins in gram-positive bacteria, *New Compr. Biochem.* **27**:217–261.

Kobe, B., and Deisenhofer, J., 1995, Proteins with leucine-rich repeats, *Curr. Opin. Struct. Biol.* **5**:409–416.

Koehler, J.E., and Tappero, J.W., 1993, Bacillary angiomatosis and bacillary peliosis in patients infected with human immunodeficiency virus, *Clin Infect. Dis.* **17**:612–624.

Kordick, D.L., and Breitschwerdt, E.B., 1995, Intraerythrocytic presence of *Bartonella henselae*, *J. Clin. Microbiol.* **33**:1655–1656.

Kostrzewski, J., 1950, The epidemiology of trench fever, *Med. Dosw. Mikrobiol.* **11**:233–263.

Koehler, J.E., Quinn, F.D., Berger, T.G., Leboit, P.E., and Tappero, J.W., 1992, Isolation of *Rochalimaea* species from cutaneous and osseous lesions of bacillary angiomatosis, *N. Engl. J. Med.* **325**:1625–1631.

Koonin, E.V., 1993, A highly conserved sequence motif defining the family of MutT-related proteins from eubacteria, eukaryotes and viruses, *Nucleic Acids Res.* **21**:4847.

Krampitz, H.E., 1962, Weitere Untersuchungen an *Grahamella* Brumpt 1911, *Zeitsch. Tropenmed. Parasit.* **13**:34–53.

Kreier, J.P., and Ristic, M., 1981, The biology of hemotrophic bacteria, *Ann. Rev. Microbiol.* **35**:325–338.

LeBoit, P.E., Berger, T.G., Egbert, B.M., Beckstead, J.H., Yen, T.S., and Stoler, M.H., 1988, Epithelioid haemangioma-like vascular proliferation in AIDS: manifestation of cat-scratch disease bacillus infection?, *Lancet*, **1**:960–963.

LeBoit, P.E., Berger, T.G., Egbert, B.M., Beckstead, J.H., Yen, T.S., and Stoler, M.H., 1989, Bacillary angiomatosis: the histopathology and differential diagnosis of a pseudoneoplastic infection in patients with human immunodeficiency virus disease, *Am J. Surg. Pathol.* **13**:909–920.

Maurin M., Eb, F., Etienne, J., and Raoult, D., 1997, Serological cross-reactions between *Bartonella* and *Chlamydia* species: implications for diagnosis, *J Clin Microbiol.* **35**:2283–2287.

McAllister, S.J., Peek, J.A., and Minnick, M.F., 1995, Identification and isolation of bundle-forming fimbriae from *Bartonella bacilliformis*. In *Abstracts of the 95th General Meeting of the American Society for Microbiology*, Abstract no. D43.

McCrary, B., 1994, Neuroretinitis in cat-scratch disease associated with the macular star, *Pediatr. Infect. Dis. J.* **13**:838–839.

McGinnis-Hill, E., Raji, A., Valenzuela, M.S., Garcia, F., and Hoover, R., 1992, Adhesion to and invasion of cultured human cells by *Bartonella bacilliformis*, *Infect. Immun.* **60**:4051–4058.

McNee, J.W., and Renshaw, A., 1916, "Trench fever": a relapsing fever occurring with the British forces in France, *Br. Med. J.* **1**:225–234.

Mehock, J.R., Greene, C.E., Gherardini, F.C., Hahn, T-W, and Krause, D.C., 1998, *Bartonella henselae* invasion of feline erythrocytes *in vitro*, *Infect. Immun.* (in press).

Mernaugh, G., and Ihler, G.M., 1992, Deformation factor: an extracellular protein synthesized by *Bartonella bacilliformis* that deforms erythrocyte membranes, *Infect. Immun.* **60**:937–943.

Merrell, B.R., Weiss, E., and Dasch, G.A., 1978, Morphological and cell association characteristics of *Rochalimaea quintana*: comparison of the vole and Fuller strains, *J. Bacteriol.* **135**:633–640.

Milam, M.W., Balerdi, M.J., Toney, J.F., Foulis, P.R., Milam, C.P., and Behnke, R.H., 1990, Epithelioid angiomatosis secondary to disseminated cat scratch disease involving the bone marrow and skin in a patient with acquired immune deficiency syndrome: a case report, *Am. J. Med.* **88**:180–183.

Miller, V.L., Bliska, J.B., and Falkow, S., 1990, Nucleotide sequence of the *Yersinia enterocolitica ail* gene and characterization of the Ail protein product, *J. Bacteriol.* **172**:1062–1069.

Minnick, M.F., 1994, Identification of outer membrane proteins of *Bartonella bacilliformis*, *Infect. Immun.* **62**:2644–2648.

Minnick, M.F., Mitchell, S.J., and McCallister, S.J., 1996, Cell entry and the pathogenesis of *Bartonella* infections, *Trends Microbiol.* **4**:343–347.

Mitchell, S.J., and Minnick, M.F., 1995, Characterization of a two-gene locus from *Bartonella bacilliformis* associated with the ability to invade human erythrocytes, *Infect. Immun.* **63**:1552–1562.

Mitchell, S.J., and Minnick, M.F., 1997a, A carboxy-terminal processing protease gene is located immediately upstream of the invasion-associated locus from *Bartonella bacilliformis*, *Microbiol.* **143**:1221–1233.

Mitchell, S.J., and Minnick, M.F., 1997b, Cloning, functional expression, and complementation analysis of an inorganic pyrophosphatase from *Bartonella bacilliformis*, *Can. J. Microbiol.* **43**:734–743.

Murakawa, G.J., Peek, J.A., Tompkins, L.S., and Falkow, S., 1996, Sequence characterization of an invasion locus in *Bartonella henselae*, in: *Abstracts of the 12th National Meeting of the American Society for Rickettsiology and Rickettsial Diseases*, Abstract no. 50.

Murakawa, G.J., 1997, Pathogenesis of *Bartonella henselae* in cutaneous and systemic disease, *J. Am. Acad. Dermatol.* **Nov.** 775–776.

Myers, W.F., Cutler, L.D., and Wisseman, C.L., 1969, Role of erythrocytes and serum in the nutrition of *Rickettsia quintana*, *J. Bacteriol.* **97**:663–666.

Palmari, J., Teysseire, N., Dussert, C., and Raoult, D., 1996, Image cytometry and topographical analysis of proliferation of endothelial cells *in vitro* during *Bartonella* (*Rochalimaea*) infection, *Anal. Cell Pathol.* **11**:13–30.

Peek, J.A., Batterman, H.J., Falkow, S., and Tompkins, L.S., 1994, Piliation of *Rochalimaea henselae* and *Rochalimaea quintana*, in: *Abstracts of the 11th National Meeting of the American Society for Rickettsiology and Rickettsial Diseases*.

Perkocha, L.A., Geaghan, S.M., Yen, T.S.B., Nishimura, S.L., Chan, S.P., Garcia-Kennedy, R., Honda, G., Stoloff, A.C., Klein, H.Z., Goldman, R.L., Van Meter, S., Ferrell, L.D., and LeBoit, P.E., 1990, Clinical and pathological features of bacillary peliosis hepatitis in association with human immunodeficiency virus infection, *New. Engl. J. Med.* **323**:1581–1586.

Raoult, D., Fournier, P.E., Drancourt, M., Marrie, T.J., Etienne, J., Cosserat, J., Cacoub, P., Poinsignon, Y., Leclercq, P., and Sefton, A.M., 1996, Diagnosis of 22 new cases of *Bartonella* endocarditis, *Ann. Intern. Med.* **125**:646–652.

Reynafarje, C., and Ramos, J., 1961, The hemolytic anemia of human bartonellosis, *Blood* **17**:562–578.

Ricketts, W.E., 1949, Clinical manifestations of Carrion's disease, *Arch. Intern. Med.* **84**:751–781.

Roberts, N.J., 1995, *Bartonella bacilliformis* (bartonellosis), in: *Principles and practice of infectious diseases*, 4th Ed. (G.L. Mandell, J.E. Bennett, and R. Dolin, eds.), Livingstone Press, New York, pp. 2209–2210.

Scherer, D.C., DeBuron-Connors, I., and Minnick, M.F., 1993, Characterization of *Bartonella bacilliformis* flagella and effect of antiflagellin antibodies on invasion of human erythrocytes, *Infect. Immun.* **61**:4962–4971.

Slater, L.N., Welch, D.F., and Min, K., 1992, *Rochalimaea henselae* causes bacillary angiomatosis and peliosis hepatitis, *Arch. Intern. Med.* **152**:602–606.

Spach, D.H., Panther, L.A., Thorning, D.R., Dunn, J.E., Plorde, J.J., and Miller, R.A., 1992, Intracerebral bacillary angiomatosis in a patient infected with human immunodeficiency virus, *Ann. Intern Med.* **116**:740–742.

Tappero, J.W., Koehler, J.E., Berger, T.G., Cockerell, C.J., Lee, T-H., Busch, M.P., Stites, D.P.,

Mohle-Boetani, J., Reingold, A.L., and LeBoit, P.E., 1993, Bacillary angiomatosis and bacillary splenitis in immunocompetent adults, *Ann. Intern. Med.* **118**:363–365.
Tyzzer, E.E., 1942, A comparative study of *Grahamella*, *Haemobartonella* and *Eperythrozoa* in small mammals, *Proc. Am. Philos. Soc.* **85**:359–398.
Urteaga, B.O., and Payne, E.H., 1955, Treatment of the acute febrile phase of Carrion's disease with chloramphenicol, *Am J. Trop. Med.* **4**:507–511.
Vinson, J.W., and Fuller, H.S., 1961, Studies on trench fever. I. Propogation of rickettsia-like microorganisms from a patient's blood, *Path. Microbiol.* **24**:152–166.
Walker, T.S., and Winkler, H.H., 1981, *Bartonella bacilliformis*: colonial types and eyrthrocyte adherence, *Infect. Immun.* **31**:480–486.
Webster, G.F., Cockerell, C.J., and Friedman-Kien, A.E., 1992, The clinical spectrum of bacillary angiomatosis, *Br. J. Dermatol.* **126**:535–541.
Weinman, D. 1965, The *bartonella* group, in: *Bacterial and mycotic infections of man* (R.J. Dubos and J.G. Hirsch, eds.), Lippincott, Philadelphia, pp. 775–785.
Wolbach, S.B., Todd, J.L., and Palfrey, F.W., 1922, *The etiology and pathology of typhus.* Harvard University Press, Cambridge, Massachusetts.
Xu, Y-H., Lu, Z-Y., and Ihler, G.M., 1995, Purification of deformin, an extracellular protein synthesized by *Bartonella bacilliformis* which causes deformation of erythrocyte membranes, *Biochim. Biophys. Acta* **1234**:173–183.

Chapter 6

Host Cell Invasion by *Streptococcus pneumoniae*

Axel Ring and Elaine Tuomanen

1. INTRODUCTION

Streptococcus pneumoniae remains one of the world's leading invasive human pathogens causing pneumonia, sepsis, and meningitis. Infants and toddlers between 18 months and 4 years of age are particularly susceptible (Tuomanen *et al.*, 1995). Virtually every child up to age 5 will experience pneumococcal otitis media while the incidence of pneumonia (1000 per 100,000 inhabitants) and meningitis (10 per 100,000 inhabitants) is much lower. However, pneumococcal meningitis continues to be a serious threat with a mortality of 25% and a morbidity of 80% in children. Despite the development of novel antibiotics, the mortality from pneumococcal meningitis has not changed significantly during the last 20 years. Taken together with increasing antibiotic resistance among pneumococci, it has become imperative to improve our understanding of the mechanisms by which this bacterium colonizes, invades and kills its victims. The development of

---

AXEL RING and ELAINE TUOMANEN Department of Infectious Diseases, St. Jude Children's Research Hospital, Memphis, Tennessee 38105, E-mail: elaine.tuomanen@stjude.org
Subcellular Biochemistry, Volume 33: Bacterial Invasion into Eukaryotic Cells, edited by Oelschlaeger and Hacker. Kluwer Academic / Plenum Publishers, New York, 2000.

vaccines and novel adjunctive drugs targeting defined pneumococcal virulence determinants is a crucial goal of current research.

2. STRUCTURAL DETERMINANTS OF INVASION

Pneumococci are low-efficiency invaders in that a maximum of 0.2% of an inoculum enters cells (Ring *et al.*, 1998; Talbot *et al.*, 1996; Cundell *et al.*, 1995). This figure is ten fold lower than other streptococci. Clinical isolates exhibit a wide variability in invasive capacity which arises from multiple factors. Invasion is affected nonspecifically by capsule and pneumolysin and specifically by cell wall and associated adhesins. Capsular polysaccharide inhibits phagocytosis and thereby increases the number of bacteria in a position to invade tissues. This factor is particularly important in enabling penetration of the blood-brain barrier (BBB). However, at the level of cell-cell contact, capsule inhibits adherence and thus invasion into eukaryotic cells up to 200 fold *in vitro* (Ring *et al.*, 1998; Talbot *et al.*, 1996; Cundell *et al.*, 1995). This may arise through the net negative charge of the polysaccharide, particularly for types 7F and 14.

Pneumolysin is an intracellular, multifunctional toxin with cytolytic and complement activating domains (Mitchel *et al.*, 1997). Pneumococci undergo autolysis and release pneumolysin at high bacterial densities. Pneumolysin enhances the transition of bacteria from lung into blood by causing host cell necrosis and detachment leading to the denuding of mucosal barriers and exposure of basement membrane constituents such as collagen, laminin and fibronectin. Binding to extracellular matrix proteins exposed by mucosal injury would be expected to enhance dissemination, particularly in environments where attachment to cell receptors is poor, e.g. in the upper respiratory tract. Localization of pneumococci to sites of exposed basement membrane under cells damaged by viruses *in vivo* (Plotkowski *et al.*, 1986) or pneumolysin *in vitro* (Ryner *et al.*, 1994) has been demonstrated. Pneumococcal binding to type IV collagen, laminin and vitronectin has been described (Kostrzynska *et al.*, 1992). In addition, pneumococci bind avidly to immobilized but not soluble fibronectin (van der Flier *et al.*, 1995). The binding site is within the C-terminal heparin binding domain, which is an unusual site for microbial targeting since *Staphylococcus aureus*, oral streptococci and some gram-negative bacteria bind to the N-terminal heparin binding domain of fibronectin (Westerlund and Korhonen, 1993). On the pneumococcal side, a 14 kDa and a 50 kDa protein may be ligands for fibronectin (T. Wizemann, A. Ring, and E. Tuomanen, unpublished). These events are important for pathogenicity since binding to extracellular matrix is only encountered under pathological conditions.

Two surface components are mediators of specific adherence and subsequent cell invasion: cell wall phosphorylcholine and choline binding protein A (Figure 1). Cell wall components such as peptidoglycan and teichoic acid possess a striking inflammatory potential when applied to the lung or subarachnoid space of animals (Tuomanen *et al.*, 1987, 1985). An unusual characteristic of pneumococcal teichoic acid is the presence of phosphorylcholine, a determinant also found in the proinflammatory chemokine, platelet-activating factor (PAF). Phosphorylcholine is a key molecular determinant of invasion, acting both as an adhesive ligand and

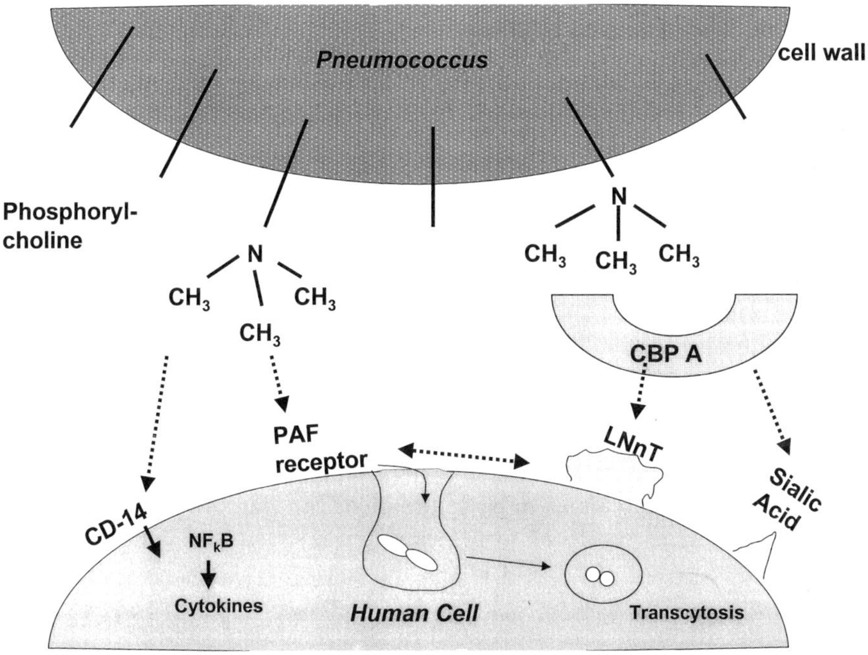

FIGURE 1. Elements of pneumococcal adherence leading to invasion and transcytosis of human cells. Pneumococcal cell wall binds CD14 leading to the activation of the nuclear transcription factor NF-kB and *de novo* expression of proinflammatory cytokines. Cytokine activated cells upregulate expression of the PAF receptor, C3, and other receptors bearing lacto-N-neotetraose or sialic acid. Pneumococci bind to these new receptors by cell wall phosphorylcholine and choline binding proteins and subsequently enter cytoplasmic vacuoles by receptor mediated endocytosis. Given sufficient time, intracellular migration of the vacuole across the cell and exit at the basolateral membrane is the predominant fate. Some intracellular bacteria are shunted back to the apical membrane (recycling). Bacteria not bearing CbpA (choline binding protein A) and not entering via the PAF (platelet activating factor) receptor, die in lysosomes.

as a docking station for protein adhesins. Intact bacteria use phosphorylcholine to tether to the PAF receptor and thereby gain access to the cytoplasmic compartment in a vacuole (Cundell *et al.*, 1995; Geelen *et al.*, 1993). This vacuole is subsequently sorted to the basolateral cell surface resulting in transcytosis of the bacterium across the cell.

The cell wall phosphorylcholine also promotes adherence by serving as a docking station for a family of CBPs (Tuomanen and Masure 1997). CBPs are surface-exposed proteins noncovalently bound to the bacterial surface by a signature choline binding domain. Well known members are the autolysin (LytA), pneumococcal surface protein A (PspA) and choline-binding protein A (CbpA) (Rosenow *et al.*, 1997; McDaniel *et al.*, 1991; Ronda *et al.*, 1987). Nine additional CBPs have been identified ranging in size from 20 to 110 kDa, but their functions are as yet unknown (Tuomanen and Masure, 1997). The expression of choline-binding proteins varies significantly consistent with the dramatic role phase variation plays in adherence to and invasion of host cells (Weiser *et al.*, 1994).

CbpA, a major pneumococcal structural adhesin, is a protein of 663 amino acids with two distinct domains (Rosenow *et al.*, 1997). The C-terminus consists of a choline-binding domain and the N-terminal domain (amino acids 1–433) contains two large repeat regions, each containing three alpha helices. Recombinant CbpA blocks pneumococcal binding to lacto-N-neotetraose, sialic acid, lung cells and nasopharyngeal cells. Wild type pneumococci bind to these substrates while CbpA deficient mutants bind much less effectively or not at all. These data suggest that the repeats constitute lectin domains. A variant of CbpA containing only one repeat and the choline-binding domain was reported to bind IgA and secretory component (Hammerschmidt *et al.*, 1997). Whether this represents a lectin type of interaction remains to be determined. Thus far, the general scheme of pneumococcal translocation across all endothelial and epithelial cells studied involves cell wall and CbpA.

3. THE PROCESS OF INVASION

Invasion of host cells by pneumococcus is believed to be a multistage process initiated by adherence. Pneumococci bind avidly to epithelial cells of the nasopharynx and lung as well as to endothelial cells from peripheral and brain microvasculature. The initial encounter of nasopharyngeal epithelial cells with pneumococci results in colonization which is found in up to 40% of healthy adults without any adverse effects (Gray and Dillon, 1986). For resting nasopharyngeal cells and type II pneumocytes, simple, primary attachment of pneumococci is mediated by as yet unidentified bacterial lectins. Inhibition studies with soluble sugars suggest the bacteria recognize host cell glycoconjugates bearing N-acetyl-D-galactosamine linked either β1-3 or β1-4 to galactose (Barthelson *et al.*, 1998; Cundell and Tuomanen, 1994; Krivan *et al.*, 1988; Andersson *et al.*, 1983). These interactions are distinct from those leading to invasion since (a) they are not dependent on quorum sensing events, (b) there is no difference between opaque and transparent phase variants and (c) they do not directly result in symptomatic disease.

In a minority of individuals, the bacteria then progress from the nasopharynx into blood, middle ear or the respiratory tract. Progression to invasive disease requires the activation of host cells and presentation of new receptor types. Glycoconjugates bearing sialic acid and lacto-N-neotetraose are expressed *de novo* on activated cells, providing new opportunities for pneumococci to tether to target cells. These determinants may be present on several human cell determinants including the PAF receptor, the third component of complement (C3), IgA or other unidentified determinants (Smith and Hostetter, 1998; Hammerschmidt *et al.*, 1997; Cundell *et al.*, 1995). CbpA is the major adhesin for activated cells. In addition to CbpA-mediated adherence, choline itself serves as a direct ligand for binding to the PAF receptor (Cundell *et al.*, 1995). The role of phosphorylcholine in pneumococcal disease has gained broader significance with the recognition that other pulmonary pathogens (e.g. *Haemophilus influenzae, Mycoplasma pneumoniae, Pseudomonas aeruginosa*) also decorate major surface virulence determinants with phosphorylcholine (Weiser *et al.*, 1997). In each case, the phosphorylcholine is subject to phase variation and modulates virulence (Weiser *et al.*, 1998).

Pneumococci do not require *de novo* synthesis of DNA, RNA or protein in order to invade lung epithelial cells as evidenced by uptake of heat-killed bacteria (Ring *et al.*, 1998; Geelen *et al.*, 1993). In both type II pneumocytes and brain microvascular endothelial cells, cytochalasin D at non-toxic concentrations totally abrogates invasion indicating the need for actin microfilament polymerization as a prerequisite to formation of the vacuole. Nocodazole, a microtubule inhibitor, decreases invasion of brain endothelial cells, but only to about 60% (Ring *et al.*, 1998), indicating that microtubules are not required for all invasion. Once intracellular, the number of viable pneumococci decreases steadily (Ring *et al.*, 1998). This is a result both of intracellular death as well as exit from the cell either by transcytosis through the cell or recycling back to the original port of entry. Intracellular replication does not appear operative during pneumococcal pathogenicity. In polarized brain microvascular endothelial cell monolayers, PAF-receptor-mediated uptake of pneumococci (70% of internalized bacteria) leads to transcytosis to the basolateral side compared to

PAF-receptor-independent entry mechanisms (30% of internalized bacteria) (Ring *et al.*, 1998). This suggests that the vesicle is driven across the cell by an as yet unknown intracellular signaling mechanism activated by pneumococcus-bound PAF receptors. PAF receptor is coupled to G proteins that activate phospholipase C (Kunz *et al.*, 1992). This eventually induces release of calcium from intracellular stores, calcium binding to calmodulin and activation of calcium-calmodulin dependent protein kinases. In contrast to PAF, PAF-receptor binding by pneumococci does not result in classical G-protein mediated signaling (Garcia Rodrigues *et al.*, 1995). However, calcium-calmodulin antagonists have a dramatic negative impact on pneumococcal invasion into brain endothelial cells raising the possibility that some but not all of the PAF receptor signalling pathway participates in pneumococcal translocation (J. Zhang and E. Tuomanen, unpublished results). The PAF receptor is known to be rapidly internalized after interaction with its natural ligand, and, after release of PAF, the receptor recycles back to the apical side of the cell (Gerard and Gerard, 1994). It is possible that the continued binding of pneumococci to the PAF receptor, even inside the vacuole, redirects the vesicle from recycling to transcytosis.

4. A MODEL OF PNEUMOCOCCAL INVASION

Transit of bacteria across a barrier can occur by either a transcellular or paracellular pathway, the latter necessitating compromising tight junctions. Using a two-chamber cell culture system of human or rat brain microvascular endothelial cells grown on a microporous polycarbonate filter, the validity of the model of pneumococcal transcytosis was tested (Ring *et al.*, 1998). The monolayer was infected with pneumococci from the apical side, and the trafficking of intracellular bacteria was quantitated after antibiotic killing of extracellular colony-forming units. Electron microscopy studies revealed intracellular pneumococci located in vacuoles but never free in the cytoplasm, a finding in concordance with evidence in type II pneumocytes and human umbilical vein endothelial cells (Talbot *et al.*, 1996; Geelen *et al.*, 1993). However, the number of invasive bacteria was greater in BBB cells compared to peripheral endothelia. The tight junctions remained intact and there was no evidence for bacteria located between cells. There were no signs of cell death or injury to the monolayers over 20 hours. Emergence of intracellular bacteria was observed first in the apical and subsequently also in the basal chamber within 4 hours. These data suggest that pneumococci cross the polarized brain endothelial cell monolayer by passing through the cells in a vacuole without affecting cell viability or tight junctions. A smaller portion of bacteria is released back to the apical side, the original port of entry, possibly establishing a holding reservoir enabling a pool of invasive bacteria to exit and re-enter the cell for transcytosis. Transcytosis was virtually completely restricted to the transparent pneumococcal phenotype and was completely absent in mutants lacking the adhesin CbpA. Invasion was significantly increased with human cell activation by cytokines. PAF-receptor antagonist blocked ~60% of transcytosis. Figure 2 summarizes a validated model of pneumococcal trafficking across the BBB.

Given that pneumococci do not directly impair tight junctions, they appear to cross the BBB by a bona fide vectorial vesical transport initiated at least in part by tethering to the PAF receptor. In contrast to pneumococci, group B streptococci, *Haemophilus influenzae* and *Neisseria meningitidis* cross the barrier by transcytosis or by inducing cytotoxicity to brain endothelial cells (Nizet *et al.*, 1997; Birkness *et al.*, 1995; Patrick *et al.*, 1992). Thus, transcytosis without a paracellular route of passage by a cytotoxic effect on tight junctions appears to be unique to pneumococci as compared to other meningeal pathogens.

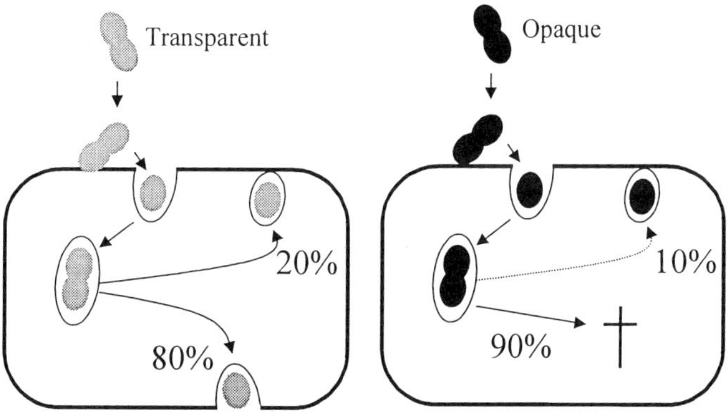

FIGURE 2. A model of pneumococcal transcytosis across brain endothelial cells in a vacuole. Opaque bacteria have a selective advantage over transparent in the bloodstream due to increased expression of polysaccharide capsule enabling them to evade phagocytosis. However, only the transparent bacteria are capable of migrating through the endothelial cells; opaque forms die intracellularly. Thus, switching of colony morphology phenotypes is likely to be a critical step in bacterial adaptation to the host environment by allowing them to survive in the blood as opaque and cross epithelia and endothelia as transparent forms. Penetration of cells by the transparent forms occurs through binding of cell wall choline to the PAF receptor and subsequent receptor-mediated endocytosis. The vacuole is driven to and released from the basolateral side of the cell. A small proportion of bacteria-laden vacuoles are recycled to the apical cell surface.

Given the very low efficiency of pneumococcal invasion *in v*itro even within the group of low-grade invaders like *Staphylococcus aureus*, other streptococci, and *E. coli*, it could be argued that the information gathered on pneumococcal invasion *in vitro* may not reflect events *in vivo*. The mechanisms underlying brain and pulmonary inflammation have been extensively studied using animal models. However, the molecular details of the process by which pneumococci encountering an intact barrier, be it the alveolar epithelium or the BBB, manage to cross it must be validated *in vivo*. In the rabbit pneumonia model, a PAF-receptor antagonist, administered at the time of intratracheal pneumococcal challenge, greatly attenuates the bacterial load in the lung and prevents bacteremia (Indapaan-Heikkila *et al.*, 1997). A comparable activity was found for lacto-N-neotetraose and a sialylated derivative, LSTc. Single doses of these compounds prevent the initial attachment to lung and nasopharyngeal cells *in vivo*, virtually abrogate the morphopathologic changes of pulmonary inflammation and reduce invasion of the bloodstream. Lacto-N-neotetraose, a constitutent of ligands present on the surface of activated lung cells, is the most active agent in the rabbit pneumonia model. The correlation between agents inhibiting pneumococcal adherence to lung cells *in vitro* and beneficial effects on disease in animal models suggests that the details of adherence and invasion described *in vitro* reasonably represent true events during real infection.

5. DYNAMIC VIRULENCE FEATURES

Pneumococcal pathogenicity is not merely a static function of virulence factor expression, but rather is subject to modulation by interdependent dynamic processes. Pneumococci spontaneously phase vary between opaque and transparent colony forms (Weiser *et al.*, 1994). Opaque colonies are selected in blood while their transparent counterparts are capable of colonization of the nasopharynx. Adherence to cytokine-stimulated endothelial cells and PAF-receptor-mediated entry into the cytoplasm of brain endothelial cells with subsequent transcytosis through the BBB are limited to transparent variants (Cundell *et al.*, 1995). This correlates with increased surface expression of phosphorylcholine and CbpA in transparent variants. These findings suggest that transparent pneumococci colonizing the nasopharynx would have to switch to opaque upon entry into the bloodstream, and switch back again to transparent when they cross the BBB to cause meningitis. No evidence for phase variation within the vacuole of human cells has been found. The frequency of spontaneous switching is between 10^{-3} and 10^{-6}, and given that pneumococcus is a low-grade invader, this would provide a rationale for the relative infrequency of meningitis compared to bacteremia.

The nature and diversity of pneumococcal virulence factors reflect the complexity of molecular events underlying pathogen-host interactions. It is now recognized that crucial processes in pneumococcal physiology are interrelated with pathogenicity in a model of programmed development (Tuomanen and masure, 1997). The key to this hypothesis is the temporal analysis of the pneumococcal growth curve (Figure 3). *In vitro*, all members of a culture are triggered to undergo major events like DNA transformation, adherence to the PAF receptor or sialylated glycoconjugates and

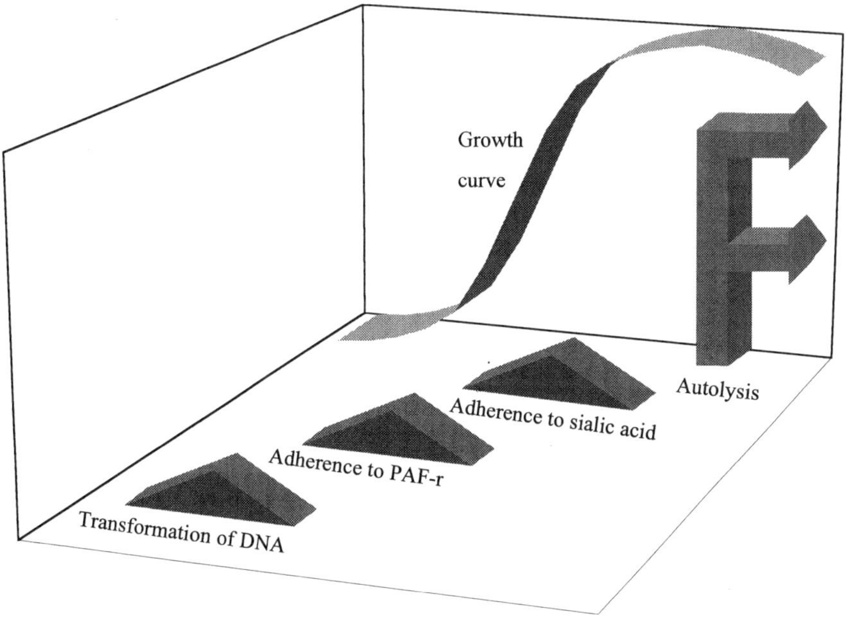

FIGURE 3. The pneumococcal life cycle. Pneumococci pass through four different growth phases in the course of their life. 1) Early in the cycle, natural DNA transformation is triggered by the 17 amino acid competence-stimulating peptide, which induces expression of over a dozen genes necessary for binding, uptake and incorporation of exogenous DNA. 2 and 3) Adherence to the PAF receptor and subsequently to sialic acid, both of which are receptors up-regulated on cytokine-activated cells, is optimal at distinct, defined points during the growth cycle. In contrast, adherence to resting cells is expressed equally throughout the entire life cycle. 4) In stationary phase, pneumococci autolyse through the activity of N-acetylmuramyl-L-alanine amidase (autolysin). The major cytotoxin of pneumococcus, pneumolysin, is released from the cytosol during this final phase.

autolysis at defined points of the pneumococcal life cycle consistent with quorum sensing events. Transformation occurs during early logarithmic phase followed by the ability to bind to the PAF receptor and subsequently to sialic acid-containing human cell carbohydrates. Eventually, autolysis occurs during the stationary phase, accompanied by release of pneumolysin from the cytosol. Supernatant fluids collected from cultures at the peak of each event can confer that activity on bacteria at other points in the life cycle. This suggests that autoinducing molecules are produced at different bacterial densities and these are in turn sensed by the pneumococci which then express multiple gene products in a coordinate fashion to elicit the respective physiological event. Quorum sensing is carried out through two-component signal transduction systems consisting of a surface-exposed histidine kinase that phosphorylates a response regulator, which subsequently alters DNA transcription (Cheng *et al.*, 1997; Havarstein *et al.*, 1996; Guenzi and hakenbeck, 1995).

6. CONCLUSIONS

Pneumococci selectively adhere to respiratory epithelial and endothelial cells. The biology of attachment to resting cells differs strikingly from adherence to cytokine-activated cells. The latter involves interaction of choline or CbpA with the PAF receptor or lacto-N-neotetraose, respectively. Adherence to cytokine-activated cells has significant pathophysiological relevance since it precedes invasion by receptor-mediated endocytosis. Pneumococci in a cytoplasmic vacuole are subject to three different fates: intracellular death, transcytosis through the cell with exocytosis at the basolateral side or recycling back to the apical side. Transcytosis is limited to transparent phase variants while opaque forms are more likely to be sorted to a lethal lysosomal compartment. Recycling might serve to keep invasive bacteria in a holding reservoir allowing them to re-enter cells and be protected from phagocytosis until transcytosis has taken place.

7. REFERENCES

Andersson, B., Dahmen, J., Frejg, T., Leffler, H., Magnusson, G., Noori, G., and Svanborg, E.C., 1983, Identification of an active disaccharide unit of a glycoconjugate receptor for pneumococci attaching to human pharyngeal epithelial cells, *J. Exp. Med.* **158**: 559–570.

Barthelson, R., Mobasseri, A., Zopf, D., and Simon, P., 1998, Adherence of *Streptococcus pneumoniae* to respiratory epithelial cells is inhibited by sialylated oligosaccharides, *Infect. Immun.* **66**:1439–1444.

Birkness, K., Swisher, B., White, E., Long, E., Ewing, E., and Quinn, F., 1995, A tissue cultre bilayer model to study the passage of *Neisseria meningitidis*, *Infect. Immun.* **63**:402–409.

Cheng, Q., Campbell, E., Naughton, A., Johnson, S., and Masure, H., 1997, The *com* locus controls genetic transformation in *Streptococcus pneumoniae*, *Mol. Microbiol.* **23**:683–692.

Cundell, D., and Tuomanen, E., 1994, Receptor specificity of adherence of *Streptococcus pneumoniae* to human type II pneumocytes and vascular endothelial cells *in vitro*, *Microb. Pathog.* **17**:361–374.

Cundell, D., Gerard, N., Gerard, C., Idanpaan-Heikkila, I., and Tuomanen, E., 1995, *Streptococcus pneumoniae* anchors to activated eukaryotic cells by the receptor for platelet activating factor, *Nature* **377**:435–438.

Garcia Rodriguez, C., Cundell, D., Tuomanen, E., Kolakowski, L.J., Gerard, C., and Gerard, N., 1995, The role of N-glycosylation for functional expression of the human platelet-activating factor receptor. Glycosylation is required for efficient membrane trafficking, *J. Biol. Chem.* **270**:25178–25184.

Geelen, S., Battacharyya, C., and Tuomanen, E., 1993, Cell wall mediates pneumococcal attachment and cytopathology to human endothelial cells, *Infect. Immun.* **61**:1538–1543.

Gerard, N., and Gerard, C., 1994, Receptor-dependent internalization of platelet-activating factor, *J. Immunol.* **152**:793–800.

Gray, B., and Dillon, H., 1986, Clinical and epidemiologic studies of pneumococcal infection in children, *Pediatr. Infect. Dis.* **5**:201–207.

Guenzi, R., and Hakenbeck, R., 1995, Genetic competence and a two component regulatory system in pneumococci, *Mol. Microbiol.* **12**:505–515.

Hammerschmidt, S., Talay, S., Brandtzaeg, P., and Chhatwal, G., 1997, SpsA, a novel pneumococcal surface protein with specific binding to secretory immunoglobulin A and secretory component, *Mol. Microbiol.* **25**:1113–1124.

Havarstein, L., Gaustad, P., Nes, I., and Morrison, D., 1996, Identification of the streptococcal competence pheromone receptor, *Mol. Mcirobiol.* **21**:965–971.

Idanpaan-Heikkila, I., Simon, P., Cahill, C., Sokol, K., and Tuomanen, E., 1997, Oligosaccharides interfere with the establishment and progression of experimental pneumococcal pneumonia, *J. Infect. Dis.* **176**:704–712.

Kostrzynska, M., and Wadstrom, T., 1992, Binding of laminin, type IV collagen, and vitronectin by *Streptococcus pneumoniae*, *Zbl. Bakt.* **277**:80–83.

Krivan, H.C., Roberts, D.D., and Ginsburg, V., 1988, Many pulmonary pathogenic bacteria bind specifically to the carbohydrate sequence GalNAcβ1-4Gal found in some glycolipids, *Proc. Natl. Acad. Sci.* USA **85**:6157–6161.

Kunz, D., Gerard, N., and Gerard, C., 1992, The human leukocyte platelet activating factor receptor, *J. Biol. Chem.* **267**:9101–9106.

McDaniel, L.S., Sheffield, J.S., Delucchi, P., and Briles, D.E., 1991, PspA, a surface protein of *Streptococcus pneumoniae*, is capable of eliciting protection against pneumococci of more than one capsular type, *Infect. Immun.* **59**:222–228.

Mitchell, T., and Andrew, P., 1997, Biological properties of pneumolysin, *Microb. Drug Resistance* **3**:19–26.

Nizet, V., Kim, K., Stins, M., Jonas, M., Chi, E.Y., Nguyen, D., and Rubens, C., 1997, Invasion of brain microvascular endothelial cells by Group B streptococci, *Infect. Immun.* **65**:5074–5081.

Patrick, D., Betts, J., Fery, E., Prameya, R., Dorovini-Zis, K., and Finaly, B., 1992, *Haemophilus influenzae* lipopolysaccharide disrupts confluent monolayers of bovine brain endothlial cells via a serum-dependent cytotoxic pathway, *J. Infect. Dis.* **165**:865–872.

Chapter 7

High Frequency Invasion of Mammalian Cells by β Hemolytic Streptococci

P. Patrick Cleary and David Cue

1. STREPTOCOCCAL PATHOGENESIS

The impact of intracellular invasion on the virulence of *S. pyogenes* (group A streptococci) and *S. agalactiae* (group B streptococci) is a rapidly expanding field of investigation. The discovery that members of these species are internalized by a variety of mammalian cells at frequencies equal to or beyond those of the more classical bacterial pathogens has piqued that interest. Investigators are attempting to relate this newfound potential to the pathophysiology of streptococcal disease. This chapter will focus on *S. pyogenes*, but where possible comparisons will be made to another important streptococcal pathogen, *S. agalactiae*. Knowledge of bacterial adhesions, is of course, essential for understanding the mechanisms by which bacteria are ingested by non-professional phagocytes; however, studies of *S. pyogenes* and *S. agalactiae* adherence mechanisms are a large body of work that is littered with debate and uncertainty and will therefore not be directly addressed here. Reviews of *S. pyogenes* (Hasty

---

P. PATRICK CLEARY and DAVID CUE Department of Microbiology, University of Minnesota, Minneapolis, Minnesota 55455.

Subcellular Biochemistry, Volume 33: Bacterial Invasion into Eukaryotic Cells, edited by Oelschlaeger and Hacker. Kluwer Academic / Plenum Publishers, New York, 2000.

et al., 1992) and *S. agalactiae* (Tamura and Rubens, 1994) adherence phenomena are cited.

β hemolytic streptococci cause a wide spectrum of human infections, however, *S. pyogenes* is the most common pathogen. In temperate climates pharyngitis is one of the most frequent reasons for a child to visit a general practitioner's office; whereas, in warmer climates impetigo, a superficial infection of the skin, can be very common among children. More recently public health authorities were reminded that this species has the capacity to cause extremely serious disease. The incidence of septicemia, toxic shock, necrotizing fasciitis, rheumatic fever, and childbed sepsis increased dramatically over the past ten years on a global scale. The pathogenesis of *S. pyogenes* infection is complex and varies within the serotype. There are more than 146 recognized serotypes. This streptococcus has mastered control of the early inflammatory response. At least five extracellular superantigens, two hemolysins, DNases, proteases, and several hyaluronidases are known to be secreted by different strains of *S. pyogenes*. Their potential to induce a shower of pro-inflammatory cytokines in experimental animals has motivated many to suggest a cause and effect relationship between the superantigenicity of these exotoxins, toxic shock and necrotizing fasciitis (Schlievert *et al.*, 1996). The SpeB exotoxin, once thought to be a superantigen is now known to be a highly active cysteine protease that is required for virulence in animal models of systemic infection (Burns *et al.*, 1998).

Activation of the alternate complement pathway and subsequent clearance of streptococci from sites of infection are initially impeded by two surface proteins, the C5a peptidase and M protein. Infiltration of professional phagocytes is slowed by specific proteolytic destruction of the early granulocyte chemotaxin, C5a. The C5a peptidase is a serine protease that cleaves the chemoattractant in the leukocyte binding site. In most strains the peptidase is anchored to the cell wall, strategically positioned to eliminate C5a at its source (Ji *et al.*, 1998).

Superimposed on the delay in accumulation of phagocytes, is the potential for M protein to restrict the deposition of C3b opsonin onto the surface of the bacterium, a requirement for efficient phagocytosis by polymorphonuclear phagocytes (PMNs). The M proteins are known adhesins (Okada *et al.*, 1994) and invasins (Cue *et al.*, 1998; Jadoun *et al.*, 1998; Dombek *et al.*, 1998) associated with resistance to phagocytosis (Fischetti, 1989). The first 25–30 amino acids of the mature form of M proteins are highly variable and responsible for serotype specificity. The carboxy two thirds of the molecule is conserved in sequence and composed of amino acid repeats that bind a variety of plasma proteins. Regardless of serotype, the extracellular portions of M proteins are primarily α-helical and exist as rod-like, coiled-coil molecules extending from the cell surface. Like the C5a peptidase, M and M-like proteins are bound to the cell wall by a common anchor sequence at their carboxy end. Vaccines for prevention of pharyngitis, composed of the C5a peptidase or M protein, are under development.

Other M-like proteins are also expressed by many serotypes. They bind various immunoglobulins and plasma proteins. Their architecture is similar to M protein, and they share extensive sequence homology. The wide distribution of these proteins and the fact that their expression is co-regulated with M protein and other virulence factors suggests an important role in the pathogenesis of streptococci. They may function as adhesins, activators of complement or sponges that inactivate antibodies however, their precise contribution to virulence is still unclear.

Clinical isolates of *S. pyogenes* express hyaluronic acid capsules of various size. The impact of these capsules on pathogenesis varies from strain to strain and with growth phase. The biosynthesis of hyaluronic acid is coded by the *has* operon; however, environmental factors that regulate capsule size have not been discovered. Capsule is the primary determinant of resistance to phagocytosis for some strains, yet is non-essential for others (Dale *et al.*, 1996).

S. agalactiae are responsible for neonatal sepsis and meningitis with an incidence of 0.3–4 cases per 1000 live births (Tamura *et al.*, 1994). Neonates are infected *in utero* or during birth from their colonized mother with devastating consequences. Surface antigens distinguish nine different serotypes, but one, serotype III, is the most common cause of disease. Virulence determinants include polysaccharide capsules, hemolysins, a surface bound C5a peptidase, and various surface proteins.

2. *S. pyogenes* EFFICIENTLY INVADES EPITHELIAL CELLS

β hemolytic streptococci were first shown to be ingested by epithelial cells by Nath (1989). The superficial nature of throat and skin infections, however, caused most in the field to consider *S. pyogenes* an extracellular pathogen. Therefore, our discovery that these streptococci are efficiently internalized by cultured human epithelial cells was a surprise (LaPenta *et al.*, 1994). Using the standard gentamicin protection assay up to 50% of a 1×10^5 CFU inoculum is internalized. This frequency exceeds that of the better-studied bacterial intracellular pathogens, such as Listeria and Salmonella. The assay has been modified by some investigators to include penicillin in addition to gentamicin (LaPenta *et al.*, 1994). Other variations have included addition of Triton X-100 (Molinari *et al.*, 1997; Greco *et al.*,

1995) to the lysis buffer. It is our experience, however, that internalized streptococci are sensitive to even very low concentrations of this detergent, and that addition of distilled water to the trypsinized monolayer more efficiently releases viable streptococci. The frequency of invasion also depends on growth phase, the bacterial strain, the size of the inoculum and the cell line. Although Greco *et al.* (1995) used early log phase streptococci in their studies LaPenta *et al.* (1994) showed that stationary phase streptococci are internalized at significantly higher frequency than log phase bacteria. This has also been shown to be true for *S. agalactiae* (Rubens *et al.*, 1992). The basis for growth phase differences has not been investigated. When the inoculum exceeds 5×10^5, CFU monolayers begin to slough off the microtiter plate and the apparent frequency of invasion is less (unpublished data). Frequencies of internalization vary from day to day and with lots of fetal calf serum (FCS) added to the culture medium. The latter variation is presumed to depend on differences in the concentration of fibronectin in the serum (see below).

Although frequencies vary, the M1 strain 90-226 can efficiently invade several epithelial cell lines, the macrophage line J774 and primary cultures of tonsil epithelial cells at relatively high frequency (Table 1) (Cleary and Cue, unpublished data). Frequencies ranged from 1–50%. Low frequency invasion of keratinocytes and high frequency invasion of J774 macrophages was independent of M proteins. Internalization of strain 90-226 by J774 cells was also independent of serum; whereas high frequency invasion of epithelial cells was always serum dependent. Significant variation has also been observed between serotypes and strains within a serotype. Serotype M12

Table I
Variety of Cultured Mammalian Cells Known to Internalize Serotype M1 *S. pyogenes*

Mammalian cells	Percent of inoculum ingested	M1 protein required	Serum required or serum factors
A549 human epithelial lung	10–50	+	+
HeLa epithelial cells	5–20	+	NT[a]
HEp2 pharyngeal epithelial cells	10–20	+	NT
Keratinocytes SCC-15 Squamous cell carcinoma	1–3	–	NT
Primary keratinized tonsillar epithelial cells	1–10	+	+
J774 mouse macrophage	20–35	–	–

[a] NT equals not tested.

strains are poorly internalized by HEp2 and A549 cells; (Molinari *et al.*, 1997; Greco *et al.*, 1995; LaPenta *et al.*, 1994). This low frequency is comparable to that of a M^- mutant of strain 90-226 (Cue *et al.*, 1998). In contrast cultured epithelial cells ingest M1, M3, M6, M18, M28 and M49 strains at high frequency.

2.1. High and Low Frequency Invasion

The existence of high and low efficiency pathways of ingestion of *S. pyogenes* is best exemplified by serotype M1 cultures. The $M1inv^+$ subclone (an example is strain 90-226) that was primarily responsible for the global increase in sepsis and toxic shock in the late 1980s and early 1990s is distinguished from other M1 strains by allelic variation in M1, streptokinase and SpeA toxin sequences; (Musser *et al.*, 1995; Cleary *et al.*, 1992), and by two chromosomal prophages, one of which encodes SpeA (Cleary, *et al.*, 1998a). Comparison of invasion frequencies of clinical isolates demonstrated that the $M1inv^+$ subclone was internalized by A549 cells at significantly higher frequency than other M1 subclones, which lacked these prophages. The source of the strain, i.e., blood or the throat of a carrier was irrelevant. Efficient internalization only correlated with genotype (LaPenta *et al.*, 1994). Efforts to transmit this property to low invasive laboratory strains by the prophage have been unsuccessful (Cleary *et al.*, 1998a). Although the M1 protein is required, highly efficient phagocytosis of these streptococci by epithelial cells probably requires additional factors since other subclones that produce nearly identical M1 proteins are poorly ingested.

2.2. Polysaccharide Capsules Impede Uptake of Streptococci by Epithelial Cells

Expression of virulence factors by *S. pyogenes* (Cleary *et al.*, 1998b) and *S. agalactiae* (Pincus *et al.*, 1992) is genetically unstable. The *vir* regulon controls the expression of several surface proteins, including the M protein (Cleary *et al.*, 1998b). In some strains expression of the *has* operon (hyaluronic acid biosynthesis) is coordinated with M protein expression by the transcriptional regulator Mga (Cleary *et al.*, 1998b). One M1 culture proved to be a mixture of encapsulated highly invasive and unencapsulated poorly invasive streptococci. The culture could be enriched for the more invasive form by multiple passage through monolayers of A549 cells. Genetic analysis of these variants revealed that encapsulated variants produced M protein, a requirement for high frequency invasion and that unencapsulated streptococci did not express M protein. Capsule seemed to be irrelevant to invasion by strain 90-226. Others have demonstrated that capsules interfere with ingestion of streptococci by epithelial cells (Schrager *et al.*, 1996; Hulse *et al.*, 1993). Schrager *et al.* (1996) showed that the hyaluronic acid capsule limited internalization of a highly encapsulated M18 strain. Elimination of the capsule dramatically increased the invasion frequency of cultured keratinocytes. The contradiction of this finding and those of Cleary *et al.* (1998a) can be explained by differences in capsule size. Although capsules of M1inv$^+$ streptococci are small relative to those expressed by the M18 culture used in the Schrager study. It is also possible that M1 strains activate a potent hyaluronidase as they enter stationary phase. A null mutation in capsule biosynthetic genes would likely further enhance ingestion of M1 streptococci by epithelial cells as well.

The type III capsule was shown to significantly block uptake of *S. agalactiae* by A549 epithelial cells (Hulse *et al.*, 1993) and endothelial cells (Nizet *et al.*, 1997). Källman and Kihlstrom (1997) did not, however, find a good correlation between capsule production by *S. agalactiae* and invasion of canine kidney cells. Although *S. agalactiae* is not known to produce enzymes that destroy the capsule, expression is genetically unstable (Pincus *et al.*, 1992) and may be limited by the acid pH of the vagina (Sellin *et al.*, 1995). Therefore, it is unlikely that capsules enhance internalization of either *S. pyogenes* or *S. agalactiae*, nor are they likely to be an absolute impediment to invasion of mammalian cells.

2.3. Ingestion Mechanisms

Transmission microscopy revealed what appeared to be polymerized actin beneath adherent streptococci, and streptococci contained in cytoplasmic vacuoles (LaPenta *et al.*, 1994). Scanning electron microscopy (SEM) suggested that streptococci are phagocytized by a zipper-like mechanism, mediated by receptor-ligand interactions (Dombek *et al.*, 1998). Neither membrane ruffling (Francis *et al.*, 1992), nor membrane coiling structures (Horwitz, 1984) were observed to be associated with uptake of streptococci. Figure 1 shows A549 cells with adherent streptococci (A) and a chain partially ingested by a HeLa cell (B). Streptococci and latex beads, coated with M1 protein were frequently observed to be in close contact with microvilli extending from the surface of either HeLa or A549 cells. The microvilli appeared to be firmly attached and stretched across the surface of bacteria or latex beads coated with M1 protein (A, C). In some cases the microvilli had broadened (Figure 1B) and appeared to be undergoing morphogenesis to form structures that resemble pseudopodia. Similar microvilli-like protrusions from the surface of HEp2 cells were seen to be in close contact with *S. agalactiae* (Valentine-Weigand *et al.*, 1997). These

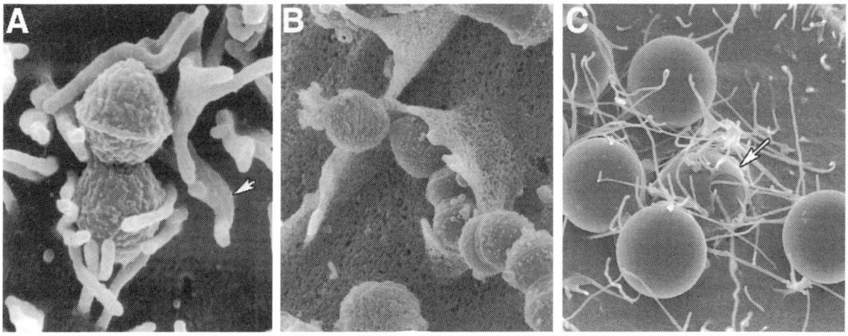

FIGURE 1. Scanning electron micrographs of A549 human lung carcinoma epithelial cells infected with *S. pyogenes* 90-226. Panel A shows microvilli of A549 cells in close contact with streptococci. The arrow indicates a microvillus that appears to be undergoing morphological change. Panel B shows a streptococcal chain about to be ingested by A549 cells. Panel C shows 3 μm latex beads coated with M1 protein. Microvilli appear to be attracted to the coated surface of the bead. The arrow marks a bead that is nearly ingested by a HeLa cell. Micrographs were provided by Dr. Priscilla Dombek, Department of Microbiology, University of Minnesota.

interactions may be similar to those reported for *Bordetella parapertussis* upon entering HeLa and respiratory epithelial cells (Ewanowich *et al.*, 1989). Although it is tempting to suggest that contacts with microvilli are an early stage of adhesion, or a prerequisite for ingestion of streptococci, definitive experiments have not been performed. It is equally possible that contacts between microvilli and streptococci are an artifact of fixation for microscopy. Co-infection of A549 cells with differentially marked, antibiotic resistant, high and low invading M1 cultures suggested that ingestion is an exclusive process. Preincubation or simultaneous incubation of highly invasive M1 streptococci with poorly invasive M1 bacteria did not increase or augment internalization of the latter (Cleary, unpublished data).

Several investigators have reported that ingestion of *S. pyogenes* by epithelial cells is inhibited by cytochalasin D (Fluckiger *et al.*, 1998; Dombek, *et al.*, 1998; Greco *et al.*, 1995; LaPenta *et al.*, 1994), suggesting that microfilaments were involved in endocytic uptake of these bacteria. Electron dense areas, beneath the plasma membrane, adjacent to attached streptococci were proposed to be actin filaments (LaPenta *et al.*, 1994). Dombek *et al.* (1998) directly confirmed that ingestion involves actin polymerization using double immunofluorescence, combined with phalloidin-Texas red labels and confocal microscopy. Aggregates and single chains of streptococci, internalized by HeLa cells, co-localized with polymerized actin. F-actin cup-like structures surrounded streptococci retained in vacuoles. An occasional streptococcal chain that spanned the plasma membrane was associated with F-actin, and was presumed to be a partially engulfed chain (Dombek *et al.*, 1998). This supported the conclusion that actin polymerization contributes to the mechanics of internalization.

The microtubulin system contributes to the uptake of *E. coli* O157:H7, Citrobacter and Campylobacter (Oelschlaeger *et al.*, 1994) and *Haemophilus influenzae* (St. Geme and Falkow, 1990). Although high concentrations of colchicine inhibited invasion of epithelial cells by 20–60%, a role for microtubules in the endocytosis of *S. pyogenes* was not further investigated (Fluckiger *et al.*, 1998; LaPenta *et al.*, 1994). Several investigators reported that inhibitors of the microtubule network impaired the ingestion of *S. agalactiae* by both epithelial and endothelial cells and macrophages (Winram *et al.*, 1998; Nizet *et al.*, 1997; Valentine-Weigland *et al.*, 1996). Valentine-Weigand *et al.* (1997), however, concluded that the inhibitory effects of nocodazal, an inhibitor of microtubulin function, on *S. agalactiae* internalization by HEp2 cells was most likely indirect, and due to its effect on DNA replication.

2.4. Adherence and Invasion are Independent

Adherence of streptococci to epithelial cells in itself is not sufficient to trigger events that lead to their ingestion. Streptococcal adherence to epithelial cells is dependent on multiple adhesins. Serotype M6 streptococci depend on both M protein and protein F for adherence to HEp2 cells (Jadoun *et al.*, 1998; Fluckiger *et al.*, 1998) and M1 streptococci depend on M protein and an unidentified adhesin. Several lines of evidence suggest that adherence in itself is not sufficient to initiate ingestion of the bacteria. Mutations in either *emm6* or *prtF* reduced both adherence to and invasion of HEp2 cells by strain JRS4, and the double $M6^- F^-$ mutation completely eliminated these interactions (Jadoun *et al.*, 1998). A mutation in the *emm1* gene reduced adherence to HeLa cells by 76%, but had a negligible effect on adherence to A549 cells. This mutation reduced the ability of strain 90-226 to invade both cell lines by more than 95% (Cue *et al.*, 1998; Dombek *et al.*, 1998). Moreover, deposition of $M1^-$ streptococci or latex beads without M1 protein onto monolayers by centrifugation does not initiate actin polymerization. On the other hand, latex beads with bound M1 protein were internalized by HeLa and A549 cells. Thus M1 protein is an invasin that is associated with cytoskeletal changes that are required for cultured epithelial cells to phagocytize streptococci (Dombek *et al.*, 1998). The M1 protein also binds fibronectin (Cue *et al.*, 1998). So it is possible that signals which induce cytoskeletal rearrangement are initiated by interaction of this extra cellular matrix (ECM) protein with integrin receptors rather than direct interaction of M1 or SfbI/F1 proteins with epithelial surfaces. (The fibronectin binding proteins, SfbI and F1, are discussed below). The signals generated by the interaction of epithelial cells with streptococci have not been investigated.

2.5. Persistence and Multiplication of Intracellular Streptococci

The fate of internalized group A streptococci has not been systematically studied. They end up in vacuoles in several epithelial cell lines that have been examined and in epithelial cells visualized in sections of human tonsils (Österlund *et al.*, 1997). By 8 hrs post-inoculation Greco *et al.* (1995) noted that some intracellular streptococci have undergone partial degradation, and demonstrated phagosomal-lysosomal fusion using HEp2 cells, pre-loaded with ferritin. Exposure of cells to NH_4Cl did not reduce the invasion frequency, suggesting that acidification of the vacuole does not occur. Vacuoles that contain *S. typhimurium* associate with lysosomal membrane proteins (Lgp), and by-pass fusion with mature endosomal vesicles. Garcia-del Portilla and Finlay (1995) suggested that these phagosomes are exocytic vesicles that are directed to the plasma membrane of infected cells. In a preliminary analysis of infected HeLa cells, phagosomes that contained streptococci were shown to contain Lamp-I lysosomal proteins. After 2 hours incubation these were the most common vacuoles associated with streptococci (Dombek *et al.*, 1998). Other Lgp markers were not investigated, so it was not possible to conclude whether these streptococci are targeted to be killed and degraded, or transported to the exterior of the HeLa cells. In contrast to most reports, Österlund and Engstrand (1995) observed streptococci in the cytoplasm of HEp2 cells. Actin tails, similar to those that form on cytoplasmic *Listeria monocytogenes* and Shigella were not seen following prolonged incubated of infected HeLa cells (unpublished data).

Whether streptococci survive and multiply in epithelial cells is still debated. Most investigators who examined this question incubated infected cells for various periods in culture media that contained gentamicin and reported significant initial reduction in viable bacteria. This prompted Schrager *et al.* (1996) to suggest that internalization by epithelial cells is a dead end for *S. pyogenes*, perhaps a defense mechanism of their host. Others do not share this opinion. Streptococci were found to persist from 4–7 days in HEp2 cells, and removal of gentamicin at anytime during the experiment resulted in re-growth of associated bacteria (Österlund and Engstrand, 1995). Although viability initially decreased, serotype M1 streptococci persisted at least 72 hours in A549 cells and 24 hours in tonsillar epithelial cells (unpublished data). Drevets *et al.* (1991) demonstrated that the decline in viability of intracellular *Listeria monocytogenes* is due to uptake of gentamicin by infected cells. Thus, the gradual demise of intracellular streptococci could be antibiotic related.

S. pyogenes have not been reported to multiply inside epithelial cells. Again, definitive experiments have not been performed. Transmission electron microscopy often reveals cells containing multiple cocci (unpublished observation) suggesting that intracellular bacteria had divided. A computer-generated image from confocal sections was purported to show a micro colony that transversed an entire cell (Dombek *et al.*, 1998). Experiments such as plaque assays and more thorough analyses by electron microscopy are required before this issue will be settled. It is more than likely that internalized streptococci face alternative fates, some destined to be killed by lysosome products, others able to take over control of the target cell and/or be transcytosed to the external environment or to adjacent cells.

3. ADHESINS, INVASINS, AND INTEGRIN RECEPTORS

Intracellular invasion by bacterial pathogens is frequently mediated by a subclass of adhesins referred to as invasins. Typically, invasins are proteins expressed on the surfaces of bacterial cells that recognize, directly or indirectly, specific host cell receptors (Finlay and Falkow 1997). Two well studied invasins are the InvA and InlA proteins of *Yersinia pseudotuberculosis* and *Listeria monocytogenes*, respectively, The *Yersinia* invasin is a 108 kDa, outer membrane protein capable of binding multiple β1 integrins (Isberg and Leong, 1990). Internalin is an 88 kDa, cell wall anchored molecule for which E-cadherin is the receptor (Mengaud *et al.*, 1996).

A number of factors contribute to whether a microbial adhesin can function as an invasin (Finlay and Falkow, 1997). In general, however, the latter are capable of inducing reorganization of the host cell cytoskeleton upon interaction with host cell receptor(s). Integrins and cadherins, which possess cytoplasmic domains capable of transmitting signals to cytoskeletal elements (Schwartz, 1995), are frequently exploited by invasive pathogens to gain entry into host cells. The cytoskeletal rearrangements resulting from engagement of these receptors by invasins, ultimately lead to endocytic uptake of adherent microbes.

S. pyogenes is known to express an array of cell surface proteins that facilitate bacterial colonization of a variety of human tissues (Courtney *et al.*, 1997). Recent studies have implicated several of these adhesins that mediate intracellular invasion, including two related fibronectin binding proteins (Jadoun *et al.*, 1998; Ozeri *et al.*, 1998; Molinari *et al.*, 1997) and four different M proteins (Jadoun *et al.*, 1998; Dombek *et al.*, 1998).

Additionally, the epithelial cell receptors for two of the streptococcal invasins have been tentatively identified (Cue *et al.*, 1998; Ozeri *et al.*, 1998).

3.1. High Affinity Fibronectin Binding Proteins, SfbI and F1 are Invasins

SfbI and F1 are two closely related streptococcal adhesins that bind the serum/ECM protein fibronectin (Fn). Approximately 70% of *S. pyogenes* isolates carry the gene encoding SfbI/F1 (Molinari *et al.*, 1997; Jaffe *et al.*, 1996). These proteins each contain two domains that are capable of Fn binding (Ozeri *et al.*, 1996; Talay *et al.*, 1994) (Figure 2). One domain, designated RD2, is comprised of a tandemly repeated amino acid sequence that shares identity with Fn binding proteins of other gram-positive bacteria (Patti *et al.*, 1994). In protein F1, the second domain is comprised of a unique 43 amino acid sequence located immediately N-terminal of the RD2 domain, and includes six amino acids of the first RD2 repeat.

Molinari *et al.* (1997) demonstrated that SfbI can mediate invasion of HEp2 cells. Internalization of streptococci can be blocked by antiserum raised against SfbI or by preincubation of HEp2 cells with recombinant SfbI (rSfbI). Latex beads that are coated with rSfbI readily adhere to and are efficiently ingested by epithelial cells, demonstrating that the interaction of SfbI with host cells is sufficient for internalization. Latex beads coated with a rSfbI peptide lacking the Fn binding domains fail to efficiently adhere to HEp2 cells. The authors hypothesized that SfbI mediated invasion occurs by a process whereby the bacterial invasin binds Fn which, in turn, binds a host cell Fn receptor.

This proposal is supported by the work of Jadoun *et al.* (1998) and Ozeri *et al.* (1998). These researchers reported that the addition of exogenous Fn can stimulate invasion of HEp2 cells by protein F1-expressing streptococci. F1-mediated invasion of HeLa cells is highly dependent on addition of serum or purified Fn (Ozeri *et al.*, 1998). In both studies anti-Fn antibody and rF1 peptides, containing at least one Fn binding domain, were demonstrated to specifically inhibit F1-mediated invasion.

Protein F1 binds to a 70 kDa N-terminal fragment of Fn encompassing the fibrin and collagen binding domains (Figure 2). The 70 kDa fragment, which lacks the portion of Fn bound by epithelial cells, inhibits Fn mediated invasion by competing with intact Fn for F1 binding. Only intact Fn is capable of facilitating bacterial invasion. Antibody directed against the integrin β1 subunit specifically blocks Fn-F1 mediated invasion of HeLa cells, suggesting that one or more β1-containing integrins are involved in bacterial internalization. In contrast, β1 integrins do not appear to partici-

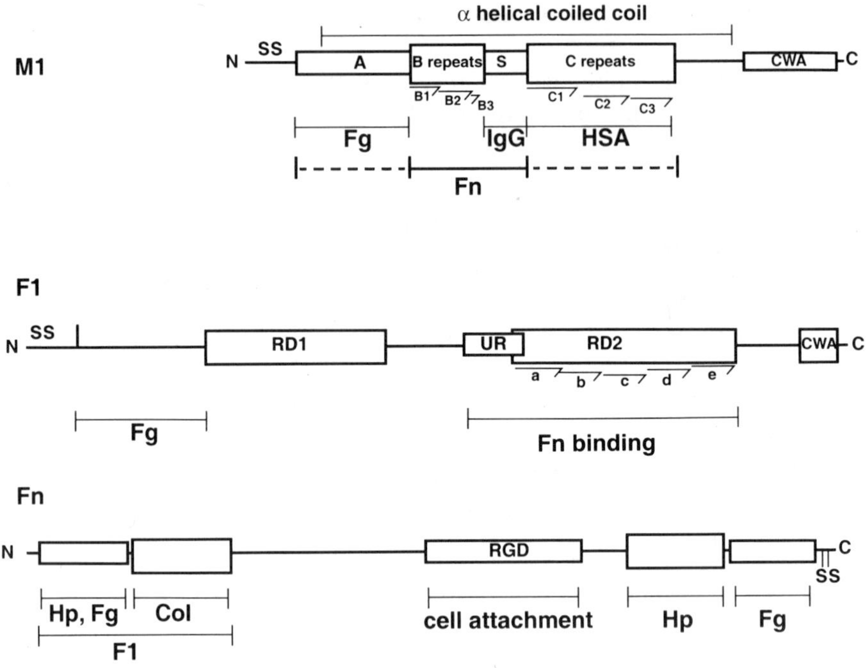

FIGURE 2. Schematic representation of M1 protein, protein F1 and human Fn. M1 depicts domains of M1 as previously described (Cedervall *et al.*, 1995; Akesson *et al.*, 1994). SS is the signal sequence; A is the N-terminal region of mature M1 protein; B repeats are 2.2 repeats (B1, 2 and 3) of a 35 amino acid sequence. S is a unique 35 amino acid sequence; C repeats are 2.7 repeats of a 42 amino acid sequence that are conserved among different serotypes of M proteins; CWA is the cell wall anchor, membrane spanning and cytoplasmic regions of M1. The brackets below the figure indicate regions of M1 responsible for binding the indicated serum proteins: Fg is fibrinogen (Cleary, unpublished); IgG is human immunoglobulin G; HAS is human serum albumin (Akesson *et al.*, 1994); Fn, fibronectin (Cleary, unpublished data). Recombinant M1 protein fragments containing the ABSC, ABS or BSC domains are all capable of Fn binding. A protein fragment containing only the Band S domains, however, does not bind Fn. This suggests that binding is a function of the BS region. The dashed line is intended to reflect these results. While uncertain at this time, the failure of the BS fragment to bind Fn may be due to the inability of the protein fragment to maintain a conformation necessary for ligand binding (Cedervall *et al.*, 1995; Nilson *et al.*, 1995). F1 (protein F1): RD1 is repeat domain 1, RD2 is repeat domain 2 containing 5 repeats of a 37 amino acid sequence. RD2 is capable of independently binding the N-terminal fibrin binding domain of Fn. UR; upstream fibronectin binding region is comprised of a unique 43 amino acid sequence and 6 amino acids of the first RD2 repeat. UR is capable of independent binding of the N-terminal region of FN containing the collagen binding domains of Fn (Ozeri *et al.*, 1996). High affinity binding of cellular (matrix) Fn requires both the UR and RD2 domains (Okada *et al.*, 1997). The fibrinogen binding domain is also (Katerov *et al.*, 1998). Fn (human fibronectin): Regions of Fn involved in binding of heparin (Hp), fibrin (Fg) and collagen (Col) are indicated. Cell attachment; region of Fn bound by integrins and containing the RGD sequence (Schwarzbauer, 1991). The portion of Fn bound by protein F1 is also indicated. The representation of Fn is not in scale with M1 and F1 proteins.

pate in invasion of GD25 (embryonic mouse stem) cells. Rather, invasion of GD25 cells appears to be mediated by integrin $\alpha v\beta 3$ the major Fn receptor of this cell line (Ozeri *et al.*, 1998). As a whole, these results are consistent with the proposal that Fn serves as a molecular bridge between F1 and host cell Fn receptors. This type of mechanism was originally proposed by Okada *et al.* (1994) to account for the stimulation of protein F mediated adherence by soluble Fn.

3.2. M Proteins Can Function as Invasins

While the studies cited above firmly established that SfbI/FI can function as an invasin, reports from other laboratories indicate that this protein is only one of several invasins expressed by *S. pyogenes*. M proteins are a large family of cell surface proteins recently shown to be involved in adherence to host tissues, and intracellular invasion (Dombek *et al.*, 1998; Jadoun *et al.*, 1998; Fischetti *et al.*, 1989). Jadoun *et al.* (1998) studied adherence and invasion by an *S. pyogenes* strain, JRS4, that expresses protein F1 and type 6 M protein. F1 functions as the major adhesin/invasin for this strain, but inactivation of the gene encoding M6, *emm6*.1, decreased adherence to HEp2 cells by 26% and invasion by 46%. Disruption of *emm6*.1 in a protein F1 mutant, resulted in further decreases in both adherence and invasion. Fluckiger *et al.* (1998) also demonstrated a role for M6 protein in invasion by showing that anti-M6 IgG can block bacterial invasion of pharyngeal epithelial cells, however, bacterial adherence was unaffected by the antibody. Expression of M6 in the noninvasive streptococcal species, *S. gordonii*, is reportedly insufficient for invasion of HeLa cells suggesting that other bacterial factors may be required for M6 mediated invasion (Greco *et al.*, 1998).

M6 protein has been proposed to mediate adherence to human keratinocytes, at least in part, via recognition of CD46, or membrane cofactor protein, a receptor expressed in a number of human tissues (Okada *et al.*, 1995). Berkower *et al.* (1998) recently investigated the possibility that *S. pyogenes* invasion can be mediated by the interaction of M6 with CD46. Their results indicate that while expression of M6 protein can facilitate invasion, M6 mediated invasion does not require and is unaffected by CD46 expression.

The studies cited above established that M6 is not absolutely required for invasion, but expression of M6 can clearly increase the efficiency of invasion. In contrast, invasion by a serotype M1 strain of *S. pyogenes*, strain 90-226, was found to be highly dependent upon M1 protein expression (Dombek *et al.*, 1998; Cue *et al.*, 1998). In these studies, antiserum directed against the mature amino terminus of M1 reduced invasion by approximately 10 fold and latex beads coated with M1 were readily ingested by HeLa cells. Inactivation of *emm1*, the gene encoding M1, decreased HeLa cell invasion by 50-fold. M1 expression is also required for invasion of A549 human lung epithelial cells, HEp2 cells and primary cultures of human tonsillar epithelial (HTE) cells, but is dispensable for invasion of a mouse macrophage cell line (J774A.1) (Cue and Cleary, unpublished). These results indicate that invasion of human epithelial cells by strain 90-226 is largely mediated by M1 protein.

Expression of M1 and the presence of Fn is sufficient for high efficiency invasion by strain 90-226. M1 Protein appears to account for most of the Fn binding by this strain, as inactivation of *emm1* reduces binding by 88%. Moreover, M1 protein is capable of binding Fn (Cue *et al.*, 1998). It was recently reported that invasion of HEp2 (Jadoun *et al.*, 1998) or HeLa (Greco *et al.*, 1998) cells by other $M1^+$ strains is very inefficient, with only 0.1–1% internalization of streptococci invasion efficiencies comparable to 90-226 in the absence of an agonist (i.e. serum, Fn or Lm). The apparent explanation for these seemingly contradictory results is that in the latter studies, invasion experiments were performed in the absence of serum. Recent unpublished results indicate that at least some $M1^+$ strains that exhibit low efficiency invasion can be efficiently internalized by cultured cells but as in the case of 90-226, invasion is serum dependent (Cue and Cleary, unpublished).

Fn binding is not a general property of M proteins and only two types, M1 and M3, have been reported to bind Fn (Cue *et al.*, 1998; Schmidt *et al.*, 1993). Interestingly, M3 protein was recently found to promote serum dependent invasion of HEp2 cells (Berkower *et al.*, 1998). Protein H, an M-like protein expressed by some *S. pyogenes* strains, also has Fn binding activity (Frick *et al.*, 1995; Akesson *et al.*, 1994). The extracellular portions of M1, M3 and protein H do not display significant similarity to protein SfbI/F1 or to the conserved Fn binding motifs present in well characterized Fn binding proteins of gram-positive bacteria (Patti *et al.*, 1994). Thus, Fn binding by M-like proteins likely occurs via a novel mechanism. The finding that protein H binds to type III Fn repeats (Frick *et al.*, 1995) whereas SfbI/F1 recognizes the heparin and collagen binding domains of Fn (Ozeri *et al.*, 1996) (Figure 2), supports this assertion.

M1, M3 and protein H have a high degree of sequence similarity in their C-terminal regions. These segments, containing the C repeats and anchoring domains, however, are apparently dispensable for Fn binding by M1 and protein H (Cue and Cleary, unpublished; Frick *et al.*, 1995). The N terminal portions of these proteins, implicated in Fn binding by M1 and protein H, share only short segments of sequence similarity. The amino terminal segments of all three proteins have been demonstrated to be, or are predicted to be, almost entirely α-helical in conformation and to exist as coiled-coil dimers on the surface of bacterial cells (Nilson *et al.*, 1995). It is possible that Fn binding by these proteins is more a function of the conformational state of M proteins, rather than being due to the interaction of specific M amino acid residues with ligand.

In addition to M1 and M6, other types of M protein have recently been found to function as invasins. Berkower *et al.* (1998) introduced the genes coding for types 3 and 18 M proteins into an *emm6*.1 deleted derivative of *S. pyogenes* JRS4, then compared the abilities of the resulting strains to invade cultural mammalian cells. These findings led these authors to suggest that M proteins of different serotypes likely recognize different receptors on the surfaces of eukaryotic cells. Collectively, these studies suggests that while several types of M proteins can facilitate invasion by *S. pyogenes*, there is no single mechanism underlying M protein mediated invasion.

3.3. Fibronectin and Laminin Independently Trigger Internalization of *S. pyogenes*

Expression of M1 protein per se, is not sufficient for invasion, however, as internalization of 90-226 is also dependent upon the exogenous serum, serum fibronectin or laminin (Lm) (Cue *et al.*, 1998). Lm, like Fn, is a high-molecular weight, extracellular glycoprotein present in the ECM of numerous tissues (Timpl and Brown, 1994). Isogenic $M1^+$ and $M1^-$ strains were tested for their responses to fetal bovine serum (FBS), Fn or Lm, with regard to invasion of A549 cells. In the absence of an agonist, invasion by either strain is very inefficient with less than 1% internalization of the inoculum. These agonists can stimulate invasion of $M1^+$ streptococci by up to 70-fold. Neither FBS or Fn promote invasion by $M1^-$ bacteria and Lm stimulates invasion by only 2-fold. Therefore, the ability of either Fn or Lm to stimulate invasion by strain 90-226 is dependent upon expression of M1 protein. Anti-Fn serum is effective at blocking invasion stimulation by FBS or Fn, but does not abrogate Lm-mediated invasion (Cue *et al.*, 1998). These results establish that Fn and Lm are distinct, M1-dependent, invasion agonists. The presence of at least one of these agonists is also required for bacterial invasion of HTE cells (Cleary, unpublished).

Strain 90-226 adheres to A549 and HTE cells independently of invasion agonists, although adherence can be stimulated approximately 2-fold by the addition of serum, Fn or Lm. Agonist independent adherence is also M1 independent, as $M1^+$ and $M1^-$ bacteria adhere equally well to epithelial monolayers in the absence of invasion agonists. In contrast to $M1^+$ bacteria, adherence by the M1$^-$ mutant is not appreciably affected by Fn or Lm. Thus, strain 90-226 possesses both factor-dependent and factor-independent adhesins for binding host cells. Epithelial cell binding via the latter is apparently insufficient for efficient internalization of bacteria. Adherence mediated by the factor-dependent adhesin, M1 protein, presumably targets bacteria to the appropriate cell receptor for efficient internalization to occur. In the case of Fn-mediated invasion, the appropriate receptor is integrin α5β1. A monoclonal antibody (mAb) that specifically blocks Fn binding by this integrin, can also block invasion of A549, HeLa and HTE cells. MAbs that react with either the α5 or β1 integrin subunits or small, nonpeptidyl α5β1 antagonists, are also effective invasion inhibitors (Cue *et al.*, 1998; Cue and Cleary, unpublished).

The inhibitory effects of α5β1 antagonists are not due to generalized effects on either bacterial or host cells as the inhibitors do not abrogate either Lm or RGD peptide (see below) mediated invasion. Rather, the inhibitory effects ofαβ1 antagonists is only observed when bacteria are exposed to either serum or purified Fn. The inefficient invasion that occurs in the absence of an invasion agonist is also largely unaffected by α5β1 antagonists. Thus, as in the case of protein F1$^+$ strains, Fn appears to function as a bridging molecule in promoting invasion by M1$^+$ bacteria directing bacterial interaction to α5β1 (Figure 3).

Another finding that is consistent with the proposal that Fn primarily functions as a bridging molecule that triggers invasion, is that invasion agonists promote ingestion of already adherent bacteria by epithelial cells. Only 0.1–1% of M1$^+$ bacteria that adhere to A549 cells in the absence of an agonist, are internalized. The addition of an invasion agonist, either prior to or subsequent to bacterial adherence, increases this value by one to two orders of magnitude (Cue *et al.*, 1998). Thus, for *S. pyogenes* 90-226, invasion appears to be dependent upon the formation of a trinary complex of M1, Fn and integrin α5β1.

3.4. β1 Integrin Receptors Specifically Mediate Internalization of *S. pyogenes*

Fn is the only known α5β1 ligand (Hynes 1992); therefore, in the context of a bridging model, antagonists of this integrin would not be anticipated to abrogate Lm-mediated invasion. However, a mAb directed against the integrin β1 subunit does block Lm mediated invasion (Cue *et al.*, 1998). This result suggests that Lm facilitates invasion by binding to one or more integrins containing the β1 subunit. Integrin α6β1 is the primary Lm receptor of A549 cells (Falcioni *et al.*, 1994) and HTE cells express the α6 subunit (Cue and Cleary, unpublished). While this suggests that the α6β1

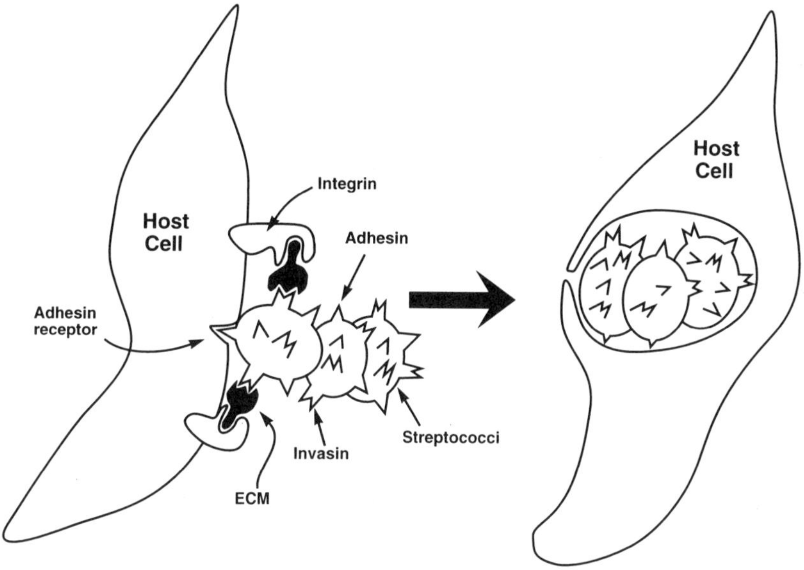

FIGURE 3. A model for extracellular matrix protein triggered invasion of epithelial cells by *S. pyogenes*. ECM are extracellular matrix proteins, laminin or fibronectin. Integrins are various β1 integrins. ^ symbolizes an unidentified adhesin that does not initiate ingestion of streptococcus. M symbolizes M protein. An ECM bridge between the streptococcal surface and a functional integrin receptor induces the cytoskeletal changes required for ingestion of the bacteria.

integrin may be the receptor involved in Lm-mediated invasion, this proposal has yet to be substantiated.

In some respects invasion by $M1^+$ streptococci is analogous *to Y. pseudotuberculosis* invasion which also exploits multiple β1 integrins for entry into mammalian cells. The Yersinia invasin, InvA is capable of binding at least four different β1 chain integrins, including the α5β1 receptor (Isberg and Leong, 1990). Interaction of invasin with α5β1 is not dependent on Fn binding. Rather, invasin binds directly to α5β1 with high affinity, and binding can be inhibited by Fn or RGD-containing peptides (van Nhieu *et al.*, 1991). This mechanism is clearly distinct from that used by $M1^+$ or $F1^+$ streptococci, for which integrin engagement is mediated by integrin ligands. It is not yet clear whether integrin recognition of M1-bound ligands is sufficient to promote bacterial entry. Direct engagement of integrins or possibly other host receptor molecules by M1, may be required for efficient internalization of bacteria. On the other hand, Fn-coated beads are efficiently internalized by both A549 (Cue and Cleary, unpublished) or HeLa (Ozeri *et al.*, 1998) cells, suggesting that the interaction of Fn with $\alpha5\beta1$ is sufficient for high efficiency invasion.

3.5. Other Potential *S. pyogenes* Invasins

S. pyogenes 90-226 encodes multiple pathways for invasion of cultured cells (Figure 4). Two pathways, described above, are M1-dependent, require the presence of exogenous Fn or Lm, and involve exploitation of multiple $\beta1$ integrins for bacterial uptake. A third invasion pathway is known to exist in this strain that is M1 protein independent and does not appear to involve the participation of $\beta1$ integrins. Activation of the latter pathway requires bacterial exposure to small peptides containing the sequence RGD (Cue *et al.*, 1998).

The tripeptide RGD sequence is present in most integrin ligands, including Fn, and is an essential determinant in the binding of ECM proteins by their cognate receptors (Hynes, 1992). Synthetic RGD peptides can compete with ECM proteins for integrin binding and it was anticipated that such peptides would inhibit Fn mediated invasion by 90-226. This does occur under the appropriate conditions, however, a variety of small (4 to 6 amino acid) synthetic peptides (e.g. GRGDTP, RGDS) are able to activate invasion by both $M1^+$ and $M1^-$ bacteria by approximately five fold. It appears that peptide mediated invasion is a specific response to RGD containing peptides, as peptides such as EHIPA and KGDS do not promote invasion (Cue and Cleary, unpublished).

RGD containing peptides increase both bacterial invasion and adherence and are not dependent on bacterial protein synthesis to exert their effects (Cue and Cleary, unpublished). It is likely that peptides capable of stimulating *S. pyogenes* invasion *in vitro* bear some degree of similarity to proteins encountered in an infected host. Thus the response to RGD peptides may be an indication of the bacterium's capacity to respond to changing conditions encountered during the course of infection. Testing of this hypothesis, however, awaits identification of the bacterial peptide receptor.

Lm mediated invasion by strain 90-226 is heavily dependent on M1 protein expression but this protein can modestly stimulate (~2 fold) invasion by an $M1^-$ derivative of 90-226. This suggests that this strain expresses an unidentified Lm receptor that facilitates invasion. The contribution of this receptor to M1-mediated invasion is uncertain at this time. To date, there is no direct evidence that M1 protein is capable of Lm binding. It is possible that M1 protein works in concert with other bacterial factors to promote bacterial internalization. Switalski *et al.* (1984) described an *S. pyogenes* Lm receptor for which ligand binding is inhibitable by fibrinogen. While this receptor was not fully characterized, its size (10^6 Da) suggests that it is not an M protein. Studies cited above indicate that *S. pyogenes* makes use of at least three pathways to invade epitheial cells (Figure 4).

Integrins and integrin ligands have been implicated in intracellular invasion by a number of diverse microbial pathogens. Vitronectin and Fn have been found to facilitate intracellular invasion by *Neisseria gonorrhoeae* (van Putten *et al.*, 1998; Duensing *et al.*, 1997; Gomez-Duarte *et al.*, 1997). Invasion by *Mycobacterium bovis* (Kuroda *et al.*, 1993) is also

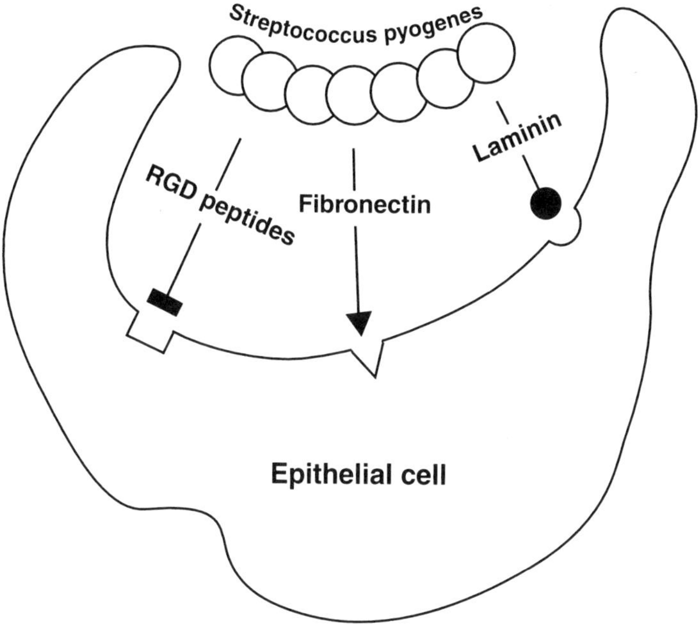

FIGURE 4. Multiple pathways for invasion of epithelial cells by *S. pyogenes*. Three pathways have been identified for serotype M1 strains 90-226 (Cue *et al.*, 1998). The black symbols indicate that each route involves a different receptor on an epithelial cell. Fibronectin can bind to M1 protein or one of the other Fn binding proteins expressed by *S. pyogenes* (Okada *et al.*, 1994; Talay *et al.*, 1994). The laminin binding protein has not been identified. In the absence of serum, RGD peptides can enhance invasion. The streptococcal receptors for these peptides have not been identified. The M6 protein of strain JRS4 can mediate invasion of HEp2 cells in the absence of serum. This may represent, yet another invasion mechanism.

stimulated by Fn. *M. leprae* invasion can be stimulated by microbial binding of Fn (Schorey *et al.*, 1995) or, as suggested by a recent report, Lm (Rambukkana, 1997). Various *S. pyogenes* strains are known to bind serum and ECM proteins, such as vitronectin, collagen and fibrinogen, all of which can bind to human cells via their cognate integrins. It is likely that as work on *S. pyogenes* invasion progresses, that one or more of these proteins will be implicated in promoting streptococcal invasion of human cells.

Streptococcal pyrogenic exotoxin B is produced by all serotypes of *S. pyogenes*. This cysteine protease has a broad specificity. It has been shown to cleave extracellular matrix proteins, fibronectin and vitronectin, and to activate human matrix metalloprotease, and IL-1β. Mutations in the *speB* gene were shown by two different research groups to alter the capacity of *S. pyogenes* to invade A549 cells. Unfortunately, they reported opposite effects on invasion. Tsai *et al.* (1998) observed a decrease in invasion frequency when the *speB* genes of M1 and M49 strains were altered; whereas Burns *et al.* (1998) saw an increased frequency by M2 and M3 strains with mutated *speB* genes. At this point there is no rational means to reconcile these differences.

4. REALITY CHECK: DOES INTRACELLULAR INVASION IMPACT ON THE VIRULENCE OR EPIDEMIOLOGY OF STREPTOCOCCAL INFECTIONS?

Although pharyngitis and superficial skin infection are most common, *S. pyogenes* has historically also been associated with sepsis and fulminate systemic infections. The mechanisms by which streptococci breach oral mucosal or epithelial barriers are unknown. Over the past 11 years, the incidence of systemic disease has increased across most continents of the world. Several serotypes have been associated with severe disease, but an M1 subclone, designated M1inv^{+}, was the cause of temporally and geographically clustered cases of sepsis, toxic shock and necrotizing fasciitis (Schlievert *et al.*, 1996; Musser *et al.*, 1995; Cleary *et al.*, 1992).

Curiously, patients with devastating soft tissue infection often reported an absence of prior trauma or disease that might initiate such invasive diseases. Wounds appeared to develop spontaneously or from bruises that did not break the skin (Stevens *et al.*, 1989). La Penta *et al.* (1994) reported that an M1inv^{+} strain invaded A549 epithelial cells at higher frequency than other M1 subclones, prompting them to suggest that high frequency intracellular invasion may account for the association of this M1inv^{+} subclone with more serious infection. A more extensive analysis of serotypes M1 strains confirmed that the M1inv$^+$ subclone invaded A549 epithelial cells at higher frequency than the M1 subclones (Cleary *et al.*, 1998a). M1inv$^+$ strains harbor two prophages, one of which encodes the SpeA erythrogenic toxin. Together these prophages augment the organism's genome by at least 70 kb of DNA. The origins of the strains, i.e. uncomplicated disease, or more invasive disease (blood and wound isolates) were irrelevant. Only the M1inv$^+$ genotype correlated with high frequency invasion. Although low intracellular invasive strains lack these prophages and the *speA* gene, there is no evidence that either influence the efficiency by which bacteria are internalized by epithelial cells. The molecular basis for different invasion frequencies within a serotype is unknown.

4.1. Intracellular Infection, Recurrent Tonsillitis, and Carriage of *S. pyogenes*

Other investigators also failed to correlate high frequency intracellular invasion of epithelial cells with systemic or invasive streptococcal diseases. Recent analyses of a variety of strains and serotypes from carriers, patients with uncomplicated pharyngitis, and septic patients came to the opposite conclusion (Molinari *et al.*, 1998; Neeman *et al.*, 1998). Strains from carriers were internalized at higher frequency by HEp2 cells than those isolated from blood cultures. A genetically unstable M1inv$^+$ clinical isolate was shown to be enriched for more invasive streptococci by *in vitro* serial passage through A549 cells (Cleary *et al.*, 1998b). The carrier state, which may cycle bacteria in and out of the host epithelium could also select for variants that are more efficiently internalized.

Schrager *et al.* (1996) also questioned the relationship between intracellular invasion and systemic disease. They reported that an unencapsulated mutant, which was readily internalized *in vitro* by cultured keratinocytes did not more frequently induce systemic infection than the parent culture in a mouse model. The relevance of this model to human disease is, however, questionable. To reconcile these observations it was suggested that high frequency intracellular invasion may have an important influence on the epidemiology of group A streptococcal disease in general (Cleary *et al.*, 1998a; Molinari *et al.*, 1998; Neeman *et al.*, 1998). High frequency intracellular invasion may increase the rate of antibiotic therapy failure and persistence of a given strain in a community. With more individuals in a community shedding these streptococci, the probability of more serious disease increases. Strains or serotypes that are less endowed with this potential may be less likely to result in carriers and to cause recurrent infections. Of course other factors, both host and bacterial, could also impact on the incidence of severe systemic infections. An epidemiological analysis of an outbreak of toxic shock and necrotizing fasciitis in southern Minnesota supported this concept (Cockerill *et al.*, 1997). Nearly 40 percent of the school children in surrounding communities were carriers of the serotype M3 clone that was responsible for systemic disease in adults. These investigators argued that school children were the reservoir for the bacteria that produce toxic shock in adults who often had underlying disease.

The most direct evidence that intracellular streptococci in the throats of carriers are a reservoir for this pathogen comes from microscopic examination of tonsils excised from individuals plagued by recurrent tonsillitis. Österlund *et al.* (1997) observed that 13/14 tonsils from such individuals harbored intracellular streptococci. Streptococci were located in epithelial cells and macrophage-like cells. Tonsils from controls who underwent surgery for other reasons were free of intracellular streptococci. They also showed that infection of cultured tonsil biopsies resulted in intracellular bacteria. Unpublished studies from our laboratory confirmed that the $M1inv^+$ strain 90-926 efficiently invades primary cultures of keratinized tonsillar epithelial cells. Moreover, internalization of these streptococci was determined to be dependent on M protein, and the same agonists and integrins that are required for invasion of A549 cells (Cue and Cleary, unpublished data). The fact that streptococci inside such cells are highly recalcitrant to penicillin may account for the frequent failure of penicillin to eliminate these bacteria from the throats of patients. More knowledge of both the streptococcal adhesins and host receptors that lead to high frequency invasion of epithelial tissue could lead to improved antibiotic therapy.

4.2. *S. agalactiae* Can Traverse Polarized Epithelium and Endothelium from the Apical Side

S. agalactiae is a primary cause of neonatal sepsis and meningitis. The fetus can be infected *in utero* from contaminated amniotic fluid or during birth by aspiration of contaminated mucosal secretions. In both situations the bacteria must penetrate several epithelial and endothelial barriers in order to reach the blood and/or brain. Barriers include the genital and respiratory epithelium, the chorioamnionic membrane, and pulmonary and brain vascular endothelium. Efforts to reconstruct pathophysiological events that lead to newborn disease have focused on intracellular invasion as a means to transcend these barriers. Serotype III strains, which account for more than a third of early onset cases and the majority of late onset cases were the subject of most studies. *S. agalactiae* are internalized by a variety of cells lines (Valentin-Weigand *et al.*, 1996, 1997; Greco *et al.*, 1995;

Valentin-Weigand and Chhatwal, 1995; Tamura and Rubens, 1994) where they survive for at least 48 hours without signs of degradation. Fifty percent of the inoculum is rapidly phagocytized by cultured J774 macrophages and internalized streptococci remained completely viable for 8 hours (Valentin-Weigand *et al.*, 1996). Bacteria that were opsonized with human serum directed against *S. agalactiae* were also efficiently phagocytized by these macrophages, but their life span was significantly shortened by the antibody. Thus, in a non-immune host macrophage could be a vehicle for systemic dissemination of the organism.

Infection of the lungs, followed by pneumonia and dissemination to the blood requires that bacteria invade from the apical side of the respiratory epithelium. In a primate model of early onset disease, *S. agalactiae* were observed to invade pulmonary epithelium and lung capillary endothelium (Rubens *et al.*, 1991). Using polarized Madin-Darby canine kidney cells (MDCK), a thoroughly characterized model of polarized epithelium, Källman and Kihlstrom (1997) showed that *S. agalactiae* can readily traverse monolayers of these cells from the apical to basolateral side. Transcytosis was time and temperature dependent, and peaked at 16 hours after infection. In that study, the capsule size of different strains did not correlate with invasion frequency, nor did it appreciably influence transcytosis. The integrity of the polarized monolayer was maintained for up to 24 hours and continued to prevent non-invasive *E. coli* and *Streptococcus gordonii* from reaching the basolateral side of the monolayer. Attempts to infect MDCK cells from the basal side of the monolayer were unsuccessful, suggesting that receptors for appropriate streptococcal invasins were only on the apical side of the monolayer. The mechanism by which these bacteria traverse the cytoplasm has not been studied to our knowledge.

The blood-brain barrier is a single layer of microvascular endothelial cells (BMEC) that are poorly phagocytic and held together by tight junctions. Transcytosis of *S. agalactiae* by immortalized cultures of these cells has also been demonstrated (Nizet *et al.*, 1997). Polarized cells, infected from the apical side with serotype III streptococci retained *S. gordonii*, but transported *S. agalactiae* to the basal side. A non-encapsulated mutant invaded BMEC twice as frequently as the parent culture, suggesting that capsule may sterically interfere with invasin-receptor interactions. At high multiplicity of infection the bacteria visibly damaged BMEC. Disruption of BMEC depended in part on the hemolysin produced by *S. agalactiae*. Thus, it is also possible that breaks in the blood-brain barrier also provide a doorway for streptococci to reach the central nervous system.

The route by which *S. agalactiae* reach the amnionic fluid of the placenta has long intrigued investigators of this infectious agent. Colonization of the vaginal mucosa with *S. agalactiae* is associated with chorioamnionitis, and is thought to precede premature rupture of placental membranes and neonatal sepsis. Streptococci have been postulated to traverse the chorioamnionic membrane, which is comprised of a chorion epithelial monolayer on the maternal side, and the amnionic monolayer on the fetus side. Winram *et al.* (1998) examined the capacity of *S. agalactiae* to invade and traverse primary cultures of each cell type. Bacteria adhere to both cell types, and invaded chorion cells *in vitro*. Intracellular streptococci also underwent limited replication during the first 9 hours of infection. Streptococci were transcytosed from the apical to basolateral side of chorion cell monolayers without disrupting tight junctions between cells. In contrast amnion cell monolayers were not penetrated by *S. agalactiae*, thus this single cell layer can remain a barrier, and is presumed to prevent streptococci from reaching the luxurious growth environment of the amnionic fluid. The authors suggested that transcytosis of *S. agalactiae* through the chorionic membrane into the proteinaceous stromal layer between it and the amnionic membrane may be the event that initiates chorioamnionitis. The resulting inflammatory response could damage the amnionic membrane and permit bacteria to reach the neonate side of the placenta.

It is clear from our research of the literature that *S. agalactiae* are well-endowed with the potential to cross the many epithelial and endothelial barriers to sterile tissue and organs. Although intracellular *S. agalactiae* have not been observed in human tissue, experiments in the *Macaca nemestrina* primate model lend considerable credibility to the above *in vitro* studies (Rubens *et al.*, 1991). A major void exists in the identification of the bacterial adhesins and invasins, required for intracellular infection. Recent advances in the application of genetic and molecular techniques to this species and completion of its genome sequence should speed the identification of these factors.

5. CONCLUSION

Our level of understanding of events that lead to the internalization of β hemolytic streptococci is elementary relative to other more studied intracellular pathogens. The fact that *S. pyogenes* have evolved at least three pathways to the interior of epithelial cells from a variety of tissues, however, suggests that intracellular infection is pivotal to the pathogenesis of this species. Interesting and important questions have emerged from these humble beginnings in just a few years (LaPenta *et al.*, 1994; Rubens *et al.*, 1992). What is the fate of intracellular streptococci? Do they multiply, persist in a dormant state, or merely use the epithelium as pipeline to other tissues? Are different cellular routes to the intracellular state brought online as the organism migrates from superficial tissue to other organs of the body? The molecular mechanisms by which Fn or Lm trigger endocytosis, including down stream signaling, is an area of research that may produce important new ideas. Are other bacterial factors required for high frequency invasion? Does interaction with the bacteria upregulate expression, or increase affinity and/or clustering of integrin receptors on target cells? Underlying high efficiency invasion, mediated by Lm and Fn, is serum independent invasion mediated by M6 protein (Jadoun *et al.*, 1998) or RGD-containing peptides (Cue *et al.*, 1998). The latter suggests that streptococci may have integrin-like molecules on their surface. It is also possible that these peptides upregulate expression of another unidentified invasin.

Even less is known about the cellular interactions that result in the internalization and transcytosis of *S. agalactiae*. Bacterial invasins have not been identified, but the capacity of this organism to traverse relevant polarized epithelial and endothelial monolayers suggests that intracellular infection plays a critical role in disease. Recent experiments have again raised interesting questions. How does the organism invade from the apical side of cultured cells? The invasins and receptors on these cells may be very different from those used by *S. pyogenes*. The transcytosis process is a complete mystery. Does the organism escape the vacuole and become mobilized by actin polymerization, or are vacuoles that contain *S. agalactiae* marked for transport to the outside or to the plasma membrane for fusion to an adjacent cell? Answers to the above questions will surely contribute to our understanding of the natural history of streptococcal infections and may reveal new targets for more effective treatments.

6. REFERENCES

Akesson, P., Schmidt, K.-H., Cooney, J., and Björck, L., 1994, M1 protein and Protein H: IgGFc- and albumin-binding streptococcal surface proteins encoded by adjacent genes, *J. Biochem.* **300**:877–886.

Burns, E.H., Jr., Lukomski, S., Rurangirwa, J., Podbielski, A., and Musser, J.M., 1998, Genetic inactivation of the extracellular cysteine protease enhances *in vitro* internalization of group A streptococci by human epithelial and endothelial cells, *Microb. Pathog.* **24**:333–339.

Caparon, M.G., Stephens, D.S., Olsen, A., and Scott, J.R., 1991, Role of M protein in adherence of group A streptococci, *Infect. Immun.* **59**:1811–1817.

Cedervall, T., Akesson, P., Stenberg, L., Herrmann, A., and Akerstrom, B., 1995, Allosteric and temperature effects on the plasma protein binding by streptococcal M protein family members, *Scand. J. Immunol.* **42**:433–441.

Cleary, P.P., LaPenta, D., Vessela, R., Lam, H., and Cue, D., 1998a, A globally disseminated M1 subclone of group A streptococcus differs from other subclones by 70 kb of prophage DNA and capacity for high frequency intracellular invasion, *Infect. Immun.* **66**:5592–5597.

Cleary, P.P., McLandsborough, L., Ikedo, L., Cue, D., Krawezak, J., and Lam, H., 1998b, High frequency intracellular infection and erythrogenic toxin A expressions undergo phase variation in M1 group A streptococci, *Mol. Microb.* **28**:157–167.

Cleary, P., Kaplan, E.L., Handley, J.P., Wlazlo, A., Kim, M.H., Hauser, A.R., and Schlievert, P.M., 1992, Clonal basis for resurgence of serious streptococcal disease in the 1980's, *Lancet* **321**:518–521.

Cockerill, F.R., MacDonald, K.L., Thompson, R.L., Robertson, F., Kohner, P.C., Besser-Wiek, J., Manahan, J.M., Musser, J.M., Schlievert, P.M., Talbot, J., Frankfort, B., Steckelberg, J.M., Wilson, W.R., and Osterholm, M.T., 1997, An outbreak of invasive group A streptococcal disease associated with high carrige rates of the invasive clone among school-aged children, *JAMA* **277**:38–43.

Courtney, H.S., Dale, J.B., and Hasty, D.L., 1997, Host cell specific adhesins of group A streptococci, *Adv. Exp. Med. Biol.* **418**:605–606.

Cue, D., Dombek, P.E., Lam, H., and Cleary, P.P., 1998, Serotype M1 *Streptococcus pyogenes* encodes multiple pathways for entry into human epithelial cells, *Infect. Immun.* **66**: 4593–4601.

Dale, J.B., Washburn, R.G., Marques, M.B., and Wessels, M.R., 1996, Hyaluronate capsule and surface M protein in resistance to opsonization of group A streptococci, *Infect. Immun.* **64**:1495–1501.

Dombek, P.E., Cue, D., Sedgewick, J., Lam, H., Ruschkowski, S., Finlay, B.B., and Cleary, P.P., 1998, High frequency intracellular invasion of epithelial cells by serotype M1 group A streptococci: M1 protein mediated invasion and cytoskeletal rearrangements, *Mol. Microbiol.* **31**: in press.

Drevets, D.A., and Campbell, P.A., 1991, Macrophage phagocytosis: use of fluorescence microscopy to distinguish between extracellular and intracellular bacteria, *J. Immunol. Methods* **142**:31–38.

Duensing, T.D., and van Putten, J.P., 1997, Vitronectin mediates internalization of *Neisseria gonorrhoeae* by Chinese hamster ovary cells, *Infect. Immun.* **65**:964–970.

Ewanowich, C.A., Sherburne, R.K., Man, S.F., and Peppler, M.S., 1989, *Bordetella parapertussis* invasion of HeLa 229 cells and human respiratory epithelial cells in primary culture, *Infect. Immun.* **57**:1240–1247.

Falcioni, R., Cimino, L., Gentileschi, M.P., D'Agnano, I., Zupi, G., Kennel, S.J., and Sacchi, A., 1994, Expression of β1, β3, β4, and β5 integrins by human lung carcinoma cells of different histotypes, *Exp. Cell Res.* **210**:113–122.

Finlay, B.B., and Falkow, S., 1997, Common themes in microbial pathogenicity revisited, *Microbiol. Molec. Biol. Rev.* **61**:136–169.

Fischetti, V.A., 1989, Streptococcal M protein: molecular design and biological behavior, *Clin. Microbiol. Rev.* **2**:285–314.

Fluckiger, U., Jones, K.F., and Fischetti, V.A., 1998, Immunoglobulins to group A streptococcal surface molecules decrease adherence to and invasion of human pharyngeal cells, *Infect. Immun.* **66**:974–979.

Francis, C.L., Starbach, M.M., and Falkow, S., 1992, Morphological and cytoskeletal changes in epithelial cells occur immediately upon interaction of *Salmonella typhimurium* grown under low-oxygen conditions, *Mol. Microbiol.* **6**:3077–3087.

Frick, I.-M., Crossin, K.L., Edelman, G.M., and Björck, L., 1995, Protein H- a bacterial surface protein with affinity for both immunoglobulin and fibronectin type III domains, *J. EMBO* **14**:1674–1679.

Garcia-del Portillo, F., and Finlay, B.B., 1995, Targeting of *Salmonella typhimurium* to vesicles containing lysosomal membrane glycoproteins bypasses compartments with mannose 6-phosphate receptors, *J. Cell Biol.* **129**:81–97.

Gomez-Duarte, O.G., Dehio, M., Guzman, C.A., Chhatwal, G.S., Dehio, C., and Meyer, T.F., 1997, Binding of vitronectin to Opa-expressing *Neisseria gonorrhoeae* mediates invasion of HeLa cells, *Infect. Immun.* **65**:3857–3866.

Greco, R., De Martino, L., Donnarumma, G., Conte, M.P., Seganti, L., and Valenti, P., 1995, Invasion of cultured human cells by Streptococcus pyogenes, *Res. Microbiol.* **146**: 5551–5560.

Greco, R., von Hunolstein, C., Orefici, G., Donnarumma, G., Nicoletti, M., and Valenti, P., 1998, Protein M and fibronectin-binding proteins are not sufficient to promote internalization of group A streptococci into HeLa cells, *International J. Immunopathology and Pharmacology* In press.

Hanski, E., and Caparon, M., 1992, Protein F, a fibronectin-binding protein, is an adhesin of the group A streptococcus *Streptococcus pyogenes*, *Proc. Natl. Acad. Sci. USA* **89**:6172–6176.

Horwitz, M.A., 1994, Phagocytosis by the Legionnaires' disease bacterium (*Legionella pneumophilia*) occurs by a novel mechanism: engulfment within a pseudopod coil, *Cell* **36**:27–33.

Hulse, M.L., Smith, S., Chi, E.Y., Pham, A., and Rubens, C.E., 1993, Effect of type III group B streptococcae capsular polysaccharide on invasion of respiratory epithelial cells, *Infect. Immun.* **61**:4835–4841.

Hynes, R.O., 1992, Integrins: versatility, modulation, and signaling in cell adhesion, *Cell* **69**:11–25.

Isberg, R.R., and Leong, J.M., 1990, Multiple β1 chain integrins are receptors for invasin, a protein that promotes bacterial penetration into mammalian cells, *Cell* **60**:861–871.

Jadoun, J., Ozeri, V., Burstein, E., Skutelsky, E., Hanski, E., and Sela, S., 1998, Protein F1 is required for efficient entry of *Streptococcus pyogenes* into epithelial cells, *J. Infect. Dis.* **178**:147–158.

Jaffe, J., Natanson-Yaron, S., Caparon, M.G., and Hanski, E., 1996, Protein F2, a novel fibronectin-binding protein from *Streptococcus pyogenes*, possesses two binding domains, *Mol. Microbiol.* **21**:373–384.

Ji, Y., Carlson, B., Kondagunta, A., and Cleary, P., 1997, Intranasal immunization with C5a peptidase prevents nasopharyngeal colonization by groupA Streptococcus, *Infect. Immun.* **65**:2080–2087.

Källman, J., and Kihlstrom, E., 1997, Penetration of group B streptococci through polarized Madin-Darby canine kidney cells, *Pediatric Research* **42**:799–804.

Katerov, V., Andreev, A., Schalen, C., and Totolian, A.A., 1998, Protein F, a fibronectin-binding protein of *Streptococcus pyogenes*, also binds human fibrinogen: isolation of the protein and mapping of the binding region, *Microbiology* **14**:119–126.

Kuroda, K., Brown, E.J., Telle, W.B., Russell, D.G., and Ratliff, T.L., 1993, Characterization of the internalization of bacillus Calmette-Guerin by human bladder tumor cells, *J. Clin. Invest.* **91**:69–76.

LaPenta, D., Rubens, C., Chi, E., and Cleary, P.P., 1994, Group A steptococci efficiently invade human respiratory epithelial cells, *Proc. Natl. Acad. Sci. USA* **91**:12115–12119.

Mengaud, J., Ohayon, H., Gounon, P., Mege, R.-M., and Cossart, P., 1996, E-cadherin is the receptor for internalin, a surface protein required for entry of *L. monocytogenes* into epithelial cells, *Cell* **84**:923–932.

Molinara, G., and Chhatwal, G.S., 1998, Invasion and survival of *Streptococcus pyogenes* in eukaryotic cells correlates with the source of the clinical isolates, *J. Infect. Dis.* **177**:1600–1607.

Molinari, G., Talay, S.R., Valentin-Weigand, P., Rohde, M., and Chhatwal, G.S., 1997, The fibronectin-binding protein of *Streptococcus pyogenes*, SfbI, is involved in the internalization of group A streptococci by epithelial cells, *Infect. Immun.* **65**:1357–1363.

Musser, J.M., Kapur, V., Szeto, J., Pan, X., Swanson, D.S., and Martin, D.R., 1995, Genetic diversity and relationships among *Streptococcus pyogenes* strains expressing serotype M1 protein: recent intercontinental spread of a subclone causing episodes of invasive disease, *Infect. Immun.* **63**:994–1003.

Nath, S.K., 1989, Invasion of HeLa cells by β-hemolytic group G streptococci, *Can. J. Microbiol.* **35**:515–517.

Neeman, R., Keller, N., Barzilai, A., Korenman, Z., and Sela, S., 1998, Prevalence of the internalization-associated gene, *prt*F1, among persisting group A streptococcus strains isolated from asymptomatic carriers, *Lancet* **352**:1974–1977.

Nilson, B.H., Frick, I.M., Akesson, P., Forsen, S., Bjorck, L., Akerstrom, B., and Wikstrom, M., 1995, Structure and stability of protein H and the M1 protein from *Streptococcus pyogenes*. Implications for other surface proteins of gram-positive bacteria, *Biochemistry* **34**:13688–13698.

Nizet, V., Kim, K.S., Stins, M., Jonas, M., Chi, E.Y., Nguyen, D., and Rubens, C.E., 1997, Invasion of brain microvascular endothelial cells by group B streptococci, *Infect. Immun.* **65**:5074–5081.

Oelschlaeger, T.A., Barrett, T.J., and Kopecko, K.J., 1994, Some structures and processes of human epithelial cells involved in uptake of enterohemorrhagic *Escherichia coli* O157:H7 strains, *Infect. Immun.* **62**:5142–5150.

Okada, N., Pentland, A.P., Falk, P., and Caparon, M.G., 1994, M protein and protein F act as important determinants of cell-specific tropism of *Streptococcus pyogenes* in skin tissue, *J. Clin. Invest.* **94**:965–977.

Okada, N., Liszewski, M.K., Atkinson, J.P., and Caparon, M., 1995, Membrane cofactor protein (CD46) is a keratinocyte receptor for the M protein of the group A streptococcus, *Proc. Natl. Acad. Sci. USA* **92**:2489–2493.

Okada, N., Watarai, M., Ozeri, V., Hanski, E., Caparon, M., and Sasakawa, C., 1997, A matrix form of fibronectin mediates enhanced binding of *Streptococcus pyogenes* to host tissue, *J. Biol. Chem.* **272**:26978–26984.

Österlund, A., Popa, R., Nikkila, T., Scheynius, A., and Engstrand, L., 1997, Intracellular reservoir of *Streptococcus pyogenes in vivo*: a possible explanation for recurrent pharyngotonsillitis, *Laryngoscope* **107**:640–647.

Österlund, A., and Engstrand, L., 1995, Intracellular penetration and survival of *Streptococcus pygenes* in respiratory epithelial cells *in vitro*, *Acta Otolaryngol* **115**:685–688.

Ozeri, V., Rosenshine, I., Mosher, D.F., Fassler, R., and Hanski, E., 1998, Roles of integrins and fibronectin in the entry of *Streptococcus pyogenes* into cells *via* protein F1, *Mol. Microbiol.* **30**:625–637.

Ozeri, V., Tovi, A., Burstein, I., Natanson-Yaron, S., Caparon, M.G., Yamada, K.M., Akiyama, S.K., Vlodavsky, I., and Hanski, E., 1996, A two-domain mechanism for group A streptococcal adherence through protein F to the extracellular matrix, *J. EMBO* **15**:989–998.

Patti, J.M., Allen, B.L., McGavin, M.J., and Hook, M., 1994, MSCRAMM-mediated adherence of microorganisms to host tissues, *Annu. Rev. Microbiol.* **48**:585–617.

Pincus, S.H., Cole, R.L., Wessels, M.R., Corwin, M.D., Kamanga-Sollo, E., Hayes, S.F., and Cieplak, W., Jr., Swanson, J., 1992, Group B streptococcal opacity variants, *J. Bacteriol.* **174**:3739–3749.

Rambukkana, A., Salzer, J.L., Yurchenco, P.D., and Tuomanen, E.I., 1997, Neural targeting of *Mycobacterium leprae* mediated by the G domain of the laminin-α2 chain, *Cell* **88**:811–821.

Rubens, C.E., Smith, S., Hulse, M., and Chi, E.Y., 1992, Respiratory epithelial cells invasion by group B streptococci, *Infect. Immun.* **60**:5157–5163.

Rubens, C.E., Raff, H.V., Jackson, J.C., and Chi, E.Y., 1991, Pathophysiology and histopathology of group B streptococcal sepsis in *Macaca hemestrina* primates induced after intra-amniotic inoculation: evidence for bacterial cellular invasion, *J. Infect. Dis.* **164**:320–333.

Schlievert, P.M., Assimacopoulos, A.P., and Cleary, P.P., 1996, Severe invasive group A streptococcal disease: clinical description and mechanisms of pathogenesis. *J. Lab. Clin. Med.* **127**:13–22.

Schmidt, K.-H., Mann, K., Cooney, J., and Kohler, W., 1993, Multiple binding of type 3 streptococcal M protein to human fibrinogen, albumin and fibronectin, *FEMS Immunol. Med. Microbiol.* **7**:135–144.

Schorey, J.S., Li, Q., McCourt, D.W., Bong-Mastek, M., Clark-Curtiss, J.E., Ratliff, T.L., and Brown, E.J., 1995, A *Mycobacterium leprae* gene encoding a fibronectin binding protein is used for efficient invasion of epithelial cells and Schwann cells, *Infect. Immun.* **63**:2652–2657.

Schrager, H.M., Rheinwald, J.G., and Wessels, M.R., 1996, Hyaluronic acid capsule and the role of streptococcal entry into keratinocytes invasive skin infection, *J. Clin. Invest.* **98**:1954–1958.

Schwartz, M.A., Schaller, M.D., and Ginsberg, M., 1995, Integrins: emerging paradigms of signal transduction, *Annu. Rev. Cell Dev. Biol.* **11**:549–559.

Schwarzbauer, J.E., 1991, Fibronectin: from gene to protein, *Curr. Opin. Cell Biol.* **3**:786–791.

Sellin, M., Hakausson, S., and Norgren, M., 1995, Phase-shift of polysaccharide capsule expression in group B streptococci, type III, *Microb. Pathog.* **18**:401–415.

Stevens, D.L., Tanner, M.H., Winship, J., Swarts, R., Ries, K., Schlievert, P., and Kapland, E., 1989, Severe group A streptococcal infections associated with a toxic shock-like syndrome and scarlet fever toxin A, *New Engl. J. Med.* **321**:1–7.

St. Geme, J.W., III, and Falkow, S., 1990, *Haemophilus influenzae* adheres to and enters cultured human epithelial cells, *Infect. Immun.* **58**:4036–4044.

Switalski, L.M., Speziale, P., Hook, M., Wadstrom, T., and Timpl, R., 1984, Binding of *Streptococcus pyogenes* to laminin, *J. Biol. Chem.* **259**:3734–3738.

Talay, S.R., Valentin-Weigand, P., Timmis, K.N., and Chhatwal, G.S., 1994, Domain structure and conserved epitopes of Sfb protein, the fibronectin-binding adhesin of *Streptococcus pyogenes*, *Mol. Microbiol.* **13**:531–539.

Tamura, G.S., and Rubens, C.E., 1994, Pathogenesis of group B streptococcal infections, *Curr. Opin. Inf. Dis.* **7**:317–322.

Timpl, R., and Brown, J.C., 1994, The laminins, *Matrix Biol.* **14**:275–281.

Tsai, P.J., Kuo, C.F., Lin, K.Y., Lin, Y.S., Lei, H.Y., Chen, F.F., Wang, J.R., and Wu, J.J., 1998, Effect of group A streptococcal cysteine protease on invasion of epithelial cells, *Infect. Immun.* **66**:1460–1466.

Valentin-Weigand, P., Benkel, P., Rohde, M., and Chhatwal, G.S., 1996, Entry and intracellular survival of group B streptococci in J774 macrophages, *Infect. Immun.* **64**:2467–2473.

Valentin-Weigand, P., and Chhatwal, G.S., 1995, Correlation of epithelial cell invasiveness of group B streptococci with clinical source of isolation, *Microb. Pathog.* **19**:83–91.

Valentin-Weigand, P., Jungnitz, H., Zock, A., Rohde, M., and Chhatwal, G.S., 1997, Characterization of group B streptococcal invasion in HEp-2 epithelial cells, *FEMS Microb. Letts.* **147**:69–74.

van Nhieu, G.T., and Isberg, R.R., 1991, The *Yersinia pseudotuberculosis* invasin protein and human fibronectin bind to mutually exclusive sites on the $\alpha5\beta1$ integrin receptor, *J. Biol. Chem.* **266**:24367–24375.

van Putten, J.P., Duensing, T.D., and Cole, R.L., 1998, Entry of OpaA+ gonococci into HEp-2 cells requires concerted action of glycosaminoglycans, fibronectin and integrin receptors, *Mol. Microbiol.* **29**:369–379.

Winram, S.B., Jonas, M., Chi, E., and Rubens, C.E., 1998, Characterization of group B streptococcal invasion into human chorion and amnion epithelial cells *in vitro*, *Infect. Immun.* **66**:4932–4941.

Chapter 8

Nocardia asteroides as an Invasive, Intracellular Pathogen of the Brain and Lungs

Blaine L. Beaman* and LoVelle Beaman

1. INTRODUCTION

1.1. *Nocardia* spp

This genus is defined morphologically as gram-positive, partially acid-fast, branching filamentous bacteria that divide by fragmentation. They are strictly aerobic, mycolic acid containing actinomycetes phylogenetically related to Mycobacterium, Corynebacterium, Rhodococcus, Gordona, and Tsukamurella (Wilson *et al.*, 1998). Twelve species comprise this genus; however, the type species, *Nocardia asteroides*, has been most frequently studied (Beaman, B. and L. Beaman, 1994). This species is phenotypically diverse, and many investigators believe that the *N. asteroides* complex consists of at least 5 to 6 distinct groups. Some of these have now been assigned

BLAINE L. BEAMAN and LOVELLE BEAMAN Department of Medical Microbiology and Immunology, University of California School of Medicine, Davis, California 95616. Corresponding Author: Blaine L. Beaman Department of Medical Microbiology and Immunology, University of California School of Medicine, Davis, California USA 95616, E-mail: blbeaman@ucdavis.edu

Subcellular Biochemistry, Volume 33: Bacterial Invasion into Eukaryotic Cells, edited by Oelschlaeger and Hacker. Kluwer Academic / Plenum Publishers, New York, 2000.

species status. Thus, the *N. asteroides* complex currently includes *N. asteroides*, *N. farcinica*, and *N. nova*. It is likely that additional species will be created from further subdivision of this taxon. Since routine laboratory identification does not differentiate these species, most reports of infections due to *N. asteroides* represent the total complex and underestimate infections due to individual members such as *N. farcinica* and *N. nova* (Workman *et al.*, 1998).

1.2. Host Specificity for Nocardial Infection

The nocardiae include organisms that have been cultured from such diverse hosts as plants (Lechevalier, 1989), dinoflagellates (Tosteson *et al.*, 1989), insects (Umunnabuike and Irokanulo, 1986), oysters (Friedman *et al.*, 1998), fish (Chen, 1992), birds (Sileo *et al.*, 1990), and numerous mammals (Marino and Jaggy, 1993; Beaman and Sugar, 1983) including humans (Beaman, B. and L. Beaman, 1994).

The diseases caused by nocardiae in fish, birds and mammals appear to be similar (Beaman, B. and L. Beaman, 1994; Beaman and Sugar, 1983). In these animals, the nocardiae are facultative intracellular pathogens invading a variety of host cells within specific regions of the body. Thus, the lungs (gills in fish) represent a dominant site for infection with frequent dissemination to the brain and skin. Primary cutaneous and subcutaneous infections including mycetomatous lesions have been reported in many of these animal species. There appears to be little or no unique host specificity for infection by members of the *N. asteroides* complex (Beaman, B. and L. Beaman, 1994).

1.3. Animal versus Human Infections

It is believed that nocardial infections are derived from exogenous, environmental sources such as soil, dust, water and vegetation. This view is supported by reports that most of the pathogenic strains of *Nocardia* have been isolated from these sources on a global scale (Khan *et al.*, 1997; Provost *et al.*, 1997). Furthermore, nocardiae have never been shown to be present normally as part of the resident microflora in any animal species except perhaps *N. asteroides* in the gut of the cockroach (Umunnabuike and Irokanulo, 1986).

In animals including humans, diseases caused by nocardiae are diverse; however, most cases can be divided into the following 6 distinct categories: 1. Pulmonary nocardiosis with infection limited to the lungs; 2. Systemic or disseminated nocardiosis with lesions at 2 or more distinct anatomical locations; 3. Primary central nervous system nocardiosis; 4. Extrapulmonary nocardiosis with infection recognized at only 1 anatomical site (i.e. ocular nocardiosis); 5 Cutaneous and lymphocutaneous nocardiosis; and 6. Actinomycetoma (Beaman, B. and L. Beaman, 1994). Thus, pulmonary nocardiosis in a horse is not significantly different from pulmonary infection in either a whale or a human; and, mycetomas caused by *Nocardia* spp. in cats and dogs appear essentially the same as those in humans and "so on" (Beaman, B. and L. Beaman, 1994; Beaman and Sugar, 1983).

1.4. Nocardiae as Facultative Intracellular Pathogens

Early studies of nocardial interactions with phagocytic cells from both humans and other animals indicate that pathogenic strains of nocardiae are not killed readily by either macrophages or polymorphonuclear neutrophils (PMNs). Indeed, virulent strains of *N. asteroides* grow rapidly within these phagocytes indicating that *N. asteroides* is a facultative intracellular pathogen (Beaman, B. and L. Beaman, 1994).

1.5. Defining Pathogenicity and Virulence

Since, many species of Nocardia are capable of causing disease in a wide range of hosts, these bacteria must be defined as potential pathogens. Nevertheless, there is great variability in how readily these organisms cause disease depending upon the host as well as the specific strain of nocardia. As compared to many "traditional" primary pathogens of mammals (i.e. *Yersinia pestis*), the pathogenic strains of nocardiae usually are not as virulent, and often they take advantage of a host with compromised defenses. Thus, nocardiae are both primary and opportunistic pathogens in mammals, and virulence is defined by an arbitrary value in an experimental model system (either *in vivo* or *in vitro*). Usage of a description such as "highly virulent strain of *N. asteroides*" must be defined within a specific model because it does not necessarily have the same meaning or ramifications as the phrase "a highly virulent strain of *Yersinia pestis*" within the same model.

Nocardial virulence frequently is defined as an LD_{50} dose within a murine model. However, several studies demonstrate that the same inoculum of nocardiae injected into different strains of mice results in different LD_{50} values. Furthermore, the route of inoculation, the culture conditions, and phase of growth affect profoundly the apparent virulence of nocardiae in these assays. Even when all of these variables are controlled, the virulence of different strains of the same species of nocardia for "normal" BALB/c female mice (for example) may vary by a factor of more that 10,000 fold (Beaman, B. and L. Beaman, 1994; Beaman and Maslan, 1978).

2. *In Vivo* ANIMAL MODELS

2.1. Dogs, Rabbits, Rats, and Guinea Pigs

Nocard first recognized nocardioform organisms causing bovine farcy in 1888 (Nocard, 1888). Pulmonary infection with dissemination to the brain caused by *Nocardia asteroides* was diagnosed in a human in 1889 (Eppinger, 1890). In 1891, Eppinger was the first investigator to use rabbits and guinea pigs as experimental models to study nocardiosis. By using these animals, he proved that *N. asteroides* was the etiology of the fatal disease "pseudotuberculosis" in humans (Eppinger, 1891). Rabe described aerobic actinomycetes associated with abscesses in dogs in 1888 (MacCallum, 1902). In 1902, MacCallum was the first investigator to demonstrate that healthy dogs were very susceptible to systemic infection by intravenous injection of *N. asteroides* (MacCallum, 1902). Since these early studies from 1902 to the mid-1970's, there were numerous attempts to establish reliable experimental animal models for investigating pulmonary, systemic and actinomycotic nocardiosis using rabbits, rats, guinea pigs and mice (Beaman, B. and L. Beaman, 1994). The results from these studies were extremely variable, and often they lead investigators to reach erroneous and contradictory conclusions. This frustrated efforts to use animals to investigate the mechanisms of nocardial pathogenesis. A major contributor to this problem was the general, but incorrect, belief that these organisms were fungi. As a consequence, the nocardiae were frequently scraped from cultures grown on a variety of solid media designed for growing fungi. Most often, these cultures were incubated for periods of days to weeks, and the nocardial masses were macerated to yield a cellular suspension. The methods were crude, and the nocardial suspension was not standardized. To quantitate the amount of inoculum injected into the animal, a wet weight determination of this crude pellet was done. Somtimes these methods resulted in a progressive, fatal infection in the animal, and at other times nothing happened. During this period, several investigators demonstrated that suspending certain fungi in either hog gastric mucin or mineral oil significantly enhanced their virulence. Therefore, these methods were adapted to the nocardiae; and indeed, these procedures enhanced the ability of most strains of nocardiae to cause disease in a variety of experimental animals. Unfortunately, this approach was considered to be artificial, and it did not permit analysis of the mechanisms of nocardial interactions with either individual host cells or immune responses to the "natural" infection (Beaman and Beaman, 1994).

In 1975, Beaman showed that the rate of growth of *Nocardia asteroides* in brain heart infusion broth (BHI-b) could be controlled (Beaman, 1975). Furthermore, it was demonstrated that homogeneous suspensions of single cells at specific stages of growth could be prepared by differential centrifugation. The cell walls of *N. asteroides* during logarithmic phase of growth were ultrastructurally and biochemically different from the cell walls from the same organisms at stationary phase of growth. These growth stage differences in the cell envelope correlated with the relative virulence of *N. asteroides* for experimental animals. It was demonstrated that nocardial virulence for mice was altered by the age of the culture; the methods used to start the culture; the medium used to grow the nocardiae; and the incubation conditions (Beaman and Maslan, 1978). These observations probably explain many of the difficulties and contradictions described above that were reported during the early studies of nocardiosis in experimental animals (Beaman, Beaman, 1994).

2.2. Primates

As noted above, a wide variety of animals, including primates, are susceptible to environmentally acquired pulmonary and systemic nocardiosis (Sakakibara *et al.*, 1984; Kessler and Brown, 1981; Mahajan *et al.*, 1977; McClure *et al.*, 1976; Boncyk, *et al.*, 1975; Al-Doory *et al.*, 1969; Jonas and Wyand, 1966). Sakakibara *et al.* (1984) reported a case of naturally acquired, disseminated nocardiosis in a cynomolgus monkey. These investigators recovered *N. asteroides* from both a mouth lesion and the lungs. They suggested that these were the initial foci of infection, and that the organisms invaded through the blood vessels at these sites. Once in the blood, the nocardiae disseminated to other regions where they caused multiple abscesses in the kidneys, heart, liver, and brain (Sakakibara *et al.*, 1984). Mahajan *et al.* (1977) noted that monkeys were very susceptible to experimental infection by *N. asteroides*. Within 2 weeks following inoculation through the canaliculus connecting to the upper airways, monkeys readily developed progressive, fatal pneumonia. These investigators stated that the rhesus monkey was an excellent model for studying pulmonary nocardiosis was the same in humans (Mahajan *et al.*, 1977). Beaman (unpublished data) demonstrated that cynomolgus monkeys were very susceptible to disseminated nocardiosis following intravenous injection of log phase *N. asteroides* GUH-2 (LD_{50} is approximately 1×10^6 CFU/3 kg monkey). Indeed, very much like in the murine model, *N. asteroides* GUH-2 displayed a remarkable degree of specificity for certain regions of the brain in the monkey (Beaman, unpublished data).

2.3. Murine Models

Many aspects of human disease can be reproduced in mice, and large numbers of animals at a reasonable cost can be studied to provide statistical evaluation of host responses. Numerous varieties of different strains of mice have been genetically engineered so that single, defined factors affecting host-pathogen interactions may be dissected. This level of genetic definition and manipulation is not available in any other animal species. Therefore, the laboratory mouse is the best experimental tool for investigating the mechanisms of nocardial pathogenesis *in vivo* (Beaman and Beaman, 1994).

The murine response to *N. asteroides* appears to be compartmentalized, and cellular responses differ from one anatomical region to another. Because of this, the route of exposure of the mouse to the nocardiae significantly affects the virulence of the organisms based on lethality (Beaman, *et al.*, 1980). For example, the LD_{50} of log phase *N. asteroides* GUH-2 injected i.v. into female Swiss Webster mice is approximately 2×10^4 CFU whereas the LD_{50} of the same culture administered intranasally is approximately 2×10^6 CFU/left lung. In contrast, the LD_{50} of log phase cells of *N. asteroides* GUH-2 injected IP into Swiss Webster mice is more than 5×10^7 CFU/mouse. Thus, there is more than a 2000 fold difference in the lethality of log phase GUH-2 for these mice depending entirely upon the route of exposure (Beaman *et al.*, 1980). The strain of the mouse (immunologically intact) alters susceptibility to nocardiae. As noted above, the approximate LD_{50} of log phase *N. asteroides* GUH-2 injected i.v. into 18–20 g female Swiss Webster mice is approximately 2×10^4 CFU/mouse. In contrast, the LD_{50} of log phase *N. asteroides* GUH-2 injected IV into 18–20 g female BALB/c mice is approximately 4×10^5 CFU/mouse; whereas, the LD_{50} of log phase GUH-2 injected i.v. into 18–20 g female C57BL/6J mice is about 2×10^6 CFU/mouse. Thus, there is at least a 100 fold difference in the lethality of log phase GUH-2 following i.v. injection depending entirely upon the murine strain (Beaman and Beaman, 1994; Beaman, unpublished data).

The stage of growth significantly affects the relative virulence of *N. asteroides* for mice. For example, log phase cells of *N. asteroides* GUH-2 are 1000 times more virulent than stationary phase cells from the same culture when injected i.v. into Swiss Webster mice (Beaman and Maslan, 1978). This association of stage of growth with virulence was repeated with more than 60 strains of *Nocardia*, representing all pathogenic species. It is important to emphasize that the methods used to grow the nocardiae affected their growth rate. Therefore, these parameters were defined and controlled. Nevertheless, the culture age altered nocardial virulence regardless of how these organisms were grown; but, the methods used to start the cultures affected the relative degree of these differences (Beaman and Beaman, 1994).

3. *In Vitro* TISSUE CULTURE MODELS

3.1. Macrophages and Monocytes

In pulmonary nocardiosis, alveolar macrophages should play a key role in the host resistance to infection; however, a mixed response was observed in cultured alveolar macrophages *in vitro*. Splino reported that guinea pig alveolar macrophages phagocytized *N. asteroides* strain Weipheld but within 8h of culture nocardial filaments were observed growing through the phagocytic cell membrane; and, this strain appeared to be cytotoxic for phagocytic cells (Splino *et al.*, 1975). Beaman found *N. asteroides* 10,905, a strain of low virulence for mice, was rapidly phagocytized and largely killed by normal rabbit alveolar macrophages except for a few organisms that survived in an altered gram-negative form. Wall-less, spheroplast-like organisms with many properties of L-forms appeared 9 days after infection with strain 10,905 both in rabbit alveolar macrophages and in murine peritoneal macrophages (Beaman and Smathers, 1976; Bourgeois and Beaman, 1974). The more virulent strain, *N. asteroides* ATCC 14,759, survived in rabbit alveolar macrophages and grew out of these phagocytes as gram-positive, branching filaments. Subsequent observations showed macrophages migrated to *N. asteroides* 14,759 infected macrophages resulting in the formation of multinucleated giant cells which slowed the growth of 14,759, however acid fast filaments grew out of the individually infected nonfused macrophages in these cultures (Beaman, 1977).

Alveolar macrophages obtained from the lungs of normal rabbits more readily phagocytized the stationary phase cells of *N. asteroides* GUH-2 than log phase organisms; but, log phase cells were more toxic to the macrophage monolayer. Nevertheless, both stationary and log phase cells of GUH-2 grew in these cultured alveolar macrophages. On the other hand, macrophages obtained from immunized rabbits inhibited the growth of stationary phase GUH-2 but only slowed the growth of log phase temporarily. If the log phase cells of GUH-2 were presensitized with specific antibody then the intracellular growth was inhibited further by alveolar macrophages from immunized rabbits (Beaman, 1979). Filice *et al.* (1980) showed that activated macrophages from mice infected with *Toxoplasma gondii* or injected with *Corynebacterium parvum* inhibited the growth of stationary phase nocardial cells. One of the more virulent nocardial strains for mice,

N. asteroides GUH-2, was found by Davis-Scibienski to be most resistant to killing by *in vitro* maintained rabbit alveolar macrophages. The less virulent *N. asteroides* 14,759 was intermediate in resistance, and strain 10,905 had little resistance to being killed by rabbit alveolar macrophages (Davis-Scibienski and Beaman, 1980a). Cells of GUH-2 inhibited phagosome-lysosome fusion and did not appear to be damaged by rabbit alveolar macrophages when observed by electron and fluorescent microscopy. In contrast, the less virulent strain 10,905 did not inhibit phagosome-lysosome fusion and exhibited considerable cellular damage in macrophage cultures (Davis-Scibienski and Beaman, 1980a). The ability of *N. asteroides* GUH-2 to inhibit phagosome-lysosome fusion in rabbit alveolar macrophages is growth stage dependent since log phase organisms are more efficient at inhibiting fusion than the stationary phase organisms from the same culture (Davis-Scibienski and Beaman, 1980b). Specifically activated rabbit alveolar macrophages incubated with primed lymph node cells plus immune serum and alveolar lining material were more able to phagocytize GUH-2, inhibit bacterial growth, and increase the amount of bacterial cell damage and nocadial killing than were macrophages lacking any one of these constituents (Davis-Scibienski and Beaman, 1980c).

Nocardia asteroides GUH-2 affects the levels of lysosomal enzymes, such as acid phosphatase, in infected macrophages as measured by means of a computer-assisted cytospectrophotometry system. Black *et al.* showed that the level of acid phosphatase activity in alveolar and peritoneal murine macrophages was decreased by infection with live cells of virulent GUH-2 (Black *et al.*, 1983). In contrast, the decrease was not observed following infection with killed GUH-2 or with the less virulent strain 10,905 (Black *et al.*, 1983). Lysosomal acid phosphatase activity was found to be an effective marker for the ability of macrophages to inhibit the growth of and kill *N. asteroides* (Black *et al.*, 1985). In these studies, kupffer cells (macrophages) from immunized mice were more effective at killing nocardiae than either peritoneal or alveolar macrophages from the same animal (Black *et al.*, 1985). The decrease in acid phosphatase activity in GUH-2 infected macrophages did not reflect an overall reduction in lysosomal enzyme levels due to degranulation or membrane leakage. In these macrophages, the levels of lysozyme and esterase-neutral protease activity remained either unchanged or increased following infection by increasing numbers of nocardial cells (Black *et al.*, 1986). These results appear to be explained by the observation that *N. asteroides* GUH-2 preferentially metabolizes acid phosphatase during growth within phagocytes. This strain of nocardia was shown to use acid phosphatase as a sole carbon source for growth (Beaman, L. *et al.*, 1988).

Black *et al.* showed that when virulent GUH-2 was phagocytized by murine peritoneal macrophages the phagosomal pH remained above pH 7 for more than 2 hours. In contrast, the normal phagosome rapidly became acidified to less than pH 5.0. This ability of virulent *N. asteroides* to prevent phagosomal acidification may be prerequisite to the survival of nocardiae within phagocytes (Black *et al.*, 1986).

Macrophages from immunized mice killed *N. asteroides* more readily than macrophages from normal mice, and these macrophages incubated with lymphocytes from immunized mice were even more activated to kill nocardiae (Black *et al.*, 1985). However, macrophages activated with interferon-gamma to kill Toxoplasma and the fungus *Coccidioides immitis*, was unable to kill *N. asteroides* GUH-2 (Beaman and Beaman, 1992; Beaman, L., 1987; Black *et al.*, 1987). Pretreatment of human monocytes with either interferon-gamma or tumor necrosis factor-alpha prior to infection resulted in enhanced nocardial growth as shown by filament elongation. *N. asteroides* apparently used acid phosphatase as a carbon source which may partially explain the increase in nocardial growth seen in activated macrophages. In addition, the effect of enhanced oxidative killing seen in interferon-gamma activated macrophages was probably neutralized by catalase and cell surface associated superoxide dismutase secreted by virulent GUH-2 (Beaman and Beaman, 1990; Beaman *et al.*, 1985; Filice, 1983).

The observations described above show that the macrophage response to nocardial infection is a multifactorial and complicated interaction between the bactericidal mechanisms of the phagocytes and the ability of nocardia to evade or neutralize these bactericidal responses. The host must mount a lymphocyte response with the production of antibody and/or lymphocyte signals which enable macrophages to kill *N. asteroides*.

3.2. Polymorphonuclear Neutrophils

Filice *et al.* found that human neutrophils and monocytes were able to kill *Listeria monocytogenes* and *Staphylococcus aureus* but were not able to kill *N. asteroides*, eventhough there was an oxidative metabolic burst by these phagocytes (Filice *et al.*, 1980). This resistance of the nocardiae to oxidative killing appears to be due partially, but not completely, to a relatively high intracytoplasmic catalase activity combined with secreted and surface associated SOD (Beaman *et al.*, 1985; Filice, 1983).

Human neutrophils killed 80% of the less virulent *N. asteroides* strain 10,905 and about 50% of the log phase cells of the more virulent strain GUH-2 within 3h incubation (Beaman *et al.*, 1985). In contrast, PMNs were not able to kill any of the early stationary phase cells of GUH-2 that contained 10 times more intracytoplasmic catalase than log phase cultures (Beaman *et al.*, 1985). In addition, *N. asteroides* GUH-2 has a cell surface associated superoxide dismutase which, with catalase, may play a major role in protection against oxidative killing by phagocytes. Thus, neutrophils killed about 50% of early stationary phase GUH-2 after treatment with specific antibody for nocardial superoxide dismutase. None were killed without pre-treatment with anti-SOD antibody. In these experiments, more than 90% of *Listeria monocytogenes* used as a control were killed by these PMNs. Both nocardial catalase and SOD added to cells of *N. asteroides* strain 10,905 completely protected them from being killed by neutrophils (Beaman *et al.*, 1985). Furthermore, catalase added to the log phase of *N. asteroides* GUH-2 (which possesses a cell surface associated SOD) blocked completely PMN nocardicidal activity (Beaman *et al.*, 1985). Apparently there is an essential role for both catalase and SOD in the resistance of *N. asteroides* GUH-2 to the microbicidal activities of human PMNs. Nevertheless, Odell and Segal (1991) believe that oxygen-dependent killing mechanisms are important in controlling nocardiae because they found two pulmonary isolates of *N. asteroides* that were resistant to oxygen-independent killing by human neutrophil granule lysates were still susceptible to intact PMNs (Odell and Segal, 1991).

Filice (1985) demonstrated that human neutrophils inhibited both filament formation of nocardiae and amino acid uptake for the first 7.5 h after infection. The addition of fresh neutrophils prolonged this inhibition (Filice, 1985). These observations suggest that neutrophils play a key role in slowing nocardial growth until cell-mediated responses involving macrophages are activated sufficiently.

3.3. Primary Tissue Culture Cells

Murine brain cultures prepared from newborn mice were enriched for microglia or astrocytes. After 13 to 15 days in culture, a population of esterase positive cells with macrophage-associated antigen were identified as microglia cells. These microglia phagocytized early stationary phase *N. asteroides* GUH-2, but after 6 hr the continued presence of coccoid cells indicated nocardial growth was inhibited. In contrast, in astroglial cells (positive for GFAP) infected with the same culture of GUH-2, nocardial filaments appeared after 6 h incubation indicating growth was not inhibited. Astroglial cultures supported the growth of nocardiae in a manner similar to peritoneal exudate cells; while in contrast, microglia inhibited nocardial growth. Scanning electron microscopy showed that the apex of filaments of GUH-2 were adhering to the astrocyte cell surface with penetration into the cytoplasm suggesting an invasion process. This was not observed with microglia; but instead, the microglial cells actively phagocytized the nocardiae (Beaman and Beaman, 1993).

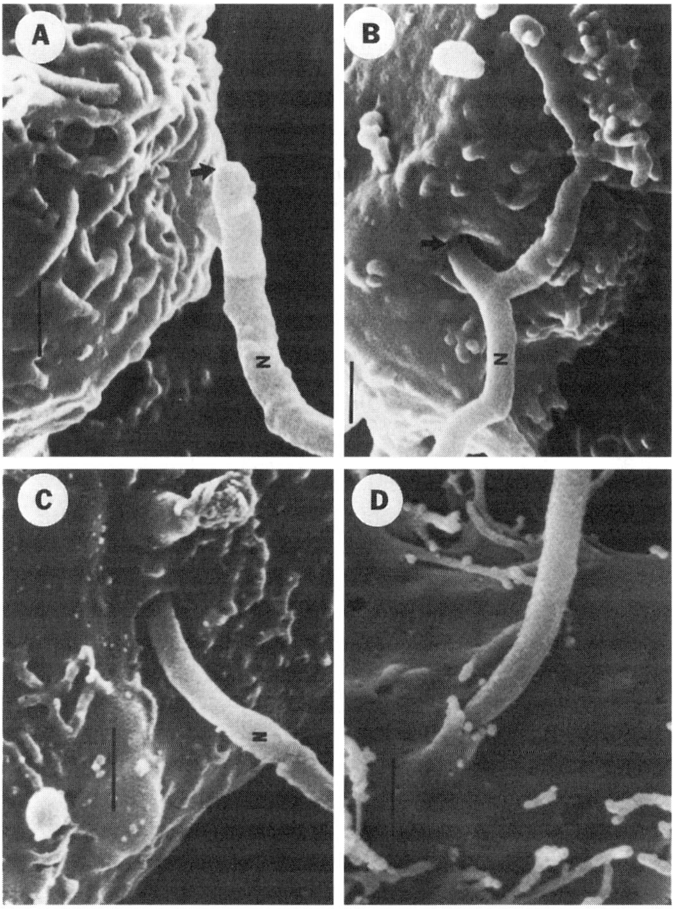

FIGURE 1. Interaction of log phase cells of *Nocardia asteroides* with tissue culture cell lines 1 hour after inoculation. (A) Filament tip adherence to the surface of Cytochalasin treated macrophage cell line J774-A.1. The arrow indicates point of attachment between host cytoplasmic membrane and nocardial surface. N indicates nocardial filament. Bar = 1.0 μm. (B) Nocardial filament penetration (arrow) into Cytochalasin treated mcarophage cell line J774-A1. Bar = 1.0 μm. (C) Nocardial filament penetration (arrow) into Cytochalasin treated human astrocytoma CCF-STTG1 cell line. Bar = 1.0 μm. (Micrographs (A, B, C) reprinted from Beaman and Beaman (1994) with permission of the publisher). (D) Log phase filament of *Nocardia asteroides* strain UC-63 penetrating into Type II pneumocyte cell line. Note the difference in the surface of the host cell in response to nocardial penetration. (Compare (B) and (C) [strain GUH-2] with (D) [strain UC-63]) Bar = 1.0 μm. These data suggest both actin mediated uptake of nocardiae as well as nocardial invasion independent of an actin driven endocytic process.

3.4. Established Tissue Culture Cell Lines

Tissue culture cell lines were infected with nocardia in an *in vitro* model for studying nocardial-host cell interactions. In macrophage cell lines J774A.l and P388D1, log phase GUH-2 infected a higher percentage of the host cells than did stationary phase nocardia (Beaman and Beaman, 1994). When macrophage cell lines infected with stationary phase GUH-2 were incubated for 6 hr, filaments appeared; thus, indicating that nocardial growth had occurred. However, only 31 to 57% of the total stationary phase coccobacillary cells phagocytized by these macrophage cell lines formed filaments. The microfilament inhibitor, cytochalasin B, prevented uptake of nocardia in cell lines J774A.l and P388D1(Beaman and Beaman, 1994).

Pulmonary artery endothelium cell line CPAE, rat glial tumor cell line C6 and human astrocytoma cell lines CCF-STTG1 and U-373 MG were infected with either log phase or stationary phase *N. asteroides* GUH-2 as described above. Scanning electron microscopy showed distinct types of phagocytosis by the different cell lines. The nocardiae grew within these astrocytoma and endothelial cell lines as shown by nocardial filament elongation. The microfilament inhibitor, cytochalasin B, reduced uptake of nocardia by all of the cell lines except by astrocytoma cell line U-373. In cytochalasin B treated astrocytoma cell line U-373, the nocardiae were shown to penetrate through the cell surface and become internalized in a manner distinct from typical phagocytosis (Beaman and Beaman, 1994). Thus, invasion of this cell line by log phase cells of *N. asteroides* GUH-2 appears to be independent of either microfilaments composed of actin or microtubules (Beaman and Beaman, 1994).

An *in vitro* model using HeLa cells demonstrated that the tip of log phase cells of GUH-2 adhered to and penetrated the surface of HeLa cells. Treatment with amikacin demonstrated that the nocardiae were internalized. Immune mouse serum absorbed with stationary phase GUH-2 labeled the tip of log phase GUH-2 and blocked attachment to and invasion of HeLa cells, indicating filament tip associated proteins are involved in attachment to and invasion of HeLa cells by *N. asteroides* GUH-2 (Beaman and Beaman, 1998).

4. ADHERENCE OF NOCARDIAE TO HOST TISSUES

The first event necessary for most pathogens to establish an infection within the host depends on adherence of the organism to an appropriate site. In humans and other mammals, members of the *N. asteroides* complex frequently cause pulmonary infections (Beaman and Beaman, 1994). The bacteria in these lesions may remain either localized within the lungs or they may disseminate. There appears to be organ specificity (tropism) during systemic spread of nocardiae from a localized site of infection (with a preference for the brain). In contrast, cells of *N. brasiliensis* tend to remain localized to cutaneous and lymphocutaneous regions. The spread of these organisms is by direct extension through the tissues, often invading adjacent muscle and bone (Beaman and Beaman, 1994).

4.1. *In Vivo* Specificity for Adherence

Inoculation of nocardiae into mice demonstrate specificity for adherence within different anatomical sites. This selectivity depends upon the route and method of inoculation, the growth stage of the organism, and the strain of *Nocardia* being studied. For example, log phase cells of different strains of *N. asteroides* vary considerably in their ability to bind to endothelial surfaces in the brain following tail vein injection into mice (Beaman, 1996).

The comparative interactions of log phase cells of *N. asteroides* strain GUH-2 (GUH-2) in the brain and lungs have been studied most extensively. The filament tip of log phase cells of GUH-2 injected i.v. into mice bind to capillary endothelial surfaces within the brain. This apical attachment is followed rapidly by penetration of the endothelial cell surface, and the bacteria become internalized (Beaman and Ogata, 1993). The nocardiae then pass through the endothelial cell and the basal lamina. Once the bacteria enter the brain parenchyma, they grow within, among, and through astroglia, neurons, and axonal extensions. Frequently, there is neither detectable damage to the integrity of the blood-brain barrier nor is there an inflammatory response (Beaman, 1993). However, neurodegenerative alterations may be observed at the site of invasion. There is specificity for this binding of GUH-2 within the brain (Ogata and Beaman, 1992a). On a gram-weight basis, GUH-2 specifically targets the substantia nigra region (Ogata and Beaman, 1992b). In monkeys, there also appears to be increased specificity for the lateral geniculate, thalamus, and hypothalamus. The nocardiae adhere poorly to endothelial cells in the white matter of the cerebral cortex, cerebellum, hippocampus, and meninges. These interactions appear to be identical following i.v. inoculation of log phase cells of GUH-2 into either mice or monkeys (Beaman, unpublished data).

Log phase GUH-2 shows specificity for adherence to pulmonary epithelial cells following intranasal (IN) administration into mice. The filament tip binds to the surface of Clara cells in bronchioles, Type II alveolar cells in the alveoli, and alveolar septa. As in the brain, the apically attached nocardiae then penetrate into these epithelial cells and they are internalized completely (Beaman and Beaman, 1998; Beaman, 1996).

The ligand on the filament tip of *N. asteroides* that binds to epithelial cells in the lung is different from the component responsible for adherence to endothelial cells in the brain (Beaman and Beaman, 1998).

Thus, some strains of *N. asteroides* bind to and invade the brain and not the lungs; whereas, other strains invade the lungs and not the brain. A few strains, such as GUH-2, adhere to and invade both the brain and lungs in mice. As should be expected, there are strains of *N. asteroides* that do not adhere to cells in either the brain or lungs (Beaman, 1996).

There are additional components on the surface of nocardiae that are associated with adherence. These substances are not restricted to the tip of the growing filament; but instead, they are present on both log phase and stationary phase cells of *N. asteroides*. These additional adhesins appear to lack specificity for either the brain or lungs. Indeed, the heart, liver, spleen, and kidneys of mice appear to respond quite differently to stationary phase nocardiae as compared to log phase cells of the same culture. These responses are distinctly different in strains that do not adhere in either the brain or lungs of mice. Therefore, the data suggest that nocardial specificity for other regions of the body is mediated by substances distinct from those involved in adherence within either the brain or lungs. There is little or no evidence that nocardiae adhere specifically to cells in other anatomical sites such as the gastrointestinal tract, urogenital tract, oral cavity, and skin (Beaman and Beaman, 1994).

4.2. *In Vitro* Specificity for Adherence

As described above, L. Beaman and B. L. Beaman (1994) studied the interactions of different strains of *N. asteroides* with tissue culture cells representing different lineages. There was specificity for differential adherence of nocardiae to these cells. For example, *N. asteroides* GUH-2 adhered dramatically to the surface of type II primary murine brain astrocytes; while in contrast, few of these bacteria attached to the surface of type I astrocytes (Beaman and Beaman, 1994). These data support the hypothesis that *N. asteroides* possess a variety of growth stage dependent adherence factors that bind differentially to the surface of host cells.

5. POSSIBLE MECHANISMS FOR SPECIFICITY FOR ADHERENCE

It is not clear why soil organisms such as *Nocardia* spp. possess adherence mechanisms for specific cells within regions of the mammalian host. The environmental pressures that maintain the genes for these substances on the nocardial surface are not known. The specific ligands responsible for attachment to and invasion of host cells are probably encoded by genes on the bacterial chromosome since analysis of their association with plasmids appear to be non-concordant (Provost *et al.*, 1996).

5.1. The Nocardial Cell Envelope

The nocardial cell wall is structurally and chemically complex (Beaman and Beaman, 1994). Furthermore, the cell envelope of nocardia shares many features with the phylogenetically related mycobacteria (Riess *et al.*, 1998). Members of both of these genera possess surface associated adherence factors, and the ability to invade a variety of eucaryotic host cells (Beaman and Beaman, 1998; Bermudez *et al.*, 1997). Some species of *Mycobacterium* are parasites of humans and other animals; however, most of them are saprophytic bacteria occupying the same environmental niche as the nocardiae. It is tempting to speculate that the linkages of these organisms may be through their evolution either to or from members who are obligate mammalian pathogens.

The basal peptidoglycan layer of the cell wall is composed of β-*N*-acetylglucosaminyl-1, 4-*N*-glycolylmuramic acid. The *N*-glycolylmuramic acid is unique to the cell walls of *Mycobacterium* and *Nocardia* (Uchida and Seino, 1997). Other mycolic acid containing bacteria appear to lack this *N*-glycolylmuramic acid; but instead, possess *N*-acetylmuramic acid. Peptide side chains consisting of either diamidated L-alanine-D-α-glutamine-*meso*-diaminopimelic acid (DAP) tripeptides or L-alanine-D-α-glutamine-*meso*-DAP-D-alanine tetrapeptides are linked to the carboxyl group of the *N*-glycolylmuramic by L-alanine. These peptidoglycan polymers are cross-linked directly by bonds between the *meso*-DAP on one side chain to the D-alanine on the adjacent polymer. This cross-linked peptidoglycan structure varies from 15–45% of the total cell wall mass depending upon the strain of nocardiae, stage of growth, and culture conditions (Beaman and Beaman, 1994).

All nocardiae possess an arabinogalactan polymer covalently bound to the peptidoglycan by a phosphodiester linkage between muramic acid and arabinose. In addition, the terminal arabinose of the arabinogalactan polymer is esterified to a mycolic acid resulting in a peptidoglycan-arabinogalactan-mycolate complex. This complex forms the basal layer of the cell envelope of all mycolic acid containing bacteria (Beaman and Beaman, 1994).

The total cell envelope of nocardiae contains a large variety of other compounds either covalently linked to or loosely associated with the basal cell wall complex described above. Thus, the nocardial envelope becomes a dynamically changing, layered structure consisting of not only the basal layer but also many complex and unique lipids, peptides, proteins, lipoproteins, peptidolipids, glycolipids, peptidoglycolipids, and polysaccharides (Beaman and Moring, 1988; and Beaman *et al.*, 1988; Beaman *et al.*, 1981; Beaman, 1975). Many of these compounds are embedded deeply within the envelope; however, several such as trehalose dimycolate (TDM), become either permanently or transiently surface associated. Some of these surface associated components are involved in providing specificity for nocardial adherence to and invasion of host cells (Beaman and Beaman, 1994).

It was established clearly that the relative virulence of *N. asteroides* for mice and cells in tissue culture is growth stage dependent (Davis-Scibienski and Beaman, 1980b; Beaman, 1979; Beaman and Maslan, 1978). Therefore, the compositional changes in cell walls from virulent and avirulent strains of *N. asteroides* during their growth cycle was studied (Beaman and Moring, 1988). Mutants were selected that were either more or less virulent for mice than the wild type strain (Beaman and Moring, 1988). These studies revealed that there are significant modifications in the structure of mycolic acids at different stages of growth, and that the fatty acid composition of the cell wall changes as well (Beaman *et al.*, 1988). Both mycolic and fatty acids are significantly more unsaturated during log phase (the most virulent form) and become saturated (with few double bonds) during stationary phase (least virulent stage). These correlations were reconfirmed using different strains and mutants of *N. asteroides* that varied considerably in their virulence for mice. In addition to the degree of saturation, the relative size of these compounds also correlated with virulence. Less virulent strains and mutants had shorter mycolic acids with a predominance of odd carbon chain length than their more virulent counterparts (Beaman and Moring, 1988). The cell envelope and the mycolic acids of a less virulent mutant (GUH-2AI) of *N. asteroides* GUH-2 was studied in considerable detail (Beaman and Moring, 1988). This mutant lacked a $C_{54:3}$ mycolic acid in the cell wall and had increased odd carbon sized mycolates. In contrast, the cell envelope of wild type GUH-2 at the same stage of growth had large amounts of $C_{54:3}$ mycolic acid and relatively small amounts of odd carbon sized mycolates. Otherwise, the cell wall composition and ultrastructure of GUH-2AI appeared to be identical to the parental, wild type strain (Vistica and Beaman, 1983). The Mutant, GUH-2AI was 10 times less virulent for mice than GUH-2, did not adhere to the brain, did not grow in the brain, did not invade host cells either *in vitro* or *in vivo*, was cleared rapidly from the lungs, did not alter phagocyte function, and did not inhibit phagosome-lysosome fusion in macrophages (Beaman and Moring, 1988; Vistica and Beaman, 1983). Both the wild type GUH-2 and the mutant GUH-2AI had identical growth curve characteristics in BHI broth as well as on all routinely used culture media for differentiating *Nocardia* spp. These data indicate that the $C_{54:3}$ mycolic acid in the cell wall of *N. asteroides* GUH-2 is critical for host interactions with this pathogen, and appears to represent a virulence factor for GUH-2. The role, if any, of the odd carbon sized mycolic acids in these host-pathogen interactions is less clearly defined (Beaman and Beaman, 1994; Beaman and Moring, 1988; Beaman *et al.*, 1988; Vistica and Beaman, 1983).

5.2. Possible Adhesins

Nocardiae grow by apical extension to form filamentous cells (Beaman and Beaman, 1998). The growing tips of these cells are ultrastructurally and antigenically distinct from the remaining body of the filament. Discrete biochemical constituents are dominant at the filament apex as compared to the longitudinal surface. These differences become apparent when log and stationary phase cells are incubated with various types of host cells both *in vitro* and *in vivo* (Beaman and Beaman, 1998; Beaman and Beaman, 1994). For example, log phase cells of *N. asteroides* GUH-2 adhere to the surface of HeLa cells by way of both the filament tip and longitudinally. In contrast, stationary phase cells from the same culture adhere longitudinally to Hela cells with little evidence of apical attachment (Beaman and Beaman, 1998). If the nocardial filaments are first incubated with antibody specific for stationary phase GUH-2, then only filament tip attachment to HeLa cells occurs. On the other hand, if the nocardial filaments are preincubated with adsorbed immune mouse serum specific for filament tip antigens, then filament tip attachment of GUH-2 to HeLa cells is blocked. Only longitudinal adherence is observed. These observations show that there are different growth stage dependent ligands on the nocardial cell for HeLa surface receptors. Those "adhesins" on the filament tip of *N. asteroides* GUH-2 are antigenically distinct from "adhesins" present along the longitudinal surface of the same bacterial cell (Beaman and Beaman, 1998).

A 43 kDa protein on the filament tip of log phase cells of *N. asteroides* GUH-2 is the probable "adhesin" for apical adherence to both murine pulmonary epithelial cells *in vivo* and HeLa cells *in vitro* (Beaman and Beaman, 1998). Data suggest that the glycolipid, threhalose dimycolate (cord factor; CF), may be the ligand along the nocardial surface that is involved in longitudinal attachment to host cells since monoclonal antibody against this CF blocks longitudinal adherence (Beaman, unpublished data). Immunoflourescent labeling combined with SDS extraction and Western immunoblot analysis using an antibody prepared against a purified 36 kDa culture filtrate antigen of GUH-2 demonstrated a filament tip associated

36 kDa protein (Beaman and Beaman, 1998). This protein appears to be involved in attachment of *N. asteroides* GUH-2 to endothelial cells in the murine brain because anti-36 kDa antibody blocks adherence of GUH-2 in the brain (Beaman, unpublished data). Furthermore, Western blot studies demonstrate that both 36 kDa and 31 kDa surface proteins bind the extracellular matrix (ECM) protein laminin but not type IV collagen and fibronectin (Beaman and Beaman, 1998). Nevertheless, the longitudinal surface of GUH-2 possesses a fibronectin binding substance. These observations suggest that ECM proteins my be involved in the adherence of *N. asteroides* GUH-2 to the surface of some host cells. Other substances related to lipoarabinomannans (LAMs) on the nocardial surface are probably involved in adherence, since these classes of compounds have been implicated in the adherence of *Mycobacterium* spp. to host cells (Schlesinger *et al.*, 1996). Unfortunately, the structure of LAMs of nocardiae have not been investigated; therefore, their roles, if any, in adherence cannot be deduced.

6. INTERNALIZATION OF NOCARDIAE WITHIN HOST CELLS

After adherence to the surface of host cells, most strains of *Nocardia* become internalized either by classical phagocytosis or by an invasion process (Beaman and Beaman, 1994). The mechanisms regulating internalization of infectious agents are distinct for each of these processes, and have been the focus of numerous investigations. However, few studies have been published regarding the mechanisms of uptake of nocardiae by either professional phagocytes or by "non-phagocytic" cells. Therefore, the mechanisms involved in cellular uptake of nocardiae must be deduced by analogy through investigation of closely related bacteria. We will summarize only briefly some of these studies since it is beyond the scope of our review to discuss the plethora of data published in other bacteria. However, this information is presented in more detail elsewhere in this book.

6.1. Uptake by Phagocytes

Schlesinger *et al.* (1990) found serum components such as the complement component C3 and the type 1 and 3 complement receptors (CR1 and CR3) on monocytes mediated adherence of virulent strains of *M. tuberculosis* to human monocytes. In the absence of serum, C3 secreted by the macrophages may also mediate phagocytosis of tubercle bacilli (Schlesinger, 1993). Modulation of the Fc receptor does not alter adherence of tubercle bacilli to human monocytes (Schlesinger *et al.*, 1990);

however, the mannose receptor on human monocytes appears to affect phagocytosis of virulent *M. tuberculosis* (strains Erdman and H37Rv) in either the presence or absence of serum (Schlesinger, 1993). It has been postulated that the mannose receptor recognizes mannose oligosaccharide "caps" on the arabinose saccharides at the terminal end of lipoarabinomannan (LAM) on the surface of *M. tuberculosis* (Schlesinger, 1993). *M. tuberculosis* H37Ra (derived from H37Rv) does not induce progressive disease in animals, and therefore, it is considered avirulent (Schlesinger, 1993; Oatway and Steenken, 1936; Steenken *et al.* 1934). These two strains have biochemical differences in LAM. The H37Rv LAM has mannosyl "caps" on the arabinose side chains which are absent on H37Ra LAM (Schlesinger *et al.*, 1994; Chatterjee *et al.*, 1992). This may result in a different response to the mannose receptor on macrophages. It is clear that both complement receptors and mannose receptors are involved in phagocytic uptake of *M. tuberculosis* (Schlesinger, 1998; Kang and Schlesinger, 1998; Schorey *et al.*, 1997). In addition, LAM stimulates a variety of immunomodulatory effects (Chan *et al.*, 1991). The effects of LAM can be duplicated with cord factors (TDM) that also reside on the surface of these organisms. It is not clear that the avirulent *M. tuberculosis* H37Ra possesses surface associated TDM. The mycolic acid composition of H37Ra is quite different from mycolates found in the virulent H37Rv strain (Beaman, unpublished data). Thus, if strain H37Ra contains TDMs, then they should be quite distinct from those of H37Rv. Recently, Silver *et al.*, (1998) reported that both H37Rv and H37Ra were phagocytised by human monocytes equally well. However, only the virulent H37Rv grew within these phagocytes; whereas, H37Ra did not. These investigators suggested that the virulence of *M. tuberculosis* correlated more with intracellular growth and induction of TNF-α than with other factors such as adherence and phagocytosis (Silver *et al.*, 1998). TDM is a well described inducer of TNF-α (Ozeki *et al.*, 1997; Silva and Faccioli, 1988). The cell envelope of *N. asteroides* GUH-2 appears to contain a lipoarabinomannan; however, except for its general chemical composition, this component has never been studied. As noted above, these organisms also possess considerable amounts of TDM on their surface. This nocardial TDM is membrane interactive, it alters membrane fluidity, membrane hydration, and membrane permeability (Crowe *et al.*, 1994). In addition, it prevents calcium dependent membrane fusion, and blocks phagosome-lysosome fusion within human monocytes (Spargo *et al.*, 1991). All of these observations suggest that there may be an interrelationship between the glycolipids, TDM and LAM, and host cell responses (Ozeki *et al.*, 1997; Barnes *et al.*, 1992; Goren, 1990; Moreno *et al.*, 1989; Silva and Faccioli, 1988).

6.2. Uptake by Non-phagocytic Cells (Invasion)

Relatively few bacterial species invade non-phagocytic, eukaryotic cells (Falkow, 1991). Nevertheless, several gram-negative organisms belonging to the genera *Escherichia, Yersinia, Shigella, Salmonella, Pseudomonas, Erwinia, Xanthomonas*, and others possess a complex set of genes for unique secretory proteins (termed the type III secretion system) that regulate invasion of both animal and plant cells (Barinaga, 1996; Jarvis *et al.*, 1995). A gram-positive counterpart to this type III secretion system has not been established. However, *Listeria monocytogenes* binds to surface proteins promoting entry into a variety of non-phagocytic cells whereby they are then surrounded by actin filaments that form a mobile tail enabling the bacteria to spread from cell to cell (Gaillard *et al.*, 1991). A number of gene products including the internalin multigene family (i.e. surface associated InlA, InlB, etc.), as well as listeriolysin O (LLO), ActA, and possibly lecithinase are involved in adherence, invasion, and intercellular spread (Portnoy *et al.*, 1992). There is some evidence that invasive strains of *N. asteroides* may share similar adherence and invasion mechanisms (Beaman and Beaman, 1998).

Most studies on adherence to and invasion of host cells by mycolic acid containing pathogens (MACPs) have focused on *Mycobacterium spp.* (Schlesinger, 1998; Kang and Schlesinger, 1998; Schorey *et al.*, 1997; Bermudez and Young, 1994; Arruda *et al.*, 1993). Studies with *M. avium* indicate that log phase cells grown at 37°C adhere to and invade cultured intestinal (HT-29) or laryngeal (HEp-2) cell lines better than the same cells in stationary phase of growth at 30°C (Bermudez and Young, 1994). This invasion is inhibited by cytochalasin B, the protein kinase inhibitors staurosporin and H7, and tyrosine protein kinase inhibitor genistein. These treatments do not prevent attachment of the bacteria to the cell lines. These observations suggest that *M. avium* adheres to specific ligands on the host cell surface triggering specific signals to cause actin in the cytoskeleton to undergo rearrangement with the resultant internalization of the bacteria (Bermudez and Young, 1994).

7. POSSIBLE MECHANISMS FOR INVASION

The mechanisms controlling the invasion of host cells by *N. asteroides* are not known. Nocardiae in log phase of growth form long, branching filamentous cells. These filaments grow by apical extension, and they divide by forming cross walls along the filament that then separate by fragmentation. During grow in broth media, this process continues until the bacterial cell density depletes nutrients with an accumulation of secondary metabolites that probably signal termination of filament formation and increased fragmentation. Thus, at stationary phase, the nocardial cells are short, pleomorphic rods and cocci (Beaman and Beaman, 1994). Log phase organisms of invasive strains of *N. asteroides* adhere to the surface of the host primarily at the filament tip which then protrudes into the cell. Once this attachment and penetration occurs, the bacteria rapidly become internalized. Apical attachment to and penetration of host tissues have not been observed with stationary phase rods and cocci of *N. asteroides*. Instead, these cells adhere longitudinally to the surface, and they are internalized by phagocytosis (Beaman and Beaman, 1998). The longitudinal adherence and uptake of stationary phase nocardiae appear to be quite different from the apical attachment to and penetration of non-phagocytic cells by log phase *N. asteroides* (Beaman and Beaman, 1994; Beaman and Beaman, 1998).

7.1. Tip Associated Proteins

Mice inoculated i.v. with alive, sublethal dose of log phase *N. asteroides* GUH-2 developed a strong IgG antibody response within 4–6 weeks. FITC fluorescence showed that this immune mouse serum (IMS) bound uniformly to the surface of both log phase and stationary phase organisms (Beaman and Beaman, 1998). However, after incubation of the IMS with several changes of cell suspensions of stationary phase GUH-2, this adsorbed IMS (ADS-IMS) had no reactivity to stationary phase nocardiae. Instead there was a very strong immunoreactivity of the ADS-IMS against the filament tip of log phase GUH-2. Western immunoblot analysis of SDS extractable proteins from log phase cells demonstrated that this ADS-IMS reacted primarily with 43 kDa and 62 kDa proteins (Beaman and Beaman, 1998). Immunofluorescence utilizing antibody specific for purified culture filtrate proteins from *N. asteroides* GUH-2 revealed that the filament tip of log phase GUH-2 had significantly enhanced immunoreactivity for both 43 kDa and 36 kDa culture filtrate proteins. None of the other culture filtrate antigens including the 62 kDa protein were localized to the filament tip. In addition, these antibodies did not label the tip of stationary phase organisms. The ADS-IMS incubated with log phase GUH-2 blocked adherence to and invasion of both HeLa cells *in vitro* and murine lung epithelial cells *in vivo*. Preincubation of log phase GUH-2 with monospecific antibodies against the culture filtrate proteins (62 kDa; 55 kDa; 43 kDa; 36 kDa; 31 kDa; and 25 kDa proteins) demonstrated that only the anti-43 kDa antibody prevented adherence to and invasion of pulmonary epithelia in mice (Beaman and Beaman,

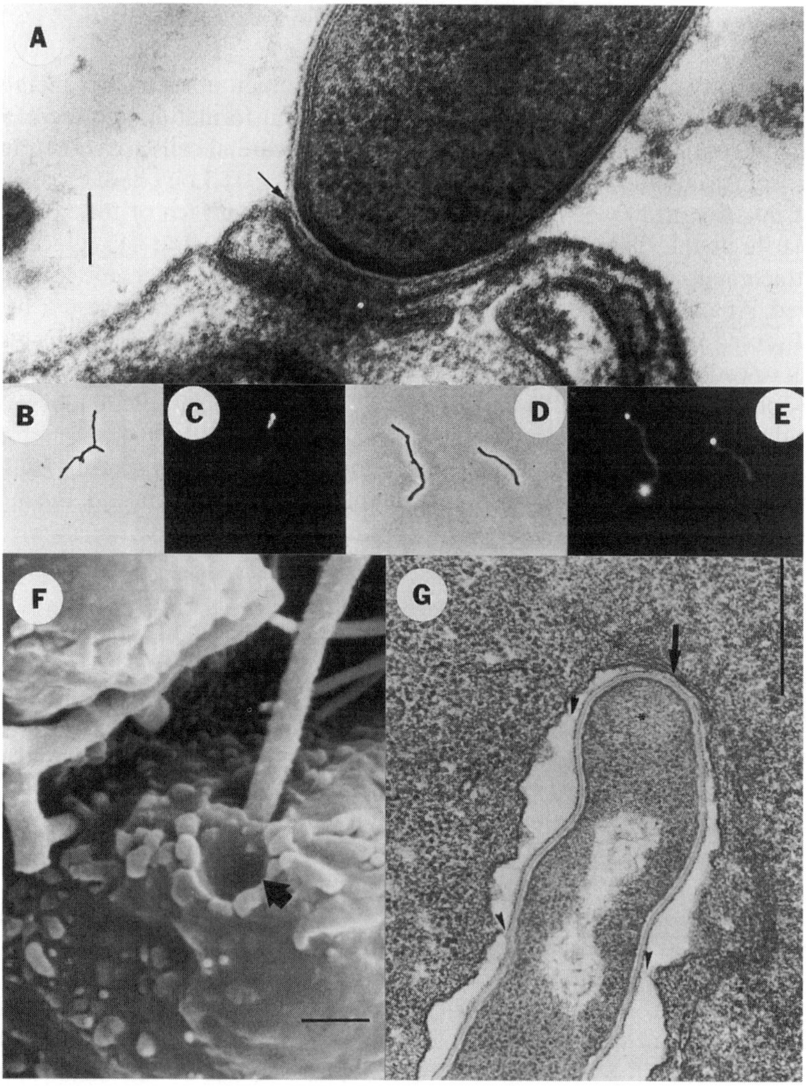

FIGURE 2. Differential adherence to and invasion of endothelial cells in the brain and epithelial cells in the lungs by *Nocardia asteroides* GUH-2. (A) Filament tip adherence to a capillary endothelial cell in the pons region of the murine brain 15 min after cardiac perfusion of a suspension of log phase *N. asteroides* GUH-2 bacteria. At the point of contact, there is a tight interaction between the outer layer of the nocardial cell wall and the cytoplasmic membrane of the host cell resulting in the formation of a "cup-like" depression that precedes penetration (arrow). Bar = 0.1 µm. (Micrograph reprinted from Beaman and Ogata (1993) with permission of the publisher.) The mechanisms controlling internalization of the nocardiae are not known. However, there are apical specific proteins on the elongated nocardial filament that appear to mediate this initial interaction which in-turn signals the invasion process. These tip associated proteins can be visualized by specific immunofluorescence microscopy (panels

1998). Furthermore, only the anti-36kDa antibody prevented adherence to and invasion of murine brain endothelial cells *in vivo*. These data suggest that a 43kDa protein on the filament tip of log phase *N. asteroides* GUH-2 is responsible for adherence to and invasion of pulmonary epithelial cells (Beaman and Beaman, 1998). In contrast, it is the 36kDa protein on the filament tip that appears to be involved in adherence to and invasion of the brain by log phase *N. asteroides* GUH-2.

7.2. Tip Associated Glycolipids

As described above, a variety of glycolipids such as LAM and TDM on the nocardial surface may be involved in adherence to the host. Monoclonal antibody (Mab) against the TDMs of *N. asteroides* GUH-2 was prepared (Beaman, unpublished data). This antibody bound to the entire nocardial cell. It did not show specificity for the filament tip. Nevertheless, log phase cells of GUH-2 preincubated with anti-TDM Mab, displayed significantly decreased attachment to and invasion of Hela cells. These observations suggest that trehalose dimycolate on the surface of nocardiae may interact with the cytoplasmic membrane of the host cell. This interaction may trigger an internal kinase resulting in internalization of the bacteria possibly through signal transduction (Beaman; unpublished data).

B, C, D, & E: Micrographs reprinted from Beaman and Beaman (1998) with permission of the publisher). (B) Phase contrast micrographs of log phase filament of *N. asteroides* GUH-2 incubated with antibody against a 43kDa culture filtrate protein secreted by GUH-2. (C) Immunofluorescence of the same bacterium shown in panel (B) localizing the 43kDa protein to the filament tip. (D) Phase contrast micrograph of log phase filaments of *N. asteroides* GUH-2 incubated with antibody against a 36kDa culture filtrate protein secreted by GUH-2. (E) Immunofluorescence of the same bacteria shown in panel D localizing the 36kDa protein to the filament tip. These data show that both a 36kDa and a 43kDa protein become associated with the filament tip of the nocardiae during growth. Antibody to the 43kDa protein reduces GUH-2 adherence to and invasion of the lungs whereas antibody against 36kDa protein reduces GUH-2 adherence within the brain suggesting distinct site specificity for these 2 apical ligands (Beaman and Beaman, 1998). (F) Scanning electron micrograph of log phase *N. asteroides* GUH-2 showing penetration of a Clara cell in the bronchiole following intranasal infection. The arrow points to a "cup like" depression in the surface of the epithelial cell left after a nocardial filament was removed by the knife during sectioning of the lung. (G) Transmission electron micrograph of a nocardial filament penetrating into a HeLa cell 1 hour after inoculation. The arrow indicates a tight association between the outer layer of the cell wall of the nocrdial filament tip and the cytoplasmic membrane of the HeLa cell. Additional pointers indicate sites of attachment between the outer bacterial cell wall and the HeLa cell membrane as this nocardial filament enters the host cell (Micrograph reprinted from Beaman and Beaman, 1998, with permission of the publisher).

7.3. Possible Cytoskeletal Rearrangements during Invasion

It is not clear how nocardiae gain entrance into nonphagocytic cells in either the lungs or brain. Many gram-positive and gram-negative pathogens enter epithelial and endothelial cells by triggering cytoskeletal rearrangement through signal transduction mechanisms. Thus, some bacterial surface ligands (adhesins) activate a kinase mediated rearrangement of either microfilaments composed of actin or microtubules (Cossart and Lecuit, 1998; Bermudez and Young, 1994; Falkow, 1991). This results in a phagocytic response; as a consequence, these bacteria become internalized.

The phagocytic process involving microfilaments composed of actin is prevented by disruption with cytochalasin. The uptake of bacteria brought about by microtubule activity is inhibited by colchicine. Therefore, the roles of different cytoskeletal rearrangements on internalization of invasive bacteria can be deduced by using these inhibitory compounds. By utilizing different tissue culture cell lines, it was found that nocardiae were internalized into host cells by at least 3, possibly 4, distinct mechanisms (Beaman and Beaman, 1994). Colchicine inhibited nocardial uptake only in phagocytic cell lines such as monocyte-macrophage derived J774A.1 and P388D1. Cytochalasin inhibited internalization of nocardiae in bovine pulmonary artery endothelial cell line CPAE, rat glial cell line C-6, and human astrocytoma cell line STTG1. Neither Colchicine nor Cytochalasin inhibited invasion of human astrocytoma cell line U-373 by *N. asteroides* GUH-2. In addition, scanning electron microscopy revealed that the surface interactions, penetration, and uptake characteristics were distinct for each cell line when incubated with log phase GUH-2 (Beaman and Beaman, 1994).

8. INTRACELLULAR GROWTH

After the nocardiae enter the host cell, several different outcomes may occur. The organism may be killed, it may be structurally and metabolically altered and then persist in a non-replicating form, its intracellular growth may be inhibited, nothing may occur, or it may grow within the host cell. All of these intracellular possibilities have been described depending upon the strain of the nocardia and the type of host cell (Beaman and Beaman, 1994). In general, invasive strains of *N. asteroides* grow as facultative intracellular pathogens in a variety of host cells.

8.1. Intracellular Modulation of Host Cell Function

The effects of nocardiae on host cell function have been studied most extensively with *N. asteroides* GUH-2 in phagocytic cells (both human and murine macrophages, monocytes and PMNs). Stationary phase cells of

GUH-2 are more readily phagocytosed by macrophages-monocytes than log phase cells from the same culture. Nevertheless, log phase cells modulate phagocyte function to a greater extent than stationary phase. For example, log phase GUH-2 inhibit phagosome-lysosome fusion, block phagosomal acidification, alter lysosomal enzyme activity, neutralize oxidative killing mechanisms, exhibit significant cytotoxicity (possible apoptotic activity), and grow in murine and human macrophages-monocytes and PMNs better than stationary phase GUH-2 (Beaman and Beaman, 1994).

9. POSSIBLE MECHANISMS FOR DISSEMINATION AFTER INVASION

It is well documented that pathogenic strains of *Nocardia* often disseminate from a site of primary infection to other regions of the body in humans and most other animals (Beaman and Beaman, 1994). The mechanisms involved in this spread are not known. However, dissemination is more frequently recognized in hosts who have any form of immunocompromise. Therefore, it is likely that completely intact innate and acquired immune responses play a critical role in keeping nocardiae sequestered within the original focal lesion, thus preventing them from disseminating.

In most mammals, including humans, the primary site of infection with *N. asteroides* and related species is the lungs. These infcctions are characteristically invasive, with bacteria extending from the bronchi to the alveoli and into the pulmonary vasculature. Once they pass through the blood vessel, the organisms can spread to all regions of the body. There is specificity for nocardial spread since most secondary sites for this blood borne dissemination are in the brain, skin, retina of the eye, joints, bone and kidneys (Beaman and Beaman, 1994). There is a uniquely high specificity for nocardial dissemination from the lungs to certain regions of the brain.

9.1. Factors Affecting Dissemination

Blocking adherence of *N. asteroides* GUH-2 within the lungs of C57BL/6 mice prevented both invasion of pulmonary epithelial cells and dissemination to the brain. Antibody against the 43 kDa tip associated protein was effective in reducing this adherence (Beaman and Beaman, 1998). However, log phase GUH-2 preincubated with anti-43 kDa antibody still bound within the brain following an i.v. inoculation. In contrast, log phase GUH-2 preincubated with anti-36 kDa antibody had decreased adherence in the brain after i.v. injection; but, there was no effect following i.n. administration. There are strains of *N. asteroides* that invade the brain following i.v. inoculation but not the lungs after i.n. administration. These organisms possess the 36 kDa tip protein but not the 43 kDa antigen (Beaman, 1996; Kjelstrom and Beaman, 1993). Antibodies to other culture filtrate antigens did not alter adherence of log phase GUH-2 in either the brain or lungs even though they bound to the nocardial surface (Beaman and Beaman, 1998). Only anti-43 kDa antibody prevented dissemination of GUH-2 from the lungs to the brain in C57BL/6 mice. Our observations suggest that this filament tip associated protein is involved somehow in adherence to pulmonary epithelial cells, invasion of cells, and dissemination to the brain. On the other hand, the 36 kDa tip associated protein appears not to be essential for dissemination of GUH-2 from the lungs to the brain even though this protein is involved in selective adherence to endothelial cells in the brain.

Preliminary data suggest that other factors are involved in nocardial dissemination. These include the surface glycolipids (TDM) and a variety of host factors such as $\gamma\delta$ T-cells. The roles of the other components on dissemination of nocardiae within the host require further study.

10. CONCLUSIONS

The genus *Nocardia* contains numerous species pathogenic for a variety of hosts. The mechanisms controlling pathogenicity are not understood fully. However, there appears to be commonality among the different species and the diverse animals susceptible to infection. In most mammals, members of the *N. asteroides* complex grow initially in cells in the lungs with a propensity for dissemination to and invasion of the brain. Therefore, *N. asteroides*, a bacterium found in the soil, is an invasive, intracellular pathogen of both the lungs and brain. Animal and tissue culture models are being utilized to elucidate the subcellular biochemistry behind nocardial invasion of eukaryotic cells.

11. REFERENCES

Al-Doory, Y., Pinkerton, M.E., Vice, T.E., and Hutchinson V., 1969, Pulmonary nocardiosis in a vervet monkey, *J. Amer. Vet. Med. Assoc.* **155**:1179–1180.

Arruda, S., Bomfim, G., Knights, R., Huima-Byron, T., and Riley, L.W., 1993, Cloning of an *M. tuberculosis* DNA fragment associated with entry and survival inside cells, *Science* **261**:1454–1456.

Barinaga, M., 1996, A shared strategy for virulence, *Science* **272**:1261–1263.

Barnes, P.F., Chatterjee, D., Abrams, J.S., Lu, S., Wang, E., Yamamura, M., Brennan, P.J., and Modlin, R.L., 1992, Cytokine production induced by *Mycobacterium tuberculosis* lipoarabinomannan, relationship to chemical structure, *J. Immunol.* **149**:541–547.

Beaman, B.L., 1975, Structural and biochemical alterations of *Nocardia asteroides* cell walls during its growth cycle, *J. Bacteriol.* **123**:1235–1253.

Beaman, B.L., 1977, The *in vitro* response of rabbit alveolar macrophages to infection with *Nocardia asteroides*, *Infect. Immun.* **15**:925–937.

Beaman, B.L., 1979, Interaction of *Nocardia asteroides* at different phases of growth with *in vitro*-maintained macrophages obtained from the lungs of normal and immunized rabbits, *Infect.Immun.* **26**:355–361.

Beaman, B.L., 1993, Ultrastructural analysis of growth of *Nocardia asteroides* during invasion of the murine brain, *Infect. Immun.* **61**:274–283.

Beaman, B.L., 1996, Differential binding of *Nocardia asteroides* in the murine lung and brain suggest multiple ligands on the nocardial surface, *Infect. Immun.* **64**:4859–4862.

Beaman, B.L., and Beaman, L., 1994, *Nocardia* species: Host-parasite relationships, *Clin. Microbiol. Rev.* **7**:213–264.

Beaman, B.L., and Beaman, L., 1998, Filament tip-associated antigens involved in adherence to and invasion of murine pulmonary epithelial cells *in vivo* and Hela cells *in vitro* by *Nocardia asteroides*, *Infect. Immun.* **66** (in press).

Beaman, B.L., Black, Ç.M., Doughty, F., and Beaman, L., 1985, Role of superoxide dismutase and catalase as determinants of pathogenicity of *Nocardia asteroides*: Importance in resistance to microbicidal activities of human polymorphonuclear neutrophils, *Infect. Immun.* **47**:135–141.

Beaman, B.L., Bourgeois, A.L., and Moring, S.E., 1981, Cell wall modification resulting from *in vitro* induction of L-phase variants of *Nocardia asteroides*, *J. Bacteriol.* **148**: 600–609.

Beaman, B.L., and Maslan, S., 1978, Virulence of *Nocardia asteroides* during its growth cycle, *Infect. Immun.* **20**:290–295.

Beaman, B.L., Maslan, S., Scates, S., and Rosen, J., 1980, Effect of route of inoculation on host resistance to *Nocardia*, *Infect. Immun.* **28**:185–189.

Beaman, B.L., and Moring, S.E., 1988, Relationship among cell wall composition, stage of growth, and virulence of *Nocardia asteroides* GUH-2, *Infect. Immun.* **56**:557–563.

Beaman, B.L., Moring, S.E., and Ioneda, T., 1988, Effect of growth stage on mycolic acid structure in cell walls of *Nocardia asteroides* GUH-2, *J. Bacteriol.* **170**:1137–1142.

Beaman, B.L., and Ogata, S.A., 1993, Ultrastructural analysis of attachment to and penetration of capillaries in the murine pons, midbrain, thalamus and hypothalamus by *Nocardia asteroides*, *Infect. Immun.* **61**:955–965.

Beaman, B.L., and Smathers, M., 1976, Interaction of *Nocardia asteroides* with cultured rabbit alveolar macrophages, *Infect. Immun.* **13**:1126–1131.

Beaman. B.L., and Sugar, A., 1983, *Nocardia* in naturally acquired and experimental infections in animals, *J. Hyg., Camb.* **91**:393–419.

Beaman, L., Paliescheskey, M., and Beaman, B.L., 1988, Acid phosphatase stimulation of the growth of *Nocardia asteroides* and its possible relationship to the modification of lysosomal enzymes in macrophages, *Infect. Immun.* **56**:1652–1654.

Beaman, L., 1987, Fungicidal activation of murine macrophages by recombinant gamma interferon, *Infect. Immun.* **55**:2951–2955.

Beaman, L., and Beaman, B.L., 1990, Monoclonal antibodies demonstrate that superoxide dismutase contributes to protection of *Nocardia asteroides* within the intact host, *Infect. Immun.* **58**:3122–3128.

Beaman, L., and Beaman, B.L., 1992, The timing of exposure of mononuclear phagocytes to recombinant interferon-gamma and recombinant turmor necrosis factor-alpha alters interactions with *Nocardia asteroides*, *J. Leukocyte Biol.* **51**:251–281.

Beaman, L., and Beaman, B.L., 1993, Interactions of *Nocardia asteroides* with murine glia cells in culture, *Infect. Immun.* **61**:343–347.

Beaman, L., and Beaman, B.L., 1994, Differences in the interactions of *Nocardia asteroides* with macrophage, endothelial, and astrocytoma cell lines, *Infect. Immun.* **62**:1787–1798.

Bermudez, L.E., Petrofsky, M., and Goodman, J., 1997, Exposure to low oxygen tension and increased osmolarity enhance the ability of *Mycobacterium avium* to enter intestinal epithelial (HT-29) cells, *Infect. Immun.* **65**:3768–3773.

Bermudez, L.E., and Young, L.S., 1994, Factors affecting invasion of HT-29 and HEp-2 epithelial cells by organisms of the *Mycobacterium avium* complex, *Infect. Immun.* **62**:2021–2026.

Black, C., Beaman, B.L., Donovan, R.M., and Goldstein, E., 1983, Effect of virulent and less virulent strains of *Nocardia asteroides* on acid-phosphatase activity in alveolar and peritoneal macrophages maintained *in vitro*, *J. Infect. Dis.* **148**:117–124.

Black, C.M., Beaman, B.L., Donovan, R.M., and Goldstein, E.M., 1985, Intracellular acid phosphatase content and ability of different macrophage populations to kill *Nocardia asteroides*, *Infect. Immun.* **47**:375–383.

Black, C.M., Paliescheskey, M., Beaman, B.L., Donovan, R.M., and Goldstein, E., 1986, Acidification of phagosomes in murine macrophages, *J. Infect. Dis.* **154**:952–958.

Black, Ç.M., Paliescheskey, M., Beaman, B.L., Donovan, R.M., and Goldstein, E., 1986, Modulation of lysosomal protease-esterase and lysozyme in Kupffer cells and peritoneal macrophages infected with *Nocardia asteroides*, *Infect. Immun.* **54**:917–919.

Black, C.M., Catterral, J.R., and Remington, J.S., 1987, *In vivo* and *in vitro* activation of alveolar macrophages by recombinant interferon-gamma, *J. Immunol.* **138**:491–495.

Boncyk, L.H., McCullough, B., Grotts, D.D., and Kalter, S.S., 1975, Localized nocardiosis due to *Nocardia caviae* in a baboon (*Papio cynocephalus*), *Lab. An. Sci.* **25**:88–91.

Bourgeois, L., and Beaman, B.L., 1974, Probable L-forms of *Nocardia asteroides* induced in cultured mouse peritoneal macrophages, *Infect. Immun.* **9**:576–590.

Chan, J., Fan, X., Hunter, S.W., Brennan, P.J., and Bloom, B.R., 1991, Lipoarabinomannan, a possible virulence factor involved in persistence of *Mycobacterium tuberculosis* within macrophages, *Infect. Immun.* **59**:1755–1761.

Chatterjee, D., Lowell, K., Rivoire, B., McNeil, M.R., and Brennan, P.J., 1992, Lipoarabinomannan of *Mycobacterium tuberculosis*: capping with mannosyl residues in some strains, *J. Biol. Chem.* **267**:6234–6236.

Chen, S.C., 1992, Study on the pathogenicity of *Nocardia asteroides* to the formosa snakehead *Channamaculata* (lacepede), and largemouth bass, *Micropterus salmoides* (lacepede), *J. Fish Dis.* **15**:47–53.

Cossart, P., and Lecuit, M., 1998, Interactions of *Listeria monocytogenes* with mammalian cells during entry and actin-based movement: bacterial factors, cellular ligands and signaling, *EMBO* **17**:3797–3806.

Crowe, L.M., Spargo, B.J., Ioneda, T., Beaman, B.L., and Crowe, J.H., 1994, Interaction of cord factor (a,a'-trehalose-6,6′-dimycolate) with phospholipids, *Biochim. Biophys. Acta* **1194**:53–60.

Davis-Scibienski, C., and Beaman, B.L., 1980a, Interaction of *Nocardia asteroides* with rabbit alveolar macrophages: Association of virulence, viability, ultrastructural damage, and phagosome-lysosome fusion, *Infect. Immun.* **28**:610–619.

Davis-Scibienski, C., and Beaman, B.L., 1980b, Interaction of *Nocardia asteroides* with rabbit alveolar macrophages: Effect of growth phase and viability on phagosome-lysosome fusion, *Infect. Immun.* **29**:24–29.

Davis-Scibienski, C., and Beaman, B.L., 1980c, Interaction of alveolar macrophages with *Nocardia asteroides*: Immunological enhancment of phagocytosis, phagosome-lysosome fusion and microbicidal activity, *Infect. Immun.* **30**:578–587.

Eppinger, H., 1890, Uber eine neue, pathogene *Cladothrix* und eine durch sie hervogerufene Pseudotuberculosis, *Wien. Klin. Wochenschr.* **3**:321–323.

Eppinger, H., 1891, Uber eine neue, pathogene *Cladothrix* und eine durch sie hervogerufene Pseudotuberculosis (cladothrichica), *Beitr. Pathol. Anat. Allg. Pathol.* **9**:287–328.

Falkow, S., 1991, Bacterial entry into eukaryotic cells, *Cell* **65**:1099–1102.

Filice, G.A., Beaman, B.L., Krick, J.A., and Remington, J.S., 1980, Effects of human neutrophils and monocytes on *Nocardia asteroides*: Failure of killing despite occurrence of the oxidative metabilic burst, *J. Infect. Dis.* **142**:432–438.

Filice, G.A., 1983, Resistance of *Nocardia asteroides* to oxygen-dependant killing by neutrophils, *J. Infect. Dis.* **148**:861–867.

Filice, G.A., 1985, Inhibition of Nocardia asteroides by neutrophils, *J. Infect. Dis.* **151**:47–56.

Friedman, C.S., Beaman, B.L., Chun, J., Goodfellow, M., Gee, A., and Hedrick, R.P., 1998, *Nocardia crassostreae* sp. nov., the causal agent of nocardiosis in Pacific oysters, *Int. J. Syst. Bacteriol.* **48**:237–246.

Gaillard, J.L., Berche, P., Frehel, C., Gouin, E., and Cossart, P., 1991, Entry of *L. monocytogenes* into cells is mediated by internalin, a repeat protein reminiscent of surface antigens from gram-positive bacteria, *Cell* **65**:1127–1141.

Goren, M.B., 1990, Mycobacterial fatty acid esters of sugars and sulfosugars, in: *Glycolipids, Phosphoglycolipids and Sulfoglycolipids. Handbook of Lipid Research*, Volume 6 (M. Kaitz, ed.), Plenum Press, New York, pp. 363–461.

Jarvis, K.G., Giron, J.A., Jerse, A.E., McDaniel, T.K., Donnenberg, M.S., and Kaper, J.B., 1995, Enteropathogenic *Escherichia coli* contains a putative type III secretion system necessary for the export of proteins involved in attaching and effacing lesion formation, *Proc. Natl. Acad. Sci.* **92**:7996–8000.

Jonas, A.M., and Wyand, D.S., 1966, Pulmonary nocardiosis in the Rhesus monkey, *Path. Vet.* **3**,588–600.

Kang, B.K., and Schlesinger, L.S., 1998, Characterization of mannose receptor-dependent phagocytosis mediated by *Mycobacterium tuberculosis* lipoarabinomannan, *Infect. Immun.* **66**:2769–2777.

Kessler, M.J., and Brown, R.J., 1981, Mycetomas in a squirrel monkey, *J. Zoo An. Med.* **12**:91–93.

Khan, Z.U., Neil, L., Chandy, R., Chugh, T.D., Al-Sayer, H., Provost, F., and Boiron, P., 1997, *Nocardia asteroides* in the soil of Kuwait, *Mycopathologia.* **137**:159–163.

Kjelstrom J.A., and Beaman, B.L., 1993, Development of a serological panel for recognition of nocardial infections in a murine model, *Diagn. Microbiol. Infect. Dis.* **16**:291–301.

Lechevalier, H.A., 1989, Nocardioform actinomycetes, in: *Bergey's Manual of Systematic Bacteriology*, Volume 4 (S.T. Williams, M.E. Sharpe, and J.G. Holt, eds.), The Williams & Wilkins Co., Baltimore, pp. 2348–2404.

MacCallum, W.G., 1902, On the life history of *Actinomyces asteroides*, *Zentralbl. Bakteriol. Parasitenkd. Infektionskr. Hy. Abt. 1 Orig.* **31**:528–547.

Mahajan, V.M., Padhy, S.C., Dayal, Y., Bhatia, I.M., and Ratnakar, K.S., 1977, Experimental pulmonary nocardiosis in monkeys, *Sabouraudia* **15**:47–50.

Marino, D.J., and Jaggy, A., 1993, Nocardiosis: A literature review with selected case reports in two dogs, *Vet. Intern. Med.* **7**:4–11.

McClure, H.M., Chang, J., Kaplan, W., and Brown, J.M., 1976, Pulmonary nocardiosis in an Orangutan, *J. Amer. Vet. Med. Assoc.* **169**:943–945.

Moreno, C.J., Taverne, A., Mehlert, C.A.W., Bate, R.J., Brealey, A., Meager, G.A.W., and Playfair, J.H.L., 1989, Lipoarabinomannan from *Mycobacterium tuberculosis* induces the production of tumor necrosis factor from human and murine macrophages, *Clin. Exp. Immunol.* **76**:240–245.

Oatway, W.H., Jr., and Steenken, W., Jr., 1936, The pathogenesis and fate of the tubercle produced by dissociated variants of tubercle bacilli, *J. Infect. Dis.* **59**:306–325.

Odell, L.W., and Segal, A.W., 1991, Killing of pathogens associated with chronic granulomatous disease by the non-oxidatiive microbicidal mechanisms of human neutrophils, *J. Med. Microbiol.* **34**:129–135.

Ogata, S.A., and Beaman, B.L., 1992a, Adherence of *Nocardia asteroides* within the murine brain, *Infect. Immun.* **60**:1800–1805.

Ogata, S.A., and Beaman, B.L., 1992b, Site specific growth of *Nocardia asteroides* within the murine brain, *Infect. Immun.* **60**:3262–3267.

Ozeki, Y., Kaneda, K., Fujiwara, N., Morimoto, M., Oka, S., and Yano, I., 1997, *In vivo* induction of apoptosis in the thymus by administration of mycobacterial cord factor (trehalose 6,6′-dimycolate), *Infect. Immun.* **65**:1793–1799.

Nocard, E., 1888, Note sur la maladie des boeufs de la Gouadeloupe connue sous le nom de farcin, *Ann. Inst. Pasteur* (Paris) **2**:293–302.

Portnoy, D.A., Chakraborty, T., Goebel, W., and Cossart, P., 1992, Molecular determinants of *Listeria monocytogenes* pathogenesis, *Infect. Immun.* **60**:1263–1267.

Provost, F., Blanc, M.V., Beaman, B.L., and Boiron, P., 1996, Occurrence of plasmids in pathogenic strains of *Nocardia*, *J. Med. Microbiol.* **45**:344–348.

Provost, F., Laurent, F., Blanc, M.V., and Boiron, P., 1997, Transmission of nocardiosis and molecular typing of *Nocardia* species: a short review, *Eur. J. Epidemiol.* **13**:235–238.

Riess, F.G., Lichtinger, T., Cseh, R., Yassin, A.F., Schaal, K.P., and Benz, R., 1998, The cell wall porin of *Nocardia farcinica*: Biochemical identification of the channel-forming protein and biophysical characterization of the channel properties, *Mol. Microbiol.* **29**:139–150.

Sakakibara, I., Sugimoto, Y., Minato, H., Takasaka, M., and Honjo, S., 1984, Spontaneous nocardiosis with brain abscess caused by *Nocardia asteroides* in a cynomolgus monkey, *J. Med. Primatol.* **13**:89–95.

Schlesinger, L.S., 1993, Macrophage phagocytosis of virulent but not attenuated strains of *Mycobacterium tuberculosis* is mediated by mannose receptors in addition to complement receptors, *J. Immunol.* **150**:2920–2930.

Schlesinger, L.S., 1998, *Mycobacterium tuberculosis* and the complement system, *Trends Microbiol.* **6**:47–49.

Schlesinger, L.S., Bellinger-Kawahara, C.G., Payne, N.R., and Horwitz, M.A., 1990, Phagocytosis of *Mycobacterium tuberculosis* is mediated by human monocyte complement receptors and complement component C3, *J. Immunol.* **144**:2771–2780.

Schlesinger, L.S., Hull, S.R., and Kaufman, T.M., 1994, Binding of the terminal mannosyl units of lipoarabinomannan from a virulent strain of *Mycobacterium tuberculosis* to human macrophages, *J. Immunol.* **152**:4070–4079.

Schlesinger, L.S., Kaufman, T.M., Iyer, S., Hull, S.R., Marchiando, L.K., 1996, Differences in mannose receptor-mediated uptake of lipoarabinomannan from virulent and attenuated strains of *Mycobacterium tuberculosis* by human macrophages, *J. Immunol.* **157**: 4568–4575.

Schorey, J.S., Carroll, M.C., and Brown, E.J., 1997, A macrophage invasion mechanism of pathogenic mycobacteria, *Science* **277**:1091–1093.

Sileo, L., Sievert, P.R., and Samuel, M.D., 1990, Causes of mortality of albatross chicks at Midway Atoll, *Wildlife Dis.* **26**:329–338.

Silva, C.L., and Faccioli, L.H., 1988, Tumor necrosis factor (cachectin) mediates induction of cachexia by cord factor from mycobacteria, *Infect. Immun.* **56**:3067–3071.

Silver, R.F., Li, Q., and Ellner, J.J., 1998, Expression of virulence of *Mycobacterium tuberculo-*

sis within human monocytes: Virulence correlates with intracellular growth and induction of tumor necrosis factor alpha but not with evasion of lymphocyte-dependent monocyte effector functions, *Infect. Immun.* **66**:1190–1199.

Spargo, B.J., Crowe, L.M., Ioneda, T., Beaman, B.L., and Crowe, J.H., 1991, Cord factor (a,a′-trehalose-6,6′-dimycolate) inhibits fusion between phospholipid vesicles, *Proc. Natl. Acad. Sci.* USA. **88**:737–740.

Splino, M., Merka, V., and Kyntera, F., 1975, Phagocytosis and intracellular proliferation of *Nocardia asteroides* (strain Weipheld) in cell structures *in vitro*. 1. Alveolar macrophages of Guinea-pigs, *Zentrabl. Bakteriol. Mikrobiol. Hyg. 1 Abt. Orig. A* **232**:334–340.

Steenken, W., Jr., Oatway, W.H., Jr., and Petroff, S.A., 1934, Biological studies of the tubercle bacillus. III. Dissociation and pathogenicity of the R and S variants of the human tubercle bacillus (H37), *J. Exp. Med.* **60**:515–540.

Tosteson, T.R., Ballantine, D.L., Tosteson, C.G., Hensley V., and Bardales, A.T., 1989, Associated bacterial flora, growth, and toxicity of cultured benthic dinoflagellates *Ostreopsis lenticularis* and *Gambierdiscus toxicus*, *Appl. Environ. Microbiol.* **55**:137–141.

Uchida, K., and Seino, A., 1997, Intra- and intergeneric relationships of various actinomycete strains based on the acyl types of the muramyl residue in cell wall peptidoglycans examined in a glycolate test, *Int. J. Syst. Bacteriol.* **47**:182–190.

Umunnabuike, A.C., and Irokanulo, E.A., 1986, Isolation of *Campylobacter* subsp. *jejuni* from Oriental and American cockroaches caught in kitchens and poultry houses in Vom, Nigeria, *Int. J. Zoonoses.* **13**:180–186.

Vistica, C.A., and Beaman, B.L., 1983, Pathogenic and virulence characterization of colonial mutants of *Nocardia asteroides* GUH-2, *Can. J. Microbiol.* **29**:1126–1135.

Wilson, R.W., Steingrube, V.A., Brown, B.A., and Wallace, R.J., Jr., 1998, Clinical application of PCR-restriction enzyme pattern analysis for rapid identification of aerobic actinomycete isolates, *J. Clin. Microbiol.* **36**:148–152.

Workman, M.R., Philpott-Howard, J., Yates, M., Beighton, D., and Casewell, M.W., 1998, Identification and antibiotic susceptibility of *Nocardia farcinica* and *N. nova* in the UK, *Med. Microbiol.* **47**:85–90.

Chapter 9

Mycoplasma Interaction with Eukaryotic Cells

Shlomo Rottem and David Yogev

1. INTRODUCTION

Mollicutes are the smallest and simplest self-replicating prokaryotes. These microorganisms lack a rigid cell wall and are bound by a single membrane, the plasma membrane (Razin *et al.*, 1998). Wall-less prokaryotes were first described 100 years ago and now over 180 species, widely distributed among humans, animals, insects and plants are known (Razin *et al.*, 1998). The lack of a cell wall is used to distinguish these microorganisms from ordinary bacteria and to include them in a separate class named Mollicutes. Most human and animal mollicutes are *Mycoplasma* and *Ureaplasma* species of the family Mycoplasmataceae. The trivial name mycoplasmas will be used by us to denote any organisms within this family. Mycoplasmas have an extremely small genome size of 0.58–1.35 mb (compared with the 4.64 mb of *E. coli*). Over the last three years the genomes of *Mycoplasma genitalium* (0.58 mb) and *Mycoplasma pneumoniae* (0.816 mb) have been

SHLOMO ROTTEM and DAVID YOGEV Department of Membrane and Ultrastructure Research, The Hebrew University-Hadassah Medical School, Jerusalem 91120, Israel; E-mail: rottem@cc.huji.ac.il

Subcellular Biochemistry, Volume 33: Bacterial Invasion into Eukaryotic Cells, edited by Oelschlaeger and Hacker. Kluwer Academic / Plenum Publishers, New York, 2000.

sequenced (Himmelreich *et al.*, 1996; Fraser *et al.*, 1995), showing only 470 and 500 protein coding regions respectively. Their small genomes impose on these organisms limited metabolic options for replication and survival (Himmelreich *et al.*, 1997; Pollack *et al.*, 1997).

Most mycoplasmas are parasitic species, exhibiting strict host and tissue specificities. The mycoplasmas enter an appropriate host in which they multiply and survive for extended periods. The major question is whether mycoplasmas cause damage to the host cells and to what extent the damage is clinically apparent. Mycoplasmas have long been refractory to detailed analyses because of complex nutritional requirements, poor growth yields, and a paucity of useful genetic tools. While questions still far outnumber answers, significant progress has been made in identifying the major players and their probable roles in the interaction of the mycoplasmas with eukaryotic host cells (Krause, 1996). This review will describe the major mechanisms by which mycoplasmas interact and damage host cells and circumvent the host immune system.

2. ADHERENCE TO HOST CELLS

Many animal mycoplasmas depend on adhesion to host tissues for colonization and infection. Adherence is regarded as the major virulence factor of mycoplasmas and adherence-deficient mutants are avirulent (Baseman and Tully, 1997; Razin and Jacobs, 1992). The best studied species are *M. pneumoniae*, the causative agent of primary atypical pneumonia in humans, which inhabits the respiratory tract and *M. genitalium* which preferentially colonize the urogenital tract. These organisms exhibit the typical polymorphism of mycoplasmas, but most prominent are flask-shaped elongated organisms. Cytadherence of these organisms to the respiratory or urogenital epithelium is an initial and essential step in tissue colonization and subsequent disease pathogenesis.

2.1. Adhesins

M. pneumoniae and *M. genitalium* have a tiny tip at one of the poles, termed the tip organelle, which functions as an attachment organelle. Two surface proteins, a 169 kDa protein designated P1 and a 30 kDa protein designated P30, have been identified as the major adhesins of *M. pneumoniae* on the basis of biochemical, immunological and ultrastructural studies (Baseman *et al.*, 1996; Krause, 1996; Baseman, 1993). The P1 and P30 genes were cloned, sequenced and characterized (Dallo *et al.*, 1990; Inamine *et al.*, 1988). The P30 adhesin shares a number of properties with P1, notably a proline-rich C-terminus, expressed also by substantial amino acid sequence homology with the C-terminus of P1. Both adhesins are clustered at the tip organelle of virulent mycoplasmas, providing polarity to the cytadherence event. Also, these adhesins elicit a strong immunological response in convalescent-phase sera from humans and experimentally infected hamsters, and anti-adhesin monoclonal antibodies block cytadherence (Baseman, 1993; Razin and Jacobs, 1992).

2.2. Accessory Proteins

The isolation and characterization of *M. pneumoniae* mutants that possess P1 and P30 yet fail to cytadhere, has led to the notion that the tip-mediated adherence of *M. pneumoniae* to eucaryotic target cells is much more complex (Krause, 1996; Razin and Jacobs, 1992). Accordingly two groups of accessory proteins, the first including proteins A- C, and the second, proteins HMW1-HMW3, were identified (Krause, 1996; Krause *et al.*, 1982). These proteins cannot be defined as adhesins *per se*, but are required for the proper functioning of the adhesins. The loss of an ac-cessory protein(s) is also associated with the inability to cytadhere, whereas reacquisition of this protein(s) is accompanied by reversion to a cytadherence-positive phenotype (Razin and Jacobs, 1992). Proteins B (90 kDa) and C (40 kDa) were intensively investigated and were found to be surface proteins, localized mainly at the tip organelle. These proteins are closely associated with P1 (Franzoso *et al.*, 1994; Layh-Schmitt and Herrmann, 1994) but are not directly involved in receptor binding. Although P1 and P30 are present on the surface of strains lacking B and/or C, they are not clustered at the tip organelle, but scattered on the surface of the mycoplasma (Krause, 1996). It appears that the mycoplasmal adhesins interact cooperatively with these accessory proteins and it has been suggested that these proteins have a scaffolding role in maintaining the proper disposition/distribution of P1 and P30 on the mycoplasma membrane (Baseman, 1993).

Also, it has been discovered that HMW1, HMW2 and/or HMW3 are missing in certain non-cytadhering mutant strains and regained in their corresponding revertants. HMW1 (112 kDa) and HMW3 (74 kDa) are members of a family of mycoplasma proteins that share acidic residues and an internal proline-rich domain in repeated motifs (Krause, 1996). The proline-rich domains probably impart an extended conformation to the protein backbone, while the hydrophobic surface provided by the proline residues may contribute to interactions with other mycoplasma proteins. Similar to their counterparts lacking A, B and C, mutant strains devoid of HMW1-HMW3 (class I mutants) are avirulent, cytadhere very poorly, and fail to cluster P1 at the tip organelle (Hahn *et al.*, 1998; Popham *et al.*, 1997).

Furthermore, their tip organelle lacks the characteristic truncated appearance seen in wild-type *M. pneumoniae*, suggesting that one or more of the HMW proteins is required both to anchor P1 at the attachment organelle and to maintain the proper architecture of the tip organelle (Stevens and Krause, 1992). Consistent with this interpretation is the observation that these proteins are components of the Triton X-100-insoluble, mycoplasma cytoskeleton and as such are thought to have a structural role in P1 localization. The deduced amino acid sequence of HMW3 shows that this protein is largely hydrophilic and results of whole-cell radio-immunoprecipitation experiments revealed that HNW3 is not exposed on the *M. pneumoniae* cell surface (Krause, 1996).

As the tip organelle in HMW3 mutants lacks the truncated tip appearance characteristic of wild-type *M. pneumoniae* (Stevens and Krause, 1992), it has been suggested that HMW3 contribute to the architecture of the tip organelle by providing a platform to which P1 and P30 are anchored (Krause, 1996). Similar to HMW3, HMW1 has an unusual subcellular distribution, localizing primarily along the leading and trailing filaments that extend from the mycoplasma cell body (Stevens and Krause, 1991). These extensions seem to form as the mycoplasma establishes contact with surfaces, moves by gliding along those surfaces and undergoes cell division (Krause, 1996). The subcellular location of HMW2 has not been established but the possibility that this protein is present near the base of the tip organelle has been brought up (Hahn *et al.*, 1998).

How do the accessory proteins promote adherence? As already mentioned, clustering of P1 and P30 at the tip organelle appears vital to attachment, as clustering provides a critical concentration of adhesion molecules required for securing a stable primary association with receptor molecules on the host cells. In the non-adhering mutants, P1 fails to cluster at the tip organelle (Baseman *et al.*, 1993). Hence, it is conceivable that one or more of the accessory proteins plays a role in the lateral movement and concentration of the adhesion molecules at the tip organelle. How this is accomplished is questionable. A plausible hypothesis is that to fulfill this role the accessory proteins must be associated with the mycoplasma cytoskeleton, which is responsible not only for the lateral movement and proper orientation of P1, but also for changes in cell shape, cell division and motility (Razin *et al.*, 1998).

2.3. Receptors

The role of host cell surface sialoglycoconjugates as receptors for mycoplasmas has long been established (Razin and Jacobs, 1992) and the carbohydrate moiety of an active glycoprotein, which has been shown to serve as the *M. pneumoniae* receptor on human erythrocytes, has been identified as having a terminal NeuAc(α2–3)Gal(β1–4)GlcNAc sequence (Roberts *et al.*, 1989). Nevertheless, neuraminidase treatment has frequently failed to abolish the ability of various eukaryotic cells to bind *M. pneumoniae* (Geary and Gabridge, 1987). A sialic acid-free glycoprotein, isolated from cultured human lung fibroblasts, which serves as a receptor for *M. pneumoniae*, has been isolated by Geary *et al.* (1990) and sulphated glycolipids containing terminal Gal($3SO_4$)β1-residues were also found to function as receptors (Krivan *et al.*, 1989). Clearly, there is more than one type of receptor for *M. pneumoniae*, and apparently for other adhering mycoplasmas as well. It is interesting to note that at the primary site of *M. pneumoniae* infection the apical microvillar border and the cilia carried the sialo-oligosaccharide type of receptors, whereas the secretory cells and mucus lacked them, favoring attachment of *M. pneumoniae* to the ciliated cells and avoiding the secreted mucus barrier (Loveless and Feizi, 1989).

2.4. Damage to Host Cells

Adherence to host cells may interfere with membrane receptors or alter transport mechanisms of the host cell. For example, disruption of the K^+ channels of ciliated bronchial epithelial cells by *Mycoplasma hyopneumoniae* is known to depolarize the cell membrane, resulting in ciliostasis (DeBey and Ross, 1994). The host cell membrane is also vulnerable to toxic materials released by the adhering mycoplasmas. Although toxins have not been associated with mycoplasmas, the production of cytotoxic metabolites and the activity of cytolytic enzymes is well established. Oxidative damage to the host cell membrane by peroxide and superoxide radicals excreted by the adhering mycoplasmas appears to be experimentally well-substantiated (Almagor *et al.*, 1986). The intimate contact of the mycoplasma with the host cell membrane may also result in the hydrolysis of host cell phospholipids catalyzed by the potent membrane-bound phospholipases present in many mycoplasma species (Shibata *et al.*, 1995). This could trigger specific signal cascades Rosenshine and Finlay, 1993) or release cytolytic lysophospholipids capable of disrupting the integrity of the host cell membrane (Salman and Rottem, 1995).

3. FUSION WITH HOST CELLS

The lack of a rigid cell wall allows direct and intimate contact of the mycoplasma membrane with the cytoplasmic membrane of the host cell. Under the appropriate conditions, such contact may lead to cell fusion.

Fusion of mycoplasmas with eukaryotic host cells has been first observed by electron-microscopic studies (Rottem and Naot, 1998). The development of energy transfer and fluorescence dequenching methods has enabled investigation of the fusion process on a quantitative basis, and various fusogenic strains have been identified. In all the species tested, fusogenicity is dependent on the unesterified cholesteral content of the cell membrane (Tarshis *et al.*, 1993). Fusogenic strains, found only among mollicutes requiring unesterified cholesterol for growth, whereas *Acholeplasma* species which do not require cholesterol were nonfusogenic. Furthermore, adaptation of a fusogenic strain of *Mycoplasma capricolum* to grow in the absence of cholesterol, resulting in a marked reduction in membrane- cholesterol content, renders the organism nonfusogenic (Tarshis *et al.*, 1993). In some mycoplasmas, fusogenicity also depends on the proton gradient across the mycoplasma cell membrane and it is markedly decreased when the proton gradient is collapsed by proton ionophores (Citovsky *et al.*, 1987).

Among *Mycoplasma* species, *Mycoplasma fermentans* is highly fusogenic, capable of fusing with a variety of cells (Franzoso *et al.*, 1992). This organism was first isolated from the lower urogenital tract of both women and men, but its role in urogenital diseases has not been established. A unique strain of *M. fermentans* (incognitus strain), isolated from patients with acquired immunodeficiency syndrome (AIDS), was suggested to be involved in the pathogenesis of the disease (Brenner *et al.*, 1996; Lo, 1992). Intensive studies on this organism led to the identification of an unusual phosphocholine-containing phosphoglycolipid in its cell membrane that was capable of enhancing fusion. This fusogenic lipid has been recently identified as 6′-O-[3″-phosphocholine-2″-amino-1″, 3″-propanediol)-α-D-glucopyranosyl]-(1′□3)—1, 2-di-acyl-*sn*-glycerol (MfGL-II) (Zahringer *et al.*, 1997).

During the fusion process, mycoplasma components are delivered into the host cell, and affect the normal functions of the cell. A whole array of potent hydrolytic enzymes has been identified in mycoplasmas, including, phospholipases, proteases and nucleases. Recently, it has been shown that *M. fermentans* contains a potent phosphoprotein phosphatase (Shibata *et al.*, 1994; Borovsky and Rottem, unpublished data). Phosphorylation of cellular constituents by interacting cascades of serine/threonine and tyrosine protein kinases and phosphatases is a major means by which a eukaryotic cells responds to exogenous stimuli (Taylor-Robinson *et al.*, 1991). The delivery of an active phosphoprotein phosphatase into the eurkaryotic cell upon fusion may interfere with the normal signal transduction cascade of the host cell. In addition to delivery of the mycoplasmal cell content into the host cell, fusion also allows insertion of mycoplasmal membrane components into the membrane of the eukaryotic host cell. This could alter receptor recognition sites, as well as affect the induction and expression of cytokines and alter the cross-talk between the various cells in an infected tissue.

4. INVASION OF HOST CELLS

Although it is believed that mycoplasmas remain attached to the surface of epithelial cells, recent reports show that several *Mycoplasma* species can survive within non-phagocytic cells (Lo, 1992; Andreev *et al.*, 1995; Taylor-Robinson *et al.*, 1991). The ability of *Mycoplasma penetrans*, recently isolated from the urogenital tract of AIDS patients (Lo, 1992), to penetrate and live within host cells has been intensively studied. This microorganism has invasive properties and localizes in the cytoplasm and perinuclear regions (Giron *et al.*, 1996; Andreev *et al.*, 1995).

In studying bacterial invasion it is essential to differentiate between microorganisms adhering to a host cell and those which have been internalized. The gentamicin protection assay has been extensively used to assess the number of intracellular bacteria (Elinghorst, 1994; Shaw and Falkow, 1988). In this assay, the extracellular bacteria are killed, while the intracellular bacteria are shielded from the antibiotic effect, due to limited penetration of gentamicin into eukcaryotic cells. *M. penetrans* cells are relatively stable against gentamicin, yet the susceptibility to the antibiotic can be increased by adding low concentrations of Triton X-100 to the medium. Thus, a combination of 200 μg gentamicin and 0.01% Triton X-100 resulted in an 8 log decrease in CFU within 1 h of incubation at 37°C (Andreev *et al.*, 1995). The low Triton X-100 concentrations affected neither the viability of the host cells nor their permeability to gentamicin. Low Triton X-100 concentrations have only a slight effect on the viability of *M. penetrans* or on the binding of *M. penetrans* to HeLa cells (Andreev *et al.*, 1995). Usually the number of intracellular bacteria is determined by washing the host cells free of the antibiotic, lysing them with mild detergents to release the bacteria and counting the colonies (Finlay and Falkow, 1988). As mycoplasmas are as susceptible to detergent lysis as the host cells, dilutions of the mycoplasma-infected host cells should be plated directly onto solid mycoplasma media without lysing them beforehand. Each mycoplasma colony thus obtained represents one infected host cell rather than a single intracellular mycoplasma (Elsinghorst, 1994).

Differential immunofluorescent staining of internalized bacteria and of those remaining on the cell surface, combined with confocal laser scanning microscopy, has been also used to demonstrate that *M. penetrans* is capable of penetrating eukaryotic cells (Borovsky *et al.*, 1998; Baseman *et al.*, 1995). This nondestructive, high-resolution method allowed infected host cells to be optically sectioned, following fixation and immunofluorescent labeling and localization of the mycoplasmas within the host cell. Single-cell imaging of infected HeLa cells revealed that invasion is both time- and temperature-dependent. Penetration of the HeLa cells has been observed as early as 20 min postinfection (Borovsky *et al.*, 1998), whereas invasion of cultured Hep-2 cells by *M. penetrans* has been shown to begin after 2 h of infection (Baseman *et al.*, 1995).

M. penetrans invasion of HeLa cells depends on the capacity of the cells to assemble actin microfilaments, as treatment with cytochalasin D inhibits invasion by *M. penetrans* (Andreev *et al.*, 1995). This conclusion is supported also by the results of confocal studies (Borovsky *et al.*, 1998). Furthermore, as taxol, a drug known to disorganize microtubules and vinblastine, which disrupts microtubules, virtually abolishes the penetration of *M. penetrans* (Borovsky *et al.*, 1998), it seems that alterations in the polymerization dynamics and stability of both the microfilaments and the microtubules have a dramatic effect on the invasion process. Drugs that disrupt microtubules may inhibit contractile intracellular processes such as the invagination and transport of membrane-bound bacteria from the plasma membrane into the cell, either by preventing their movement along the microtubules or by inhibiting the necessary actin-mediated contractile forces (Guzman *et al.*, 1994; Finlay *et al.*, 1991). Consistent with previous observations (Baseman *et al.*, 1995), adherence of *M. penetrans* to HeLa cells is not inhibited by cytochalasin D, vinblastine or taxol (Borovsky *et al.*, 1998).

It has been suggested that the invading mycoplasma generate uptake signals which cause the assembly of highly organized cytoskeletal structures composed of elements such as actin filaments, tubulin and α-actinin in the host cells (Giron *et al.*, 1996). Yet, the nature of these signals and the mechanisms used to transduce them are not fully understood. It has been shown that invasion of HeLa cells by *M. penetrans* is associated with tyrosine-phosphorylation of a 145 kDa host cell protein, and it is possible that this protein is a phospholipase which is activated by phosphorylation (Andreev *et al.*, 1995). More recently, changes in host cell lipid turnover as a result of *M. penetrans* binding and/or invasion of Molt-3 lymphocytes have been observed (Salman *et al.*, 1998). These changes included the accumulation of diacylglycerol (DAG), a well-established second messenger released by activated phospholipase C, and the release of unsaturated fatty acids, predominantly long chain polyunsaturated ones such as docosahexanoic acid ($C_{22:6}$, Salman *et al.*, 1998). These observations support the notion that *M. penetrans* stimulates host phospholipases to cleave membrane phospholipids, thereby initiating the signal transmission cascade. This, and possible other factors, then trigger cytoskeletal rearrangement, enabling the internalization of *M. penetrans*. The role of these signals in the penetration, survival and proliferation of the mycoplasma within host cells, as well as the involvement of the lipid intermediates in the pathobiological alterations taking place in the host cells, merit further investigation.

Bacterial invasion may lead to cytopathic effects (Kelly, 1990). In the case of *M. penetrans*, vacuolation of HeLa cells infected by *M. penetrans* has been demonstrated as early as 4 h postinfection. The vacuoles appeared to be empty, differing from the described membrane-bound vesicles containing clusters of bacteria (Lo *et al.*, 1993). The number and size of the vacuoles depended on duration of infection. As vacuolation is not obtained with *M. penetrans* cell fractions or with the growth medium remains after harvesting the microorganisms (Borovsky *et al.*, 1998), it is unlikely that a necrotizing cytotoxin is involved in the generation of the cellular lesions. A possible mechanism that might lead to vacuolation is the accumulation of organic peroxides upon invasion of HeLa cells by *M. penetrans*. Indeed, when HeLa cells were grown with the antioxidant α-tocopherol, the level of accumulated organic peroxides has been very low and vacuolation was almost completely abolished (Borovsky *et al.*, 1998).

5. COMPETITION FOR BIOSYNTHETIC PRECURSORS

Genetic analyses of *M. genitalium* and *M. pneumoniae* (Himmelreich *et al.*, 1996; Fraser *et al.*, 1995) have revealed the limited biosynthetic capabilities of these microorganisms (Himmelreich *et al.*, 1997). Genomic analysis has revealed that as they evolved, the mycoplasmas apparently lost almost all the genes involved in the biosynthesis of amino acids and most of the genes responsible for cofactor biosynthesis. They thus depend on the host microenvironment to supply the full spectrum of essential amino acids and vitamins (Himmelreich *et al.*, 1997). Mycoplasmas also require fatty acids for growth, and most of them have an absolute requirement for sterols (Pollack *et al.*, 1997). Competition for nutrients or biosynthetic precursors by mycoplasmas may disrupt host cell integrity and alter host cell function. The most pronounced effect on host cells is from the competition for arginine by non-fermenting mycoplasmas that utilize the arginine dihydrolase pathway for generating ATP (Pollack *et al.*, 1997). The arginine-utilizing mycoplasmas rapidly deplete the host's arginine reserves, thereby altering protein synthesis and affecting host cell division and growth (Rottem and Barile, 1993). Certain strains of arginine-utilizing *Mycoplasma* species were found to induce chromosomal aberrations in host cells, most commonly chromosomal breakage, multiple translocations, a reduction in chromosome number and the appearance of new and/or additional chromosome varieties (Rottem and Barile, 1993). As histones are rich in arginine, it has been suggested that arginine utilization by mycoplasmas inhibits histone synthesis and causes chromasomal damage (Rottem and Barile, 1993). Other mechanisms might also be involved including competition for nucleic acid precursors or degradation of host cell DNA by mycoplasma nucleases (Razin *et al.*, 1998).

6. MODULATING THE IMMUNE SYSTEM

It is increasingly recognized that for many mycoplasmas, induction of cytokines is a major virulence mechanism (Henderson *et al.*, 1996). The induced cytokines have a wide range of effects on the eukaryotic host cell and are recognized as important mediators of tissue pathology in infectious diseases (Henderson *et al.*, 1996). It appears that although mycoplasmas circumvent phagocytosis (Marshall *et al.*, 1995), they interact with mononuclear and polymorphonuclear phagocytes, suppressing or stimulating them by a combination of direct and indirect cytokine-mediated effects. These immunomodulatory influences depend on both the immune cells and the *Mycoplasma* species involved. Until fairly recently, the only bacterial component known to stimulate cytokine synthesis was LPS. Only in the past decade has it been demonstrated that other bacterial components, mainly those associated with the cell wall, such as peptidoglycan fragments, lipoteichoic acid and murein lipoproteins, have the capacity to stimulate mammalian cells to produce a diverse array of cytokines (Henderson *et al.*, 1996). With the exception of *Mycoplasma arthritidis*, a strain that synthesizes a well-characterized protein with superantigen properties (Cole *et al.*, 1996), the biochemical nature of mycoplasma-derived putative macrophage activators is not clear. Recent attempts to identify mycoplasmal cytokine-inducing moieties have revealed that membrane lipoproteins of several mycoplasmas are able to induce the synthesis of proinflammatory cytokines (Brenner *et al.*, 1996; Mühlradt and Frisch, 1994). Whereas in most eubacteria the number of lipoproteins is limited, lipoproteins are extremely abundant in mollicutes. In *M. pneumoniae*, for example, out of an estimated 150 membrane proteins, 46 open reading frames encoding putative lipoprotein genes have been identified (Himmelreich *et al.*, 1996). Yet, it is not certain whether all naturally occurring mollicute membrane lipoproteins are potent macrophage activators. Mühlradt *et al.* (1996) showed that a lipopeptide with an O-acylated S-(2,3-dihydroxypropyl) cysteine terminus is the active component in *M. fermentans* (Mühlradt *et al.*, 1996). Information on the functionally important lipopeptide moieties has been obtained by synthesizing and assaying various analogs. Thus, the presence of ester-bound fatty acids is a prerequisite for biological activity, whereas the amide-bound fatty acid has been found to be dispensable (Metzger *et al.*, 1995). It is likely that the extent of stimulation is also determined by the amino acid composition of the lipopeptide, although little information is available on the specific sequence required for optimal activity. Cytokine-inducing activity is also detected in the polar lipid fraction of *M. fermentans* (Salman *et al.*, 1994). The active lipid moiety has been identified as a choline-containing phosphoglycolipid and its structure was recently elucidated (Zahringer *et al.*, 1997). These observations suggest that a single mycoplasma species may possess several different active molecules.

7. CIRCUMVENTING THE HOST IMMUNE SYSTEM

On epithelial surfaces, the main antibacterial immune defense of the host is the protection afforded by secretory IgA. Once the epithelial surfaces have been penetrated, however, the major immune defenses of humeral and cellular immunity are encountered. To meet this challenge, microorganism populations must, on the whole, possess mechanisms and strategies allowing them to bypass or overcome the immune defenses of the host, which contribute to the virulence of the microbe and the pathology of the disease. Only microorganisms able to exhibit environmentally responsive and adaptive molecular traits, facilitating them, adherence and replication within the host, will survive. Thus, successful bacterial pathogens are those which have evolved molecular mechanisms to deal with the rigors of the host immune response and the need to be transferred and reestablished in a new host. Such mechanisms include mimicry of host antigens, survival within professional phagocytes, and generation of phenotypic plasticity.

Phenotypic plasticity has been defined as the ability of a single genotype to produce more than one alternative form of morphology, physiological state and/or behavior in response to environmental conditions. One of the most common means for achieving phenotypic plasticity is antigenic variation. The term "antigenic variation" or as it is also known, "phenotypic switching" refers to the ability of a microbial species to alter the antigenic character of its surface components, including flagella, pili, outer membrane proteins and capsules that enhance colonization of host tissues and evade phagocytosis. (Markham *et al.*, 1994; Dong *et al.*, 1992). These surface organelles are the major targets for the host antibody response. Therefore, the ability of a microorganism to rapidly change the surface antigenic repertoire, and consequently vary the immunogenicity of these structures, allows effective avoidance of immune recognition.

Surface antigenic variation, or phenotypic switching can be accomplished by two distinct means: microbial pathogens may use signal transduction pathways to sense signals in the host environment and respond accordingly by expressing virulence-gene products necessary for survival in the host (Robertson and Meyer, 1992). Alternatively, the microbial population may spontaneously and randomly generate distinct cell populations with a variety of antigenic phenotypes, "heterotypes", that will survive the specific host response capable of eliminating the predominant "homotypes". The frequency of occurrence of such genetic variants is strikingly high (10^{-4}–10^{-2} per cell per generation, compared with 10^{-6}–10^{-8} for a normally occurring mutation). Thus, the presence of a large repertoire of genetic variants may provide the pathogen with the desired escape variant needed for survival in case of a sudden environmemetal change, or when confronting the host response. Notably, the molecular switching events leading to the generation of these heterotypes are reversible, and the escape variants produced through random genetic variation must inherit the ability to produce, at a high frequency, a wide range of antigenic phenotypes. A considerable evolutionary dividend to the microbial pathogen of such random phenotypic switching can be achieved even before the onset of a specific immune response. For example, by "fine tuning" of the specificities of variant receptors or adhesion factors throughout the cell population, there is a better chance that a given variant will succeed in finding the preferred receptors on the mosaic of different tissues displayed by the host. It may also provide the pathogen, during the course of infection, the flexibility to reach and adapt to different niches within the host where distinctive receptors may be required for colonization.

7.1. The Use of Large Gene Families for the Generation of Surface Diversity

Despite the very limited genetic information that mycoplasma contain, the number of genes in mycoplasmas involved in diversifying the antigenic nature of their cell surface is unexpectedly high. The utilization of multiple variable genes organized as gene families, allowing the generation of an extensive repertoire of antigenic variants is a common theme in pathogenic bacteria and parasites for maintaining surface variability (Robertson and Meyer, 1992; Swanson *et al.*, 1992). By oscillating at a high frequency, these genes allow numerous combinatorial antigenic repertoires to be generated.

7.2. The pMGA Family of *Mycoplasma gallisepticum*

The most remarkable example, in terms of the amount of genetic material used for antigenic variation, is found in the avian pathogen *Mycoplasma gallisepticum.* This species possesses a large family of related genes, designated the pMGA gene family, encoding variant copies of a major surface lipoprotein (Baseggio *et al.*, 1996; Glew *et al.*, 1995; Markham *et al.*, 1994). The estimated number of pMGA gene copies in the *M. gallisepticum* genome varies from 32 in strain F to 70 in strain R. Assuming that all members of the pMGA family are similar in length to those which have already been characterized, a minimum of 79 kb or 7.7% of the genome of strain F and 168 kb or 16% of the genome of strain R are dedicated exclusively to the generation of variants of the same surface antigen (Baseggio *et al.*, 1996). Moreover, it has been shown that despite the presence of multiple copies of the pMGA gene, individual genes are expressed one at a time in a given strain (Glew *et al.*, 1995). These striking findings place the pMGA gene family of *M. gallisepticum* as probably the largest known family of translatable mollicute genes.

7.3. The Vsp Family of *Mycoplasma bovis*

M. bovis, an important pathogen of cattle, presents another interesting example. A family of genes encoding membrane surface lipoproteins and undergoing noncoordinate high- frequency phase variation, as well as size-variation (designated Vsps), is utilized to achieve extensive antigenic variation (Lysnyansky *et al.*, 1996; Behrens *et al.*, 1994; Rosengarten *et al.*, 1994; Lysnyansky and Yogev, unpublished data). Thirteen Vsp open reading frames, not all similarly oriented, were identified within a 16 kb genomic fragment carrying the *vsp* locus. These ORFs predicted features typical of the authentic Vsp products characterized so far. Several striking aspects of Vsp structural similarity, sequence divergence and variability are associated with specific regions of these proteins (Figure 1). Each *vsp* gene is flanked at its 5′ end by a highly conserved 190 bp non-coding region containing a putative ribosome binding site. The first 29 amino acids represent also a highly homologous N-terminal portion of the Vsp proteins which contains a positively-charged amino terminal region and a central hydrophobic region. The N-terminal portion ends in a cysteine residue at the predicted acylation site and point of membrane anchorage of a mature processed procaryote lipoprotein (Sutcliffe and Russell, 1995). A small block of eight amino acids localized immediately after the cysteine residue is present in all *vsp* genes. Examination of the deduced amino acid Vsp sequences has revealed an unusual structure. About 80% of the Vsp molecules are

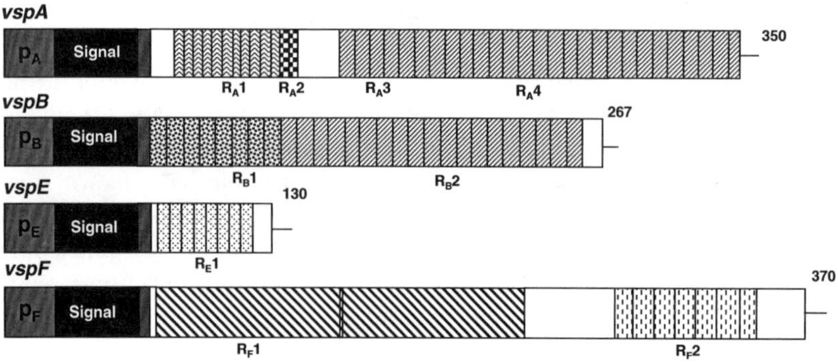

FIGURE 1. Schematic representation structural features and comparison of four representative *Vsp* proteins (VspA, VspB, VspE and VspF) of *M. bovis*. Vsp ORFs are shown as a rectangle consisting of internal blocks delineating various features of the Vsp proteins. The solid block, labeled signal, contains 25 amino acids of a putative lipoprotein signal peptide. A gray box, labeled P, represents a highly conserved promoter region common to all *vsp* genes. Different hatched blocks represent in-frame repetitive regions encoding distinctive periodic amino acid sequences.

composed of reiterated sequences of different length, extending from the N-terminus to the C-terminus of the Vsp proteins. Between one to four distinct internal regions of repetitive sequences, usually organized as tandem in-frame blocks, have been identified. These regions create a periodic polypeptide structure spanning the entire length of the Vsp molecules (Figure 1).

Actually, the potential of *M. bovis* to produce a wide spectrum of Vsp-antigenic phenotypes is more complex than that described above. Genetic analysis of the *vsp* gene family in a few *M. bovis* field- and clinical-isolates revealed that different *M. bovis* strains possess a modified *vsp* gene complex (Kotzer *et al.*, unpublished data). In these strains, major and extensive sequence alterations have occurred throughout and in all the *vsp* structural genes. These changes were localized mainly in *vsp* gene regions encoding reiterated sequences of different length, extending from the N-terminus to the C-terminus portion of the Vsp proteins. Notably, the 5′-*vsp* promotor region and the N-terminal part encoding the lipoprotein signal peptide have remained unaltered and are highly conserved in all *vsp* genes. The finding that *M. bovis* isolates possess different versions of the same *vsp* gene family, thereby leading to an amplified array of phenotypic variants, demonstrates a remarkable and efficient way by which mycoplasmas utilize their limited genetic material to increase their adaptive capability.

7.4. The Vlp Family of *Mycoplasma hyorhinis*

One of the well documented examples of a gene family providing an impressive surface variation system is the *vlp*-gene family of the swine pathogen *M. hyorhinis.* This system encodes a set of variable lipoproteins (Vlp) that constitute the major coat protein of this mycoplasma (Rosengarten and Wise, 1991, 1990). By combinatorial expression and high-frequency phase variation and size variation of the Vlps, an extensive array of antigenic variants can be generated. These lipoproteins are products of multiple and related but divergent *vlp* genes, which occur as single chromosomal gene copies organized in a cluster (Yogev *et al.*, 1991). Structurally, the *vlp* genes are divided into four domains: (i) a highly conserved promotor region; (ii) a highly homologous N-terminal region containing a typical prokaryotic signal peptide sequence, consistent with a prokaryotic lipoprotein signal peptidase recognition site (Sutcliffe and Russell, 1995); (iii) a region of considerable sequence divergence that contains several short blocks of homologous amino acid sequences recurring at variable locations, within or among different Vlps; and (iv) an external C-terminus domain containing reiterated sequences in the form of tandem, in-frame units encoding 12–13 amino acids, which undergoes size variation by loss or gain of these repetitive intragenic elements. Interestingly, size variants of a particular Vlp were associated with a characteristic degree of colony opacity, serving as a useful marker in their isolation (Rosengarten and Wise, 1990). Each *vlp* gene is subject to noncoordinate phase, as well as size-variation. The finding that different *M. hyorhinis* strains carry variable numbers of *vlp* genes (Yogev *et al.*, 1995) provides additional evidence for the vastly expanded potential of structural diversity spawned modulation of the *vlp* repertoire.

The ability to modulate the repertoire of the *vsp* and the *vlp* genes affords another important dimension of antigenic variation capability in mycoplasmas. The extensive phenotypic switching observed in mycoplasmas is achieved not only by the cluster of genes undergoing ON/OFF switching, but also as a result of an expanded reservoir of genes present in different strains. It is intriguing to speculate that gene clusters, such as the *vlp* or the *vsp* gene family, may be in dynamic flux among propagating individual cells of a certain species, leading to the exchange of genetic material and giving rise to alternative coding sequences.

7.5. The Vsa Family of *Mycoplasma pulmonis*

Another gene family mediating antigenic variation and displaying features characteristic of the *vlp* and the *vsp* gene systems is that of the murine pathogen *M. pulmonis* (Simmons *et al.*, 1996; Bhugra *et al.*, 1995). *M. pulmonis* possesses a genomic region, designated the *vsa* locus, containing multiple *vsa* genes and encoding a variable protein (V-1), that has been implicated as one of the virulence factors in *M. pulmonis*-induced murine respiratory disease. V-1 antigens undergo high-frequency phase switching, and selective patterns of V-1 expression have been correlated with variation in colony opacity. Most of the *vsa* locus from *M. pulmonis* UAB 6510 has been cloned, characterized, and shown to contain at least seven distinct *vsa* genes. However, only one *vsa* gene is expressed, while the other *vsa* genes are transcriptionally silent (Bhugra *et al.*, 1995). Structurally, the expressed *vsa* gene displays features similar to those of the *vlp* and *vsp* genes (Lysnyansky *et al.*, 1996; Yogev *et al.*, 1991). The expressed *vsa* gene is divided into three regions: (i) encoding a putative lipoprotein signal sequence; (ii) the region encoding the conserved N-terminus of the mature Vsa protein; and (iii) a variable region containing the 3′ repetitive elements (Simmons *et al.*, 1996; Bhugra *et al.*, 1995). In contrast, the silent *vsa* genes are truncated and lack regions (i) and (ii).

Comparative analysis of the gene structure of the mycoplasmal gene families responsible for antigenic variation demonstrates remarkable conservation of the 5′ flanking regions and of the N-terminal portion encoding the lipoprotein signal peptide. In contrast, other parts of the structural genes exhibit considerable sequence divergence. Conservation of the 5′ region may offer compelling evidence for the conceit that during evolution, mycoplasmas marshaled their antigenic variation through mechanisms of gene duplication, while the distinctive host selective pressure could be the driving force for the variable nature of the external repetitive domains. In summary, a common theme in generating surface diversity in mycoplasmas is based on the utilization of a cluster of several related but divergent variable lipoprotein genes. The genes encode a conserved N-terminal region and divergent external domains comprising reiterated sequences undergoing contraction or expansion, generating size variants (Figure 1).

7.6. Genetic Mechanisms Generating Mycoplasma Antigenic Variation

The apparent scarcity in mycoplasmas of regulatory genes functioning as sensors to environmental stimuli and of genes encoding transcriptional factors suggests, but does not rule out, that adaptation of mycoplasmas to the changing environment is not *per se* a response to signals. In other words, the major survival strategy of the mycoplasmas seems to depend on random and stochastic processes, comprising various mutational mechanisms that give rue to high-frequency phenotypic switching (Rainey *et al.*, 1993; Wise, 1993; Robertson and Meyer, 1992; Wise and Rosengarten, 1992; DiRita and

Mekalanos, 1989). Genetic mechanisms of antigenic variation emerging from the mycoplasma studies can be broadly divided into three categories: (i) variation by homopolymeric repeats; (ii) variation by chromosomal rearrangements; (iii) variation by reiterated coding sequence domains.

7.6.1. Variation by Homopolymeric Repeats

The essence of this type of variation is the presence of small regions containing reiterated bases (homopolymeric repeats) or oligonucleotide repeats. These "hot spots" provide favorite targets for frequent insertion or deletions of nucleotides and are quite commonly used by several pathogenic bacteria to switch genes ON and OFF (Rainey *et al.*, 1993; Robertson and Meyer, 1992). Loss or gain of nucleotides is thought to occur due to transient misalignment during DNA replication by a process termed slipped-strand mispairing (Levinson and Gutman, 1987). The location of these homopolymeric repeats within the regulatory region, or within the structural coding region of the corresponding variable genes, determines the level of regulation.

The conserved promoter of the *vlp* genes in *M. hyorhinis* contains a tract of contiguous adenine residues (poly-A region) immediately upstream of the TATAAT box and downstream of a –35 site, or of a two directly repeated structure (DR-1) that may accommodate a regulatory DNA binding protein (Yogev *et al.*, 1991). This homopolymeric tract is subject to frequent mutation, altering its length in exact correspondence with the ON and OFF expression states of individual *vlp* genes (Figure 2). These mutations, affecting the length of the poly-A, from 17 and 18 residues in the ON state to longer stretches of up to 20A in the OFF state, is the only sequence change detected during phase transition, and correlates well with the expression state of multiple *vlp* genes. The frequent mutations apparently affect the spacing, or secondary structure, between the -10 site and the –35 box or the DR-1 structure, thus influencing the optimal positioning of RNA polymerase in relation to the promoter. These findings strongly argue in favor of the control of Vlp phase variation, is at least in part, at the transcriptional level (Citti and Wise, 1995; Yogev *et al.*, 1991).

In contrast to the *M. hyorhinis vlp* gene family, the variable adherence-associated (Vaa) antigen of *Mycoplasma hominis* is controlled at the translational level. The Vaa antigen, which is an abundant surface lipoprotein adhesin that apparently mediates interactions of this mycoplasma with its human host, is subject to high-frequency phase variation in expression, precisely correlating with the ability of *M. hominis* to adhere to cultured human cells. It has been shown that an oscillating mutation involving a single nucleotide deletion or insertion in a short homopolymeric tract of adenine residues correlates with the Vaa expression state (Zhang and Wise, 1997). The poly-A tract has been localized near the 5′ end of the mature Vaa coding sequences, creating a translational frameshift that results in either complete Vaa ORF or an in-frame UAG stop codon immediately downstream of the poly-A tract (Figure 2).

Similar translational control of variable surface lipoproteins has been demonstrated in the human pathogen, *M. fermentans* (Figure 2). A putative ABC type transport operon encoding four gene products has been recently identified in this mycoplasma (Theiss and Wise, 1997). The 3′ distal gene encoding P78, a known surface-exposed antigen undergoing high-frequency phenotypic switching, is the proposed substrate-binding lipoprotein of the ABC transporter. The P78 gene is subject to localized hypermutation in a short homopolymeric tract of adenine residues located at the N-terminal coding region of the mature product. Reversible, high-frequency insertion/deletion frameshift mutations lead to selective phase variation in P78 expression, whereas the P63 protein, encoded by the gene located at the 5′ end of the operon, is continually expressed.

A hot spot for frequent mutations within a structural gene has been also found in the major cytadhesin gene (P1) of *M. pneumoniae.* Reversible and spontaneous cytaderence- negative and cytadherence-positive mutants were associated with the insertion or deletion of a single nucleotide in a strech of seven adenines. These mutations resulted in a frameshift and the generation of a termination codon causing the premature termination of P1 translation (Su *et al.*, 1989).

An interesting issue, not yet fully understood, concerns the molecular mechanism mediating high-frequency switching of the major surface lipoprotein pMGA of *M. gallisepticum*. A distinct hot spot region in the form of a trinucleotide GAA repeat motif, $(GAA)_{n,}$ where n represents the number of GAA repeats, has been identified in the 5′ region of all members of the *pMGA* gene family, 21 bases upstream from the –35 box of the promotor (Figure 2). The number of GAA repeats range from 10–16 (Baseggio *et al.*, 1996). Genetic analysis of phenotypically switched clonal isolates, representing ON or OFF expression states of the pMGA product from a number of *pMGA* genes in *M. gallisepticum* strain R, revealed that *pMGA* genes always contain a $(GAA)_{12}$ motif. Different numbers of the (GAA) repeats were correlated with the OFF expression state of *pMGA* (Glew, 1997). How does the $(GAA)_{12}$ motif function to initiate *pMGA* transcription? According to an intriguing model proposed by Glew (1997), the $(GAA)_{12}$ motif serves as a linker or a spacer for two flanking binding sites necessary for the initiation of *pMGA* transcription. The first binding site, localized downstream of the $(GAA)_{12}$ motif, is a well-defined promotor site for RNA polymerase. The second site, upstream of the $(GAA)_{12}$ motif,

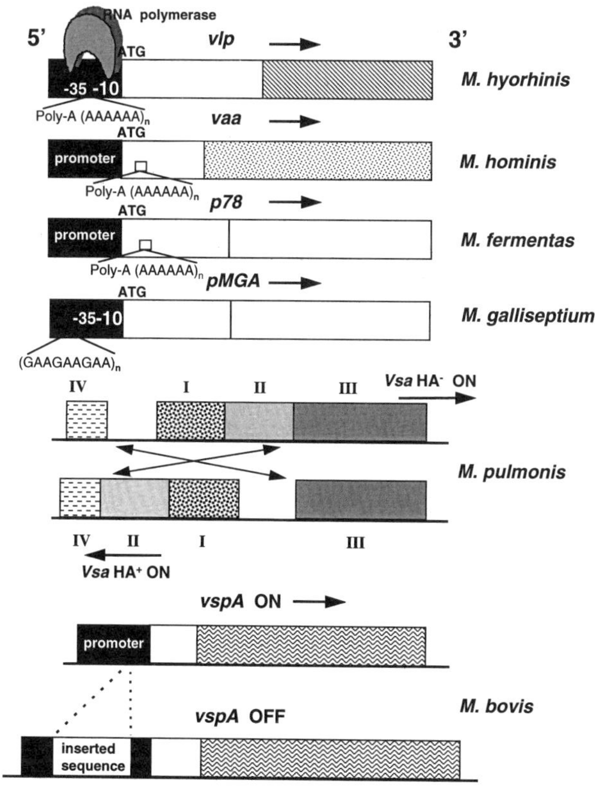

FIGURE 2. Schematic representation of regulatory and structural features of antigenic variation systems in mycoplasmas. Representative genes of distinct antigenic variation systems of six mycoplasma species are shown (not on scale) as rectangles consisting of internal blocks aligned from the 5′- end to the 3′- end of each gene. System designation and the corresponding mycoplasma species are indicated above the corresponding gene and on the right, respectively. The solid black block, labeled promoter, represents the 5′ upstream region of each gene. Different hatched blocks represent system-specific in-frame reiterated sequences. The location of a homopolymeric tract of contiguous adenine (Poly-A) or of oligonucleotide repeats $(GAA)_n$, within the promoter region (*vlp*, *pMGA*) or within the coding gene region (*vaa*, *P78*) is shown. Two *vsp* genes from two *M. bovis* clonal isolates, exhibiting ON or OFF expression states of the variable surface lipoprotein VspA, are presented. An insertion element that was inserted within the *vspA* promoter region, leading to its OFF expression state, is shown by a box labelled inserted sequence. Two *vsa* genes isolated from two *M. pulmonis* variants displaying the vsaHA– or the vsaHA+ phenotypes are shown. A chromosomal fragment that inverted during phase transition (vsaHA– vsaHA+) is indicated by arrows. The direction of expression of the *vsa* gene from each variant is marked by an arrow.

contains a region of sequence homology present in all known *pMGA* genes. These two binding sites presumably allow initiation of transcription only when a putative activator protein is in the correct spatial alignment with the RNA polymerase bound to the *pMGA* promoter region. The correct alignment in this case would correspond to the $(GAA)_{12}$ repeat acting as a linker between the two binding sites. Interestingly, this model is analogous to activation of transcription by the cyclic-AMP receptor protein (CRP) for class I promoters in *E. coli*.

7.6.2. Variation by Chromosomal Rearrangements

DNA inversions, gene conversions, duplications or deletions of tandem homologous blocks of DNA, as well as the movement of transposable elements, are widely used to regulate expression of phase variable surface antigens in bacteria. Most of these chromosomal rearrangements are usually considered random or spontaneous. Homologous recombination is a major mechanism allowing genetic variation, depending on RecA function, which promotes the annealing of single-stranded DNA to any complementary sequence in double-stranded DNA, with as little as 20–100 bp homology (Dybvig, 1993). Therefore, large gene families, such as those involved in generating surface diversity, which contain homologous sequences common to members of a particular gene family, are favorite targets for recombinatorial events. As a result, an extraordinary collection of phenotypic variants can be produced. For example, homologous recombination within the *pil* gene family of *Neisseria gonorrhoeae* results in the production of over a million combinations of antigenically variant pili (Swanson *et al.*, 1992). It is, therefore, reasonable to expect, that variable mycoplasmal gene families, containing regions of significant sequence similarity, would be subject to recombinative events regulating the high-frequency ON/OFF switching. So far, however, regulation of variable genes through genomic rearrangements has been demonstrated in only two mycoplasma species. The study of chromosomal rearrangements associated with phenotypic switching requires the isolation of a lineage of clonal isolates representing successive generations and exhibiting ON/OFF switching of only the antigen of interest.

Phenotypic switching of the variable surface lipoprotein in *M. bovis* has been shown to involve chromosomal rearrangements (Lysnyansky *et al.*, 1996). The nature of these rearrangement events was revealed when a lineage of *M. bovis* PG45 clonal isolates, representing successive generations and exhibiting oscillating phase transitions of only VspA, *i.e.*, ON→OFF→ON, have been analyzed. The *vspA* gene, a representative variable surface lipoprotein of the Vsp family, has been found to be subject to high-frequency reversible DNA rearrangements during the oscillating phase transition of VspA (Lysnyansky *et al.*, 1996). Such genomic rearrangements resulted in the disappearance of a 1.5 kb *Hind*III genomic fragment carrying the vspA ON gene, and in the generation of a restriction fragment of 2.3 kb, present only in the clonal variant in which the *vsp*A gene was turned OFF. Sequence analysis of the 2.3 kb *Hind*III fragment carrying the *vspA* gene, in the OFF expression state, revealed the insertion of a fragment of about 800 bp, 70 bp upstream of the initiation codon of the *vspA* gene, giving rise to the observed size variation during *vspA* ON/OFF switching (Lysnyansky and Yogev, unpublished data). The deduced amino acid sequences of this insertion fragment suggests an uninterrupted ORF containing a region of reiterated sequences containing 10 repeats each encoding 11 aa. These repeats were not found in any member of the *vsp* gene family (Figure 2).

The precise mechanism of this unique transposition event is not clear as yet. However, the finding that this mobile genetic element exists as a single chromosomal copy and undergoes precise excision and insertion during ON/OFF switching of the *vspA* gene (Lysnyansky and Yogev, unpublished data), argues against a replicative mechanism of transposition. The position of the insertion site within the promotor region of the *vspA* gene also indicates that *vspA* ON/OFF switching is regulated at the transcriptional level. An exciting recent finding (Lysnyansky and Yogev, unpublished data) has localized adjacent to the *vsp* locus, a gene family displaying significant homology with several bacterial insertion sequence (IS) elements, including the mobile genetic element IS*30* of *E. coli* (61% amino acid homology), the transposase for insertion sequence element IS*4351* transposon Tn*4551* (58.6%), IS*1161* (57.9%) and IS*1086* (56%) (Dong *et al.*, 1992). The possibility that transposase enzymes or other regulatory proteins play a role in dictating the genomic rearrangement event responsible for the VspA phase transition, in which environmental signals are involved, is intriguing although not experimentally proven.

The second known example in which chromosomal rearrangements regulate the expression of variable surface proteins is that of the V-1 antigens of *M. pulmonis* (Simmons *et al.*, 1996; Bhugra *et al.*, 1995; Bhugra and Dybvig, 1992). Identification of *M. pulmonis* isolates exhibiting variations in V-1 expression is fairly simple, as expression of the V-1 antigen has been found to correlate with the ability of the mycoplasma colonies to hemadsorb (HA). Sequence analysis of the V-1 gene from two *M. pulmonis vsa* HA^+ and *vsa* HA^- phenotypes indicated that these two genotypes arose as a result of high-frequency chromosomal inversion (Figure 2). Silent *vsa* genes lack the 5′ end region containing the promoter and ribosome binding site present in the expressed gene. During DNA rearrangements, gene expression is regulated by reassorting the 5′ end region from an expressed gene with the 3′ end region from a previously silent gene. All *vsa* rearrangements identified so far are site-specific DNA inversions occurring between copies of a specific 34 bp sequence that is conserved among *vsa* genes (Simmons *et al.*, 1996; Bhugra *et al.*, 1995).

It is worthwhile to compare the frequencies of the rearrangements described in the mycoplasmas with those in other microbial genomes. The frequency of the *vspA* transposition mechanism, as calculated from the frequency of colonies oscillating between the ON and OFF expression state of the VspA, VspB and VspC lipoproteins, is 2×10^{-3} per cell per generation (Rosengarten *et al.*, 1994). A rate of 10^{-2}–10^{-3} has been calculated for the site-specific DNA inversion regulating the switch of the V-1 antigen in *M. pulmonis* (Simmons *et al.*, 1996; Bhugra *et al.*, 1995). Such high frequencies of genomic rearrangements place the mycoplasma chromosome among the most variable genomes known, and underscore the efficiency with which mycoplasmas utilize their limited genomic material.

7.6.3. Variation by Reiterated Coding Sequences Domains

Surface antigens with tandem repetitive domains are being increasingly identified as molecules involved in pathogen-host interactions in a variety of pathogenic microorganisms. The role of repetitive units within an antigen is directly associated with ligand binding involved in generating antigenic variation, and having an impact on protein conformation that may facilitate pathogen-host interaction (Kehoe, 1994; Dramsi *et al.*, 1993; Hollingshead *et al.*, 1987). The presence of multiple in-tandem repetitive domains within the antigen-encoding gene generates a highly mutable module, subject to frequent contraction or expansion of these repetitive intragenic coding sequences, and resulting in the expression of a distinct size variant of the corresponding protein. Molecular mechanisms capable of precisely deleting or inserting repetitive sequences, include homologous recombination and slipped-strand mispairing (Levinson and Gutman, 1987). It is, therefore, not surprising that the majority of mycoplasmal variable surface antigens identified so far fit this pattern, i.e., contain significant regions of reiterated sequences. These include: the Vlps of *M. hyorhinis* (Rosengarten and Wise, 1991, 1990), the Vsps of *M. bovis* (Lysnyansky *et al.*, 1996; Behrens *et al.*, 1994; Rosengarten *et al.*, 1994), the MB antigen of *Ureaplasma urealyticum* (Zheng *et al.*, 1995), the Vaa adhesin and the Lmp1 antigen of *M. hominis* (Zhang and Wise, 1996), the p30 and P1 of *M. pneumoniae* (Dallo *et al.*, 1990) and the V-1 antigen of *M. pulmonis* (Rainey *et al.*, 1993). Notably, in a few phase transitions of distinct Vlps in *M. hyorhinis* or Vsps in *M. bovis*, variation at the DNA level occurred without expression of the corresponding protein (Yogev *et al.*, 1991; Lysnyansky and

Yogev, unpublished data). This indicates that deletion or addition of repetitive coding sequences in mycoplasmas can be independent of phase variation. It has been also established that variation in the number of the reiterated sequences can occur in various locations within the coding region. This is clearly demonstrated in the Vsp system of *M. bovis*, where deletions or additions of repetitive sequences were identified not only within the C-terminal repetitive structures-R_A-4 (see Figure 2) but also within the N-terminal repetitive domain-R_A-1 (Rosengarten *et al.*, 1994; Lysnyansky and Yogev, unpublished data).

Perhaps the most intriguing recent finding associated with antigenic variation in general and size-variation in particular, in terms of its functional role in mycoplasmas, has been described in *M. hyorhinis* (Citti *et al.*, 1997). The Vlp system plays an important role in modulating susceptibility to mycoplasma growth inhibition by host antibodies, thus providing a mutational framework in which a propagating population can escape host antibody inhibitory activity. Variants expressing longer versions of Vlps were resistant to host antibodies, whereas variants expressing shorter allelic versions of each Vlp were susceptible. This study has also shown that the emergence of a prevalent, protective Vlp phenotype in antibody- selected populations is not due to expression of a particular Vlp, but rather results from optional mutational pathways leading to expression of any long Vlp product. Selection of mutational pathways appear to be determined by the most favorable mutation available to generate a long Vlp in a given genetic background.

8. PROSPECTS FOR FUTURE RESEARCH

Significant developments have occurred over the last decade that help us to understand the strategy employed by a mycoplasma while interacting with an host eukaryotic cell. The identification of membrane components that act as adhesins will direct future efforts towards understanding the molecular organization of adhesion associated proteins and to further identify mammalian membrane receptors for mycoplasmas and mycoplasma products. The finding that some mycoplasmas can reside intracellularly open up new horizons to study the role of mycoplasma and host surface molecules in invasion and the signaling mechanisms involved in this process. The fusion of mycoplasmas with eukaryotic host cells observed *in vitro* raises exciting questions of how microinjection of mycoplasma components into eukaryotic cells subvert and damage the host cells and what is the significance *in vivo* of this process. The observations of diverse interactions of mycoplasmas with host immune cells and the cytokines that they trigger will allow to characterize the cytokine inducing mycoplasma components, to identify the cells that secrete proinflammatory cytokines *in vivo* and to study the significance *in vivo*, of the different host cells responses observed *in vitro*. The role of homologous sequences and cross reactive epitopes that exist in mycoplasmas and their host cells in evasion of host defense mechanisms and/or induction of autoimmune manifestations needs to be further explored.

The discovery of genetic systems that enable the mycoplasma cell to rapidly change its antigenic characteristics has been one of the major developments in mycoplasma research over the past decade. It is now clear that these minute wall-less microorganisms possess an impressive capability of maintaining a surface architecture that is antigenically and functional versatile. These variable surface antigens undoubtedly contributes to the capability of the mycoplasmas to adapt to a large range of habitats and to cause diseases which are often chronic in nature. Future molecular characterization of genetic systems present in different mycoplasma species will provide an experimenal basis to explore genes potentially involved in immune evasion and in pathogen-host interaction during mycoplasma infection.

9. REFERENCES

Almagor, M., Kahane, I., Gilon, C., and Yatziv, S., 1986, Protective effects of the glutathione redox cycle and vitamin E on cultured fibroblasts infected by *Mycoplasma pneumoniae*, *Infect. Immun.* **52**:240–244.

Andreev J., Borovsky, Z., Rosenshine, I., and Rottem, S., 1995, Invasion of HeLa cells by-*Mycoplasma penetrans* and the induction of tyrosine phosphorylation of a 145 kDa host cell protein, *FEMS Lett.* **132**:189–194.

Baseggio, N., Glew, M.D., Markham, P.F., Whithear, K.G., and Browning, G.F., 1996, Size and genomic location of the *pMGA* multigene family of *Mycoplasma gallisepticum*, *Microbiology* **142**:1429–1435.

Baseman, J.B., 1993, The cytadhesins of *Mycoplasma pneumoniae* and *Mycoplasma genitalium*, in *Subcellular Biochemistry*, Volume 20 (S. Rottem, and I. Kahane, eds), Plenum Press, New York, pp. 243–259.

Baseman, J.B., Lange, M., Criscimagna, N.L., Giron, J.A., and Thomas, C.A., 1995, Interplay between mycoplasmas and host target cells, *Microb. Pathog.* **19**:105–116.

Baseman, J.B., Reddy, S.P., and Dallo, S.F., 1996, Interplay between mycoplasma surface proteins, airway cells, and the protein manifestations of mycoplasma-mediated human infections, *Am. J. Resp. Crit. Care Med.* **154**:S137–S144.

Baseman, J.B., and Tully, J.G., 1997, Mycoplasmas: sophisticated, reemerging, and burdened by their notoriety, *Emerg. Infect. Dis.* **3**:21–32.

Behrens, A., Heller, M., Kirchhoff, H., Yogev, D., and Rosengarten, R., 1994, A family of phase- and size-variant membrane surface lipoprotein antigens (Vsps) of *Mycoplasma bovis*, *Infect. Immun.* **62**:5075–5084.

Bhugra, B., and Dybvig, K., 1992, High-frequency rearrangements in the chromosome of *Mycoplasma pulmonis* correlate with phenotypic switching, *Mol. Microbiol.* **6**:1149–1154.

Bhugra, B., Voelker, L.L., Zou, N., Yu, H., and Dybvig, K., 1995, Mechanism of antigenic variation in *Mycoplasma pulmonis*: interwoven, site-specific DNA inversions, *Mol. Microbiol.* **18**:703–714.

Borovsky, Z., Tarshis, M., Zhang, P., and Rottem, S., 1998, *Mycoplasma penetrans* invasion of HeLa cells induces protein kinase C activation and vacuolation in the host cells, *J. Med. Microbiol.* **47**:915–922.

Brenner, C., Neyrolles, O., and Blanchard, A., 1996, Mycoplasmas and HIV infection: From epidemiology to their interaction with immune cells, *Front. Biosci.* **1**:42–54.

Citti, C., and Wise, K.S., 1995, *Mycoplasma hyorhinis vlp* gene transcription: critical role in phase variation and expression of surface lipoproteins, *Mol. Microbiol.* **18**:649–660.

Citti, C., Kim, M.F., and Wise, K.S., 1997, Elongated versions of Vlp surface lipoproteins protect *Mycoplasma hyorhinis* escape variants from growth-inhibiting host antibodies, *Infect. Immun.* **65**:1773–1785.

Citovsky, V., Rottem, S., Nussbaum, O., Rott, R., and Loyter, A., 1987, Animal viruses are able to fuse with prokaryotic cells: fusion between Sendai or Influenza virions and *Mycoplasma*, *J. Biol. Chem.* **263**:461–468.

Cole, B.C., Knudtson, K.L., Oliphant, A., Sawitzke, A.D., Pole, A., Manohar, M., Benson, L.S., Ahmed, E., and Atkin, C.L., 1996, The sequence of the *Mycoplasma arthritidis* superantigen, MAM: identification of functional domains and comparison with microbial superantigens and plant lectin mitogens, *J. Exp. Med.* **183**:1105–1110.

Dallo, S.F., Chavoya, A., and Baseman, J.B., 1990, Characterization of the gene for a 30kDa adhesin-related protein of *Mycoplasma pneumoniae, Infect. Immun.* **58**:4163–4165.

DeBey, M.C., and Ross R.F., 1994, Ciliostasis and loss of cilia induced by *Mycoplasma hyopneumoniae* in porcine tracheal organ cultures, *Infect. Immun.* **62**:5312–5318.

DiRita, V.J., and Mekalanos, J.J., 1989, Genetic regulation of bacterial virulence, *Annu. Rev. Genet.* **23**:455–482.

Dong, Q., Sadouk, A., van der Lelie, D., Taghavi, A., Ferhat, A., Mutten, A., Borremans, B., Mergeay, M., and Toussaint, A., 1992, Cloning and sequencing of IS*1086*, an *Alcaligenes eutrophus* insertion element related to IS*30* and IS*4351*, *J. Bacteriol.* **174**:8133–8138.

Dramsi, S., Dehoux, P., and Cossart, P., 1993, Common features of gram-positive bacterial proteins involved in cell recognition, *Mol. Microbiol.* **9**:1119–1121.

Dybvig, K., 1993, DNA rearrangements and phenotypic switching in prokaryotes, *Mol. Microbiol.* **10**:465–471.

Elsinghorst, E.A., 1994, Measurement of invasion by gentamicin resistance, *Methods Enzymol.* **236**:405–420.

Finlay, B.B., and Falkow, S., 1988, Comparison of the invasion strategies used by *Salmonella choleraesuis. Shigella flexneri* and *Yersina enterocolitica* to enter cultured animal cells: endosomic acidification is not required for bacterial invasion or intracellular replication, *Biochimie* **80**:248–254.

Finlay, B.B., Ruschkowski, S., and Dedhar, S., 1991, Cytoskeletal rearrangements accompanying *Salmonella* entry into epithelial cells, *J. Cell Sci.* **99**:283–296.

Franzoso, G., Dimitrov, D.S., Blumenthal, R., Barile, M.F., and Rottem, S., 1992, Fusion of *M. fermentans*, strain incognitus, with T-lymphocytes, *FEBS Lett.* **303**:251–254.

Franzoso, G., Hu, P.-C., Meloni, G., and Barile, M.R., 1994, Immunoblot analysis of chimpanzee sera after infection and after immunization and challenge with *Mycoplasma pneumoniae*, *Infect. Immun.* **62**:1008–1014.

Fraser, C.M., Gocayne, J.D., White, O., Adams, M.D., Clayton, R.A., Fleischmann, R.D., Bult, C.J., Kerlavage, A.R., Sutton, G., Kelly, J.M., Fritchman, J.L., Weidman, J.F., Small, K.V., Sandusky, M., Fuhrmann, J., Nguyen, D., Utterback, T.R., Saudek, D.M., Phillips, C.A., Merrick, J.M., Tomb, J.-F., Dougherty, B.A., Bott, K.F., Hu, P.-C., Lucier, T.S., Petterson, S.N., Smith, H.O., Hutchison, C.A. III, and Venter, J.C., 1995, The minimal gene complement of *Mycoplasma genitalium*, *Science* **270**:397–403.

Geary, S.J., and Gabridge, M.G., 1987, Characterization of a human lung fibroblast receptor site for *Mycoplasma pneumoniae*, *Isr. J. Med. Sci.* **23**:462–468.

Geary, S.J., Gabridge, M.G., Intres, R., D.L., Draper, R., and Gladd, M.F., 1990, Identification of mycoplasma binding proteins utilizing a 100 kilodalton lung fibroblast receptor, *J. Rec. Res.* **9**:465–478.

Girón, J.A., Lange, M., and Baseman, J.B., 1996, Adherence, fibronectin binding, and induction of cytoskeleto reorganization in cultured human cells by *Mycoplasma penetrans*, *Infect. Immun.* **64**:197–208.

Glew, M.D., Markham, P.F., Browning, G.F., and Walker, I.D., 1995, Expression studies on four members of the pMGA multigene family in *Mycoplasma gallisepticum* S6, *Microbiology* **141**:3005–3014.

Glew, M.D., 1997, Gene expression studies of pMGA, A surface protein of *Mycoplasma gallisepticum*, Ph. D. Thesis, The University of Melbourne.

Guzman, C.A., Rohde, M., and Timmis, K.N., 1994, Mechanisms involved in uptake of *Bordetella bronchiseptica* by mouse dendritic cells, *Infect. Immun.* **62**:5538–5544.

Hahn, T.-W., Willby, M.J., and Krause, D.C., 1998, HMW1 is required for cytadhesin P1 trafficking to the attachment organelle in *Mycoplasma pneumoniae*, *J. Bacteriol.* **180**:1270–1276.

Henderson, B., Poole, S., and Wilson, M., 1996, Bacterial modulins: a novel class of virulence factors which cause host tissue pathology by inducing cytokine synthesis, *Microbiol. Rev.* **60**:316–341.

Himmelreich, R., Hilbert, H., Plagens, H., Pirkl, E., Li, B.-C., and Herrmann, R., 1996, Complete sequence analysis of the genome of the bacterium *Mycoplasma pneumoniae*, *Nucleic Acids Res.* **24**:4420–4449.

Himmelreich, R., Plagens, H., Hilbert, H., Reiner, B., and Herrman, R., 1997, Comparative analysis of the genomes of the bacteria *Mycoplasma pneumoniae* and *Mycoplasma genitalium*, *Nucleic Acids Res.* **25**:701–712.

Hollingshead, S.K., Fischetti, V.A., and Scott, J.R., 1987, Size variation in group A streptococcal M protein is generated by homologous recombination between intragenic repeats, *Mol. Gen. Genet.* **207**:196–203.

Inamine, J.M., Denny, T.P., Loechel, S., Schaper, U., Huang, C.-H., Bott, K.F., and Hu, P.-C., 1988, Nucleotide sequence of the P1 attachment-protein gene of *Mycoplasma pneumoniae*, *Gene* **64**:217–229.

Kehoe, M.A., 1994, Cell-wall-associated proteins in gram-positive bacteria, in: *Bacterial cell wall* (J-M. Ghuysen, and R. Hakenbeck, eds.), Elsevier Biomedical Press, Amsterdam, pp. 217–261.

Kelly, R.B., 1990, Microtubules, membrane traffic, and cell organization, *Cell* **61**:5–7.

Krause, D.C., Leith, D.K., Wilson, R.M., and Baseman, J.B., 1982, Identification of *Mycoplasma pneumoniae* proteins associated with hemadsorption and virulence, *Infect. Immun.* **35**:809–817.

Krause, D.C., 1996, *Mycoplasma pneumoniae* cytadherence: unravelling the tie that binds, *Mol. Microbiol.* **20**:247–253.

Krivan, H.C., Olson, L.D., Barile, M.F., Ginsburg, V., and Roberts, D.D., 1989, Adhesion of *Mycoplasma pneumoniae* to sulfated glycolipids and inhibition by dextran sulfate, *J. Biol. Chem.* **264**:9283–9288.

Layh-Schmitt, G., and Herrmann, R., 1994, Spatial arrangement of gene products of the P1 operon in the membrane of *Mycoplasma pneumoniae*, *Infect. Immun.* **62**:974–979.

Levinson, G., and Gutman, G.A., 1987, Slipped-strand mispairing: A major mechanism for DNA sequence evolution, *Mol. Biol. Evol.* **4**:203–221.

Lysnyansky, I., Rosengarten, R., and Yogev, D., 1996, Phenotypic switching of variable surface lipoproteins in *Mycoplasma bovis* involves high-frequency chromosomal rearrangements, *J. Bacteriol.* **178**:5395–5401.

Lo, S.C., 1992, Mycoplasmas in AIDS, in *Mycoplasmas: molecular biology and pathogenesis* (J. Maniloff, R.N. McElhaney, L.R. Finch, and J.B. Baseman, eds.), American Society for Microbiology, Washington, D.C., pp. 525–548.

Lo, S.C., Hayes, M.M., and Kotani, H., Pierce, P.F., Wear, D.J., Newton, P.B., Tully, J.G., and Shih, J.W-K., 1993, Adhesion onto and invasion into mammalian cells by *Mycoplasma penetrans*: a newly isolated mycoplasma from patients with AIDS., *Mod. Pathol.* **6**:276–280.

Loveless, R.W., and Feizi, T., 1989, Sialo-oligosaccharide receptors for *Mycoplasma pneumoniae* and related oligosaccharides of poly-*N*-acetyllactosamine series are polarized at the cilia and apicalmicrovillar domains of the ciliated cells in human bronchial epithelium, *Infect. Immun.* **57**:1285–1289.

Markham, P.F., Glew, M.D., Sykes, J.E., Bowden, T.R., Pollocks, T.D., Browning, G.F., Whithear, K.G., and Walker, I.D., 1994, The organization of the multigene family which encodes the major cell surface protein, pMGA, of *Mycoplasma gallisepticum*, *FEBS Lett.* **352**:347–352.

Marshall, A.J., Miles, R.J., and Richards, L., 1995, The phagocytosis of mycoplasmas, *J. Med. Microbiol.* **43**:239–250.

Metzger, J.W., Beck-Sickinger, A.G., Loleit, M., Eckert, M., Bessler, W.G., Jung, G., 1995, Synthetic S-(2,3-dihydroxypropyl)-cysteinyl peptides derived from the N-terminus of the cytochrome subunit of the photoreaction centre of *Rhodopseudomonas viridis* enhance murine splenocyte proliferation, *J. Pept. Sci.* **3**:184–190.

Mühlradt, P.F., and Frisch, M., 1994, Purification and partial biochemical characterization of a *Mycoplasma fermentans* derived substance that activates macrophages to release nitric oxide, tumor necrosis factor, and interleukin-6, *Infect. Immun.* **62**:3801–3807.

Mühlradt, P.F., Meyer, H., and Jansen, R., 1996, Identification of S-(2,3-dihydroxypropyl) cystein in a macrophage-activating lipopeptide from *Mycoplasma fermentans*, *Biochemistry* **35**:7781–7786.

Pollack, J.D., Williams, M.V., and McElhaney, R.N., 1997, The comparative metabolism of the mollicutes (mycoplasmas): The utility for taxonomic classification and the relationship of putative gene annotation and phylogeny to enzymatic function, *Crit. Rev. Microbiol.* **23**:269–354.

Popham, P.L., Hahn, T.-W., Krebes, K.A., and Krause, D.C., 1997, Loss of HMW1 and HMW3 in noncytadhering mutants of *Mycoplasma pneumoniae* occurs posttranslationally, *Proc. Natl. Acad. Sci. USA* **94**:13979–13984.

Rainey, P.B., Moxon, E.R., and Thompson, I.P., 1993, Intraclonal polymorphism in bacteria, *Advan. Microb. Ecol.* **13**:263–300.

Razin, S., and Jacobs, E., 1992, Mycoplasma adhesion, *J. Gen. Microbiol.* **138**:407–422.

Razin, S., Yogev, D., and Naot, Y., 1998, Molecular biology and pathogenicity of mycoplasmas, *Microbiol. Rev.* **62**:1094–1156.

Roberts, D.D., Olson, L.D., Barile, M.F., Ginsburg, V., and Krivan, H.C., 1989, Sialic acid-dependent adhesion of *Mycoplasma pneumoniae* to purified glycoproteins, *J. Biol. Chem.* **264**:9289–9293.

Robertson, B.D., and Meyer, T.F., 1992, Genetic variation in pathogenic bacteria, *Trends Genet.* **8**:422–427.

Rosengarten, R., and Wise, K., 1990, Phenotypic switching in mycoplasmas: Phase variation of diverse surface lipoproteins, *Science* **247**:315–318.

Rosengarten, R., and Wise, K., 1991, The Vlp system of *Mycoplasma hyorhinis*: combinatorial expression of distinct size variant lipoproteins generating high-frequency surface antigenic variation, *J. Bacteriol.* **173**:4782–4793.

Rosengarten, R., Behrens, A., Stetefeld, A., Heller, M., Ahrens, M., Sachse, K., Yogev, D., and Kirchhoff, H., 1994, Antigen heterogeneity among isolates of *Mycoplasma bovis* is generated by high-frequency variation of diverse membrane surface proteins, *Infect. Immun.* **62**:5066–5074.

Rosenshine, I., and Finlay, B.B., 1993, Exploitation of host signal transduction pathways and cytoskeletal functions by invasive bacteria, *BioEssays* **15**:17–24.

Rottem, S., and Barile, M.F., 1993, Beware of mycoplasmas, *Trends Biotechnol.* **11**:143–151.

Rottem, S., and Naot, Y., 1998, Subversion and exploitation of host cells by mycoplasmas, *Trends Microbiol.* **6**:436–440.

Salman, M., Deutsch, I., Tarshis, M., Naot, Y., and Rottem, S., 1994, Membrane lipids of *Mycoplasma fermentans*, *FEMS Microbiol. Lett.* **123**:255–260.

Salman, M., and Rottem, S., 1995, The cell membrane of *Mycoplasma penetrans*: Lipid composition and phospholipase A_1 activity, *Biochim. Biophys. Acta* **1235**:369–377.

Salman, M., Borovsky, Z., and Rottem, S., 1998, *Mycoplasma penetrans* invasion of Molt-3 lymphocytes induces changes in the lipid composition of host cells, *Microbiology* **144**:3447–3454.

Shaw, J.H., and Falkow, S., 1988, Model for invasion of human tissue culture cells by *Neisseria ghonorrhoeae*, *Infect. Immun.* **56**:1625–1632.

Shibata, K.-I., Mamoru, N., Yoshihiko, S., and Tsuguo, W., 1994, Acid phosphatase purified from *Mycoplasma fermentans* has protein tyrosine phosphatase-like activity, *Infect. Immun.* **62**:313–315.

Shibata, K.-I., Sasaki, T., and Watanabe, T., 1995, AIDS-Associated mycoplasmas possess phospholipases C in the membrane, *Infect. Immun.* **63**:4174–4177.

Simmons, W.L., Zuhua, C., Glass, J.I., Simecke, J.W., Cassell, G.H., and Watson, H.L., 1996, Sequence analysis of the chromosomal region around and within the V-1-encoding gene of *Mycoplasma pulmonis*: Evidence for DNA inversion as a mechanis for V-1 variation, *Infect. Immun.* **64**:472–479.

Stevens, M.K., and Krause, D.C., 1991, Localization of the *Mycoplasma pneumoniae* cytadherence-accessory proteins HMW1 and HMW4 in the cytoskeleton like triton shell, *J. Bacteriol.* **173**:1041–1050.

Stevens, M.K., and Krause, D.C., 1992, *Mycoplasma pneumoniae* cytadherence phase-variable protein HMW3 is a component of the attachment organelle, *J. Bacteriol.* **174**:4265–4274.

Su, C.J., Chavoya, A., and Baseman, J.B., 1989, Spontaneous mutation results in loss of the cytadhesin (P1) of *Mycoplasma pneumoniae*, *Infect. Immun.* **57**:3237–3239.

Sutcliffe, I.C., and Russell, R.R., 1995, Lipoproteins of gram-positive bacteria, *J. Bacteriol.* **177**:1123–1128.

Swanson, J., Belland, R.J., and Hill., S.A., 1992, Neisserial surface variation: how and why?, *Curr. Opin. Genet. Develop.* **2**:805–811.

Tarshis, M., Salman, M., and Rottem, S., 1993, Cholesterol is required for the fusion of single unilamellar vesicles with *M. capricolum.*, *Biophys. J.* **64**:709–715.

Taylor-Robinson, D., Davies, H.A., Sarathchandra, P., and Furr, P.M., 1991, Intracellular location of mycoplasmas in cultured cells demonstrated by immunocytochemistry and electron microscopy, *Int. J. Exp. Pathol.* **72**:705–714.

Theiss, P., and Wise, K.S., 1997, Localized frameshift mutation generates selective, high-frequency phase variation of a surface lipoprotein encoded by a mycoplasma ABC transporter operon, *J. Bacteriol.* **179**:4013–4022.

Wise, K.S., Yogev, D., and Rosengarten, R., 1992, Antigenic variation, in: *Mycoplasmas: molecular biology and pathogenesis* (J. Maniloff, R.N. McElhaney, L.R. Finch, and J.B. Baseman, eds.), American Society for Microbiology, Washington, D.C., pp. 473–489.

Wise, K.S., 1993, Adaptive surface variation in mycoplasmas, *Trends. Microbiol.* **1**:59–63.

Yogev, D., Rosengarten, R., Watson-McKown, R., and Wise, K.S., 1991. Molecular basis of mycoplasma surface antigenic variation: a novel set of divergent genes undergo spontaneous mutation of periodic coding regions and 5′ regulatory sequences, *EMBO J.* **10**:4069–4079.

Yogev, D., Watson, M.R., Rosengarten, R., Im, J., and Wise, K.M., 1995, Increased structural and combinatorial diversity in an extended family of genes encoding Vlp surface proteins of *Mycoplasma hyorhinis, J. Bacteriol.* **177**:5636–5643.

Zähringer, U., Wagner, F., Rietschel, E.Th., Ben-Menachem, G., Deutsch, J., and Rottem, S., 1997, Primary structure of a new phosphocholine-containing glycoglycerolipid of *Mycoplasma fermentans, J. Biol. Chem.* **272**:26262–26270.

Zhang, Q., and Wise, K.S., 1996, Molecular basis of size and antigenic variation of a *Mycoplasma hominis* adhesin encoded by divergent *vaa* genes, *Infect. Immun.* **64**: 2737–2744.

Zhang, Q., and Wise, K.S., 1997, Localized reversible frameshift mutation in an adhesin gene confers a phase-variable adherence phenotype in mycoplasma, *Mol. Microbiol.* **25**:859–869.

Zheng, X., Teng, L.J., Watson, H.L., Glass, J.I., Blanchard, A., and Cassell, G.H., 1995, Small repeating units within the *Ureaplasma urealyticum* MB antigen gene encode serovar specificity and are associated with antigen size variation, *Infect. Immun.* **63**:891–898.

Part II

Professional Facultative Intracellular Bacteria

Chapter 10

Mycobacterial Invasion of Epithelial Cells

Luiz E. Bermudez* and Felix J. Sangari

1. INTRODUCTION

Infections caused by organisms of the genera mycobacteria are responsible for a great percentage of yearly deaths dues to infectious diseases. Tuberculosis is a widespread infection caused by *Mycobacterium tuberculosis*. It is transmitted mainly as an airborne infection. *M. bovis* is found in cattle and it is transmitted to humans via contaminated milk and dairy products. Infections caused by *Mycobacterium avium* complex can cause localized disease usually in patients with underlying lung disease as well as disseminated disease in individuals with immunosuppression or Acquired Immunodeficiency Syndrome (AIDS). According to the World Health Organization, infections caused by mycobacteria will rise to 12 million cases annually by the year 2000.

Our mistake has always been to seek to control infectious diseases based on the empirical discovery of novel drugs. While this approach may work initially, ultimately it will certainly fail due to the continuous evolu-

LUIZ E. BERMUDEZ and FELIX J. SANGARI Kuzell Institute for Arthritis and Infectious Diseases, California Pacific Medical Center Research Institute, San Francisco, CA 94115.
*To whom correspondence should be addressed.

Subcellular Biochemistry, Volume 33: Bacterial Invasion into Eukaryotic Cells, edited by Oelschlaeger and Hacker. Kluwer Academic / Plenum Publishers, New York, 2000.

tion of pathogens. Therefore, only understanding of the fundamental factors that microbes use to cause disease will lead to better means for control of infectious diseases.

Despite the clear importance of tuberculosis and disseminated infection by *M. avium* to public health, very little is known about the specific mechanisms by which those organisms infect the host. How the bacterium establishes the first contact with the host, and whether a minimal number of organisms is needed to initiate infection and disease. Assuming that studies carried out in experimental animal models are correct, just a few bacteria (both *M. tuberculosis* or *M. avium*) are necessary to cause disease.

For many years it was assumed that monocytes and tissue macrophages were the only cells that had significance for the pathogenesis of mycobacterial infection, since pathogenic mycobacteria as an intracellular pathogen interacts with mononuclear phagocytic cells and ultimately conquer this arm of the immune system. A number of reviews have been written addressing this interaction (Benson and Ellner 1993; Inderlied *et al.*, 1993; Horsburgh Jr, 1991).

The advent of AIDS brought our attention to the manner by which *M. avium* interacts with the host. As an environmental microorganism, present in natural as well as urban sources of water and soil, more often *M. avium* is ingested by the host as opposed to contact by aerosol with the respiratory tract. The bacterium has the ability to resist acidic conditions of the stomach although the mechanisms are not fully understood (Bodmer *et al.*, Submitted). Different from Salmonella or Shigella that require a pre-adaptation period to become resistant to acid, *M. avium* apparently encodes this property without the requrement of pre-adaptation. Current evidence suggests that exposure to hypoosmolar conditions increase the acid resistance indicating that life in water prior to ingestion by the human host is the ideal condition to resist the presence of acid in the stomach (Bodmer *et al.*, Submitted). Gastrointestinal colonization was then observed preceding the appearance of disseminated disease in some patients, although in a small percentage of individuals the source of *M. avium* infection could never be established (Torriani *et al.*, 1995; Jacobson *et al.*, 1991). It then led us to the study of the intestinal mucosa as the protective barrier against *M. avium* infection and the mechanisms used by the bacterium to overcome the intestinal barrier.

A similar rationale can be used regarding the entry of *M. tuberculosis* into the host. Alveolar epithelial cells outnumber macrophages in the alveolar space, and therefore by a question of chance are more likely to interact with an aerosolized inoculum of *M. tuberculosis* even before there is contact between bacteria and macrophages. Then, if one assumes this as a fact, questions such as what is the outcome when *M. tuberculosis* invades epithelial cells of the respiratory tract or are these cells a simple side trip for the organism without any importance or is the infection of respiratory tract epithelial cells a stage of *M. tuberculosis* pathogenesis? become very relevant for *M. tuberculosis* pathogenesis.

This chapter will review succinctly what is currently known about the mechanisms of interaction of both *M. avium* and *M. tuberculosis* with mucosal cells.

2. *Mycobacterium avium* INTERACTION WITH INTESTINAL EPITHELIAL CELLS

To cause infection through the gastrointestinal tract both binding and invasion of the intestinal epithelial lining are required. The intestinal epithelium consists mainly of enterocytes and M cells. Enterocytes, the most abundant cell type, present a differentiated apical surface to the lumen with a well developed brush border, and separated from the basolateral area by intercellular tight junctions. M cells are localized in Peyer's patches, contain few microvillae, and are specialized epithelial cells responsible for sampling the lumen, transporting antigens and organisms to the underlying macrophages and lymphocytes. A number of bacteria are known to use M cells for gaining access to the internal medium, including *M. bovis*, BCG, *Streptococcus pneumoniae*, *Salmonella typhimurium*, *Yersinia enterocolitica*, and *Shigella flexneri* (Jepson and Clark, 1998). In *M. avium*, using a mouse intestinal loop model of infection, as well as ingestion of bacteria orally, we have been able to see bacteria primarily invading enterocytes, but failed to show them invading M cells.

2.1. Uptake of *M. avium* by Polarized Intestinal Cells

Intestinal cells *in vivo* form a polarized monolayer with a gradient of differentiation from the cripts where new enterocytes are formed to the tip of the villi. Polarization is characterized by the presence of tight junctions that seal the internal medium from the exterior, separating the apical surface of the cell, in contact with the lumen, from the basolateral surface. Differentiation consist of a maturation process that results in both physiological and morphological changes such as the formation of an apical brush border.

To study the entry process of *M. avium* in an assay model that resembles the *in vivo* process, we used both HT-29 and Caco-2 cell lines, both of them able to form polarized cell monolayers when grown to confluence.

Different bacterial species interact with different domains of intestinal epithelial cells, and this interaction is believed to have profound implications on the pathogenesis of bacterial disease. For instance, Salmonella enters polarized Caco-2 cells through the apical surface, whereas *Shigella* spp. invade them through the basolateral surface (Finlay and Ruschkowski, 1991). In contrast, *Listeria monocytogenes* enters through the entire surface of nonpolarized Caco-2 cells but through the basolateral surface when they form polarized monolayers.

To evaluate the *M. avium* mechanism of entry, confluent monolayers were treated with Ca^{++}-free medium, supplemented or not with EGTA, a process that opens the intercellular junctions, allowing access of the bacteria to the basolateral surface of the cells. In both HT-29 and Caco-2 cell monolayers, the level of *M. avium* invasion was similar in all the treatment and control groups regardless of the level of exposure of the basolateral surface. Therefore, it can be assumed that *M. avium* is capable of entering intestinal epithelial cells by the apical surface but cannot enter through the basolateral membrane. The results were subsequently confirmed by using anti-*M. avium* antibody conjugated with FITC or Texas Red that labeled bacteria entering the cells by the apical surface which does not express E-cadherin (a marker to basolateral membrane) (Neutra *et al.*, 1996).

2.2. Communication between the Bacterium and the Mucosal Epithelial Cell

Invasion of mammalian cells by pathogens has been shown to be a process that depends on the participation of both the host and the pathogen (Hueck, 1998; Neutra *et al.*, 1996). Gram-negative bacteria encode a pathogenicity mechanism termed type III secretion system. Type III secretion enables gram-negative bacteria to secrete and inject pathogenicity-related proteins into the cytosol of eukaryotic host cells (reviewed in Hueck, 1998; Lee, 1977). The type III secretion apparatus is conserved in very distinct pathogens such as Yersinia and Erwinia, but similar machinery was not identifed in *M. avium* (Bermudez LE, Tan T, personal communication). Electron micrographs of intestinal mucosa of mice given *M. avium* orally demonstrates that the interaction between the bacteria and the mucosa involves cytoskeleton rearrangement (Hsu *et al.*, 1996). Interestingly, actin polymerization in the mammalian cell can be observed before contact with the bacterium, suggesting that somehow the bacterium is capable of triggering that specific response in the host cell prior to establishing contact. Incubation of *M. avium* with polarized intestinal epithelial cells *in vitro* is associated with the secretion of at least seven different proteins in the supernatant, which are not observed in supernatant of bacteria cultured alone or HT-29 cells alone (Tan *et al.*, Submitted). Release of these proteins can occur as early as within one hour of contact and increase significantly over time, up to four hours of exposure to intestinal epithelial cells. The sequence of the N-terminal amino acid of five purified proteins revealed novel proteins with sequences not present in the data base.

Supernatant of bacteria and HT-29 cells in contrast to supernatant of bacteria alone or supernatant of HT-29 cells alone can trigger cytoskeleton rearrangement in polarized HT-29 cells which supports the hypothesis that those proteins interact with intestinal epithelial cells triggering a response in the host cell that is probably necessary for *M. avium* internalization (Tan *et al.*, Submitted).

An effort to clone the genes encoding for these secreted proteins is currently underway.

2.3. Participation of Cytoskeleton in the Internalization Process

M. avium internalization by epithelial cells is an active process like the process of internalization of almost all bacteria, and requires the participation of the host's cell. As mentioned above, all the evidence indicates that *M. avium* uptake is preceded by cytoskeleton rearrangement in the mammalian cell. Microscopic studies staining the monolayer with phalloidin at several time points after adding *M. avium* confirmed that the bacterium's uptake by HEp-2 and HT-29 epithelial cells is accompanied by actin polymerization. Interestingly, actin bundles surrounding the bacterium is not seen 18 hours after uptake indicating that the process is only required for the initial interaction. The use of either cytochalasin B or cytochalasin D inhibit *M. avium* internalization in both HEp-2 and HT-29 cells, confirming that actin polymerization is a requirement of *M. avium* uptake (Bermudez and Goodman, Submitted; Bermudez and Young, 1994).

Other cytoskeleton proteins such as α-actinin are also involved in the internalization process. Use of anti- α-actinin antibodies clearly shows accumulation of the protein around *M. avium* following uptake and α-actinin is present in close proximity to the plasma membrane, forming a direct link between the cytoplasmic domain of the receptors and F-actin filaments (Bermudez and Goodman, Submitted). In contrast, talin and tubulin do not appear to participate in the process of *M. avium* internalization by intestinal epithelial cells, suggesting a specific pathway within HT-29 cells.

Cytoskeleton reorganization is observed upon entry of *M. avium* into the intestinal enterocytes and the cytoplasmic vacuole is surrounded by polymerized actin.

2.4. Recognition of Binding Sites

Once bacteria get to the intestinal lumen the difficulties faced by the microorganism enhance. Enteropathogens have fimbria which facilitate the binding and localization of the bacteria in specific sites on the mucosa.

A prerequisite for the development of any bacterial disease is localization of the bacteria to a niche that is suitable for growth. In a mammalian intestine attachment is critical to avoid displacement from the preferred site by the continuous flow of the intestinal contents. The initial step in bacterial attachment to the host epithelium is usually mediated by fimbria. Because *M. avium* does not have fimbria, the bacteria must rely on a mechanism that allows for binding on the mucus (glycocalix) and/or the microvilli. One consequence of this strategy is the impaired efficiency of the process and only a small percentage of the bacterial inoculum enters intestinal epithelial cells over time (Bermudez and Young, 1994).

We recently identified lectins present in the intestinal villi that are recognized by *M. avium*. Use of either wheat germ agglutinin or *Ulex europhaeus* agglutinin I is able to significantly block *M. avium* from binding to polarized intestinal epithelial cells (HT-29 cells) suggesting that *M. avium* can recognize sialic acid (Neu5Ac(GlcNAc)2-5 or Neu5Ac,Neu5Glc) and fucose (fuc α 1-2Gal) on the surface of the enteric cell. In fact, double staining techniques show that *M. avium* only binds to differentiated HT-29 or Caco2 cells on sites containing the two carbohydrates epitopes. Treatment of epithelial cells with neuramidase reduced binding and invasion of *M. avium* supporting the possibility that sialic acid residues are recognized by *M. avium*. These findings are in agreement with the hypothesis that *M.* avium needs to recognize antigens on intestinal villi to be able to anchor, and would certainly require further studies to determine the adhesin on the bacteria capable of recognizing glycoconjugates. In addition, very likely *M. avium* recognizes antigens lcoated on the tip of the brush border or the glycocalix. This hypothesis is further supported by studies using electron microscopy *in vivo* which have shown that *M. avium* like *Salmonella typhimurium* initially adheres to the tips of microvilli, and subsequently induces reorganization of the brush border membrane (with efficacing in the case of *M. avium*) and cytoskeleton. The epithelial cell's response might expose additional binding sites and make them accessible to the bacteria. It may explain why a subpopulation of intestinal mucosal cells is targeted by the bacteria (those cells that express the *M. avium* binding sites) while other cells are not.

2.5. Other Putative Adhesins

Invasion tactics probably ensure a protected cellular niche for the bacterium to replicate and persist. Cultured eukaryotic cell lines are commonly used to study bacterial invasion although few closely mimic the original site from where they were isolated. Generally, invasive organisms adhere to host cells by using a class of adhesin molecules. A small number of organisms, such as *Rickettsia prowazeckii*, appear to forcefully enter into the host cells by a local enzymatic digestion of the host cell membrane (Walker *et al.*, 1983). Following the initial contact of *M. avium* with the intestinal mucosal cell, the uptake process appears to be rapid. Usually about 15 to 30 minutes, after adding the bacterium to an intestinal cell monolayer, is sufficient for uptake to occur. *M. avium* entry into eukaryotic cells appears to be a very complex phenomenon requiring a number of bacterial and host factors. Currently, there is no information how the bacteria gain access to the intracellular environment, although two adhesins to epithelial cells have been identified. A 55 kDa protein, fibronectin attachment protein, has been identified and the gene encoding for the protein has been cloned and shown to be important for the uptake of *M. avium* by bladder epithelium cells *in vitro* (Schorey *et al.*, 1996b). The protein is supposed to bind fibronectin on the mucosa and uses it as a bridge to enter the cell. All studies were carried out, however, with soluble fibronectin and no data is available about the binding to insoluble (membrane) fibronectin. The same gene has been identified on *M. leprae* although *M. avium* and *M. leprae* do not seem to share the port of entry into the host (Schorey *et al.*, 1996a). At least one other fibronectin binding protein, 30–31 kDa, has been characterized in mycobacteria (Abou-Zeid *et al.*, 1988), but its role in invasion of mammalian cells has not been investigated. In addition, no explanation exists about why one and not the other fibronectin binding protein would bind to mucosal epithelial cells.

Another protein of 27 kDa has been identified and the gene cloned (Bermudez *et al.*, 1995). This protein is a *M. avium* adhesin that participates in the binding to epithelial cells. Subsequently, it was shown that latex beads covered by the *M. avium* 27 kDa protein bind to intestinal epithelial cells but are not internalized, indicating that this specific adhesin very likely is associated only with binding but not invasion.

Therefore, no definitive information exists at this time about adhesins and binding sites for *M. avium* on epithelial cells, but it is possible that *M. avium*, similar to other bacteria, utilize a number of pathways to enter cells.

2.6. Effect of Environmental Conditions on Binding and Uptake

Intestinal pathogens must adapt to a variety of adverse conditions even before getting near to an invasion site in the intestinal lumen. *M. avium* is an environmental organism encountered in natural sources of water as well as in urban water reservoirs (Falkinham III, 1996; Glover *et al.*, 1994). Therefore, the bacterium must be able to adapt to diverse environments.

When ingested by the host, the bacterium faces continuous changes, not only in temperature but also pH and osmolarity of the environment, added to the intense competition with the intestinal biota for nutrients. The adaptation process is certainly complex and requires rapid changes, although for several years it was assumed that mycobacteria were not able to synthesize proteins rapidly. More recent studies, however, have shown that the synthesis of total RNA in slow-growing mycobacteria can happen quickly, following a shift in environmental conditions (Bermudez *et al.*, 1997b). This observation can explain the findings that short exposure to environmental variables encountered in the intestinal lumen, such as hyperosmolarity and low tension of oxygen, increase significantly (by several fold) the efficiency with which *M. avium* can invade mucosal intestinal cells (Bermudez *et al.*, 1997b). Then this essential step in the pathogenesis of *M. avium* is regulated by the conditions found within the intestinal environment. In fact, the shift in protein synthesis upon exposure to different conditions that mimic the intestine can be abrogated by suppressing *de novo* synthesis of proteins with sub-inhibitory concentration of amikacin, an antibiotic (Bermudez *et al.*, 1997b).

2.7. Putative Participation of Environmental Amoeba in the Invasion Process

Environmental bacteria like *M. avium* are constantly exposed to other water organisms such as fresh water amoeba. In the past, a number of studies have shown that fresh water amoeba can be the host of *Legionella pneumophila*, a human pathogen, and that it has a significant impact on *Legionella's* pathogenicity (Cirillo *et al.*, 1994; Barker *et al.*, 1993), including invasion of macrophages by a different pathway than Legionella grown on media.

Recent studies have shown that the same is probably true for *M. avium* as well. *M. avium* can survive and replicate within fresh water amoeba independent of the temperature the amoeba is exposed to (Cirillo *et al.*, 1997). This characteristic of *M. avium* differs from Legionella which can only survive in amoeba when in an environment with high temperature. It has been observed that the amoeba environment triggers an invasive phenotype on *M. avium*, which can then enter intestinal epithelial cells both *in vitro* and *in vivo* with an increased efficiency (Cirillo *et al.*, 1997). The mechanism by which *M. avium* change phenotypically is currently unknown, but similar changes can be observed in *M. avium* phenotype when the bacterium infects macrophages (Bermudez *et al.*, 1997a).

The role of environmental amoeba in the epidemiology of *M. avium* disease has not been proved, but the observations are provocative enough to suggest a possible role for environmental amoeba in *M. avium* pathogenesis. Amoeba would not only protect the intracellular bacteria from the dangers of the environment, but also release it in the intestines as a well adapted bacterial strain capable of invading intestinal mucosal cells with increased efficiency.

2.8. Inhibition of Chemokine Release by Epithelial Cells

No longer considered as just a physical barrier between the external and internal media, epithelial cells of different origins are known to respond to invasive bacterial pathogens by secreting molecules known as chemokines. Chemokines initiate the mucosal inflammatory response in the earliest phases of microbial invasion, attracting and activating inflammatory cells in an effort to control the process. Secretion of IL-1, IL-6, IL-8, MCP-1, RANTES, GM-CSF and TNF-α have been demonstrated in a number of studies *in vitro* and *in vivo* following infection of mucosal epithelial cells *in vitro* with Salmonella, Shigella, Helicobacter, Yersinia or Listeria among other pathogenic microorganisms (Rasmussen *et al.*, 1997; Hedges *et al.*, 1995; Huang *et al.*, 1995; Jung *et al.*, 1995). As mentioned, in AIDS patients, in contrast to non-AIDS patients where the pulmonary infection is more frequent, the gastrointestinal tract appears to be the primary route of *M. avium* infection. *M. avium* is capable of invading intestinal and laryngeal epithelial cells *in vitro*, and the intestinal mucosa of healthy mice *in vivo*. In mice invasion is followed by little inflammatory response, with infilration of neutrophils into the Peyer's patches observed approximately one week after infection. This was the first indication that *M. avium* could induce a different response in epithelial cells than that observed with the majority of the enteric pathogens.

Subsequent assays *in* vitro using HT-29 (intestinal) or HEp-2 (laryngeal) cells showed that following infection with different strains of *M. avium*, no increase of IL-8 or RANTES was observed, compared to non-infected monolayers. In contrast, Salmonella-infected controls triggered a rapid release of great amounts of IL-8 and RANTES. Infection of HT-29 cells for up to seven days did not result in increased production of any of the chemokines analyzed (IL-8, IL-1α, IL-1β, IL-6), even with a high multiplicity of infection. *M. avium*-infected HEp-2 cells, although not releasing chemokines intiailly, showed an increase in IL-8 and RANTES production by 72 h. These results suggest that *M. avium* is able to interfere in some way with the transduction pathway that results in the augmented production and release of proinflammatory cytokines. The process appears to be more intense or effective in epithelial cells of intestinal origin rather than laryngeal origin, although more cell lines should be studied to confirm this dichotomy. It is tempting to speculate that this inhibition of the chemokine release allows some extra time for the bacteria before being detected by the host defense cells, thus increasing the probability of a successful infection of the intestinal mucosa.

The mechanism(s) by which *M. avium* suppresses chemokine production in epithelial cells is currently unknown. Some pathogens, such as *Cryptosporidium* inhibit IL-8 production by host cells for 24h after entry (Laurent *et al.*, 1997), but the suppresion of IL-8 and RANTES in HT-29 cells when invaded by *M. avium* is significantly longer. Therefore, whatever mechanism of inhibition, probably in signal transduction, it persists for long periods of time.

2.9. Fate of Intracellular Bacteria

The majority of bacteria that are phagocytosed by macrophages are killed; however, several pathogens, including mycobacteria, devise successful strategies that enable them to survive and replicate within potentially lethal cells. In macrophages, *M. avium* live within specialized vacuoles which show characteristics of early endosome stage, with no proton pump needed for vesicle acidification, no mannose-6-phosphate receptor-mediated delivery and which contain lysosomal markers such as cathepsin D and glycoprotein LAMP-1 (Sturgill-Koszycki *et al.*, 1994).

Studies looking at *M. avium* fate after entering epithelial cells show that bacteria remain vacuole-bound during all the intracellular life. Vacuoles that contain several bacteria initially undergo segmentation and ultimately become a vacuole with a single bacterium. There is no evidence that *M. avium* ever leaves the vacuole, as well as that the vacuole ever fuses with lysosomes. The pH of intracellular vacuoles within epithelial cells ranges between 6.8 and 6.5; very similar to the pH in macrophage vacuoles (Sturgill-Koszycki *et al.*, 1994).

We have neither knowledge about the mechanism that *M. avium* uses to escape the mucosal intestinal cell nor about the intracellular course of the infection, although there is preliminary evidence that either apoptosis or cytotoxicity are responsible for the release of *M. avium*. It is also plausible that infected epithelial cells end up being ingested or lysed by activated macrophages, which would provide a fresh host cell reservoir for the intracellular organism.

2.10. Invasion of the Intestinal Mucosa (the "Real Thing")

A large number of AIDS patients with disseminated *M. avium* infection have been shown to have the gastrointestinal tract colonized with *M.*

avium sometimes months prior to the onset of bacteremia (Torriani *et al.*, 1995; Damsker and Bottone, 1985). In some studies, it was demonstrated by DNA fingerprinting that the *M. avium* strain later isolated from the blood was present among the strains cultured from the intestinal tract confirming the source of infection.

Despite the epidemiologic evidence, *M. avium* cannot be isolated from the intestinal tract of several individuals with AIDS (Mazurek *et al.*, 1997), perhaps due to the lack of sensitivity of the methods for culture, but nonetheless suggesting that in some patients the route of infection could be the respiratory tract (Mazurek *et al.*, 1997; Jacobson *et al.*, 1991). As an alternative, since *M. avium* is an environmental bacterium and serum antibodies against the organism are found in a large percentage of the population, it is plausible to hypothesize that the infecting stage had occurred prior to the AIDS onset, and the bacteria would stay dormant in regional lymph nodes probably for years.

During the early years of the epidemic, it was assumed that *M. avium* bacteria present in the intestinal tract translocated the intestinal mucosa through lesions produced by other intestinal pathogens such as HIV-1, cytomegalovirus or intestinal parasites, and that *M. avium* by itself was not able to cross the intestinal mucosa. The strongest evidence that *M. avium* is capable of adhering and translocating through the intestinal mucosa and is not an opportunistic pathogen that gains access to the intestinal mucosa/sub-mucosa through intestinal microulcerations caused by other pathogens came from experimental studies in which the bacterium fed orally to otherwise healthy mice would ultimately cause disseminated disease in 100% of the mice after a determined period of time (Bermudez *et al.*, 1992). In these studies, although the percentage of mice with bacteremia would vary with the *M. avium* strain used, even less virulent strains, such as an opaque morphotype variant of *M. avium* strain 101, caused disseminated infection. Additionally, it was observed that the terminal ileum was preferentially targeted by the bacteria with a large number of bacteria found associated with the mucosa and submucosa. Bacteria were also observed in association with the mucosa of the ascending colon but were mostly absent from the other segments of the gut (Bermudez *et al.*, 1992). As noted above, different strains are associated with different degrees of infection in tissues, confirming the heterogenicity of *M. avium* strains (Bermudez *et al.*, 1992). In fact, subsequent studies using the closed intestinal loop model, and a number of different *M. avium* strains, showed a large variability in efficiency to invade the intestinal mucosa, confirming the previous results in the oral model of infection (Hsu *et al.*, 1996).

Recent studies demonstrated that *M. avium* interaction with the intestinal mucosa results in necrotic lesions (Kim *et al.*, 1998). While the initial contact does not induce a rapid inflammatory response with the infiltration of immune cells, by 3 to 4 weeks after *M. avium* ingestion, lymphocyte infiltration and edema of the intestinal villi can be observed. After 5 to 6 weeks, the majority of the mice show segmental necrosis of the intestinal mucosa. Whether open lesions represent a mechanism by which large numbers of *M. avium* bacteria gain access to the submucosa remains to be shown. Later stages of infection both in mice as well as in humans (Kim *et al.*, 1998; Inderlied *et al.*, 1993) show bacilli-laden macrophages in the lamina propria.

Assays using electron microscopy as well as immunohistochemistry to investigate the interaction between *M. avium* and the intestinal mucosa have shown that the bacterium adheres preferentially to enterocytes of the terminal ileum inside and outside of the Peyer's patches than M cells. M cells are cells present in the Peyer's patches but also in other sites of the body. They are specialized in the uptake of antigens and have been shown to be associated with translocation of a number of enteric pathogens such as Salmonella and Yersinia (Pepe *et al.*, 1995; Jones *et al.*, 1994). In addition, it has been shown that *M. bovis* uses the M cell pathway to gain access to the intestinal mucosa (Fujimura, 1986). The reason why *M. avium* interacts preferentially with enterocytes is currently unknown, but as described above it may be related to the bacterium's ability to avoid early influx of immune cells to the site of infection. Therefore, by entering enterocytes instead of M cells the bacterium avoids contact with immune cells before establishing a niche. This period within enterocytes may be crucial for the survival of a microorganism that is not as virulent as Salmonella or *M. tuberculosis*.

Close observation of the uptake process confirms the findings *in vitro* showing that cytoskeleton rearrangement in the mammalian cell accompanies invasion *in vivo* (Hsu *et al.*, 1996). In addition, internalized bacteria are always surrounded by a dense, dark halo indicative of actin polymeralization (Hsu *et al.*, 1996).

Cytoskeleton rearrangement parallels the uptake of a number of bacteria such as Salmonella and Shigella by epithelial cells (Adam *et al.*, 1995; Finlay and Ruschkowski, 1991), and the ability to trigger actin polymerization in the host cell has been shown to be a virulence feature of pathogenic microorganisms. In addition, electron micrographs documenting the interaction of *M. avium* with intestinal mucosal cells in the Peyer's patches also show that effacing of the target cells in a manner similar to the effacing observed with enteropathogenic *Escherichia coli* and *Helicobacter pylori* (Segal *et al.*, 1996; Rosenshine *et al.*, 1992). The mechanisms associated with *M. avium*-dependent effacing are currently being investigated.

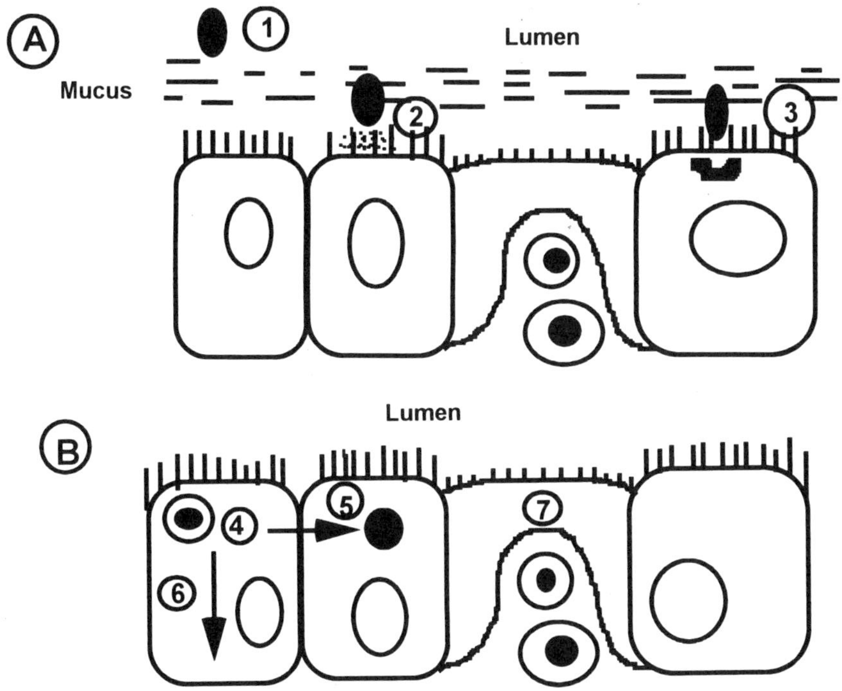

FIGURE 1. Interaction of *Mycobacterium avium* with enterocytes.
1. Recognition of a structure to anchor.
2. Secretion of proteins.
3. "Effacing" and cytoskeleton reorganization. Entry by the apical surface.
4. Intracellular, intravacuolar *M. avium*. Organisms are immobile.
5. There is no evidence of spread from cell to cell through the basolateral surface. Suppression of chemokine release by epithelial cells.
6. Translocation within cell. Exit by the basal surface.
7. M cell.

3. *Mycobacterium tuberculosis* AND ALVEOLAR EPITHELIAL CELLS

M. tuberculosis is an important pathogen by medical and epidemiological standards. *M. tuberculosis* establishes infection after inhalation of the bacilli into the alveolar space in the lungs (Bloom, 1994). It is largely accepted that during this process, *M. tuberculosis* is ingested by or enters alveolar macrophages (Bloom, 1994).

Although it is assumed that *M. tuberculosis* infection occurs primarily as described above, it is plausible to hypothesize that the aerosolized bacterium can bind and invade epithelial cells as well upon its arrival in the alveolar space. Type II alveolar epithelial cells line 85% of the epithelial surface of the alveolar space and certainly contact between these cells and *M. tuberculosis* gets established. Passage of the organism through the alveolar epithelium and into the lymphatics or capillaries is a prerequisite to the establishment of infection. The alveolar epithelium is a barrier that contains tight junctions and the integrity of the barrier depends on the tight junctions.

3.1. Interaction with Alveolar Epithelial Cells

There are two possible pathways that could explain *M. tuberculosis'* migration through the alveolar epithelial barrier: (1) invasion of epithelial cells; and/or (2) reduction of the bioelectrical properties of the epithelial barrier.

The ability of mycobacteria to bind to and enter epithelial cells was initially demonstrated in studies by Shepard (1957). Recent work from two laboratories has demonstrated that *M. tuberculosis* interacts and enters alveolar epithelial cells by a receptor-mediated mechanism (Zhang *et al.*, 1997; Bermudez and Goodman, 1996). The interaction is dependent on the presence of components of the integrin family (Bermudez and Goodman, 1996) and is more efficient at 37°C than at 30°C (Bermudez and Goodman, 1996). Current evidence rules out pinocytosis as a mechanism for invasion (Bermudez and Goodman, 1996). Once within the cell, *M. tuberculosis* can replicate rapidly, demonstrating its adaptation to the intracellular environment. Although H37Rv and H37Ra invade alveolar epithelial cells at comparable fashion, H37Rv replicates intracellularly faster than H37Ra (Zhang *et al.*, 1997; Bermudez and Goodman, 1996). H37Rv (virulent strain) and H37Ra (attenuated strain) are derived from a single parent strain and became phenotypically and genotypically different (H37Ra contains a number of deletions in the chromosome) over the years.

It has also been shown that *M. tuberculosis* is capable of altering the bioelectric properties of the alveolar epithelium *in vitro* (Zhang *et al.*, 1997), causing a significant decrease in resistance across the monolayer. Strains H37Rv as well as H37Ra were associated with a similar effect on the membrane potential. It was further shown that *M. tuberculosis* triggers the release of TNF-α from the epithelial cells by the apical membrane and TNF-α was responsible for reducing the epithelial cell barrier resistance (Zhang *et al.*, 1997).

Therefore, it is possible that *M. tuberculosis* would both enter alveolar epithelial cells and then decrease the permeability of the epithelial barrier, allowing large numbers of bacteria to translocate through the epithelium.

This mechanism of infection would allow the bacterium to gain access to both lymphatics and blood capillaries, with subsequent dissemination.

It has also been shown that invasion of alveolar epithelial cells by *M. tuberculosis* increases the virulence of the bacteria which once released from the epithelial cells would invade and lyse macrophages (McDonough *et al.*, 1993). It is tempting to speculate that *M. tuberculosis* released from epithelial cells can efficiently lyse macrophages; however, this hypothesis does not explain thus far what would be the advantage of this mechanism for the invasion and survival of the organism.

For many years it has been argued that *M. tuberculosis* expresses hemolysis (Mehta *et al.*, 1996) and therefore would lyse the membrane of the intracellular vacuoles releasing the bacteria into the cytoplasma of the infected cell. More recently it was shown (although controversial at this point) that *M. tuberculosis* ingested by epithelial cells can be released from vacuole compartments and live in the cytoplasm (McDonough and Kress, 1995). This work waits for confirmation.

3.2. Receptors and Mechanisms of Invasion

M. tuberculosis uptake by alveolar epithelial cells (A549 cells) depend on both microfilaments and microtubules (Bermudez and Goodman, 1996). Treatment of A549 cells with cytochalasin D, colchicine or nocodazole has been shown to significantly inhibit the internalization of *M. tuberculosis*, indicating that the process, as demonstrated with a number of other bacteria, require active participation of the eukaryotic cell. Why both pathways (actin-dependent and microtubule-dependent) seem to be important for the uptake of *M. tuberculosis* by type II alveolar cells is currently unknown, but dependence on both pathways has been described before for *Candida albicans*, *Campylobacter jejuni*, and *Citrobacter freundii* (Oelschlaeger *et al.*, 1993).

Bacteria utilize a number of different receptors to be internalized by cells, an indication of the degree of adaptation to the environment in the host. It has been demonstrated that blocking β_1 integrin and the vitronectin receptors on the membrane of alveolar epithelial cells results in a significant inhibition (approximately 80%) of binding and entry. These receptors link the membrane to intracellular pathways of signal transduction. Other receptors, such as transferin receptors were shown to have no role in the internalization process by epithelial cells.

3.3. Chemokine Production

M. tuberculosis infection of human alveolar epithelial cells elicits production of chemokines, among them MCP-1 and IL-8 through upregulation of expression of their respective mRNA (Lin *et al.*, 1998). Chemokine production in response to *M. tuberculosis* was not dependent on production of inflammatory cytokines, TNF-α, IL-1 and IL-6. In contrast, work by Lin and colleagues (1998) showed that A549 cells infected with *M. avium* do not induce chemokine production, a finding similar to the one observed when intestinal cells (*in vitro* and *in vivo*) were infected by different strains of *M. avium* (Bermudez and Goodman, 1996). Therefore, it seems clear that the pathogenesis of *M. tuberculosis* at this stage of the infection differs from *M. avium*'s pathogenesis. As a less virulent bacterium, *M. avium* tends to be more quiet and ultimately "hide" from the immune system, a characteristic that does not seem necessary for *M. tuberculosis*.

4. CONCLUSIONS

Historically, mononuclear phagocytes were considered the only host cell that would get infected by mycobacteria during infection. Recent studies have shown, however, that both *M. avium* and *M. tuberculosis* can efficiently infect epithelial mucosal cells. With the advent of AIDS, the gastrointestinal tract was determined to be the main route of *M. avium* infection. Therefore, *M. avium* behaves more like an enteric pathogen than a lung pathogen.

M. tuberculosis was always considered a "macrophage pathogen" and it is assumed to be ingested by macrophages when within the alveolar space. However, in this case, too, little by little evidence acumulates showing that *M. tuberculosis* invade alveolar epithelial cells and that port of entry appears to be important to the pathogenesis of the infection.

The future will show us what is the role of the intestinal and respiratory mucosis in mycobacteria infection.

5. REFERENCES

Abou-Zeid, C., Ratliff, T.L., Wiker, H.G., Harboe, M., Bennedsen, J., and Rook, G.A.W., 1988, Characterization of fibronectin-binding antigens released by *Mycobacterium tuberculosis* and *Mycobacterium bovis*, *Infect. Immun.* **56**:3046–3051.

Adam, T., Arpim, M., Prevost, M.C., Gounon, P., and Sansonetti, P.J., 1995, Cytoskeletal rearrangements and the functional role of T-plastin during entry of *Shigella flexneri* into HeLa cells, *J. Cell Biol.* **129**:367–381.

Barker, J., Lambert, P.A., and Brown, M.R.W., 1993, Influence of intramoebic and other growth conditions on the surface properties of *Legionella pneumophila*, *Infect. Immun.* **62**:3254–3261.

Benson, C.A., and Ellner, J.J., 1993, *Mycobacterium avium* complex infection and AIDS: Advances in theory and practice, *Clin. Infect. Dis.* **17**:7–20.

Bermudez, L.E., and Goodman, J., 1996, *Mycobacterium tuberculosis* invades and replicates within type II alveolar cells, *Infect. Immun.* **64**:1400–1406.

Bermudez, L.E., and Goodman, J., 1998, *Mycobacterium avium* attachment and invasion of intestinal epithelial cells is associated with cytoskeleton rearrangement, submitted.

Bermudez, L.E., Parker, A., and Goodman, J., 1997a, Growth within macrophages increases the efficiency of *Mycobacterium avium* to invade other macrophages by complement receptor independent pathway, *Infect. Immun.* **65**:1916–1922.

Bermudez, L.E., Petrofsky, M., and Goodman, J., 1997b, Exposure to low oxygen tension and increased osmolarity enhance the ability of *Mycobacterium avium* to enter intestinal epithelial (HT-29) cells, *Infect. Immun.* **65**:3768–3772.

Bermudez, L.E., Petrofsky, M., Kolonoski, P., and Young, L.S., 1992, An animal model of *Mycobacterium avium* complex disseminated infection after colonization of the intestinal tract, *J. Infect. Dis.* **165**:75–79.

Bermudez, L.E., Shelton, K., and Young, L.S., 1995, Comparison of the ability of *M. avium*, *M. smegmatis*, and *M. tuberculosis* to invade and replicate within HEp-2 epithelial cells, *Tubercle Lung Dis.* **76**:240–247.

Bermudez, L.E., and Young, L.S., 1994, Factors affecting invasion of HT-29 and HEp-2 epithelial cells by organisms of the *Mycobacterium avium* complex, *Infect. Immun.* **62**:2021–2026.

Bloom, B.R., 1994, *Tuberculosis: Pathogenesis, protection and control, ASM Press*. Washington, D.C.

Bodmer, T., Miltzer, E., and Bermudez, L.E., 1998, *Mycobacterium avium* resists exposure to acidic conditions of the stomach, a property that is enhanced by re-adaptation to hypoosmolarity, Submitted.

Cirillo, J., Falkow, S., Tompkins, L., and Bermudez, L.E., 1997, Growth of *Mycobacterium avium* within environmental amoeba enhances virulence, *Infect. Immun.* **65**:3759–3769.

Cirillo, J.D., Falkow, S., and Tompkins, L.S., 1994, Growth of *Legionella pneumophila* in *Acanthamoeba castellani* enhances invasion, *Infect. Immun.* **62**:3254–3261.

Damsker, B., and Bottone, E.J., 1985, *Mycobacterium avium-Mycobacterium intracellulare* from the intestinal tracts of patients with the acquired immunodeficiency syndrome: concepts regarding acquisition and pathogenesis, *J. Infect. Dis.* **151**:179–180.

Falkinham III, J.O., 1996, Epidemiology of infection by nontuberculous mycobacteria, *Clin. Microbiol. Rev.* **9**:177–215.

Finlay, B.B., and Ruschkowski, S., 1991, Cytoskeletal rearrangements accompanying Salmonella entry into epithelial cells, *J. Cell Sci.* **99**:283–296.

Fujimura, Y., 1986, Functional morphology of microfold cells (M cells) in Peyer's Patches—phagocytosis and transport of BCG by M cells in rabbit Peyer's Patches, *Gastroenterol. Jap.* **21**:325–330.

Glover, N., Holzman, A., Aronson, T., Proman, B., Berlin, G.W., Dominguez, P., Konzel, K.A., Overturf, G., Stelma, G., Smith, C., and Yakrus, M., 1994, The isolation and identification of *Mycobacterium avium* complex recovered from Los Angeles potable water, a possible source of infection in AIDS patients, *Int. J. Environ. Health Res.* **4**:63–72.

Hedges, S.R., Agace, W.W., and Svanborg, C., 1995, Epithelial cytokine responses and mucosal cytokine networks, *Trends Microbiol.* **3**:266–270.

Horsburgh Jr, C.R., 1991, *Mycobacterium avium* complex in the acquired immunodeficiency syndrome (AIDS), *N. Engl. J. Med.* **324**:1332–1338.

Hsu, N., Goodman, J.R., Young, L.S., and Bermudez, L.E., 1996, Interaction between *Mycobacterium avium* complex and intestinal mucosal cells *in vivo*, in: *36th Interscience Conference on Antimicrobial Agents and Chemotherapy*, American Society for Microbiology, Abstract #B25.

Huang, J., O'Toole, P.W., Doig, P., and Trust, T.J., 1995, Stimulation of interleukin-8 production in epithelial cell lines by *Helicobacter pylori*, *Infect. Immun.* **63**:1732–1738.

Hueck, C.J., 1998, Type III protein secretion systems in bacterial pathogens of animals and plants, *Microbiol. Mol. Biol. Rev.* **62**:379–433.

Inderlied, C.B., Kemper, C.A., and Bermudez, L.E., 1993, The *Mycobacterium avium* complex, *Clin. Microbiol. Rev.* **6**:266–310.

Jacobson, M.A., Hopewell, P.C., Yajko, D.M., Hadley, W.K., Lazarus, E., Mohanty, P.K., Modin, G.W., Feigal, D.W., Cusick, P.S., and Sande, M.A., 1991, Natural history of disseminated *Mycobacterium avium* complex infection in AIDS, *J. Infect. Dis.* **164**:994–998.

Jepson, M.A., and Clark, M.A., 1998, Studying M cells and their role in infection, *Trends Microbiol.* **6**:359–365.

Jones, B.D., Ghori, N., and Falkow, S., 1994, *Salmonella typhimurium* initiates murine infection by penetrating and destroying the specialized epithelial M cells of the Peyer's Patches, *J. Exp. Med.* **180**:15–23.

Jung, H.C., Eckmann, L., Yang, S.K., Panja, A., Fierer, J., Wroblewska, E., and Kagnoff, M.F., 1995, A distinct array of proinflammatory cytokines is expressed in human colon epithelial cells in respone to bacterial invasion, *J. Clin. Invest.* **95**:55–65.

Kim, S.Y., Goodman, J.R., Petrofsky, M., and Bermudez, L.E., 1998, *Mycobacterium avium* infection of the gut mucosa in mice is associated with later inflammatory response and ultimately results in areas of intestinal cell necrosis, *J. Med. Microbiol.* **47**:725–731.

Laurent, F., Eckmann, L., Savidge, T.C., Morgan, G., Theodos, C., Naciri, M., and Kagnoff, M.F., 1997, *Cryptosporidium parvum* infection of human intestinal epithelial cells induces the polarized secretion of C-X-C chemokines, *Infect. Immun.* **65**:5067–5073.

Lee, C.A., 1977, Type III secretion systems: machines to deliver bacterial proteins into eukaryotic cells? *Trends Microbiol.* **5**:148–156.

Lin, Y., Zhang, M., and Barnes, P., 1998, Chemokine production by a human alveolar epithelial cell line in response to *Mycobacterium tuberculosis*, *Infect. Immun.* **66**:1121–1126.

Mazurek, G.H., Chin, D.P., Hartman, S., Reddy, V., Horsburgh Jr., R.C., Green, T.A., Yajko, D.M., Hopewell, P.C., Reingold, A.L., and Crawford, J.T., 1997, Genetic similarity among *Mycobacterium avium* isolates from blood, stool and sputum of persons with AIDS, *J. Infect. Dis.* **176**:976–983.

McDonough, K., and Kress, Y., 1995, Cytotoxicity for lung epithelial cells is a virulence-associated phenotype of *Mycobacterium tuberculosis*, *Infect. Immun.* **63**:4802–4811.

McDonough, K.A., Kress, Y., and Bloom, B.R., 1993, Pathogenesis of tuberculosis: Interaction with macrophages, *Infect. Immun.* **61**:2763–2773.

Mehta, P.K., King, C.H., White, E.H., Murtagh Jr., J.J., and Quinn, R.D., 1996, Comparison of *in vitro* models for the study of *Mycobacterium tuberculosis* invasion and intracellular replication, *Infect. Immun.* **64**:2673–2679.

Neutra, M.R., Frey, A., and Kraehenhuhl, J.P., 1996, Epithelial M Cells: Gatesways for mucosal infection, *Cell* **86**:345–348.

Oelschlaeger, T.A., Guerry, P., and Kopecko, D.J., 1993, Unusual microtubule-dependent endocytosis mechanisms triggered by *Campylobacter jejuni* and *Citrobacter freundii*, *Proc. Natl. Acad. Sci. USA* **90**:6884–6888.

Pepe, J.C., Wachtel, M.R., Wager, E., and Miller, V.L., 1995, Pathogenesis of the defined invasion mutants of *Yersinia enterocolitica* in a BALB/C mouse model of infection, *Infect. Immun.* **63**:4837–4848.

Rasmussen, S.J., Eckmann, L., Quayle, A.J., Shen, L., Zhang, Y.X., Anderson, D.J., Fierer, J., Stephens, R.S., and Kagnoff, M.F., 1997, Secretion of proinflammatory cytokines by epithelial cells in response to Chlamydia infection suggests a central role for epithelial cells in chlamydial pathogenesis, *J. Clin. Invest.* **99**:77–87.

Rosenshine, I., Donnenberg, M.S., Karper, J.B., and Finlay, B.B., 1992, Signal transduction between enteropathogenic *Escheria coli* (EPEC) and epithelial cells: EPEC induces tyrosine phosphorylation of host cell proteins to initiate cytoskeletal rearrangement and bacterial uptake, *EMBO J.* **11**:3551–3560.

Schorey, J.F., Li, Q., McCourt, D., Bong-Mastek, M., Clark-Curtiss, J., Ratliff, T.L., and Brown, E.J., 1996a, A *Mycobacterium leprae* gene encoding a fibronectin binding protein is used for efficient invasion of epithelial cells and Schwann cells, *Infect. Immun.* **63**:2652–2657.

Schorey, J.S., Holsti, M.D., Ratliff, T.L., Allen, P.M., and Brown, E.J., 1996b, Characterization of fibronectin-attachment protein of *Mycobacterium avium* reveals a fibronectin-binding motif conserved among mycobacteria, *Mol. Microbiol.* **21**:321–330.

Segal, E.D., Falkow, S., and Tompkins, L.S., 1996, *Helicobacter pylori* attachment to gastric cells induces cytoskeletal rearrangements and tyrosine phosphoylation of host cell proteins, *Proc. Natl. Acad. Sci. USA* **93**:1259–1264.

Shepard, C.C., 1957, Growth characteristics of tubercle bacilli and certain other mycobacteria in HeLa cells, *J. Exp. Med.* **105**:39–48.

Sturgill-Koszycki, S., Schlesinger, P.H., Chakraborty, P., Haddix, P.L., Collins, H.L., Fok, A.K., Allen, R.D., Gluck, S.L., Heuser, J., and Russell, D.G., 1994, Lack of acidification in Mycobacterium phagosomes produced by exclusion of the vesicular proton-ATPase, *Science* **263**:678–681.

Tan, T., Sangari, F., Petrofsky, M., and Bermudez, L., 1998, *M. avium* secretes proteins upon contact with intestinal cells *in vitro*, Submitted.

Torriani, F., Maslow, J.N., Kornbluth, R., Arbeit, R.D., McCutchan, J.A., Hasegawa, P., Keays, L., and Havlir, D., 1995, Analysis of *Mycobacterium avium* complex isolates infecting multiple organs of AIDS patients using pulsed field gel eletrophoresis, in: *Interscience Conference on Antimicrobial Agents and Chemotherapy*, San Francisco, American Society for Microbiology. Abstract #I-91.

Walker, D.H., Firth, W.T., Ballard, J.G., and Hegarty, B.C., 1983, Role of phospholipase-associated penetration mechanism in cell injury by *Rickettsia rickettsii*, *Infect. Immun.* **40**:840–842.

Zhang, M., Kim, K.J., Iyer, D., Lin, Y., Belisle, J., McEnery, K., Crandall, E.D., and Barnes, P.F., 1997, Effects of *Mycobacterium tuberculosis* on the bioelectric properties of the alveolar epithelium, *Infect. Immun.* **65**:692–698.

Chapter 11

Invasion of Epithelial Cells by Bacterial Pathogens

The Paradigm of Shigella

Kirsten Niebuhr and Philippe J. Sansonetti

1. INTRODUCTION

A necessary step in the successful colonization and production of disease by microbial pathogens is their ability to persist in the host. Pathogenic bacteria have achieved this goal by developing mechanisms to adhere to mucosal surfaces or, going even further, by entering into and surviving within eukaryotic cells. Shigella is a well-studied example of a facultative intracellular bacterium that is able to enter into non-professional phagocytes and disseminate in the infected tissue. Thus, Shigella infection can be used as a model system to study the complex interplay between host and pathogen that occurs during the process of disease. In this review, we will discuss how the unraveling of the molecular mechanisms underlying this lifestyle can help to understand and combat the disease caused by this

KIRSTEN NIEBUHR and PHILIPPE J. SANSONETTI Unité de Pathogénie Microbienne Moléculaire Institut Pasteur, F-75724 Paris Cédex 15, France.
Subcellular Biochemistry, Volume 33: Bacterial Invasion into Eukaryotic Cells, edited by Oelschlaeger and Hacker. Kluwer Academic / Plenum Publishers, New York, 2000.

pathogen and provide insights into basic host cell functions that are exploited by the bacterium.

Shigella flexneri is a gram-negative bacillus belonging to the family of *Enterobactericeae* and it is closely related to *Escherichia coli*. In fact, *Shigella* species have more than 80% nucleotide sequence homology with the *E. coli* chromosome (Brenner *et al.*, 1973). Enteroinvasive *E. coli* (EIEC) are considered as an intermediate in the evolution between *E. coli* and Shigella, but although the disease symptoms are very similar, EIEC are less virulent in humans than Shigella. The infectious dose of EIEC that causes disease in human volunteers is 10^6 cfus compared to 100 cfus for Shigellae (DuPont *et al.*, 1989), perhaps reflecting a better adaptation of Shigellae for the successful colonisation of their host.

Shigellosis, also called bacillary dysentery, is a severe diarrhoeal disease. The only natural hosts of Shigella are humans and monkeys, and the transmission occurs via the fecal-oral route or by contaminated food and water. During infection, the invasive process remains localized to the colonic and rectal mucosa, causing a severe inflammation that leads to mucosal destruction. The disease is characterized by initial watery diarrhea rapidly followed by fever, violent intestinal cramps and emission of mucopurulent and bloody stools. In general, it is a disease of the poor, primarily affecting young children in the developing world where about five million cases require hospitalization and 600,000 persons die each year. The highly infectious nature of the pathogen and the rapid increase of strains that display multiple resistance to antibiotics make prevention and treatment of shigellosis a difficult task; thus, the development of a vaccine against Shigella is a high priority for the World Health Organization (Kotloff *et al.*, 1999). The understanding of the molecular mechanisms underlying this disease is a prerequisite for the design of novel vaccination strategies.

2. THE MODEL OF SHIGELLA INFECTION

After oral uptake, the bacteria progress along the intestinal tract until they reach the mucosal surface of the colon and rectum, which they invade, causing inflammation and tissue destruction. In contrast to Salmonella and enteropathogenic *E. coli* (EPEC)/Shiga toxin producing *E. coli*, which are able to interact with their host cells from the apical side, Shigella entry into polarized epithelial cells occurs most efficiently from the basolateral side (Mounier *et al.*, 1992). Thus, the bacteria must cross the epithelium before they can invade the cells from subepithelial tissues. Experiments carried out in macaque monkeys and in the rabbit ligated intestinal loop model have indicated that Shigella primarily translocates through the epithelial lining via the M cells (Sansonetti *et al.*, 1996; 1991; Wassef *et al.*, 1989). These are specialized cells devoid of brush border which belong to the follicle-associated epithelium (FAE) that covers the lymph nodes associated with the mucosa. After ingestion by M cells and translocation to the dome region of the FAE, the bacteria are taken up by macrophages, which are particularly abundant in this area. Surprisingly, Shigellae are not killed and degraded by macrophages, but they induce apoptosis of these cells (Zychlinsky *et al.*, 1996; 1992). This not only results in the release of the bacteria, but also that of massive amounts of cytokines (Zychlinsky *et al.*, 1994a), which will be discussed in detail later. The subsequent inflammatory reaction results from the influx of polymorphonuclear leukocytes (PMN) and monocytes and favours further translocation of bacteria by destabilizing the epithelial integrity (Perdomo *et al.*, 1994a,b). Having crossed the epithelial barrier, Shigella invades the enterocytes by triggering its uptake in a process that resembles macropinocytosis and requires massive cytoskeletal rearrangements (Clerc and Sansonetti, 1987). The bacteria subsequently lyse the vacuole and escape into the cytoplasm (Sansonetti *et al.*, 1986), where they multiply and spread from cell to cell (Sansonetti *et al.*, 1994; Bernardini *et al.*, 1989; Makino *et al.*, 1986), thus disseminating efficiently through the host cell epithelium. The infectious process is schematically summarized in Figure 1; we will discuss in detail the progress that has been made during the last two decades in unravelling the molecular cross-talk between Shigella and its host cells accounting for entry, intracellular movement and cell-to-cell spread, as well as the apoptotic process in macrophages.

3. ENTRY INTO EPITHELIAL CELLS: MOLECULES AND SIGNALS

A key factor in the pathogenesis of shigellosis is the capacity of the bacteria to invade epithelial cells. Early studies made use of *in vitro*-cultured cells to study the capacity of Shigella to penetrate into cells which are not professional phagocytes (LaBrec *et al.*, 1964). Since then, the tissue-culture model of infection has been refined to become quantitative. For example, the "gentamicin protection assay" is now commonly used to quantitate the invasiveness. Gentamicin cannot penetrate the cell membrane and remains extracellularly thereby killing those bacteria that have not entered cells, and preserving intracellular bacteria. The "plaque assay" (Oaks *et al.*, 1986) and the "infectious focus assay" (Sansonetti *et al.*, 1994) are both used to study cell-to-cell spread on confluent cell monolayers. Combined with genetic analyses of bacterial pathogenicity, these tests have permitted the identification of the bacterial factors involved in entry, escape into the

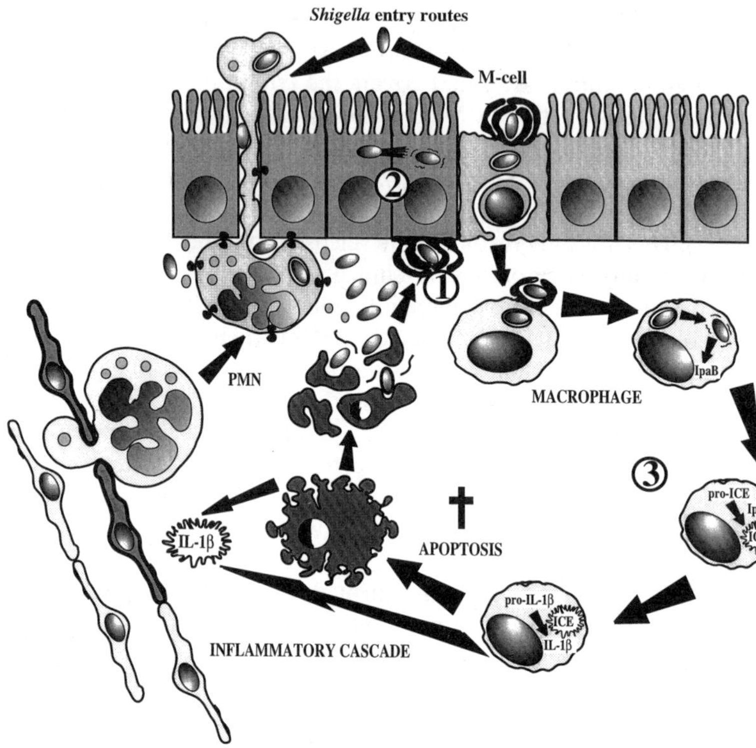

FIGURE 1. The vicious circle. This scheme summarizes the early steps of *Shigella* infection: the crossing of the follicular-associated epithelium and the initiation of inflammation (i.e. mostly influx of PMNs) following induction of macrophage apoptosis (3). The destabilization of the integrity of the epithelial barrier enhances bacterial invasion (1), and the process is subsequently amplified by the capacity of Shigella to spread from cell to cell (2). These three points will be the focus of this review.

cytoplasm, intracellular motility, and cell-to-cell spread. Moreover, they have also been used to identify major eukaryotic cell components supporting these processes, particularly the actin cytoskeleton (Goldberg and Sansonetti, 1993). These approaches have contributed to the establishment of the concept of "cellular microbiology", which analyses microbial pathogenesis at its intersection between the prokaryotic and eukaryotic worlds (Cossart *et al.*, 1996).

Shigella invade the enterocytes, in other words the bacteria induce their uptake into cells that under normal circumstances are non-phagocytic. They share this ability with a variety of pathogenic bacteria, like Salmonella, Listeria, Mycobacterium and several other pathogens that are the subject of different chapters of this volume. Different strategies of entry into non-professional phagocytes have been adopted by enteroinvasive pathogens: the "zipper"-like mechanism, which entails the zippering of the host cell membrane around the bacterium as it enters (Mengaud *et al.*, 1996; Isberg and Leong, 1990), and the "triggering" mechanism, which results in a dramatic response at the epithelial cell surface and bacterial uptake via membrane ruffles, a process resembling macropinocytosis (Adam *et al.*, 1995; Francis *et al.*, 1993; Finlay and Falkow, 1990).

An example for uptake by zippering is entry of the gram-positive bacterium *Listeria monocytogenes* into host cells. This process is mediated by the internalins (Gaillard *et al.*, 1991) and best examplified by the interaction of Internalin A and E-cadherin (Mengaud *et al.*, 1996). The most cited example of the zippering mechanism, however, is the entry of *Yersinia pseudotuberculosis* into eukaryotic cells. The interaction of the bacterial surface and the host cell membrane is mediated by a classical receptor-ligand interaction; the bacterial surface ligand being "invasin", which binds with high affinity to its host cell receptor β_1 integrin (Isberg and Leong, 1990). However, pathological examinations of experimentally infected animals indicate that yersiniae remain largely extracellular, but firmly affixed to the host cell surface, and several virulence factors of Yersinia have been found to paralyze phagocytotic cells, thereby preventing bacterial uptake (Fallman *et al.*, 1995; Rosqvist *et al.*, 1991; Simonet *et al.*, 1990).

The zippering mechanism of entry appears to be a straightforward ligand-receptor mediated process and is characterized by cytoskeletal rearrangements that are not dramatic and disappear within a few minutes of bacterial entry (Young *et al.*, 1992). In contrast, the triggering mechanism is a complex process involving many bacterial factors and an important reorganization of the cytoskeleton at the site of bacterial entry (Figure 2). This mode of entry is employed by Salmonella and Shigella, both of which secrete a set of invasion proteins via a specialized type III secretory apparatus upon contact with their cellular targets to stimulate entry.

3.1. Shigella Effector Molecules Involved in Entry

The invasive phenotype of Shigella (and EIEC) is encoded by a large virulence plasmid of 200 kb that harbors the bacterial factors necessary for bacterial uptake and intracellular motility (Sansonetti *et al.*, 1982, 1981). A 30 kb-region of this virulence plasmid is necessary and sufficient to confer the invasive ability to an *E. coli* K-12 strain (Sasakawa *et al.*, 1988; Venkatesan *et al.*, 1988; Maurelli *et al.*, 1985). This "entry region" is divided into two operons, that are transcribed in opposite directions (Figure 3) and encode, in one orientation, a specialized secretion apparatus, the Mxi-Spa

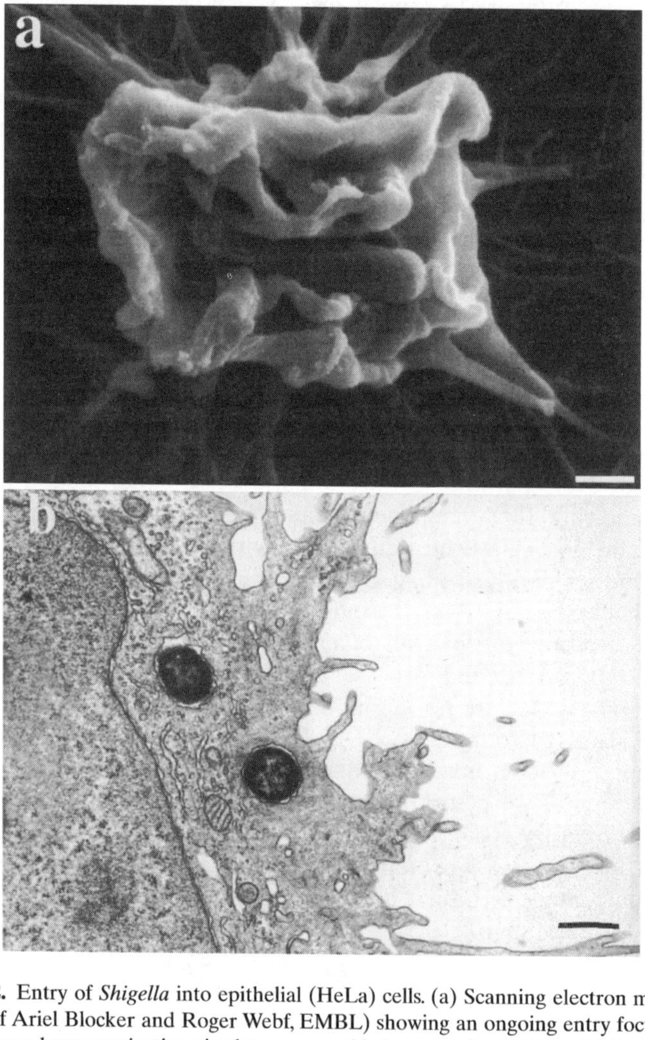

FIGURE 2. Entry of *Shigella* into epithelial (HeLa) cells. (a) Scanning electron microscopy (courtesy of Ariel Blocker and Roger Webf, EMBL) showing an ongoing entry focus characterized by membrane projections in the process of being organized into multiple ruffles that eventually form a macropinocytic vacuole. (b) Transmission electron microscopy (P. Gounon and P. Sansonetti, Station Centrale de Microscopie Electronique, Institut Pasteur) showing the section of an entry focus with multiple projections characterized by massive rearrangements of the cell subcortical cytoskeleton. Bars = 1 μm.

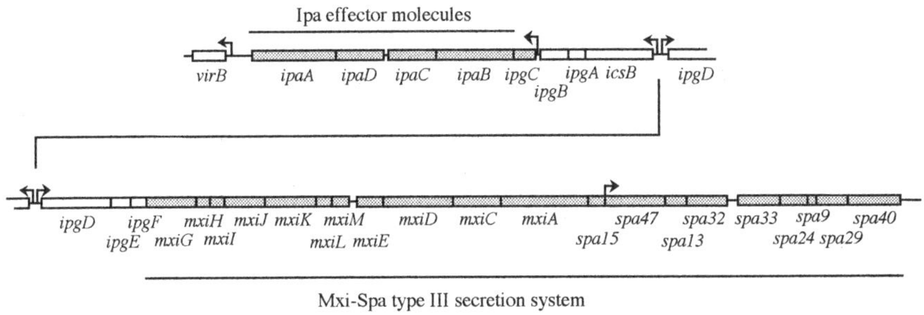

FIGURE 3. The "entry region" on the *Shigella* virulence plasmid. Genetic map of the *Shigella flexneri* 5 entry locus. This sequence of 30 kb is located on the large virulence plasmid (pWR100) of this species. It comprises two sub-loci transcribed in the opposite directions. On top, the *ipa* operon which encodes the entry effectors is shown in grey. At the bottom, the *mxi* and *spa* operons encoding the components of the type III secretory apparatus are also shown in grey.

translocon, and, in the other, the secreted virulence factors and their cytoplasmic molecular chaperones.

3.1.1. The Mxi-Spa Secretion System

Only in the last few years have type III secretion systems been identified as a novel secretion system that plays an active role in the secretion of virulence factors in many gram-negative human, animal and plant pathogens (Van Gijsegem *et al.*, 1993). The *mxi* (*m*embrane e*x*cretion of *I*pa)—*spa* (*s*urface *p*resentation of invasion plasmid *a*ntigens) locus of Shigella comprises about 20 genes coding for such a type III secretion apparatus (Sasakawa *et al.*, 1993; Allaoui *et al.*, 1993a,b; Venkatesan *et al.*, 1992; Andrews *et al.*, 1991), which is expressed and assembled at 37°C, and activated upon contact with the target cells (Watarai *et al.*, 1995; Ménard *et al.*, 1994a).

Little is known about the localization of the Mxi and Spa proteins in the bacterial cell. MxiJ and MxiM are lipoproteins and may be anchored to the outer membrane by their terminal lipid moiety (Allaoui *et al.*, 1992). MxiA has been detected in the inner membrane and the N-terminal domain has six putative transmembrane segments, whereas the C-terminal domain is thought to be cytoplasmic (Andrews and Maurelli, 1992; Andrews *et al.*, 1991). The presence of hydrophobic regions in Spa9, Spa15, Spa24, Spa29, and Spa40 suggest that these proteins might also be located in the inner membrane (Sasakawa *et al.*, 1993; Venkatesan *et al.*, 1992). MxiD was tentatively localized in the outer membrane (Allaoui *et al.*, 1993a) and MxiG was shown to associate with both the inner and the outer membranes (Allaoui *et al.*, 1995). It is thought that the Mxi-Spa secretion apparatus forms a continuous channel across the inner and outer bacterial membrane, but this has not yet been formally proven. However, striking homologies have been found between some genes of type III secretion systems and genes involved in flagellar assembly (Carpenter *et al.*, 1993), suggesting a relationship between these two structures. Much excitement has been generated recently by the electron microscopic detection of the type III apparatus of *Salmonella typhimurium* (Kubori *et al.*, 1998), which indeed resembles the basal body of a flagellum with a needle- like structure on top.

The components of the type III secretion systems of different pathogens show a high degree of homology among these species. The secreted effector molecules do not show classical signal sequences for export, but some of them require specific cytoplasmic chaperones (see below). Apparently, the chaperones recognize a portion of the secreted protein that is critical for mediating its export (usually the amino-terminal region). Interchangeability of secreted effector molecules has been shown in Shigella, Salmonella, and Yersinia, thus reflecting the high degree of functional conservation of these systems among these pathogens. For example, when certain effector molecules are expressed in the presence of their proper chaperones in a different pathogen containing a type III system, heterologous secretion occurs (Hermant *et al.*, 1995; Rosqvist *et al.*, 1995; Groisman *et al.*, 1993).

The activity of type III secretion systems is tightly regulated. In broth culture under normal growth conditions, *Shigella flexneri* secretes hardly any effector proteins into the medium. Although these proteins are produced and stored in the bacterial cytoplasm, they are only released upon certain triggers (Watarai *et al.*, 1995; Ménard *et al.*, 1994a). Host cell activators of the Mxi-Spa secretory system are generally unknown although fibronectin has been proposed as a possible candidate (Watarai *et al.*, 1995). In addition, dyes such as Congo red and some of its steric equivalents are able to induce the secretion of the Ipa proteins and may provide some information about the structure of the components that are able to activate secretion (Bahrani *et al.*, 1997). The ability to artificially induce the release of the Shigella effector molecules provides a means for further characterizing these proteins. As mentioned above, the type III system is closed under normal conditions, i.e., no protein translocation takes place unless it is activated. Mutants in components of the *mxi-spa* genes are unable to secrete under activating conditions, suggesting that the integrity of the translocation machinery is disturbed rendering it non-functional (Allaoui *et al.*, 1995; 1993; 1992; Venkatesan *et al.*, 1992; Andrews *et al.*, 1991). In contrast, two mutants in genes coding for secreted virulence factors (*ipaB* and *ipaD*) show constitutive protein secretion, suggesting that a complex formed by the IpaB and IpaD proteins regulates the secretion of the Ipa proteins and is required to keep the channel closed (Ménard *et al.*, 1994a).

The best characterized type III secretion system is the Yersinia Ysc secreton (for reviews see Cornelis and Wolf-Watz, 1997; Cornelis, 1998). Purified secreted Yops, the Yersinia effector molecules, do not show a cytotoxic effect on cultured cells, even though live extracellular bacteria have such an activity (Rosqvist *et al.*, 1990). These findings led to the hypothesis that the effectors must be directly injected by the bacteria into the host cell's cytosol in order to exert their effect. This hypothesis was corroborated in 1994 by two independent approaches. The first approach involved the use of immunofluorescence and confocal microscopy to detect translocated Yops in the cytoplasm of eukaryotic cells (Rosqvist *et al.*, 1994), and in the second approach, a reporter enzyme strategy was used to demonstrate the direct delivery of a bacterial factor into the host cell cytoplasm (Sory and Cornelis, 1994).

The discovery of the Salmonella needle structure (Kubori *et al.*, 1998) also supports the hypothesis that the factors are injected by an "organelle" which strikingly resembles the flagellar basal body and so the term "injectisome" has recently been suggested for this specialized secretion-translocation system (Cornelis, 1998). Given the similarity of the type III components and the interchangeability of the effector molecules between Shigella, Salmonella, and Yersinia, it is very likely that some Shigella proteins are also injected into the host cell with the help of the Mxi-Spa system (Figure 4). However, this is just a hypothesis and the structure and the elements of this injectisome remain to be identified.

3.1.2. The Bacterial Effector Molecules

Shigella secretes approximately 15 proteins via the Mxi-Spa system, with the four Ipas (*i*nvasion *p*lasmid *a*ntigen) A, B, C and D and Ipg (*i*nvasion *p*lasmid *g*ene) D being the most abundant. The analysis of nonpolar deletion mutants showed that IpaB, IpaC and IpaD are essential for Shigella entry into HeLa cells and for virulence *in vivo*, as assayed by the induction of keratoconjunctivitis in guinea pigs (Ménard *et al.*, 1993). These proteins were identified as the key players from the bacterial side. The *ipa* operon encodes IpaB (62 kDa), IpaC (42 kDa), IpaD (37 kDa), IpaA (70 kDa) and IpgC, a 18 kDa cytoplasmic protein (Figure 3). Subcellular localization and coimmunoprecipitation studies revealed that, under non-secreting conditions, IpaB and IpaC accumulate in the bacterial cytoplasm and independently associate with IpgC (Ménard *et al.*, 1994b). A mutant in

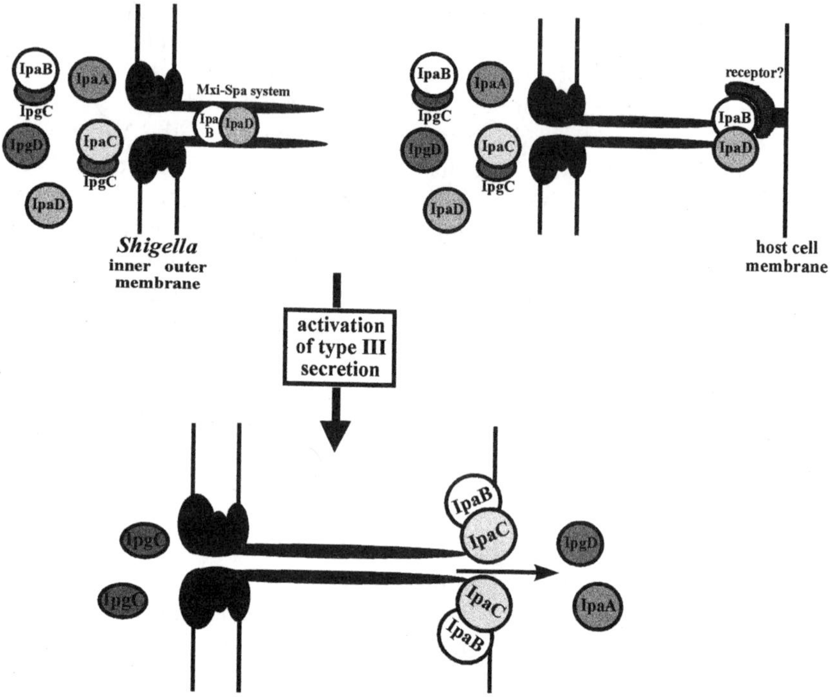

FIGURE 4. Injectisome. Hypothetical scheme showing activation of the Mxi-Spa apparatus in the presence of the target eukaryotic cell membrane, formation of a pore by the IpaB-IpaC complex, and injection of proteins such as IpaA and IpgD. Before activation, the injectisome is closed by a "plug" composed of IpaB and IpaD. This plug could block the needle structure from inside, or it could be located at the tip of the needle and interact with receptors on the host cell surface.

the *ipgC* gene is avirulent and only shows low levels of IpaB and IpaC expression, suggesting that IpgC serves as a molecular chaperone by preventing IpaB and IpaC aggregation and proteolytic degradation (Ménard *et al.*, 1994b). Once secreted, however, IpaB and IpaC form a soluble complex in the extracellular medium; latex beads coated with the IpaB-C complex are internalized by HeLa cells, indicating that this molecular complex acts as an entry effector (Ménard *et al.*, 1996). However, the surface rearrangements observed during internalization of the beads are much less dramatic than the cytoskeletal reorganization induced by Shigella during entry, suggesting that the entry mechanism is only partially reproduced by the IpaB/C complex.

The mode of action of the IpaB/C complex in entry is still unclear. The Ipa complex has been shown to bind the fibronectin receptor $\alpha_5\beta_1$ integrin

(Watarai *et al.*, 1996). Integrins do not appear to be the exclusive receptors for the IpaB-C complex since another cell surface receptor, the hyaluronate receptor CD44, also interacts with IpaB [Skoudy *et al.*, manuscript submitted]. It is possible that these interactions contribute to a transient adherence stage which allows the IpaB-IpaC complex to insert into the epithelial cell membrane, thus forming a pore or translocator structure which allows subsequent injection of other effector proteins into the cell. This hypothesis is based largely on the analogies between the Mxi-Spa system of Shigella and the Ysc type III secretion system of Yersinia. The pore-forming capacity of the complex is illustrated by the ability of IpaC to insert into and destabilize lipid bilayers (De Geyter *et al.*, 1997) and on the requirement of both IpaB and IpaC to cause hemolysis of red blood cells (High *et al.*, 1992; Barzu *et al.*, 1997). The insertion of this IpaB-IpaC complex may have two different effects: (i) the induction of nucleation and polymerization of actin filaments initiating the cytoskeletal rearrangements resulting in bacterial uptake, and (ii) the translocation of the other molecules into the host cell cytoplasm through the putative injectisome discussed above (compare Figure 4).

Undoubtedly, IpaB, IpaC and IpaD are indispensible for Shigella entry into host cells. Since mutations in IpaB and IpaD cause constitutive protein secretion, these proteins are presumably required to keep the injectisome closed. This allows for the accumulation of effector molecules in the bacterial cytoplasm and permits the massive release when they are needed, i.e., upon encountering a target cell. It is not yet clear whether this is the only function of IpaD, or whether IpaD exerts additional effects inside the eukaryotic cell. The fact that the *ipaD* mutant has a very basic defect, i.e. the integrity of the system required for translocation of the bacterial effectors makes elucidation of its particular role more difficult to address. In principle, the same is also true for IpaB, but it is obvious that IpaB is a multifunctional molecule. It does not only act as a plug for the injectisome, but, as mentioned above, it also seems to play a role in establishing contact with the host cell via cell surface receptors. Moreover, it was already observed in 1992 that an *ipaB* mutant failed to lyse the vacuole of infected macrophages and to efficiently kill these cells as observed with wild type Shigella (High *et al.*, 1992). As it turns out, IpaB is the bacterial factor responsible for inducing apoptosis in macrophages (Zychlinsky *et al.*, 1994b; 1992), a process that will be discussed in detail later. With respect to type III secretion, the *ipaC* mutant shows similar levels of secreted proteins as that observed with wild type Shigella, suggesting that the non-invasive phenotype is due to other defects. A functional analysis of IpaC by insertional mutagenesis led to the construction of *ipaC* mutants that still form a complex with IpaB but are affected in entry and hemolytic membrane disruption (Barzu *et al.*, 1997), suggesting that IpaC plays an active role in these processes.

All entry-related functions described in this review so far for IpaB, IpaC and IpaD performed from the outside of the host cell. They are most likely also involved in the translocation of bacterial effector molecules from the bacterial into the eukaryotic cytosol and with the release of the bacteria from the vacuole after being ingested by the host cell. However, it has yet to be demonstrated that these molecules themselves are injected into the cell or that they actively participate in cytoskeletal rearrangements by triggering signalling pathways or interacting with cytoskeletal elements. The reporter enzyme strategy introduced for Yersinia (Sory and Cornelis, 1994) would be the method of choice to address this question, but the intracytoplasmic lifestyle of Shigella makes the detection of translocated proteins more complicated, and, in addition, the amounts of these proteins translocated into the eukaryotic cell are extremely low. So with respect to this question, there is still a way to go.

In contrast to the other Ipas, the inactivation of the *ipaA* gene only leads to a partial decrease of invasion (Tran Van Nhieu *et al.*, 1997b). Although this molecule is not absolutely required for the induction of the entry structures, it does play a role with respect to the "fine tuning" of this process. As seen by immunofluorescence microscopy during infection, the mutant is defective in recruiting the cytoskeletal proteins vinculin and α-actinin into the entry foci (see below). A direct interaction between IpaA and vinculin was demonstrated *in vitro* by blot overlay experiments as well as *in vivo* by coimmunoprecipitation of an IpaA-vinculin complex from Shigella infected cells (Tran Van Nhieu *et al.*, 1997b). These findings strongly implicate IpaA as one of the effector molecules translocated by the injectisome. The fact, however, that IpaA is not absolutely required for entry implies that Shigella entry into host cells is a multi-step process, utilizing signals mediated by different invasins and activating alternative pathways in parallel.

3.1.3. Other Type III Secreted Proteins from Shigella

As mentioned above, not only the Ipa effector molecules are secreted via the Mxi-Spa type III secretion system; a set of about 15 proteins can be detected in the supernatant after activation of secretion (Parsot *et al.*, 1995). *IpgD*, for example, is the first gene of the *mxi-spa* operon (Allaoui *et al.*, 1993b), but the encoded protein is not involved in type III secretion. Instead it is found secreted in amounts comparable to those of the Ipa proteins (Niebuhr *et al.*, unpublished results). The functional role of this protein in the pathogenesis of Shigella remains to be elucidated. This is also the case for IpgA, B, E and F, the genes of which are all located in the Shigella "entry region" of the virulence plasmid.

Recent findings indicate that the type III machinery in Shigella is responsible not only for protein secretion, but that it is also involved in the control mechanisms of transcription of other target genes located on the virulence plasmid (Demers *et al.*, 1998). Analysis of supernatants of constitutively secreting Shigella mutants (Δ*ipa*B, Δ*ipa*D or Δ*ipa*A,B,C,D) identified secreted proteins of 46 and 60 kDa, the products of *virA* and *ipaH9.8*, respectively. Using *lacZ* transcriptional fusions, it was found that, unlike the Ipa proteins, these proteins were not constitutively synthesized and stored in the bacterial cytoplasm; their expression was markedly increased after initial activation of the secretion system. The characterization of a *virA* mutant indicated that VirA is not required for entry into epithelial cells. The differential expression of secreted proteins might reflect differences in the function of these proteins during infection. It is possible that this regulation enables the secretion of a second "wave" of effector molecules that are important for later stages in the infection process, after bacterial entry into the host cell has been achieved.

3.2. Host Cell Proteins Involved in Shigella Entry

3.2.1. Cytoskeletal Proteins

Shigella can enter several cell lines *in vitro*, regardless of their species and organ of origin, and so it appears that the bacteria manipulate very basic processes within the eukaryotic cell for the purpose of invasion. Shigella-induced phagocytosis is characterized by the formation of cellular extensions at the site of bacterial interaction with the cell membrane, which are very similar to the structures induced by Salmonella (Adam *et al.*, 1995; Francis *et al.*, 1993). The membrane-ruffles induced by the bacteria seem to initiate as microspikes and form a blossom-like structure that extends around the bacterial body (see Figure 2). These projections finally merge and engulf the bacterial body, which is a process that lasts five to ten minutes (Tran Van Nhieu *et al.*, 1997a; Adam *et al.*, 1995).

Actin polymerization is essential for bacterial entry, since treatment of cells with cytochalasins completely inhibits Shigella uptake (Clerc and Sansonetti, 1987). Consistent with the dynamic role of actin, observation by electron microscopy after S1-myosin decoration at the early stage of infection shows progressive accumulation of dense dots underneath the cytoplasmic membrane of the epithelial cell at the site of interaction with the bacteria. These dots correspond to actin nucleation zones from which filaments further extend. During the course of the entry process, the membrane ruffles induced by the bacteria contain tightly bundled actin filaments organized in a parallel orientation with their barbed ends oriented towards the tip of the cell extension (Adam *et al.*, 1995). As maturation of this entry focus proceeds, a dense meshwork of polymerized actin accumulates around the bacteria at the base of the ruffle, forming a coat—or cup-like structure—around the nascent bacterial phagosome. This process seems to represent an important step for the internalization of the bacteria (Tran Van Nhieu *et al.*, 1997b).

Immunolocalization studies performed in order to characterize the composition of the membrane ruffles identified actin-binding proteins, such as plastin (Adam *et al.*, 1995), α-actinin (Tran Van Nhieu *et al.*, 1997b) and talin (Watarai *et al.*, 1997), as well as proteins that are considered as linkers between the cytoskeleton and membranes, like myosin I (Clerc and Sansonetti, 1987), cortactin (Dehio *et al.*, 1995), and ezrin [Skoudy *et al.*, manuscript submitted]. Transfection experiments with transdominant negative forms of plastin and ezrin showed a significant inhibition of Shigella entry into these cells, indicating that these proteins play an important role in the integrity and stability of these bacteria-induced cellular extensions (Adam *et al.*, 1995; Skoudy *et al.*, manuscript submitted).

With respect to the cytoskeletal-associated proteins seen in the actin cups that form at the base of the Shigella induced ruffle, there are striking analogies between the composition of these structures and focal adhesion complexes. In cultured fibroblasts or epithelial cells, the focal adhesion complexes, or focal contacts, are the regions where the plasma membrane of the cell is attached to components of the extracellular matrix that have become absorbed onto the culture dish. These complexes serve to anchor the cell to the substratum and are also the sites where the ends of actin stress fibers attach to the plasma membrane. Thus, focal contacts form a link between the actin cytoskeleton and the extracellular matrix, and, as such, have been referred to as "signal transduction organelles" (Lo and Chen, 1994). The main transmembrane linker proteins of focal contacts are members of the integrin family. Proteins of this family contain an external domain that binds to an extracellular matrix component while the cytoplasmic domain is indirectly linked to actin stress fibers via multiple attachment proteins like talin and vinculin (for a review see Burridge and Chrzanowska-Wodnicka, 1996). Like focal contacts, the cup-like cytoskeletal structures formed around Shigella during entry contain several of the characteristic components including the structural proteins α-actinin, $\alpha_5\beta_1$ integrin, paxillin, talin and vinculin (Tran Van Nhieu *et al.*, 1997b; Watarai *et al.*, 1996), and the signalling molecules Src (Dehio *et al.*, 1995) and RhoA (Adam *et al.*, 1996). In addition, the bacterial protein IpaA can be localized in these pseudo-focal adhesions; the IpaA-vinculin interaction is obviously a critical step in the formation of this structure. The observation that the Ipa-complex can bind to β_1 integrins (Watarai *et al.*, 1996) together with the fact that in the absence of these cups bacterial uptake is much less efficient (Tran Van Nhieu *et al.*, 1997b), makes it tempting to speculate that this structure serves to fix and focus the bacteria in the entry structure until uptake is complete, i.e., a phagosome has formed.

Multiple events are therefore orchestrated by Shigella during entry: IpaA-vinculin binding which is responsible for the formation of this focal-adhesion-like structure at the base of the cup, and other bacterial signals, which remain to be identified, that are responsible for the induction of the massive cytoskeletal reorganisation resulting in membrane ruffles at the bacterial entry site.

3.2.2. Role of Signalling Molecules

The signals that are triggered in the host cell and lead to the formation of entry structures are the result of a complex cross talk between Shigella and its target cell. Over the last few years, several pieces of this mosaic have been identified. A useful approach for the identification of signalling pathways involved in bacterial uptake has been the analysis of host cell proteins that are phosphorylated on tyrosine residues upon Shigella entry. Using a specific antibody, it has been shown by immunofluorescence that tyrosyl-phosphorylated proteins are recruited to the site of entry. Western blot analysis and immunoprecipitation experiments identified the actin-binding protein cortactin as the major protein that becomes phosphorylated on tyrosine residues (Dehio *et al.*, 1995). Cortactin has been shown to be a substrate for the Src tyrosine kinase (Wu *et al.*, 1991), suggesting that Src is activated during entry. Recently, it has indeed been demonstrated that expression of a dominant negative form of Src in epithelial cells inhibits cortactin phosphorylation as well as the efficiency of Shigella entry. A significant reduction in actin rearrangements at the entry focus is observed in such cells (Duménil *et al.*, 1998). This indicates that Src activation represents an important step in a signalling pathway that is implicated in this process. Besides cortactin, two other cytoskeletal proteins have been shown to be tyrosine-phosphorylated upon Shigella entry (Watarai *et al.*, 1996): the integrin-directed focal adhesion kinase (FAK), which also contains a binding site for talin (Chen *et al.*, 1995), and paxillin, a component of focal adhesion plaques that contains binding sites for FAK and vinculin (Turner and Miller, 1994). Interestingly, there is evidence that sequential steps in the formation of signalling complexes at focal adhesion sites are (i) the autophosphorylation of FAK thus creating a Src homology (SH) 2 binding site for Src, and (ii) further phosphorylation of FAK as well as paxillin (Calalb *et al.*, 1995; Schaller *et al.*, 1994). This shows that the Shigella entry "cup" and focal contacts do not only reveal similarities with respect to the composition of their structural elements, as discussed above, but also at the level of signalling molecules that are recruited to and involved in the formation of these structures. A summary of the components so far localized in the Shigella entry structure is provided in Figure 5.

Src is not the only important player in the formation of focal adhesion complexes; the small GTPases of the Rho family also play an important role in this phenomenon. Rho can be activated by the addition of extracellular ligands, and its activation results in the assembly of stress fibers and of associated focal adhesion complexes (Ridley and Hall, 1992). Furthermore, it has been shown that Rho controls the translocation of Src to the cell periphery (Finchan *et al.*, 1996). Taken together, these results led to the conclusion that Rho acts as a molecular switch controlling a signal transduction pathway that links membrane receptors to the cytoskeleton. Activation of Rac, another member of the Rho family, leads to the assembly of a meshwork of actin filaments at the cell periphery and the formation of lamellipodia and membrane ruffles (Ridley *et al.*, 1992), and Cdc42, the third Rho subfamily member, induces actin-rich surface protrusions called filopodia (Nobes and Hall, 1995; Kozma *et al.*, 1995). The fact that in all three cases the cytoskeletal changes are associated with distinct, integrin-based adhesion complexes, and that there is crosstalk between Cdc42, Rac

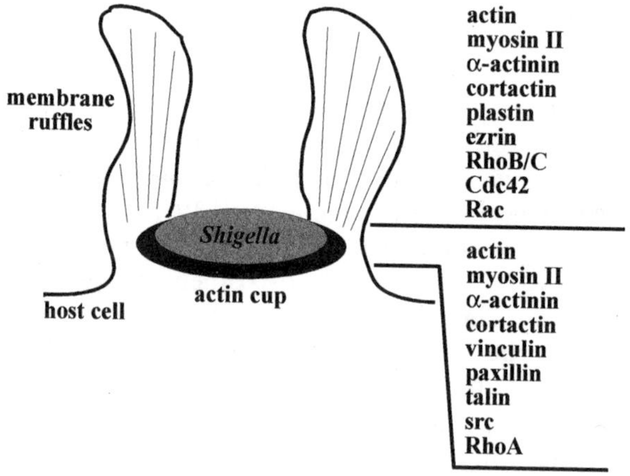

FIGURE 5. Components of the entry structure. Localization of structural, motility/scaffolding and signaling-related cytoskeletal proteins according to sub-domains of the Shigella entry focus.

and Rho resulting in mutual activation suggests that members of the Rho GTPase family are key regulatory molecules that link surface receptors to the organisation of the actin cytoskeleton.

Members of the Rho family act as molecular switches, and cycle between their active-GTP bound form and their inactive GDP bound form. Investigation of the functions of these GTPases was facilitated by the fact that they can be specifically inhibited by over-expressing a GTPase inactive, dominant negative form bearing a mutation in the GTPase domain. First clues to their biological functions, however, came from the use of the C3 exoenzyme of *Clostridium botulinum*, which specifically inactivates Rho by ADP-ribosylation on a critical residue (Aktories *et al.*, 1989), and through TcdB of *Clostridium difficile*, which inactivates Rho, Rac and Cdc42 by monoglucosylation within their effector domain (Just *et al.*, 1995).

The first demonstration that Rho plays a key role in Shigella entry came from the observation that C3 treatment of cells before infection inhibits Shigella entry by 90% (Watarai *et al.*, 1997; Adam *et al.*, 1996). Transfection of cells with epitope-tagged Rho isoforms (RhoA, B and C) showed a recruitment of Rho to entry structures, and, interestingly, the three isoforms showed a differential localization relative to the membrane structure: RhoA preferentially accumulated in close vicinity to the cup-like pseudo focal adhesion plaque formed around bacteria during the process of entry, whereas RhoB and RhoC were highly concentrated in the tips of the cellular projections, where intense actin polymerization was also observed (Adam *et al.*, 1996). In samples treated with C3, actin nucleation zones were still observed around the bacteria suggesting that Rho is essential for the elongation and bundling of actin filaments rather than for the initial actin nucleation (Adam *et al.*, 1996). This is consistant with observations made in quiescent Swiss3T3 cells, where microinjection of fluorescently labelled actin revealed that Rho activation leads to the elongation and bundling of preexisting actin filaments rather than to actin nucleation (Machesky and Hall, 1997).

Using the approach of transfection with constitutively active/inactive small GTPases, we have recently shown that Cdc42 and Rac also play an important role in the entry of Shigella into HeLa cells [Mounier *et al.*, manuscript submitted]. Transient overexpression of a dominant negative form of Rac or Cdc42 led to a dramatic decrease in actin polymerization at the level of entry foci, resulting in an inhibition of Shigella entry by ~70%. When looking at the stepwise formation of Shigella entry structures, it is tempting to speculate that during this process a signalling cascade via Cdc42, Rac and Rho is triggered. Shigella entry begins with the nucleation of actin in the area around and underneath the bacteria, followed by the projection of membrane protrusions that grow and form membrane ruffles engulfing the bacterium. Finally, an actin cup forms around the phagosome. The Rho GTPases could play a role in each of these steps. For example, Shigella entry may involve Cdc42-mediated actin polymerization, followed by Rac-mediated dynamic remodeling of the actin-made entry focus which ends up in the development of a Rho-mediated mature structure (i.e. focal-plaque like) able to efficiently internalize the bacterial body.

The pathways by which Shigella effector molecules activate the three GTPases and Src are not yet clear. As outlined above, up to 15 different proteins are secreted by Shigella upon activation of the type III system (Parsot *et al.*, 1995) and they represent potential candidates as bacterial effectors acting inside the host cell. Although some of these products might not be directly relevant for bacterial entry, it is possible that many of them play a role in fine-tuning of the cellular response and modulate it at different levels of regulatory steps of cytoskeletal organization. The identification of the molecular functions of these different Shigella virulence factors and the elucidation of the role that they play during infection is a challenge for the future and should provide further insights into the host cell functions that are successfully exploited by bacterial pathogens.

4. BACTERIAL MOTILITY AND CELL-TO-CELL SPREAD

As outlined in the last paragraphs, the entry process represents a concerted action of numerous bacterial factors and host cell molecules, ranging from the proper assembly and regulation of the injectisome of Shigella to the release of effector molecules upon receiving a certain signal, and to the translocation of the effector molecules into the host cytoplasm where they trigger a complex response involving different signalling pathways and dramatic reorganisation of the host cell cytoskeleton. In contrast, the ability of Shigella to subvert the actin cytoskeleton once they have reached the host cell cytoplasm, can be attributed to one single bacterial factor: IcsA (VirG).

4.1. The Shigella Virulence Factor IcsA (VirG)

Many bacterial pathogens are able to enter host cells by inducing cytoskeletal rearrangements; some bacteria remain in a vacuole (for example Salmonella or Mycobacteria) while others gain access to the cytosol and subvert the host cytoskeleton in order to colonize the entire tissue. Shigella, Listeria and Rickettsia are the only three bacterial genera found so far that are able to escape from the phagocytic vacuole and to use cytoplasmic cytoskeletal components in order to achieve movement inside infected cells, to reach the cell membrane and, finally, to induce cellular protrusions.

Engulfment of these cellular extensions containing moving bacteria by neighbouring cells leads to cell-to-cell spread of the pathogen (for reviews see Dramsi and Cossart, 1998; Theriot, 1995), which helps them to disseminate throughout the tissue without ever leaving the host cell cytoplasm. Among these three bacteria, Shigella was the first pathogen identified as an intracellularly spreading bacterium inside infected cells. Using time-lapse video microscopy, Ogawa and colleagues (1968) showed that intracellular Shigellae are able to move independently of cellular organelles and to induce formation of cellular extensions at the cell surface. Since this pioneering work, it has been shown that Shigella intracellular motility is associated with the formation of actin tails at one pole of the bacterium and that it is the virulence factor IcsA, also named VirG (Lett *et al.*, 1988; Makino *et al.*, 1986), which accounts for this ability (Bernardini *et al.*, 1989). Using heterologous expression of IcsA, it has been shown that the presentation of IcsA on the bacterial surface is necessary and sufficient to induce this actin polymerization. Expression of IcsA in a Δ*ompT E. coli* K-12 mutant induces actin comet tail formation and bacterial motility in cell-free Xenopus eggs cytoplasmic extracts, thus identifying IcsA as the sole Shigella factor required to induce this process (Goldberg and Theriot, 1995; Kocks *et al.*, 1995).

Like most of the genes involved in bacterial entry into cells, the *icsA* gene is present on the virulence plasmid of *S. flexneri.* It is located outside of the entry locus, 15 kb downstream of the *spa* operon (Demers *et al.*, 1998). Interestingly, the *icsA* gene shows a significantly higher GC content than the genes of the entry region (41% compared to 34%), which suggests an independent acquisition of the *icsA* gene by Shigella. *In vivo* assays showed that the ability to move intracellularly and to spread from cell to cell is critical for Shigella virulence, and Δ*ics*A mutants have been used as candidates for the development of live attenuated Shigella vaccine strains (Sansonetti *et al.*, 1991).

IcsA is a surface protein anchored in the outer membrane of Shigella. The IcsA polypeptide is composed of 1102 amino acids (116 kDa) and it consists of three distinctive domains: the N-terminal signal sequence (amino acids 1–52), the 706 amino acids α-domain (53–758), and the 344 amino acids C-terminal β-core. The β-core is embedded in the outer membrane, whereas the α-domain is presented on the surface of Shigella to interact with the host cell cytoskeleton (Suzuki *et al.*, 1995). The translocation of IcsA across the two membranes and its surface anchoring are not dependent on the Mxi-Spa type III secretion apparatus, but are mediated by an autotransporter secretion pathway, similar to the IgA-protease secretion mechanism (Klausner *et al.*, 1993). It seems that the C-terminal domain (IcsAβ) serves as an autotransporter domain that possibly forms a β-barrel-like structure composed of anti-parallel hydrophobic stretches allowing translocation and anchorage of the N-terminal IcsAα domain at the outer membrane (Suzuki *et al.*, 1995). The surface distribution of IcsA is unusual among bacterial proteins. In wild type bacteria, IcsA is asymmetrically distributed, being present exclusively at the bacterial pole opposite to the septation furrow in dividing bacteria (Goldberg *et al.*, 1993). Inside infected cells, IcsA colocalizes with the base of the actin tail and determines the site of actin assembly and direction of movement. A yet uncharacterized intrinsic property of IcsA may direct most of the protein to the bacterial pole, although a significant proportion is still expressed over the entire cell surface. It has been noted that in Shigella cultures, about 50% of the total amount of IcsA is cleaved at the junction between the IcsAα and the IcsAβ domain, resulting in the release of IcsAα in the culture medium (Fukuda *et al.*, 1995). Cleavage occurs on a sequence previously shown by *in vitro* experiments to be the target of PKA mediated-phosphorylation -S_{756}SRRASS_{762}-(d'Hauteville *et al.*, 1996; 1992), and recently the bacterial surface protease SopA (IcsP) has been shown to be involved in the cleavage of IcsA (Egile *et al.*, 1997; Shere *et al.*, 1997). Like *icsA*, the gene for SopA has also been found on the Shigella virulence plasmid, and this protein has been identified as a member of the OmpT/OmpP family of serine proteases. A Δ*sopA* mutant is unable to polarize IcsA and to induce actin comet tail formation, showing that processing of IcsA by SopA is important for bacterial motility (Egile *et al.*, 1997). SopA may contribute to the polar distribution of IcsA by pronounced cleavage of the IcsA fraction that is expressed on the lateral side of the bacterial surface.

The α-domain of IcsA (Icsα) shares no homology with any known protein, neither with cytoskeletal proteins, nor with ActA, its functional counterpart in the gram-positive pathogen *Listeria monocytogenes*. Over the last few years, the ActA protein has become the model system to study actin-based motility of intracellular bacteria. Apart from the N-terminal signal sequence and a C-terminal membrane anchor, the sequence of ActA exhibits a striking central proline-rich region with a motif that is repeated four times (Domann *et al.*, 1992; Kocks *et al.*, 1992). This motif is also present in the cytoskeletal protein vinculin, and so it was postulated that this region of the molecule might be involved in interactions of Listeria with the host cell cytoskeleton. As it turned out later, this particular repeated proline-rich motif represents a specific ligand for proteins that are implicated in microfilament dynamics (Niebuhr *et al.*, 1997; Gertler *et al.*, 1996), and so Listeria subvert the host cell cytoskeleton by mimicking this ligand motif on their surface.

IcsAα of Shigella also contains a series of repeats of 32 to 34 residues in the N-termal part, but these are not rich in proline, but in glycine (Goldberg *et al.*, 1993). However, no homologies between these repeats and known protein sequences have been found so far. The expression of recombinant IcsA carrying internal in-frame deletions on the surface of Δ*icsA* bacteria has allowed a dissection of the functional domains of this molecule (Suzuki *et al.*, 1997; 1996). Analysis of cells infected with these mutant Shigella strains revealed that the N-terminal two-thirds of the α-domain, which contains the glycine-rich repetitive sequences and the following ~80 amino acids, is necessary for F-actin assembly on the bacterial surface (Suzuki *et al.*, 1996). It was also noticed that the C-terminal ~220 amino acids of IcsAα (509–729) are essential for the unipolar distribution of IcsA on the bacterial surface (Suzuki *et al.*, 1996). A strain harboring this deletion displays a nonpolar distribution of IcsA and although it still elicits F-actin assembly on its surface, this mutant fails to form actin tails and to spread intercellularly. An explanation could be that this internal deletion leads to an inaccessability of the SopB cleavage site (-S_{756}SRRASS_{762}-) thus hampering its proper processing. However, it has been reported that drastic bacterial surface modifications, such as the lack of expression of O-side chains of LPS, also impair polar localization of IcsA, secretion of IcsAα, and actin-based motility (Sandlin *et al.*, 1995; Rajakumar *et al.*, 1994). Therefore, it seems as if several factors are implicated in the proper polar localization of IcsA on the bacterial surface.

4.2. Host Cell Molecules Implicated in Bacterial Movement

As mentioned above, IcsA is the sole factor of Shigella that is required for bacterial movement and intracellular spread. Neither a direct interaction with actin, nor actin nucleacting properties have been shown for IcsA, and so it is likely that the recruitment of a cytosolic protein complex to the bacterial vicinity creates the proper conditions for actin polymerization and comet tail formation. The steps necessary for this process are (i) actin nucleation, (ii) elongation of actin filaments, (iii) capping of the filaments' pointed ends and bundling of the filaments necessary for the proper structure of the comet.

Thermodynamic studies of bacterial intracellular motility were originally done with Listeria and the important processes identified also hold true for Shigella. Using video microscopy and microinjection of labelled actin monomers, it has been shown that the actin tail remains stationary in the cytoplasm, trailing behind the moving bacterium, and that the rate of incorporation of actin monomers correlates directly with the rate of bacterial movement (Theriot *et al.*, 1992). These findings suggest that continuous actin polymerization at one pole of the bacterium is itself sufficient to generate the motile force. Shigella movement inside the cytoplasm is random and rapid (6–60 μm/min) (Zeile *et al.*, 1996), and it occurs optimally at the stage of bacterial division (Goldberg *et al.*, 1994). Ultrastructural analysis of Shigella actin comet tails by transmission electron microscopy after S1 myosin decoration revealed that intracellular comets are composed of short, cross-linked, randomly organized filaments with their fast growing (barbed) ends always oriented towards the bacterial surface. The density of actin filaments is high in the vicinity of the bacterial body and decreases at the distal part of the comet [C. Egile, P.J. Sansonetti and P. Cossart, manuscript in preparation].

Several cytoskeletal proteins found in lamellipodial structures have been detected in Shigella actin tails by immunofluorescence analysis. These proteins include the actin-bundling proteins T-plastin and α-actinin (Prévost *et al.*, 1992), vinculin (Laine *et al.*, 1997; Kadurugamuwa *et al.*, 1991), the focal adhesion protein VASP (Chakraborty *et al.*, 1995), and the *a*ctin-*r*elated *p*roteins Arp2/3 [Egile and Sansonetti, unpublished results]. The only direct binding partners for IcsA that have been identified so far are vinculin and N-WASP (Suzuki *et al.*, 1998; 1996). They have both been found to bind to the glycine-rich repeat region of IcsA that is essential for F-actin assembly, suggesting that these proteins play an important role in this process.

Vinculin is a structural protein involved in the cross-linking of the actin cytoskeleton to the cell membrane via the focal adhesion complex, and, as outlined earlier, it is also implicated in Shigella entry into cells. Immunofluorescence microscopy revealed a colocalization of IcsA and vinculin, and coprecipitation experiments demonstrated a direct interaction between the vinculin head portion and IcsA *in vitro* (Suzuki *et al.*, 1996). As vinculin is considered to be a linker between actin filaments and focal contacts, a model was proposed in which the interaction of the vinculin head with IcsA and interaction of the tail with actin filaments may serve as a similar link between the bacterial surface and the actin comet (Suzuki *et al.*, 1996). The relevance of the IcsA-vinculin interaction for Shigella actin-based motility is still debated, as it has been reported that Shigella motility is unaffected in a vinculin-deficient murine cell line (Goldberg, 1997). However, these cells might not have been completely vinculin negative, since another group detected several truncated vinculin head polypeptides by Western blot analysis as well as messenger RNA spanning the entire head region of vinculin (Laine *et al.*, 1997). An alternative explanation could be that in this cell line another, yet unidentified functional homolog of vinculin might be recruited by IcsA.

Evidence for a functional role for vinculin in Shigella motility was provided by microinjection experiments that were based on the knowledge of the actin-based motility of Listeria. A dissection of the Listeria ActA molecule revealed that there are two separable domains required for interaction with the actin cytoskeleton: an N-terminal region that is essential for actin filament nucleation, and a central, proline-rich repeat region required for efficient actin recruitment (Lasa *et al.*, 1995; Pistor *et al.*, 1995). The proline-rich motif, which is also found in vinculin, represents a specific binding site for proteins of the Ena-VASP family (Niebuhr *et al.*, 1997; Gertler *et al.*, 1996). These proteins in turn recruit the actin monomer binding protein profilin. This observation led to the formation of a model in which ActA attracts profilin-bound actin to the interface of the bacterium and its actin tail by recruiting proteins from the Ena-VASP family, thereby enhancing local actin polymerization and enabling rapid actin-based movement (reviewed by Pollard, 1995). In cells infected with Shigella, microinjection of such a proline-rich ActA/vinculin peptide or of a $(GPPPPP)_3$ peptide representing the profilactin binding site of Ena-VASP family proteins inhibits bacterial motility (Zeile *et al.*, 1996). From these observations, it was concluded that Shigella recruits the proteins from the Ena-VASP family indirectly via vinculin, instead of directly mimicking the ligand motif on their surface like Listeria. In subsequent studies by the same authors, microinjection of the vinculin head portion in Shigella-infected cells led to a 3-fold increase of the bacterial rate of movement (Laine *et al.*, 1997). Based on these results, it was proposed that IcsA binding to vinculin may unmask the VASP interaction motif present in the vinculin head portion leading to recruitment of the VASP-profilin-actin complex and, in turn, actin polymerization (Laine *et al.*, 1997). These observations suggest that Listeria and Shigella have evolved a similar motility mechanism involving VASP recruitment either directly in the case of Listeria or indirectly by an ActA analog, such as vinculin, in the case of Shigella.

The second direct host cell binding partner identified so far for IcsA is the neural Wiskott-Aldrich syndrome protein (N-WASP) (Suzuki *et al.*, 1998). The WASP family proteins, including WASP, the protein deficient in patients suffering of the Wiskott-Aldrich Syndrome (Derry *et al.*, 1994), N-WASP (Miki *et al.*, 1996), Dictyostelium SCAR protein (Bear *et al.*, 1998) and its human homolog WAVE (Miki *et al.*, 1998), are required for normal cytoskeletal function in both humans and yeast and are most likely involved in transmitting signals from receptors to the cytoskeleton. All these proteins are organized into several modular domains: a central region with polyproline stretches that function as SH3 ligands (Rivero-Lezcano *et al.*, 1995), an actin binding motif in the carboxy-terminal part, known as the WH2 or verprolin homology domain, and a conserved series of acidic residues. WASP and N-WASP bind to activated Cdc42 and Rac through their N-terminal CRIB domain, a Cdc42/Rac small GTPase binding motif (Aspenström *et al.*, 1996; Symons *et al.*, 1996) and it has been shown that N-WASP enhances Cdc42-induced formation of filopodia (Miki *et al.*, 1998).

Immunofluorescence analysis revealed that during Shigella infection N-WASP is recruited to the interface of the bacterial pole and the actin comet tail (Suzuki *et al.*, 1998). Immunodepletion of N-WASP in Xenopus extracts abolishes bacterial motility and add-back of recombinant N-WASP restores the actin nucleation step on the bacterial surface. Unexpectedly, N-WASP-mediated nucleation during Shigella motility does not seem to require Cdc42, as inactivation of Rho GTPases by TcdB toxin in infected cells or in Xenopus extracts has no effect on bacterial motility [Mounier *et al.*, manuscript submitted]. Coprecipitation assays using truncated forms of IcsA identified the glycine-rich repeats of IcsA (residues 103–433) as the binding site for N-WASP. This finding underlines the importance of this IcsA domain, as it is also the region of vinculin interaction (residues 53–506) and is essential for F-actin assembly (Suzuki *et al.*, 1998). Very recently it has been found that the C-terminal acidic domain of the WASP family proteins directly interacts with the p21-Arc subunit of the Arp complex (Machesky and Insall, 1998), which can nucleate, cap and crosslink actin filaments (Mullins *et al.*, 1998; Welch *et al.*, 1998). This finding provides a link between the WASP familiy proteins and their cytoskeletal function.

The direct interaction of a bacterial surface molecule with vinculin and N-WASP is a specific feature of the actin-based motility of Shigella that has not been found for Listeria. Since the Arp2/3 complex has been shown to be absolutely essential for actin nucleation on the Listeria surface (Welch *et al.*, 1998; 1997), it seems that ActA directly interacts with the Arp2/3 complex and proteins from the Ena/VASP family. IcsA, in contrast, recruits these proteins indirectly via N-WASP and vinculin. The comparison of the different, albeit so similar, mechanisms that Shigella and Listeria have evolved to hijack the actin machinery of the host cell is a fascinating example of how bacterial pathogens have evolved strategies to exploit physiological processes within a host cell for their own purposes.

4.3. Intercellular Spread

Following bacterial entry, lysis of the phagocytic vacuole, multiplication and intracellular spread, the bacteria form protrusions which extend from the infected cells and are endocytosed by the adjacent cell (Sansonetti *et al.*, 1994; Prévost *et al.*, 1992). Bacteria were observed to move along the actin ring of the perijunctional area of polarized Caco-2 cells, and cell-to-cell spread seems to predominantly occur in the region of the eukaryotic intermediate junctions (Vasselon *et al.*, 1992). A variety of host cell components are implicated in this process, including actin, constituents of the intermediate junctions and, quite likely, host cell signalling molecules. The bacterial- induced protrusions appear as extensions of cellular junctions, as

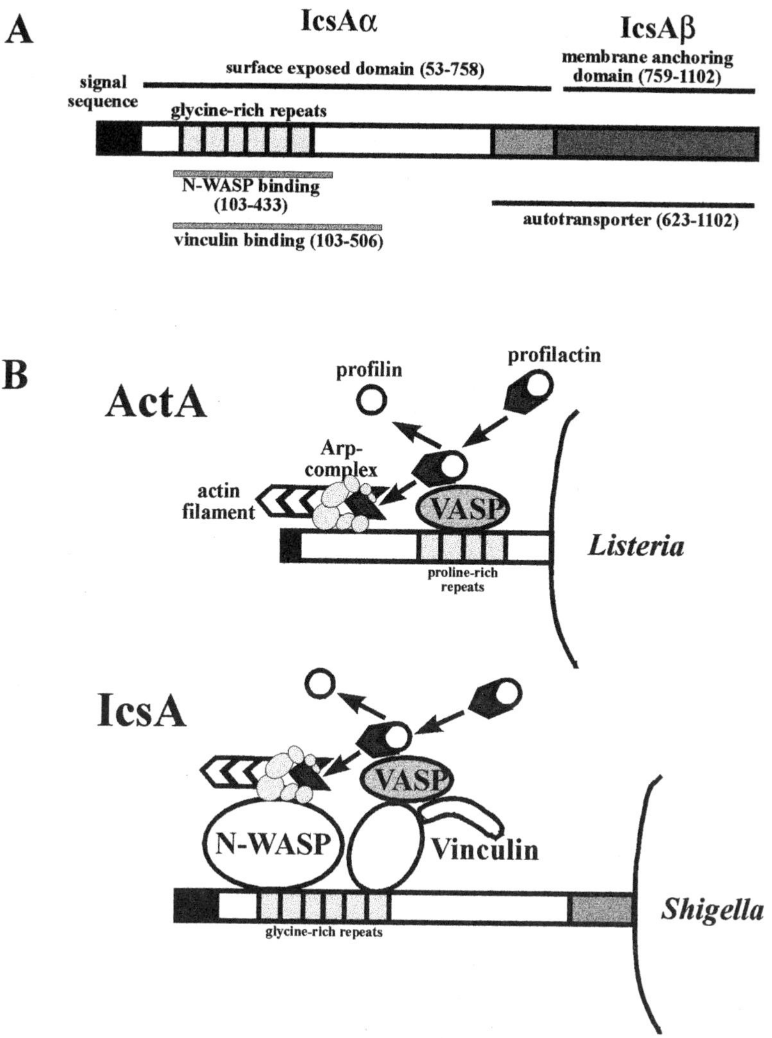

FIGURE 6. IcsA-model/comparison of IcsA with ActA. (A) Schematic drawing of the IcsA protein and its functional domains. (B) Hypothetic model of IcsA mediated actin polymerization on the surface of Shigella and a comparison with the current model on actin-based motility. The Listeria ActA protein seems to recruit the Arp-complex and VASP directly, whereas Shigella IcsA recruits N-WASP and vinculin as adaptor molecules.

α-catenin, β-catenin, α-actinin and vinculin, the major components of this specialized structure, colocalize with the protrusion (Sansonetti *et al.*, 1994; Kadurugamuwa *et al.*, 1991). Using a mouse fibroblastic sarcoma cell line that does not produce cell adhesion molecules and the same cell line transfected with cadherins, it was demonstrated that cadherin expression dramatically facilitates cell-to-cell spread of *S. flexneri*. In these transfected cells, the cadherins were shown to be necessary for the proper structure and dynamics of the protrusions, as well as for their efficient internalization by adjacent cells (Sansonetti *et al.*, 1994). Intermediate junctions are believed to participate in cell signalling and communication between cells, and so the elucidation of the signalling events triggered by Shigella during cell-to-cell spread is another challenge for future studies.

Once endocytosed by the adjacent cell, the bacteria are trapped inside a pocket surrounded by a double membrane. The Shigella gene *icsB* has been shown to be necessary for lysis of these two membranes that surround the bacterium. *IcsB* is located on the virulence plasmid, approximately 1.5 kb upstream of the *ipa* genes. A strain containing a mutation in this gene forms protrusions similar to those of the wild type, but it is unable to lyse the double membrane and remains trapped within this pocket (Allaoui *et al.*, 1992). The exact function of IcsB in this process remains to be identified.

In summary, in the context of an epithelial lining, the invasive phenotype of Shigella leads to an efficient process of intracellular colonization. This pathogen has the capacity to induce entry into non phagocytic cells, escape to the cytoplasm, grow intracellularly, move within the infected cell and spread to adjacent cells. This multifactorial process is a spectacular example of integration of several defined steps leading to progression of infection in an enclosed sanctuary that is relatively protected from humoral and cellular effectors of the innate and adaptative immune response.

5. HOST CELL KILLING

As Shigella fail to enter polarized epithelial cells from the apical pole, a successful outcome of infection is dependent on their access to the basolateral pole from subepithelial tissues and, as discussed earlier, involves entry through M cells and transport through the follicular cells underneath. This is where Shigella encounter macrophages, the resident professional phagocytes that are ready to engulf and destroy any invading microorganism (Soestayo *et al.*, 1990; Mahida *et al.*, 1989). Although Shigella are phagocytosed by this macrophage population, wild type strains, unlike their non-invasive mutants, escape from the phagosome and induce apoptotic death of the host macrophage (Zychlinsky *et al.*, 1992). The observation of programmed macrophage cell death during Shigella infection has been confirmed *in vivo* in infected Peyer's patches in the rabbit ligated loop model (Zychlinsky *et al.*, 1996) and in the rectal mucosa of human dysenteric patients (Islam *et al.*, 1997). Background levels of apoptosis are detected in this area when infection is carried out with a non invasive mutant, whereas numerous apoptotic cells can be seen if an invasive Shigella is used. Induction of apoptosis has been found for all invasive clinical isolates belonging to the different species of *Shigella* so far (Guichon and Zychlinsky, 1997). Shigella-induced apoptosis seems to remain restricted to monocytes/macrophages, as it does not occur during epithelial cell infection (Mantis *et al.*, 1996).

Apoptosis is an essential cellular function in the development and homeostasis of multicellular organisms; it represents death from within, where the dying cell provides the weapons for its own destruction. One of the main features of such programmed cell death is the expression of specific surface receptors that allow phagocytes to recognize and engulf cellular cadavers, thus avoiding spillage of intracellular contents that could cause tissue destruction, inflammation and presentation of intracellular antigens (Savill *et al.*, 1993; Arends and Wyllie, 1991). By inducing macrophage apoptosis, Shigella kills the most effective bactericidal cell in the tissue and thereby avoids its own death. Unlike normal apoptotic cells that die in "immunological anonymity", this programmed cell death results in an acute inflammation causing tissue damage, which in turn permits further bacterial invasion. The signalling pathways leading to Shigella-induced apoptosis in macrophages have been elucidated over the last few years, and the studies have facilitated the understanding of how so few bacteria can provoke such fast and severe inflammation as seen during shigellosis.

The Shigella IpaB protein is required for apoptotic killing of macrophages, provided that the bacteria are phagocytosed and the vacuoles disrupted. Direct cytotoxicity of IpaB has been demonstrated both by a genetic approach (Zychlinsky *et al.*, 1994b) and more directly by microinjecting this purified protein into macrophages (Chen *et al.*, 1996). The mode of action of IpaB appears to reside in its capacity to bind to pro-caspase-1. Caspase-1 is the interleukin-1β (IL-1β)-converting enzyme (ICE) that belongs to the growing family of cysteine proteases which act in a proteolytic cascade involving initiator and effector caspases. The latter enzymes in this cascade achieve the degradation of various substrates and lead to programmed cell death (Thornberry and Lazebnik, 1998). All caspases are translated as proenzymes which are subsequently cleaved into two polypeptides and dimerize or tetramerize to form the active enzyme. Caspase-1 belongs to a group of pro-inflammatory caspases whose role appears to be the cleavage of pro-IL-1β and IL-18. Macrophages synthesize IL-1β as a biologically inactive 30 kDa protein that lacks a secretion signal sequence and accumulates in the cytoplasm. The binding of IpaB to caspase 1 initiates the apoptotic process, which is accompanied by the massive release of mature IL-1β *in vitro* (Zychlinsky *et al.*, 1994a). The key role of IL-1 in the *in vivo* situation has been clearly demonstrated; inflammation is prevented by pretreating animals with IL-1 receptor antagonist (IL-ra), a macrophage product that binds to the IL-1 receptor without initiating signal transduction, before infection (Sansonetti *et al.*, 1995).

Macrophages isolated from caspase-1 knock-out mice are not susceptible to Shigella-induced phagocytosis, although they respond normally to other apoptotic stimuli (Hilbi *et al.*, 1998). This effect was specific for caspase-1, as macrophages from other caspase knock-out mice were efficiently killed by Shigella. These findings suggest that Shigella-induced apoptosis is distinct from other forms of apoptosis and seems to be uniquely dependent on caspase-1, which is directly activated through IpaB, bypassing signal transduction events and caspases upstream of caspase-1 (Hilbi *et al.*, 1998). The direct activation of caspase-1 by IpaB, the release of IL-1 by infected macrophages, and the importance of IL-1 in the inflammatory response during shigellosis all suggest that IL-1β functions as an "emergency" proinflammatory cytokine. One can envision that during infection with invasive agents that cause havoc in a short period of time, the only way for the host to prevent the infection from spreading is to initiate an intense inflammatory response as fast as possible. Although uncontrolled growth of Shigella would destroy the entire colonic epithelium in a few hours, dysentery is a self-limiting disease in patients that do not succumb to early local or systemic complications. Hence, in spite of its deleterious effects, the massive inflammatory response mounted by the host acts to control the infection.

6. CONCLUDING REMARKS

With the combination of both *in vitro* and *in vivo* approaches, the complex mechanisms involved in the pathogenesis of shigellosis are progressively being unraveled. The major contributions in this area over the last two decades have been (i) the establishment of the genetic basis of epithelial cell invasion by Shigella and its regulation by environmental cues, (ii) the description of the infectious cycle of the bacteria in epithelial cells and the recognition of the molecular cross-talk accounting for entry, intracellular movement and cell-to-cell spread, (iii) the description of apoptotic killing of macrophages with its dual implication on bacterial survival and elicitation of inflammation, and finally (iv) the description of the role of inflammation in facilitating mucosal invasion. These concepts have also led to the development of a series of promising live vaccine candidates against shigellosis.

It is clear that further studies are required in order to get a more complete picture of the pathogenesis of shigellosis, since several questions remain. It is not clear, for example, why Shigella is so highly adapted for infecting the human and primate colonic epithelium and why only such a low infective dose is required to cause disease. Also, we are only at the beginning of deciphering the signalling pathways that lead to entry into epithelial cells, to programmed cell death of macrophages and to the particularly severe inflammation observed during shigellosis. It also needs to be confirmed that translocation through the epithelium occurs essentially via M cells, since inflammation occuring at distant sites may also facilitate entry. Another gap in our understanding is the bases of immune protection against the disease. Hopefully the elucidation of these aspects of Shigella pathogenesis will lead to the development of effective attenuated vaccine candidates or the design of novel therapeutic strategies for blocking specific interactions between the pathogen and the host target components.

ACKNOWLEDGMENTS. The authors wish to thank Claude Parsot, Dana Philpott and Michelle Rathman for many helpful comments on the manuscript, and Ariel Blocker for communication of unpublished results.

7. REFERENCES

Adam, T., Arpin, M., Prévost, M.C., Gounon, P., and Sansonetti, P.J., 1995, Cytoskeletal rearrangements and the functional role of T-plastin during entry of *Shigella flexneri* into HeLa cells, *J. Cell Biol.* **129**:367–381.

Adam, T., Giry, M., Boquet, P., and Sansonetti, P.J., 1996, Rho-dependent membrane folding causes Shigella entry into epithelial cells, *EMBO J.* **15**:3315–3321.

Allaoui, A., Sansonetti, P.J., and Parsot, C., 1992, MxiJ, a lipoprotein involved in secretion of Shigella Ipa invasin, is homologous to YscJ, a secretion factor of the Yersinia Yop proteins, *J. Bacteriol.* **174**:7661–7669.

Allaoui, A., Sansonetti, P.J., and Parsot, C., 1993a, MxiD: an outer membrane protein necessary for the secretion of the *Shigella flexneri* Ipa invasins, *Mol. Microbiol.* **7**:59–68.

Allaoui, A., Sansonetti, P.J., and Parsot, C., 1993b, Characterization of the *Shigella flexneri ipgD* and *ipgF* genes, which are located in the proximal part of the *mxi* locus, *Infect. Immun.* **61**:1707–1714.

Allaoui, A., Sansonetti, P.J., Ménard, R., Barzu, S., Mourinier, J., Phalipon, A., and Parsot, C., 1995, MxiG, a membrane protein required for secretion of Shigella invasins: involvement in entry into epithelial cells and intracellular dissemination, *Mol. Microbiol.* **17**:461–470.

Aktories, K., Braun, S., Rösener, S., Just, I., and Hall, A., 1989, The *rho* gene product expressed in *E. coli* is a substrate for botulinum ADP-ribosyltransferase C3, *Biochem. Biophys. Res. Commun.* **158**:209–213.

Andrews, G.P., and Maurelli, A.T., 1992, *mxiA* of *Shigella flexneri* 2a, which faciliates export of invasion plasmid antigens, encodes a homolog of the low-calcium response protein, LcrD, of *Yersinia pestis, Infect. Immun.* **60**:3287–3295.

Andrews, G.P., Horomockyj, A.E., Coker, C., and Maurelli, A.T., 1991, Two novel virulence loci, *mxiA* and *mxiB*, in *Shigella flexneri* 2a faciliate excretion of invasion plasmid antigens, *Infect. Immun.* **59**:1997–2005.

Aspenström, P., Lindberg, U., and Hall, A., 1996, Two GTPases, Cdc42 and Rac, bind directly to a protein implicated in the immunodeficiency disorder Wiskott-Aldrich syndrome, *Curr. Biol.* **6**:70–75.

Arends, M.J., and Wyllie, A.H., 1991, Apoptosis: mechanisms and role in pathology, *Int. Rev. Exp. Pathol.* **32**:223–254.

Bahrani, F.K., Sansonetti, P.J., and Parsot, C., 1997, Secretion of Ipa proteins by *Shigella flexneri*: inducing molecules and kinetics of activation, *Infect. Immun.* **65**:4005–4010.

Barzu, S., Benjelloun-Touimi, Z., Phalipon, A., Sansonetti, P.J., and Parsot, C., 1997, Functional analysis of the *Shigella flexneri* IpaC invasin by insertional mutagenesis, *Infect. Immun.* **65**:1599–1605.

Bear, J.E., Rawls, J.F., and Saxe, C. III., 1998, SCAR, a WASP-related protein, isolated as a suppressor of receptor defects in late Dictyostelium development, *J. Cell. Biol.* **142**:1325–1335.

Bernardini, M.L., Mounier, J., d'Hauteville, H., Coquis-Rondon, M., and Sansonetti, P.J., 1989, Identification of *ics*A, a plasmid locus of *Shigella flexneri* which governs bacterial intra- and intercellular spread through interaction with F-actin, *Proc. Natl. Acad. Sci. USA* **86**:3867–3871.

Brenner, D.J., Fanning, G.R., Miklos, G.V., and Steigerwalt, A.G., 1973, Polynucleotide sequence relatedness among *Shigella* species, *Int. J. Syst. Bacteriol.* **23**:1–7.

Burridge, K., and Chrzanowska-Wodnicka, M., 1996, Focal adhesions, contractility and signalling, *Annu. Rev. Cell Dev. Biol.* **12**:463–519.

Calalb, M.B., Polte, T.R., and Hanks, S.K., 1995, Tyrosine phosphorylation of focal adhesion kinase at sites in the catalytic domain regulates kinase activity: a role for the Src family kinases, *Mol. Cell Biol.* **15**:954–963.

Carpenter, P.B., Zuberi, A.R., and Ordal, G.W., 1993, *Bacillus subtilis* flagellar proteins FliP, FliQ, FliR and FlhB are related to *Shigella flexneri* virulence factors, *Gene* **137**:243–245.

Chakraborty, T., Ebel, F., Domann, E., Niebuhr, K., Gerstel, B., Pistor, S., Temm-Grove, C.J., Jockush, B.M., Reinhard, M., Walter, U., and Wehland, J., 1995, A focal adhesion factor directly linking intracellularly motile *Listeria monocytogenes* and *Listeria ivanovii* to the actin based cytoskeleton of mammalian cells, *EMBO J.* **14**:1314–1321.

Chen, H.C., Appeddu, P.A., Parsons, J.T., Hildebrand, J.D., Schaller, M.D., and Guan, J.L., 1995, Interaction of focal adhesion kinase with the cytoskeletal protein talin, *J. Biol. Chem.* **270**:16995–16999.

Chen, Y., Smith, M.R., Thirumalai, K., and Zychlinsky, A., 1996, A bacterial invasin induces macrophage apoptosis by binding directly to ICE, *EMBO J.* **15**:3853–3860.

Clerc, P., and Sansonetti, P.J., 1987, Entry of *Shigella flexneri* into HeLa cells: evidence for directed phagocytosis involving actin polymerization and myosin accumulation, *Infect. Immun.* **55**:2681–2688.

Cornelis, G.R., 1998, The Yersinia deadly kiss, *J. Bacteriol.* **180**:5495–5504.

Cornelis, G.R., and Wolf-Watz, H., 1997, The Yersinia Yop virulon, a bacterial system for subverting eukaryotic cells, *Mol. Microbiol.* **23**:861–867.

Cossart, P., Boquet, P., Normark, S., and Rappuoli, R., 1996, Cellular microbiology emerging, *Science* **271**(5247):315–316.

De Geyter, C., Vogt, B., Benjelloun-Touimi, Z., Sansonetti, P.J., Ruysschaert, J.-M., Parsot, C., and Cabiaux, V., 1997, Interaction of IpaC, a protein involved in entry of *S. flexneri* into epithelial cells, with lipid membranes, *FEBS Letter* **400**:149–154.

Dehio, C., Prévost, M.C., and Sansonetti, P.J., 1995, Invasion of epithelial cells by *Shigella flexneri* induces tyrosine phosphorylation of cortactin by a pp60$^{c\text{-}src}$ mediated signalling pathway, *EMBO J.* **14**:2471–2482.

Derry, J.M.J., Ochs, H.J., and Francke, U., 1994, Isolation of a novel gene mutated in Wiskott-Aldrich Syndrome, *Cell* **78**:635–644.

Duménil, G., Olivo, J.C., Pellegrini, S., Fellous, M., Sansonetti, P.J., and Tran Van Nhieu, G., 1998, Interferon α inhibits a Src-mediated pathway necessary for Shigella-induced cytoskeletal rearrangements in epithelial cells, *J. Cell Biol.* **143**:1–10.

Demers, B., Sansonetti, P.J., and Parsot, C., 1998, Induction of type III secretion in *Shigella flexneri* is associated with differential control of transcription of genes encoding secreted proteins, *EMBO J.* **17**:2894–2903.

Domann, E., Wehland, J., Rohde, M., Pistor, S., Hartl, M., Goebel, W., Leimeister-Wächter, M., Wuenscher, M., and Chakraborty, T., 1992, A novel bacterial virulence gene in *Listeria monocytogenes* required for host cell microfilament interaction with homology to the proline-rich region of vinculin, *EMBO J.* **11**:1981–1990.

DuPont, H.L., Levine, M.M., Hornick, R.B., and Formal, S.B., 1989, Inoculum size in shigellosis and impliations for expected mode of transmission, *J. Infect. Dis.* **159**:1126–1128.

Dramsi, S., and Cossart, P., 1998, Intracellular pathogens and actin cytoskeleton, *Ann. Rev. Cell Dev. Biol.* **14**:137–166.

Egile, C., d'Hauteville, H., Parsot, C., and Sansonetti, P.J., 1997, SopA, the outer membrane protease responsible for polar localization of IcsA in *Shigella flexneri*, *Mol. Microbiol.* **23**:1063–1074.

Fallman, M., Andersson, K., Hakansson, S., Magnusson, K.E., Stendahl, O., and Wolf-Watz, H., 1995, *Yersinia pseudotuberculosis* inhibits Fc receptor-mediated phagocytosis in J774 cells, *Infect. Immun.* **63**:3117–3124.

Finchan, V.J., Unlu, M., Bruntou, V.G., Pitts, J.D., Wyke, J.A., and Frame, M.C., 1996, Translocation of *src* kinase to the cell periphery is mediated by the actin cytoskeleton under the control of the *rho* family of small G proteins, *J. Cell Biol.* **135**:1551–1564.

Finlay, B.B., and Falkow, S., 1990, Salmonella interactions with polarized human intestinal Caco-2 epithelial cells, *J. Infect. Dis.* **162**:1096–1106.

Francis, C.L., Ryan, T.A., Jones, B.D., Smith, S.J., and Falkow, S., 1993, Ruffles induced by Salmonella and other stimuli direct micropinocytosis of bacteria, *Nature* **364**:639–642.

Fukuda, I., Suzuki, T., Munakata, H., Hayashi, N., Katayama, E., Yoshikawa, M., and Sasakawa, C., 1995, Cleavage of Shigella surface protein VirG occurs at a specific site, but the secretion is not essential for intracellular spreading, *J. Bact.* **177**:1719–1726.

Gaillard, J.L., Berche P., Frehel, C., Gouin, E., and Cossart, P., 1991, Entry of *L. monocytogenes* into cells is mediated by internalin, a repeat protein reminiscent of surface antigens from gram-posistive cocci, *Cell* **65**:1127–1141.

Gertler, F.B., Niebuhr, K., Reinhard, M., Wehland, J., and Soriano, P., 1996, Mena, a relative of VASP and Drosophila Enabled, is implicated in the control of microfilament dynamics, *Cell* **87**:227–239.

Goldberg, M.B., 1997, Shigella actin-based motility in the absence of vinculin, *Cell. Motil. Cytoskeleton* **37**:44–53.

Goldberg, M.B., and Sansonetti, P.J., 1993, Shigella subversion of the cellular cytoskeleton: a strategy for epithelial colonization, *Infect. Immun.* **61**:4941–4946.

Goldberg, M.B., and Theriot, A.J., 1995, *Shigella flexneri* surface protein IcsA is sufficient to direct actin-based motility, *Proc. Natl. Acad. Sci. USA* **92**:6572–6576.

Goldberg, M.B., Barzu, O., Parsot, C., and Sansonetti, P.J., 1993, Unipolar localization and ATPase activity of IcsA, a *Shigella flexneri* protein involved in intracellular movement, *Infect. Immun.* **175**:2189–2196.

Goldberg, M.B., Theriot, J.A., and Sansonetti, P.J., 1994, Regulation of surface presentation of IcsA, a Shigella protein essential to intracellular movement and spread, is growth phase dependent, *Infect. Immun.* **62**:5664–5668.

Groisman, E.A., and Ochman, H., 1993, Cognate gene clusters govern invasion of host epithelial cells by *Salmonella typhimurium* and *Shigella flexneri*, *EMBO J.* **12**:3779–3787.

Guichon, A., and Zychlinsky, A., 1997, Clinical isolates of *Shigella* species induce apoptosis in macrophages, *J. Infect. Dis.* **175**:470–473.

d'Hauteville, H., and Sansonetti, P.J., 1992, Phosphorylation of IcsA by cAMP-dependent protein kinase and its effect on intercellular spread of *Shigella flexneri*, *Mol. Microbiol.* **6**:833–841.

d'Hauteville, H., Dufourcq-Lagelouse, R., Nato, F., and Sansonetti, P.J., 1996, Lack of cleavage of IcsA in *Shigella flexneri* causes aberrant movement and allows demonstration of a cross reactive eukaryotic protein, *Infect. Immun.* **64**:511–517.

Hermant, D., Ménard, R., Arricau, N., Parsot, C., and Popoff, M.Y., 1995, Functional conservation of the Salmonella and Shigella effectors in entry into epithelial cells, *Mol. Microbiol.* **17**:781–789.

High, N., Mounier, J., Prévost, M.C., and Sansonetti, P.J., 1992, IpaB of *Shigella flexneri* causes entry into epithelial cells and escape from the phagocytic vacuole, *EMBO J.* **11**:1991–1999.

Hilbi, H.J., Moss, E., Hersh, D., Chen, Y., Arondel, J., Banerjee, R.A., Flavell, J., Yuan, J., Sansonetti, P.J., and Zychlinsky, A., 1998, Shigella-induced apoptosis is dependent on Caspase-1 which binds to IpaB, *J. Biol. Chem.* **273**:32864.

Isberg, R.R., and Leong, J.M., 1990, Multiple β1 chain integrins are receptors for invasin, a protein that promotes bacterial penetration into mammalian cells, *Cell* **60**:861–871.

Islam, D., Veress, B., Bardhan, P.K., Lindberg, A.A., and Christensson, B., 1997, *In situ* characterization of inflammatory responses in the rectal mucosae of patients with shigellosis, *Infect. Immun.* **65**:739–749.

Just, I., Selzer, J., Wilm, M., von Eichel-Streiber, C., and Aktories, K., 1995, Glucosylation of Rho proteins by *Clostridium difficile* toxin B, *Nature* **375**:500–503.

Klausner, T., Pohlner, J., and Meyer, T.F., 1993, The secretion pathway of IgA protease-type proteins in gram-negative bacteria, *Bioessays* **15**:799–805.

Kocks, C., Gouin, E., Tabouret, M., Berche, P., Ohayon, H., and Cossart, P., 1992, *Listeria monocytogenes*-induced actin assembly requires the *actA* gene product, a surface protein, *Cell* **68**:521–531.

Kocks, C., Marchand, J.B., Gouin, E., d'Hauteville, H., Sansonetti, P.J., Carlier, M.F., and Cossart, P., 1995, The unrelated surface proteins ActA of *Listeria monocytogenes* and IcsA of *Shigella flexneri* are sufficient to confer actin-based motility to *L. innocua* and *E. coli* respectively, *Mol. Microbiol.* **18**:413–423.

Kotloff, K.L., Winickoff, J.P., Ivanoff, B., Clemens, J.D., Swerdlow, D.L., Sansonetti, P.J., Adak, G.K., and Levine, M.M., Global burden of Shigella infections: implication for vaccine development and implementation, *WHO Bulletin*. (in press).

Kozma, R., Ahmed, S., Best, A., and Lim, L., 1995, The Ras-related protein Cdc42Hs and bradykinin promote formation of peripheral actin microspikes and filopodia in Swiss3T3 fibroblasts, *Mol. Cell Biol.* **15**:1942–1949.

Kubori, T., Matsushima, Y, Nakamura, D., Uralil, J., Lara-Tejero, M., Sukhan, A., Galán, J.E., and Aizawa, S.-I., 1998, Supermolecular structure of the *Salmonella typhimurium* type III protein secretion system, *Science* **280**:602–605.

LaBrec, E.H., Schneider, H., Magnani, T.J., and Formal, S.B., 1964, Epithelial cell penetration as an essential step in the pathogenesis of bacillary dysentery, *J. Bacteriol.* **88**:1503–1518.

Laine, R.O., Zeile, W., Kang, F., Purich, D.L., and Southwick, F.S., 1997, Vinculin proteolysis unmasks an ActA homolog for actin-based Shigella motility, *J. Cell. Biol.* **138**:1255–1264.

Lasa, I., David, V., Gouin, E., Marchand, J.-B., and Cossart, P., 1995, The N-terminal part of ActA is critical for the actin-based motility of *Listeria monocytogenes*; the central proline-rich region acts as a stimulator, *Mol. Microbiol.* **18**:425–426.

Lett, M.C., Sasakawa, C., Okada, N., Sakai, T., Makino, S., Yamada, M., Komatsu, K., and Yoshikawa, M., 1988, *vir*G, a plasmid-coded virulence gene of *Shigella flexneri*: identification of the VirG protein and determination of the complete coding sequence, *J. Bacteriol.* **171**:353–359.

Lo, S.H., and Chen, L.B., 1994, Focal adhesion as a signal transduction organelle, *Cancer Metastasis Rev.* **13**:9–24.

Machesky, L.M., and Hall, A., 1997, Role of actin polymerization and adhesion to extracellular matrix in Rac- and Rho-induced cytoskeletal reorganization, *J. Cell Biol.* **138**:913–926.

Machesky, L.M., and Insall, R.H., 1998, Scar1 and the related Wiskott-Aldrich syndrome protein, WASP, regulate the actin cytoskeleton through the Arp2/3 complex, *Curr. Biol.* **8**:1347–1356.

Mahida, Y.R., Patel, S., Gionchetti, P., Vaux, D., and Jewell, D.P., 1989, Macrophage subpopulations in lamina propria of normal and inflamed colon and terminal ileum, *Gut* **30**:826–834.

Makino, S., Sasakawa, C., Kamata, K., Kurata, T., and Yoshikawa, M., 1986, A genetic determinant required for continuous reinfection of adjacent cells on large plasmid in *S. flexneri* 2a, *Cell* **46**:551–555.

Mantis, N., Prévost, M.C., and Sansonetti, P.J., 1996, Anaysis of epithelial cell stress responses during infection by *Shigella flexneri*, *Infect. Immun.* **64**:2474–2482.

Maurelli, A.T., Baudry, B., d'Hauteville, H., Hale, T.L., and Sansonetti, P.J., 1985, Cloning of plasmid DNA sequences involved in invasion of HeLa cells by *Shigella flexneri*, *Infect. Immun.* **49**:164–171.

Ménard, R., Sansonetti, P.J., and Parsot, C., 1993, Non polar mutagenesis of the *ipa* genes defines IpaB, IpaC and IpaD as effectors of *Shigella flexneri* entry into epithelial cells, *J. Bacteriol.* **175**:5899–5906.

Ménard, R., Sansonetti, P.J., and Parsot, C., 1994a, The secretion of the *Shigella flexneri* Ipa invasins is induced by the epithelial cell and controlled by IpaB and IpaD, *EMBO J.* **13**:5293–5302.

Ménard, R., Sansonetti, P.J., Parsot, C., and Vasselon, T., 1994b, Extracellular association and cytoplasmic partitioning of the IpaB and IpaC invasins of *Shigella flexneri*, *Cell* **79**:515–525.

Ménard, R., Prévost, M.C., Gounon, P., Sansonetti, P.J., and Dehio, C., 1996, The secreted Ipa complex of *Shigella flexneri* promotes entry into mammalian cells, *Proc. Natl. Acad. Sci. USA* **93**:1254–1258.

Mengaud, J., Ohayon, H., Gounon, P., Mège, R.M., and Cossart, P., 1995, E-cadherin is the receptor for internalin, a surface protein required for entry of *L. monocytogenes* into epithelial cells, *Cell* **84**:923–932.

Miki, H., Miura, K., and Takenawa, T., 1996, N-WASP, a novel actin-depolymerizing protein regulates the cortical cytoskeletal rearrangements in a PIP2-dependent manner downstream of tyrosine kinases, *EMBO J.* **15**:5326–5335.

Miki, H., Suetsugu, S., and Takenawa, T., 1998, WAVE, a novel WASP-family protein involved in actin reorganization induced by Rac, *EMBO J.* **17**:6932–6941.

Miki, H., Sasaki, T., Takai, Y., and Takenawa, T., 1998, Induction of filopodium formation by a WASP-related actin-depolymerizing protein N-WASP, *Nature* **391**:93–96.

Mounier, J., Vasselon, T., Hellio, R., Lesourd, M., and Sansonetti, P.J., 1992, *Shigella flexneri* enters human colonic Caco-2 epithelial cells through their basolateral pole, *Infect. Immun.* **60**:237–248.

Mullins, R.D., Heuser, J.A., and Pollard, T.D., 1998, The interaction of Arp2/3 complex with actin: nucleation, high affinity pointed end capping, and formation of branching networks of filaments, *Proc. Natl. Acad. Sci. USA* **95**:6181–6186.

Niebuhr, K., Ebel, F., Ronald, F., Reinhard, M., Domann, E., Carl, U.D., Ulrich, W., Gertler, F.B., Wehland, J., and Chakraborty, T., 1997, A novel proline-rich motif present in ActA of *Listeria monocytogenes* and cytoskeletal proteins is the ligand for the EVH1 domain, a protein module present in the Ena/VASP family, *EMBO J.* **16**:2793–2802.

Nobes, C.D., and Hall, A., 1995, GTPases regulate the assembly of multimolecular focal complexes associated with actin stress fibers, lamellipodia and filopodia, *Cell* **81**:53–62.

Oaks, E.V., Hale, T.L., and Formal, S.B., 1986, Serum immune response to Shigella protein antigens in rhesus monkeys and humans infected with *Shigella* spp., *Infect. Immun.* **53**:57–63.

Ogawa, H., Nakamura, A., and Nakaya, R., 1968, Cinemicrographic study of tissue cell cultures infected with *Shigella flexneri, Jpn. J. Med. Sci. Biol.* **21**:259–273.

Parsot, C., Ménard, R., Gounon, P., and Sansonetti, P.J., 1995, Enhanced secretion through the *Shigella flexneri* Mxi-Spa translocon leads to assembly of extracellular proteins into macromolecular structures, *Mol. Microbiol.* **16**:291–300.

Perdomo, J.J., Gounon, P., and Sansonetti, P.J., 1994a, Polymorphonuclear leukocyte transmigration promotes invasion of colonic epithelial monolayer by *Shigella flexneri*, *J. Clin. Invest.* **93**:633–643.

Perdomo, J.J., Cavaillon, J.M., Huerre, M., Ohayon, H., Gounon, P., and Sansonetti, P.J., 1994b, Acute inflammation causes epithelial invasion and mucosal destruction in experimental shigellosis, *J. Exp. Med.* **180**:1307–1319.

Pistor, S., Chakraborty, T., Walter, U., and Wehland, J., 1995, The bacterial actin nucleator protein ActA of *Listeria monocytogenes* contains multiple binding sites for host microfilament proteins, *Curr. Biol.* **5**:517–525.

Pollard, T.D., 1995, Missing link for intracellular bacterial movement? *Curr. Biol.* **5**:837–840.

Prévost, M.C., Lesourd, M., Arpin, M., Vernel, F., Mounier, J., Hellio, R., and Sansonetti, P.J., 1992, Unipolar reorganisation of F-actin layer at bacterial division and bundling of actin filaments by plastin correlate with movement of *Shigella flexneri* within Hela cells, *Infect. Immun.* **60**:4088–4099.

Rajakumar, K., Jost, B.H., Sasakawa, C., Okada, N., Yoshikawa, M., and Adler, B., 1994, Nucleotide sequence of the rhamnose biosynthetic operon of *Shigella flexneri* 2a and role of lipopolysaccharide in virulence, *J. Bacteriol.* **176**:2362–2373.

Ridley, A.J., and Hall, A., 1992, The small GTP-binding protein Rho regulates the assembly of focal adhesives and actin stress fibers in response to growth factors, *Cell* **70**:389–399.

Ridley, A.J., Paterson, H.F., Johnston, C., and Hall, A., 1992, The small GTP-binding protein Rac regulates growth factor induced membrane ruffling, *Cell* **70**:401–410.

Rivero-Lezcano, O.M., Macilla, A., Sameshima, J.H., and Robbins, K.C., 1995, Wiskott-Aldrich syndrome protein physically associates with Nck through Src homology domains, *Mol. Cell Biol.* **15**:5725–5731.

Rosqvist, R., Forsberg, A., Rimpiläinen, M, and Wolf-Watz, H., 1990, The cytotoxic protein YopE of Yersinia obstructs the primary host defence, *Mol. Microbiol.* **4**:657–667.

Rosqvist, R., Forsberg, A., and Wolf-Watz, H., 1991, Intracellular targeting of the Yersinia YopE cytotoxin in mammalian cells induces actin microfilament disruption, *Infect. Immun.* **59**:4562–4569.

Rosqvist, R., Magnusson, K.E., and Wolf-Watz, H., 1994, Target cell contact triggers expression and polarized transfer of Yersinia YopE cytotoxin into mammalian cells, *EMBO J.* **13**:964–972.

Rosqvist, R., Hakansson, S., Forsbery, A., and Wolf-Watz, H., 1995, Functional conservation of the secretion and translocation machinery for virulence proteins of yersiniae, salmonellae and shigellae, *EMBO J.* **14**:4187–4195.

Sandlin, R.C., Lampel, K.A., Keasler, S.P., Goldberg, M.B., Stolzer, A.L., and Maurelli, A.T., 1995, Avirulence of rough mutants of *Shigella flexneri*: requirement of O antigen for correct unipolar localization of IcsA in the bacterial outer membrane, *Infect. Immun.* **63**:229–237.

Sansonetti, P.J., Kopecko, D.J., and Formal, S.B., 1982, Involvement of a large plasmid in the invasive ability of *Shigella flexneri*, *Infect. Immun.* **35**:852–860.

Sansonetti, P.J., Ryter, A., Clerc P., Maurelli, A.T., and Mounier, J., 1986, Multiplication of *Shigella flexneri* within HeLa cells: lysis of the phagocytic vacuole and plasmid-mediated contact hemolysis, *Infect. Immun.* **51**:461–469.

Sansonetti, P.J., Arondel, J., Fontaine, A., d'Hauteville, H., and Bernardini, M.L., 1991, *omp*B (osmo-regulation) and *ics*A (cell to cell spread) mutants of *Shigella flexneri*: vaccine candidates and probes to study the pathogenesis of shigellosis, *Vaccine* **9**:416–422.

Sansonetti, P.J., Mounier, J., Prévost, M.C., and Mège, R.M., 1994, Cadherin expression is required for the spread of *Shigella flexneri* between epithelial cells, *Cell* **76**:829–839.

Sansonetti, P.J., Arondel, J., Cavaillon, J.M., and Huerre, M., 1995, Role of IL-1 in the pathogenesis of experimental shigellosis, *J. Clin. Invest.* **96**:884–892.

Sansonetti, P.J., Arondel, J., Cantey, R.J., Prévost, M.C., and Huerre, M., 1996, Infection of rabbit Peyer's patches by *Shigella flexneri*: effect of adhesive or invasive bacterial phenotypes on follicular-associated epithelium., *Infect. Immun.* **64**:2752–2764.

Sasakawa, C., Kamata, K., Sakai, T., Makino, S.I., Yamada, M., Okada, N., and Yoshikawa, M., 1988, Virulence associated genetic regions comprising 30 kilobases of the 230-kilobase plasmid in *Shigella flexneri* 2a, *J. Bacteriol.* **170**:2480–2484.

Sasakawa, C., Komatsu, K., Tobe, T., Suzuki, T., and Yoshikawa, M., 1993, Eight genes in region 5 that form an operon are essential for invasion of epithelial cells by *Shigella flexneri* 2a, *J. Bacteriol.* **175**:2334–2346.

Savill, J., Fadok, V., Henson, P., and Haslett, C., 1993, Phagocyte recognition of cells undergoing apoptosis, *Immunol. Today* **14**:131–136.

Schaller, M.D., Hildebrand, J.D., Shannon, J.D., Fox, J.W., Vines, R.R., and Parsons, J.T., 1994, Autophosphorylation of the focal adhesion kinase, $pp125^{FAK}$, directs SH2-dependent binding of $pp60^{src}$, *Mol. Cell. Biol.* **14**:1680–1688.

Shere, K.D., Sallustio, S., Manessis, A., d'Aversa, T.G., and Golberg, M.B., 1997, Disruption of IcsP, the major *Shigella* protease that cleaves IcsA, accelerates actin-based motility, *Mol. Microbiol.* **25**:451–462.

Simonet, M., Richard, S., and Berche, P., 1990, Electronmicroscopic evidence for *in vivo* extracellular localization of *Yersinia pseudotuberculosis* harboring the pYV plasmid, *Infect. Immun.* **58**:841–845.

Soestatyo, M., Biewenga, J., Kraal, G., and Sminia, T., 1990, The localization of macrophage subsets and dendritic cells in the gastrointestinal tract of the mouse with special reference to the presence of high endothelial venules. An immuno- and enzyme-histochemical study, *Cell Tissue Res.* **259**:587–593.

Sory, M.P., and Cornelis, G.R., 1994, Translocation of a hybrid YopE-adenylate cyclase from *Yersinia enterocolitica* into Hela cell, *Mol. Microbiol.* **14**:583–594.

Suzuki, T., Lett, M.C., and Sasakawa, C., 1995, Extracellular tansport of VirG protein in *Shigella*, *J. Biol. Chem.* **270**:30874–30880.

Suzuki, T., Shinsuke, S., and Sasakawa, C., 1996, Functional analysis of *Shigella* VirG domains essential for interaction with vinculin and actin-based motility, *J. Chem. Biol.* **271**:21878–21885.

Suzuki, T., Miki, H., Takenawa, T., and Sasakawa, C., 1998, Neural Wiskott-Aldrich Syndrome Protein is implicated in the actin-based motility of *Shigella flexneri*, *EMBO J.* **17**:2767–2776.

Symons, M., Derry, J.M.J., Karlak, B., Jiang, S., Lemahieu, V., McCormick, F., Francke, U., and Abo, A., 1994, Wiskott-Aldrich Syndrome Protein, a novel effector for the GTPases Cdc42Hs, is implicated in actin polymerization, *Cell* **84**:723–734.

Theriot, J.A., 1995, The cell biology of infection by intracellular bacterial pathogens, *Annu. Rev. Cell. Dev. Biol.* **11:**213–239.

Theriot, J.A., Mitchison, T.J., Tilney, L.G., and Portnoy, D.A., 1992, The rate of actin-based motility of intracellular *Listeria monocytogenes* equals the rate of actin polymerization, *Nature* **357**:257–260.

Thornberry, N.A., and Lazebnik, Y., 1998, Caspases: enemies within, *Science* **281**:1312–1316.

Tran Van Nhieu, G., Adam, T., Dehio, C., Ménard, R., Skoudy, A., Mounier, J., Hellio, R., Gounon, P., and Sansonetti, P.J., 1997a, Shigella-induced cytoskeletal reorganisation during host cell invasion, in: *Molecular aspects of host-pathogen interaction*, (M.A. Mc Crae, J.R. Saunders, C.J.Smyth, and N.D. Stow; eds.), Cambridge University press, pp. 237–252.

Tran Van Nhieu, G., Ben Ze'ev, A., and Sansonetti, P.J., 1997b, Modulation of bacterial entry in epithelial cells by association between vinculin and the Shigella IpaA invasin, *EMBO J.* **16**:2717–2729.

Turner, C.E., and Miller, J.T., 1994, Primary sequenceof paxillin contains putative SH2 and SH3 domain binding motifs and multiple LIM domains: Identification of a vinculin and pp125^{FAK}-binding region, *J. Cell Science* **107**:1583–1591.

Van Gisjegem, F., Génin, S., and Boucher, C., 1993, Conservation of secretion pathways for pathogenicity determinants of plant and animal bacteria, *Trends Microbiol.* **1**:175–180.

Vasselon, T., Mounier, J., Hellio, R., and Sansonetti, P.J., 1992, Movement along actin filaments of the perijunctional area and *de novo* polymerization of cellular actin are required for *Shigella flexneri* colonization of epithelial Caco-2 cell monolayers, *Infect. Immun.* **60**:1031–1040.

Venkatesan, M., Buysse, J.M., and Kopecko, D.J., 1988, Characterization of invasion plasmid antigen (*ipaBCD*) genes from *Shigella flexneri*. DNA sequence analysis and control of gene expression, *Proc. Natl. Acad. Sci. USA* **85**:9317–9321.

Venkatesan, M.M., Buysse, J.M., and Oaks, E.V., 1992, Surface presentation of *Shigella flexneri* invasion plasmid antigen requires the products of the *spa* locus, *J. Bacteriol.* **174**:1990–2001.

Wassef, J., Keren D.F., and Mailloux, J.L., 1989, Role of M cells in initial bacterial uptake and in ulcer formation in the rabbit intestinal loop model in shigellosis, *Infect. Immun.* **57**:858–863.

Watarai, M., Tobe, T., Yoshikawa, M., and Sasakawa, C., 1995, Contact of Shigella with host cells triggers release of Ipa invasins and is an essential function of invasiveness, *EMBO J.* **14**:2461–2470.

Watarai, M., Funato, S., and Sasakawa, C., 1996, Interaction of Ipa proteins of *Shigella flexneri* with alpha5beta1 integrin promotes entry of the bacteria into mammalian cells, *J. Exp. Med.* **183**:991–999.

Watarai, M., Kamata, Y., Kozaki, S., and Sasakawa, C., 1997, Rho, a small GTP-binding protein, is essential for Shigella invasion of epithelial cells, *J. Exp. Med.* **185**:281–292.

Welch, M.D., Iwamatsu, A., and Mitchison, T.J., 1997, Actin polymerization is induced by the Arp2/3 complex at the surface of *Listeria monocytogenes*, *Nature* **385**:265–269.

Welch, M.D., Rosenblatt, J., Skoble, J., Portnoy, D.A., and Mitchison, T.J., 1998, Interaction of Arp2/3 complex and the *Listeria monocytogenes* ActA protein in actin filament nucleation, *Science* **281**:105–108.

Wu, H., Reynolds, A., Kanner, S., Vines, R., and Parsons, J., 1991, Identification and characterization of a novel cytoskeleton-associated pp60src substrate, *Mol. Cell. Biol.* **11**:5113–5123.

Young, V.B., Falkow, S., and Schoolnik, G.K., 1992, The invasin protein of *Yersinia enterolitica*: internalization of invasin-bearing bacteria by eukaryotic cells is associated with reorganization of the cytoskeleton, *J. Cell Biol.* **116**:197–207.

Zeile, W.L., Purich, D.L., and Southwick, F.S., 1996, Recognition of two classes of oligoproline sequences in profilin-mediated acceleration of actin based Shigella motility, *J. Cell Biol.* **133**:49–59.

Zychlinsky, A., Prévost, M.C., and Sansonetti, P.J., 1992, *Shigella flexneri* induces apoptosis in infected macrophages, *Nature* **358**:167–169.

Zychlinsky, A., Fitting, C., Cavaillon, J.M., and Sansonetti, P.J., 1994a, Interleukin-1 is released by macrophages during apoptosis induced by *Shigella flexneri*, *J. Clin. Invest.* **94**:1328–1332.

Zychlinsky, A., Kenny, B., Ménard, R., Prévost, M.C., Holland, I.B., and Sansonetti, P.J., 1994b, IpaB mediates macrophage apoptosis induced by *Shigella flexneri*, *Mol. Microbiol.* **11**:619–627.

Zychlinsky, A., Thirumalai, K., Arondel, J., Cantey, J.R., Aliprantis, A.O., and Sansonetti, P.J., 1996, *In vivo* apoptosis in *Shigella flexneri* infections, *Infect. Immun.* **64**:5357–5365.

Chapter 12

Salmonella Invasion of Non-Phagocytic Cells

Lisa M. Schechter and Catherine A. Lee

1. INTRODUCTION

1.1. Salmonella Classification and Host Range

The gram-negative, facultatively anaerobic bacteria of the genus *Salmonella* are able to infect a wide range of animal hosts and produce a variety of clinical manifestations. The ability of salmonellae to cause such a spectrum of diseases is attributable to the genetic diversity of this genus. Based on multilocus enzyme electrophoresis analysis, DNA hybridization studies, and gene sequencing, the genus *Salmonella* contains two species: *bongori* and *enterica* (Selander *et al.*, 1996). *Salmonella enterica* is further subdivided into six subspecies, designated by the roman numerals I, II, IIIa, IIIb, IV, VI, and VII. Each subspecies is additionally divided into serovars based on the variability of surface antigens. Prior to the determination of Salmonella phylogenetic relationships by modern molecular methods, each Salmonella serovar was considered to be a unique species. Salmonella serovars are still

LISA M. SCHECHTER and CATHERINE A. LEE Department of Microbiology and Molecular Genetics, Harvard Medical School, Boston, Massachusetts 02115.

Subcellular Biochemistry, Volume 33: Bacterial Invasion into Eukaryotic Cells, edited by Oelschlaeger and Hacker. Kluwer Academic / Plenum Publishers, New York, 2000.

commonly referred to as different species, although this practice is now taxonomically incorrect.

Serovars of *S. bongori* and *S. enterica* subspecies II, IIIa, IIIb, IV, VI, and VII are mainly isolated from reptiles, whereas serovars of *S. enterica* subspecies I are primarily isolated from warm-blooded vertebrates. Subspecies I contains almost all serovars pathogenic to humans, and accounts for 99% of clinical isolates (Bäumler, 1997). *S. enterica* subspecies I serovars usually only infect a specific host. These host-adapted *Salmonella* subspecies I serovars include the human pathogen *S. enterica* serovar Typhi (*S. typhi*), the avian pathogens *S. pullorum* and *S. gallinarum*, the swine pathogen *S. choleraesuis*, and the bovine pathogen *S. dublin*. Other serovars, such as *S. typhimurium* and *S. enteriditis*, have a broad host range, causing disease in many different hosts. Host-adapted *Salmonella* serovars generally cause systemic diseases, and are more virulent than *Salmonella* serovars that lack host specificity. For example, *S. typhi* causes typhoid fever in humans, while *S. typhimurium* and *S. enteriditis* usually only cause gastroenteritis in humans. However, *S. typhimurium* and *S. enteriditis* may cause systemic diseases in young or immunocompromised animal hosts.

1.2. General Course of Salmonella Infection

Despite their genetic heterogeneity, all salmonellae are thought to infect their hosts by a similar route. Salmonella infections are usually contracted by ingestion of contaminated food or water, followed by passage of the bacteria through the stomach to the small intestine. Studies in mice have shown that in the distal ileum, *S. enteritidis* crosses the intestinal epithelium to gain access to deeper tissues (Carter and Collins, 1974). However, the *Salmonella* entry site may vary depending on the host and the bacterial serovar (Galán and Sansonetti, 1996). In many hosts, *S. typhimurium* invades enterocytes (Watson *et al.*, 1995; Reed *et al.*, 1986; Wallis *et al.*, 1986; Takeuchi, 1967), although in mice, this serovar preferentially invades M cells (Pascopella *et al.*, 1995; Clark *et al.*, 1994; Jones *et al.*, 1994; Kohbata *et al.*, 1986). M cells are specialized epithelial cells that transport antigens to underlying lymphoid follicles. After salmonellae transcytose through enterocytes or M cells to the lamina propria, they encounter macrophages and lymphocytes. Non-typhoidal salmonellae may additionally encounter polymorphonuclear leukocytes, which are often seen infiltrating the small intestine at the site of bacterial entry (Wallis *et al.*, 1986; Smith, 1967; Takeuchi and Sprinz, 1967). Depending on the host and the bacterial serovar, Salmonella may establish a local infection in the lamina propria that typically leads to diarrhea, dysentery, and fever (Miller *et al.*, 1995). Alternatively, host-specific serovars may travel through the lymphatic system and the blood stream to infect systemic sites such as the liver and spleen. This dissemination may occur via infected macrophages, as Salmonella survives and replicates inside macrophages (Gulig, 1996).

During the course of an infection, Salmonella must overcome a variety of stresses and barriers. First, Salmonella must survive the acidic pH of the stomach to reach the small intestine. In the small intestine, Salmonella must overcome the actions of pancreatic enzymes, bile salts, defensins (small cationic antimicrobial peptides), secretory IgA, and mucus to reach the intestinal wall (Miller *et al.*, 1995). Salmonella may also have to compete with the normal flora of the intestine to gain access to the intestinal epithelium. Finally, Salmonella must breach the intestinal epithelial barrier and evade the host immune system. Because epithelial cells are normally non-phagocytic, Salmonella actively induces its own uptake into these host cells to traverse the intestinal epithelium. In this chapter, the host responses and bacterial factors involved in Salmonella entry into non-phagocytic cells will be reviewed.

2. EXPERIMENTAL SYSTEMS TO STUDY SALMONELLA INTERACTIONS WITH NON-PHAGOCYTIC CELLS

2.1. Epithelial Cell Culture

Tissue culture models have allowed the study of Salmonella invasion *in vitro* and have greatly facilitated the identification of bacterial genes and host cell factors required for invasion. Salmonella invasion *in vitro* was first demonstrated in HeLa cells, and bacterial entry into a number of other cell lines has subsequently been demonstrated (Giannella *et al.*, 1973b). Salmonella interactions with epithelial monolayers have been studied using polarized cell lines, such as MDCK, Caco-2, or T84, which can form tight junctions (McCormick *et al.*, 1993; Finlay and Falkow, 1990; Finlay *et al.*, 1988a). In contrast to non-polarized cultured cells, these epithelial monolayers more closely resemble the enterocyte brush border *in vivo*. T84 cells have also been used to study PMN migration through epithelial monolayers (McCormick *et al.*, 1993). Bacterial invasion into cultured cells can be quantitated using methods that differentiate internal and external bacteria (Tang *et al.*, 1993). For example, tissue culture cells containing internalized Salmonella can be incubated with antibiotics that cannot freely diffuse across the plasma membrane and thus preferentially kill extracellular bacteria. After removal of the antibiotic and lysis of the tissue culture cells, viable intracellular bacteria can be quantitated by determining colony-forming units.

2.2. Animal Models

The Balb/c mouse, which contracts a typhoid-like disease when orally infected with *S. typhimurium*, is the best characterized animal model for studying the importance of Salmonella interactions with intestinal cells. Many *S. typhimurium* invasion-defective mutants that were identified using tissue culture systems *in vitro* were also found to be defective in their ability to penetrate the murine intestinal barrier *in vivo* (Penheiter *et al.*, 1997; Jones and Falkow, 1994; Behlau and Miller, 1993; Galán and Curtiss, 1989). It has been estimated that 4% of the *S. typhimurium* genome is required for systemic infection of mice, and that many of the genes encoded by this 4% are required during the early stages of a Salmonella infection (Bowe *et al.*, 1998). Unfortunately, detailed analyses of the initial stages of bacterial invasion through the intestinal barrier are complicated by the fact that salmonellae encounter many barriers and stresses before reaching the small intestine, and very few bacteria from an oral inoculum may actually enter the intestinal epithelium. For this reason, ligated loops, which are tied segments of intestine in anesthetized animals that are directly inoculated with bacteria, have been used to specifically study Salmonella interactions with the intestinal epithelium *in vivo*. Epithelial cell damage induced by Salmonella has been analyzed in murine, rabbit, bovine, and porcine ligated loops (Jones *et al.*, 1995; Watson *et al.*, 1995; Reed *et al.*, 1986; Giannella *et al.*, 1973a). Rabbit ligated loops have additionally been used to study aspects of Salmonella-induced gastroenteritis such as fluid accumulation and PMN influx (Wallis *et al.*, 1986; Giannella *et al.*, 1973a).

3. HOST CELL RESPONSES TO SALMONELLA INVASION

3.1. Host Cell Morphological Changes

Takeuchi was the first to microscopically observe Salmonella interactions with enterocytes in the guinea pig small intestine (1967), and his observations have subsequently been confirmed using other animal models and tissue culture cells. Figure 1 shows the progression of a Salmonella infection in polarized cells. Soon after coming in contact with the epithelial brush border, Salmonella induces distortions in the host cell membrane that resemble the membrane ruffling caused by growth factors, hormones, or oncogenic stimuli (Ridley, 1994). However, unlike membrane ruffling induced by other stimuli, membrane distortions induced by Salmonella are localized to the site of bacterial-host cell interaction. The stimulation of membrane ruffling by Salmonella leads to an increase in macropinocytosis,

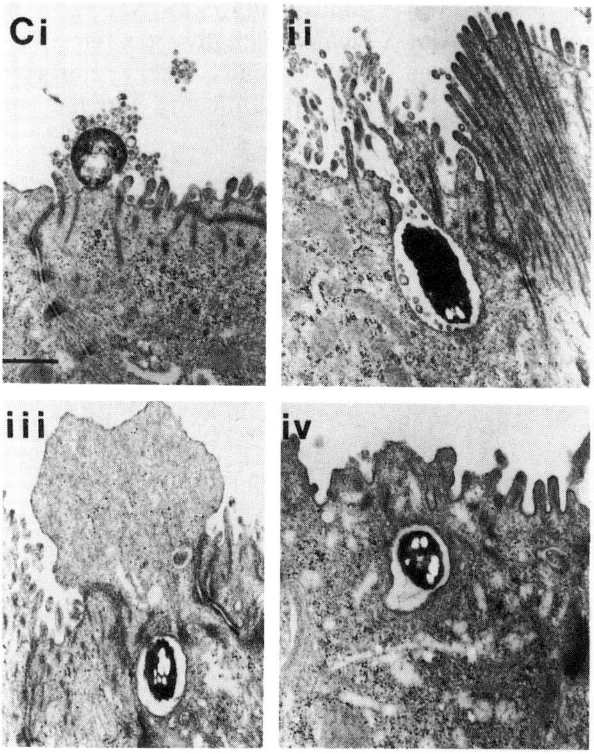

FIGURE 1. Salmonella invasion of polarized T84 cells (reproduced from McCormick *et al.*, 1993 by copytight permission of The Rockefeller University Press). Upon contact with host cells (Ci), Salmonella induces cytoskeletal rearrangements that allow bacterial entry (ii), and cause apical protrusion of the host cell plasma membrane (iii). After Salmonella has been internalized, the host cell morphology returns to normal (iv).

which directly facilitates the internalization of cell surface-associated bacteria within a membrane bound vacuole (Garcia-del Portillo and Finlay, 1994; Francis *et al.*, 1993). Destruction of the epithelial brush border is transient and reversible, as microvilli are reconstituted following bacterial entry.

In mice, Salmonella preferentially invades M cells to gain access to deeper tissues (Jones *et al.*, 1995). Although the reason for this preference is not known, it is possible that these cells express surface receptors for Salmonella, or that they are more accessible due to the lack of both a dense glycocalyx layer and closely packed microvilli on their apical surface (Neutra and Kraehenbuhl, 1993). *S. typhimurium* invades murine M cells by triggering the same morphological changes seen during enterocyte entry

(Jones *et al.*, 1994). However, unlike enterocytes, M cells do not appear to return to normal following *S. typhimurium* invasion. Bacterial invasion can lead to M cell extrusion and death, causing adjacent enterocytes to be shed and allowing lumenal bacteria to pass directly into the lamina propria (Jones *et al.*, 1994).

3.2. Host Factors and Signaling Pathways Required for Salmonella Invasion

The alteration of host cell morphology during Salmonella invasion is correlated with changes in the host cell cytoskeleton. Many proteins, including actin, α-actinin, tropomyosin, and talin, accumulate at the site of bacterial entry (Finlay *et al.*, 1991). In fact, actin rearrangements are required for Salmonella invasion because bacterial entry is inhibited when epithelial cells are pretreated with cytochalasin D, a drug that disrupts actin polymerization (Finlay *et al.*, 1991). Although the mechanisms by which Salmonella induces changes in the host cell cytoskeleton are not well understood, several signaling pathways have been identified that may be involved in this process (see Figure 2).

One way that Salmonella may induce cytoskeletal rearrangements in cultured epithelial cells is by stimulating an increase in free intracellular calcium levels (Pace *et al.*, 1993; Ruschkowski *et al.*, 1992). Calcium ions activate a number of actin-binding proteins that are involved in the disassembly of actin filaments (Stossel, 1993). Therefore, Salmonella-induced increases in $[Ca^{2+}]_i$ may stimulate depolymerization of subcortical actin filaments, which could supply actin monomers for use in the assembly of new cytoskeletal structures associated with bacterial entry. Although increases in free intracellular calcium are required for Salmonella internalization, the mechanisms underlying this calcium flux differ in particular cell types (Galán, 1994). In Henle-407 cells, Salmonella entry is reduced in the

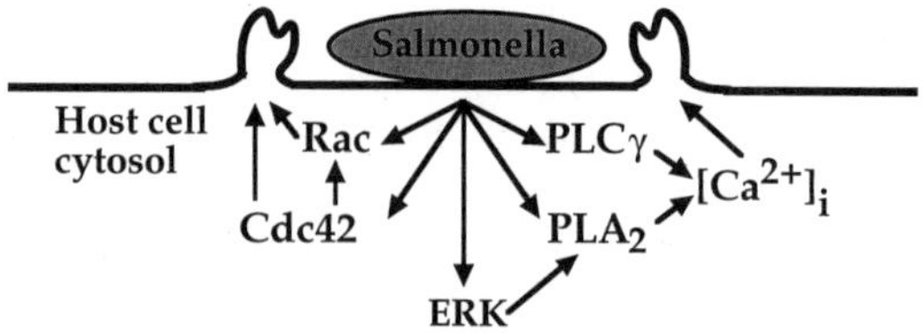

FIGURE 2. Host cell signaling pathways involved in Salmonella invasion. Salmonella may induce cytoskeletal rearrangements by increasing free intracellular calcium levels $[Ca^{2+}]_i$ via either phospholipase A_2 (PLA_2) or phospholipase C_γ (PLC_γ). Salmonella also induces cytoskeletal rearrangements via Cdc42 and Rac. Although not shown, it is possible that Cdc42 or Rac could activate the phospholipase pathways.

presence of calcium channel antagonists, or inhibitors of phospholipase A_2 and 5-lipooxygenase (Pace, 1993). These data support a pathway in which interactions with Salmonella stimulates phospholipase A_2 in the host cell, causing the release of arachidonic acid from membrane phospholipids. 5-lipooxygenase metabolizes arachidonic acid to the leukotriene D_4 (LTD_4), which directly or indirectly activates calcium channels, leading to an influx of extracellular calcium. In HeLa cells, Salmonella appears to increase $[Ca^{2+}]_i$ by stimulating release of calcium from intracellular stores (Ruschkowski *et al.*, 1992). This calcium flux may be mediated by phospholipase Cγ, because Salmonella entry into HeLa cells stimulates inositol trisphosphate production, and is reduced by phospholipase Cγ inhibitors (Galán, 1994; Ruschkowski *et al.*, 1992).

Salmonella may also trigger cytoskeletal reorganization by activating small GTP-binding proteins in the Rho family. When activated in response to extracellular signals, these GTPases induce morphological changes, including membrane ruffling (Zigmond, 1996). Expression of a dominant negative form of Cdc42 in COS-1 cells prevents Salmonella entry and bacterial-induced cytoskeletal rearrangements, suggesting that Cdc42 mediates Salmonella invasion into COS-1 cells (Chen *et al.*, 1996a). Expression of a dominant negative Racl in COS-1 cells also reduces Salmonella-induced cytoskeletal rearrangements and bacterial entry, although to a lesser extent than the dominant negative Cdc42 (Chen *et al.*, 1996a). It should be noted that inhibition of Racl and Rho in Swiss 3T3 fibroblasts, MDCK cells, and HEp-2 cells did not prevent the induction of membrane ruffling by Salmonella (Jones *et al.*, 1993). However, the effects of Cdc42, Racl, and Rho inhibition on Salmonella invasion into these cell lines were not tested. The contrasting results obtained by different laboratories could be due to the different techniques used to inhibit the GTPases, or the use of different tissue culture cell lines.

Little information is known about how Salmonella stimulates the phospholipase and GTPase signaling pathways. It was initially believed that Salmonella might stimulate membrane ruffling by activating the EGF receptor. This idea, based on evidence that the EGF receptor is phosphorylated in Henle-407 cells in response to Salmonella infection, was attractive for several reasons (Galán *et al.*, 1992b). Like Salmonella, EGF is known to stimulate membrane ruffling, actin rearrangements, and an increase in free intracellular calcium levels. Furthermore, both EGF receptor activation and Salmonella invasion stimulate phosphorylation of the mitogen-activated protein (MAP) kinases ERK 1 and 2 (Hobbie *et al.*, 1997; Rosenshine *et al.*, 1994; Pace *et al.*, 1993). The MAP kinases are known to phosphorylate phospholipase A_2, thus providing a link between EGF receptor activation and Salmonella stimulation of a calcium influx in Henle-407 cells.

Despite the similarities between the host cell responses to EGF and Salmonella invasion, several observations indicate that the EGF receptor may not have a primary role in Salmonella internalization. First, Salmonella can invade cell lines that do not express the EGF receptor (Francis *et al.*, 1993; Galán *et al.*, 1992b). Second, tyrosine kinase inhibitors do not reduce Salmonella entry into many different cell types (Reed *et al.*, 1996; Rosenshine *et al.*, 1994). Third, removing the EGF receptor from the surface of Henle-407 cells does not reduce *S. typhmurium* internalization (Jones *et al.*, 1993). Finally, *S. typhimurium* penetrates the intestinal mucosa of EGF receptor-defective and wild-type mice equally well (McNeil *et al.*, 1995). These results suggest that either the EGF receptor does not directly participate in Salmonella internalization, or that Salmonella can also stimulate another signaling pathway that is functionally redundant. Additional studies are needed to determine exactly how *Salmonella* manipulates host cell signaling pathways to achieve internalization. As will be discussed later in this chapter, a recent study indicates that proteins secreted by *S. typhimurium* into the host cell cytosol may directly activate Cdc42 and Rac1 (Hardt *et al.*, 1998a).

Interestingly, the cystic fibrosis transmembrane conductance regulator (CFTR), a chloride ion transporter, appears to be required for *S. typhi* but not *S. typhimurium* entry into intestinal epithelial cells (Pier *et al.*, 1998). *S. typhi* enters cells expressing wild-type CFTR more efficiently than those containing the ΔF508 CFTR mutation, the most common CFTR allele found in cystic fibrosis patients (Pier *et al.*, 1998). An increased resistance to *S. typhi* infection conferred by the ΔF508 CFTR mutation may explain why mutant CFTR alleles are so frequent in certain human populations.

3.3. Salmonella Trafficking Inside Epithelial Cells

Once inside epithelial cells, salmonellae replicate within membrane bound vacuoles and appear to subvert the classical endocytic pathway. Normally, endocytosed material is delivered to early endosomes, which mature into late endosomes. Late endosomes, which contain mannose 6-phosphate receptors (M6PRs), lysosomal membrane glycoproteins (Lgps), and other integral membrane proteins, fuse with secondary lysosomes in a microtubule-dependent manner (Mellman, 1996). In contrast, experiments in HeLa cells indicate that Salmonella-containing vacuoles do not fuse with early endosomes, and only partially fuse with secondary lysosomes in a microtubule-independent manner (Garcia-del Portillo and Finlay, 1995). Over time, Salmonella-containing vacuoles acquire Lgps, but unlike late endosomes, these vacuoles do not contain M6PRs (Garcia-del Portillo and Finlay, 1995; Garcia-del Portillo *et al.*, 1993). Several hours after bacterial internalization, unusual lgp-rich filamentous structures are formed that extend from the vacuole membrane (Mills and Finlay, 1994; Garcia-del Portillo *et al.*, 1993). The appearance of these Salmonella-induced filaments (Sif) in host cells correlates with the onset of bacterial replication, and requires vacuolar acidification and an intact microtubule network (Garcia-del Portillo *et al.*, 1993). *S. typhimurium*-induced filament formation also requires a bacterial gene, *sifA*, although the role that this novel gene plays in filament development has not yet been elucidated (Stein *et al.*, 1996).

3.4. Mucosal Immune Responses to Salmonella Invasion

When infecting humans, *S. typhimurium* and other non-adapted salmonellae often provoke an acute intestinal inflammatory response that leads to symptoms of gastroenteritis. This inflammatory response may be triggered by the apical attachment of Salmonella to the intestinal wall, which causes epithelial cells to secrete cytokines that then attract polymorphonuclear neutrophils (PMNs) to the site of bacterial infection (Miller *et al.*, 1995). *S. typhimurium* stimulates the secretion of many proinflammatory cytokines, including TNF-α, IL-1, and IL-6, into the lumen of rabbit ileal ligated loops (Klimpel *et al.*, 1995; Arnold *et al.*, 1993). However, it is difficult to distinguish *in vivo* whether epithelial cells or other cells are the source of these cytokines. Therefore, an *in vitro* system using polarized T84 epithelial monolayers and isolated human peripheral PMNs has been developed to study how mucosal immune responses are elicited by Salmonella interactions (McCormick *et al.*, 1993). Soon after bacterial attachment to T84 monolayers, epithelial cells synthesize and secrete cytokines, including IL-8, TNF-α, monocyte chemotactic protein-1, and GM-CSF (Jung *et al.*, 1995; McCormick *et al.*, 1993). *S. typhimurium* induction of IL-8 synthesis and secretion is particularly intriguing, as this cytokine is a chemoattractant for PMNs. It appears that *S. typhimurium* induces IL-8 expression by activating the host cell transcription factors NF-κB and AP-1 (Hobbie *et al.*, 1997). IL-8 is released from epithelial cells into the basolateral space, where it binds to glycosaminoglycans of the subepithelial matrix and forms a chemotactic gradient that is resistant to washout by fluid (McCormick *et al.*, 1995a; McCormick *et al.*, 1993). This IL-8 gradient may function *in vivo* to stimulate PMN passage through the endothelium of blood vessels into the lamina propria. Other cytokines secreted into the intestinal lumen may then recruit PMNs to traverse the intestinal barrier. One such cytokine, termed pathogen-elicited chemoattractant (PEEC), has been semi-purified from the apical supernatant of epithelial cells colonized with *S. typhimurium* (McCormick *et al.*, 1998). This 1–3 kDa bioactivity is released from epithelial cells in a polarized apical fashion following *S. typhimurium* infection, and elicits the transepithelial migration of PMNs (McCormick *et al.*, 1998). Thus, *S. typhimurium* may induce the secretion and differential localization of cytokines within the intestinal mucosa and submucosa to provoke inflammation.

The distinct diseases caused by salmonellae may result, at least in part, from differences in the initial mucosal immune responses to these pathogens. For example, *S. typhimurium* and *S. enteritidis* induce PMN migration in the T84 cell model, but the host-specific serovars *S. pullorum*, *S. typhi*, and *S. paratyphi* do not (McCormick *et al.*, 1995b). Therefore, the ability of Salmonella to cause gastroenteritis in a host may depend on its ability to recruit PMNs. Distinct host cytokine responses to specific Salmonella serovars may also contribute to the different pathogenic properties of these organisms. Although there is little experimental evidence available to support this idea, it has been shown that *S. typhi* stimulates the production and secretion of significantly more IL-6 than *S. dublin* or *S. typhimurium* in human intestinal epithelial cells (Weinstein *et al.*, 1998).

Several studies in rabbit ileal ligated loops indicate that the fluid secretion induced by *S. typhimurium* requires bacterial invasion, and correlates with the appearance of PMNs in the intestinal mucosa (Wallis *et al.*, 1989; Wallis *et al.*, 1986; Giannella *et al.*, 1973a). The mechanism by which *S. typhimurium* induces transepithelial fluid secretion remains unclear. It is possible that PMNs recruited to the lamina propria release factors that stimulate epithelial secretory responses (Madara *et al.*, 1993). Adenylate cyclase may play a role in Salmonella-induced fluid secretion, as cAMP levels in intestinal epithelial and mucosal cells increase upon *S. typhimurium* invasion (Peterson *et al.*, 1983; Giannella *et al.*, 1975). Although there are several possible ways that *S. typimurium* could activate adenylate cyclase in intestinal cells, a pathway involving host cell leukopeptides and prostaglandins is favored. This idea is supported by observations that *S. typhimurium* invasion leads to increases in epithelial cell levels of leukopeptides and prostaglandins, and that an inhibitor of prostaglandin synthesis reduces Salmonella-dependent cAMP production and fluid secretion (Pace *et al.*, 1993; Duebbert and Peterson, 1985; Giannella *et al.*, 1975). It is also possible that *S. typhimurium* encodes enterotoxins or cytotoxins that induce fluid secretion. A heat-labile, cholera toxin-like Salmonella enterotoxin (Stn) appears to possess biological activities similar to cholera toxin when it is expressed in *E. coli* (Prasad *et al.*, 1990). However, both *S. typhimurium* and *S. dublin stn* mutants were not impaired in their abilities to induce fluid secretion and PMN migration in bovine ileal ligated loops (Watson *et al.*, 1998). These findings indicate that Stn does not play a primary role in Salmonella-induced enteritis. Recently, proteins secreted into host cells by Salmonella have been identified, and one or more of these factors may contribute to transepithelial PMN migration and fluid secretion. These proteins will be discussed in Section 4.1.2 of this chapter.

4. BACTERIAL FACTORS REQUIRED FOR SALMONELLA INVASION OF NON-PHAGOCYTIC CELLS

4.1. The Invasion-Associated Type-III Secretion Pathway

Salmonella proteins that manipulate host cell signal transduction cascades appear to be the central mediators of bacterial invasion. These proteins are directly translocated into host cells via a specialized bacterial export system, called a type-III secretion apparatus. Type-III secretion pathways are important for the virulence of a variety of gram-negative bacterial pathogens. Pathogenic properties such as Yersinia inhibition of phagocytosis, enteropathogenic *E. coli* (EPEC) effacement of enterocytes, and Shigella invasion of epithelial cells all require the participation of type-III secretion pathways (Hueck, 1998). In addition to the invasion-associated type-III secretion pathway, salmonellae (excluding *S. bongori*) even encode a second type-III secretion system that is required for survival in macrophages (Cirillo *et al.*, 1998; Hensel *et al.*, 1998; Ochman *et al.*, 1996). In all of these type-III secretion systems, an export apparatus, which is usually composed of greater than 20 proteins and spans the inner and outer bacterial membranes, directs the secretion of proteins that do not contain classical *sec*-dependent signal sequences (Lee, 1997). In fact, the secretion signals of type-III exported factors may not be protein encoded, and may reside in their mRNA (Anderson and Schneewind, 1997). As shown in Figure 3, the type-III secretion apparatus required for Salmonella invasion is encoded by a large cluster of genes located near minute 63 of the Salmonella chromosome. This 40 kb region of DNA, termed Salmonella pathogenicity island-1 (SPI-1), also encodes secreted proteins, accessory factors, and regulators.

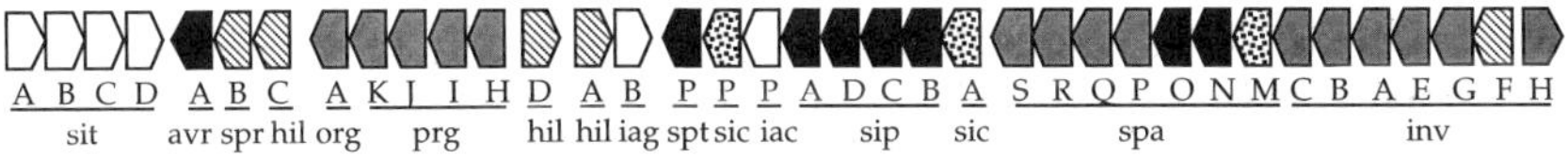

FIGURE 3. Salmonella Pathogenicity Island-1 (SPI-1). The open reading frames on SPI-1 (not drawn to scale) are shown as arrows pointing in the direction of transcription, and are shaded according to their function: secretion apparatus components (grey), secreted proteins (black), chaperones (polka-dotted), and regulators (hatched). The *sit* genes encode an iron transport system, *iacP* encodes an acyl-carrier protein, and the function of *iagB* is unknown.

4.1.1. The Type-III Secretion Apparatus

4.1.1.1. Structure and Function of the Secretion Apparatus The Salmonella invasion-associated type-III secretion apparatus is predicted to be composed of the Inv proteins, (except for InvF), the Spa proteins (except for SpaM/InvI), the Prg proteins, and OrgA (Eichelberg *et al.*, 1994; Jones and Falkow, 1994; Kaniga *et al.*, 1994; Altmeyer *et al.*, 1993; Groisman *et al.*, 1993; Galán *et al*, 1992a; Ginnochio *et al.*, 1992). As shown in Figure 4, transmission electron microscopy indicates that the invasion-associated type-III secretion apparatus is a needle-like structure that protrudes from the bacterial surface (Kubori *et al.*, 1998). The base of the apparatus contains two rings associated with the inner membrane, and single rings associated with the peptidoglycan and outer membrane. Loss-of-function mutations affecting individual components of the apparatus (excluding *invB*) all cause severe defects in Salmonella invasion, suggesting that each component is crucial for the structure or function of the secretion machinery (Collazo and Galán, 1997b).

Because most secretion machinery components have not yet been biochemically characterized, the location of individual proteins in this needle-like structure is not well defined. However, the positions and functions of some Salmonella apparatus proteins can be predicted based on sequence homologies with components of the better studied Yersinia and Shigella type-III secretion systems. The base of the SPI-1 type-III secretion apparatus is most likely composed of the InvC, InvA, SpaP, SpaQ, SpaR, SpaS, and PrgH proteins. The cytoplasmic protein InvC is similar to the β-subunit of the F_0F_1 ATPase, indicating that this protein may energize the secretion

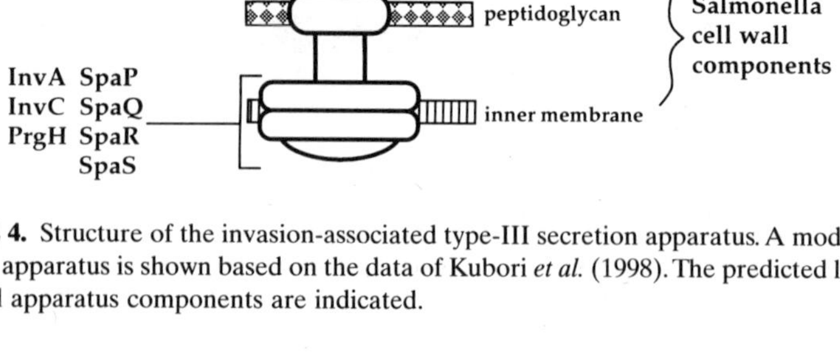

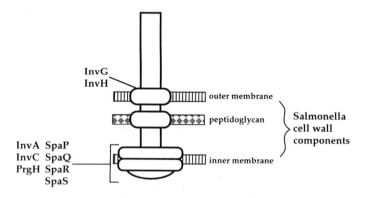

FIGURE 4. Structure of the invasion-associated type-III secretion apparatus. A model of the secretion apparatus is shown based on the data of Kubori *et al.* (1998). The predicted locations of several apparatus components are indicated.

process. In support of this hypothesis, a point mutation in the predicted nucleotide-binding region of InvC abolishes its ATPase activity *in vitro*, and its ability to complement an *invC* null mutation *in vivo* (Eichelberg *et al.*, 1994). As shown by subcellular fractionation and Tn*phoA* analysis, InvA is an inner membrane protein containing two domains: a N-terminal region possessing 8 potential membrane spanning segments, and a C-terminal cytoplasmic domain (Collazo and Galán, 1997b; Galán *et al.*, 1992a). While the N-terminal membrane spanning domains of InvA and its Yersinia homolog LcrD are interchangeable and highly conserved, the C-terminal domains of these proteins are more divergent (Ginocchio and Galán, 1995). It is thought that InvA may function as the inner membrane channel, and that the cytoplasmic portion of this protein may interact with secretion substrates. The SpaP, SpaQ, SpaR, and SpaS proteins, which all contain putative transmembrane segments, are additionally predicted to be in the inner membrane and are similar to type-III secretion apparatus components in other systems (Groisman and Ochman, 1993). According to immunoelectron microscopy studies, PrgH may also be a component of the apparatus base (Kubori *et al.*, 1998). PrgH was originally thought to be a lipoprotein because it contains a consensus lipoprotein signal sequence (Pegues *et al.*, 1995). However, lipoproteins are cleaved prior to acylation by *N*-acyltransferase, and PrgH isolated from purified apparatus preparations was not processed, suggesting that PrgH is not a lipoprotein (Kubori *et al.*, 1998).

The outer membrane ring of the secretion channel may be composed of InvG. This hypothesis is based on InvG homology to the PulD family of outer membrane proteins, and the observation that purified InvG forms oligomeric ring structures *in vitro*. (Crago and Koronakis, 1998; Pugsley, 1993). InvG stability and localization appears to depend on interactions with the outer membrane lipoprotein InvH (Crago and Koronakis, 1998; Daefler and Russel, 1998). Another SPI-1 encoded lipoprotein, PrgK, does not appear to be required for InvG localization (Daefler and Russel, 1998). InvH has been repeatedly identified in screens for invasion mutants (Leclerc *et al.*, 1998; Lodge *et al.*, 1995; Altmeyer *et al.*, 1993; Stone *et al.*, 1992). Because *invH* mutants are less adherent to host cells, unlike other SPI-1 invasion mutants, it was originally proposed that InvH was an adhesin (Altmeyer *et al.*, 1993). Although InvH may play a role in adherence, it has recently been shown that an *invH* mutation abolishes secretion of a SPI-1 encoded effector protein, confirming that InvH is part of the invasion-associated type-III secretion apparatus (Daefler and Russel, 1998; Watson *et al.*, 1998). Interestingly, other type-III secretion systems do not contain homologs of InvH (Hueck, 1998).

Filamentous appendages have also been seen to protrude from the Salmonella surface by high resolution scanning electron microscopy (Ginocchio *et al.*, 1994). These appendages are observed shortly after bacterial contact with host cells, and are shed upon the induction of host cell membrane ruffling (Ginocchio *et al.*, 1994). Although it is unknown whether the appendages are part of the needle-like structures or if they are important for invasion, the secretion apparatus may play a role in the formation of appendages. Salmonella *invC* and *invG* mutants do not produce these appendages, and *invA* and *invE* mutants produce aberrant longer filaments that are not shed (Ginocchio *et al.*, 1994). However, Reed *et al.* recently found that appendage formation was not affected by SPI-1 mutations, and the reason for these conflicting observations is unclear (1998).

4.1.1.2. Virulence Defects of Secretion Apparatus Mutants Loss-of-function mutations in SPI-1 secretion apparatus components (excluding *invB*) severely reduce Salmonella entry into cultured cells, enterocytes, and M cells (Collazo and Galán, 1997b; Penheiter *et al.*, 1997; Watson *et al.*, 1995; Jones *et al.*, 1994). The invasion defects of SPI-1 secretion mutants are attributed to their inability to induce membrane ruffling, calcium fluxes, or actin rearrangements in host cells (Galán *et al.*, 1992a; Ginocchio *et al.*, 1992). However, invasion defects can be rescued *in vitro* if mutants are coinfected with wild-type bacteria, or if host cell membrane ruffling is stimulated by other means, such as by addition of EGF (Francis *et al.*, 1993; Galán *et al.*, 1992b; Ginocchio *et al.*, 1992). Therefore, type-III secretion is essential for activation of the host cell signaling pathways and membrane ruffling that mediate bacterial internalization. Because *S. typhimurium* SPI-1 secretion apparatus mutants cannot efficiently penetrate the intestinal epithelium, they exhibit virulence defects in orally infected mice (Penheiter *et al.*, 1997; Jones and Falkow, 1994; Behlau and Miller, 1993; Galán and Curtiss, 1989). However, these mutants are fully virulent when inoculated intraperitoneally, indicating that SPI-1 secretion apparatus genes may not be required for Salmonella infection once bacteria have breached the intestinal barrier. It should be noted that *S. typhimurium* SPI-1 secretion mutants are not completely avirulent in mice, and that orally inoculated bacteria are able to gain access to systemic sites by another unknown mechanism.

In addition to its role in intestinal epithelial invasion, the SPI-1 encoded type-III export system and its secreted products may be important for other aspects of Salmonella pathogenesis. *S. typhimurium* SPI-1 secretion apparatus mutants are unable to induce transepithelial PMN migration *in vitro*, suggesting a role for the SPI-1 type-III secretion system in the induction of the host inflammatory response (McCormick *et al.*, 1995b). In fact, a mutation in *invH* reduces the ability of *S. typhimurium* to induce PMN migration and fluid secretion in bovine ligated loops (Watson *et al.*,

1998). *S. typhimurium* SPI-1 secretion apparatus mutants are also defective in their ability to kill macrophages *in vitro* (Chen *et al.*, 1996b; Monack *et al.*, 1996). It is known that bacterial interactions with macrophages play an important role in systemic infection. However, the relevance of SPI-1 in bacterial interactions with macrophages *in vivo* is unclear, as SPI-1 secretion apparatus mutations do not affect intraperitoneal LD_{50} values (Penheiter *et al.*, 1997; Jones and Falkow, 1994; Behlau and Miller, 1993; Galán and Curtiss, 1989).

4.1.2. Secreted Factors

When grown in Luria-Bertani media, Salmonella secretes many proteins into the culture supernatant via the invasion-associated secretion apparatus (Wood *et al.*, 1996; Pegues *et al.*, 1995). Most of these secreted proteins are encoded on SPI-1, but at least two are encoded elsewhere on the bacterial chromosome. The secreted proteins can be divided into two categories: delivery factors that facilitate the translocation of other secreted proteins into host cells, and effector proteins that are translocated into host cells and may directly affect cellular functions. Certain secreted factors, namely SipB, SipC, and SipD, may have dual roles.

Effector proteins that are predicted to be translocated into host cells via the invasion-associated secretion apparatus include the SPI-1 encoded factors SptP, SipA, and AvrA, and the non-SPI-1 encoded factors SopE and SopB (Galyov *et al.*, 1997; Hardt and Galán, 1997; Wood *et al.*, 1996; Kaniga *et al.*, 1996; Hueck *et al.*, 1995; Kaniga *et al.*, 1995a). These effector proteins are generally thought to share two common features: they are not involved in the secretion or translocation of other proteins (Collazo and Galán, 1997a; Hueck *et al.*, 1995; Kaniga *et al.*, 1995a), and they all presumably modify host cell functions to facilitate Salmonella-induced host cell responses. Surprisingly, mutations in most of the genes encoding these factors do not significantly reduce invasion *in vitro* or virulence in mice. A possible explanation for this finding is that effector proteins have redundant functions.

The activities of the effector proteins in host cells are not yet well understood. SptP contains two modular domains—a C-terminal region similar to the catalytic domain of the YopH protein tyrosine phosphatase, and an N-terminal region similar to the YopE and ExoS cytotoxins. SptP exhibits phosphatase activity *in vitro*, suggesting that its C-terminal domain may dephosphorylate host cell signaling proteins *in vivo* (Kaniga *et al.*, 1996). Interestingly, either domain of SptP can induce cytoskeletal rearrangements in cultured cells (Fu and Galán, 1998b). The function of SipA is unknown, and its translocation into host cells is undetectable by immunofluorescence microscopy (Collazo and Galán, 1997a). Based on the similarity of SipA to IpaA from Shigella, SipA may play a role in the formation of bacteria-induced focal adhesion-like structures in host cells (Tran Van Nhieu *et al.*, 1997). It has also been difficult to define the activity of AvrA. AvrA is similar to YopJ from Yersinia, which is required for Yersinia-induced apoptosis of macrophages (Mills *et al.*, 1996; Monack *et al.*, 1997). However, an *S. typhimurium avrA* mutant is not less cytotoxic to J774 macrophages (Hardt and Galán, 1997). The SopB protein is required for *S. dublin* to fully induce the inflammatory response in bovine ileal ligated loops, and may also play a role in *S. typhimurium* invasion of epithelial cells (Galyov *et al.*, 1997; Hong and Miller, 1998). Finally, the best characterized effector protein is SopE, which can induce membrane ruffling and cytoskeletal rearrangements in a Cdc42 and Rac1 dependent manner when transiently transfected into COS-1 cells. SopE specifically binds to and stimulates GDP/GTP nucleotide exchange on Rac-1 and Cdc42 *in vitro*, suggesting that SopE may function as a GEF (guanine nucleotide exchange factor) *in vivo* (Hardt *et al.*, 1998a).

Secreted proteins that appear to be involved in the delivery of effector proteins into host cells include SpaN (InvJ), SpaO, SipB, SipC, and SipD. Loss-of-function mutations in the genes encoding these factors severely reduce Salmonella invasion of tissue culture cells (Collazo *et al.*, 1995; Hueck *et al.*, 1995; Kaniga *et al.*, 1995a; Kaniga *et al.*, 1995b; Groisman and Ochman, 1993). SpaN and SpaO are required for export of the Sip proteins, as well as other secreted proteins (Hardt *et al.*, 1998b; Hardt and Galán, 1997; Collazo and Galán, 1996). Both SpaN and SpaO are only weakly similar to their counterparts in other bacterial type-III secretion systems, and they appear to have functions unique to the Salmonella SPI-1 export apparatus. Although these proteins are secreted into the culture supernatant during growth in rich media, when bacteria are pelleted, washed, and resuspended in tissue culture medium, SpaN is only secreted in the presence of live epithelial cells or serum (Collazo *et al.*, 1995; Li *et al.*, 1995; Zierler and Galán, 1995). This finding implies that SpaN secretion can be triggered by host cell contact. Mutations in *invG* and *invC*, but not *invE*, abolish contact-dependent secretion of SpaN (Collazo *et al.*, 1995; Zierler and Galán, 1995). These results correlate with the observations that *invG* and *invC* mutants, but not *invE* mutants, lack Salmonella surface filamentous appendages (Ginocchio *et al.*, 1994). Thus, it is tempting to speculate that SpaN could be a component of the appendages that are shed upon host cell contact.

In contrast to SpaN and SpaO, the SipB, SipC, and SipD proteins are not required for the secretion of other proteins into the culture supernatant, but are instead necessary for the translocation of other factors into host cells (Fu and Galán, 1998b; Collazo and Galán, 1997a; Wood *et al.*, 1996). Although the specific functions of SipB, SipC, and SipD have not yet been determined, predictions can be made based on their homology to the Shigella Ipa proteins. By analogy to IpaD, SipD may be involved in the modulation of the secretion process. A regulatory role for SipD is supported by the finding that *sipD* mutants export greater quantities of other Sips into the culture supernatant (Kaniga *et al.*, 1995a). It has also been suggested that the SipB homolog IpaB is a pore-forming protein, and may form a secretion channel in the host cell membrane (High *et al.*, 1992). In addition to facilitating translocation of proteins into host cells, the Shigella Ipa proteins have several effector functions (Hueck, 1998). The observation that SipB and SipC are translocated into host cells suggests that these proteins may also have effector functions (Collazo and Galán, 1997a).

4.1.3. Chaperones

SPI-1 appears to encode at least three chaperones, SicP, SicA, and SpaM (InvI), which are thought to bind to type-III secreted factors in the bacterial cytoplasm, and maintain them in a competent state for secretion (Fu and Galán, 1998a; Collazo *et al.*, 1995; Kaniga *et al.*, 1995b; Groisman *et al.*, 1993). All three proteins are predicted to be small (14–19 kDa) and highly charged. SicP has been shown to specifically bind to SptP *in vitro*, and increase SptP stability *in vivo* (Fu and Galán, 1998a). Although *sicA* and *spaM* mutants cannot efficiently enter cultured epithelial cells, chaperone functions have not been experimentally demonstrated for the SicA and SpaM proteins (Collazo *et al.*, 1995; Kaniga *et al.*, 1995b). SicA is a homolog of the Shigella protein IpgC, a chaperone for IpaB and IpaC, suggesting that SicA may serve as a chaperone for SipB and SipC (Ménard *et al.*, 1994).

4.2. Factors Mediating Salmonella Contact with Host Cells

Translocation of type-III effectors requires direct contact between Salmonella and host cells. Heterologous expression of an *E. coli* afimbrial adhesin in *S. typhimurium* increases bacterial invasion, suggesting that factors that promote bacterial adhesion may allow more efficient translocation of SPI-1 effector proteins, and thus enhance invasion (Francis *et al.*, 1992). *S. typhimurium* may encode adherence factors that function independently of SPI-1 invasion factors, as SPI-1 secretion mutants can still associate with epithelial cells. In contrast to *S. typhimurium*, mutations in the *S. typhi* SPI-1 secretion apparatus reduce both adherence and invasion, suggesting that *S. typhi* can only associate with host cells by entering them (Leclerc *et al.*, 1998).

S. typhimurium encodes several fimbrial adhesins, including long polar (LP) and type 1 fimbriae, that mediate bacterial attachment to host cells (Bäumler *et al.*, 1997b). *In vitro*, mutations in the *lpf* (LP fimbriae) and *fim* (type 1 fimbriae) operons reduce entry of *S. typhimurium* into tissue culture cells (Bäumler *et al.*, 1996a; Ernst *et al.*, 1990). In ligated loops, LP-fimbriae appear to mediate *S. typhimurium* attachment and invasion into murine M cells, because *lpfC* mutants cannot invade or destroy these cells (Bäumler *et al.*, 1996b). However, mutations in the *lpf*, or *fim* genes have little to no effect on *S. typhimurium* virulence in mice, indicating that multiple fimbriae may be involved in bacterial colonization *in vivo* (Bäumler *et al.*, 1996b; Lockman and Curtiss, 1992). Although LP fimbriae may enhance SPI-1-dependent invasion of intestinal cells, these fimbriae must have another role in infection, because combining *lpfC* and *invA* mutations reduces *S. typhimurium* virulence more than when either mutation is present alone (Bäumler *et al.*, 1997c).

Salmonella invasion can additionally be influenced by motility, which may allow the bacteria to actively move towards host cells. Nonflagellate and nonmotile mutants of *S. typhi* and *S. typhimurium* are unable to enter tissue culture cells (Betts and Finlay, 1992; Khoramian *et al.*, 1990; Liu *et al.*, 1988). In some cases, the ability of non-motile mutants to enter cultured cells can be increased by using centrifugation to assist bacterial-host cell contact (Jones *et al.*, 1992). However, centrifugation does not suppress the invasion defect in *S. typhi*, suggesting that motility may also be important to maintain or enhance bacterial contact with or invasion into host cells. Consistent with this idea, *che* mutations that cause smooth swimming phenotypes actually increase *S. typhimurium* invasion, while mutations that lead to perpetual tumbling reduce bacterial invasion (Jones *et al.*, 1992; Lee *et al.*, 1992; Khoramian *et al.*, 1990).

Lipopolysaccharide (LPS) may also be important for invasion by certain *Salmonella* serovars. For example, disruption of the *S. typhi* and *S. choleraesuis* O-side chains significantly reduces bacterial invasion (Mroczenski-Wildey *et al.*, 1989; Finlay *et al.*, 1988b). These mutants are also non-adherent, suggesting that LPS directly or indirectly assists in promoting bacterial interactions with the host cell surface. However, the contribution of LPS to invasion is not universal because *S. typhimurium* rough mutants appear to enter tissue culture cells as well as smooth strains (Gahring *et al.*, 1990; Giannella *et al.*, 1973b).

5. EVOLUTIONARY ASPECTS OF SALMONELLA INVASION

Many virulence traits of Salmonella are encoded on chromosomal elements that are absent from the genome of closely related, non-pathogenic strains of *E. coli*. These chromosomal elements, termed pathogenicity islands, are often located at prophage integration sites, and usually have a GC/AT base pair composition that differs from that of the rest of the Salmonella chromosome, suggesting that they were horizontally transferred from distantly related organisms (Lee, 1996). Several pathogenicity islands in Salmonella appear to contribute to bacterial entry into non-phagocytic cells. Although Salmonella also contains a large virulence plasmid associated with systemic pathogenesis, this plasmid does not appear to be required for invasion (Gulig and Curtiss, 1987).

Salmonella pathogenicity island-1 (SPI-1) contains the large cluster of genes that encode the invasion-associated type-III secretion apparatus and some of its secreted factors (see Section 4.1 and Figure 3). This 40 kb region of DNA, at centisome 63 of the Salmonella chromosome, is located between the *fhlA* and *mutS* genes, which are contiguous in the corresponding region of the *E. coli* K-12 chromosome (Mills *et al.*, 1995). Because SPI-1 is present in *S. bongori* and all subspecies of *S. enterica*, SPI-1 must have been acquired soon after the divergence of Salmonella from other enteric genera. Interestingly, the regions at the ends of SPI-1 do not appear to encode genes required for Salmonella invasion. Deletion of the *sitABCD* (Salmonella iron transport) region, which encodes components of an ABC iron transport system, does not effect Salmonella entry into cultured epithelial cells (Zhou *et al.*, 1997). It has also been suggested that sequences between *invH* and *mutS* are not required for Salmonella invasion (Galán, 1996). Pathogenicity islands are often flanked by repeated sequences or insertion sequence (IS) elements. There are no repeated sequences at the ends of SPI-1 in *S. typhimurium*, although sequences similar to IS*3* have been detected downstream of *invH* in *S. Choleraesuis* (Mills *et al.*, 1995; Altmeyer *et al.*, 1993). Further analysis of the ends of SPI-1 in diverse Salmonella species and serovars may shed light on how SPI-1 was acquired and why the *sit* genes are located on this element.

The secreted proteins SopE and SopB are also encoded by genes unique to Salmonella. SopE is encoded on a P2-like cryptic bacteriophage near centisome 61 of the Salmonella chromosome (Hardt *et al.*, 1998b). The *sopB* gene is located on Salmonella pathogenicity island-5 (SPI-5), near centisome 20 of the Salmonella chromosome (Wood *et al.*, 1998). The gene located directly downstream from *sopB*, *pipC* (*sigE* in *S. typhimurium*), appears to encode a chaperone for SopB (Hong and Miller, 1998). Mutations in some of the other genes on SPI-5 (*pipA*, *pipB*, and *pipD*) reduce the ability of *S. dublin* to produce an inflammatory response in bovine ileal ligated loops (Wood *et al.*, 1998).

The differential acquisition or loss of invasion genes by Salmonella serovars may contribute to the diverse pathogenic effects and host specificities of these organisms. Although SPI-1 is present in all Salmonella, the SPI-1 *avrA* gene is replaced by another open reading frame in *S. typhi* and *S. choleraesuis* (Hardt and Galán, 1997). In addition, the *sopE* gene, as well as certain fimbrial operons, are only found in a subset of Salmonella serovars (Bäumler *et al.*, 1997a; Hardt *et al.*, 1998b). Detailed genomic comparisons between Salmonella serovars may reveal more information about how genetic diversity contributes to the differential pathogenic characteristics of these organisms.

6. REGULATION OF SALMONELLA INVASION

6.1. Conditions that Regulate Salmonella Invasion

Salmonellae encounter diverse environments during infection of their hosts. Following passage through the acidic stomach, bacteria enter the intestinal lumen, which is thought to contain a low oxygen level, a high osmolarity, a neutral pH, and a high cation concentration. If salmonellae are grown *in vitro* under low oxygen and high osmolarity conditions, they enter tissue culture cells much more efficiently than when grown aerobically or in low osmolarity media (Tartera and Metcalf, 1993; Schiemann and Shope, 1991; Galán and Curtiss, 1990; Lee and Falkow, 1990). Salmonella are also less invasive in the presence of glucose or other utilizable carbohydrates (Schiemann, 1995). These observations have led to the hypothesis that Salmonella senses its environment during infection, and expresses invasion determinants when in the intestinal lumen. Although it was proposed that the expression of bacterial invasion factors only occurs after Salmonella adheres to host cells, subsequent studies have shown that salmonellae grown under oxygen-limiting conditions induce morphological and cytoskeletal changes in epithelial cells immediately upon bacterial contact (Francis *et al.*, 1992; Finlay *et al.*, 1989). Moreover, *de novo* bacterial protein synthesis is not required for invasion (Lee and Falkow, 1990). However, if salmonellae are grown to stationary phase, or subjected to prolonged inhibition of protein synthesis prior to invasion, they cannot efficiently enter host cells (MacBeth and Lee, 1993; Lee and Falkow, 1990). These findings indicate that invasion factors are not stable, and the ability to enter host cells is transient. Tight control of the expression and activity of invasion factors may allow Salmonella to avoid potential detrimental effects that these proteins might impart during other stages of bacterial infection.

6.2. Regulatory Factors that Control Invasion Gene Expression

Many of the conditions that regulate Salmonella invasion appear to do so by controlling the expression of SPI-1 invasion genes. For example, the expression of *invF*, *prgH*, *prgK*, *orgA*, *sipA*, and *sipC* in *S. typhimurium* is coordinately controlled by oxygen, osmolarity, and pH (Bajaj *et al.*, 1996). These findings have led to the speculation that all of the genes encoding the type-III secretion apparatus and secreted factors are regulated in a similar manner. Although the transcriptional organization of SPI-1 is not well characterized, it appears that expression of the secretion genes requires at least three promoters. One promoter upstream of *prgH* directs the transcription of the *prgHIJK* and *orgA* genes (Pegues *et al.*, 1995; B. Jones, personal communication). Another promoter directs the expression of *invH*. Finally, the presence of short intergenic distances and overlapping coding regions in the *invF-spaS* region suggests that these genes may all be part of an operon and transcribed from a promoter upstream of *invF*. Further studies will be required to determine exactly how these and other genes encoded on SPI-1 are transcribed.

The coordinate regulation of invasion gene expression requires the participation of regulatory factors. As shown in Figure 5, SPI-1 encodes

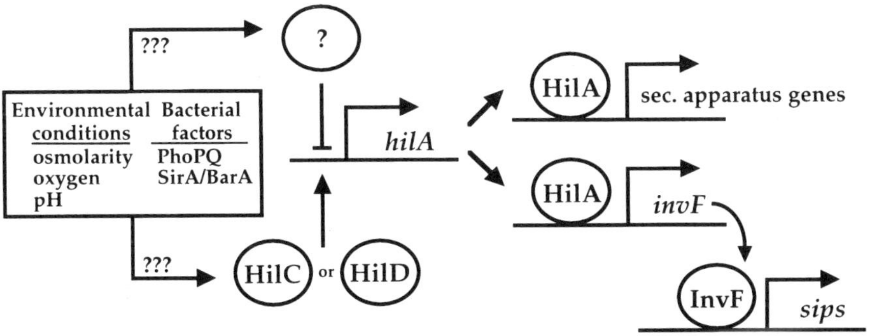

FIGURE 5. Regulation of invasion gene expression in *S. typhimurium*. *S. typhimurium* invasion genes are regulated by several environmental conditions and sensor/effector pathways. Under conditions that are unfavorable for invasion, expression of the SPI-1 encoded regulator HilA is repressed by a factor or factors that have not yet been identified. Under conditions that promote invasion, *hilA* expression is derepressed by the SPI-1 encoded HilC and HilD proteins. HilA activates expression of SPI-1 secretion apparatus genes as well as the *sip* genes via InvF, another SPI-1 encoded regulator.

several regulatory factors that appear to act in a sequential manner to control invasion gene expression. InvF, a member of the AraC/XylS family of transcriptional regulators, is required for expression of *sipC* (Kaniga *et al.*, 1994; Eichelberg *et al.*, 1996). The long intergenic space present between *spaS* and *sicA* suggests that a promoter may be located in this region that mediates the *invF*-dependent expression of the *sic-sip* genes. The expression of *invF* and *prgH* is directly controlled by HilA (IagA), a member of the OmpR/ToxR family of transcriptional regulators (Arricau *et al.*, 1998; Bajaj *et al.*, 1996; Bajaj *et al.*, 1995). Under conditions that are not optimal for Salmonella invasion, *hilA* expression appears to be repressed by an as yet unidentified repressor that is not encoded on SPI-1 (Schechter *et al.*, in press). When conditions are optimal for Salmonella invasion, the derepression of *hilA* expression can be independently elicited by either HilC or HilD, two members of the AraC/XylS family that are each encoded on SPI-1 (Schechter *et al.*, in press).

Little is known about the bacterial signaling pathways that are presumably involved in transmitting environmental signals to the SPI-1 encoded factors that directly control invasion gene expression. One two-component system implicated in controlling invasion gene expression is PhoPQ, which appears to simultaneously activate intramacrophage survival genes and repress invasion genes in response to low extracellular cation levels (Groisman, 1998). Because cation levels are thought to be low inside host cells, it is believed that the PhoPQ system represses invasion gene expression after Salmonella invasion has occurred. Another factor that regulates SPI-1 invasion gene expression is SirA, a member of the FixJ family of response regulators (Johnston *et al.*, 1996). SirA also regulates the expression of *sigD* (*sopB*), a secreted factor that is not encoded on SPI-1 (Hong and Miller, 1998). SirA activity may be controlled by the sensor kinase BarA, but the environmental signal that activates BarA has not yet been identified (R. Maurer, personal communication). In *S. typhi*, invasion genes are negatively controlled by the transcriptional regulator RcsB (Arricau *et al.*, 1998). RcsB regulates the expression of capsular polysaccharide biosynthesis enzymes in both *E. coli* and *S. typhi*, but the environmental signal that RcsB responds to is unknown (Gottesman, 1995). Finally, invasion gene expression is also affected by certain mutations in SPI-2, a pathogenicity island that encodes another type-III secretion system required for Salmonella to survive intracellularly in macrophages (Deiwick *et al.*, 1998). The mechanisms by which these mutations affect invasion gene expression are not understood. However, this finding raises the interesting possibility that factors encoded on a pathogenicity island involved in another stage of Salmonella infection may regulate expression of SPI-1 invasion genes.

The complex regulation of SPI-1 invasion genes by several environmental and regulatory factors suggests that it is important to limit the expression of invasion genes to the appropriate times during Salmonella infection. Further studies may reveal exactly how environmental signals in the small intestine are sensed by Salmonella, and how bacterial regulatory circuits control the SPI-1 regulatory cascade.

7. CONCLUSIONS

Salmonellae are bacterial pathogens that have evolved and diversified in order to infect and cause disease in many hosts. One important feature of Salmonella pathogenesis is bacterial entry into intestinal epithelial cells. Genetic, biochemical, and cell biology studies have revealed that Salmonella invasion involves many bacterial and host cell factors. The bacteria produce a complex secretion apparatus that can translocate bacterial proteins directly into the cytosol of host cells to stimulate signaling pathways. Bacterial adherence factors and host cell receptors also contribute to invasion. Elucidation of the activities of each bacterial factor and their host cell targets is needed to fully understand how Salmonella induces cytoskeletal changes and cytokine production in epithelial cells. More detailed analyses of Salmonella infection *in vivo* is needed to understand the contribution of bacterial invasion and host cell responses to disease. Finally, comparison of bacterial invasion factors and their regulation in diverse *Salmonella* species and serovars may shed light on the different host specificities of salmonellae, and different diseases caused by salmonellae.

ACKNOWLEDGMENTS. We thank Beth McCormick, Alison Criss, and Peter Juo for critical reading of the manuscript and helpful discussions, Aaron Kelly for providing Figure 4, and Brad Jones and Russ Maurer for sharing information prior to publication.

8. REFERENCES

Atmeyer, R.M., McNern, J.K., Bossio, J.C., Rosenshine, I., Finlay, B.B., and Galán, J.E., 1993, Cloning and molecular characterization of a gene involved in *Salmonella* adherence and invasion of cultured epithelial cells, *Mol. Microbiol.* **7**:89–98.

Anderson, D.M., and Schneewind, O., 1997, A mRNA signal for the type III secretion of Yop proteins by *Yersinia enterocolitica*, *Science* **278**:1140–1143.

Arnold, J.W., Niesel, D.W., Annable, C.R., Hess, C.B., Asuncion, M., Cho, Y., Peterson, J.W., and Klimpel, G.R., 1993, Tumor necrosis factor-a mediates the early pathology in Salmonella infection of the gastrointestinal tract, *Microb. Pathog.* **14**:217–227.

Arricau, N., Hermant, D., Waxin, H., Ecobichon, C., Duffey, P.S., and Popoff, M.Y., 1998, The RcsB-RcsC regulatory system of *Salmonella typhi* differentially modulates the expression of invasion proteins, flagellin and Vi antigen in response to osmolarity, *Mol. Microbiol.* **29**:835–850.

Bajaj, V., Hwang, C., and Lee, C.A., 1995, *hilA* is a novel *ompR/toxR* family member that activates the expression of *Salmonella typhimurium* invasion genes, *Mol. Microbiol.* **18**:715–727.

Bajaj, V., Lucas, R.L., Hwang, C., and Lee, C.A., 1996, Co-ordinate regulation of *Salmonella typhimurium* invasion genes by environmental and regulatory factors is mediated by control of *hilA* expression, *Mol. Microbiol.* **22**:703–714.

Bäumler, A.J., 1997, The record of horizontal gene transfer in *Salmonella*, *Trends Microbiol.* **5**:318–322.

Bäumler, A.J., Tsolis, R.M., and Heffron, F., 1996a, Contribution of fimbrial operons to attachment to and invasion of epithelial cell lines by *Salmonella typhimurium*, *Infect. Immun.* **64**:1862–1865.

Bäumler, A.J., Tsolis, R.M., and Heffron, F., 1996b, The *lpf* fimbrial operon mediates adhesion of *Salmonella typhimurium* to murine Peyer's patches, *Proc. Natl. Acad. Sci. USA* **93**:279–283.

Bäumler, A.J., Gilde, A.J., Tsolis, R.M., van der Velden, A.W., Ahmer, B.M., and Heffron, F., 1997a, Contribution of horizontal gene transfer and deletion events to development of distinctive patterns of fimbrial operons during evolution of *Salmonella* serotypes, *J. Bacteriol.* **179**:317–322.

Bäumler, A.J., Tsolis, R.M., and Heffron, F., 1997b, Fimbrial adhesins of *Salmonella typhimurium*. Role in bacterial interactions with epithelial cells, *Adv. Exp. Med. Biol.* **412**:149–158.

Bäumler, A.J., Tsolis, R.M., Valentine, P.J., Ficht, T.A., and Heffron, F., 1997c, Synergistic effect of mutations in *invA* and *lpfC* on the ability of *Salmonella typhimurium* to cause murine typhoid, *Infect. Immun.* **65**:2254–2259.

Behlau, I., and Miller, S.I., 1993, A PhoP-repressed gene promotes *Salmonella typhimurium* invasion of epithelial cells, *J. Bacteriol.* **175**:4475–4484.

Betts, J., and Finlay, B.B., 1992, Identification of *Salmonella typhimurium* invasiveness loci, *Can. J. Microbiol.* **38**:852–857.

Bowe, F., Lipps, C.J., Tsolis, R.M., Groisman, E., Heffron, F., and Kusters, J.G., 1998, At least four percent of the *Salmonella typhimurium* genome is required for fatal infection of mice, *Infect. Immun.* **66**:3372–3377.

Carter, P.B., and Collins, F.M., 1974, The route of enteric infection in normal mice, *J. Exp. Med.* **139**:1189–1203.

Chen, L.M., Hobbie, S., and Galán, J.E., 1996a, Requirement of CDC42 for *Salmonella*-induced cytoskeletal and nuclear responses, *Science* **274**:2115–2118.

Chen, L.M., Kaniga, K., and Galán, J.E., 1996b, *Salmonella* spp. are cytotoxic for cultured macrophages, *Mol. Microbiol.* **21**:1101–1015.

Cirillo, C.M., Valdivia, R.H., Monack, D.M., and Falkow, S., 1998, Macrophage-dependent induction of the *Salmonella* pathogenicity island 2 type III secretion system and its role in intracellular survival, *Mol. Microbiol.* **30**:175–188.

Clark, M.A., Jepson, M.A., Simmons, N.L., and Hirst, B.H., 1994, Preferential interaction of *Salmonella typhimurium* with mouse Peyer's patch M cells, *Res. Microbiol.* **145**:543–552.

Collazo, C.M., and Galán, J.E., 1996, Requirement for exported proteins in secretion through the invasion-associated type III system of *Salmonella typhimurium*, *Infect. Immun.* **64**:3524–3531.

Collazo, C.M., and Galán, J.E., 1997a, The invasion-associated type III system of *Salmonella typhimurium* directs the translocation of Sip proteins into the host cell, *Mol. Microbiol.* **24**:747–756.

Collazo, C.M., and Galán, J.E., 1997b, The invasion-associated type-III protein secretion system in *Salmonella*–a review, *Gene* **192**:51–59.

Collazo, C.M., Zierler, M.K., and Galán, J.E., 1995, Functional analysis of the *Salmonella typhimurium* invasion genes *invl* and *invJ* and identification of a target of the protein secretion apparatus encoded in the *inv* locus, *Mol. Microbiol.* **15**:25–38.

Crago, A.M., and Koronakis, V., 1998, *Salmonella* InvG forms a ring-like multimer that requires the InvH lipoprotein for outer membrane localization, *Mol. Microbiol.* **30**:47–56.

Daefler, S., and Russel, M., 1998, The *Salmonella typhimurium* InvH protein is an outer membrane lipoprotein required for the proper localization of InvG, *Mol. Microbiol.* **28**:1367–1380.

Deiwick, J., Nikolaus, T., Shea, J.E., Gleeson, C., Holden, D.W., and Hensel, M., 1998, Mutations in *Salmonella* Pathogenicity Island 2 (SPI2) genes affecting transcription of SPI1 genes and resistance to antimicrobial agents, *J. Bacteriol.* **180**:4775–4780.

Duebbert, I.E., and Peterson, J.W., 1985, Enterotoxin-induced fluid accumulation during experimental salmonellosis and cholera: involvement of prostaglandin synthesis by intestinal cells, *Toxicon* **23**:157–172.

Eichelberg, K., Ginocchio, C.C., and Galán, J.E., 1994, Molecular and functional characterization of the Salmonella typhimurium invasion genes *invB* and *invC*: homology of InvC to the F0F1 ATPase family of proteins, *J. Bacteriol.* **176**:4501–4510.

Eichelberg, K., Kaniga, K., and Galán, J.E., 1996, Transcriptional regulation of *Salmonella* secreted virulence determinants, in: *The 96th General Meeting of the American Society for Microbiology*, New Orleans, ASM Press, Washington D.C., p. 161.

Ernst, R.K., Dombroski, D.M., and Merrick, J.M., 1990, Anaerobiosis, type 1 fimbriae, and growth phase are factors that affect invasion of HEp-2 cells by Salmonella typhimurium, *Infect. Immun.* **58**:2014–2016.

Finlay, B.B., and Falkow, S., 1990, *Salmonella* interactions with polarized human intestinal Caco-2 epithelial cells, *J. Infect. Dis.* **162**:1096–1106.

Finlay, B.B., Gumbiner, B., and Falkow, S., 1988a, Penetration of *Salmonella* through a polarized Madin-Darby canine kidney epithelial cell monolayer, *J. Cell Biol.* **107**:221–230.

Finlay, B.B., Starnbach, M.N., Francis, C.L., Stocker, B.A., Chatfield, S., Dougan, G., and Falkow, S., 1988b, Identification and characterization of Tn*phoA* mutants of *Salmonella* that are unable to pass through a polarized MDCK epithelial cell monolayer, *Mol. Microbiol.* **2**:757–766.

Finlay, B.B., Heffron, F., and Falkow, S., 1989, Epithelial cell surfaces induce *Salmonella* proteins required for bacterial adherence and invasion, *Science* **243**:940–943.

Finlay, B.B., Ruschkowski, S., and Dedhar, S., 1991, Cytoskeletal rearrangements accompanying Salmonella entry into epithelial cells, *J. Cell. Sci.* **99**:283–296.

Francis, C.L., Starnbach, M.N., and Falkow, S., 1992, Morphological and cytoskeletal changes in epithelial cells occur immediately upon interaction with *Salmonella typhimurium* grown under low-oxygen conditions, *Mol. Microbiol.* **6**:3077–3087.

Francis, C.L., Ryan, T.A., Jones, B.D., Smith, S.J., and Falkow, S., 1993, Ruffles induced by *Salmonella* and other stimuli direct macropinocytosis of bacteria, *Nature* **364**:639–642.

Fu, Y., and Galán, J.E., 1998a, Identification of a specific chaperone for SptP, a substrate of the centisome 63 type III secretion system of *Salmonella typhimurium*, *J. Bacteriol.* **180**:3393–3399.

Fu, Y., and Galán, J.E., 1998b, The *Salmonella typhimurium* tyrosine phosphatase SptP is translocated into host cells and disrupts the actin cytoskeleton, *Mol. Microbiol.* **27**:359–368.

Gahring, L.C., Heffron, F., Finlay, B.B., and Falkow, S., 1990, Invasion and replication of *Salmonella typhimurium* in animal cells, *Infect. Immun.* **58**:443–448.

Galán, J.E., 1994, *Salmonella* entry into mammalian cells: different yet converging signal transduction pathways, *Trends Cell Biol.* **4**:196–199.

Galán, J.E., 1996, Molecular genetic bases of *Salmonella* entry into host cells, *Mol. Microbiol.* **20**:263–271.

Galán, J.E., and Curtiss, R.III, 1989, Cloning and molecular characterization of genes whose products allow *Salmonella typhimurium* to penetrate tissue culture cells, *Proc. Nat. Acad. Sci. USA* **86**:6383–6387.

Galán, J.E., and Curtiss, R.III, 1990, Expression of *Salmonella typhimurium* genes required for invasion is regulated by changes in DNA supercoiling, *Infect. Immun.* **58**:1879–1885.

Galán, J.E., and Sansonetti, P.J., 1996, Molecular and cellular bases of *Salmonella* and *Shigella* interactions with host cells, in: *Escherichia coli and Salmonella typhimurium: Cellular and Molecular Biology*, (F.C. Neidhardt, R.I. Curtiss, J.L. Ingraham, E.C.C. Lin, K.B. Low, B. Magasanik, W.S. Reznikoff, M. Schaechter, and E.H. Umbarger, eds.) ASM Press, Washington D.C., pp. 2757–2773.

Galán, J.E., Ginocchio, C., and Costeas, P., 1992a, Molecular and functional characterization of the *Salmonella* invasion gene *invA*: homology of InvA to members of a new protein family, *J. Bacteriol.* **174**:4338–4349.

Galán, J.E., Pace, J., and Hayman, M.J., 1992b, Involvement of the epidermal growth factor receptor in the invasion of cultured mammalian cells by *Salmonella typhimurium*, *Nature* **357**:588–589.

Galyov, E.E., Wood, M.W., Rosqvist, R., Mullan, P.B., Watson, P.R., Hedges, S., and Wallis, T.S., 1997, A secreted effector protein of *Salmonella dublin* is translocated into eukaryotic cells and mediates inflammation and fluid secretion in infected ileal mucosa, *Mol. Microbiol.* **25**:903–912.

Garcia-del Portillo, F., and Finlay, B.B., 1994, *Salmonella* invasion of nonphagocytic cells induces formation of macropinosomes in the host cell, *Infect. Immun.* **62**:4641–4645.

Garcia-del Portillo, F., and Finlay, B.B., 1995, Targeting of *Salmonella typhimurium* to vesicles containing lysosomal membrane glycoproteins bypasses compartments with mannose 6-phosphate receptors, *J. Cell. Biol.* **129**:81–97.

Garcia-del Portillo, F., Zwick, M.B., Leung, K.Y., and Finlay, B.B., 1993, *Salmonella* induces the formation of filamentous structures containing lysosomal membrane glycoproteins in epithelial cells, *Proc. Nat. Acad. Sci. USA* **90**:10544–10548.

Giannella, R.A., Formal, S.B., Dammin, G.J., and Collins, H., 1973a, Pathogenesis of salmonellosis. Studies of fluid secretion, mucosal invasion, and morphologic reaction in the rabbit ileum, *J. Clin. Invest.* **52**:441–453.

Giannella, R.A., Washington, O., Gemski, P., and Formal, S.B., 1973b, Invasion of HeLa cells by *Salmonella typhimurium*: a model for study of invasiveness of *Salmonella*, *J. Infect. Dis.* **128**:69–75.

Giannella, R.A., Gots, R.E., Charney, A.N., Greenough, W.B., and Formal, S.B., 1975, Pathogenesis of *Salmonella*-mediated intestinal fluid secretion. Activation of adenylate cyclase and inhibition by indomethacin, *Gastroenterology* **69**:1238–1245.

Ginocchio, C., Pace, J., and Galán, J.E., 1992, Identification and molecular characterization of a *Salmonella typhimurium* gene involved in triggering the internalization of salmonellae into cultured epithelial cells, *Proc. Natl. Acad. Sci. USA* **89**:5976–5980.

Ginocchio, C.C., Olmsted, S.B., Wells, C.L., and Galán, J.E., 1994, Contact with epithelial cells induces the formation of surface appendages on *Salmonella typhimurium*, *Cell* **76**:717–724.

Ginocchio, C.C., and Galán, J.E., 1995, Functional conservation among members of the *Salmonella typhimurium* InvA family of proteins, *Infect. Immun.* **63**:729–732.

Gottesman, S., 1995, Regulation of capsule biosynthesis: Modification of the two-component paradigm by an accessory unstable regulator, in *Two-Component Signal Transduction.* (J.A. Hoch, and T.J. Silhavy, eds.), ASM Press, Washington D.C., pp. 253–262.

Groisman, E.A., 1998, The ins and outs of virulence gene expression: Mg^{2+} as a regulatory signal, *BioEssays* **20**:96–101.

Groisman, E.A., and Ochman, H., 1993, Cognate gene clusters govern invasion of host epithelial cells by *Salmonella typhimurium* and *Shigella flexneri*, *EMBO J.* **12**:3779–3787.

Gulig, P.A., 1996, Pathogenesis of systemic disease, in: *Escherichia coli and Salmonella typhimurium: Cellular and Molecular Biology*, (F.C. Neidhardt, R.I. Curtiss, J.L. Ingraham, E.C.C. Lin, K.B. Low, B. Magasanik, W.S. Reznikoff, M. Schaechter, and E.H. Umbarger, eds.) ASM Press, Washington D.C., pp. 2774–2787.

Gulig, P.A., and Curtiss, R.III, 1987, Plasmid-associated virulence of *Salmonella typhimurium*, *Infect. Immun.* **55**:2891–2901.

Hardt, W.-D., and Galán, J.E., 1997, A secreted *Salmonella* protein with homology to an avirulence determinant of plant pathogenic bacteria, *Proc. Natl. Acad. Sci. USA* **94**:9887–9892.

Hardt, W.-D., Chen, L.-M., Schuebel, K.E., Bustelo, X.R., and Galán, J.E., 1998a, *S. typhimurium* encodes an activator of Rho GTPases that induces membrane ruffling and nuclear responses in host cells, *Cell* **93**:815–826.

Hardt, W.-D., Urlaub, H., and Galán, J.E., 1998b, A substrate of the centisome 63 type III protein secretion system of *Salmonella typhimurium* is encoded by a cryptic bacteriophage, *Proc. Natl. Acad. Sci. USA* **95**:2574–2579.

Hensel, M., Shea, J.E., Waterman, S.R., Mundy, R., Nikolaus, T., Banks, G., Vazquez-Torres, A., Gleeson, C., Fang, F.C., and Holden, D.W., 1998, Genes encoding putative effector proteins of the type III secretion system of Salmonella pathogenicity island 2 are required for bacterial virulence and proliferation in macrophages, *Mol. Microbiol.* **30**:163–174.

High, N., Mounier, J., Prévost, M.-C., and Sansonetti, P.J., 1992, IpaB of *Shigella flexneri* causes entry into epithelial cells and escape from the phagocytic vacuole, *EMBO J.* **11**:1991–1999.

Hobbie, S., Chen, L.M., Davis, R.J., and Galán, J.E., 1997, Involvement of mitogen-activated protein kinase pathways in the nuclear responses and cytokine production induced by *Salmonella typhimurium* in cultured intestinal epithelial cells, *J. Immunol.* **159**:5550–5559.

Hong, K.H., and Miller, V.L., 1998, Identification of a novel *Salmonella* invasion locus homologous to *Shigella* IpgDE, *J. Bacteriol.* **180**:1793–1802.

Hueck, C.J., 1998, Type III protein secretion systems in bacterial pathogens of animals and plants, *Microbiol. Mol. Biol. Rev.* **62**:379–433.

Hueck, C.J., Hantman, M.J., Bajaj, V., Johnston, C., Lee, C.A., and Miller, S.I., 1995, *Salmonella typhimurium* secreted invasion determinants are homologous to *Shigella* Ipa proteins, *Mol. Microbiol.* **18**:479–490.

Johnston, C., Pegues, D.A., Hueck, C.J., Lee, A., and Miller, S.I., 1996, Transcriptional activation of *Salmonella typhimurium* invasion genes by a member of the phosphorylated response-regulator superfamily, *Mol. Microbiol.* **22**:715–727.

Jones, B.D., and Falkow, S., 1994, Identification and characterization of a *Salmonella typhimurium* oxygen-regulated gene required for bacterial internalization, *Infect. Immun.* **62**:3745–3752.

Jones, B.D., Lee, C.A., and Falkow, S., 1992, Invasion by *Salmonella typhimurium* is affected by the direction of flagellar rotation, *Infect. Immun.* **60**:2475–2480.

Jones, B.D., Paterson, H.F., Hall, A., and Falkow, S., 1993, *Salmonella typhimurium* induces membrane ruffling by a growth factor-receptor-independent mechanism, *Proc. Nat. Acad. Sci. USA* **90**:10390–10394.

Jones, B.D., Ghori, N., and Falkow, S., 1994, *Salmonella typhimurium* initiates murine infection by penetrating and destroying the specialized epithelial M cells of the Peyer's patches, *J. Exp. Med.* **180**:15–23.

Jones, B., Pascopella, L., and Falkow, S., 1995, Entry of microbes into the host: using M cells to break the mucosal barrier, *Curr. Opin. Immunol.* **7**:474–478.

Jung, H.C., Eckmann, L., Yang, S.K., Panja, A., Fierer, J., Morzycka-Wroblewska, E., and Kagnoff, M.F., 1995, A distinct array of proinflammatory cytokines is expressed in human colon epithelial cells in response to bacterial invasion, *J. Clin. Invest.* **95**:55–65.

Kaniga, K., Bossio, J.C., and Galán, J.E., 1994, The *Salmonella typhimurium* invasion genes *invF* and *invG* encode homologues of the AraC and PulD family of proteins, *Mol. Microbiol.* **13**:555–568.

Kaniga, K., Trollinger, D., and Galán, J.E., 1995a, Identification of two targets of the type III protein secretion system encoded by the *inv* and *spa* loci of *Salmonella typhimurium* that have homology to the *Shigella* IpaD and IpaA proteins, *J. Bacteriol.* **177**:7078–7085.

Kaniga, K., Tucker, S., Trollinger, D., and Galán, J.E., 1995b, Homologs of the *Shigella* IpaB and IpaC invasins are required for *Salmonella typhimurium* entry into cultured epithelial cells, *J. Bacteriol.* **177**:3965–3971.

Kaniga, K., Uralil, J., Bliska, J.B., and Galán, J.E., 1996, A secreted protein tyrosine phosphatase with modular effector domains in the bacterial pathogen *Salmonella typhimurium*, *Mol. Microbiol.* **21**:633–641.

Khoramian, F.T., Harayama, S., Kutsukake, K., and Pechere, J.C., 1990, Effect of motility and chemotaxis on the invasion of *Salmonella typhimurium* into HeLa cells, *Microb. Pathog.* **9**:47–53.

Klimpel, G.R., Asuncion, M., Haithcoat, J., and Niesel, D.W., 1995, Cholera toxin and *Salmonella typhimurium* induce different cytokine profiles in the gastrointestinal tract, *Infect. Immun.* **63**:1134–1137.

Kohbata, S., Yokoyama, H., and Yabuuchi, E., 1986, Cytopathogenic effect of *Salmonella typhi* GIFU 10007 on M cells of murine ileal Peyer's Patches in ligated ileal loops: An ultrastructural study, *Microbiol. Immunol.* **30**:1225–1237.

Kubori, T., Matsushima, Y., Nakamura, D., Uralil, J., Lara-Tejero, M., Sukhan, A., Galán, J.E., and Aizawa, S.-I., 1998, Supramolecular structure of the *Salmonella typhimurium* type III protein secretion system, *Science* **280**:602–605.

Leclerc, G.J., Tartera, C., and Metcalf, E.S., 1998, Environmental regulation of *Salmonella typhi* invasion-defective mutants, *Infect. Immun.* **66**:682–691.

Lee, C.A., 1996, Pathogenicity islands and the evolution of bacterial pathogens, *Inf. Agents Dis.* **5**:1–7.

Lee, C.A., 1997, Type III secretion systems: machines to deliver bacterial proteins into eukaryotic cells, *Trends Microbiol.* **5**:148–156.

Lee, C.A., and Falkow, S., 1990, The ability of *Salmonella* to enter mammalian cells is affected by bacterial growth state, *Proc Natl Acad Sci USA* **87**:4304–4308.

Lee, C.A., Jones, B.D., and Falkow, S., 1992, Identification of a *Salmonella typhimurium* invasion locus by selection for hyperinvasive mutants, *Proc Natl Acad Sci USA* **89**:1847–1851.

Li, J., Ochman, H., Groisman, E.A., Boyd, E.F., Solomon, F., Nelson, K., and Selander, R.K., 1995, Relationship between evolutionary rate and cellular location among the Inv/Spa invasion proteins of *Salmonella enterica*, *Proc. Natl. Acad. Sci. USA* **92**:7252–7256.

Liu, S.-L., Ezaki, T., Miura, H., Matsui, K., and Yabuuchi, E., 1988, Intact motility as a *Salmonella typhi* invasion-related factor, *Infect. Immun.* **56**:1967–1973.

Lockman, H.A., and Curtiss, R.I., 1992, Virulence of non-type 1-fimbriated and nonfimbriated nonflagellated *Salmonella typhimurium* mutants in murine typhoid fever, *Infect. Immun.* **60**:491–496.

Lodge, J., Douce, G.R., Amin, I.I., Bolton, A.J., Martin, G.D., Chatfield, S., Dougan, G., Brown, N.L., and Stephen, J., 1995, Biological and genetic characterization of Tn*phoA* mutants of *Salmonella typhimurium* TML in the context of gastroenteritis, *Infect. Immun.* **63**:762–769.

MacBeth, K.J., and Lee, C.A., 1993, Prolonged inhibition of bacterial protein synthesis abolishes *Salmonella* invasion, *Infect. Immun.* **61**:1544–1546.

Madara, J.L., Patapoff, T.W., Gillece-Castro, B., Colgan, S.P., Parkos, C.A., Delp, C., and Mrsny, R.J., 1993, 5′-Adenosine Monophosphate is the neutrophil-derived paracrine factor that elicits chloride secretion from T84 intestinal epithelial cell monolayers, *J. Clin. Invest.* **91**:2320–2325.

McCormick, B.A., Colgan, S.P., Delp-Archer, C., Miller, S.I., and Madara, J.L., 1993, *Salmonella typhimurium* attachment to human intestinal epithelial monolayers: transcellular signalling to subepithelial neutrophils, *J. Cell Biol.* **123**:895–907.

McCormick, B.A., Hofman, P.M., Kim, J., Carnes, D.K., Miller, S.I., and Madara, J.L., 1995a, Surface attachment of *Salmonella typhimurium* to intestinal epithelia imprints the subepithelial matrix with gradients chemotactic for neutrophils, *J. Cell Biol.* **131**:1599–1608.

McCormick, B.A., Miller, S.I., Carnes, D., and Madara, J.L., 1995b, Transepithelial signaling to neutrophils by salmonellae: a novel virulence mechanism for gastroenteritis, *Infect. Immun.* **63**:2302–2309.

McCormick, B.A., Parkos, C.A., Colgan, S.P., Carnes, D.K., and Madara, J.L., 1998, Apical secretion of a pathogen-elicited epithelial chemoattractant activity in response to surface colonization of intestinal epithelia by *Salmonella typhimurium*, *J. Immunol.* **160**:455–466.

McNeil, A., Dunstan, S.J., Clark, S., and Strugnell, R.A., 1995, *Salmonella typhimurium* displays normal invasion of mice with defective epidermal growth factor receptors, *Infect. Immun.* **63**:2770–2772.

Mellman, I., 1996, Endocytosis and molecular sorting, *Annu. Rev. Cell Dev. Biol.* **12**:575–625.

Ménard, R., Sansonetti, P.J., Parsot, C., and Vasselon, T., 1994, Extracellular association and cytoplasmic partitioning of the IpaB and IpaC invasins of *S. flexneri*, *Cell* **79**:515–525.

Miller, S.I., Hohmann, E.L., and Pegues, D.A., 1995, *Salmonella* (including *Salmonella typhi*), in: *Principles and practice of infectious disease*, (G. L. Mandell, J.E. Bennett, and R. Dolin, eds.) Churchill Livingstone, New York, pp. 2013–2033.

Mills, D.M., Bajaj, V., and Lee, C.A., 1995, A 40 kb chromosomal fragment encoding *Salmonella typhimurium* invasion genes is absent from the corresponding region of the *Escherichia coli* K-12 chromosome, *Mol. Microbiol.* **15**:749–759.

Mills, S.D., and Finlay, B.B., 1994, Comparison of *Salmonella typhi* and *Salmonella typhimurium* invasion, intracellular growth and localization in cultured human epithelial cells, *Microb. Pathog.* **17**:409–423.

Mills, S.D., Boland, A., Sory, M.P., van der Smissen, P., Kerbourch, C., Finlay, B.B., and Cornelis, G.R., 1996, *Yersinia enterocolitica* induces apoptosis in macrophages by a process requiring functional type III secretion and translocation mechanisms and involving YopP, presumably acting as an effector protein, *Proc. Natl. Acad. Sci. USA* **94**:12638–12643.

Monack, D.M., Raupach, B., Hromockyj, A.E., and Falkow, S., 1996, *Salmonella typhimurium* invasion induces apoptosis in infected macrophages, *Proc. Natl. Acad. Sci. USA* **93**:9833–9838.

Monack, D.M., Mecsas, J., Ghori, N., and Falkow, S., 1997, *Yersinia* signals macrophages to undergo apoptosis and YopJ is necessary for this cell death, *Proc. Natl. Acad. Sci. USA* **94**:10385–10390.

Mroczenski-Wildey, M.J., Di Fabio, J.L., and Cabello, F.C., 1989, Invasion and lysis of HeLa cell monolayers by *Salmonella typhi*: the role of lipopolysaccharide, *Microb. Pathog.* **6**:143–152.

Neutra, M.R., and Kraehenbuhl, J.P., 1993, The role of transepithelial transport by M cells in microbial invasion and host defense, *J. Cell Sci.* **17**:209–215.

Ochman, H., Soncini, F.C., Solomon, F., and Groisman, E.A., 1996, Identification of a pathogenicity island required for *Salmonella* survival in host cells, *Proc. Natl. Acad. Sci. USA* **93**:7800–7804.

Pace, J., Hayman, M.J., and Galán, J.E., 1993, Signal transduction and invasion of epithelial cells by *S. typhimurium*, *Cell* **72**:505–514.

Pascopella, L., Raupach, B., Ghori, N., Monack, D., Falkow, S., and Small, P.L., 1995, Host restriction phenotypes of *Salmonella typhi* and *Salmonella gallinarum*, *Infect. Immun.* **63**:4329–4435.

Pegues, D.A., Hantman, M.J., Behlau, I., and Miller, S.I., 1995, PhoP/PhoQ transcriptional repression of *Salmonella typhimurium* invasion genes: evidence for a role in protein secretion, *Mol. Microbiol.* **17**:169–181.

Penheiter, K.L., Mathur, N., Giles, D., Fahlen, T., and Jones, B.D., 1997, Non-invasive *Salmonella typhimurium* mutants are avirulent because of an inability to enter and destroy M cells of ileal Peyer's patches, *Mol. Microbiol.* **24**:697–709.

Peterson, J.W., Molina, N.C., Houston, C.W., and Fader, R.C., 1983, Elevated cAMP in intestinal epithelial cells during experimental cholera and salmonellosis, *Toxicon* **21**:761–775.

Pier, G.B., Grout, M., Zaidi, T., Meluleni, G., Mueschenborn, S.S., Banting, G., Ratcliff, R., Evans, M.J., and Colledge, W.H., 1998, *Salmonella typhi* uses CFTR to enter intestinal epithelial cells, *Nature* **393**.

Prasad, R., Chopra, A.K., Peterson, J.W., Pericas, R., and Houston, C.W., 1990, Biological and immunological characterization of a cloned cholera toxin-like enterotoxin from *Salmonella typhimurium*, *Microb. Pathog.* **9**:315–329.

Pugsley, T., 1993, The complete general secretory pathway in gram-negative bacteria, *Microbiol. Rev.* **57**:50–108.

Reed, K.A., Booth, T.A., Hirst, B.H., and Jepson, M.A., 1996, Promotion of *Salmonella typhimurium* adherence and membrane ruffling in MDCK epithelia by staurosporine, *FEMS Microbiol. Lett.* **145**:233–238.

Reed, K.A., Clark, M.A., Booth, T.A., Hueck, C.J., Miller, S.I., Hirst, B.H., and Jepson, M.A., 1998, Cell-contact-stimulated formation of filamentous appendages by *Salmonella typhimurium* does not depend on the type III secretion system encoded by *Salmonella* pathogenicity island 1, *Infect. Immun.* **66**:2007–2017.

Reed, W.M., Olander, H.J., and Thacker, H.L., 1986, Studies on the pathogenesis of *Salmonella typhimurium* and *Salmonella choleraesuis* var *kunzendorf* infection in weanling pigs, *Am. J. Vet. Res.* **47**:75–83.

Ridley, A.J., 1994, Membrane ruffling and signal transduction, *BioEssays* **16**:321–327.

Rosenshine, I., Ruschkowski, S., Foubister, V., and Finlay, B.B., 1994, *Salmonella typhimurium* invasion of epithelial cells: role of induced host cell tyrosine protein phosphorylation, *Infect. Immun.* **62**:4969–4974.

Ruschkowski, S., Rosenshine, I., and Finlay, B.B., 1992, *Salmonella typhimurium* induces an inositol phosphate flux in infected epithelial cells, *FEMS Microbiol. Lett.* **95**:121–126.

Schechter, L.M., Damrauer, S.M., and Lee, C.A., Two AraC/XylS family members can independently counteract the effect of repressing sequences upstream of the *hilA* promoter, *Mol. Microbiol.*, in press.

Schiemann, D.A., 1995, Association with MDCK epithelial cells by *Salmonella typhimurium* is reduced during utilization of carbohydrates, *Infect. Immun.* **63**:1462–1467.

Schiemann, D.A., and Shope, S.R., 1991, Anaerobic growth of *Salmonella typhimurium* results in increased uptake by Henle-407 and mouse peritoneal cells *in vitro* and repression of a major outer membrane protein, *Infect. Immun.* **59**:437–440.

Selander, R.K., Li, J., and Nelson, K., 1996, Evolutionary genetics of *Salmonella enterica*, in: *Escherichia coli and Salmonella typhimurium: Cellular and Molecular Biology*, (F.C. Neidhardt, R.I. Curtiss, J.L. Ingraham, E.C.C. Lin, K.B. Low, B. Magasanik, W.S. Reznikoff, M. Schaechter, and E.H. Umbarger, ed.) ASM Press, Washington D.C., pp. 2691–2707.

Smith, H.W., 1967, Observations on experimental oral infection with *Salmonella dublin* in calves and *Salmonella choleraesuis* in pigs, *J. Pathol. Bacteriol.* **93**:141–156.

Stein, M.A., Leung, K.Y., Zwick, M., Garcia-del Portillo, F., and Finlay, B.B., 1996, Identification of a *Salmonella* virulence gene required for formation of filamentous structures containing lysosomal membrane glycoproteins within epithelial cells, *Mol. Microbiol.* **20**:151–164.

Stone, B.J., Garcia, C.M., Badger, J.L., Hassett, T., Smith, R.I., and Miller, V.L., 1992, Identification of novel loci affecting entry of *Salmonella enteritidis* into eukaryotic cells, *J. Bacteriol.* **174**:3945–3952.

Stossel, T.P., 1993, On the crawling of animal cells, *Science* **262**:1086–1094.

Takeuchi, A., 1967, Electron microscope studies of experimental *Salmonella* infection. I. Penetration into the intestinal epithelium by *Salmonella typhimurium*, *Am. J. Pathol.* **50**:109–136.

Takeuchi, A., and Sprinz, H., 1967, Electron-microscope studies of experimental *Salmonella* infection in hte preconditioned guinea pig. II. Response of the intestinal mucosa to the invasion by *Salmonella typhimurium*, *Am. J. Pathol.* **51**:137–161.

Tang, P., Foubister, V., Pucciarelli, M.G., and Finlay, B.B., 1993, Methods to study bacterial invasion, *J. Microbiol. Meth.* **18**:227–240.

Tartera, C., and Metcalf, E.S., 1993, Osmolarity and growth phase overlap in regulation of *Salmonella typhi* adherence to and invasion of human intestinal cells, *Infect. Immun.* **61**:3084–3089.

Tran Van Nhieu, G., Ben-Ze'ev, A., and Sansonetti, P.J., 1997, Modulation of bacterial entry into epithelial cells by association between vinculin and the *Shigella* IpaA protein, *EMBO J.* **16**:2717–2729.

Wallis, T.S., Starkey, W.G., Stephen, J., Haddon, S.J., Osborne, M.P., and Candy, D.C., 1986, The nature and role of mucosal damage in relation to *Salmonella typhimurium*-induced fluid secretion in the rabbit ileum, *J. Med. Microbiol.* **22**:39–49.

Wallis, T.S., Hawker, R.J., Candy, D.C., Qi, G.M., Clarke, G.J., Worton, K.J., Osborne, M.P., and Stephen, J., 1989, Quantification of the leucocyte influx into rabbit ileal loops induced by strains of *Salmonella typhimurium* of different virulence, *J. Med. Microbiol.* **30**:149–156.

Watson, P.R., Paulin, S.M., Bland, A.P., Jones, P.W., and Wallis, T.S., 1995, Characterization of intestinal invasion by *Salmonella typhimurium* and *Salmonella dublin* and effect of a mutation in the *invH* gene, *Infect. Immun.* **63**:2743–2754.

Watson, P.R., Galyov, E.E., Paulin, S.M., Jones, P.W., and Wallis, T.S., 1998, Mutation of *invH*, but not *stn*, reduces *Salmonella*-induced enteritis in cattle, *Infect. Immun.* **66**:1432–1438.

Weinstein, D.L., O'Neill, B.L., Hone, D.M., and Metcalf, E.S., 1998, Differential early interactions between *Salmonella enterica* serovar Typhi and two other pathogenic *Salmonella* serovars with intestinal epithelial cells, *Infect. Immun.* **66**:2310–2318.

Wood, M.W., Rosqvist, R., Mullan, P.B., Edwards, M.H., and Galyov, E.E., 1996, SopE, a secreted protein of *Salmonella dublin*, is translocated into the target eukaryotic cell via a sip-dependent mechanism and promotes bacterial entry, *Mol. Microbiol.* **22**:327–338.

Wood, M.W., Jones, M.A., Watson, P.R., Hedges, S., Wallis, T.S., and Galyov, E.E., 1998, Identification of a pathogenicity island required for *Salmonella* enteropathogenicity, *Mol. Microbiol.* **29**:883–891.

Zhou, D., Hardt, W.-D., and Galán, J.E., 1997, Identification and characterization of a putative iron transport system in the pathogenicity island in *Salmonella typhimurium*, in: *The 97th General Meeting of the American Society for Microbiology*, Miami, ASM Press, p. 77.

Zierler, M.K., and Galán, J.E., 1995, Contact with cultured epithelial cells stimulates secretion of *Salmonella typhimurium* invasion protein InvJ, *Infect. Immun.* **63**:4024–4028.

Zigmond, S.H., 1996, Signal transduction and actin filament organization, *Curr. Opin. Cell Biol.* **8**:66–73.

Chapter 13

Salmonella Interactions with Professional Phagocytes

Robert A. Kingsley and Andreas J. Bäumler

1. INTRODUCTION

Salmonella enterica serotype Typhimurium (*S. typhimurium*) initiates infection by adhering to and invading the intestinal mucosa. Bacterial multiplication in the gut-associated lymphoid tissue (GALT) is accompanied by spread to the regional lymph nodes where macrophages that line the lymphatic sinuses form a first effective barrier to prevent further spread. In humans, pigs or cattle, this host defense mechanism successfully limits bacterial expansion and the infection remains localized to the intestine and the GALT. This localized disease is termed Salmonella gastroenteritis and the prominent symptom or sign of disease is acute diarrhea (Figure 1). However, if the macrophages located in the draining lymph nodes fail to limit bacterial spread, *S. typhimurium* can cause a systemic illness. Both the immune status of the host and the innate susceptibility of the host species determine whether infection results in localized or systemic disease.

ROBERT A. KINGSLEY and ANDREAS J. BÄUMLER Department of Medical Microbiology and Immunology, College of Medicine, Texas A&M University, College Station, Texas 77843-1114, E-mail: abaumler@tamu.edu

Subcellular Biochemistry, Volume 33: Bacterial Invasion into Eukaryotic Cells, edited by Oelschlaeger and Hacker. Kluwer Academic / Plenum Publishers, New York, 2000.

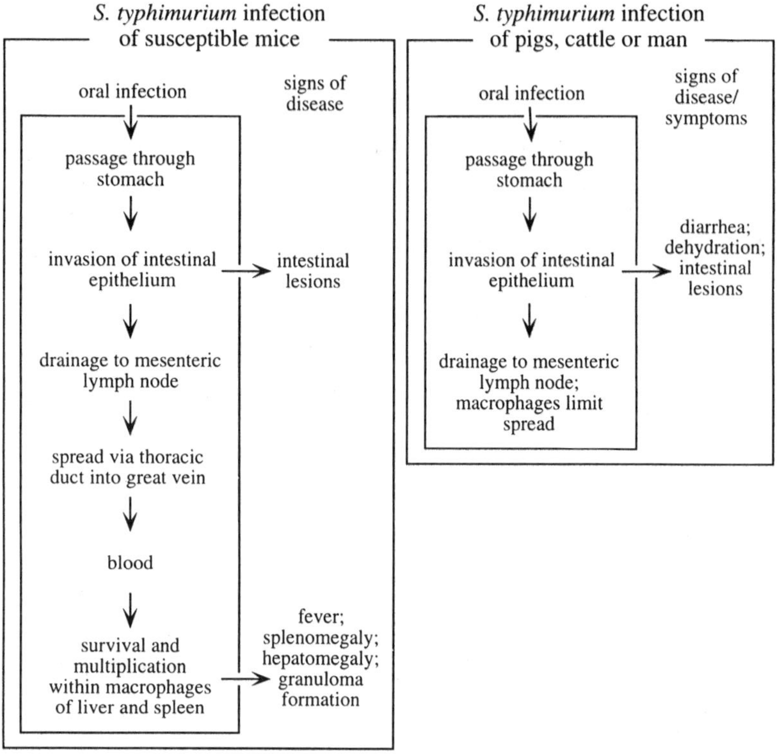

FIGURE 1. Course of infection characteristic of the two main disease syndromes caused by *S. typhimurium*. Murine typhoid (left half) is a systemic illness with signs of disease that differ from those developing in pigs, cattle or man infected with *S. typhimurium* (right half).

For instance, in children, elderly or immunocompromised patients *S. typhimurium* infections are more frequently complicated by bacteremia and systemic illness, than in immunocompetent individuals (Miller *et al.* 1995). In contrast, in immunocompetent, genetically susceptible mice, *S. typhimurium* causes a systemic disease called murine typhoid (Figure 1). During this infection *S. typhimurium* spreads from the GALT by way of the efferent lymphatics and the thoracic duct into the vena cava. Bacteria are removed from circulation by macrophages of the reticuloendothelial system (RES) located in the capillary systems of liver and spleen, thus focusing infection to liver and spleen, resulting in hepatomegaly and splenomegaly. Murine typhoid closely resembles typhoid fever, a systemic disease caused by the strictly human-adapted *S. enterica* serotype Typhi (*S. typhi*) in man (Carter and Collins 1974). Diarrhea does not develop during *S. typhimurium* infection of mice, and similarly, diarrhea only develops in about one third of typhoid fever patients and then only after several days following the onset of fever (Miller *et al.* 1995).

In order to cause murine typhoid, *S. typhimurium* must be able to breach the local defense formed by macrophages of regional lymph nodes and subsequently resist attack by macrophages of liver and spleen, the principle sites for bacterial multiplication during systemic infection (Figure 1). Apparently, the ability of *Salmonella* serotypes to cause systemic diseases, such as murine typhoid or typhoid fever, requires an adaptation to survival within mononuclear phagocytes of a given host species. This is illustrated by the ability of the human-adapted *S. typhi* to survive better in human macrophages than *S. typhimurium*, while only the latter survives well in murine macrophages (Vladoianu *et al.* 1990). Thus it appears that mononuclear phagocytes are an important barrier restricting the host range of those *Salmonella* serotypes which cause typhoid fever or related systemic illness (Barrow *et al.* 1994). The role of macrophage survival during Salmonella gastroenteritis, on the other hand, remains unclear.

2. *Salmonella* AS A PATHOGEN RESIDING WITHIN MACROPHAGES

Whether or not *S. typhimurium* is a facultative intracellular pathogen has been a subject of debate. The controversy arose since most arguments supporting the idea that *S. typhimurium* multiplies intracellularly were based on indirect evidence. For instance, it has been shown that, (i) macrophages of mice which are susceptible to *S. typhimurium* have a defective Nramp 1 protein, which is normally expressed primarily in macrophages and involved in their ability to kill intracellular bacteria (Vidal *et al.*, 1993; Lissner *et al.*, 1983, O'Brien *et al.*, 1982), (ii) *S. typhimurium* mutants that have a reduced capacity to survive in murine macrophages *in vitro* are unable to cause systemic disease in mice (Fields, 1986), and (iii) bacteria in the spleen are resistant to antibiotics that do not reach intracellular compartments (Gulig and Doyle, 1993; Dunlap *et al.*, 1991). However, attempts to provide direct evidence by microscopic analysis of infected organs produced contradicting results concerning the location of *S. typhimurium* within liver or spleen. Implicated sites of localization included macrophages, hepatocytes, polymorphnuclear phagocytes (PMN), and in the extracellular space (Hsu 1993; Conlan and North, 1992; Dunlap *et al.*, 1992; Nnalue *et al.*, 1992; Lin *et al.*, 1987). Since it is technically difficult to visualize small numbers of bacteria within an organ by microscopy, experiments relied on artificially high inocula or on analysis of infected organs after extensive multiplication has already occurred (>5 days post infection). Therefore, these microscopic observations may not accurately depict host pathogen interactions which occur when only a few organisms colonize and multiply within liver and spleen during a naturally acquired infection. Recent advances in technology have helped to overcome this problem, allowing the fate of <100 bacteria reaching the liver during the initial stages of infection to be observed. Using confocal laser scanning microscopy of immunostained thick sections of murine liver, Finlay and coworkers provided direct evidence that *S. typhimurium* resides intracellularly within macrophages (Richter-Dahlfors *et al.*, 1997). These data suggest that bacterial multiplication at systemic sites of infection occurs intracellularly within macrophages. As discussed later in this chapter, intracellular bacteria eventually kill the macrophage, and upon release, are exposed to PMN. Depletion of PMN exacerbates *S. typhimurium* infections in mice, suggesting that bacterial killing mediated by these host cells is an important host defense mechanism during murine typhoid (Conlan and North, 1992). Thus, the adaptations which allow *Salmonella* serotypes to enter and multiply within macrophages could be seen as a mechanism to evade the non-specific immune defense formed by PMN (Vassiloyanakopoulos *et al.*, 1998).

3. UPTAKE AND TRAFFICKING

3.1. Uptake by Macrophages

The mechanisms used by *Salmonella* serotypes to enter macrophages have not been exhaustively investigated and a definitive model is yet to emerge. Phagocytosis via the complement receptor CR1 presumably following opsonization with complement components C3B and C4B has been reported to account for approximately 1/3 of the observed phagocytosis of *S. typhimurium* and *S. typhi* by murine macrophages and human monocyte-derived macrophages respectively, resulting in subsequent intracellular multiplication (Ishibashi and Arai, 1996, 1990). In contrast, phagocytosis of *S. typhimurium* by human macrophage and *S. typhi* by murine macrophages was found to be mediated by receptor CR3 and correlated with efficient killing of the bacteria by the macrophage, suggesting that subsequent events may be influenced by the route of entry and that this is a possible mechanism for host specificity. These data indicate that phagocytosis via complement receptors represents one possible route of entry. At least a proportion of the remaining uptake may be accounted for by entry via macropinocytosis. Alpuche-Aranda *et al.* (1994) reported macropinocytosis of opsonized and non-opsonized Salmonella by a process which is distinct in morphology from that observed during Salmonella invasion of epithelial cells and dependent upon one or more PhoP-repressed genes. While entry into epithelial cells occurs following adherence and by a mechanism dependent upon formation of relatively tight localized ruffles, Salmonella induced ruffles over a large portion of the macrophage was resulted, in some cases, without overt bacterial adherence. The latter observation prompted the authors to implicate a possible rôle for a soluble induction factor. However, a number of investigators have reported entry into macrophages by macropinocytosis mediated by a contact dependent secretion system encoded on Salmonella pathogenicity island 1 (SPI1) (Chen *et al.*, 1996; Monack *et al.*, 1996). It remains to be seen if these *in vitro* studies accurately reflect *in vivo* mechanisms since it is not clear whether environmental factors necesary for induction of the invasion locus prevail following passage across the epithelium.

Several other mechanisms for uptake by macrophages have been described in other pathogens; however their contribution to entrance by *S. typhimurium* has thus far not been investigated (reviewed recently by Ernst, 1998). For instance, opsonization with the serum component surfactant protein A (Sp-A) results in uptake of *Mycobacterium* spp. via the Sp-A receptor. A number of non-opsonic mechanisms which involve direct binding of ligands present on the bacterial surface by macrophage receptors have also been implicated. These include the mannose receptor which binds lipoarabinomannan of *M. tuberculosis*, scavenger receptors which bind polyanionic macromolecules such as lipopolysaccharides (LPS) of gram-negative bacteria and lipoteichoic acid of gram-positive bacteria, and CD14, a high affinity receptor for LPS.

3.2. Trafficking of *Salmonella* containing Vacuoles

The trafficking pathway followed by Salmonella containing vacuoles (SCV's) is unclear since various investigators report differing ability of SCV's to fuse with other endocytic vesicles and acquisition patterns of endocytic markers. These differences may be accounted for by the use of different macrophage cell lines, their activation state, or growth state of the bacterial challenge. However, a consensus has emerged indicating that SCV's exhibit a unique trafficking pattern and that the ability of Salmonella to induce this phenomenon is essential for pathogenesis, presumably because of a requirement for modification of the micro-environment in such a way that it is favourable for survival and multiplication.

Observations using light microscopy and time-lapse video microscopy (Alpuche-Aranda *et al.*, 1995) indicate that the majority of SCV's appear as structures described as spacious phagosomes; 2–6 μm vacuoles which are indistinguishable from neighboring macropinosomes. These structures form primarily following entry via the ruffling mechanism described above, but are also observed following phagocytosis of bacteria opsonized with anti-Salmonella IgG. The latter entry mechanism initially results in smaller phagosomes which, within minutes of formation, enlarge by fusion with macropinosomes to form structures indistinguishable from spacious phagosomes. At later time points (40 minutes +) two populations of SCV's are observed; tight and spacious phagosomes, indicating that a subset of SCV's maintains a large volume by an unknown mechanism. The formation of tight and spacious phagosomes appears to result in the presence of at least two populations of intracellular bacteria, one static and the other rapidly dividing (Abshire and Neidhardt, 1993). Two lines of evidence suggest that the ability to form spacious phagosomes is essential for progression of disease. First, a constitutive *phoP* mutant, which is attenuated for virulence is also impaired in formation of spacious phagosomes (Alpuche-Aranda *et al.*, 1994), and second, spacious phagosomes formation correlates with serotype specific survival in murine macrophages and lethality in mice; *S. typhi*, *S. pullorum* and *S. arizonae* all exhibited reduced spacious phagosome formation and are avirulent in mice (Alpuche-Aranda *et al.*, 1995).

Morphological descriptions of the SCV contributes little to the intriguing question concerning the nature of the SCV. A number of investigators have defined compartments of the endosome-lysosome pathway on the basis of immunological markers. In order to define the unique compartment in which Salmonella resides it is useful to briefly reflect upon the normal trafficking of the macropinosome in non-infected murine macrophages. Determination of the antigenic profile of maturing macropinosomes has been assessed using phase microscopy and immunofluorescence to identify markers specific for a number of compartments (Desjardins *et al.*, 1994; Racoosin and Swanson, 1993). Early macropinosomes are generally positive for the transferrin receptor (TfR) which is rapidly lost as a result of recycling. Maturation initiates following the acquisition of rab-5 (marker for early endosome) and rab-7 (marker for late endosome). The lysosomal glycoprotein A (lgp-A) is detected in early endosomes and all subsequently formed compartments (Racoosin and Swanson, 1993). Finally, the fusion of the endosome with tubular lysosomes coincides with detection of cathepsin L and the cation-dependent mannose-6-phosphate receptor (CI-MPR) (Figure 2). Initial observations of SCV's using electron microscopy indicated that these fuse with lysosomes at a low frequency compared to *E. coli* or heat killed *S. typhimurium* (Buchmeier and Heffron, 1991; Ishibashi and Arai, 1990). Using confocal immunofluorescence these observations were

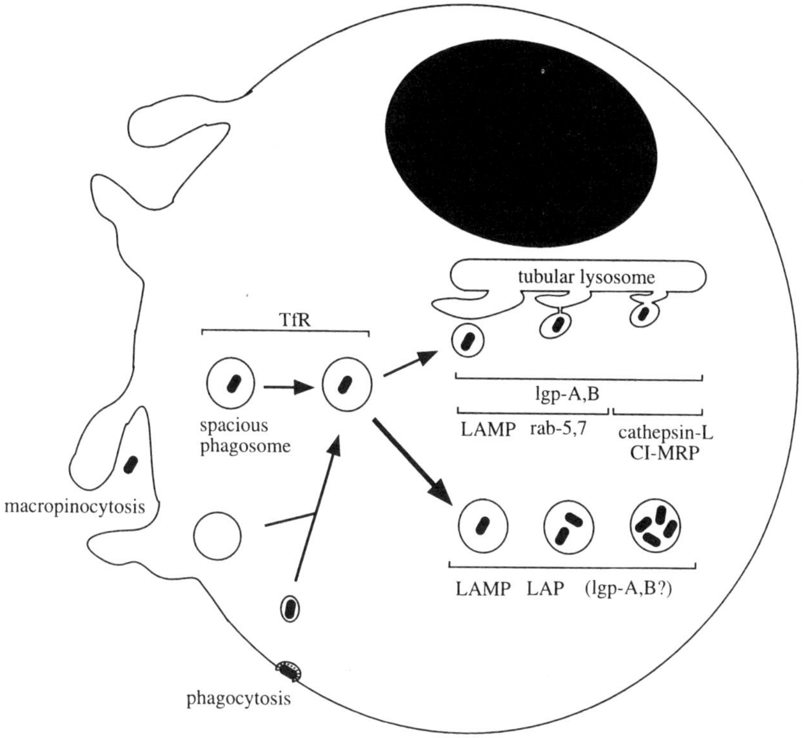

FIGURE 2. Current model of the trafficking pathway of the SCV in macrophages. Following entry by phagocytosis or macropinocytosis, partial inhibition of fusion with lysosomes results in two populations of SCV. SCV which do not label with lysosomal markers contain replicating bacteria. In contrast, bacteria are presumably killed in SCV which fuse with the tubular lysosome. For further explanation see text.

confirmed by showing that SCV's colocalized at a considerably lower frequency with the lysosome specific markers CI-M6PR and cathepsin L, when compared to vesicles containing latex beads or heat killed *S. typhimurium* (Rathman *et al.*, 1997). Conflicting observations (Oh *et al.*, 1996), also made using imunofluorescence microscopy, may be explained by the presence of two populations of intracellular Salmonella in macrophages which are distinct in growth rate and compartment morphology as described previously, or by discrepancy in the working defininition of fused vesicles used in each study.

Normal progression along the endocytic continuum is known to be accompanied by a drop in pH mediated by a vacuolar-type H^+-ATPase in murine macrophages (Lukacs *et al.*, 1990). There is aggreement among investigators that the SCV does indeed acidify and, furthermore, that this is required for induction of virulence genes (Rathman *et al.*, 1996; Alpuche-Aranda *et al.*, 1995). However, while Alpuche Aranda *et al.* report that acidification of the SCV is both delayed and attenuated following infection with live *S. typhimurium* compared with a heat killed control, Rathman *et al.* report acidification to be comparable regardless of whether the inoculum was administered live, heat killed or formalin fixed prior to infection.

Although all aspects regarding uptake and trafficking of Salmonella in macrophages have by no means been elucidated, a model is emerging (Figure 1). Salmonella may enter macrophages by a number of routes including a Salmonella-induced macropinocytosis and macrophage mediated phagocytosis. Either of these routes leads to formation of the nascent vacuole described as a spacious phagosome. Two subsequent outcomes may result, (i) normal trafficking of the SCV along the endosomal continuum accompanied by shrinking of the vacuole, limited or no bacterial multiplication and ultimately fusion with tubular lysosomes, or (ii) subversion of this pathway, signified by maintainance of the spacious vacuole, significant multiplication and failure to fuse with lysosomes (Rathman *et al.*, 1997; Buchmeier and Heffron, 1991; Ishibashi and Arai, 1990). The bacterial encoded determinants responsible for these events are not well characterized. However, a type III secretion system encoded on SPI2 may be involved since it is required for macrophages survival (Ochman *et al.*, 1996), presumably as a result of interaction of effector proteins following injection across the vacuolar membrane.

4. THE PARASITOPHOROUS VACUOLE

4.1. Nutritional Immunity

It is difficult to study the nature of the intra-macrophage microenvironment occupied by *S. typhimurium* directly. One method which provides information on the intracellular environment is to study regulation of genes expressed within the SCV. Recently developed screens, such as *in vivo* expression technology (IVET) and differential fluorescence induction (DFI) have identified a large number of genes induced in certain organs or host cells (Valdivia and Falkow, 1997; Mahan *et al.*, 1995; Mahan, 1993). These developments have been reviewed comprehensively and the reader is referred to recent articles for more detail (Conner *et al.*, 1998; Falkow, 1997; Heithoff *et al.*, 1997). Information regarding the conditions encountered within macrophages can be further inferred from mutant analysis.

Although this approach provides only indirect evidence, it has resulted in accumulation of a large body of data on environmental parameters present in the SCV.

S. typhimurium replicates in the liver and spleen at a net growth rate of 0.5–1.5 log/day (Hormaeche, 1980; Maw and Meynell, 1968). In order to obtain enough nutrients to support this considerable growth rate, bacteria must overcome nutritional immunity, a non-specific host defense mechanism which limits bacterial multiplication by withholding compounds that cannot be synthesized by microbes, for example metal ions. The best studied example of nutritional immunity is the iron-withholding defense which prevents bacterial growth in extracellular compartments (Weinberg, 1984). Iron is essential for bacterial multiplication since it is a cofactor of enzymes involved in important biosynthetic pathways, such as in DNA biosynthesis (ribonucleoside reductase). The host limits the availability of this trace element in blood using the high affinity Fe(III)-binding protein transferrin (Morgan, 1981). In addition, upon stimulation, PMN release an additional iron-binding protein, lactoferrin, which further reduces the availability of Fe(III) at sites of inflammation. Furthermore, inflammatory cytokines, such as IL-1, IL-6 and TNFα induce physiological changes, which reduce the serum iron content to about 30% of its normal level, while an increased amount of iron is stored in cells of the liver and spleen (Brock, 1989). Thus infection leads to an hypoferremia, thereby limiting the availability of iron for bacterial growth in blood and other extracellular environments (Konijn and Hershko, 1989).

Extracellular pathogens possess mechanisms to obtain host iron from Fe(III)-lactoferrin, Fe(III)-transferrin, heme or hemoglobin by binding these compounds directly or indirectly, by releasing low molecular weight Fe(III)-chelators, designated siderophores (Cornelissen and Sparling, 1994; Wooldridge and Wiliams, 1993; Otto *et al.*, 1992; Braun and Hantke, 1991; Crosa, 1989). In gram-negative bacteria, these Fe(III) uptake mechanisms depend on TonB, an inner membrane protein which, in concert with substrate specific outer membrane receptor proteins, mediates transport across the outer membrane of Fe(III)-siderophores, heme, heme released from hemoglobin, and Fe(III) released from transferrin or lactoferrin (Postle, 1990). A mutation in *tonB* attenuates extracellular pathogens, such as *Heamophilus influenzae* (Jarosik *et al.*, 1994). However, a *S. typhimurium tonB* mutant exhibited wild type levels of multiplication in the liver and spleen of mice (Tsolis *et al.*, 1996). Similarly, the ability of *S. typhimurium* to produce the siderophore enterochelin is not essential for virulence in mice (Tsolis *et al.*, 1996; Benjamin *et al.*, 1985). These data suggest that *S. typhimurium* uses a different mechanism to overcome the iron-withholding defense of the host.

An important function of resident macrophages of the liver and spleen is the removal of senescent erythrocytes from the circulation (Kay, 1989). The large amount of heme-associated iron internalized during this process is released into the blood and transported to the bone-marrow for haemopoiesis. During infection, macrophages decrease the amount of Fe(III) released into the bloodstream and instead increase the amount of iron stored intracellularly in ferritin and hemosiderin. Thus, while the serum iron level decreases, the amount of the intracellular iron storage is increased in liver and spleen during infection. Compared to extracellular body fluids, the iron availability within macrophages may thus be more favorable for bacterial growth, although the iron level within the SCV is currently unknown. However, the finding that *S. typhimurium* does not require functional high affinity iron-uptake mechanisms for multiplication in liver and spleen of mice suggest that this nutrient is not limiting for growth in the SCV (Tsolis *et al.*, 1996). It could therefore be argued that mutiplication within macrophages of liver and spleen may be a strategy of *S. typhimurium* to evade the iron-withholding defense.

Recent data from Groisman and coworkers suggest that *S. typhimurium* encounters an environment within macrophages in which magnesium is a nutrient limiting bacterial multiplication. *S. typhimurium* senses changes in external magnesium concentration using PhoPQ, a two component regulatory system (Vescovi *et al.*, 1997; Vescovi *et al.*, 1996). Upon entry, PhoPQ activates expression of several *S. typhimurium* genes within macrophages and this activation can be mimicked *in vitro* during growth in low Mg^{2+} (10 μM) minimal medium (Soncini *et al.*, 1996; Buchmeier and Heffron, 1990). Expression of PhoPQ activated genes appears to be required for intracellular growth, since *S. typhimurium phoPQ* mutants are recovered in reduced numbers from infected macrophage *in vitro* and are avirulent for mice (Miller *et al.*, 1989; Fields, 1986). A growth defect of a *S. typhimurium phoP* mutant can also be demonstarted by mimicking growth conditions in the SCV using low Mg^{2+} (10 μM) minimal medium, suggesting that the SCV is a Mg^{2+}-limiting environment (Blank-Potard and Groisman, 1997). In response to the Mg^{2+}-limiting environment of the SCV, PhoPQ activates genes involved in Mg^{2+} uptake, including *mgtA* and *mgtCB* (Soncini *et al.*, 1996). The idea that Mg^{2+} uptake is required during infection is supported by the finding that a mutation in *mgtC* reduces mouse virulence and that this attenuation is further increased in a strain harboring mutations in both *mgtA* and *mgtCB* genes. Furthermore, growth defects of *phoP* and *mgtCB* mutants in macrophages can be restored by addition of Mg^{2+} to the host cells *in vitro* (Blank-Potard and Groisman, 1997). Mutational analysis of *S. typhimurium* has therefore identified an important new mechanism for nutritional immunity, which may be referred to as Mg^{2+}-

withholding defense. Growth conditions encountered by intracellular pathogens are apparently drastically different from those of extracellular bacteria. While organisms which multiply extracellularly have to overcome the iron-withholding defense, intracellular pathogens require mechanisms to overcome the Mg^{2+}-withholding defense encountered in macrophages. However, additional limitation of other nutrients in the intracellular compartment cannot be discounted.

A *S. typhimurium mgtB mgtA* mutant is not as attenuated as a *S. typhimurium phoP* mutant, suggesting that PhoPQ controls expression of additional virulence genes required for survival and/or multiplication within the SCV. PhoPQ has recently been shown to activate expression of *pmrAB*, two genes encoding a second two-component regulatory system (Gunn and Miller, 1996; Soncini and Groisman, 1996). In turn, PmrAB leads to an increased substitution of phosphates with 4-amino-4-deoxy-L-arabinose in both the core oligosaccharide and the lipid A components of LPS. This process may compensate for the lack of divalent cations present in the SCV, since during growth in Ca^{2+} and Mg^{2+} replete conditions, these cations stabilize the outer membrane by neutralizing the negative charge of phosphate groups, thus bridging adjacent LPS molecules (Vaara, 1992). Therefore, under the growth conditions encountered in the SCV, PmrAB induced modification of LPS composition may be vital to maintainance of outer membrane integrity. Furthermore, these structural changes in LPS result in increased resistance to bactericidal/permeability-increasing protein (BPI), a cationic antibacterial protein that is released by human PMN during inflammation (Helander *et al.*, 1994). Collectively, these recent reports indicate that much of the PhoPQ regulated response may be an adaptation of *S. typhimurium* to the low Mg^{2+} environment encountered inside macrophages.

4.2. Resistance against Killing Mechanisms

A major bacteriocidal mechanisms of host phagocytes is the release of reactive oxygen intermediates (ROI) and reactive nitrogen intermediates (RNI). Bacterial resistance mechanisms against host induced oxidative and nitrosative damage have been reviewed comprehensively (Laval, 1996; DeGroote and Fang, 1995; Hassett and Cohen, 1989; Beaman and Beaman, 1984) and we will only provide a brief overview of aspects related to pathogenesis.

Superoxide radicals ($\cdot O_2^-$) and nitric oxide radicals ($NO\cdot$) can react to form peroxynitrite ($OONO^-$), a compound which has antimicrobial activity. Possibly as a result of this synergy between ROI and RNI, many bacterial defense mechanisms are active against both bacteriocidal mechanisms

(Fang, 1997). For instance, the function of superoxide dismutases is not only to scavenge superoxide radicals but also to prevent formation of peroxynitrite, thereby protecting the bacterial cell from both oxidative and nitrosative damage (DeGroote *et al.*, 1997). Similarly, *S. typhimurium* mutants that are unable to synthesize glutathione are more susceptible to hydrogen peroxide (H_2O_2), peroxynitrite and nitric oxide radicals. In addition, RecBCD is involved in repair of DNA damage caused by both ROI (Buchmeier *et al.*, 1993) and RNI (DeGroote *et al.*, 1995).

Not all mechanisms to prevent or repair ROI or RNI mediated damage are equally important during infection. Among the effector genes involved in detoxifying superoxide radicals or hydrogen peroxide, only *sodC* is essential for virulence in the mouse (Figure 3) (DeGroote *et al.*, 1997; Farrant *et al.*, 1997). SodC is a periplasmic space-localized superoxide dismutase suggesting that during growth in liver and spleen the bacterial surface is exposed to toxic levels of ROI, RNI or both. Homocystein is a low molecular weight compound which scavenges RNI. A mutation in *metL*, an enzyme involved in homocysteine biosynthesis, results in hypersusceptibility to macrophages *in vitro* and attenuation in the mouse (DeGroote *et al.*, 1996). In addition to mechanisms for detoxification, *S. typhimurium* requires proteins which repair oxidative or nitrosative injury. HtrA and Prc, two proteases involved in degradation of misfolded proteins in the periplasm are required for macrophage survival and full mouse virulence (Figure 4) (Bäumler *et al.*, 1994). Furthermore, RecABC mediated DNA repair is essential to survive the ROI or RNI induced injury during growth of *S. typhimurium* within mice (DeGroote *et al.*, 1995; Buchmeier *et al.*, 1993). Although these examples suggest that *S. typhimurium* is exposed to ROI and RNI during infection, it is not clear whether these bacteriocidal mechanisms are encountered during contact with macrophages or PMN. In addition, there are a number of genes (*sodA*, *katG*, *ahpC*, *soxS*, and *OxyR*) which are required for defense against ROI or RNI *in vitro*, but are not essential for full virulence of *S. typhimurium* in mice (Taylor *et al.*, 1998; Fang, 1997; Fang *et al.*, 1997; Tsolis *et al.*, 1995; Papp-Szabò *et al.*, 1994).

5. MACROPHAGE DEATH

Salmonella induced apoptosis was first described in macrophages infected with *S. enterica* serotype Choleraesuis (*S. choleraesuis*) in the presence of anti IL-10 monoclonal antibody approximately 6 hours post-infection (Arai *et al.*, 1995), suggesting that IL-10 plays a protective role in preventing apoptosis during infection. Subsequently, however, a number of investigators have reported cytotoxic activity of *S. typhimurium* in the

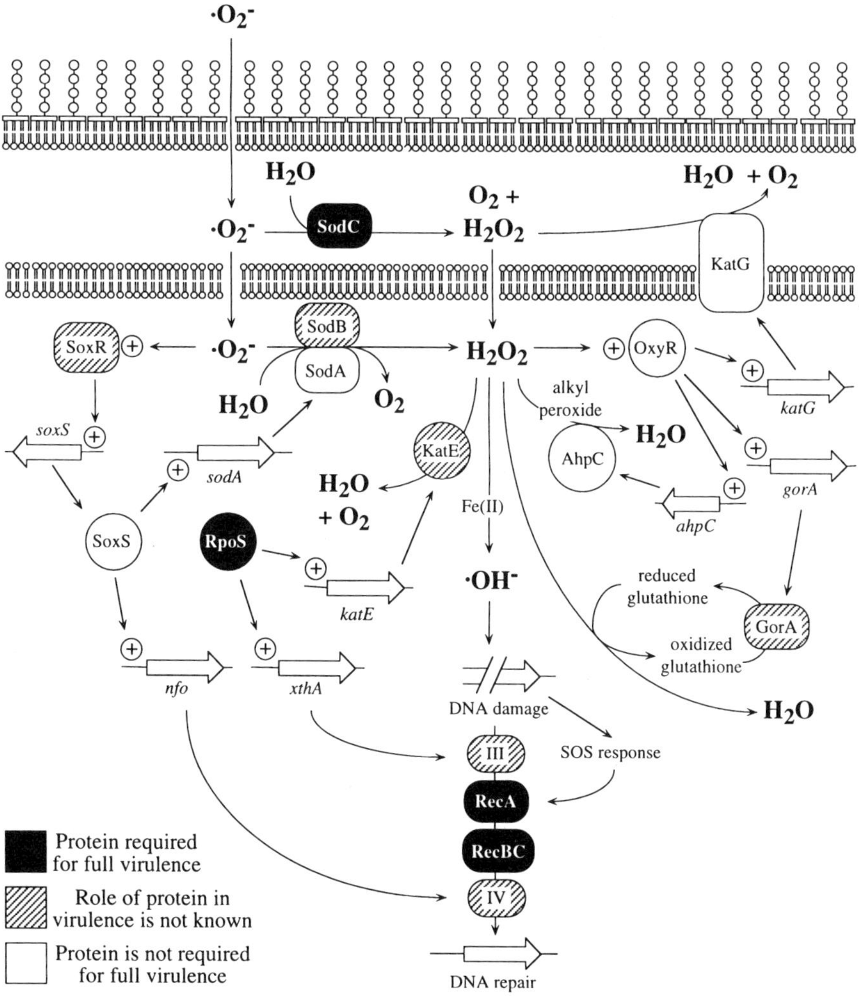

FIGURE 3. Detoxification of RNI and ROI and repair of DNA damage in *S. typhimurium*. Mutations in *sodC*, *rpoS* and *recABC* reduce mouse virulence of *S. typhimurium* (DeGroote *et al.*, 1997; Farrant *et al.*, 1997; Buchmeier *et al.*, 1993; Fang *et al.*, 1992). Changes in gene expression in response to oxidative stress mediated by SoxRS and OxyR have been reviewed previously (Rosner and Storz, 1997; Storz and Altuvia, 1994).

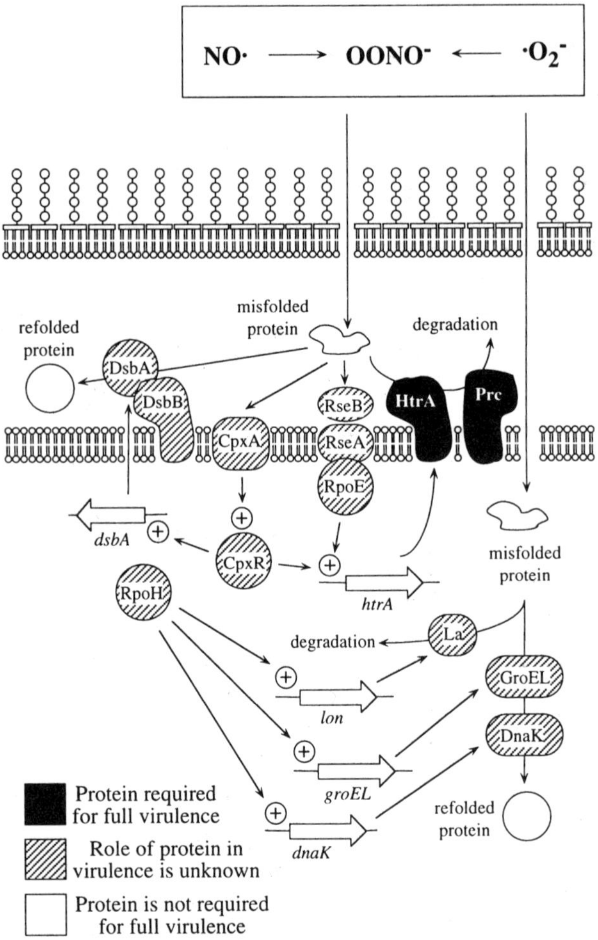

FIGURE 4. Repair of protein damage in *S. typhimurium*. Contribution of HtrA and Lon to degradation of aggregated proteins has been reported previously (Laskowska *et al.*, 1996). Signal transduction pathways controlling *htrA* expression have been reviewed by Pallen and Wren (Pallen and Wren, 1997). Synergy in the antimicrobicidal activity of RNI and ROI has been proposed by Fang and coworkers (1997).

absence of IL-10 neutralizing antibodies. As a result of these reports considerable uncertainty surrounds the mechanism of apoptosis in Salmonella infections and its possible role(s) in pathogenesis. In part at least, the uncertainty may be explained by the fact that at least two mechanisms appear to result in macrophage cytotoxicity and furthermore this may be a consequence of apoptosis or necrosis. The major difference between these two mechanisms is the time after initial entry at which cytotoxic events are observed. In common with the cytotoxicity of macrophages infected with *S. choleraesuis* described previously, Lindgren *et al.* (1996) described cytotoxicity which initiated six hours post infection resulting in gross morphological changes and death after approximately 12–14 hours. Cytotoxicity was not dependent on invasion via SPI-1 encoded determinants, since an *invA* mutant was still cytotoxic, but may require Salmonella entry, since a *nagC* mutant which is taken up poorly by macrophages was not cytotoxic. In contrast two laboratories have described necrosis occuring within two hours following infection which was further distinguished from the late onset events since it was absolutely dependent upon SPI-1 encoded determinants (Chen *et al.*, 1996; Monack *et al.*, 1996). However, these investigators disagree over the requirement for entry into the macrophage in order to exert their cytotoxic effects. By treatment of macrophages with cytochalasin D, which prevents entry via the ruffling mechanism mediated by SPI-1 encoded determinants, Chen *et al.* (1996) reported no affect on cytotoxicity while Monack *et al.* (1996) reported a significant reduction.

The early mechanism for cytotoxicity (Chen *et al.*, 1996; Monack *et al.*, 1996) could play a role during interaction with macrophages encountered shortly after invasion of the epithelium, since SPI-1 is required during this phase of infection. Recently, macrophage apoptosis has been observed in sections of liver from mice infected with *S. typhimurium* (Richter-Dahlfors *et al.*, 1997) using confocal laser scanning microscopy. Since mutations in *ompR*, but not SPI-1 encoded determinants, affects the systemic phase of infection the apoptosis observed in the RES *in vivo* may result from OmpR-mediated late cytotoxicity described previously (Lindgren *et al.*, 1996). However, since *ompR* mutations are pleiotropic, attenuation of virulence may not be entirely accounted for by reduction in macrophage cytotoxicity. At a first approximation, these data may be interpreted as Salmonella actively causing death of the macrophage, a mechanism which may facilitate spread within the organ. On the other hand, the macrophage is a safe site for multiplication, and therefore macrophage apoptosis could be viewed as a host defence mechanism depriving infecting bacilli of this privileged intracellular niche. Indeed, such activity would expose Salmonella to killing mechanisms of PMN's. Accordingly, attenuation of virulence due to an *ompR* mutation may simply result in failure to trigger this defence mechanism (Vassiloyanakopoulos *et al.*, 1998).

6. EVOLUTION OF INTRACELLULAR PARASITISM

The genus Salmonella and *E. coli* contain closely related pathogens that share a common ancestor some 100 million years ago. Since their divergence, both lineages acquired multiple virulence determinants by way of phage or plasmid mediated horizontal gene transfer events. A recent estimate suggests that new genetic material was introduced at a rate of approximately 16 kb per million years per lineage (Lawrence and Ochman, 1998). Many genes acquired in this way are lost subsequently although determinants which confer a selective advantage persist in the genome. As a result of horizontal transfer, the *E. coli* K-12 genome is estimated to contain at least 755 genes that were received after branching from the Salmonella lineage (Lawrence and Ochman, 1998). Virulence determinants introduced horizontally may lead to punctuated evolution of pathogens suggesting that these genes encode mechanisms that are key factors which facilitate occupation of the preferred niche in vertebrate hosts (Bäumler, 1997). Virulence determinants which are present in the genus Salmonella but absent from *E. coli* could thus be viewed as niche-specific genes which contribute to the differences in colonization observed between these two groups of pathogens. Indeed, the prefered niche of these two closely related pathogens differs considerably; pathogenic *E. coli* establish themselves in the intestinal luman (enterotoxigenic *E. coli*, enteropathogenic *E. coli*, enterohemorrhagic *E. coli*) or invade the epithelial lining of the alimentary tract, without subsequent systemic involvement (enteroinvasive *E. coli*, *Shigella* spp.). In contrast, upon penetration of the intestinal epithelium, Salmonella serotypes colonize the underlying lymphoid tissue and the mesenteric lymph node where they encounter professional phagocytes. Survival within macrophages may have evolved as an adaptation to this environment and results in the ability of some *Salmonella* serotypes to cause systemic infections such as typhoid fever (Bäumler, 1997).

There is good evidence that niche-specific virulence determinants required for growth of *S. typhimurium* in the liver and spleen have been obtained by horizontal transfer (Bäumler, 1997). These include SPI-2, a large pathogenicity island encoding a type III excretion system (Ochman *et al.*, 1996; Shea *et al.*, 1996), and the *spv* (Salmonella plasmid virulence) operon (Gulig *et al.*, 1993). SPI-2 is inserted in a t-RNA gene, implicating phage mediated transfer as mechanism for its acquisition since t-RNA genes can serve as phage attachment sites (Hensel *et al.*, 1997). SPI2 was introduced after the Salmonella lineage branched into two species, designated *S. enterica* and *S. bongori*. As a consequence, SPI-2 is present in all subspecies of *S. enterica* but is absent from *S. bongori* serotypes (Hensel *et al.*, 1997; Ochman and Groisman, 1996). Similarly, the *spv* operon was obtained by *S. enterica* after its divergence from *S. bongori* (Boyed and Hartl, 1998). These data suggest that adaptations to professional phagocytes developed soon after or during formation of the species *S. enterica*. In addition to SPI-2 and the *spv* operon, Salmonella serotypes acquired other niche-specific virulence determinants required for occupation of an intracellular compartment, including a high affinity Mg^{2+} uptake system ecoded by *mgtC* and a periplasmic superoxide dismutase encoded by *sodC* (Blank-Potard and Groisman, 1997; DeGroote *et al.*, 1997). While the function of *sodC* and *mgtC*, during intracellular growth, is known to be detoxification and nutrient uptake respectively, the role of *spv* and SPI2 has not been ellucidated to date. However, the *spv* locus functions to increase multiplication of *S. typhimurium* in the liver and spleen of mice (Gulig and Doyle, 1993) and is likely required for interaction with both macrophages and PMN (Vassiloyanakopoulos *et al.*, 1998). Similarly, SPI2 is essential for colonization of murine liver and spleen and survival within macrophages *in vitro* (Ochman *et al.*, 1996; Shea *et al.*, 1996; Hensel *et al.*, 1995).

These data support the concept that acquisition of SPI2, the *spv* locus, *sodC* and *mgtC* (and likely other virulence genes) by horizontal transfer was essential for the development of the facultative intracellular lifestyle of *S. enterica* serotypes. Further analysis of horizontally acquired, niche-specific virulence genes will be a key to the understanding of Salmonella pathogenesis.

7. REFERENCES

Abshire, K.Z., and Neidhardt, F.C., 1993, Growth rate paradox of *Salmonella typhimurium* within host macrophages, *J Bacteriol*, **175**:3744–3748.

Alpuche-Aranda, C.M., Berthiaume, E.P., Mock, B., Swanson, J.A., and Miller, S.I., 1995, Spacious phagosome formation within mouse macrophages correlates with *Salmonella* serotype pathogenicity and host susceptibility, *Infec. Immun.*, **63**:4456–4462.

Alpuche-Aranda, C.M., Racoosin, E.L., Swanson, J.A., and Miller, S.I., 1994, *Salmonella* stimulate macrophage macropinocytosis and persist within spacious phagosomes, *J Exp Med*, **179**:601–608.

Arai, T., Hiromatsu, K., Nishimura, H., Kimura, Y., Kobayashi, N., Ishida, H., Nimura, Y., and Yoshikai, Y., 1995, Endogenous interleukin 10 prevents apoptosis in macrophages during *Salmonella* infection, *Biochemical & Biophysical Research Communications*, **213**:600–607.

Barrow, P.A., Huggins, M.B., and Lovell, M.A., 1994, Host specificity of *Salmonella* infection in chickens and mice is expressed *in vivo* primarily at the level of the reticuloendothelial system, *Infect. Immun.*, **62**:4602–4610.

Bäumler, A.J., 1997, The record of horizontal gene transfer in *Salmonella*, *Trends Microbiol.*, **5**:318–322.

Bäumler, A.J., Kusters, J.G., Stojiljkovic, I., and Heffron, F., 1994, *Salmonella typhimurium* loci involved in survival within macrophages, *Infect. Immun.*, **62**:1623–1630.

Beaman, L., and Beaman, B.L., 1984, The role of oxygen and its derivatives in microbial pathogenesis, *Ann. Rev. Microbiol.*, **38**:27–48.

Benjamin, W.H., Turnbough, C.L., Posey, B.S., and Briles, D.E., 1985, The ability of *Salmonella typhimurium* to produce the siderophore enterobactin is not a virulence factor in mouse typhoid, *Infect. Immun.*, **50**:392–397.

Blank-Potard, A.-B., and Groisman, E.A., 1997, The *Salmonella selC* locus contains a pathogenicity island mediating intramacrophage survival, *EMBO J.*, **16**:5376–5385.

Boyed, F.E., and Hartl, D.L., 1998, *Salmonella* virulence plasmid: modular acquisition of the *spv* virulence region by an F-plasmid in *Salmonella enterica* subspecies I and insertion into the chromosome in subspecies II, IIIa, IV, and VII isolates, *Genetics*, **149**:1183–1190.

Braun, V., and Hantke, K., 1991, Genetics of Bacterial Iron Transport, in: *Handbook of Microbial Iron Chelates* (G. Winkelmann, ed.), CRC Press; Boca Raton, FL, pp. 107–138.

Brock, J.H., 1989, Iron and cells of the immune system, in: *Iron in immunity, cancer and inflammation* (J.H. Brock, ed.), Jon Wiley & Sons, Inc., New York, pp. 81–108.

Buchmeier, N.A., and Heffron, F., 1990, Induction of *Salmonella* stress proteins upon infection of macrophages, *Science*, **248**:730–2.

Buchmeier, N.A., and Heffron, F., 1991, Inhibition of macrophage phagosome-lysosome fusion by *Salmonella typhimurium*, *Infect. Immun.* **59**:2232–2238.

Buchmeier, N.A., Lipps, C.J., So, M.Y., and Heffron, F., 1993, Recombination-deficient mutants of *Salmonella typhimurium* are avirulent and sensitive to the oxidatove burst of macrophages., *Mol. Microbiol.* **7**:933–936.

Carter, P.B., and Collins, F.M., 1974, The route of enteric infection in normal mice, *J. Exp. Med.* **139**:1189–1203.

Chen, L.M., Kaniga, K., and Galan, J.E., 1996, *Salmonella* spp are cytotoxic for cultured macrophages, *Mol. Microbiol.* **21**:1101–1115.

Conlan, J.W., and North, R.J., 1992, Early pathogenesis of infection in the liver with the facultative intracellular bacteria *Listeria monocytogenes, Francisella tularensis,* and *Salmonella typhimurium* involves lysis of infected hepatocytes by leucocytes, *Infect. Immun.* **60**:5164–5171.

Conner, C.P., Heithoff, D.M., and Mahan, M.J., 1998, *In vivo* gene expression: contributions to infection, virulence, and pathogenesis, *Curr. Top. Microbiol. Immunol.* **225**: 1–12.

Cornelissen, C.N., and Sparling, P.F., 1994, Iron piracy: acquisition of transferrin-bound iron by bacterial pathogens, *Mol. Microbiol.* **14**:843–850.

Crosa, J.H., 1989, Genetics and molecular biology of siderophore-mediated iron transport in bacteria, *Microbiol. Rev.* **53**:517–530.

DeGroote, M.A., and Fang, F.C., 1995, NO inhibitions: antimicrobial properties of nitric oxide, *Clin. Infect. Dis.* **92**:6399–6403.

DeGroote, M.A., Granger, D., Xu, Y., Campbell, G., Prince, R., and Fang, F.C., 1995, Genetic and redox determinants of nitric oxide cytotoxicity in a *Salmonella typhimurium* model, *Proc. Natl. Acad. Sci. USA*, **92**:6399–6403.

DeGroote, M.A., Ochsner, U.A., Shiloh, M.U., Nathan, C., McCord, J.M., Dinauer, M.C., Libby, S.J., Vazquez, T.A., Xu, Y., and Fang, F.C., 1997, Periplasmic superoxide dismutase protects Salmonella from products of phagocyte NADPH-oxidase and nitric oxide synthase, *Proc. Natl. Acad. Sci. USA* **94**:13997–4001.

DeGroote, M.A., Testerman, T., Xu, Y., Stauffer, G., and Fang, F.C., 1996, Homocysteine antagonism of nitric oxide-related cytostasis in *Salmonella typhimurium*, *Science*, **272**:414–7.

Desjardins, M., Huber, L.A., Parton, R.G., and Griffiths, G., 1994, Biogenesis of phagolysosomes proceeds through a sequential series of interactions with the endocytic apparatus, *J. Cell Biol.* **124**:677–88.

Dunlap, N.E., Benjamin Jr., W.H., McGall, R.D., Tilden, A.B., and Briles, D.E., 1991, A "safe-site" for *Salmonella typhimurium* is within splenic cells during the early phase of infection in mice, *Microb. Pathogen.* **10**:297–310.

Dunlap, N.E., Benjamin, W.J., Berry, A.K., Eldridge, J.H., and Briles, D.E., 1992, A "safe-site" for *Salmonella typhimurium* is within splenic polymorphonuclear cells, *Microbial Pathogenesis*, **13**:181–90.

the activation of the ERK1/2 (ERK: extracellular signal-regulated kinase), p38 and JNK MAPKs (MAPK: mitogen-activated protein kinase; JNK: Jun N-terminal kinase); in addition, this action on MAPKs strictly correlates with a reduction in TNF-α release (Ruckdeschel *et al.*, 1997a). A functional secretion machinery is required for the phenomenon to occur and a strain secreting only YopB, YopD, YopN, YopE, YopH and LcrV does not impair TNF-α release, indicating that these proteins are not, or at least not solely, responsible for the Yersinia suppressive effect on TNF-α release (Ruckdeschel *et al.*, 1997a). Similar studies, performed with either *Y. enterocolitica* or *Y. pseudotuberculosis*, showed that a functional translocation machinery is required in addition to type III secretion to observe the inhibition of TNF-α release; moreover, the translocated YopP/YopJ protein was shown to be responsible for this effect (Boland and Cornelis, 1998; Palmer *et al.*, 1998). Contrary to the previous results (Nakajima *et al.*, 1995; Beuscher *et al.*, 1995) but in agreement with the results of Ruckdeschel *et al.*, the YopB and LcrV proteins play only an indirect role in this phenomenon (Boland and Cornelis, 1998), since they are both involved in the formation of the translocation apparatus (Håkansson *et al.*, 1996b; Sarker *et al.*, 1998a). YopP/YopJ is also responsible for the downregulation of the ERK2, p38 and JNK MAPKs (Figure 4) (Boland and Cornelis, 1998; Palmer *et al.*, 1998). It is not known at what level of the MAPK cascade YopP/YopJ acts, but it seems that the inhibition of the MAPKs activities occurs at least partially via a decrease in the activities of upstream kinases such as Raf-1 (Ruckdeschel *et al.*, 1997a). All together, these results suggest that Yersinia impairs the normal production of TNF-α by acting on the MAPKs cascades. In addition, the ability of Yersinia to impair NF-κB activation also contributes to the inhibition of TNF-α production (Ruckdeschel *et al.*, 1998; Schesser *et al.*, 1998). The blockade of TNF-α production by interference with two distinct signaling pathways suggests that Yersinia probably acts at the level of a common upstream cascade component (Figure 4). Proteins of the Ras superfamily such as Cdc42, Rac-1 and Rho seem to be good candidates since they are implicated in the activation of both NF-κB and MAPKs (Perona *et al.*, 1997; Symons, 1996). The fact that YopP/YopJ is responsible for both the induction of apoptosis and the inhibition of TNF-α production could lead to the conclusion that one phenomenon is a consequence of the other, e.g. the moribund macrophage would be incapable of synthesizing TNF-α. However, the observation that inhibition of Yersinia-induced apoptosis by a broad spectrum caspase inhibitor does not affect the level of TNF-α released argues against this hypothesis (Ruckdeschel *et al.*, 1998) and rather suggests that YopP/YopJ acts at the level of an upstream event common to several signaling cascades (Figure 4).

5. CONCLUSION

Yersinia has the capacity to invade non-phagocytic cells. However, this does not seem to be a major element in their virulende. Instead, interaction of Yersinia with eukaryotic cells appears as a new type of bacteria-eukaryotic cell interaction. It leads to delivery of bacterial effectors into the cytosol of the target eukaryotic cells by extracellular bacteria adhering at the host cell surface. These effectors subvert their target cells by very fine mechanisms, most of which have not been unravelled yet. It is remarkable that two of these effector proteins are enzymes related to eukaryotic counterparts (YopO, YopH), which favors the hypothesis that Yersinia acquired these virulence factors from an eukaryotic source by horizontal transfer and suggests that the Yop virulon is an assembly of elements of bacterial and eukaryotic origins. Although a coherent picture is now emerging, the Yop virulon is far from being completely understood. In particular, the mechanisms underlying bacteria-target cell recognition, deployment of the translocation apparatus and control of Yop delivery remain to be clarified. Expression of this integrated anti-host system during an infection leads to impaired phagocytosis by professional phagocytes and inhibition of the respiratory burst; in addition, macrophages undergo apoptosis and the normal release of TNF-α is altered. All these effects probably allow the bacteria to survive in the quite hostile environment that are the Peyer's patches and other lymphoid tissues.

ACKNOWLEDGMENTS. We thank A. Boyd for critical reading of the manuscript. Our work on Yersinia was supported by the Belgian Fonds National de la Recherche Scientifique Médicale (Convention 3.4595.97), the Direction Générale de la Recherche Scientifique-Communauté Française de Belgique (Action de Recherche Concertée, 94/99-172), and by the Interuniversity Poles of Attraction Program-Belgian State, Prime Minister's Office, Federal Office for Scientific, Technical, and Cultural Affairs (PAI 4/03). A. B. is supported as a research assistant from the Belgian Fonds National de la Recherche Scientifique.

6. REFERENCES

Allaoui, A., Woestyn, S., Sluiters, C., and Cornelis, G.R., 1994, YscU, a *Yersinia enterocolitica* inner membrane protein involved in Yop secretion, *J. Bacteriol.* **176**:4534–4542.

Allaoui, A., Scheen, R., Lambert de Rouvroit, C.L., and Cornelis, G.R., 1995a, VirG, a *Yersinia enterocolitica* lipoprotein involved in Ca^{2+} dependency, is related to ExsB of *Pseudomonas aeruginosa*, *J. Bacteriol.* **177**:4230–4237.

Allaoui, A., Schulte, R., and Cornelis, G.R., 1995b, Mutational analysis of the *Yersinia enterocoliticia virC* operon: characterization of *yscE, F, G, I, J, K* required for Yop secretion and *yscH* encoding YopR, *Mol. Microbiol.* **18**:343–355.

Anderson, D.M., and Schneewind, O., 1997, A mRNA signal for the type III secretion of Yop proteins by *Yersinia enterocolitica*, *Science* **278**:1140–1143.

Andersson, K., Carballeira, N., Magnusson, K.E., Persson, C., Stendahl, O., Wolf-Watz, H., and Fällman, M., 1996, YopH of *Yersinia pseudotuberculosis* interrupts early phosphotyrosine signalling associated with phagocytosis, *Mol. Microbiol.* **20**:1057–1069.

Autenrieth, I.B., and Firsching, R., 1996, Penetration of M cells and destruction of Peyer's patches by *Yersinia enterocolitica*: an ultrastructural and histological study, *J. Med. Microbiol.* **44**:285–294.

Autenrieth, I.B., and Heesemann, J., 1992, *In vivo* neutralization of tumor necrosis factor-alpha and interferon-gamma abrogates resistance to *Yersinia enterocolitica* infection in mice, *Med. Microbiol. Immunol. Berl.* **181**:333–338.

Balligand, G., Laroche, Y., and Cornelis, G., 1985, Genetic analysis of virulence plasmid from a serogroup 9 *Yersinia enterocolitica* strain: role of outer membrane protein P1 in resistance to human serum and autoagglutination, *Infect. Immun.* **48**:782–786.

Bergman, T., Erickson, K., Galyov, E.E., Persson, C., and Wolf-Watz, H., 1994, The *lcrB (yscN/U)* gene cluster of *Yersinia pseudotuberculosis* is involved in Yop secretion and shows high homology to the *spa* gene clusters of *Shigella flexneri* and *Salmonella typhimurium*, *J. Bacteriol.* **176**:2619–2626.

Beuscher, H.U., Rodel, F., Forsberg, Å., and Rollinghoff, M., 1995, Bacterial evasion of host immune defense: *Yersinia enterocolitica* encodes a suppressor for tumor necrosis factor alpha expression, *Infect. Immun.* **63**:1270–1277.

Black, D.S., and Bliska, J.B., 1997, Identification of $p130^{Cas}$ as a substrate of Yersinia YopH (Yop51), a bacterial protein tyrosine phosphatase that translocates into mammalian cells and targets focal adhesions, *EMBO J.* **16**:2730–2744.

Bliska, J.B., 1995, Crystal structure of the Yersinia tyrosine phosphatase, *Trends. Microbiol.* **3**:125–127.

Bliska, J.B., and Black, D.S., 1995, Inhibition of the Fc receptor-mediated oxidative burst in macrophages by the *Yersinia pseudotuberculosis* tyrosine phosphatase, *Infect. Immun.* **63**:681–685.

Bliska, J.B., and Falkow, S., 1992, Bacterial resistance to complement killing mediated by the Ail protein of *Yersinia enterocolitica*, *Proc. Natl. Acad. Sci. USA* **89**:3561–3565.

Bliska, J.B., Guan, K.L., Dixon, J.E., and Falkow, S., 1991, A mechanism of bacterial pathogenesis: Tyrosine phosphate hydrolysis of host proteins by an essential Yersinia virulence determinant, *Proc. Natl. Acad. Sci. USA* **88**:1187–1191.

Bliska, J.B., Clemens, J.C., Dixon, J.E., and Falkow, S., 1992, The Yersinia tyrosine phosphatase: specificity of a bacterial virulence determinant for phosphoproteins in the J774A.1 macrophage. *J. Exp. Med.* **176**:1625–1630.

Bliska, J.B., Copass, M.C., and Falkow, S., 1993, The *Yersinia pseudotuberculosis* adhesin YadA mediates intimate bacterial attachment to and entry into HEp-2 cells, *Infect. Immun.* **61**:3914–3921.

Boland, A., and Cornelis, G.R., 1998, Role of YopP in suppression of tumor necrosis factor alpha release by macrophages during Yersinia infection, *Infect. Immun.* **66**:1878–1884.

Boland, A., Sory, M.-P., Iriarte, M., Kerbourch, C., Wattiau, P., and Cornelis, G.R., 1996, Status of YopM and YopN in the Yersinia Yop virulon: YopM of *Y. enterocolitica* is internalized inside the cytosol of PU5-1.8 macrophages by the YopB, D, N delivery apparatus, *EMBO J.* **15**:5191–5201.

Bölin, I., and Wolf-Watz, H., 1988, The plasmid encoded Yop2b protein of *Yersinia pseudotuberculosis* is a virulence determinant regulated by calcium and temperature at the level of transcription, *Mol. Microbiol.* **2**:237–245.

Bölin, I., Norlander, L., and Wolf-Watz, H., 1982, Temperature-inducible outer membrane protein of *Yersinia pseudotuberculosis* and *Yersinia enterocolitica* is associated with the virulence plasmid, *Infect. Immun.* **37**:506–512.

Boyd, A.P., Sory, M.-P., Iriarte, M., and Cornelis, G.R., 1998, Heparin interferes with translocation of Yop proteins into HeLa cells and binds to LcrG, a regulatory component of the Yersinia Yop apparatus, *Mol. Microbiol.* **27**:425–436.

Caron, E., Gross, A., Liautard, J.-P., and Dornand, J., 1996, *Brucella* species release a specific, protease-sensitive, inhibitor of TNF-α expression, active on human macrophage-like cells, *J. Immunol.* **156**:2885–2893.

Cavanaugh, D.C., and Randall, R., 1959, The role of multiplication of *Pasteurella pestis* in mononuclear phagocytes in the pathogenesis of flea-borne plague. *J. Immunol.* **83**:348–363.

Charnetzky, W.T., and Shuford, W.W., 1985, Survival and growth of *Yersinia pestis* within macrophages and an effect of the loss of the 47-megadalton plasmid on growth in macrophages, *Infect. Immun.* **47**:234–241.

Cheng, L.W., Anderson, D.M., and Schneewind, O., 1997, Two independent type III secretion mechanisms for YopE in *Yersinia enterocolitica*, *Mol. Microbiol.* **24**:757–765.

China, B., Michiels, T., and Cornelis, G.R., 1990, The pYV plasmid of Yersinia encodes a lipoprotein, YlpA, related to TraT, *Mol. Microbiol.* **4**:1585–1593.

China, B., Sory, M.-P., N'Guyen, B.T., de Bruyere, M., and Cornelis, G.R., 1993, Role of the YadA protein in prevention of opsonization of *Yersinia enterocolitica* by C3b molecules, *Infect. Immun.* **61**:3129–3136.

China, B., N'Guyen, B.T., de Bruyere, M., and Cornelis, G.R., 1994, Role of YadA in resistance of *Yersinia enterocolitica* to phagocytosis by human polymorphonuclear leukocytes, *Infect. Immun.* **62**:1275–1281.

Clark, M.A., Hirst, B.H., and Jepson, M.A., 1998, M-cell surface β1 integrin expression and invasin-mediated targeting of *Yersinia pseudotuberculosis* to mouse Peyer's patch M cells, *Infect. Immun.* **66**:1237–1243.

Cole, S.P., Guiney, D.G., and Corbeil, L.B., 1993, Molecular analysis of a gene encoding a serum-resistance-associated 76 kDa surface antigen of *Haemophilus somnus*. *J. Gen. Microbiol.* **139**:2135–2143.

Collazo, C.M., and Galan, J.E., 1997, The invasion-associated type III system of *Salmonella typhimurium* directs the translocation of Sip proteins into the host cell, *Mol. Microbiol.* **24**:747–756.

Cornelis, G.R., Sluiters, C., Delor, I., Geib, D., Kaniga, K., Lambert de Rouvroit, C.L., Sory, M.-P., Vanooteghem, J.-C., and Michiels, T., 1991, *ymoA*, a *Yersinia enterocolitica* chromosomal gene modulating the expression of virulence functions, *Mol. Microbiol.* **5**:1023–1034.

Cornelis, G.R., Iriarte, M., and Sory, M.-P., 1995, Environmental control of virulence functions and signal transduction in *Yersinia enterocolitica*, in: *Signal transduction and bacterial virulence* (R. Rappuoli, V. Scarlato, and B. Arico, eds.), R.G. Landes company, Austin. pp. 95–110.

Cornelis, G.R., Boland, A., Boyd, A.P., Geuijen, C., Iriarte, M., Neyt, C., Sory, M.-P., and Stainier, I., 1998, The virulence plasmid of Yersinia, an antihost genome, *Micro. Mol. Biol. Rev.*, in press.

Cover, T.L., and Aber, R.C., 1989, *Yersinia enterocolitica*, *N. Engl. J. Med.* **321**:16–24.

De Marco, L., Mazzucato, M., Masotti, A., and Ruggeri, Z.M., 1994, Localization and characterization of an α-thrombin-binding site on platelet glycoprotein Ibα, *J. Biol. Chem.* **269**:6478–6484.

Deibel, C., Krämer, S., Chakraborty, T., and Ebel, F., 1998, EspE, a novel secreted protein of attaching and effacing bacteria, is directly translocated into infected host cells, where it appears as a tyrosine-phosphorylated 90 kDa protein, *Mol. Microbiol.* **28**:463–474.

Delor, I., and Cornelis, G.R., 1992, Role of *Yersinia enterocolitica* Yst toxin in experimental infection of young rabbits, *Infect. Immun.* **60**:4269–4277.

Demuth, A., Goebel, W., Beuscher, H.U., and Kuhn, M., 1996, Differential regulation of cytokine and cytokine receptor mRNA expression upon infection of bone marrow-derived macrophages with *Listeria monocytogenes*, *Infect. Immun.* **64**:3475–3483.

Dunlevy, J.R., and Couchman, J.R., 1993, Controlled induction of focal adhesion disassembly and migration in primary fibroblasts, *J. Cell Sci.* **105**:489–500.

Emödy, L., Heesemann, J., Wolf-Watz, H., Skurnik, M., Kapperud, G., O'Toole, P., and Wadstrom, T., 1989, Binding to collagen by *Yersinia enterocolitica* and *Yersinia pseudotuberculosis*: evidence for *yopA*-mediated and chromosomally encoded mechanisms, *J. Bacteriol.* **171**:6674–6679.

Fällman, M., Andersson, K., Håkansson, S., Magnusson, K.E., Stendahl, O., and Wolf-Watz, H., 1995, *Yersinia pseudotuberculosis* inhibits Fc receptor-mediated phagocytosis in J774 cells, *Infect. Immun.* **63**:3117–3124.

Fällman, M., Persson, C., and Wolf-Watz, H., 1997, Yersinia proteins that target host cell signaling pathways, *J. Clin. Invest.* **99**:1153–1157.

Fields, K.A., Plano, G.V., and Straley, S.C., 1994, A low-Ca^{2+} response (LCR) secretion (*ysc*) locus lies within the *lcrB* region of the LCR plasmid in *Yersinia pestis*, *J. Bacteriol.* **176**:569–579.

Forsberg, Å., and Wolf-Watz, H., 1988, The virulence protein Yop5 of *Yersinia pseudotuberculosis* is regulated at transcriptional level by plasmid-plB1-encoded trans-acting elements controlled by temperature and calcium, *Mol. Microbiol.* **2**:121–133.

Forsberg, Å., Viitanen, A.M., Skurnik, M., and Wolf-Watz, H., 1991, The surface-located YopN protein is involved in calcium signal transduction in *Yersinia pseudotuberculosis*, *Mol. Microbiol.* **5**:977–986.

Foster, R., Thorner, J., and Martin, G.S., 1989, Nucleotidylation, not phosphorylation, is the major source of the phosphotyrosine detected in enteric bacteria, *J. Bacteriol.* **171**:272–279.

Frithz-Lindsten, E., Du, Y., Rosqvist, R., and Forsberg, Å., 1997, Intracellular targeting of exoenzyme S of *Pseudomonas aeruginosa* via type III-dependent translocation induces phagocytosis resistance, cytotoxicity and disruption of actin microfilaments, *Mol. Microbiol.* **25**:1125–1139.

Fu, Y., and Galan, J.E., 1998, The *Salmonella typhimurium* tyrosine phosphatase SptP is translocated into host cells and disrupts the actin cytoskeleton, *Mol. Microbiol.* **27**:359–368.

Galyov, E.E., Håkansson, S., Forsberg, Å., and Wolf-Watz, H., 1993, A secreted protein kinase of *Yersinia pseudotuberculosis* is an indispensable virulence determinant, *Nature* **361**:730–732.

Galyov, E.E., Håkansson, S., and Wolf-Watz, H., 1994, Characterization of the operon encoding the YpkA Ser/Thr protein kinase and the YopJ protein of *Yersinia pseudotuberculosis*, *J. Bacteriol.* **176**:4543–4548.

Gemski, P., Lazere, J.R., and Casey, T., 1980a, Plasmid associated with pathogenicity and calcium dependency of *Yersinia enterocolitica*, *Infect. Immun.* **27**:682–685.

Gemski, P., Lazere, J.R., Casey, T., and Wohlhieter, J.A., 1980b, Presence of a virulence-associated plasmid in *Yersinia pseudotuberculosis*, *Infect. Immun.* **28**:1044–1047.

Goguen, J.D., Walker, W.S., Hatch, T.P., and Yother, J., 1986, Plasmid-determined cytotoxicity in *Yersinia pestis* and *Yersinia pseudotuberculosis*, *Infect. Immun.* **51**:788–794.

Gralnick, H.R., Williams, S., McKeown, L.P., Hansmann, K., Fenton II, J.W., and Krutzsch, H., 1994, High-affinity α-thrombin binding to platelet glycoprotein Ibα: identification of two binding domains, *Proc. Natl. Acad. Sci. USA* **91**:6334–6338.

Green, S.P., Hartland, E.L., Robins Browne, R.M., and Phillips, W.A., 1995, Role of YopH in the suppression of tyrosine phosphorylation and respiratory burst activity in murine macrophages infected with *Yersinia enterocolitica*, *J. Leukoc. Biol.* **57**:972–977.

Grützkau, A., Hanski, C., Hahn, H., and Riecken, E.O., 1990, Involvement of M cells in the bacterial invasion of Peyer's patches: a common mechanism shared by *Yersinia enterocolitica* and other enteroinvasive bacteria, *GUT* **31**:1011–1015.

Guan, K.L., and Dixon, J.E., 1990, Protein tyrosine phosphatase activity of an essential virulence determinant in Yersinia, *Science* **249**:553–556.

Guan, K.L., and Dixon, J.E., 1991, Evidence for protein-tyrosine-phosphatase catalysis proceeding via a cysteine-phosphate intermediate, *J. Biol. Chem.* **266**:17026–17030.

Haimovich, B., Regan, C., DiFazio, L., Ginalis, E., Ji, P., Purohit, U., Rowley, R.B., Bolen, J., and Greco, R., 1996, The FcγRII receptor triggers $pp125^{FAK}$ phosphorylation in platelets, *J. Biol. Chem.* **271**:16332–16337.

Håkansson, S., Bergman, T., Vanooteghem, J.-C., Cornelis, G., and Wolf-Watz, H., 1993, YopB and YopD constitute a novel class of Yersinia Yop proteins, *Infect. Immun.* **61**: 71–80.

Håkansson, S., Galyov, E.E., Rosqvist, R., and Wolf-Watz, H., 1996a, The Yersinia YpkA Ser/Thr kinase is translocated and subsequently targeted to the inner surface of the HeLa cell plasma membrane, *Mol. Microbiol.* **20**:593–603.

Håkansson, S., Schesser, K., Persson, C., Galyov, E.E., Rosqvist, R., Homble, F., and Wolf-Watz, H., 1996b, The YopB protein of *Yersinia pseudotuberculosis* is essential for the translocation of Yop effector proteins across the target cell plasma membrane and displays a contact dependent membrane disrupting activity, *EMBO J.* **15**:5812–5823.

Han, Y.W., and Miller, V.L., 1997, Reevaluation of the virulence phenotype of the *inv yadA* double mutants of *Yersinia pseudotuberculosis*, *Infect. Immun.* **65**:327–330.

Hanski, C., Kutschka, U., Schmoranzer, H.P., Naumann, M., Stallmach, A., Hahn, H., Menge, H., and Riecken, E.O., 1989, Immunohistochemical and electron microscopic study of interaction of *Yersinia enterocolitica* serotype 0:8 with intestinal mucosa during experimental enteritis, *Infect. Immun.* **57**:673–678.

Hartland, E.L., Green, S.P., Phillips, W.A., and Robins Browne, R.M., 1994, Essential role of YopD in inhibition of the respiratory burst of macrophages by *Yersinia enterocolitica*, *Infect. Immun.* **62**:4445–4453.

Heesemann, J., and Grüter, L., 1987, Genetic evidence that the outer membrane protein Yop1 of *Yersinia enterocolitica* mediates adherence and phagocytosis resistance to human epithelial cells, *FEMS Microbiol. Letts.* **40**:37–41.

Holmström, A., Pettersson, J., Rosqvist, R., Håkansson, S., Tafazoli, F., Fällman, M., Magnusson, K.E., Wolf-Watz, H., and Forsberg, Å., 1997, YopK of *Yersinia pseudotuberculosis* controls translocation of Yop effectors across the eukaryotic cell membrane, *Mol. Microbiol.* **24**:73–91.

Iriarte, M., and Cornelis, G.R., 1995, MyfF, an element of the network regulating the synthesis of fibrillae in *Yersinia enterocolitica*, *J. Bacteriol.* **177**:738–744.

Iriarte, M., and Cornelis, G.R., 1998, YopT, a new Yersinia Yop effector protein, affects the cytoskeleton of host cells. *Mol. Microbiol.* in press.

Iriarte, M., Vanooteghem, J.-C., Delor, I., Diaz, R., Knutton, S., and Cornelis, G.R., 1993, The Myf fibrillae of *Yersinia enterocolitica*, *Mol. Microbiol.* **9**:507–520.

Iriarte, M., Sory, M.-P., Boland, A., Boyd, A.P., Mills, S.D., Lambermont, I., and Cornelis, G.R., 1998, TyeA, a protein involved in control of Yop release and in translocation of Yersinia Yop effectors, *EMBO J.* **17**:1907–1918.

Isberg, R.R., and Falkow, S., 1985, A single genetic locus encoded by *Yersinia pseudotuberculosis* permits invasion of cultured animal cells by *Escherichia coli* K-12. *Nature* **317**:262–264.

Isberg, R.R., and Leong, J.M., 1990, Multiple β_1 chain integrins are receptors for invasin, a protein that promotes bacterial penetration into mammalian cells, *Cell* **60**:861–871.

Isberg, R.R., and Tran Van Nhieu, G., 1995, The mechanism of phagocytic uptake promoted by invasin-integrin interaction, *Trends Cell Biol.* **5**:120–124.

Isberg, R.R., Voorhis, D.L., and Falkow, S., 1987, Identification of invasin: a protein that allows enteric bacteria to penetrate cultured mammalian cells, *Cell* **50**:769–778.

Isberg, R.R., Swain, A., and Falkow, S., 1988, Analysis of expression and thermoregulation of the *Yersinia pseudotuberculosis inv* gene with hybrid proteins, *Infect. Immun.* **56**:2133–2138.

Kapperud, G., Namork, E., and Skarpeid, H.J., 1985, Temperature-inducible surface fibrillae associated with the virulence plasmid of *Yersinia entercolitica* and *Yersinia pseudotuberculosis*, *Infect. Immun.* **47**:561–566.

Kobe, B., and Deisenhofer, J., 1994, The leucine-rich repeat: a versatile binding motif, *Trends Biochem. Sci.* **19**:415–420.

Koster, M., Bitter, W., de Cock, H., Allaoui, A., Cornelis, G.R., and Tommassen, J., 1997, The outer membrane component, YscC, of the Yop secretion machinery of *Yersinia enterocolitica* forms a ring-shaped multimeric complex, *Mol. Microbiol.* **26**:789–798.

Lachica, R.V., Zink, D.L., and Ferris, W.R., 1984, Association of fibril structure formation with cell surface properties of *Yersinia enterocolitica*, *Infect. Immun.* **46**:272–275.

Lambert de Rouvroit, C.L., Sluiters, C., and Cornelis, G.R., 1992, Role of the transcriptional activator, VirF, and temperature in the expression of the pYV plasmid genes of *Yersinia enterocolitica*, *Mol. Microbiol.* **6**:395–409.

Lee, C., 1997, Type III secretion systems: machines to deliver bacterial proteins into eukaryotic cells? *Trends Microbiol.* **5**:148–155.

Lee, V.T., Anderson, D.M., and Schneewind, O., 1998, Targeting of Yersinia Yop proteins into the cytosol of HeLa cells: one-step translocation of YopE across bacterial and eukaryotic membranes is dependent on SycE chaperone, *Mol. Microbiol.* **28**:593–601.

Leong, J.M., Fournier, R.S., and Isberg, R.R., 1990, Identification of the integrin binding domain of the *Yersinia pseudotuberculosis* invasin protein, *EMBO J.* **9**:1979–1989.

Leong, J.M., Morrissey, P.E., and Isberg, R.R., 1993, A 76-amino acid disulfide loop in the *Yersinia pseudotuberculosis* invasin protein is required for integrin receptor recognition, *J. Biol. Chem.* **268**:20524–20532.

Leong, J.M., Morrissey, P.E., Marra, A., and Isberg, R.R., 1995, An aspartate residue of the *Yersinia pseudotuberculosis* invasin protein that is critical for integrin binding, *EMBO J.* **14**:422–431.

Leung, K.Y., and Straley, S.C., 1989, The *yopM* gene of *Yersinia pestis* encodes a released protein having homology with the human platelet surface protein GPIbα, *J. Bacteriol.* **171**:4623–4632.

Leung, K.Y., Reisner, B.S., and Straley, S.C., 1990, YopM inhibits platelet aggregation and is necessary for virulence of *Yersinia pestis* in mice, *Infect. Immun.* **58**:3262–3271.

Lian, C.J., and Pai, C.H., 1985, Inhibition of human neutrophil chemiluminescence by plasmid-mediated outer membrane proteins of *Yersinia enterocolitica*, *Infect. Immun.* **49**:145–151.

Lian, C.J., Hwang, W.S., and Pai, C.H., 1987, Plasmid-mediated resistance to phagocytosis in *Yersinia enterocolitica*, *Infect. Immun.* **55**:1176–1183.

Lindler, L.E., Klempner, M.S., and Straley, S.C., 1990, *Yersinia pestis* pH 6 antigen: genetic, biochemical, and virulence characterization of a protein involved in the pathogenesis of bubonic plague, *Infect. Immun.* **58**:2569–2577.

Mantle, M., Basaraba, L., Peacock, S.C., and Gall, D.G., 1989, Binding of *Yersinia enterocolitica* to rabbit intestinal brush border membranes, mucus, and mucin, *Infect. Immun.* **57**:3292–3299.

Marra, A., and Isberg, R.R., 1997, Invasin-dependent and invasin-independent pathways for translocation of *Yersinia pseudotuberculosis* across the Peyer's patch intestinal epithelium, *Infect. Immun.* **65**:3412–3421.

Michiels, T., and Cornelis, G., 1988, Nucleotide sequence and transcription analysis of *yop51* from *Yersinia enterocolitica* W22703, *Microb. Pathog.* **5**:449–459.

Michiels, T., and Cornelis, G.R., 1991, Secretion of hybrid proteins by the Yersinia Yop export system, *J. Bacteriol.* **173**:1677–1685.

Michiels, T., Wattiau, P., Brasseur, R., Ruysschaert, J.M., and Cornelis, G., 1990, Secretion of Yop proteins by Yersiniae, *Infect. Immun.* **58**:2840–2849.

Michiels, T., Vanooteghem, J.-C., Lambert de Rouvroit, C.L., China, B., Gustin, A., Boudry, P., and Cornelis, G.R., 1991, Analysis of *virC*, an operon involved in the secretion of Yop proteins by *Yersinia enterocolitica*, *J. Bacteriol.* **173**:4994–5009.

Miller, V.L., and Falkow, S., 1988, Evidence for two genetic loci in *Yersinia enterocolitica* that can promote invasion of epithelial cells, *Infect. Immun.* **56**:1242–1248.

Mills, S.D., Boland, A., Sory, M.-P., Van der Smissen, P., Kerbourch, C., Finlay, B.B., and Cornelis, G.R., 1997, *Yersinia enterocolitica* induces apoptosis in macrophages by a process requiring functional type III secretion and translocation mechanisms and involving YopP, presumably acting as an effector protein, *Proc. Natl. Acad. Sci. USA* **94**:12638–12643.

Mittler, R., and Lam, E., 1996, Sacrifice in the face of foes: pathogen-induced programmed cell death in plants, *Trends. Microbiol.* **4**:10–15.

Monack, D.M., Mecsas, J., Ghori, N., and Falkow, S., 1997, Yersinia signals macrophages to undergo apoptosis and YopJ is necessary for this cell death, *Proc. Natl. Acad. Sci. USA* **94**:10385–10390.

Nakajima, R., and Brubaker, R.R., 1993, Association between virulence of *Yersinia pestis* and suppression of gamma interferon and tumor necrosis factor alpha, *Infect. Immun.* **61**:23–31.

Nakajima, R., Motin, V.L., and Brubaker, R.R., 1995, Suppression of cytokines in mice by protein A-V antigen fusion peptide and restoration of synthesis by active immunization, *Infect. Immun.* **63**:3021–3029.

Nilles, M.L., Williams, A.W., Skrzypek, E., and Straley, S.C., 1997, *Yersinia pestis* LcrV forms a stable complex with LcrG and may have a secretion-related regulatory role in the low-Ca^{2+} response, *J. Bacteriol.* **179**:1307–1316.

Palmer, L.E., Hobbie, S., Galan, J.E., and Bliska, J.B., 1998, YopJ of *Yersinia pseudotuberculosis* is required for the inhibition of macrophage TNFα production and downregulation of the MAP kinases p38 and JNK, *Mol. Microbiol.* **27**:953–965.

Pepe, J.C., and Miller, V.L., 1993, *Yersinia enterocolitica* invasin: a primary role in the initiation of infection, *Proc. Natl. Acad. Sci. USA* **90**:6473–6477.

Pepe, J.C., Badger, J.L., and Miller, V.L., 1994, Growth phase and low pH affect the thermal regulation of the *Yersinia enterocolitica inv* gene, *Mol. Microbiol.* **11**:123–135.

Pepe, J.C., Wachtel, M.R., Wagar, E., and Miller, V.L., 1995, Pathogenesis of defined invasion mutants of *Yersinia enterocolitica* in a BALB/c mouse model of infection, *Infect. Immun.* **63**:4837–4848.

Perona, R., Montaner, S., Saniger, L., Sanchez Perez, I., Bravo, R., and Lacal, J.C., 1997, Activation of the nuclear factor-kB by Rho, CDC42, and Rac-1 proteins, *Genes. Dev.* **11**:463–475.

Perry, R.D., and Brubaker, R.R., 1983, Vwa+ phenotype of *Yersinia enterocolitica*, *Infect. Immun.* **40**:166–171.

Persson, C., Nordfelth, R., Holmström, A., Håkansson, S., Rosqvist, R., and Wolf-Watz, H., 1995, Cell-surface-bound Yersinia translocate the protein tyrosine phosphatase YopH by a polarized mechanism into the target cell, *Mol. Microbiol.* **18**:135–150.

Persson, C., Carballeira, N., Wolf-Watz, H., and Fällman, M., 1997, The PTPase YopH inhibits uptake of Yersinia, tyrosine phosphorylation of p130^{Cas} and FAK, and the associated accumulation of these proteins in peripheral focal adhesions, *EMBO J.* **16**:2307–2318.

Petruzzelli, L., Takami, M., and Herrera, R., 1996, Adhesion through the interaction of lymphocyte function-associated antigen-1 with intracellular adhesion molecule-1 induces tyrosine phosphorylation of p130^{cas} and its association with c-CrkII, *J. Biol. Chem.* **271**:7796–7801.

Pettersson, J., Nordfelth, R., Dubinina, E., Bergman, T., Gustafsson, M., Magnusson, K.E., and Wolf-Watz, H., 1996, Modulation of virulence factor expression by pathogen target cell contact, *Science* **273**:1231–1233.

Plano, G.V., and Straley, S.C., 1993, Multiple effects of *lcrD* mutations in *Yersinia pestis*, *J. Bacteriol.* **175**:3536–3545.

Plano, G.V., and Straley, S.C., 1995, Mutations in *yscC*, *yscD*, and *yscG* prevent high-level expression and secretion of V antigen and Yops in *Yersinia pestis*, *J. Bacteriol.* **177**:3843–3854.

Plano, G.V., Barve, S.S., and Straley, S.C., 1991, LcrD, a membrane-bound regulator of the *Yersinia pestis* low-calcium response, *J. Bacteriol.* **173**:7293–7303.

Portnoy, D.A., and Falkow, S., 1981, Virulence-associated plasmids from *Yersinia enterocolitica* and *Yersinia pestis*, *J. Bacteriol.* **148**:877–883.

Reisner, B.S., and Straley, S.C., 1992, *Yersinia pestis* YopM: thrombin binding and overexpression, *Infect. Immun.* **60**:5242–5252.

Roggenkamp, A., Neuberger, H.R., Flugel, A., Schmoll, T., and Heesemann, J., 1995, Substitution of two histidine residues in YadA protein of *Yersinia enterocolitica* abrogates collagen binding, cell adherence and mouse virulence, *Mol. Microbiol.* **16**:1207–1219.

Roggenkamp, A., Ruckdeschel, K., Leitritz, L., Schmitt, R., and Heesemann, J., 1996, Deletion of amino acids 29 to 81 in adhesion protein YadA of *Yersinia enterocolitica* serotype O:8 results in selective abrogation of adherence to neutrophils, *Infect. Immun.* **64**:2506–2514.

Rosenshine, I., Duronio, V., and Finlay, B.B., 1992, Tyrosine protein kinase inhibitors block invasin-promoted bacterial uptake by epithelial cells, *Infect. Immun.* **60**:2211–2217.

Rosqvist, R., Bölin, I., and Wolf-Watz, H., 1988, Inhibition of phagocytosis in *Yersinia pseudotuberculosis*: a virulence plasmid-encoded ability involving the Yop2b protein, *Infect. Immun.* **56**:2139–2143.

Rosqvist, R., Forsberg, Å., Rimpilainen, M., Bergman, T., and Wolf-Watz, H., 1990, The cytotoxic protein YopE of Yersinia obstructs the primary host defence, *Mol. Microbiol.* **4**:657–667.

Rosqvist, R., Forsberg, Å., and Wolf-Watz, H., 1991, Intracellular targeting of the Yersinia YopE cytotoxin in mammalian cells induces actin microfilament disruption, *Infect. Immun.* **59**:4562–4569.

Rosqvist, R., Magnusson, K.E., and Wolf-Watz, H., 1994, Target cell contact triggers expression and polarized transfer of Yersinia YopE cytotoxin into mammalian cells, *EMBO J.* **13**:964–972.

Ruckdeschel, K., Roggenkamp, A., Schubert, S., and Heesemann, J., 1996, Differential contribution of *Yersinia enterocolitica* virulence factors to evasion of microbicidal action of neutrophils, *Infect. Immun.* **64**:724–733.

Ruckdeschel, K., Machold, J., Roggenkamp, A., Schubert, S., Pierre, J., Zumbihl, R., Liautard, J.-P., Heesemann, J., and Rouot, B., 1997a, *Yersinia enterocolitica* promotes deactivation of macrophage mitogen-activated protein kinases extracellular signal-regulated kinase-1/2, p38, and c-Jun NH_2-terminal kinase, *J. Biol. Chem.* **272**:15920–15927.

Ruckdeschel, K., Roggenkamp, A., Lafont, V., Mangeat, P., Heesemann, J., and Rouot, B., 1997b, Interaction of *Yersinia enterocolitica* with macrophages leads to macrophage cell death through apoptosis, *Infect. Immun.* **65**:4813–4821.

Ruckdeschel, K., Harb, S., Roggenkamp, A., Hornef, M., Zumbihl, R., Köhler, S., Heesemann, J., and Rouot, B., 1998, *Yersinia enterocolitica* impairs activation of transcription factor NF-κB: involvement in the induction of programmed cell death and in the suppression of the macrophage tumor necrosis factor α production, *J. Exp. Med.* **187**:1069–1079.

Salmond, G.P., and Reeves, P.J., 1993, Membrane traffic wardens and protein secretion in gram-negative bacteria, *Trends Biochem. Sci.* **18**:7–12.

Saltman, L.H., Lu, Y., Zaharias, E.M., and Isberg, R.R., 1996, A region of the *Yersinia pseudotuberculosis* invasin protein that contributes to high affinity binding to integrin receptors, *J. Biol. Chem.* **271**:23438–23444.

Sarker, M.R., Neyt, C., Stainier, I., and Cornelis, G.R., 1998a, The Yersinia Yop virulon: LcrV is required for extrusion of the translocators YopB and YopD, *J. Bacteriol.* **180**:1207–1214.

Sarker, M.R., Sory, M.-P., Boyd, A.P., Iriarte, M., and Cornelis, G.R., 1998b, LcrG is required for efficient internalization of Yersinia Yop effector proteins into eukaryotic cells, *Infect. Immun.* **66**:2976–2979.

Schesser, K., Frithz-Lindsten, E., and Wolf-Watz, H., 1996, Delineation and mutational analysis of the *Yersinia pseudotuberculosis* YopE domains which mediate translocation across bacterial and eukaryotic cellular membranes, *J. Bacteriol.* **178**:7227–7233.

Schesser, K., Spiik, A.-K., Dukuzumuremyi, J.-M., Neurath, M.F., Pettersson, S., and Wolf-Watz, H., 1998, The *yopJ* locus is required for Yersinia-mediated inhibition of NF-κB activation and cytokine expression: YopJ contains a eukaryotic SH2-like domain that is essential for its repressive activity, *Mol. Microbiol.* **28**:1067–1079.

Schulze-Koops, H., Burkhardt, H., Heesemann, J., von der Mark, K., and Emmrich, F., 1992, Plasmid-encoded outer membrane protein YadA mediates specific binding of enteropathogenic Yersiniae to various types of collagen, *Infect. Immun.* **60**:2153–2159.

Schulze-Koops, H., Burkhardt, H., Heesemann, J., Kirsch, T., Swoboda, B., Bull, C., Goodman, S., and Emmrich, F., 1993, Outer membrane protein YadA of enteropathogenic Yersiniae mediates specific binding to cellular but not plasma fibronectin, *Infect. Immun.* **61**:2513–2519.

Shi, L., Mai, S., Israels, S., Browne, K., Trapani, J.A., and Greenberg, A.H., 1997, Granzyme B (GraB) autonomously crosses the cell membrane and perforin initiates apoptosis and GraB nuclear localization, *J. Exp. Med.* **185**:855–866.

Siebers, A., and Finlay, B.B., 1996, M cells and the pathogenesis of mucosal and systemic infections, *Trends. Microbiol.* **4**:22–29.

Simonet, M., Richard, S., and Berche, P., 1990, Electron microscopic evidence for *in vivo* extracellular localization of *Yersinia pseudotuberculosis* harboring the pYV plasmid, *Infect. Immun.* **58**:841–845.

Simonet, M., Riot, B., Fortineau, N., and Berche, P., 1996, Invasin production by *Yersinia pestis* is abolished by insertion of an IS*200*-like element within the *inv* gene, *Infect. Immun.* **64**:375–379.

Skrzypek, E., and Straley, S.C., 1993, LcrG, a secreted protein involved in negative regulation of the low-calcium response in *Yersinia pestis*, *J. Bacteriol.* **175**:3520–3528.

Skurnik, M., and Wolf-Watz, H., 1989, Analysis of the *yopA* gene encoding the Yop1 virulence determinants of *Yersinia* spp, *Mol. Microbiol.* **3**:517–529.

Skurnik, M., Bölin, I., Heikkinen, H., Piha, S., and Wolf-Watz, H., 1984, Virulence plasmid-associated autoagglutination in *Yersinia* spp, *J. Bacteriol.* **158**:1033–1036.

Skurnik, M., el Tahir, Y., Saarinen, M., Jalkanen, S., and Toivanen, P., 1994, YadA mediates specific binding of enteropathogenic *Yersinia enterocolitica* to human intestinal submucosa. *Infect. Immun.* **62**:1252–1261.

Smith, G.L., 1994, Virus strategies for evasion of the host response to infection, *Trends. Microbiol.* **2**:81–88.

Sory, M.-P., and Cornelis, G.R., 1994, Translocation of a hybrid YopE-adenylate cyclase from *Yersinia enterocolitica* into HeLa cells, *Mol. Microbiol.* **14**:583–594.

Sory, M.-P., Boland, A., Lambermont, I., and Cornelis, G.R., 1995, Identification of the YopE and YopH domains required for secretion and internalization into the cytosol of macrophages, using the *cyaA* gene fusion approach, *Proc. Natl. Acad. Sci. USA* **92**:11998–12002.

Stainier, I., Iriarte, M., and Cornelis, G.R., 1997, YscM1 and YscM2, two *Yersinia enterocolitica* proteins causing down regulation of *yop* transcription, *Mol. Microbiol.* **26**:833–843.

Straley, S.C., and Bowmer, W.S., 1986, Virulence genes regulated at the transcriptional level by Ca^{2+} in *Yersinia pestis* include structural genes for outer membrane proteins, *Infect. Immun.* **51**:445–454.

Symons, M., 1996, Rho family GTPases: the cytoskeleton and beyond, *Trends. Biochem. Sci.* **21**:178–181.

Tamm, A., Tarkkanen, A.M., Korhonen, T.K., Kuusela, P., Toivanen, P., and Skurnik, M., 1993, Hydrophobic domains affect the collagen-binding specificity and surface polymerization as well as the virulence potential of the YadA protein of *Yersinia enterocolitica*, *Mol. Microbiol.* **10**:995–1011.

Tang, X., Frederick, R.D., Zhou, J., Halterman, D.A., Jia, Y., and Martin, G.B., 1996, Initiation of plant disease resistance by physical interaction of AvrPto and Pto kinase, *Science* **274**:2060–2063.

Tertti, R., Skurnik, M., Vartio, T., and Kuusela, P., 1992, Adhesion protein YadA of *Yersinia* species mediates binding of bacteria to fibronectin, *Infect. Immun.* **60**:3021–3024.

Van Antwerp, D.J., Martin, S.J., Verma, I.M., and Green, D.R., 1998, Inhibition of TNF-α induced apoptosis by NF-κB, *Trends Cell Biol.* **8**:107–111.

Van den Ackerveken, G., Marois, E., and Bonas, U., 1996, Recognition of the bacterial avirulence protein AvrBs3 occurs inside the host plant cell, *Cell* **87**:1307–1316.

Vassalli, P., 1992, The pathophysiology of tumor necrosis factors, *Annu. Rev. Immunol.* **10**:411–452.

Visser, L.G., Annema, A., and van Furth, R., 1995, Role of Yops in inhibition of phagocytosis and killing of opsonized *Yersinia enterocolitica* by human granulocytes, *Infect. Immun.* **63**:2570–2575.

Wachtel, M.R., and Miller, V.L., 1995, *In vitro* and *in vivo* characterization of an *ail* mutant of *Yersinia enterocolitica*, *Infect. Immun.* **63**:2541–2548.

Wattiau, P., and Cornelis, G.R., 1993, SycE, a chaperone-like protein of *Yersinia enterocolitica* involved in the secretion of YopE, *Mol. Microbiol.* **8**:123–131.

Wattiau, P., Bernier, B., Deslee, P., Michiels, T., and Cornelis, G.R., 1994, Individual chaperones required for Yop secretion by Yersinia, *Proc. Natl. Acad. Sci. USA* **91**:10493–10497.

Woestyn, S., Allaoui, A., Wattiau, P., and Cornelis, G.R., 1994, YscN, the putative energizer of the Yersinia Yop secretion machinery, *J. Bacteriol.* **176**:1561–1569.

Woestyn, S., Sory, M.-P., Boland, A., Lequenne, O., and Cornelis, G.R., 1996, The cytosolic SycE and SycH chaperones of Yersinia protect the region of YopE and YopH involved in translocation across eukaryotic cell membranes, *Mol. Microbiol.* **20**:1261–1271.

Wolff, C., Nisan, I., Hanski, E., Frankel, G., and Rosenshine, I., 1998, Protein translocation into host epithelial cells by infecting enteropathogenic *Escherichia coli*, *Mol. Microbiol.* **28**:143–155.

Yang, Y., and Isberg, R.R., 1997, Transcriptional regulation of the *Yersinia pseudotuberculosis* pH6 antigen adhesin by two envelope-associated components, *Mol. Microbiol.* **24**:499–510.

Yang, Y., Merriam, J.J., Mueller, J.P., and Isberg, R.R., 1996, The *psa* locus is responsible for thermoinducible binding of *Yersinia pseudotuberculosis* to cultured cells, *Infect. Immun.* **64**:2483–2489.

Young, V.B., Falkow, S., and Schoolnik, G.K., 1992, The invasin protein of *Yersinia enterocolitica*: internalization of invasin-bearing bacteria by eukaryotic cells is associated with reorganization of the cytoskeleton, *J. Cell Biol.* **116**:197–207.

Young, V.B., Miller, V.L., Falkow, S., and Schoolnik, G.K., 1990, Sequence, localization and function of the invasin protein of *Yersinia enterocolitica*, *Mol. Microbiol.* **4**:1119–1128.

Zaleska, M., Lounatmaa, K., Nurminen, M., Wahlstrom, E., and Makela, P.H., 1985, A novel virulence-associated cell surface structure composed of 47 kDa protein subunits in *Yersinia enterocolitica*, *EMBO J.* **4**:1013–1018.

Zhang, Z.Y., Clemens, J.C., Schubert, H.L., Stuckey, J.A., Fischer, M.W.F., Hume, D.M., Saper, M.A., and Dixon, J.E., 1992, Expression, purification, and physicochemical characterization of a recombinant Yersinia protein tyrosine phosphatase, *J. Biol. Chem.* **267**:23759–23766.

Zychlinsky, A., and Sansonetti, P., 1997, Apoptosis in bacterial pathogenesis, *J. Clin. Invest.* **100**:493–496.

Chapter 15

Invasion of Mammalian and Protozoan Cells by *Legionella pneumophila*

Yousef Abu Kwaik

1. BACKGROUND

The first recognized outbreak of pneumonia due to *Legionella pneumophila* occurred in Philadelphia in July of 1976 among 180 persons attending the 56th annual American Legion Convention. Twenty nine patients died, and the disease became known as Legionnaires' disease (Fraser *et al.*, 1977). Guinea pigs were infected with postmortem lung tissue from the patients with fatal Legionnaires' disease, and embryonated yolk sacs were inoculated with spleen homogenates from the infected guinea pigs. In January of 1977, a gram-negative bacterium was isolated and designated *L. pneumophila* (McDade *et al.*, 1977). Antisera were subsequently generated which facilitated identification of many previous outbreaks of febrile respiratory illness of unknown etiology that occurred since 1965. The source of the infection during the Legionnaires' convention was later found to be the air conditioning system in the hotel. It has been documented that the hallmark of Legionnaires' disease is the intracellular replication of *L.*

YOUSEF ABU KWAIK Department of Microbiology and Immunology, University of Kentucky, Chandler Medical Center, Lexington, Kentucky 40536-0084.

Subcellular Biochemistry, Volume 33: Bacterial Invasion into Eukaryotic Cells, edited by Oelschlaeger and Hacker. Kluwer Academic / Plenum Publishers, New York, 2000.

pneumophila in the alveolar spaces. At least another 39 species of *legionellae* have been identified, some of which are associated with disease while others are environmental isolates and whether they can cause disease is not known. *L. pneumophila* is responsible for more than 80% of cases of Legionnaires' disease, and among the 13 serogroups of *L. pneumophila*, serogroup 1 is responsible for more than 95% of Legionnaires' disease cases. It is estimated that *L. pneumophila* is responsible for at least 25,000 cases of pneumonia/year in the US. In 1980, Rowbotham described the ability of *L. pneumophila* to multiply intracellularly within protozoa (Rowbotham, 1980). Since then, *L. pneumophila* has been described to multiply in many species of protozoa, and this host-parasite interaction is central to the pathogenesis and ecology of *L. pneumophila*. Intracellular replication of *L. pneumophila* within mammalian and protozoan cells has been shown to occur in a ribosome-studded phagosome that does not fuse to lysosomes. Fields had hypothesized that the *L. pneumophila* phagosome fuses to the rough endoplasmic reticulum (RER) (Fields, 1993). Immunocytochemistry has proven this prediction by demonstrating the presence of an RER-specific chaperon, the Bip protein, in the ribosome-studded phagosome within macrophages (Swanson and Isberg, 1995), and protozoa (Abu Kwaik, 1996). Based on these characteristics the *L. pneumophila* phagosome may be accurately described as endosomal maturation-blocked (EMB) phagosome.

2. INTRODUCTION

Intracellular bacterial pathogens can be classified into three groups based on their intracellular compartment (Garcia-del Portillo and Finlay, 1995b; Theriot, 1995). One group constitutes the cytoplasmic pathogens that are able to escape from the phagosome into the cytoplasm following lysis of the phagosome by bacterial enzymes or toxins. *Listeria monocytogenes* and *Shigella flexneri* are classical examples of this group (Theriot, 1995). In addition, these two pathogens have been shown to induce polymerization of actin at one pole of the bacteria, which allows them to move in the cytoplasm and invade neighboring cells. These pathogens are free to replicate within a nutritionally rich environment away from the unfavorable conditions that may be encountered within the endocytic pathway. The second group constitutes the pathogens that reside in a phagosome that exhibits some degree of maturation through the endosomal-lysosomal pathway (Garcia-del Portillo and Finlay, 1995b). For example, *Salmonella typhimurium*, *Coxiella burnetii*, and *Mycobacterium tuberculosis* are localized to late acidified phagosomes (see the *S. typhimurium* and *M. tubercu-*

losis chapter), while *Leishmania* is localized to phagolysosomes. These pathogens have the capacity to modulate their gene expression in response to their intracellular niches, which allows them to adapt to these microenvironments. In contrast, the third group of pathogens are localized to phagosomes that are blocked from maturation through the endosomal pathway, such as *L. pneumophila*, *Chlamydia trachomatis*, and *Toxoplasma gondii*. It is important to note that the nature of the phagosomes for these pathogens is quite heterogeneous. For example, *C. trachomatis* is localized in a phagosome that is part of the exocytic rather than the endocytic pathway while *S. typhimurium* is localized in a late endosome within macrophages. The limited endosomal markers used to group these pathogens allows classification of their compartments into early, intermediate, and late endosomes. However, further characterization of the endocytic pathway will probably yield more heterogeneity in the nature of the phagosomes occupied by these pathogens. Pathogens that reside in a vacuole face the challenge of obtaining nutrients from the cytoplasm into their phagosomes. In the case of *Toxoplasma*, the phagosomal membrane has been shown to be permeable to small molecules less than 1300 dalton. It remains a mystery of how other pathogens that are localized to phagosomes obtain their nutrients.

3. INITIAL INTERACTIONS BETWEEN *L. pneumophila* AND ITS PRIMITIVE PROTOZOAN HOSTS

Legionellae are ubiquitous in the environment. The bacteria can be found extracellularly or intracellular within protozoa, but the bacteria do not multiply extracellularly. There are at least 13 species of amoebae and 2 species of ciliated protozoa that support intracellular replication of *L. pneumophila* (Fields, 1996).

One of the most predominant amoebae in water sources are the nonpathogenic amoebae hartmannellae, which have been isolated from water sources associated with Legionnaires' disease outbreaks (Fields *et al.*, 1990). *H. vermiformis* is one of the most common protozoan species used in host-parasite interaction studies with *L. pneumophila*.

In general, initial interactions between intracellular pathogens and host cells is mediated through attachment of a bacterial ligand to a receptor on the surface of the host cell. Pili are extracellular appendages on the surface of many bacteria composed of a polymer of the major protein subunit designated pilin, and mediate attachment of many bacteria to their host cells. Approximately 18 years after the first description of the presence of pili on the surface of *L. pneumophila*, a genetic evidence for the expression of at least two distinct pili on the surface of *L. pneumophila* has been recently provided (Stone and Abu Kwaik, 1998a). One of these pili is a type IV pilus, designated as CAP (competence and adherence-associated pili) (Stone and Abu Kwaik, 1998b). Type IV pilins have been described for few bacteria and are characterized by a conserved positively charged leader peptide that is processed at a conserved cleavage site by a peptidase prior to polymerization of the pilin to assemble the pili. In addition, the processed pilin contains a conserved hydrophobic amino terminal domain and the N-terminal amino acid is composed of a conserved modified amino acid (*N*-methylphenylalanine). Mutants of *L. pneumophila* defective in expression of the CAP pilus manifest reduced attachment to protozoan cells, but are not affected in their intracellular replication (Stone and Abu Kwaik, 1998a). Thus, the CAP pili of *L. pneumophila* are involved in adherence to protozoan cells. The CAP pili may provide *L. pneumophila* with a selective advantage in adherence to surfaces and biofilms in the environment. The host cell receptor to which the CAP pili bind is not known, but it is possible that the newly described lectin receptor on *H. vermiformis* (described below) to which *L. pneumophila* adhere is the receptor for the CAP pili.

Bacterial attachment to *H. vermiformis* is mediated by adherence to a protozoan receptor that has been described to be a galactose/N-acetylgalactosamine (Gal/GalNAc) lectin with similarity to the galactose lectin of the pathogenic protozoan *Entamoebae histolytica* (Harb *et al.*, 1998; Venkataraman *et al.*, 1997; Adler *et al.*, 1995; Petri *et al.*, 1987). The Gal/GalNAc lectin is involved in attachment of *E. histolytica* to colonic mucosa. It is intriguing that this lectin serves as a ligand for adherence of *E. histolytica* to mammalian cells while it serves as a receptor that mediates attachment of *L. pneumophila* to *H. vermiformis*. The Gal/GalNAc lectin of *E. histolytica* is similar to a β2 integrin of mammalian cells with potential tyrosine phosphorylation motifs within its cytoplasmic domain (Adams *et al.*, 1993; Mann *et al.*, 1991). Integrins are heterodimeric protein tyrosine kinase receptors that undergo tyrosine phosphorylation upon engagement to ligands, which subsequently results in recruitment and rearrangements of the cytoskeleton at the site of contact with the ligand or with the extracellular matrix. Interestingly, attachment of *L. pneumophila* to the Gal/GalNAc of *H. vermiformis* triggers signal transduction events in *H. vermiformis* that are manifested in dramatic tyrosine dephosphorylation of the lectin receptor and other proteins (Venkataraman *et al.*, 1997). Similar observations of tyrosine dephosphorylation of host proteins have been obtained upon infection of *H. vermiformis* by another species of *legionellae*, *L. micdadei* (Abu Kwaik *et al.*, 1998b). Among the *L. pneumophila*-induced tyrosine dephosphorylated proteins in *H. vermiformis* are the cytoskeletal proteins paxillin, vinculin, and focal adhesion kinase (Venkataraman *et al.*, 1998). This is the opposite response expected in the protozoa upon attachment to *L. pneumophila,* since phagocytosis and subsequent cytoskeletal rearrangements in mammalian cells require tyrosine phosphorylation of the integrin receptor and many cytoskeletal proteins upon engagement to ligand. Tyrosine phosphatases have been shown to disrupt the cytoskeleton in mammalian cells. It is thought that the bacteria induces a tyrosine phosphatase activity within *H. vermiformis* either by activating a protozoan tyrosine phosphatase or by direct translocation of a bacterial tyrosine phosphatase into *H. vermiformis*. This may be similar to the tyrosine phosphatase of *Yersinia* that is vectorially translocated into mammalian cells upon bacterial attachment (see the Yersinia chapter). The induced tyrosine phosphatase activity in *H. vermiformis* is probably manifested in disruption of the protozoan cytoskeleton to facilitate entry through a cytoskeleton-independent receptor-mediated endocytosis (Venkataraman *et al.*, 1998; King *et al.*, 1991). These results are consistent with the observations that uptake of *L. pneumophila* by *H. vermiformis* is not affected by cytochalasin D (which disrupts the cytoskeleton) but is inhibited by methylamine (which is an inhibitor of receptor-mediated endocytosis). Interestingly, in addition to these manipulations of the signal transduction of *H. vermiformis* by *L. pneumophila*, bacterial invasion is also associated with specific induction of gene expression in protozoa, and inhibition of this gene expression blocks entry of the bacteria (Abu Kwaik *et al.*, 1994). These observations show a dramatic and rather sophisticated adaptation of *L. pneumophila* to attach to and invade protozoa. Following this initial host-parasite interaction, uptake of *L. pneumophila* by protozoan cells occurs by conventional and coiling phagocytosis (in which the bacterium is surrounded by a multilayer coil-like structure) (Venkataraman *et al.*, 1998; Abu Kwaik, 1996; Bozue and Johnson, 1996).

4. INTRACELLULAR REPLICATION WITHIN PROTOZOA

After entry, the bacterium is enclosed in a phagosome surrounded by mitochondria and host cell vesicles during the first 60 minutes. The bacterial phagosome is blocked from fusion to the lysosomes (Bozue and Johnson, 1996). In addition, by 4 h post-infection, the phagosome is surrounded by a multilayer membrane derived from the rough endoplasmic reticulum (RER) (Abu Kwaik, 1996). Based on these characteristics, the *L. pneumophila* phagosome will be designated as EMB, for Endosomal Maturation-Blocked phagosome. The molecular and biochemical mechanisms of these manipulations of the host cell by the bacteria are not understood.

Following formation of the EMB phagosome, bacterial replication is initiated. The 4h period prior to initiation of intracellular replication may be the time required to recruit these host cell organelles that may be required for replication. Alternatively, the 4h period may be a lag phase of metabolic and environmental adjustment of the bacteria to a new niche. Bacterial replication starts following formation of the *L. pneumophila* EMB phagosome, with a generation time of approximately 1.8h, in contrast to the 2.4h generation time in rich media *in vitro* (Abu Kwaik *et al.*, 1998a). Following this prolific replication, the bacteria kill the protozoan host by an unknown mechanism and the bacteria are subsequently released.

5. ROLE OF PROTOZOA IN LEGIONNAIRES' DISEASE

Intracellular replication of *L. pneumophila* within protozoa is central to the persistence of the bacteria in the environment, bacterial resistance to and protection from harsh conditions (e.g. temperature fluctuations), and is required for bacterial amplification in the environment. Viable but nonculturable *L. pneumophila* that are unable to grow in rich media *in vitro*, can be resuscitated by growth within protozoa (Steinert *et al.*, 1997). Five species of bacteria that are related to *Legionella* but can only be isolated and grown within amoebae, have been identified and designated Legionella-Like Amoebal pathogens (LLAP), many of which have been associated with Legionnaires' disease (Adeleke *et al.*, 1996; Britles *et al.*, 1996; Fields, 1996).

It has been proposed that the infectious particle for Legionnaires' disease is amoebae infected with the bacteria (Rowbotham, 1980). Although this has not yet been proven, there are many lines of evidence to suggest that protozoa play major roles in transmission of *L. pneumophila*. First, many protozoan hosts have been identified that allow intracellular bacterial replication, the only means of bacterial amplification in the environment (Fields, 1996). Second, in outbreaks of Legionnaires' disease, amoebae and bacteria have been isolated from the same source of infection and the isolated amoebae support intracellular replication of the bacteria (Fields *et al.*, 1990). Third, following intracellular replication within protozoa, *L. pneumophila* exhibit a dramatic increase in resistance to harsh conditions including high temperature, acidity, and high osmolarity, which may facilitate bacterial survival in the environment (Abu Kwaik *et al.*, 1997). Fourth, intracellular *L. pneumophila* within protozoa are more resistant to chemical disinfection and biocides compared to *in vitro*-grown bacteria (Barker *et al.*, 1995; Barker *et al.*, 1993; Barker *et al.*, 1992). Fifth, protozoa have been shown to release vesicles of respirable size that contain numerous *L. pneumophila*, and the vesicles are resistant to freeze-thawing and sonication and the bacteria within the vesicles are highly resistant to biocides (Berk *et al.*, 1998). Sixth, following their release from the protozoan host, the bacteria exhibit a dramatic increase in their infectivity to mammalian cells *in vitro* (Cirillo *et al.*, 1994). In addition, it has been demonstrated in mice that intracellular bacteria within *H. vermiformis* are dramatically more infectious and are highly lethal (Brieland *et al.*, 1997). Seventh, the number of bacteria isolated from the source of infection of Legionnaires' disease is usually very low or undetectable, and thus, enhanced infectivity of intracellular bacteria within protozoa may compensate for the low infectious dose (O'Brein and Bhopal, 1993). Eighth, viable but non-culturable *L. pneumophila* can be resuscitated by co-culture with protozoa (Steinert *et al.*, 1997). This observation may suggest that failure to isolate the bacteria from environmental sources of infection may due to this "dormant" phase of the bacteria that can not be recovered on artificial media. Ninth, there has been no documented case of bacterial transmission between individuals. The only source of transmission is environmental droplets generated from many man-made devices such as shower heads, water fountains, whirlpools, and cooling towers of air conditioning systems (Fields, 1996).

6. ATTACHMENT AND ENTRY TO MAMMALIAN CELLS

Invasion and intracellular replication of *L. pneumophila* within pulmonary cells in the alveoli is the hallmark of Legionnaires' disease. These alveolar cells include macrophages and type I and II epithelial cells. Monocytes are recruited to the alveolar space due to the inflammatory response. Uptake of *L. pneumophila* into monocytes has been shown to be mediated, at least in part, through attachment of complement-coated bacteria to the complement receptor (Payne and Horwitz, 1987), but non-complement mediated uptake also occurs (Rodgers and Gibson III, 1993; Elliott and Winn, 1986), and most *in vitro* studies of bacterial attachment and entry in tissue culture assay systems are performed in heat-inactivated sera (Gao *et al.*, 1998; Stone and Abu Kwaik, 1998a; Gao *et al.*, 1997b). The host cell receptor involved in non-complement-mediated uptake in macrophages and epithelial cells is not known. Until the recent discovery of the type IV CAP pili of *L. pneumophila* (Stone and Abu Kwaik, 1998a), the bacterial ligands involved in non-complement-mediated attachment to macrophages and epithelial cells were not known.

The CAP pilus of *L. pneumophila* is involved in attachment to macrophages and alveolar epithelial cells, since CAP mutants that do not express the CAP pili manifest reduced attachment to these cells (Stone and Abu Kwaik, 1998a). Thus, the CAP pilus is one of the ligands involved in non-complement mediated attachment to mammalian macrophages and alveolar epithelial cells. The cell receptor on macrophages and epithelial cells involved in attachment to the CAP pilus is unknown. Mutants with severe defects in non-complement attachment to mammalian macrophages and alveolar epithelial cells have been isolated, but the defective ligands in these mutants is not the CAP pilus (Gao *et al.*, 1998; Gao *et al.*, 1997b). Therefore, multiple ligands are involved in attachment of *L. pneumophila* to mammalian macrophages and epithelial cells.

Uptake of complement-opsonized *L. pneumophila* by monocytes has been shown to occur through coiling phagocytosis (Horwitz, 1984). However, it is unlikely, that the complement receptor is involved in this process. Heat-killed or formalin-killed bacteria are also taken up by coiling phagocytosis. Moreover, some of the uptake by protozoa, which presumably do not express a complement receptor, occurs by coiling phagocytosis (Venkataraman *et al.*, 1997a; Bozue and Johnson, 1996). In addition, uptake of other intracellular pathogens that are taken up through the complement receptor, such as *Mycobacterium tuberculosis*, occurs by conventional and not coiling phagocytosis. Moreover, intracellular pathogens such as *Chlamydia psittaci*, *Leishmania donovoni*, *Borrelia burgdorferi*, and *Trypanosoma brucei* are taken up by complement-independent coiling phagocytosis. Since heat-killed and formalin-killed *L. pneumophila* are also taken up by coiling phagocytosis but are targeted to the lysosomes, this mode of uptake may not play a role in subsequent fate of the bacteria.

Many clinical isolates of *L. pneumophila* have been shown to be taken up exclusively by conventional phagocytosis (Rechnitzer and Blom, 1989; Elliott and Winn, 1986). In addition, other species of *legionellae* such as *L. micdadei*, the second most common species of *legionellae* that causes Legionnaires' disease, is taken up exclusively by "conventional" but not coiling phagocytosis (Dowling *et al.*, 1992; Rechnitzer and Blom, 1989; Weinbaum *et al.*, 1984). The bacterial ligand that mediates the coiling mode of phagocytosis is not known but it is clear that it is heat resistant and is not denatured by formalin, suggesting a remarkably denaturationresistant protein or that the lipopolysaccharide (LPS) of *L. pneumophila* is the ligand mediating the trigger for the coiling phagocytosis. The LPS of *L. pneumophila* is rather unique in its biochemical structure in which the O-side chain is made up of homopolymer repeats of a single sugar residue designated legionaminic acid, and the core region is highly *O*-acetylated providing hydrophobic characteristic on the LPS structure (Knirel *et al.*, 1996). No such highly *O*-acetylated core LPS has been described for other gram-negative bacteria. If mutants of *L. pneumophila* with truncated LPS structure can be isolated it will be possible to test this hypothesis.

7. INTRACELLULAR SURVIVAL AND REPLICATION WITHIN MACROPHAGES

Macrophages derived from alveolar lavage and peripheral blood monocytes obtained from human volunteers support intracellular replication of *L. pneumophila* (Nash *et al.*, 1984). In addition, examination of histopathological lung tissue samples from patients with Legionnaires' disease have shown intracellular replication of the bacteria within cells with morphological features of macrophages and monocytes (Winn and Myerowitz, 1981). Therefore, macrophages and monocytes have been the major focus in studying intracellular replication and pathogenesis of *L. pneumophila*. Although alveolar epithelial cells, which constitute more than 95% of the alveolar surface, have been shown to allow intracellular replication of *L. pneumophila*, their role in the pathogenesis has been largely overlooked.

Similar to the protozoan infection, following entry of the bacteria into macrophages and monocytes, the *L. pneumophila* EMB phagosome is surrounded by host cell organelles such as mitochondria, vesicles, and the RER (Swanson and Isberg, 1995; Horwitz, 1983). Similar to the trafficking of *L. pneumophila* within protozoa, the EMB phagosome within mammalian macrophages does not fuse to lysosomes (Roy *et al.*, 1998; Fields, 1996). The role of the RER in the intracellular infection is not known, but the RER is not required as a source of proteins for the bacteria (Abu Kwaik, 1998). Interestingly, examination of the intracellular infection of macrophages, alveolar epithelial cells, and protozoa by another *legionellae* species, *L. micdadei*, showed that within all of these host cells, the bacteria were localized to RER-free phagosomes (Abu Kwaik *et al.*, 1998a). Whether other *legionellae* species replicate within RER-free phagosomes is still to be determined.

The remarkable similarities in the intracellular infection of the two evolutionarily distant host cells (macrophages and protozoa) suggest that *L. pneumophila* may utilize similar molecular mechanisms to manipulate host cell processes of macrophages and protozoa (Gao *et al.*, 1997b). This may not be surprising, since macrophages and protozoa, although evolutionarily distant, both are phagocytic cells. It has been hypothesized that *L. pneumophila* has evolved as a protozoan parasite in the environment and its adaptation to this primitive phagocytic unicellular host was sufficient to allow the bacteria to survive and replicate within the biologically similar phagocytic cells of the more evolved mammalian host (Abu Kwaik, 1996; Cianciotto and Fields, 1992). It is thought that Legionnaires' disease became a threat to humans after our industrialization and production of man-made devices that generate aerosols, which allows transmission of the bacteria to humans.

8. TRAFFICKING OF THE *L. pneumophila* PHAGOSOME

Following entry of the bacteria into macrophages, monocytes, or alveolar epithelial cells, the *L. pneumophila* EMB phagosome is surrounded by host cell organelles such as mitochondria, vesicles, and is not fused to lysosomes. Mutants of *L. pneumophila* (e.g. *dotA*) defective in intracellular replication are localized into a phagosome that is trafficked through the endosomal lysosomal pathway within 5–15 minutes of entry (Roy *et al.*, 1998). This is manifested through the acquisition of the phagosome of Rab7 (required for fusion of early endosomes to late endosomes) and LAMP-1 (a late endosome/lysosomal marker) (Roy *et al.*, 1998). In contrast, phagosomes containing wild type *L. pneumophila* are excluded from these markers. Thus, the block in trafficking of the *L. pneumophila* EMB phagosome through the endosomal-lysosomal degradation pathway occurs upon bacterial entry or immediately after bacterial internalization. Interestingly, this block in endosomal maturation is not a generalized effect on other vesicles in the same host cell since co-infection of macrophages by wild type and an attenuated mutant (*dotA*) does not result in a rescue of the mutant if it was localized in a different phagosome within the same cell (C. R. Roy, personal communication). Thus, blockage of maturation of the *L. pneumophila* phagosome through the endosomal lysosomal pathway is mediated by a cis-acting element that does not affect vesicular fusions to other endosomes within the same cell.

9. MOLECULAR ASPECTS OF INTRACELLULAR REPLICATION

In order to test the hypothesis that the molecular bases of the infection of mammalian macrophages and protozoa by *L. pneumophila* are similar, a collection of 5200 miniTn*10*::*kan* insertion mutants of *L. pneumophila* have been isolated and examined for their replication within macrophages and protozoa (Gao *et al.*, 1998; Gao *et al.*, 1997b). It was reasoned that if the molecular bases of the intracellular infection of macrophages and protozoa are similar, defective mutants should exhibit similar phenotypes within both evolutionarily distant host cells. Among 121 distinct insertion mutants with varying degrees of defects in survival and replication within macrophages, 89 exhibit a very similar phenotypic defect within both macrophages and *H. vermiformis*, and the loci have been designated *pmi* (for protozoa and macrophage infectivity) (Gao *et al.*, 1997b). These observations showed that many of the molecular aspects of the intracellular infection of macrophages and protozoa are similar. However, 32 mutants with varying degrees of defects within macrophages exhibit wild type phenotypes within protozoa, and the defective loci have been designated *mil* (for macrophage-specific infectivity loci) (Gao *et al.*, 1998). Importantly, many of the *mil* mutants have been tested in peripheral blood monocytes, A/J mice-derived macrophages, and other protozoa. The macrophage-defective and protozoa-wild type phenotypes of the *mil* mutants are consistent. Thus, the *mil* loci are species-specific. These data showed that *L. pneumophila* possess genetic loci that are not required for infectivity of protozoa. Therefore, it has been hypothesized that *L. pneumophila* evolved in the environment as a protozoan parasite but acquired the *mil* loci that have allowed the bacteria to adapt to the intracellular environment of macrophages (Gao *et al.*, 1998). It is also possible that ecological co-evolution with protozoa has allowed *L. pneumophila* to possess multiple redundant mechanisms to parasitize protozoa and that some of these mechanisms are essential for survival within macrophages. These speculations suggest that there must have been pathogenic evolution in *L. pneumophila* through acquisition of the *mil* loci during its adaptation within protozoa (Gao *et al.*, 1998). The recent discoveries that *L. pneumophila* is naturally competent for DNA transformation, which is associated with expression of the type IV CAP pili (Stone and Abu Kwaik, 1998b), and the ability of *L. pneumophila* to conjugate DNA (Vogel *et al.*, 1998; Jacob *et al.*, 1994), support these speculations. Further characterization of the *mil* loci may yield interesting information that may help to elucidate these hypotheses.

Many loci of *L. pneumophila* designated *dot* (defect in organelle trafficking) and *icm* (intracellular multiplication) have been also shown to be required for intracellular replication, but it is not known whether they are required for infectivity of protozoa (Vogel *et al.*, 1998; Jacob *et al.*, 1994). Although the function of the *dot* and *icm* loci during the intracellular infection are not known, some of the gene products are thought to be involved in the assembly of a secretion apparatus. This potential secretion apparatus has been identified based on homology of a few of the Dot and Icm proteins to other proteins involved in the transfer of conjugative plasmids. The secreted products through this potential apparatus, and their role in the intracellular infection are still to be determined.

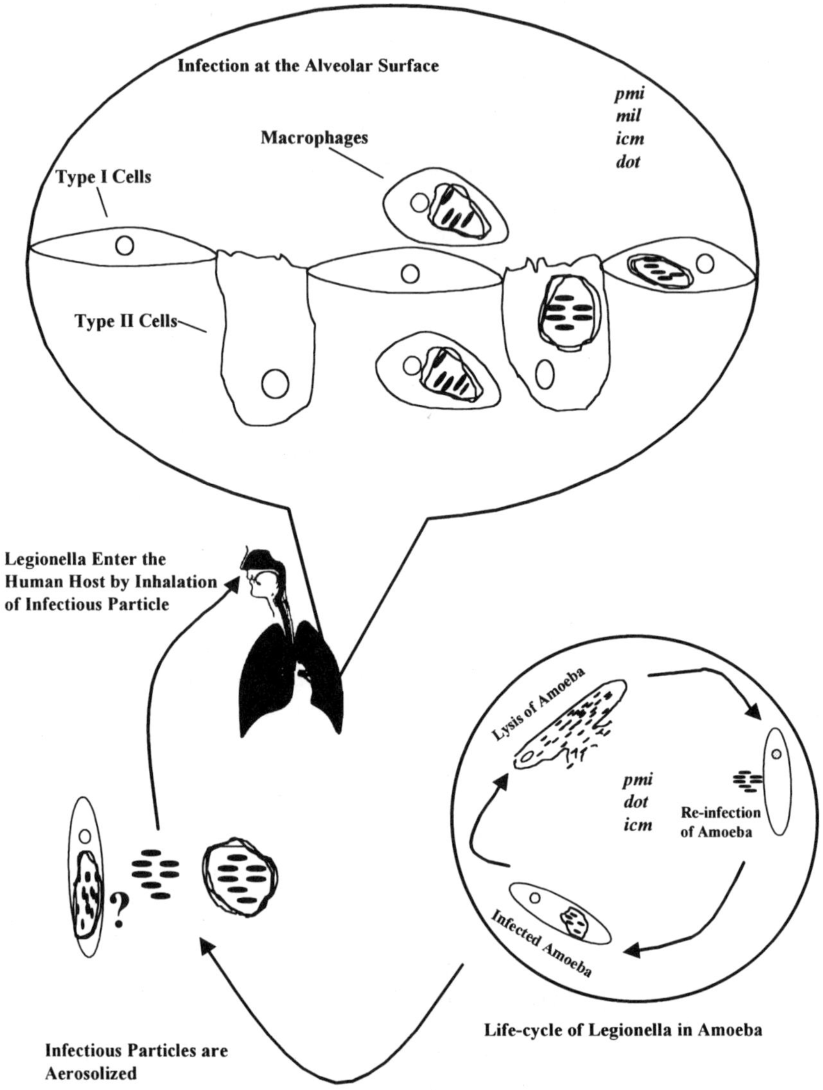

FIGURE 1. The life cycle of *L. pneumophila*. In the environment, the bacteria invade and replicate within amoebae, a process that culminates in lysis of the protozoan host and also in the release of vesicles containing bacteria. The *pmi* and probably some of the *dot/icm* loci are required for replication within protozoa. The infectious particle is not known but it could be bacteria that have been released from the amoebae and are highly infectious, *L. pneumophila*-infected amoebae, or bacteria contained within vesicles of respirable sizes that have been released from amoebae. The bacteria reach the alveolar spaces where macrophages and type I and II alveolar epithelial cells are encountered. Although bacterial replication within macrophages and monocytes that are recruited to the site of inflammation is well documented, the role of epithelial cells is not well characterized. The *dot/icm*, *pmi* and *mil* loci are all required for infectivity of the mammalian macrophages, but whether the *dot/icm, pmi*, and *mil* loci are required for infectivity of alveolar epithelial cells is still be determined.

Preliminary partial sequence analysis of the insertion junction of many of the *pmi* and *mil* mutants showed that the insertions are within loci with no similarity to virulence genes of other pathogens, suggesting that many of the *pmi*, *mil*, *dot,* and *icm* loci are novel. This may not be surprising considering 1) although few other human pathogens exhibit slight multiplication within protozoa, *L. pneumophila* is the only documented pathogen that can invade and replicate efficiently within mammalian cells and protozoa; and 2) the unique characteristics of *L. pneumophila* EMB phagosome at the ultrastructural and biochemical levels within both mammalian and protozoan cells (Abu Kwaik, 1996; Horwitz, 1983).

One of the first characterized genes that is partially required for the intracellular infection is the *mip* gene (Wintermeyer *et al.*, 1995; Cianciotto and Fields, 1992; Cianciotto *et al.*, 1989). Mutants in the *mip* gene are partially defective in early survival in macrophages, epithelial cells, and protozoa. The mutant is also partially attenuated in guinea pigs. The Mip protein is similar to a class of proteins designated as peptidyl-prolyl *cis/trans* isomerase (PPIase), and has been shown to possess this enzymatic activity (Wintermeyer *et al.*, 1995). The conserved amino acids in the PPIase catalytic domain of Mip have been shown to be conserved among 35 *legionellae* species. PPIases have been found in other intracellular pathogenic bacteria as well as nonpathogenic bacteria. Using site-directed point mutations to alter the conserved PPIase catalytic domain, it has been demonstrated that the PPIase activity of Mip is not required during the intracellular infection (Wintermeyer *et al.*, 1995).

10. NaCl^r PHENOTYPE OF *L. pneumophila* MUTANTS AND POTENTIAL ATTENUATION

It has been shown that continuous passage of *L. pneumophila* on artificial media may result in the isolation of spontaneous mutants that are NaCl^r and are attenuated for intracellular survival within macrophages, and for virulence in guinea pigs (McDade and Shepard, 1979). Using this criteria, many investigators isolated spontaneous mutants of *L. pneumophila* that are NaCl^r and are potentially attenuated by plating bacteria on NaCl-containing media. In addition, many of the transposon insertion or chemically-generated mutants that are defective in intracellular replication are NaCl^r (Swanson and Isberg, 1996; Vogel *et al.*, 1996; Sadosky *et al.*, 1993), as are many of the *pmi* mutants. However, none of the *mil* mutants are NaCl^r, indicating that the phenomenon of NaCl^r and the function of the *mil* loci are independent (Gao *et al.*, 1998). There is no explanation yet for the relationship of NaCl^r to attenuation. It is thought that the potential secretion apparatus assembled by the *dot* and *icm* loci is leaky to NaCl, which renders the wild type strain sensitive to NaCl. Upon disruption of this channel, mutants that may become defective in the secretion apparatus also become $NaCl^r$. This is still a speculation but future characterization of the potential secretion apparatus will help in explaining this phenomenon. These observation suggest that the concentration of NaCl within the EMB phagosome is very low.

11. ROLE OF ALVEOLAR EPITHELIAL CELLS IN LEGIONNAIRES' DISEASE

The ability of *L. pneumophila* to cause pneumonia is dependent on its ability to multiply intracellularly, and the bacteria do not multiply in the extracellular space within lung tissues. The alveoli are the sites of the intracellular replication. The alveolar surface is composed of resident macrophages, and type I and type II epithelial cells. Type II epithelial cells are cuboidal, cover ~5% of the alveolar surface, and secret surfactant, which is the alveolar fluid lining material. They can also differentiate into type I epithelial cells to reconstitute the alveolar surface during normal turnover or repair due to acute injury to the alveolar surface. Type I epithelial cells have thin cytoplasmic extensions that cover ~95% of the alveolar surface and cannot produce surfactant.

Most tissue materials from biopsies and autopsies from Legionnaires' disease patients showed inflammation of the alveolar space and intracellular replication of the bacteria within cells including monocytes that had been recruited to the alveolar space in response to inflammatory signals (Winn and Myerowitz, 1981). Macrophages derived from alveolar lavage and peripheral blood monocytes obtained from human volunteers support intracellular replication of *L. pneumophila in vitro* (Nash *et al.*, 1984). Although the role of alveolar epithelial cells in Legionnaires' disease has not been addressed, it has been concluded that macrophages and recruited monocytes are the sites of intracellular replication in the alveoli. This conclusion was also complicated by the limited information derived from the advanced cases of the Legionnaires' disease patients from whom tissue materials were available for histopathological studies, and thus, the early stages of infection in humans has not been studied. Thus, the role of alveolar epithelial cells in Legionnaires' disease has not been extensively studied. In contrast, *L. pneumophila* has been shown to replicate in a EMB phagosome within many epithelial cells *in vitro* including type I and II alveolar epithelial cells. Therefore, the role of epithelial cells in Legionnaires' disease may have been overlooked and underestimated. There is no reason why *L. pneumophila* is not expected to replicate at the foci of infection within alveolar epithelial cells, particularly since these constitute most of the alveolar surface, where the foci of infection are recognized. In addition, these cells are potential sites of intracellular replication during activation of macrophages by IFN-γ, which inhibits intracellular replication of *L. pneumophila* within monocytes and macrophages (Byrd and Horwitz, 1989). The importance of IFN-γ for the activation of macrophages and the resolution of infection by *L. pneumophila* has been demonstrated by treatment of A/J mice with anti-IFN-γ antibody which enhanced bacterial replication and disease progression (Brieland *et al.*, 1994). Levels of IFN-γ in intratracheally-inoculated A/J mice are elevated within 6h and peak at 24h post-infection by *L. pneumophila* (Susa *et al.*, 1998; Brieland *et al.*, 1994). These observations create a paradox because intracellular *L. pneumophila* exhibit exponential growth rate at 24h and reaches the peak number in the lungs at 48h post-inoculation of A/J mice (Susa *et al.*, 1998; Brieland *et al.*, 1994). The sites of intracellular replication of *L. pneumophila* during this period of time, when macrophages are presumably activated, are not known. Freshly recruited monocytes to the foci of infection may be potential sites of intracellular replication. Alveolar epithelial cells are a primary potential candidate, particularly since they constitute more than 95% of the alveolar surface, and may allow intracellular bacterial replication despite activation of the host immune response. It is very tempting to speculate that in infected patients or experimental animals, *L. pneumophila* preferentially replicates within both inactivated alveolar macrophages and epithelial cells. Although subsequent activation of alveolar macrophages would limit intracellular replication of *L. pneumophila*, replication of the bacterium within alveolar epithelial cells and freshly recruited monocytes may continue, and this would contribute significantly to the development of Legionnaires' disease. Immunocytochemical identification of infected alveolar cells at several stages of Legionnaires' disease will answer many of these speculations.

12. DIFFERENT FATES OF *pmi* AND *mil* MUTANTS WITHIN MACROPHAGES AND ALVEOLAR EPITHELIAL CELLS

The majority of the *mil* and *pmi* mutants (91/121) that are defective in macrophages have been shown to exhibit a wild type phenotype within type I and II alveolar epithelial cells, and also in HeLa cells (Gao *et al.*, 1997a). These observations showed that the mutants have different fates within macrophages and alveolar epithelial cells. These observation may not be surprising. Although compared to *L. pneumophila*, *Salmonella* is localized in a different phagosomal compartment within macrophages, similar observations have been made about this organism. Many of the *S. typhimurium* genes that are essential for survival within macrophages are not required for infection of epithelial cells (Behlau and Miller, 1993; Alpuche-Aranda *et al.*, 1992). The *phoP/phoQ* is a two component regulon of *S. typhimurium* that is activated and is required for survival within macrophages (Vescovi *et al.*, 1996; Miller and Mekalanos, 1990; Miller *et al.*, 1989). However, this regulon is neither activated nor is it required for survival of the bacterium in epithelial cells (Alpuche-Aranda *et al.*, 1992; Miller *et al.*, 1992; Miller, 1991). In addition, while *S. typhimurium* phagosomes are co-localized with late endosomes or lysosomes within macrophages (Oh *et al.*, 1996), they are separated from the endocytic route within epithelial cells (Garcia-del Portillo and Finlay, 1995a). These observations indicate that 1) within the same tissue or organ, certain bacterial genetic loci may be required for survival within a certain host cell but not others (Behlau and Miller, 1993; Alpuche-Aranda *et al.*, 1992), and 2) the phagosomal microenvironment for the same pathogen is probably different in macrophages and epithelial cells, and thus, alterations in bacterial gene expression during bacterial adaptation to these microenvironments is probably different (Alpuche-Aranda *et al.*, 1992; Miller *et al.*, 1992; Miller, 1991).

The *pmi* and *mil* mutants with drastically different phenotypes in macrophages and epithelial cells provide wonderful genetic tools to start dissecting the roles of alveolar macrophages and epithelial cells in Legionnaires' disease. A/J mice have been inoculated with a mutant defective in both macrophages and alveolar epithelial cells (Mq^-, Epi^-) or with a mutant defective in macrophages but exhibited a wild type phenotype within alveolar epithelial cells (Mq^-, Epi^+) (Gao *et al.*, 1998; Gao *et al.*, 1997a). Both of these mutants are also defective in intracellular replication in A/J mice-derived macrophages. The Mq^-, Epi^- mutant was defective in intrapulmonary replication in A/J mice, and was gradually killed. In contrast, the Mq^-, Epi^+ was indistinguishable from the wild type strain in intrapulmonary replication. These observations provide the first genetic evidence that attenuation *in vivo* correlates with a defect in epithelial cells *in vitro*. These data showed that a defect in macrophages *in vitro* may not correlate with attenuation *in vivo*. Future characterization of the *pmi* and *mil* loci and their differential requirements for bacterial survival within different host cells will enable characterization of the molecular and biochemical bases of the requirement for the different loci within macrophages and epithelial cells. Importantly, these observations indicate a major role for epithelial cells in the intracellular infection during Legionnaires' disease.

13. ROLE OF IRON IN THE INTRACELLULAR INFECTION

Iron is an essential nutrient for all living organisms. *L. pneumophila* requires relatively high concentrations of iron for growth *in vitro*. However, within the intracellular environment, iron is either bound to transferrin, complexed with ferritin, or sequestered in the labile iron and heme pools. How *L. pneumophila* obtain iron during intracellular growth in the EMB phagosome is not known. Several lines of evidence indicate that iron is required for intracellular replication of *L. pneumophila*. First, peripheral blood monocytes treated with several iron chelators do not support intracellular bacterial replication, a condition reversed by the addition of ferric iron (Byrd and Horwitz, 1989). IFN-γ inhibits intracellular replication of *L. pneumophila* within monocytes by reducing the amounts of intracellular iron, and the inhibitory effect of this cytokine is reversed by supplementation of iron (Byrd and Horwitz, 1989).

Fur is a conserved protein that functions as a repressor of factors involved in iron uptake in several bacteria. A *fur* gene and Fur-regulated genes have been described in *L. pneumophila* (Hickey and Cianciotto, 1997). One of these genes is a homologue of the aerobactin synthetase, raising the possibility that *L. pneumophila* utilizes siderophores to acquire iron (Hickey and Cianciotto, 1997). Importantly, a mutant defective in expression of the aerobactin synthetase is defective in intracellular replication within macrophages. In addition, mutants defective in iron acquisition and assimilation have been also isolated through transposon mutagenesis, and many of the mutants are defective in intracellular replication within macrophages and protozoa, but the functions of the defective genes are not yet known (Pope *et al.*, 1996). Further characterization of the iron uptake and assimilation systems in *L. pneumophila* will yield interesting information about how these systems are utilized within the phagosome.

14. GENE EXPRESSION BY INTRACELLULAR BACTERIA

Pathogenic bacteria such as *L. pneumophila* and *S. typhimurium* respond and adapt to the various local environmental conditions they encounter by coordinate regulation of gene expression (Abu Kwaik, 1998; Susa *et al.*, 1996; Abu Kwaik *et al.*, 1993). Upon phagocytosis and during intracellular survival, facultative intracellular pathogens are exposed to a complex mixture of stimuli, and they respond to these stimuli by a profound alteration in gene expression (Fernandez *et al.*, 1996; Abu Kwaik *et al.*, 1993; Abshire and Neidhardt, 1993). These phenotypic modulations allow intracellular bacteria to survive and adapt to environmental conditions that may be encountered intracellularly including nutrient limitation, temperature fluctuations, oxidative injury, osmolarity, pH, and other stimuli. In contrast to intracellular pathogens that reside in phagosomes and manifest induction in expression of many genes that are also induced by *in vitro* stress stimuli, cytoplasmic pathogens do no exhibit this phenomenon. These observations are not surprising since in contrast to the phagosomal microenvironment, the cytoplasmic environment is not nutrient-limited and is not exposed to contents of endosomes that fuse to phagosomes. The phenotypic alteration by intracellular pathogens is controlled by multiple regulons, and is most probably a simultaneous response to a complex combination of stimuli. Characterization of the bacterial macrophage-induced (MI) genes and examination of the kinetics of their expression during the intracellular infection will facilitate characterization of the phagosomal microenvironment that the organisms are exposed to within the host cell.

Intracellular *L. pneumophila* manifest a dramatic phenotypic modulation in gene expression in response to the intracellular environment (Abu Kwaik, 1998; Fernandez *et al.*, 1996; Susa *et al.*, 1996; Abu Kwaik *et al.*, 1993). These alterations are manifested through the induction or repression of expression of 35 and 32 genes, respectively. Many of the MI genes are also induced in response to one or more *in vitro* stress stimuli, which indicates that intracellular *L. pneumophila* is exposed to stress stimuli *in vivo* (Fernandez *et al.*, 1996; Abu Kwaik *et al.*, 1993). In order to examine the molecular aspects of the MI genes, many strategies have been utilized to clone the MI genes, including reverse genetics (Abu Kwaik, 1998; Abu Kwaik and Engleberg, 1994), antigenic detection by antisera (Susa *et al.*, 1996), and most recently by differential display PCR (Abu Kwaik and Pederson, 1996). A recent strategy of selective radiolabeling of proteins of intracellular bacteria has been developed for *S. typhimurium*. This strategy is based on the use of radioactive diaminopimelic acid (DAP), which is a major component of peptidoglycan, and is also a precursor for lysine. Bacterial auxotrophs for DAP are used to infect the host cell in presence of radioactive DAP. DAP is decarboxylated into lysine by the bacteria, which subsequently radiolabels the bacterial proteins selectively, since DAP can not be utilized by mammalian cells. In contrast to the successful use of DAP auxotrophs to selectively radiolabel proteins of intracellular *S. typhimurium*, DAP auxotroph of *L. pneumophila* can't transport DAP into the *L. pneumophila* EMB phagosome, and thus, this strategy can not be used in *L. pneumophila* (Harb and Abu Kwaik, 1998). These observations indicate differences in the permeability of the EMB phagosome of *L. pneumophila* and the phagosome occupied by *S. typhimurium*.

One of the MI genes that has been cloned by reverse genetics is the global stress gene (*gspA*) of *L. pneumophila* that is induced in response to *in vitro* stress stimuli and is also induced throughout the intracellular infection period (Abu Kwaik *et al.*, 1997; Abu Kwaik and Engleberg, 1994). Transcription of *gspA* is regulated by two promoters, one of which is a σ^{32}-regulated promoter. Intracellular *L. pneumophila* manifest differential expression of *gspA* by the σ^{32}-regulated promoter throughout the intracellular infection, which indicates continuous exposure of the bacteria to stress stimuli throughout the intracellular infection (Abu Kwaik *et al.*, 1997). A mutation in *gspA* has no effect on intracellular survival of *L. pneumophila* within mammalian macrophages and protozoan cells (Abu Kwaik *et al.*, 1997). However, the mutant exhibits a dramatic increase in susceptibility to stress stimuli *in vitro*. Interestingly, intracellular wild type strain derived from macrophages or from *H. vermiformis* exhibit a dramatic increase in their resistance to *in vitro* stress stimuli (Abu Kwaik *et al.*, 1997). The intracellular *gspA* mutant is similar to the wild type strain in being equally resistant to *in vitro* stress stimuli (Abu Kwaik *et al.*, 1997). These observations suggest that expression of other stress-induced genes during the intracellular infection may compensate for the loss of *gspA*.

The inorganic pyrophosphatase gene of *L. pneumophila* has been also shown to be induced within macrophages (Abu Kwaik, 1998). This enzyme is required for macromolecular biosynthesis, and thus, its induction within the host cell is consistent with the faster replication *in vivo* compared to that in rich medium *in vitro* (Abu Kwaik, 1998). In addition, some of the virulence genes of *L. pneumophila* may be specifically expressed or their expression is induced within the EMB phagosome. An example of these genes is the early macrophage induced locus (*eml*), the function of which is not known (Abu Kwaik and Pederson, 1996).

Flagellar expression by *L. pneumophila* is induced during late stages of the infection, and the bacteria become motile. The environmental signals that induce the expression of flagella during a specific phase of intracellular growth are not known. Many of the *dot*, *icm*, *pmi*, and *mil* mutants fail to express flagella, but the flagella are unlikely to contribute to the intracellular defect of these mutants (Gao *et al.*, 1997b). Some mutants that do not express flagella are not attenuated for intracellular survival and replication within macrophages or amoebae (Pruckler *et al.*, 1995). Thus, flagella by themsleves do not contribute to the intracellular infection.

During intracellular logarithmic growth expression of many virulence-related phenotypes of *L. pneumophila* is repressed and is subsequently induced during post-exponential growth (Byrne and Swanson, 1998). Only upon entry to post-exponential growth but not during intracellular logarithmic growth does *L. pneumophila* become sodium sensitive, osmotically resistant, flagellated, cytotoxic, infectious, and capable of evading phagolysosomal fusion (Byrne and Swanson, 1998). Expression of the above virulence-related phenotypes is a response to starvation since it can be induced upon transfer of bacteria from logarithmic growth into a broth of stationary phase cultures, but not when the broth of stationary phase cultures is supplemented with amino acids (Byrne and Swanson, 1998). The response by *L. pneumophila* to amino acid starvation may be coordinated by a conserved mechanism. The *E. coli* stringent response is mediated by RelA, a (p)ppGpp synthetase. When log phase cultures of *L. pneumophila* were induced to express the *E. coli relA* gene, the cells stopped replicating and became cytotoxic, sodium-sensitive, and flagellated (B. Hammer and M. Swanson, personal communication). Thus, (p)ppGpp appears to signal expression of some virulence-related phenotypes by *L. pneumophila* upon starvation for amino acids. Therefore, when nutrients are available, *L. pneumophila* expresses replication-related genes maximally. Upon nutrient depletion the bacteria express effector molecules to lyse the spent host cell, to disperse in the environment, to survive environmental stresses, and to reestablish a replication niche protected from lysosomal degradation (Byrne and Swanson, 1998). It would be interesting to identify the effector molecules that allow the bacteria to exhibit this dramatic phenotypic transition.

15. KILLING OF THE HOST CELL

It is well documented that intracellular replication of *L. pneumophila* within macrophages is associated with cytopathogenicity or loss of viability of the cells, as measured by dyes that detect the metabolic activity of the cells. In addition, at high multiplicity of infection (MOI), the bacteria are cytotoxic to mammalian cells (Husmann and Johnson, 1994). The bacterial factors responsible for this effect on the host cells are not known. The mechanisms by which *L. pneumophila* kills the mammalian host cell are just starting to emerge, but nothing is known about the mechanisms by which *L. pneumophila* kills the protozoan host. Müller et al (Müller *et al.*, 1996) have shown that within 24–48 h post infection of HL-60 macrophage-like cells by *L. pneumophila* at a MOI of 10-100, the infected cells underwent apoptosis, or programmed cell death, as shown by chromatin condensation and fragmentation of the chromosomal DNA into 180–200 bp fragments.

Apoptosis is a strictly regulated suicide program within the dying cell exhibiting distinct morphological and biochemical features (Anderson, 1997), which include fragmentation of chromosomal DNA into multimers of internucleosomal linker segments of approximately 180–200 bp (Wyllie, 1980). Activation of a family of cysteine proteases (caspases) that specifically cleave proteins after aspartate (Asp) residues is required for DNA fragmentation to occur (Salvesen and Dixit, 1998). A cascade mechanism for transmission of diverse apoptotic signals into a common apoptotic effector pathway by a network of caspases has been well-demonstrated, which subsequently leads to entry of the caspase-activated deoxyribonuclease (CAD) into the nucleus and degradation of chromosomal DNA (Sakahira *et al.*, 1998). A number of intracellular pathogens manipulate host cell apoptotic pathways. Several viruses, such as cowpox virus (Ray *et al.*, 1992) and baculovirus (Bertin *et al.*, 1996) and the obligate intracellular bacterium *Chlamydia trachomatis* (Fan *et al.*, 1998) inhibit host cell apoptosis, which allows these organisms to grow and persist intracellularly. The facultative intracellular pathogenic bacterium *Shigella flexneri* induces apoptosis in macrophages (Zychlinsky and Sansonetti, 1997). A single bacterial factor, the IpaB protein secreted by *S. flexneri,* is sufficient to induce apoptosis in macrophages. Secreted by *S. flexneri* upon bacterial escape from the phagosome into the cytoplasm, IpaB binds directly to caspase-1/ICE (Zychlinsky and Sansonetti, 1997). Activated ICE cleaves the precursor IL-1β into mature IL-1 to be released by infected macrophages, which promotes inflammation (Zychlinsky and Sansonetti, 1997). Although the biochemical mechanisms by which ICE activation promotes apoptosis in *Shigella*-infected macrophages are not known, these findings have established involvement of the activity of caspase-1/ICE in apoptosis induced by an intracellular bacterial pathogen. The mechanisms by which other intracellular bacterial pathogens induce apoptosis are not known. It is possible that *L. pneumophila* induces apoptosis through the direct activation of a specific caspase by a bacterial product. Alternatively, the host cell may undergo apoptosis as a result of a physical injury mediated by a bacterial product (such as a toxin) or due to the bacterial burden. Whether *L. pneumophila* induces apoptosis earlier than 24 h in HL-60 macrophages is not known (Müller *et al.*, 1996). It is also not known whether intracellular *L. pneumophila* are required to induce apoptosis or if extracellular bacteria are also capable of this induction. In addition, the relationship between induction of apoptosis and the loss of cell viability (cytopathogenicity) is still to be determined. Potential role of a specific caspase in *L. pneumophila*-mediated apoptosis will facilitate studies to elucidate the relationship between cytopathogenicity and apoptosis through the use of specific inhibitors that block the involved caspase and examine whether this blockage reduces or abolishes cytopathogenicity.

Kirby *et al.* (1998) have recently shown that *L. pneumophila* induces the formation of a pore in bone marrow-derived macrophages from A/J mice when the cells are infected at a MOI of 500, which also results in necrosis of the cells within 20–60 minutes. Apoptosis of the cells was not detected during this period under these conditions, but the investigators used one criteria (transmission electron microscopy) to detect signs of apoptosis. However, in many systems of apoptosis the apoptotic effect may not be detected by one criteria. In addition, the execution of the apoptotic pathway and its subsequent manifestation in the terminal stages of fragmentation and condensation of DNA may require more than 60 min. It is also possible that the response of A/J mice macrophages to infection by *L. pneumophila* is different from that of human-derived macrophages. Nevertheless, based on the observations made by Müller *et al.* (1996) and Kirby *et al.* (1998), it was proposed that a bi-phasic mode of killing of the host cell is mediated by *L. pneumophila* (Kirby *et al.*, 1998). These investigators proposed a rapid necrotic cell death during early stages of high MOI, and a later second phase of apoptotic cell death (Kirby *et al.*, 1998). This is a highly unlikely reflection of the natural infection in which the infectious dose has been repeatedly shown to be very low (O'Brein and Bhopal, 1993). In addition, the high MOI is most likely to be obtained later in the infection after release of intracellular bacteria subsequent to their replication, and their necrotic effect may be subsequently manifested at neighboring cells. Moreover, ATP is an absolute requirement for the induction of apoptosis (Leist *et al.*, 1997), and therefore, necrotic cells within the first 1 h of infection are highly unlikely to be able to provide the required energy for the execution of apoptosis at 24–48 h post-infection. The cells that undergo necrosis within 20–60 min of the infection may not be able to support intracellular bacterial replication for 24–48 h. It is most likely, therefore, that some necrosis occurs during late stages of the natural infection.

Why does *L. pneumophila* induce apoptosis in macrophages, which support intracellular replication of the bacteria? First, it is possible that *L. pneumophila*-induced apoptosis during early stages of the infection and that is associated with alteration of the host cell endocytic pathway. Second, *L. pneumophila* has been shown to suppress the oxidative burst in monocytes (Jacob *et al.*, 1994), and therefore, induction of apoptosis may enable *L. pneumophila* to down-regulate the bactericidal activity of macrophages. This may enable the bacteria to survive, in what is otherwise considered, a harsh environment for microorganisms. Third, cell death by apoptosis may reduce inflammation at the infection foci, which would allow further proliferation of the bacteria at the sites of infection. It is possible that apoptotic macrophages containing *L. pneumophila* are taken up by other macrophages. It would be intriguing to examine whether this process results in elimination of the bacteria through lysosomal degradation of the engulfed infected macrophage or that the bacteria are able to escape this potentially fatal fate and replicate in the new host cell.

16. CONCLUDING REMARKS

The ability of *L. pneumophila* to survive and replicate within the EMB phagosome in two evolutionarily distant hosts (mammals and protozoa) is quite remarkable. It will be intriguing to characterize the *mil* loci and their potential genetic transfer into *L. pneumophila* as a mode of pathogenic evolution of a protozoan parasite that may have not simply become a human pathogen due to generation of aerosols.

Further characterization of the function of the *pmi*, *mil*, *dot*, and *icm* loci in the intracellular infection and their role in alteration of endocytic trafficking of the bacteria will also help microbiologists to understand manipulation of host cell processes by a proficient intracellular pathogen. Understanding the functions of these loci and their roles in blocking maturation of the *L. pneumophila* EMB phagosome through the endosomal-lysosomal degradation pathway will allow both microbiologists and cell biologists to exploit them as tools to study endocytic trafficking and vesicular fusion. Thus, besides its dark side of being a human pathogen, the bright side of *L. pneumophila* is that it will contribute dramatically to our understanding of the cell biology of endocytic trafficking and vesicular fusion.

17. REFERENCES

Abshire, K.Z., and Neidhardt, F.C., 1993, Analysis of proteins synthesized by *Salmonella typhimurium* during growth within host macrophage, *J. Bacteriol.* **175**:3734–3743.

Abu Kwaik, Y., Eisenstein, B.I., and Engleberg, N.C., 1993, Phenotypic modulation by *Legionella pneumophila* upon infection of macrophages, *Infect. Immun.* **61**:1320–1329.

Abu Kwaik, Y., and Engleberg, N.C., 1994, Cloning and molecular characterization of a *Legionella pneumophila* gene induced by intracellular infection and by various *in vitro* stress stimuli, *Mol. Microbiol.* **13**:243–251.

Abu Kwaik, Y., Fields, B.S., and Engleberg, N.C., 1994, Protein expression by the protozoan *Hartmannella vermiformis* upon contact with its bacterial parasite *Legionella pneumophila*, *Infect. Immun.* **62**:1860–1866.

Abu Kwaik, Y., 1996, The phagosome containing *Legionella pneumophila* within the protozoan *Hartmanella vermiformis* is surrounded by the rough endoplasmic reticulum, *Appl. Environ. Microbiol.* **62**:2022–2028.

Abu Kwaik, Y., and Pederson, L.L., 1996, The use of differential display-PCR to isolate and characterize a *Legionella pneumophila* locus induced during the intracellular infection of macrophages, *Mol. Microbiol.* **21**:543–556.

Abu Kwaik, Y., Gao, L-Y., Harb, O.S., and Stone, B.J., 1997, Transcriptional regulation of the macrophage-induced gene (*gspA*) of *Legionella pneumophila* and phenotypic characterization of a null mutant, *Mol. Microbiol.* **24**:629–642.

Abu Kwaik, Y., 1998, Induced expression of the *Legionella pneumophila* gene encoding a 20-kilodalton protein during intracellular infection, *Infect. Immun.* **66**:203–212.

Abu Kwaik, Y., Gao, L-Y., Ticac, B., Elhage, N., and Susa, M. 1998a, Invasion and intracellular replication of *Legionella micdadei* in a ribosome-free phagosome within human-derived macrophages and alveolar epithelial cells, and within protozoa, Unpublished data.

Abu Kwaik, Y., Venkataraman, C., Gao, L-Y., and Harb, O.S., 1998b, Signal transduction in the protozoan host *Hartmannella vermiformis* upon attachment and invasion by its bacterial parasite, the Legionnaires' disease agent, *Legionella micdadei*, *Appl. Environ. Microbiol.* In Press.

Adams, S.A., Robson, S.C., Gathiram, V., et al, 1993, Immunological similarity between the 170 kDa amoebic adherence glycoprotein and human β2 integrins, *Lancet* **341**:17–19.

Adeleke, A., Pruckler, J., Benson, R., Rowbotham, T., Halablab, M., and Fields, B.S., 1996, *Legionella*-like amoebal pathogens-phylogenetic status and possible role in respiratory disease, *Emerg. Infect. Dis.* **2**:225–229.

Adler, P., Wood, S.J., Lee, Y.C., Lee, R.T., Petri, W.A. Jr., and Schnaar, R.L., 1995, High affinity binding of the *Entamoeba histolytica* lectin to polyvalent *N*-Acetylgalactosaminides, *J. Biol. Chem.* **270**:5164–5171.

Alpuche-Aranda, C.M., Swanson, J.A., Loomis, W.P., and Miller, S.I., 1992, *Salmonella typhimurium* activates virulence gene transcription within acidified macrophage phagosomes, *Proc. Natl. Acad. Sci. USA*. **89**:10079–10083.

Anderson, P., 1997, Kinase cascades regulating entry into apoptosis, *Microbiol. Mol. Biol. Rev.* **61**:33–46.

Barker, J., Brown, M.R.W., Collier, P.J.; Farrell, I., and Gilbert, P., 1992, Relationships between *Legionella pneumophila* and *Acanthamoebae polyphaga*: physiological status and susceptibility to chemical inactivation, *Appl. Environ. Microbiol.* **58**:2420–2425.

Barker, J., Lambert, P.A., and Brown, M.R.W., 1993, Influence of intra-amoebic and other growth conditions on the surface properties of *Legionella pneumophila*, *Infect. Immun.* **61**:3503–3510.

Barker, J., Scaife, H., and Brown, M.R.W., 1995, Intraphagocytic growth induces an antibiotic-resistant phenotype of *Legionella pneumophila*, *Antimicrob. Agents Chemother.* **39**:2684–2688.

Behlau, I., and Miller, S.I., 1993, A PhoP-repressed gene promotes *Salmonella typhimurium* invasion of Epithelial cells, *J. Bacteriol.* **175**:4475–4484.

Berk, S.G., Ting, R.S., Turner, G.W., and Ashburn, R.J., 1998, Production of respirable vesicles containing live *Legionella pneumophila* cells by two Acanthamoeba spp, *Appl. Environ. Microbiol.* **64**:279–286.

Bertin, J., Mendrysa, S.M., LaCount, D.J., et al, 1996, Apoptotic suppression by baculovirus P35 involves cleavage by and inhibition of a virus-induced CED-3/ICE-like protease, *J. Virol.* **70**:6251–6259.

Bozue, J.A., and Johnson, W., 1996, Interaction of *Legionella pneumophila* with *Acanthamoeba catellanii*: uptake by coiling phagocytosis and inhibition of phagosome-lysosome fusion, *Infect. Immun.* **64**:668–673.

Brieland, J., Freeman, P., kunkel, R., et al, 1994, Replicative *Legionella pneumophila* lung infection in intratracheally inoculated A/J mice: A murine model of human Legionnaires' disease, *Am. J. Pathol.* **145**:1537–1546.

Brieland, J.K., Fantone, J.C., Remick, D.G., LeGendre, M., McClain, M., and Engleberg, N.C., 1997, The role of *Legionella pneumophila*-infected *Hartmanella vermiformis* as an

infectious particle in a murine model of Legionnaires' disease, *Infect. Immun.* **65**:4892–4896.
Britles, R.J., Rowbotham, T.J., Raoult, D., and Harrison, T.G., 1996, Phylogenetic diversity of intra-amoebal legionellae as revealed by 16S rRNA gene sequence comparison, *Microbiol.* **142**:3525–3530.
Byrd, T.F., and Horwitz, M.A., 1989, Interferon gamma-activated human monocytes downregulate transferrin receptors and inhibit the intracellular multiplication of *Legionella pneumophila* by limiting the availability of iron, *J. Clin. Invest.* **83**:1457–1465.
Byrne, B., and Swanson, M.S., 1998, Expression of *Legionella pneumophila* virulence traits in response to growth conditions, *Infect. Immun.* (In Press)
Cianciotto, N.P., Eisenstein, B.I., Mody, C.H., Toews, G.B., and Engleberg, N.C., 1989, A *Legionella pneumophila* gene encoding a species-specific surface protein potentiates initiation of intracellular infection, *Infect. Immun.* **57**:1255–1262.
Cianciotto, N.P., and Fields, B.S., 1992, *Legionella pneumophila mip* gene potentiates intracellular infection of protozoa and human macrophages, *Proc. Natl. Acad. Sci. USA.* **89**:5188–5191.
Cirillo, J.D., Tompkins, L.S., and Falkow, S., 1994, Growth of *Legionella pneumophila* in *Acanthamoeba castellanii* enhances invasion, *Infect. Immun.* **62**:3254–3261.
Dowling, J.N., Saha, A.K., and Glew, R.H., 1992, Virulence factors of the family *Legionellaceae*, *Microbiol. Rev.* **56**:32–60.
Elliott, J.A., and Winn, W.C. Jr., 1986, Treatment of alveolar macrophages with cytochalasin D inhibits uptake and subsequent growth of *Legionella pneumophila*, *Infect. Immun.* **51**:31–36.
Fan, T., Lu, H., Shi, L., et al, 1998, Inhibition of apoptosis in Chlamydia-infected cells: Blockade of mitochondrial cytochrome c release and caspase activation, *J. Exp. Med.* **187**: 487–496.
Fernandez, R.C., Logan, S., Lee, S.H.S., and Hoffman, P.S., 1996, Elevated levels of *Legionella pneumophila* stress protein Hsp60 early in infection of human monocytes and L929 cells correlated with virulence, *Infect. Immun.* **64**:1968–1976.
Fields, B.S., Nerad, T.A., Sawyer, T.K., et al, 1990, Characterization of an axenic strain of *Hartmannella vermiformis* obtained from an investigation of nosocomial legionellosis, *J. Protozool.* **37**:581–583.
Fields, B.S., 1993, Legionella and protozoa: interaction of a pathogen and its natural host, in *Legionella; current status and emerging perspectives*, Washington, D.C.: American Society of Microbiology, pp. 129–136.
Fields, B.S., 1996, The molecular ecology of legionellae, *Trends. Microbiol* **4**:286–290.
Fraser, D.W., Tsai, T.R., Orenstein, W., et al, 1977, Legionnaires' disease: description of an epidemic of pneumonia, *N. Engl. J. Med.* **297**:1189–1197.
Gao, L-Y., Gutzman, M., Brieland, J.K., and Abu Kwaik, Y., 1997a, Unupublished data.
Gao, L-Y., Harb, O.S., and Abu Kwaik, Y., 1997b, Utilization of similar mechanisms by *Legionella pneumophila* to parasitize two evolutionarily distant hosts, mammalian and protozoan cells, *Infect. Immun.* **65**:4738–4746.
Gao, L-Y., Harb, O.S., and Abu Kwaik, Y., 1998, Identification of macrophage-specific infectivity loci (*mil*) of *Legionella pneumophila* that are not required for infectivity of protozoa, *Infect. Immun.* **66**:883–892.
Garcia-del Portillo, F., and Finlay, B., 1995a, Targeting of *Salmonella typhimurium* to vesicles containing lysosomal membrane glycoproteins bypasses compartments with mannose-6-phosphate receptors, *J. Cell. Biol.* **129**:81–97.
Garcia-del Portillo, F., and Finlay, B.B., 1995b, The varied life styles of intracellular pathogens within eukaryotic vacuolar compartments, *Trends. Microbiol* **3**:373–380.

Harb, O.S., and Abu Kwaik, Y., 1998, Identification of the aspartate-β-semiadehyde dehydrogenase gene of *Legionella pneumophila* and characterization of a null mutant, *Infect. Immun.* **66**:1898–1903.

Harb, O.S., Venkataraman, C., Haack, B.J., Gao, L-Y., and Abu Kwaik, Y., 1998, Heterogeniety in the attachment and uptake mechanisms of the Legionnaires' disease bacterium, *Legionella pneumophila*, by protozoan hosts, *Appl. Environ. Microbiol.* **64**:126–132.

Hickey, E.K., and Cianciotto, N.P., 1997, An iron- and Fur-repressed *Legionella pneumophila* gene that promotes intracellular infection and encodes a protein with similarity to the *Escherichia coli* aerobactin synthetase, *Infect. Immun.* **65**:133–143.

Horwitz, M.A., 1983, Formation of a novel phagosome by the Legionnaires' disease bacterium (*Legionella pneumophila*) in human monocytes, *J. Exp. Med.* **158**:1319–1331.

Horwitz, M.A., 1984, Phagocytosis of the Legionnaires' disease bacterium (*Legionella pneumophila*) occurs by a novel mechanism: engulfment within a pseudopod coil, *Cell* **36**: 27–33.

Husmann, L.K., and Johnson, W., 1994, Cytotoxicity of extracellular *Legionella pneumophila*, *Infect. Immun.* **62**:2111–2114.

Jacob, T., Escallier, J.C., Sanguedolce, M.V., et al, 1994, *Legionella pneumophila* inhibits superoxide generation in human monocytes via the down-modulation of a and b protein kinase C isotypes, *J. Leukoc. Biol.* **55**:310–312.

King, C.H., Fields, B.S., Shotts, E.B.,Jr., and White, E.H., 1991, Effects of cytochalasin D and methylamine on intracellular growth of *Legionella pneumophila* in amoebae and human monocyte-like cells, *Infect. Immun.* **59**:758–763.

Kirby, J.E., Vogel, J.P., Andrews, H.L., and Isberg, R.R., 1998, Evidence for pore-forming ability by *Legionella pneumophila*, *Mol. Microbiol.* **27**:323–336.

Knirel, Y.A., Moll, H., and Zahringer, U., 1996, Structural study of a highly *O*-acetylated core of *Legionella pneumophila* serogroup 1 lipopolysaccharide, *Carbohydr. Res.* **293**:223–234.

Leist, M., Single, B., Castoldi, A.F., Kuhnle, S., and Nicotera, P., 1997, Intracellular adenosine triphosphate (ATP) concentration: A switch in the decision between apoptosis and necrosis, *J. Exp. Med.* **185**:1481–1486.

Mann, B.J., Torian, B.E., Vedvick, T.S., and Petri, W.A. Jr., 1991, Sequence of a cysteine-rich galactose-specific lectin of *Entamoeba histolytica*, *Proc. Natl. Acad. Sci. USA.* **88**:3248–3252.

McDade, J.E., Shepard, C.C., Fraser, D.W., Tsai, T.R., Redus, M.A., and Dowdle, W.R., 1977, Legionnaires' disease: isolation of a bacterium and demonstration of its role in other respiratory disease, *N. Engl. J. Med.* **297**:1197–1203.

McDade, J.E., and Shepard, C.C., 1979, Virulent to avirulent conversion of Legionnaires' disease bacterium (*Legionella pneumophila*) its effect on isolation techniques, *J. Infect. Dis.* **139**:707–711.

Miller, S.I., Kukral, A.M., and Mekalanos, J.J., 1989, A two-component regulatory system (*phoP phoQ*) controls *Salmonella typhimurium* virulence, *Proc. Natl. Acad. Sci. USA.* **86**: 5054–5058.

Miller, S.I., and Mekalanos, J.J., 1990, Constitutive expression of PhoP regulon attenuates *Salmonella* virulence and survival within macrophages, *J. Bacteriol.* **172**:2485–2490.

Miller, S.I., 1991, PhoP/PhoQ: Macrophage-specific modulators of *Salmonella* virulence? *Mol. Microbiol.* **5**:2073–2078.

Miller, V.L., Beer, K.B., Loomis, W.P., Olson, J.A., and Miller, S.I., 1992, An unusual pagC::Tn*phoA* mutations leads to an invasion- and virulence-defective phenotype in salmonellae, *Infect. Immun.* **60**:3763–3770.

Müller, A., Hacker, J., and Brand, B.C., 1996, Evidence for apoptosis of human macrophage-like HL-60 cells by *Legionella pneumophila* infection, *Infect. Immun.* **64**:4900–4906.

Nash, T.W., Libby, D.M., and Horwitz, M.A., 1984, Interaction between the legionnaires' disease bacterium (*Legionella pneumophila*) and human alveolar macrophages. Influence of antibody, lymphokines, and hydrocortisone, *J. Clin. Invest.* **74**:771–782.

O'Brein, S.J., and Bhopal, R.S., 1993, Legionnaires' disease: the infective dose paradox, *Lancet* **342**:5–6.

Oh, Y-K., Alpuche-Aranda, C., Berthiaume, E., Jinks, T., Miller, S.I., and Swanson, J.A., 1996, Rapid and complete fusion of macrophages lysosomes with phagosomes containing *Salmonella typhimurium*, *Infect. Immun.* **64**:3877–3883.

Payne, N.R., and Horwitz, M.A., 1987, Phagocytosis of *Legionella pneumophila* is mediated by human monocyte complement receptors, *J. Exp. Med.* **166**:1377–1389.

Petri, W.A. Jr., Smith, R.D., Schlesinger, P.H., Murphy, C.F., and Ravdin, J.I., 1987, Isolation of the galactose-binding lectin that mediates the *in vitro* adherence of *Entamoeba histolytica*, *J. Clin. Invest.* **80**:1238–1244.

Pope, C.D., O'connell, W.A., and Cianciotto, N.P., 1996, *Legionella pneumophila* mutants that are defective for iron acquisition and assimilation and intracellular infection, *Infect. Immun.* **64**:629–636.

Pruckler, J.M., Benson, R.F., Moyenuddin, M., Martin, W.T., and Fields, B.S, 1995, Association of flagellum expression and intracellular growth of *Legionella pneumophila*, *Infect. Immun.* **63**:4928–4932.

Ray, C.A., Black, R.A., Kronheim, et al, 1992, Viral inhibition of inflammation: cowpox virus encodes an inhibitor of the interleukin-1 beta converting enzyme, *Cell* **69**:597–604.

Rechnitzer, C., and Blom, J., 1989, Engulfment of the Philadelphia strain of *Legionella pneumophila* within pseudopod coils in human phagocytes. Comparison with the other *Legionella* strains and species, *Acta Pathol. Microbiol. Immunol. Scand. [B]* **97**:105–114.

Rodgers, F.G., and Gibson III, F.C., 1993, Opsonin-independent adherence and intracellular development of *Legionella pneumophila* within U-937 cells, *Can. J. Microbiol.* **39**:718–722.

Rowbotham, T.J., 1980, Preliminary report on the pathogenicity of *Legionella pneumophila* for freshwater and soil amoebae, *J. Clin. Pathol.* **33**:1179–1183.

Roy, C.R., Berger, K.H., and Isberg, R.R., 1998, *Legionella pneumophila* DotA protein is required for early phagosome trafficking decisions that occur within minutes of bacterial uptake, *Mol. Microbiol.* **28**:663–674.

Sadosky, A.B., Wiater, L.A., and Shuman, H.A., 1993, Identification of *Legionella pneumophila* genes required for growth within and killing of human macrophages, *Infect. Immun.* **61**: 5361–5373.

Sakahira, H., Enari, M., and Nagata, S., 1998, Cleavage of CAD inhibitor in CAD activation and DNA degradation during apoptosis, *Nature* **391**:96–99.

Salvesen, G.S., and Dixit, V.M., 1998, Caspases: Intracellular signaling by proteolysis, *Cell* **91**:443–446.

Steinert, M., Emody, L., Amann, R., and Hacker, J., 1997, Resuscitation of viable but nonculturable *Legionella pneumophila* Philadelphia JR32 by *Acanthamoeba castellanii*, *Appl. Environ. Microbiol.* **63**:2047–2053.

Stone, B.J., and Abu Kwaik, Y., 1998a, Expression of multiple pili by *Legionella pneumophila*: Identification and characterization of a type IV pilin gene and its role in adherence to mammalian and protozoan cells , *Infect. Immun.* **66**:1768–1775.

Stone, B.J., and Abu Kwaik, Y., 1998b, Natural competency for DNA uptake by *Legionella pneumophila* and its association with expression of type IV pili, Unupublished data.

Susa, M., Hacker, J., and Marre, R., 1996, *De novo* synthesis of *Legionella pneumophila* antigens during intracellular growth in phagocytic cells, *Infect. Immun.* **64**:1679–1684.

Susa, M., Ticac, T., Rukavina, T., Doric, M., and Marre, R., 1998, *Legionella pneumophila* infection in intratracheally inoculated T cell depleted or non-depleted A/J mice, *J. Immunol.* **160**:316–321.

Swanson, M.S., and Isberg, R.R., 1995, Formation of the *Legionella pneumophila* replicative phagosome, *Infect. Agents. Dis* **2**:269–271.

Swanson, M.S., and Isberg, R.R., 1996, Identification of *Legionella pneumophila* mutants that have abarrant intracellular fates, *Infect. Immun.* **64**:2585–2594.

Theriot, J.A., 1995, The cell biology of infection by intracellular bacterial pathogens, *Annu. Rev. Cell. Dev. Biol.* **11**:213–239.

Venkataraman, C., Gao, L-Y., Bondada, S., and Abu Kwaik, Y., 1998, Identification of putative cytoskeletal protein homologues in the protozoan *Hartmannella vermiformis* as substrates for induced tyrosine phosphatase activity upon attachment to the Legionnaires' disease bacterium, *Legioenlla pneumophila*, *J. Exp. Med.* (In Press)

Venkataraman, C., Haack, B.J., Bondada, S., and Abu Kwaik, Y., 1997, Identification of a Gal/GalNAc lectin in the protozoan *Hartmannella vermiformis* as a potential receptor for attachment and invasion by the Legionnaires' disease bacterium, *Legionella pneumophila*, *J. Exp. Med.* **186**:537–547.

Vescovi, E.C., Soncini, F.C., and Groisman, E.A., 1996, Mg^{2+} as an extracellular signal: Environmental regulation of Salmonella virulence, *Cell* **84**:165–174.

Vogel, J.P., Roy, C., and Isberg, R.R., 1996, Use of salt to isolate *Legionella pneumophila* mutants unable to replicate in macrophages, *Ann. NY. Acad. Sci.* **797**:271–272.

Vogel, J.P., Andrews, H.L., Wong, S.K., and Isberg, R.R., 1998, Conjugative transfer by the virulence system of *Legionella pneumophila*, *Science* **279**:873–876.

Weinbaum, D.L., Benner, R.R., Dowling, J.N., Alpern, A., Pasculle, A.W., and Donowitz, G.R., 1984, Interaction of *Legionella micdadei* with human monocytes, *Infect. Immun.* **46**:68–73.

Winn, W.C. Jr., and Myerowitz, R.L., 1981, The pathology of the Legionella pneumonias. A review of 74 cases and the literature, *Hum. Pathol.* **12**:401–422.

Wintermeyer, E., Ludwig, B., Steinert, M., Schmidt, B., Fischer, G., and Hacker, J., 1995, Influence of site specifically altered Mip proteins on intracellular survival of *Legionella pneumophila* in eukaryotic cells, *Infect. Immun.* **63**:4576–4583.

Wyllie, A.H., 1980, Glucocorticoid-induced thymocyte apoptosis is associated with endogeneous endonuclease activation, *Nature* **284**:555–556.

Zychlinsky, A., and Sansonetti, P.J., 1997, Apoptosis as a proinflammatory event: what we can learn from bacteria-induced cell death, *Trends. Microbiol* **5**:201–204.

Chapter 16

Internalization of *Listeria monocytogenes* by Nonprofessional and Professional Phagocytes

Michael Kuhn and Werner Goebel

1. INTRODUCTION

Listeria monocytogenes was originally described by Murray *et al.* (1926), who named it *Bacterium monocytogenes* due to the characteristic monocytosis found in the blood of experimentally infected animals. In 1940 it was renamed into its present name. *L. monocytogenes* is a gram-positive, rod-shaped, non-spore forming, facultative anaerobic bacillus which grows in a wide range of temperatures. *L. monocytogenes* is catalase positive, oxidase negative, CAMP positive, and expresses a hemolysin producing characteristic zones of lysis on blood agar plates. *L. monocytogenes* is motile when grown between 20 and 25°C, but the synthesis of flagellin is repressed at 37°C and the bacteria are then nonmotile. The cell wall of *L. monocytogenes* contains peptidoglycan, teichoic acid, and lipoteichoic acid and shows all the charateristics of a typical gram-positive cell wall. While the teichoic acids are covalently linked to the peptidoglycan, the lipoteichoic acids are

MICHAEL KUHN and WERNER GOEBEL Lehrstuhl für Mikrobiologie, Universität Würzburg, 97074 Würzburg, Germany; E-Mail: goebel@biozentrum.uni-wuerzburg.de

Subcellular Biochemistry, Volume 33: Bacterial Invasion into Eukaryotic Cells, edited by Oelschlaeger and Hacker. Kluwer Academic / Plenum Publishers, New York, 2000.

anchored to the cytoplasmic membrane via a glycolipid moiety. *L. monocytogenes* is wide spread in nature and is found on plants, in soil, and in freshwater, but also in silage, wastewater, and in human and animal feces (reviewed in Schuchat *et al.*, 1991).

L. monocytogenes causes serious infections in immunocompromised individuals, newborns, pregnant women, and older people. The major symptomatic manifestations of Listeria infections are stillbirth, septicemia, encephalitis, and meningoencephalitis. Most cases of human listeriosis appear sporadically and the source of infection remains unclear. The tracing of recent epidemics to food contaminated with *L. monocytogenes*, however, suggests that the natural route of infection is the gastrointestinal tract (reviewed in Farber and Peterkin, 1991).

The interaction of facultative intracellular bacteria with their mammalian host cells was studied extensively in recent years (reviewed in Finlay and Falkow, 1997). These studies, which used simple *in vitro* cell culture systems of infection together with the development of genetic tools, allowed the identification and analysis of the genetic and molecular determinants of bacterial pathogenesis. The molecular mechanisms of entry into professional and nonprofessional phagocytes was mainly studied in Shigella, Salmonella, Yersinia, and Listeria. Whereas Shigella and Salmonella use complex type III secretion systems to deliver proteins into the host cell which results in dramatic changes in cell membrane morphology (ruffling) accompanied by the invasion process, Yersinia and Listeria have developed simple one-protein mechanisms in which binding of the bacterium to a cellular receptor leads to invasion by a zipper mechanism without extensive membrane processes. Once inside the host cell intracellular bacteria either stay inside a membrane bound vacuole (Legionella, Salmonella, Mycobacteria) which can undergo substantial modification by the pathogen or escape into the host cell cytoplasm (Shigella, Listeria) where extensive replication takes place.

This review focuses on our present view of the entry process of *L. monocytogenes* into mammalian cells. The whole intracellular life cycle of *L. monocytogenes* was the topic of several recent reviews (Kuhn and Goebel, 1998, 1995; Sheehan *et al.*, 1994) and will only be discussed shortly in the next chapter.

2. THE INTRACELLULAR LIFE CYCLE OF *L. MONOCYTOGENES*

Most studies on the cell biology of *L. monocytogenes* infections used epithelia-like and macrophage-like cell lines (Mounier *et al.*, 1990; Tilney and Portnoy, 1989; Gaillard *et al.*, 1987). Macrophages actively ingest *L. monocytogenes*, but the internalization of the bacteria by normally nonphagocytic cells is triggered by *L. monocytogenes*-specific products. Besides the internalization step, the intracellular life cycle of the listeriae in phagocytes or normally nonphagocytic mammalian cells is very similar. The bacteria are first seen in a vacuole, which is subsequently lysed by most of the ingested bacteria allowing them to escape into the cytoplasm. There they start to replicate, whereas those which remain trapped in the phagosome are killed and digested. Concomitant with the onset of intracellular replication, *L. monocytogenes* induces the nucleation of host actin filaments which form a cloud around the bacterial cell. The actin filaments are then rearranged to a polar tail which consists of short actin filaments and other host actin binding proteins which stabilize this structure. The formation of the actin tail at one pole of the bacterial cell produces the propulsive force which moves the bacteria through the cytoplasm of the host cell. This bacterial movement requires continuous *de novo* actin polymerization. Bacteria which reach the surface of the infected host cell induce the formation of pseudopod-like structures with the bacterium at the tip and the actin tail behind. These pseudopods are taken up by neighbouring cells. The bacteria entering the neighbouring cells in that fashion are within a vacuole that is surrounded by a double membrane which is subsequently lysed to release the bacteria into the cytoplasm of the new host cell (Figure 1).

Most of the known virulence genes whose products are involved in the intracellular life cycle of *L. monocytogenes* are clustered on the chromosome in the so-called PrfA-dependent virulence gene cluster. The cluster

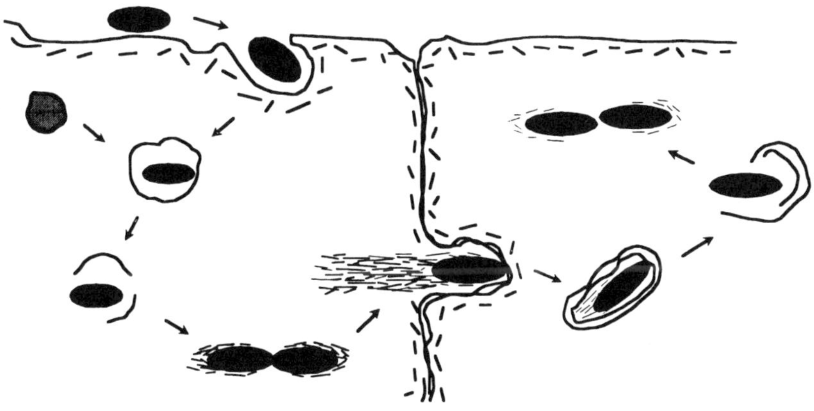

FIGURE 1. The intracellular life cycle of *L. monocytogenes*. See text for details. (Adapted from Brehm *et al.*, 1996 and Tilney and Portnoy, 1989 and kindly provided by J. Kreft).

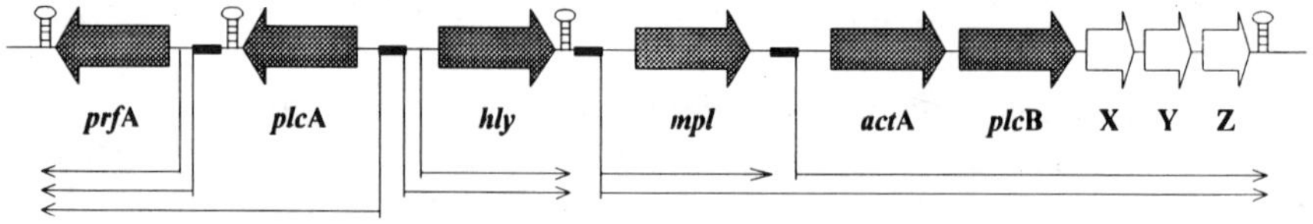

FIGURE 2. The virulence gene cluster from *L. monocytogenes*. Black boxes represent "PrfA-boxes", the arrows represent transcripts. (kindly provided by F. Engelbrecht).

comprises six well-characterized genes, *prfA*, *plcA*, *hly*, *mpl*, *actA*, and *plcB* (Figure 2) and three small ORFs of unknown functions downstream of *plcB*, called *orf* X, Y, and Z. The ends of the gene cluster are defined by genes coding for house-keeping enzymes. The products of these virulence genes are: listeriolysin (encoded by *hly*), a phosphatidylinositol-specific phospholipase C (*plcA*), a phosphatidylcholine-specific phospholipase C (*plcB*), a metalloprotease (*mpl*), ActA, a protein involved in actin polymerization (*actA*) and the positive regulatory factor PrfA (*prfA*).

3. LISTERIAL PROTEINS INVOLVED IN INVASION INTO NONPROFESSIONAL PHAGOCYTIC CELLS

With the finding that *L. monocytogenes* invades fibroblasts and epithelial cells *in vitro* the search began for the bacterial factors mediating the interactions with the mammalian cell which finally lead to invasion. The first listerial factor described to be involved in invasion of fibroblasts was the *L. monocytogenes* extracellular protein p60 (Kuhn and Goebel, 1989), followed by internalin which was shown to be needed primarily for epithelial cell invasion (Gaillard *et al.*, 1991) and InlB which is crucial for the invasion of a growing number of cell types including hepatocytes and endothelial cells (Parida *et al.*, 1998; Dramsi *et al.*, 1995). Recently, data were published showing that the listerial surface protein ActA might also be involved in invasion of mammalian cells (Alvarez-Dominguez *et al.*, 1997).

3.1. p60

Rough mutants of *L. monocytogenes* expressing reduced amounts of a 60 kDa extracellular protein, termed p60 show significantly reduced uptake by 3T6 fibroblast cells (Kuhn and Goebel, 1989). These "p60 mutants" (also referred to as R-mutants because of their rough colony appearance) form long cell chains which possess double septa between the individual cells. Treatment of *L. monocytogenes* R-mutants with partially purified p60 protein deaggregates the cell chains to normal-sized single bacteria which are again invasive for the fibroblasts. Ultrasonication which leads to physical disruption of the cell chains produces similar single cells which are, however, noninvasive. Upon treatment with wild-type p60, these ultrasonicated mutant cells are again able to invade the fibroblasts (Bubert *et al.*, 1992; Kuhn and Goebel, 1989). The reduced invasiveness of the p60 mutants is only observed with certain mammalian host cells. Cell chains of p60 mutants adhere normally to Caco-2 epithelial cells and are perfectly invasive upon disruption of the bacterial cell chains by ultrasonication without the addition of p60 (Bubert *et al.*, 1992). Protein p60 is a major secreted protein of all *L. monocytogenes* isolates (Bubert *et al.*, 1992; Kuhn and Goebel, 1989) but is also found on the cell surface of *L. monocytogenes* (Ruhland *et al.*, 1993). In contrast to other virulence factors p60 is also an essential enzyme for the cell metabolism of *L. monocytogenes* since it possesses murein hydrolase activity which appears to be involved in a late step of cell division (Wuenscher *et al.*, 1993). The gene coding for this obviously bifunctional protein, called *iap* (invasion associated protein), was cloned from a *L. monocytogenes* gene bank through a selection procedure using an anti-p60-antiserum and sequenced (Köhler *et al.*, 1990). The amino acid sequence of p60 predicts an extremely basic protein of 484 amino-acids with a 27 amino acid signal sequence, and an extended repeat domain consisting of 19 threonine-asparagine units which are separated by a proline-serine-lysine motif. A single cysteine found in the C-terminal part of p60 is probably essential for its enzymatic activity (Wuenscher *et al.*, 1993; Köhler *et al.*, 1990).

3.2. ActA

Analysis of the invasive capacity of an *inlA* deletion mutant and mutant PKP-1 without the virulence gene cluster genes (Engelbrecht *et al.*, 1996) complemented with multiple copies of PrfA strongly suggested that PrfA-dependent proteins from the virulence gene cluster may cause invasion of epithelial cells in the absence of InlA (Kuhn *et al.*, 1997). A recent study by Alvarez-Dominguez *et al.* (1997) identified the listerial surface protein ActA, a major virulence factor primarily involved in actin-based motility (Kocks *et al.*, 1992; Domann *et al.*, 1992) as contributing to internalin-independent uptake of *L. monocytogenes* by epithelial cells. ActA is a 639 amino acid surface protein containing a central proline-rich repeat region and a N-terminal region which are both essential for efficient host cell actin polymerization and intracellular movement (reviewed in Cossart and Lecuit, 1998). Whether these structures are also involved in mediating ActA-dependent invasion is not known. This ActA-dependent invasion was shown to involve interaction of ActA with host cell heparan sulfate proteoglycan receptors which are present on the cell surface of both professional and nonprofessional phagocytes. Removal of heparan sulfate from target cells with heparinase III treatment significantly impairs listerial adhesion and invasion.

3.3. The Internalins

Transposon mutagenesis using the conjugative transposon Tn*1545* and an appropriate *in vitro* invasion assay with Caco-2 epithelial cells resulted in the identification of the *inlAB* locus of *L. monocytogenes* coding for the proteins internalin (InlA) and InlB (Gaillard *et al.*, 1991). The three mutants originally identified exhibit a drastically reduced invasive capacity compared to the wild-type strain when tested on Caco-2, HT-29, HeLa, and HEp-2 cells. In addition to the reduced invasiveness, the mutants also have a lower ability to adhere to Caco-2 cells. The transposon insertions of all three mutants were mapped and shown to have occurred into a chromosomal region upstream of two open reading frames representing the *inlA* and *inlB* genes. Expression of *inlA* in *L. innocua*, a noninvasive *Listeria* species closely related to *L. monocytogenes*, renders this species invasive for Caco-2 cells showing that the *inlA* gene product is necessary and sufficient to mediate invasion of the Caco-2 epithelial cell line. Furthermore, it was also shown that a genetically engineered InlA-expressing *Enterococcus faecalis* strain as well as InlA-coated latex beads enter Caco-2 cells (Lecuit *et al.*, 1997). The deduced amino acid sequence of the *inlA* gene identified internalin as an acidic protein of 800 amino acids (Dramsi *et al.*, 1993b; Gaillard *et al.*, 1991) which possesses two extended repeat domains. Domain A consists of 15 leucine-rich repeats (LRR) of 22 amino acids each, identifying internalin as a member of the family of leucine-rich repeat proteins which were shown to be involved in protein-protein interactions (Kobe and Deisenhofer, 1995). Domain B consists of 3 repeats of about 70 amino acids each (Figure 3). Treatment with antibodies directed against the leucine-rich repeat region of internalin block entry of *L. monocytogenes* into Caco-2 cells, thereby underlining the importance of the repeat regions of internalin for its function as an invasin (Mengaud *et al.*, 1996a). A deletion analysis further showed that the LRR and the inter-repeat region of InlA are sufficeint to mediate entry, as long as the protein remains bound to the bacterial cell surface (Lecuit *et al.*, 1997). The InlA protein has a typical N-terminal transport signal sequence and in the C-terminal part a cell wall-spanning region followed by a hydrophobic sequence which represents a putative membrane anchor. Directly N-terminal to the hydrophobic sequence is a region encoding a LPTTGD hexapeptide which serves, upon cleavage, for covalent binding of internalin to the peptidoglycan of the cell wall. Internalin is therefore mainly a *L. monocytogenes* surface protein but small amounts of the protein are also found in the supernatant (Lebrun *et al.*, 1996; Gaillard *et al.*, 1991). The surface location of InlA is necessary to mediate entry of *L. monocytogenes* into epithelial cells by facilitating direct contact between the bacterium and the host cell. InlA is present on the cell surface in a polarized distribution which is also discussed for another listerial surface protein, ActA (Lebrun *et al.*, 1996).

InlB, a basic 630 amino acid protein also carries a N-terminal transport signal sequence, six leucine-rich repeats, a large inter-repeat region and a domain B with 3 repeats. The InlB protein is found mainly on the bacterial surface, but significant amounts are also secreted into the supernatant. InlB lacks the typical C-terminal cell wall anchor sequence present in InlA (Braun *et al.*, 1997; Gaillard *et al.*, 1991). Its surface association is mediated by the last 232 amino acids which form a repeat sequence also found in other gram-positive surface-associated proteins. Interestingly, surface association of InlB also occurs when purified InlB is added externally to a *L. monocytogenes* mutant. This finding and the relatively high amounts of the protein found in the supernatant suggest that the association of InlB with the listerial cell surface may occur after secretion (Braun *et al.*, 1997).

InlA and InlB were purified to homogenicity (Müller *et al.*, 1998; Mengaud *et al.*, 1996b) which allowed subsequent isolation of the mammalian receptor (as done for InlA) and the demonstration of the ability of the purified proteins to promote invasion. This was achieved for InlA by coating noninvasive *Entercoccus faecalis* and latex beads and demonstration of invasion of these particles (Lecuit *et al.*, 1997). Purified InlB was added externally to noninvasive *L. innocua* and associates with the surface and confers high level invasiveness for Vero and HeLa cells (Müller *et al.*, 1998). The availability of purified native and invasion-competent InlA and InlB should allow the analysis of the molecular basis of InlA- and InlB-mediated invasion into different cell lines as well as the characterization of internalin-receptor interactions at a molecular level.

The exact roles of either *inlA* or *inlB* in invasion of different cell lines *in vitro* were analyzed by constructing in frame deletion mutants in both genes and testing their invasive capacity in standard invasion assays. Additionally, double mutants lacking both internalin genes were complemented with either *inlA* or *inlB*. While it is unequivocally proven that *inlA* and *inlB* are needed for invasion of most mammalian cells tested, the separate role of each gene, especially for epithelial cell invasion, is still under debate. Lingnau and coworkers (1995) postulated that invasion of different epithelial cells (including Caco-2, HeLa, PtK2, HEp-2, Henle, and A549) is dependent on the presence of *inlA* and *inlB*. In contrast, Dramsi *et al.* (1995)

showed that Caco-2 invasion by *L. monocytogenes* is only dependent on *inlA*, as already suggested by Gaillard *et al.* (1991), and totally independent of *inlB*. Furthermore, these authors showed that invasion of HeLa, HEp-2, CHO, and Vero cells (all epithelia-like) is only dependent on *inlB*. A recent paper by Braun *et al.* (1998) further showed that InlB allone is sufficient to promote internalization of normally noninvasive bacteria or even latex beads by Vero, HEp-2, and HeLa cells, underlining the dominant role of InlB in invasion of these epithelial like cells. Whereas three fibroblast cell lines were found to be invaded albeit with low efficiency independently of *inlA* (S180, WI-38, and HEL 299) or InlB (WI-38 and HEL 299) (Mengaud *et al.*, 1996b; Lingnau *et al.*, 1995), hepatocyte invasion was clearly shown several times to be mainly promoted by InlB (Gaillard *et al.*, 1996; Gregory *et al.*, 1996; Dramsi *et al.*, 1995). The *inlB* gene product was found to mediate invasion of the murine TIB 73 hepatocyte-like cell line even in the absence of *inlA* (Dramsi *et al.*, 1995) and evidence for the role of *inlB* in the invasion of primary mouse hepatocytes was also presented (Gregory *et al.*, 1996). Invasion of the human hepatocyte-like cell line HepG-2 is, however, obviously promoted by a synergism of the proteins InlA and InlB (Dramsi *et al.*, 1995).

Recently, the role of *inlA* and *inlB* in endothelial cell invasion was investigated by several authors (Parida *et al.*, 1998; Greiffenberg *et al.*, 1997; Drevets *et al.*, 1995; Greiffenberg *et al.*, in revision). Drevets *et al.* (1995) speculated that *inlA* might play a role in invasion of human umbillical vein endothelial cells (HUVEC) due to the facts that an *inlAB* mutant showed reduced invasive capacity and that an anti-internalin antiserum inhibited binding of *L. monocytogenes* to HUVEC. However, the authors could not exclude the posibility that the anti-InlA antibodies used are cross-reactive with InlB. Parida *et al.* (1998) presented convincing evidence, that *inlB* plays a critical role in HUVEC invasion since a Δ*inlB* mutant invaded HUVEC 13 fold less than the wild type strain. However, in their experiments *inlA* contributed obviously to HUVEC invasion since a double Δ*inlAB* mutant was about 90 fold less invasive than the parental strain. Furthermore, the authors could demonstrate direct binding of purified InlB to HUVEC and showed that addition of InlB to a noninvasive Δ*inlAB* mutant, triggers invasion of this strain. In contrast, InlA alone neither bound to HUVEC nor did it promote invasion of the noninvasive mutant. The role of *inlA* and *inlB* in *L. monocytogenes* invasion of human brain microvascular endothelial cells (HBMEC) was also recently investigated (Greiffenberg *et al.*, in revision). In the absence of serum, *L. monocytogenes* efficiently invades HBMEC in an *inlB*-dependent manner. Whereas a deletion in *inlA* had no effect on invasion, a deletion in *inlB* reduced the invasive capacity of the *L. monocytogenes* strain by more that 200 fold and no further reduction was achieved with an *inlAB* double mutant. Taken together, these data demonstrate the outstanding role of *inlB* in *in vitro* invasion of endothelial cells of different origin. The experiments mentioned above were performed in the absence of serum. In two other studies *L. monocytogenes* invasion of HUVEC was analyzed in the presence of either 10% or 20% human serum. Under such conditiones binding to and invasion of HUVEC by *L. monocytogenes* is largely *inlAB*-independent. Drevets *et al.* (1995) using a HUVEC cell line found *inlAB* independent binding in the presence of 10% normal human serum, whereas Greiffenberg *et al.* (1997) found, using primary cultures of HUVEC and 20% pooled human serum that invasion was largely *inlA*- or *inlB*-independent with a slight effect (40% reduction) only upon deletions in *inlA* and *inlB*.

The high sequence similiarity of *inlA* and *inlB* indicated that the two genes originate from a common ancestral gene and may represent members of a gene family in *L. monocytogenes*. One additional member of this gene family, called *inlC*, was cloned and characterized and shown to encode a small (297 amino acids), secreted protein which is mainly expressed at the later stage of the intracellular life cycle and obviously not involved in the entry process into epithelial cells (Engelbrecht *et al.*, 1996) (Figure 3). InlC contains only the A repeat and no B repeat region and no membrane anchor. The *inlC* gene is located on the chromosome between the *rplS* and *infC* genes both of which share homology with housekeeping genes (Engelbrecht *et al.*, 1998a). The gene *inlC* was identified independently by Domann *et al.* (1997) and called *irpA*. A gene homologous to the *L. monocytogenes inlC* gene was cloned from *L. ivanovii* and called i-*inlC*. The i-*inlC* gene is located next to i-*inlD*, which encodes an additional small secreted internalin, and both genes map to a chromosomal location different from that of *inlC* in *L. monocytogenes* (Engelbrecht *et al.*, 1998a). Two more small internalins were recently found to be encoded by a novel chromosomal locus of *L. ivanovii* harboring the genes i-*inlE* and i-*inlF* (Engelbrecht *et al.*, 1998b). Homologs of these *L. ivanovii* genes are not present in *L. monocytogenes* and therefore these small internalins may represent factors related to the different virulence properties of these two closely related species. Meanwhile six additional members (*inlC2*, *inlD*, *inlE*, *inlF*, *inlG*, *inlH*) of the internalin multigene family were identified in *L. monocytogenes* all belonging to the large internalins with two repeat regions and the InlA-like membrane anchor (Raffelsbauer *et al.*, 1998; Dramsi *et al.*, 1997) (Figure 3). The newly identified internalin genes are located on the chromosome either clustered (*inlC2DE* or *inlGHE*) or dispersed (*inlC* and *inlF*). The gene clusters *inlC2DE* and *inlGHE* were identified in two different *L. monoytogenes* strains and are localized to the same position on the chromosome since they are flanked by the same housekeeping genes.

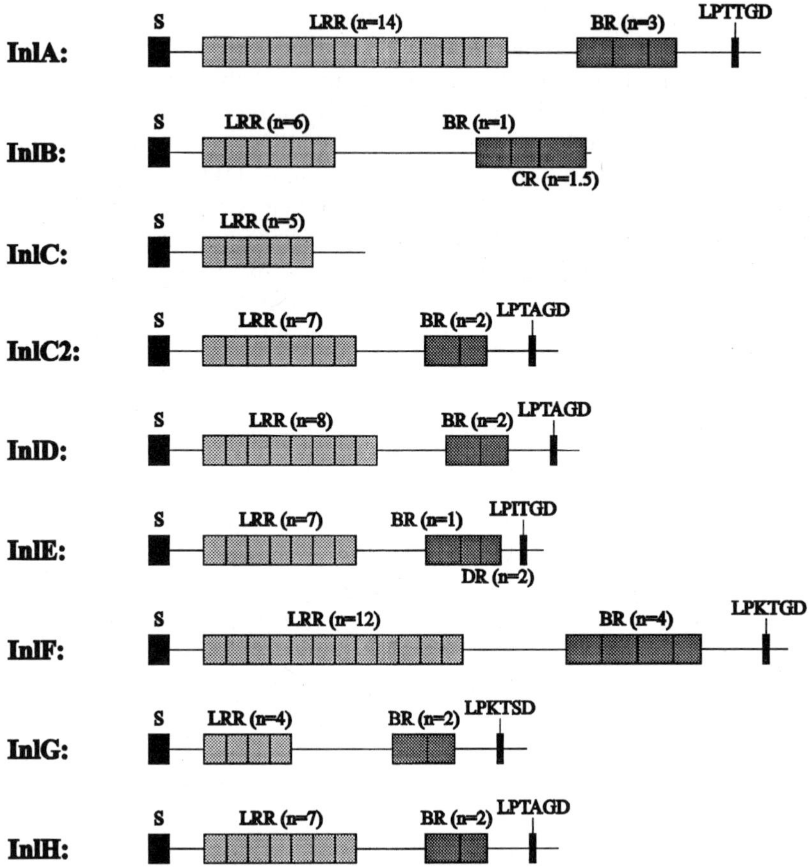

FIGURE 3. Members of the internalin multigene family found in *L. monocytogenes*. S: signal peptide; LRR: Leucine-rich repeat; BR: B-repeat; CR: C-repeat; DR: D-repeat. (kindly provided by F. Engelbrecht).

Sequence comparison of the two *inl* gene clusters indicated that *inlG* is an additional new internalin gene, while *inlH* is generated by site specific recombination leading to an in-frame deletion which removed the 3′ end of *inlC2* and the 5′ part of *inlD* (Raffelsbauer *et al.*, 1998). *In vitro* invasion assays with mutants harboring single or multiple in frame deletions in the *inlC2DE* cluster and different cell lines could not identify a role of these internalins in *L. monocytogenes* invasion *in vitro* (Dramsi *et al.*, 1997) and a deletion removing *inlGHE* did also not affect invasion, intracellular growth and cell-to-cell spread of the mutant strain upon infection of Caco-2 epithelial cells (Raffelsbauer *et al.*, 1998).

4. CELLULAR RECEPTORS, HOST CELL SIGNALING, AND THE MECHANISM OF INVASION

The mechanisms of *L. monocytogenes* invasion of cultured mammalian cells were most intensively studied with epithelial cell lines (in most cases Caco-2 cells). Using gentamicin-based infection assays, invasion of *L. monocytogenes* was studied under a wide range of experimental conditions. Especially the culture conditions of the mammalian target cells were found to have a major impact on the efficiency and the mode of invasion. Adhesion to and invasion into semiconfluent monolayers of Caco-2 and LLC-PK1 cells (pig kidney epithelial cells) was shown to occur most efficiently at the edges of cell islets where the basolateral side of the membrane is exposed (Mengaud *et al.*, 1996b; Temm-Grove *et al.*, 1994; Bubert *et al.*, 1992). Transmission and scanning electron microscopy of *L. monocytogenes* during different stages of the invasion process showed that the bacteria enter the cells efficiently with only very localized membrane perturbations in contrast to the dramatic morphological changes of the cell surface observed during entry of Salmonella and Shigella. Whereas *L. monocytogenes* enters non-differentiated Caco-2 cells in a zipper-like process during which the bacteria are successively covered by a thin fold of plasma membrane, invasion of differentiated Caco-2 cells from the apical side is accompanied by the interaction of microvilli with the bacterial surface and fusion processes of individual microvilli probably generate lamellipodia which cover the bacteria (Karunasagar *et al.*, 1994). This second mechanism is obviously much less efficient since growing Caco-2 cells beyond confluency is paralleled with a decrease in *L. monocytogenes* invasion (Gaillard and Finlay, 1996). Velge and coworkers (1994a), reported on the unexpected finding that cell immortalization of different epithelial cells through SV-40 transformation dramatically enhances *L. monocytogenes* invasion. However, the authors could not present any data concerning the molecular basis of their finding and only speculate that a putative differential expression of a relevant receptor might account for the differences in listerial uptake. This finding on highly reduced uptake capabilities of primary epithelial cells versus immortilized decendants is especially important in the light of recent *in vivo* findings which question the role of enterocyte invasion during the early stages of *L. monocytogenes* infections (see below).

The invasion process of *L. monocytogenes* is sensitive to treatment with cytochalasin D for all cell types tested so far underlining the importance of microfilaments for the uptake mechanism (Wells *et al.*, 1998; Kuhn, 1998; Kuhn *et al.*, 1988; Gaillard *et al.*, 1987). However, a dramatic accumulation of F-actin underneath the invading bacteria, as described for Shigella and Salmonella invasion which is also cytochalasin sensitive, was not observed

(Mengaud *et al.*, 1996b). In addition to multiple reports on the cytochalasin sensitivity of the invasion process in many different cell types, several recent reports demonstrated that microtubules also play some role in invasion. Uptake of *L. monocytogenes* by mouse dendritic cells, mouse bone marrow-derived macrophages, and human brain endothelial cells is sensitive to nocodazole treatment which disrupts microtubules (Kuhn, 1998; Guzmann *et al.*, 1995; Greiffenberg *et al.*, in revision).

The interaction of listerial proteins which mediate uptake by non-phagocytic cells with their eukaryotic receptors is only poorly understood. The receptor for InlA was identified as E-cadherin by a biochemical approach using matrix bound purified InlA to isolate the InlA ligand from epithelial membrane proteins (Mengaud *et al.*, 1996). E-cadherin, a member of the cadherin family is mainly expressed at the basolateral side of enterocytes (Cepek *et al.*, 1994) and was shown to promote InlA-dependent binding and invasion of *L. monocytogenes*. Expression of E-cadherin in S180 fibroblasts which are poorly invaded by *L. monocytogenes*, results in a 10-fold increase in bacterial internalization. E-cadherin binds InlA directly and its location on the basolateral membrane of epithelial cells is in line with previous observations suggesting the basolateral membrane as entry site for *L. monocytogenes* (Temm-Grove *et al.*, 1994) and the recent demonstration that disruption of the tight junctions and concomitant exposure of the basolateral membrane enhances invasion (Gaillard and Finlay, 1997). The cascade of signaling events triggered by the InlA-E-cadherin interaction and the intracellular proteins that are associated with listerial entry into normally nonphagocytic cells remain, however, to be identified.

The mammalian receptor for InlB, which triggers *L. monocytogenes* uptake by different cell types, is not known, but it was shown that InlB-mediated uptake of *L. monocytogenes* by Vero cells is associated with the activation of the lipid kinase p85/p110 (PI-3 kinase) (Ireton *et al.*, 1996). This activation requires in addition to the InlB protein tyrosine phosphorylation in the host cell, since uptake is inhibited by genistein treatment which blocks tyrosine specific protein kinases. The mechanism by which the PI-3 kinase mediates uptake is presently unknown, but one might speculate that the products of the enzymatic activity of PI-3 kinase, phosphoinositide-3,4-bisphosphate and phosphoinositide-3,4,5-trisphosphate directly interfere with the actin cytoskeleton by uncapping barbed ends of actin filaments and thereby influencing cytoskeleton dynamics as previously shown in another cellular system (Hartwig *et al.*, 1995). In addition to phosphoinositide-phosphates, leukotrienes are also suspected to be involved in *L. monocytogenes* signaling during invasion of epithelial cells. Nordihydroguaretic acid, an inhibitor of 5-lipoxygenase activity which generates leukotrienes from arachidonic acid, also inhibits *L. monocytogenes* invasion of HEp-2 cells (Mecsas *et al.*, 1998). However, it is not known at present at which step of the putative signaling cascade leukotrienes could be involved and whether the leukotriene signal is initiated by InlA- or InlB-receptor interactions. The study by Mescas *et al.* (1998) elegantly used *Yersinia pseudotuberculosis* preinfection to further demonstrate that signaling pathways which are disrupted by the delivery of the yersinial proteins YopE and YopH inhibit *L. monocytogenes* invasion. The cellular targets of the Yops are not exactly known, but it is evident that phosphatase YopH (Andersson *et al.*, 1996) could interfere with many signaling pathways. YopE causes disruption of the host cell cytoskeleton, including the collapse of actin stress fibers (Rosqvist *et al.*, 1991) and it was suggested that it might interfere with the cytoskeleton through inhibition of the small GTPase Rho (Kulich *et al.*, 1994).

Treatment of various cells with the protein kinase C inhibitor genistein (or staurosporine) inhibits invasion of *L. monocytogenes* suggesting that protein kinase C activity is critical for regulating the cytoskeletal alterations necessary for listerial uptake (Kuhn *et al.*, 1997; Guzman *et al.*, 1995; Velge *et al.*, 1994b; Tang *et al.*, 1994). Protein tyrosine phosphorylation during infection of mammalian cells was first demonstrated by Tang and coworkers (1994), who also showed that the phosphorylated proteins are the 42 and 44 kDa isoforms (ERK-1 and ERK-2) of the mitogen activated protein kinase (MAP kinase). Later it was, however, shown that this MAP kinase phosphorylation is not connected to bacterial uptake, but instead triggered by the pore-forming protein listeriolysin (Weiglein *et al.*, 1997; Tang *et al.*, 1996). In a recent paper (Tang *et al.*, 1998), it was reported that *L. monocytogenes* invasion of HeLa cells activates, in addition to ERK-1 and ERK-2, two other MAP kinases called p38 and JNK and the MAP kinase kinase MEK-1. ERK-2 was also phosphorylated upon invasion of listeriolysin-negative mutants and is a downstream target of MEK-1. Taken together, these data suggest that MEK-1/ERK-2 activation is one step in the signaling during *L. monocytogenes* uptake into host epithelial cells. ERK-2 activation is not sensitive to wortmannin treatment showing that it is not a downstream event of PI-3 kinase activation. Obviously, the PI-3 kinase and the MEK-1/ERK-2 pathways may form two components of converging or independent signal transduction systems required for *L. monocytogenes* invasion.

Another protein tyrosine kinase obviously activated during *L. monocytogenes* entry of epithelial cells is $pp60^{c\text{-}src}$ since specific inhibition of $pp60^{c\text{-}src}$ by herbimycin blocks *inlAB*-dependent entry (Van Langendonck *et al.*, 1998). $pp60^{c\text{-}src}$ activation was also found during *Shigella flexneri* entry where it phosphorylates cortactin (reviewed in Menard *et al.*, 1996). Whether cortactin is also phosphorylated during *L. monocytogenes* entry is not known and the exact role of pp60$^{c\text{-}src}$ in listerial invasion remains to be established.

The mechanisms by which the *L. monocytogenes* surface protein ActA mediates listerial uptake is even less understood. ActA-promoted attachment and invasion of CHO epithelia-like cells as well as IC-21 murine macrophages is obviously associated with the interaction of ActA with a heparan-sulfate proteoglycan receptor (Alvarez-Dominguez *et al.*, 1997). It is supposed that electrostatic interactions between heparan sulfate and positively charged residues in the N-terminal part of ActA could lead to low-stringency binding to cell surface proteoglycan receptors which are widely distributed in mammalian cells. Whether the proposed low-stringency binding of *L. monocytogenes* to heparan sulfate proteoglycan receptors triggers uptake directly or results in adequate presentation of other bacterial factors to the host cell membrane which ultimately lead to phagocytosis remains to be clarified.

5. UPTAKE BY PROFESSIONAL PHAGOCYTES

Macrophages of different origin were used in *in vitro* studies to analyse the mechanisms of *L. monocytogenes* uptake and it is generally assumed that macrophages take up *L. monocytogenes* by conventional phagocytosis. As professional phagocytic cells, all macrophages are characterized by their ability to ingest a wide variety of particles including bacteria and even inert latex beads. Additionally, macrophages dispose of specific receptors for opsonized material like complement-coated bacteria. It is therefore not surprising that *L. monocytogenes* is taken up by macrophages very efficiently.

In the absence of opsonins, mouse peritoneal macrophages or J774A.1 macrophage-like cells efficiently phagocytose *L. monocytogenes* as demonstrated by gentamicin-based invasion assays and electron-microscopy studies (Pierce *et al.*, 1996; Kuhn *et al.*, 1988; Portnoy *et al.*, 1988). The extent by which listerial proteins or surface molecules further stimulate phagocytosis is still under debate. Using bone marrow-derived macrophages, InlA was found to have only a slight effect on uptake of *L. monocytogenes*, since an *ΔinlAB* mutant still showed more than 60% invasion when compared with the wild-type strain (Hess *et al.*, 1995). Uptake of *L. monocytogenes* by the mouse macrophage-like cell line J774A.1 was shown to be inhibited by at least 50% by the pretreatment of *L. monocytogenes* with rabbit polyclonal anti-InlA antibodies or $F(ab')_2$ fragments of mouse monoclonal anti-InlA antibodies. Additionally recombinant InlA specifically binds to the macrophages supporting the assumption that InlA plays some role in facilitating the uptake of *L. monocytogenes* by macrophages (Sawyer *et al.*, 1996). Recently it was suggested that the listerial protein p60 might enhance *L. monocytogenes* phagocytosis by macrophages. A *Salmonella typhimurium* strain, constructed to express and secrete p60 using the *Escherichia coli* hemolysin secretion system is more invasive for phagocytic cells (mouse bone marrow-derived macrophages) but not for Caco-2 enterocytes. In line with this assumption is the finding that pretreatment of *L. monocytogenes* with a polyclonal anti-p60 antiserum substantially inhibits the uptake of the listeriae by a macrophage-like cell line (Hess *et al.*, 1995). As mentioned above the listerial surface protein ActA obviously stimulates listerial uptake, at least by the IC-21 macrophage-like cell line (Alvarez-Dominguez *et al.*, 1997). Another factor that could be involved in attachment and invasion of *L. monocytogenes* in macrophages is the listerial cell wall polymer lipoteichoic acid which was shown to be a ligand of the macrophage scavenger receptor. *L. monocytogenes* binds strongly to the macrophage scavenger receptor (Greenberg *et al.*, 1996; Dunne *et al.*, 1994) but it is not known whether this interaction may trigger receptor mediated phagocytosis. Taken together, the data mentioned above suggest that most of the listerial surface molecules known to be involved in entering non-professional phagocytes like fibroblasts or epithelial cells also play some role in the uptake of *L. monocytogenes* by macrophages, at least in the absence of serum factors.

The role of complement factors in phagocytosis of *L. monocytogenes* by macrophages was mainly studied by Drevets and coworkers (Drevets *et al.*, 1996, 1993, 1992; Drevets and Campbell, 1991). In the first series of experiments, the binding of complement component 3 (C3) to *L. monocytogenes* was demonstrated. Furthermore, it was shown that the complement receptor 3 (CR3) plays a significant role in phagocytosis of C3 opsonized bacteria by mouse peritoneal macrophages since anti-CR3 anibodies reduced phagocytosis to a level similar to that obtained without opsonization (Drevets and Campbell, 1991). Direct binding to *L. monocytogenes* and proteolysis of C3 was also demonstrated independently by Croize *et al.* (1993) giving direct evidence for the activation of the alternative pathway by *L. monocytogenes*. Further studies not only confirmed the dominant role of CR3 in mediating phagocytosis of opsonized *L. monocytogenes* but also showed that (i) listericidial macrophages specifically phagocytose *L. monocytogenes* through C3-CR3 interaction, (ii) that CR3 involvement is essential for efficient killing of *L. monocytogenes* by macrophages, and (iii) that the TNF-α and IFN-γ induced upregulation of CR3 expression also increases phagocytic activity and leads to enhanced killing of *L. monocy-*

togenes (Drevets *et al.*, 1996, 1993, 1992). Besides C3, the complement component C1q was also shown to directly bind to *L. monocytogenes* and to increase the uptake of the coated bacteria by IC-21 macrophage-like cells several fold (Alvarez-Dominguez *et al.*, 1993). C1q is present in body fluids in relatively high concentrations and C1q receptors are abundant on macrophages suggesting that the *in vitro* finding of C1q-C1R mediated phagocytosis of *L. monocytogenes* might also have *in vivo* relevance.

Opsonin-dependent and opsonin-independent adhesion of *L. monocytogenes* to mouse peritoneal macrophages can be inhibited by treatment with N-acetylneuramic acid or by oxidation of listerial surface carbohydrates through sodium metaperiodate treatment (Maganti *et al.*, 1998) suggesting that sialic acids are involved in the attachment of *L. monocytogenes* to murine macrophages.

The kinetics of uptake of *L. monocytogenes* by macrophages was determined using different methods (Raybourne *et al.*, 1994; Davies *et al.*, 1983). Macrophages ingest *L. monocytogenes* very rapidly and intracellular killing starts shortly after phagocytosis and leads to the destruction of most of the ingested bacteria. In a single macrophage, both killed bacteria inside acidified phagosomes and phagolysosomes and growing bacteria that escaped into the cytoplasm, are detected (De Chastelier *et al.*, 1994). These findings suggest a competition between phagosome-lysosome fusion followed by killing of the bacteria and escape of the bacteria from the acidified phagosome before phagosome-lysosome fusion occurs. The result of this competition is a population of cytoplasmic bacteria able to grow inside the macrophage.

The cellular processes underlying phagocytosis of *L. monocytogenes* by macrophages are not understood. Phagocytosis, but not binding, is cytochalasin sensitive demonstrating the importance of actin filament dynamics (Kuhn, 1998; Demuth *et al.*, 1996). Nocodazole and colchicine which depolymerize microtubules were recently shown to inhibit *L. monocytogenes* uptake by P388D1 macrophages more than ten fold demonstrating that microtubules play some role in the uptake process in certain macrophage types. The signaling cascades activated during phagocytosis are largely unknown. A MAP kinase phosphatase (MKP-1) was discussed as being involved in regulating *L. monocytogenes* uptake of J774 macrophages since overexpression of MKP-1 reduced listerial uptake (Kügler *et al.*, 1997). MAP kinases are also activated in J774 macrophages during *L. monocytogenes* entry, but this activation is obviously triggered by listeriolysin (Tang *et al.*, 1996). Finally, lipoteichoic acid which efficiently binds to the macrophage scavenger receptor also triggers a rapid and transient activation of transcription factor NF-κB (Hauf *et al.*, 1997), but it is not known whether this activation is connected to bacterial uptake.

6. REGULATION OF THE EXPRESSION OF INVASION-ASSOCIATED GENES

The regulation of the expression of the invasion associated genes *inlA*, *inlB*, *actA*, and *iap* is complex and far from being understood. Whereas *actA*, *inlA*, and *inlB* are, at least in part, expressed in a PrfA-dependent manner (for review see Brehm *et al.*, 1996; Kreft *et al.*, 1995), *iap* expression is totally PrfA-independent and controlled on the posttranscriptional level by unknown mechanisms (Köhler *et al.*, 1991). The gene *actA* is only weakly transcribed when *L. monocytogenes* is grown in rich culture media like brain heart infusion broth (BHI) but expression is induced, together with the expression of *hly*, *mpl*, *plcA*, *plcB*, when the bacteria are shifted into starvation conditions (minimal essential medium) (Sokolovic *et al.*, 1993). The expression of *actA* is also induced while *L. monocytogenes* resides in the host cell cytoplasm (Bohne *et al.*, 1994) where ActA functions in triggering actin polymerization and intracellular movement (Kocks *et al.*, 1992; Domann *et al.*, 1992). The low level of *actA* expression under extracellular conditions and the inducible *actA* expression inside the host cell cytosol argue against an important role for ActA-mediated invasion *in vivo*. However, this does not rule out the possibility that specific *in vivo* situations my exist in which ActA could function as an invasin.

The regulation of *inlA* and *inlB* expression under different growth conditions is complex and partially controversial results have been published (Bohne *et al.*, 1996; Lignau *et al.*, 1995; Dramsi *et al.*, 1993a). The genes of the *inlAB* operon are transcribed from multiple promotors upstream from *inlA* which result in 2.9 and 5.0 kb monocistronic and bicistronic transcripts and from an internal promotor located between *inlA* and *inlB* which gives rise to a monocistronic transcript. The positive regulatory factor PrfA (for review see Brehm *et al.*, 1996; Kreft *et al.*, 1995) contributes to *inlAB* expression since one of the promotors upstream of *inlA* contains the palindromic sequence motive found in all PrfA-controlled promotors and termed "PrfA-box". However, the PrfA-box in front of *inlA* is rather incomplete (Dramsi *et al.*, 1993a) and PrfA-dependent expression of *inlA* seems to be regulated inversibly to the other known PrfA-dependent genes since *inlA* expression is reduced upon shift of *L. monocytogenes* to MEM and under intracellular conditions (Bohne *et al.*, 1996; Bubert *et al.*, submitted). Probably PrfA, which obviously senses intracellular conditions, is, in the case of *inlA* used to downregulate expression of genes necessary only under extracellular conditions. The expression of *inlA* is also modulated by the growth state (maximum expression during log phase) and by temperature. At 30°C *inlA* expression is low and increases upon shift to 37°C. The expression of *inlB* was recently investigated by Lingnau *et al.* (1995) and shown to be dependent on the presence of PrfA and temperature. This is also in line with the recent finding that HBMEC invasion which is strictly InlB-dependent is also PrfA-dependent (Greiffenberg *et al.*, in revision). The multiple promotors of the *inlAB* operon should allow complex expression patterns which are possibly connected to the need of sequential or tissue specific expression of *inlA* and *inlB* during the course of infection.

7. *In Vivo* SIGNIFICANCE OF THE RESULTS OBTAINED WITH CELL CULTURE SYSTEMS

The *in vivo* relevance of listerial factors that give phenotypes in cell culture assays was either tested by measuring LD_{50} values in mice or by following the growth of the bacteria in organs of infected mice or by using the ligated ileal loop model.

Since InlA is necessary for high level invasion of Caco-2 epithelial cells it was thought that it should also be critical for the passage of the intestinal epithelium during the early stage of listeriosis. This is, however, not the case as it was demonstrated by Pron *et al.* (1998) that *inlAB* as well as *hly* and *actA* mutants and even the nonpathogenic species *L. innocua* translocate similarly to wild type *L. monocytogenes* to deep organs upon innoculation of ligated intestinal loops. Furthermore, the listeriae were only very rarely associated with the intestinal epithelium and no clear signs of epithelial invasion and interacellular growth were seen upon microscopic examination. In contrast, *L. monocytogenes* multiplied preferentially in the follicular tissue of Peyer's patches.

When measuring bacterial numbers in spleen and liver of mice infected with an *inlAB* mutant surprisingly equal numbers of wild type and mutant bacteria were detected throughout the duration of the experiment in the spleen. Only at day three postinfection the number of Δ*inlAB* bacteria in the liver was reduced upon infection when compared with the wild type strain (Dramsi *et al.*, 1995), suggesting that the *inlAB* locus may play a role in hepatocyte invasion *in vivo*. This suggestion was further sustained by the finding of Gaillard *et al.* (1996) who demonstrated that the *inlAB* locus of *L. monocytogenes* mediates entry into hepatocytes *in vivo*. Since the authors used a mutant defective in *inlA* and *inlB* expression, it is not definitively proven that InlB, shown to trigger hepatocyte invasion *in vitro*, is also the critical protein for hepatocyte invasion *in vivo*. These results were, however, questioned by independent investigations showing that the *inlAB* locus is no factor influencing rapid clearence of intravenously injected bacteria from the bloodstream und uptake in the liver. Additionally the absence of *inlAB* did not affect the distribution of *L. monocytogenes* among hepatocytes and nonparenchymal cells in the liver (Gregory *et al.*, 1996). The reason for the controversial findings of these studies are unclear at present and further studies are needed to elucidate the role of either *inlA* or *inlB* for invasion of liver cells *in vivo*.

The role of *inlA*, at least during the earlier stages of listeriosis was further questioned by the recent finding that the *L. monocytogenes* strain LO28 which is fully virulent in the mouse model (Cossart *et al.*, 1989) expresses a truncated InlA protein which is released from the bacterial surface (Jonquiers *et al.*, 1998). This strain is also less invasive for Caco-2 cells than strain *L. monocytogenes* EGD since the truncated and soluble protein is obviously unable to mediate E-cadherin dependent invasion. A limited search for *L. monocytogenes* strains expressing variants of InlA revealed that from 26 isolates, 4 strains expressed truncated forms of InlA. Three of the 4 strains were food isolates and one was a clinical isolate. In respect to the high virulence of strain LO28 in the mouse model it is important to note that it is not clear whether the mouse model of experimental listeriosis reproduces all aspects of human listeriosis e.g. meningitis or meningoencephalitis.

Whether InlA plays a role in the manifestation of these late steps of human listeriosis remains to be investigated. In several recent reports endothelial cell invasion which is regarded as a prerequisite for the development of meningitis and menignoencephalitis was found to be strictly dependent on the presence of InlB (Parida *et al.*, 1998; Greiffenberg *et al.*, in revision). Nevertheless it is possible that InlB and InlA might synergize during crossing of the blood brain barrier at the level of the choroid plexus. In order to establish an infection of the cerebral liquor the crossing of this structure, formed by an endothelial cell layer followed by an epithelium might require both, endothelial invasion triggered by InlB and epithelial invasion from the basolateral side triggered by InlA. During bacteremia which is required for invasion of the central nervous system in the mouse model (Berche, 1995), *L. monocytogenes*, if free in the blood, encounters high concentrations of human serum. The finding that invasion of endothelial cells is less InlB-dependent in the presence of human serum (Greiffenberg *et al.*, 1997; Drevets *et al.*, 1995), however, questions the *in vivo* relevance of the *in vitro* findings on the role of *inlB* for endothelial cell invasion. It is now necessary to establish the role of *inlB* in endothelial cell invasion under conditions resembling more closely the *in vivo* situation.

The *in vivo* role of the newly identified members of the internalin multigene fanily is even less clear. Mutants harboring single or multiple in frame deletions in the *inlC2DE* cluster did not show any reduction in viable counts in livers and spleens upon intravenous infection of mice (Dramsi *et al.*, 1997). However, a deletion removing the entire *inlGHE* cluster resulted in a significant loss of virulence since the numbers of viable bacteria inside the livers and spleens of orally infected C57BL/6 mice were more than two logs lower upon infection with the mutant than those found upon infection with the wid type stain.

8. CONCLUSIONS

The last decade has seen an enormous increase in our understanding of the molecular basis of infectious diseases. Knowledge of the genes and proteins determining virulence of *L. monocytogenes* and the role of specific virulence gene products in the infection of host cells is rapidly expanding. However, many problems concerning the virulence of *L. monocytogenes* still remain unsolved.

For instance, the role of the newly identified members of the internalin-multigene family in the infection process is still a mystery and more sophisticated cell culture models are needed to elucidate the function of these molecules. Furthermore, only one cellular receptor for an internalin has so far been identified further hindering the understanding of *L. monocytogenes* tissue tropism probably mediated by the different internalins as well as the analysis of the host cell signaling pathways activated by ligand-receptor interactions during invasion. Despite some interesting discoveries (PI-3 kinase, ERK-2 and $pp60^{c\text{-}src}$ activation) our picture of the signaling events during *L. monocytogenes* invasion of nonprofessional phagocytic cells is still fragmentary and the regulation of listerial uptake by macrophages is even less understood. Finally, our knowledge of the *in vivo* relevance of the results obtained using cell culture assays is still limited and much more work needs to be done in order to elucidate the contributions of the known virulence genes during the different steps of a *L. monocytogenes* infection.

ACKNOWLEDGMENTS. We thank J. Daniels and J. Kreft for critical reading of the manuscript and F. Engelbrecht and J. Kreft for providing figures. This work was supported by the Deutsche Forschungsgemeinschaft through Sonderforschungsbereich SFB 165-B4 and SFB 479-B1.

9. REFERENCES

Alvarez-Dominguez, C., Carrasco-Marin, E., and Leyva-Cobian, F., 1993, Role of complement component C1q in phagocytosis of *Listeria monocytogenes* by murine macrophage-like cell lines, *Infect. Immun.* **61**:3664–3672.

Alvarez-Dominguez, C., Vazquez-Boland, J.-A., Carrasco-Marin, E., Lopez-Mato, P., and Leyva-Cobian, F., 1997, Host cell heparan sulfate proteoglycans mediate attachment and entry of *Listeria monocytogenes*, and the listerial surface protein ActA is involved in heparan sulfate receptor recognition. *Infect. Immun.* **65**:78–88.
Andersson, K., Carballeira, N., Magnusson, K.E., Persson, C., Stendahl, O., Wolf-Watz, H., and Fallman, M., 1996, YopH of *Yersinia pseudotuberculosis* interrupts early phosphotyrosine signalling associated with phagocytosis. *Mol. Microbiol.* **20**:1057–1069.
Berche, P., 1995, Bacteremia is required for invasion of the murine central nervous system by *Listeria monocytogenes*. *Microbial Path.* **18**:323–336.
Bohne, J., Kestler, H., Uebele, C., Sokolovic, Z., and Goebel, W., 1996, Differential regulation of the virulence genes of *Listeria monocytogenes* by the transcriptional activator PrfA. *Mol. Microbiol.* **20**:1189–1198.
Bohne, J., Sokolovic, Z., and Goebel, W., 1994, Transcriptional regulation of *prfA* and PrfA-regulated virulence genes in *Listeria monocytogenes*. *Mol. Microbiol.* **11**:1141–1150.
Braun, L., Dramsi, S., Dehoux, P., Bierne, H., Lindahl, G., and Cossart, P., 1997, InlB: an invasion protein of *Listeria monocytogenes* with a novel type of surface association. *Mol. Microbiol.* **25**:285–294.
Braun, L., Ohayon, H., and Cossart, P., 1998, The InlB protein of *Listeria monocytogenes* is sufficient to promote entry into mammalian cells. *Mol. Microbiol.* **27**:1077–1087.
Brehm, K., Kreft, J., Ripio, M.T., and Vazquez-Boland, J.-A., 1996, Regulation of virulence gene expression in pathogenic *Listeria*. *Microbiologia Sem.* **12**:219–236.
Bubert, A., Kuhn, M., Goebel, W., and Köhler, S., 1992, Structural and functional properties of the p60 proteins from different *Listeria* species. *J. Bacteriol.* **174**:8166–8171.
Bubert, A., Chun, S.-K., Papatheodorou, L., Simm, A., Goebel, W., and Sokolovic, Z., Differential virulence gene expression by *Listeria monocytogenes* growing within host cells. *Mol. Gen. Genet.* (submitted for publication).
Cepek, K.L., Shaw, S.K., Parker, C.M., Russel, G.J., Morrow, J.S., Rimm, D.L., and Brenner, M.B., 1994, Adhesion between epithelial cells and T lymphocytes mediated by E-cadherin and the a_Eb_7 integrin. *Nature* **372**:190–193.
Cossart, P., and Lecuit, M., 1998, Interaction of *Listeria monocytogenes* with mammalian cells during entry and actin-based movement: bacterial factors, cellular ligands and signaling. *EMBO J.* **17**:3797–3806.
Cossart, P., Vicente, M.F., Mengaud, J., Baquero, F., Perez-Diaz, J.C., and Berche, P., 1989, Listeriolysin O is essential for virulence of *Listeria monocytogenes*: direct evidence obtained by gene complementation. *Infect. Immun.* **57**:3629–3636.
Croize, J., Arvieux, J., Berche, P., and Colomb, M.G., 1993, Activation of the human complement alternative pathway by *Listeria monocytogenes*: evidence for direct binding and proteolysis of the C3 component on bacteria. *Infect. Immun.* **61**:5134–5139.
Davies, W.A., 1983, Kinetics of killing of *Listeria monocytogenes* by macrophages: rapid killing accompanying phagocytosis. *J. Reticuloendothel. Soc.* **34**:131–141.
De Chastellier, C., and Berche, P., 1994, Fate of *Listeria monocytogenes* in murine macrophages: evidence for simultaneous killing and survival of intracellular bacteria. *Infect. Immun.* **62**:543–553.
Demuth, A., Goebel, W., Beuscher, H.U., and Kuhn, M., 1996, Differential regulation of cytokine and cytokine receptor mRNA expression upon infection of bone marrow-derived macrophages with *Listeria monocytogenes*. *Infect. Immun.* **64**:3475–3483.
Domann, E., Wehland, J., Rohde, M., Pistor, S., Hartl, M., Goebel, W., Leimeister-Wächter, M., Wuenscher, M., and Chakraborty, T., 1992, A novel bacterial virulence gene in *Listeria monocytogenes* required for host cell microfilament interaction with homology to the proline-rich region of vinculin. *EMBO J.* **11**:1981–1990.

Domann, E., Zechel, S., Lingnau, A., Hain, T., Darji, A., Nichterlein, T., Wehland, J., and Chakraborty, T., 1997, Identification and characterization of a novel PrfA-regulated gene in *Listeria monocytogenes* whose product, IrpA, is highly homologous to internalin proteins, which contain leucin-rich repeats. *Infect. Immun.* **65**:101–109.

Dramsi, S., Kocks, C., Forestier, C., and Cossart, P., 1993a, Internalin-mediated invasion of epithelial cells by *Listeria monocytogenes* is regulated by the bacterial growth state, temperature and the pleiotropic activator *prfA*. *Mol. Microbiol.* **9**:931–941.

Dramsi, S., Dehoux, P., and Cossart, P., 1993b, Common features of gram-positive bacterial proteins involved in cell recognition. *Mol. Microbiol.* **9**:1119–1122.

Dramsi, S., Biswas, I., Maguin, E., Braun, L., Mastroeni, P., Cossart, P., 1995, Entry of *Listeria monocytogenes* into hepatocytes requires expression of InlB, a surface protein of the internalin multigen family. *Mol. Microbiol.* **16**:251–261.

Dramsi, S., Dehoux, P., Lebrun, M., Goossens, P.L., and Cossart, P., 1997, Identification of four new members of the internalin multigene family of *Listeria monocytogenes* EGD. *Infect. Immun.* **65**:1615–1625.

Drevets, D.A., and Campbell, P.A., 1991, Roles of complement and complement receptor type 3 in phagocytosis of *Listeria monocytogenes* by inflammatory mouse peritoneal macrophages. *Infect. Immun.* **59**:2645–2652.

Drevets, D.A., Canono, B.P., and Campbell, P.A., 1992, Listericidal and nonlistericidal mouse macrophages differ in complement receptor type 3-mediated phagocytosis of *L. monocytogenes* and in preventing escape of the bacteria into the cytoplasm. *J. Leukoc. Biol.* **52**:70–79.

Drevets, D.A., Leenen, P.J.M., and Campbell, P.A., 1993, Complement receptor type 3 (CD11b/CD18) involvement is essential for killing of *Listeria monocytogenes* by mouse macrophages. *J. Immunol.* **151**:5431–5439.

Drevets, D.A., Sawyer, R.T., Potter, T.A., and Campbell, P.A., 1995, *Listeria monocytogenes* infects human endothelial cells by two distinct mechanisms. *Infect. Immun.* **63**:4268–4276.

Drevets, D.A., Leenen, P.J., and Campbell, P.A., 1996, Complement receptor type 3 mediates phagocytosis and killing of *Listeria monocytogenes* by a TNF-α- and IFN-γ-stimulated macrophage precursor hybrid. *Cell Immunol.* **169**:1–6.

Dunne, D.W., Resnick, D., Greenberg, J., Krieger, M., and Joiner, K.A., 1994, The type I macrophage scavenger receptor binds to gram-positive bacteria and recognizes lipoteichoic acid. *Proc. Natl. Acad. Sci.* USA **91**:1863–1867.

Engelbrecht, F., Chun, S.-K., Ochs, C., Hess, J., Lottspeich, F., Goebel, W., and Sokolovic, Z., 1996, A new PrfA-regulated gene of *Listeria monocytogenes* encoding a small, secreted protein which belongs to the family of internalins. *Mol. Microbiol.* **21**:823–837.

Engelbrecht, F., Dickneite, C., Lampidis, R., Götz, M., DasGupta, U., and Goebel, W., 1998a, Sequence comparison of the chromosomal regions encompassing the internalin C genes (*inlC*) of *Listeria monocytogenes* and *L. ivanovii*. *Mol. Gen. Genet.* **257**:186–197.

Engelbrecht, F., Dominguez-Bernal, G., Dickneite, C., Hess, J., Greiffenberg, L., Lampidis, R., Raffelsbauer, D., Kaufmann, S.H.E., Kreft, J., Vazquez-Boland, J.-A., and Goebel, W., 1998b, A novel PrfA-regulated chromosomal locus of *Listeria ivanovii* encoding two small, secreted internalins is essential for virulence in mice. *Mol. Microbiol.* in press.

Farber, J.M., and Peterkin, P.I., 1991, *Listeria monocytogenes*, a food-borne pathogen. *Microbiol. Rev.* **55**:476–511.

Finlay, B.B., and Falkow, S., 1997. Common themes in microbial pathogenicity revisited. *Microbiol. Mol. Biol. Rev.* **61**:136–169.

Gaillard, J.L., and Finlay, B.B., 1996, Effect of cell polarization and differentiation on entry of *Listeria monocytogenes* into the enterocyte-like Caco-2 cell line. *Infect. Immun.* **64**:1299–1308.

Gaillard, J.L., Berche, P., Mounier, J., Richard, S., and Sansonetti, P.J., 1987, *In vitro* model of penetration and intracellular growth of *Listeria monocytogenes* in the human enterocyte-like cell line Caco-2. *Infect. Immun.* **55**:2822–2829.

Gaillard, J.L., Berche, P., Frehel, C., Gouin, E., and Cossart, P., 1991, Entry of *Listeria monocytogenes* into cells is mediated by internalin, a repeat protein reminiscent of surface antigens from gram-positive cocci. *Cell* **65**:1127–1141.

Gaillard, J.L., Jaubert, F., and Berche, P., 1996, The *inlAB* locus mediates the entry of *Listeria monocytogenes* into hepatocytes *in vivo*. *J. Exp. Med.* **183**:359–369.

Greenberg, J.W., Fischer, W., and Joiner, K.A., 1996, Influence of lipoteichoic acid structure on recognition by the macrophage scavenger receptor. *Infect. Immun.* **64**:3318–3325.

Gregory, S.H., Sagnimeni, A.J., and Wing, E.J., 1996, Expression of the *inlAB* operon by *Listeria monocytogenes* is not required for entry into hepatic cells *in vivo*. *Infect. Immun.* **64**:3983–3986.

Greiffenberg, L., Sokolovic, Z., Schnittler, H.-J., Spory, A., Böckmann, R., Goebel, W., and Kuhn, M., 1997, *Listeria monocytogenes*-infected human umbilical vein endothelial cells: internalin-independent invasion, intracellular growth, movement, and host cell responses. *FEMS Microbiol. Lett.* **157**:163–170.

Greiffenberg, L., Goebel, W., Kim, K.S., Weiglein, I., Bubert, A., Engelbrecht, F., Stins, M., and Kuhn, M., Interaction of *Listeria monocytogenes* with human brain microvascular endothelial cells: InlB-dependent invasion, long-term intracellular growth and spread from macrophages to endothelial cells. *Infect. Immun.* (in revision).

Guzman, C., Rhode, M., Chakraborty, T., Domann, E., Hudel, M., Wehland, J., and Timmis, K., 1995, Interaction of *Listeria monocytogenes* with mouse dendritic cells. *Infect. Immun.* **63**:3665–3673.

Hartwig, J.H., Bokoch, G.M., Carpenter, C.L., Janmey, P.A., Taylor, L.A., Toker, A., and Stossel, T.P., 1995, Thrombin receptor ligation and activated Rac uncap actin filament barbed ends through phosphoinositide synthesis in permeabilized human platelets. *Cell* **82**:643–653.

Hauf, N., Goebel, W., Fiedler, F., Sokolovic, Z., and Kuhn, M., 1997, *Listeria monocytogenes* infection of $P388D_1$ macrophages results in a biphasic NF-κB (RelA/p50) activation induced by lipoteichoic acid and bacterial phospholipases and mediated by IκBα and IκBβ degradation. *Proc. Natl. Acad. Sci.* USA **94**:9394–9399.

Hess, J., Gentschev, I., Szalay, G., Ladel, C., Bubert, A., Goebel, W., and Kaufmann, S.H.E., 1995, *Listeria monocytogenes* p60 supports host cell invasion by and *in vivo* survival of attenuated *Salmonella typhimurium*. *Infect. Immun.* **63**:2047–2053.

Ireton, K., Payrastre, B., Chap, H., Ogawa, W., Sakaue, H., Kasuga, M., and Cossart, P., 1996, A role for phosphoinositide 3-kinase in bacterial invasion. *Science* **274**:780–782.

Jonquieres, R., Bierne, H., Mengaud, J., and Cossart, P., 1998, The *inlA* gene of *Listeria monocytogenes* LO28 harbors a nonsense mutation resulting in release of internalin. *Infect. Immun.* **66**:3420–3422.

Karunasagar, I., Senghaas, B., Krohne, G., and Goebel, W., 1994, Ultrastructural study of *Listeria monocytogenes* entry into cultured human colonic epithelial cells. *Infect. Immun.* **62**:3554–3558.

Kobe, B., and Deisenhofer, J., 1995, Proteins with leucine-rich repeats. *Curr. Opin. Struct. Biol.* **5**:409–416.

Kocks, C., Gouin, E., Tabouret, M., Berche, P., Ohayon, H., and Cossart, P., 1992, *Listeria monocytogenes*-induced actin assembly requires the *actA* gene product, a surface protein. *Cell* **68**:521–531.

Köhler, S., Leimeister-Wächter, M., Chakraborty, T., Lottspeich, F., and Goebel, W., 1990, The gene coding for protein p60 of *Listeria monocytogenes* and its use as a species specific probe for *Listeria monocytogenes*. *Infect. Immun.* **58**:1943–1950.

Köhler, S., Bubert, A., Vogel, M., and Goebel, W., 1991, Expression of the *iap* gene coding for protein p60 in *Listeria monocytogenes* is controlled on the posttranscriptional level. *J. Bacteriol.* **173**:4668–4674.

Kreft, J., Bohne, J., Gross, R., Kestler, H., Sokolovic, Z., and Goebel, W., 1995, Control of *Listeria monocytogenes* virulence by the transcriptional regulator PrfA. In: *Signal Transduction and Bacterial Virulenc* (R. Rappuoli, V. Scarlato, and B. Arico. R.G., eds.), Landes Company, Austin, Tex. pp. 129–142.

Kügler, S., Schüller, S., and Goebel,W., 1997, Involvement of MAP-kinases and -phosphatases in uptake and intracellular replication of *Listeria monocytogenes* in J774 macrophage cells. *FEMS Microbiol. Lett.* **157**:131–136.

Kuhn, M., 1998, The microtubule depolymerizing drugs nocodazole and colchicine inhibit the uptake of *Listeria monocytogenes* by $P388D_1$ macrophages. *FEMS Microbiol. Lett.* **60**:87–90.

Kuhn, M., and Goebel, W., 1989, Identification of an extracellular protein of *Listeria monocytogenes* possibly involved in the intracellular uptake by mammalian cells. *Infect. Immun.* **57**:55–61.

Kuhn, M., and Goebel, W., 1995, Molecular studies on the virulence of *Listeria monocytogenes*. *Genet. Eng.* **17**:31–51.

Kuhn, M., and Goebel, W., 1998, Pathogenesis of *Listeria monocytogenes*. In: *Listeria, Listeriosis and Food Safet* (E.T. Ryser and E.H. Marth. M., eds.), Dekker Inc. pp. 97–130.

Kuhn, M., Kathariou, S., and Goebel, W., 1988, Hemolysin supports survival but not entry of the intracellular bacterium *Listeria monocytogenes*. *Infect. Immun.* **56**:79–82.

Kuhn, M., Engelbrecht, F., Sokolovic, Z., Kügler, S., Schüller, S., Bubert, A., Karunasagar, I., Böckmann, R., Hauf, N., Demuth, A., Kreft, J., and Goebel, W., 1997, Interaction of intracellular bacteria with mammalian host cells and host cell responses. *Nova Acta Leopoldina* NF 75, **301**:207–221.

Kulich, S.M., Yahr, T.L., Mende-Mueller, L.M., Barbieri, J.T., and Frank, D.W., 1994, Cloning the structural gene for the 49 kDa form of exoenzyme S (*exoS*) from *Pseudomonas aeruginosa* strain 388. *J. Biol. Chem.* **269**:10431–10437.

Lebrun, M., Mengaud, J., Ohayon, H., Nato, F., and Cossart, P., 1996, Internalin must be on the bacterial surface to mediate entry of *Listeria monocytogenes* into epithelial cells. *Mol. Microbiol.* **21**:579–592.

Lecuit, M., Ohayon, H., Braun, L., Mengaud, J., and Cossart, P., 1997, Internalin of *Listeria monocytogenes* with an intact leucine-rich repeat region is sufficient to promote internalization. *Infect. Immun.* **65**:5309–5319.

Lingnau, A., Domann, E., Hudel, M., Bock, M., Nichterlein, T., Wehland, J., and Chakraborty, T., 1995, Expression of the *Listeria monocytogenes* EGD *inlA* and *inlB* genes, whose products mediate bacterial entry into tissue culture cell lines, by PrfA-dependent and -independent mechanisms. *Infect. Immun.* **63**:3896–3903.

Maganti, S., Pierce, M.M., Hoffmaster, A., Rodgers, F.G., 1998, The role of sialic acid in opsonin-dependent and opsonin-independent adhesion of *Listeria monocytogenes* to murine peritoneal macrophages. *Infect. Immun.* **66**:620–626.

Mecsas, J., Raupach, B., and Falkow, S., 1998, The *Yersinia* Yops inhibit invasion of *Listeria, Shigella* and *Edwardsiella* but not *Salmonella* into epithelial cells. *Mol. Microbiol.* **28**:1269–1281.

Menard, R., Dehio, C., and Sansonetti, P.J., 1996. Bacterial entry into epithelial cells: the paradigm of *Shigella*. *Trends Microbiol.* **4**:220–226.

Mengaud, J., Lecuit, M., Lebrun, M., Nato, F., Mazie, J.-C., and Cossart, P., 1996a, Antibodies to the leucin-rich repeat region of internalin block entry of *Listeria monocytogenes* into cells expressing E-cadherin. *Infect. Immun.* **64**:5430–5433.

Mengaud, J., Ohayon, H., Gounon, P., Mege, R.-M., and Cossart, P., 1996b, E-cadherin is the receptor for internalin, a surface protein required for entry of *Listeria monocytogenes* into epithelial cells. *Cell* **84**:923–932.

Mounier, J., Ryter, A., Coquis-Rondon, M., and Sansonetti, P.J., 1990, Intracellular and cell-to-cell spread of *Listeria monocytogenes* involves interaction with F-actin in the enterocyte like cell line Caco-2. *Infect. Immun.* **58**:1048–1058.

Müller, S., Hain, T., Pashalidis, P., Lingnau, A., Domann, E., Chakraborty, T., Wehland, J., 1998, Purification of the *inlB* gene product of *Listeria monocytogenes* and demonstration of its biological activity. *Infect. Immun.* **66**:3128–3133.

Murray, E.G.D., Webb, R.A., and Swann, M.B.R., 1926, A disease of rabbits characterized by large mononuclear leucocytosis, caused by a hitherto indescribed bacillus, *Bacterium monocytogenes* (n. sp.). *J. Pathol. Bacteriol.* **29**:407–439.

Parida, S.K., Domann, E., Rohde, M., Müller, S., Darji, A., Hain, T., Wehland, J., and Chakraborty, T., 1998, Internalin B is essential for adhesion and mediates the invasion of *Listeria monocytogenes* into human endothelial cells. *Mol. Microbiol.* **28**:81–93.

Pierce, M.M., Gibson, R.E., and Rodgers, F.G., 1996, Opsonin-independent adherence and phagocytosis of *Listeria monocytogenes* by murine peritoneal macrophages. *J. Med. Microbiol.* **45**:258–262.

Portnoy, D.A., Jacks, P.S., and Hinrichs, D.J., 1988, Role of hemolysin for the intracellular growth of *Listeria monocytogenes*. *J. Exp. Med.* **167**:1459–1471.

Pron, B., Boumaila, C., Jaubert, F., Sarnacki, S., Monnet, J.P., Berche, P., and Gaillard, J.L., 1998, Comprehensive study of the intestinal stage of listeriosis in a rat ligated ileal loop system. *Infect. Immun.* **66**:747–755.

Raffelsbauer, D., Bubert, A., Engelbrecht, F., Scheinpflug, J., Simm, A., Hess, J., Kaufmann, S.H.E., and Goebel., W., 1998, The gene cluster *inlC2DE* of *Listeria monocytogenes* contains additional new internalin genes and is important for virulence in mice. *Mol. Gen. Genet.* (in press).

Raybourne, R.B., and Bunning, V.K., 1994, Bacterium-host cell interaction on the cellular level: fluorescent labeling of the bacteria and analysis of short-term bacterium-phagocyte interaction by flow cytometry. *Infect. Immun.* **62**:665–672.

Rosqvist, R., Forsberg, A., and Wolf-Watz, H., 1991, Intracellular targeting of the *Yersinia* YopE cytotoxin in mammalian cells induces actin microfilament disruption. *Infect. Immun.* **59**:4562–4569.

Ruhland, G.J., Hellwig, M., Wanner, G., and Fiedler, F., 1993, Cell-surface location of *Listeria*-specific protein p60—detection of *Listeria* cells by indirect immunofluorescence. *J. Gen. Microbiol.* **139**:609–616.

Sawyer, R.T., Drevets, D.A., Campbell, P.A., and Potter, T.A., 1996, Internalin A can mediate phagocytosis of *Listeria monocytogenes* by mouse macrophage cell lines. *J. Leukoc. Biol.* **60**:603–610.

Schuchat, A., Swaminathan, B., and Broome, C.V., 1991, Epidemiology of human listeriosis. *Clin Microbiol. Rev.* **4**:169–183.

Sheehan, B., Kocks, C., Dramsi, S., Gouin, E., Klarsfeld, A.D., Mengaud, J., and Cossart, P., 1994, Molecular and genetic determinants of the *Listeria monocytogenes* infectious process. *Curr. Top. Microbiol. Immunol.* **192**:187–216.

Sokolovic, Z., Riedel, J., Wuenscher, M., and Goebel, W., 1993, Surface associated, PrfA-regulated proteins of *Listeria monocytogenes* synthesized under stress conditiones. *Mol. Microbiol.* **8**:219–227.

Tang, P., Rosenshine, I., and Finlay, B.B., 1994, *Listeria monocytogenes*, an invasive bacterium, stimulates MAP kinase upon attachment to epithelial cells. *Mol. Biol. Cell* **5**:455–464.

Tang, P., Rosenshine, I., Cossart, P., and Finlay, B.B., 1996, Listeriolysin O activates mitogen-activated protein kinase in eukaryotic cells. *Infect. Immun.* **64**:2359–2361.

Tang, P., Sutherland, C.L., Gold, M.R., and Finlay, B.B., 1998, *Listeria monocytogenes* invasion of epithelial cells requires the MEK-1/ERK-2 mitogen-activated protein kinase pathway. *Infect. Immun.* **66**:1106–1112.

Temm-Grove, C.T., Jokusch, B., Rohde, M., Niebuhr, K., Chakraborty, T., and Wehland, J., 1994, Exploitation of microfilament proteins by *Listeria monocytogenes*: microvillus-like composition of the comet tails and vectorial spreading in polarized epithelial sheets. *J. Cell Sci.* **107**:2951–2960.

Tilney, L.G., and Portnoy, D.A., 1989, Actin filaments and the growth, movement, and spread of the intracellular bacterial parasite, *Listeria monocytogenes*. *J. Cell Biol.* **109**:1597–1608.

Van Langendonck, N., Velge, P., and Bottreau, E., 1998, Host cell protein tyrosine kinases are activated during the entry of *Listeria monocytogenes*. Possible role of pp60$^{c\text{-}src}$ family protein kinases. *FEMS Microbiol. Lett.* **162**:169–176.

Velge, P., Bottreau, E., Kaeffer, B., and Pardon, P., 1994a, Cell immortalization enhances *Listeria monocytogenes* invasion. *Med. Microbiol. Immunol.* **183**:145–158.

Velge, P., Bottreau, E., Kaeffer, B., Yurdusev, N., Pardon, P., and Van Langendonck, N., 1994b, Protein tyrosine kinase inhibitors block the entries of *Listeria monocytogenes* and *Listeria ivanovii* into epithelial cells. *Mircobial Path.* **17**:37–50.

Weiglein, I., Goebel, W., Troppmair, J., Rapp, U.R., Demuth, A., and Kuhn, M., 1997, *Listeria monocytogenes* infection of HeLa cells results in LLO mediated transient activation of the Raf-MEK-MAP kinase pathway. *FEMS Microbiol. Lett.* **148**:189–195.

Wells, C.L., van de Westerlo, E.M., Jechorek, R.P., Haines, H.M., Erlandsen, S.L., 1998, Cytochalasin-induced actin disruption of polarized enterocytes can augment internalization of bacteria. *Infect. Immun.* **66**:2410–2419.

Wuenscher, M.D., Köhler, S., Bubert, A., Gerike, U., and Goebel, W., 1993, The *iap* gene of *Listeria monocytogenes* is essential for cell viability and its gene product, p60, has bacteriolytic activity. *J. Bacteriol.* **175**:3491–3501.

Chapter 17

Host-Plant Invasion by Rhizobia

V. Viprey, X. Perret, and W.J. Broughton

1. PROCESS OF INFECTION

Colonization of legume roots by compatible soil bacteria of the genera *Azorhizobium*, *Bradyrhizobium*, *Mesorhizobium*, *Rhizobium* and *Sinorhizobium* (collectively known as rhizobia) leads to the formation of specialized nitrogen-fixing organs called nodules. Signals produced by both partners control specificity. Flavonoids found in root exudates trigger the expression of the rhizobial genes (*nod*, *nol*, *noe*) required for nodulation (Fellay *et al.*, 1995). Many *nod* loci are involved in the synthesis and secretion of Nod-factors, a family of complex lipo-chito-oligosaccharides (Hanin *et al.*, 1998*b*; Dénarié *et al.*, 1996). Nod-factors initiate division of nodule meristems and permit entry of rhizobia into the host-plant (Relic *et al.*, 1994).

Rhizobial attachment to root-hairs precedes root-hair deformation (Had) which is the first morphologically distinguishable step in the infection process. Extreme (>360°) curling of the root-hair (Hac), which is an essential prelude to bacterial entry then follows. Localized lysis of the plant cell wall followed by the invagination of the plasma membrane entraps bac-

V. VIPREY, X. PERRET, and W.J. BROUGHTON Laboratoire de Biologie Moléculaire des Plantes Supérieures (LBMPS), Université de Genève, 1292 Chambéry/Genève, Switzerland.

Subcellular Biochemistry, Volume 33: Bacterial Invasion into Eukaryotic Cells, edited by Oelschlaeger and Hacker. Kluwer Academic / Plenum Publishers, New York, 2000.

teria in the curled region at the tip of the hair. New cell wall materials are deposited around the site of invagination, and result in the formation of an infection thread. As bacteria penetrate the plant, this tubular structure extends through layers of cells, so conducting the invading rhizobia to the root cortex. Concomitantly, the target root cortical-cells divide to form nodule primordia. Upon reaching the meristem, the growing infection threads release rhizobia into the cytoplasm of dividing plant cells where they enlarge and differentiate into nitrogen-fixing bacteroids (Cohn *et al.*, 1998; Mylona *et al.*, 1995; Dénarié *et al.*, 1992; Fischer and Long, 1992; Hirsch, 1992; Kijne *et al.*, 1992).

In fully functional nodules, bacteroids provide assimilated forms of nitrogen to the host plant, which in turn fuel the symbionts with carbohydrates. Two morphologically distinguishable types of nodules exist. In legumes that form indeterminate nodules, cell division occurs in the inner cortex, whereas primordia of determinate nodules develop from outer cortical cells. Hence, rhizobial invasion of plants possessing indeterminate nodules requires sustained growth of the infection-thread to reach the nodule meristem (Stacey *et al.*, 1991). In addition to Nod factors, a variety of prokaryotic signals are known to control successive stages of the plant infection process. Amongst these are surface and extra-cellular polysaccharides (EPS), as well as several proteins exported via type I or type III secretion systems.

2. ROLE OF BACTERIAL CELL-SURFACE COMPONENTS

Nod-factors induce the host-specific activation of cortical-cell division and root-hair curling. Later steps, such as infection-thread formation, propagation, and bacterial release into the cytoplasm of infected cells, require specific constituents of the rhizobial cell wall. These include extra-cellular- and lipo-polysaccharides (EPS, LPS), capsular polysaccharides (CPS, KPS), and cyclic β-glucans (summarized in Figure 1).

2.1. Extra-Cellular- and Lipo-Polysaccharides

Rhizobial EPS and surface polysaccharides form a complex macromolecular structure at the bacterial-plant interface. EPS either remain associated with the bacterial cell surface and thus form a bound capsule layer, or are secreted into the medium as extra-cellular slime (Whitfield and Valvano, 1993; Figure 1). Mutant strains of *Rhizobium meliloti*, deficient in EPS (Exo^-) production, form empty, non nitrogen-fixing nodules on their natural host *Medicago sativa* (Leigh *et al.*, 1985).

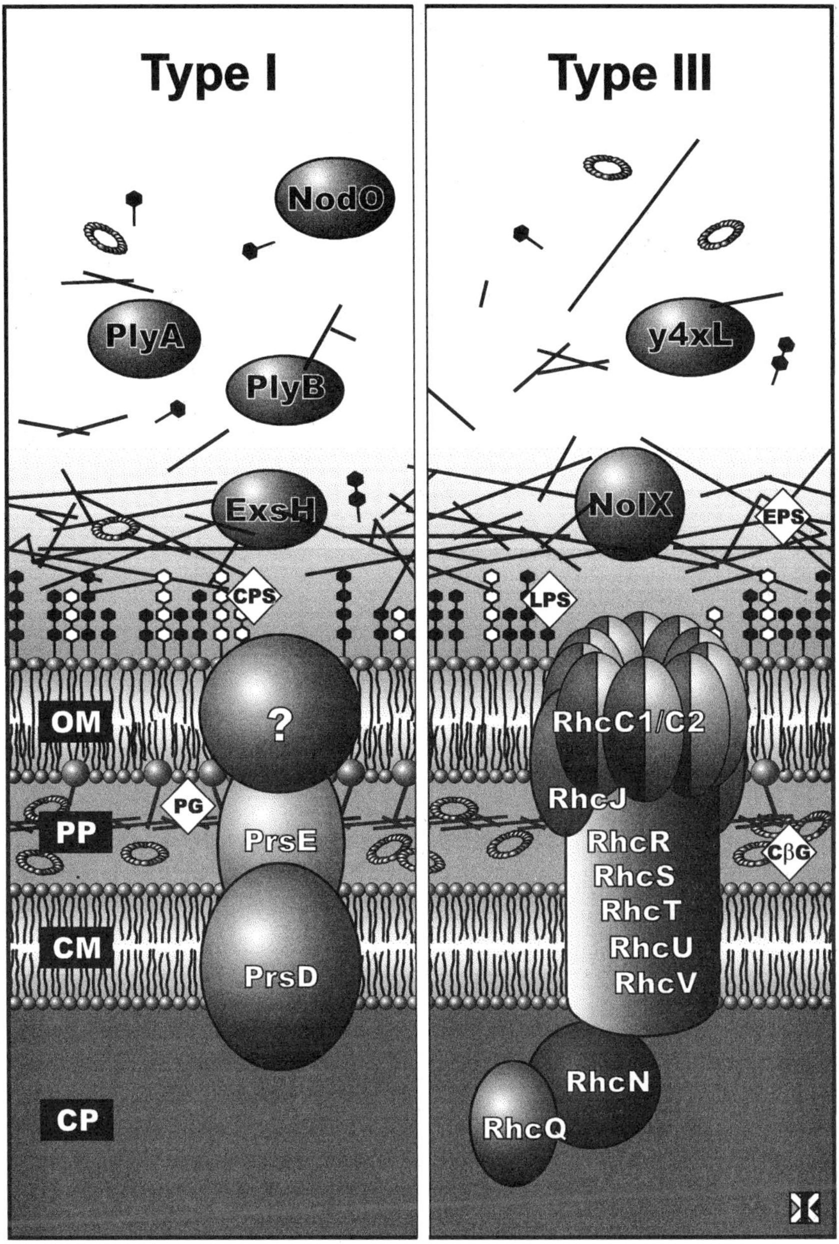

FIGURE 1. Cross-section of the rhizobial cell surface and its major constituents. Predicted positions of the proteins which form the type I and type III secretion systems of *R. leguminosarum* and NGR234 respectively, are shown in the cell cytoplasm (CP) as well as in the cytoplasmic (CM) and outer membranes (OM). Type I secreted proteins include NodO, PlyA and PlyB of *R. leguminosarum*, as well as ExsH of *R. meliloti*. Proteins secreted via the type III machinery of NGR234 include NolX and y4xL. Also drawn are other constituents, such as the cyclic-β-glucans (CβG) and peptidoglycans (PG) found in the periplasm (PP), as well as the exo-polysaccharides (EPS), lipo-polysaccharides (LPS) and capsular polysaccharides (CPS) of the cell surface.

Yang *et al.* (1992*a*) showed that inoculation of *M. sativa* with *R. meliloti* Exo$^-$ mutants, resulted in delayed root-hair curling, and that infection-threads aborted within peripheral cells of the nodule. Similarly, Exo$^-$ mutants of *Rhizobium* sp. NGR234 (Chen *et al.*, 1985), *R. leguminosarum* bv. *trifolii* (Chakravorty *et al.*, 1982), and *R. leguminosarum* bv. *viciae* (Borthakur *et al.*, 1986), are ineffective on their respective hosts, *Leucaena leucocephala*, *Trifolium repens* and *Pisum sativum*. In contrast, Exo$^-$ mutants of *R. leguminosarum* bv. *phaseoli* (Diebold and Noel, 1989) and *R. fredii* (Kim *et al.*, 1989), are fully effective on *Phaseolus vulgaris* and *Glycine max*. Although EPS$^-$ mutants of *R. loti* still nodulate the determinate host *Lotus pedunculatus*, they only form small, ineffective nodules on *L. leucocephala* (Hotter and Scott, 1991).

Taken together, these data suggest that EPS are necessary for effective nodulation of plants that form indeterminate nodules. Various exceptions to this rule exist however. Parniske *et al.* (1994) showed that an EPS-defective mutant of *B. japonicum* was altered in its ability to nodulate the determinate hosts *Glycine soja* and *G. max*. This mutant induced the formation of empty and ineffective nodule-like structures on *G. soja*. But, despite a significant delay in nodulation and a dramatic loss in competitiveness, it still formed nitrogen-fixing nodules on *G. max,* (Parniske *et al.*, 1994, 1993). Furthermore, a mutant of *Rhizobium* strain TAL1145, deficient in EPS synthesis nodulated various hosts, independently of their nodule type (Parveen *et al.*, 1997).

R. meliloti strain SU47 produces both high and low molecular weight (HMW and LMW) acidic EPS's (Leigh and Lee, 1988; Zevenhuizen and van Neerven, 1983). These oligo-saccharides, referred to as EPS I, are polymers of an octa-saccharide repeating unit of succinoglycan (Figure 2). Each sub-unit is composed of one galactose and seven glucose residues, with acetyl, succinyl, and pyruvyl modifications (Reuber and Walker, 1993a; Aman *et al.*, 1981). An *exoH* mutant of *R. meliloti* which lacks LMW but produces a succinoglycan lacking the succinyl substituent (Leigh and Lee, 1988; Leigh *et al.*, 1987), forms ineffective nodules on *M. sativa*, suggesting that specific succinoglycans are required for successful invasion of nodules. Also, addition of a fraction containing tetrameric LMW succinoglycans to *M. sativa* plants inoculated with *exo*$^-$ mutants, restores nitrogen fixation (Battisti *et al.*, 1992).

Genes responsible for the biosynthesis of EPS I are part of a 25 kb cluster located on the second mega-plasmid, pSym*b,* of *R. meliloti*. The octa-saccharide subunits of EPS I are assembled on a lipid carrier, starting with the single galactose, followed by glucose residues and various other substituents (Reuber and Walker, 1993a). *exoY* and *exoF* are responsible for

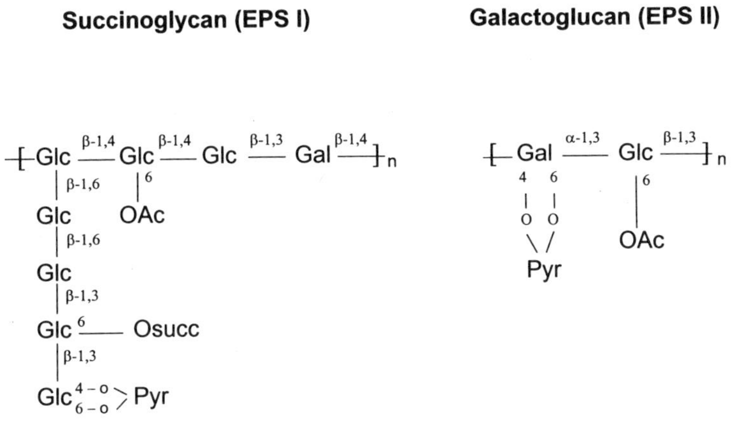

FIGURE 2. Chemical structures of succinoglycan and galactoglucan from *R. meliloti* (Leigh and Walker, 1994). Abbreviations are: Glc, glucose; Gal, galactose, OAc, acyl group; Osucc, succinyl group; Pyr, pyruvyl group.

the addition of the first galactose, whereas ExoA, ExoL, ExoM, ExoO, ExoU, and ExoW proteins are required for subsequent additions of sugars. Products of *exoH*, *exoZ* and *exoV* add the acetyl, succinyl, and pyruvyl substitutions, while ExoP, ExoQ and ExoT are required for polymerisation and secretion (Reuber and Walker, 1993a, 1993b; Leigh *et al.*, 1987). In addition, the products of the *exoB*, *exoC* and *exoN* genes are responsible for the synthesis of the sugar precursors (Leigh and Walker, 1994). Finally, ExoK specifically depolymerizes nascent succinoglycan chains, generating the LMW succinoglycan fraction (York and Walker, 1998; 1997; Becker *et al.*, 1993; Glucksmann *et al.*, 1993).

Mutants of *R. meliloti* SU47 deficient in EPS I synthesis, still produce another exo-polysaccharide (EPS II, a galactoglucan; Figure 2) if a second mutation in either of the chromosomal loci *expR101* or *mucR* results in derepression of its synthesis (Her *et al.*, 1990; Glazebrook and Walker, 1989; Zhan *et al.*, 1989). Only the *expR101* mutant produces LMW EPS II capable of restoring nodule infection on *M. sativa*, suggesting that EPS II can substitute for the lack of EPS I during nodule invasion (Gonzàles *et al.*, 1996). Biosynthesis of EPS II in *R. meliloti* is encoded by the 32 kb *exp* gene cluster (Becker *et al.*, 1997; Glazebrook and Walker, 1989) which is also carried on pSym*b*.

Thus, in *R. meliloti*, the LMW but not HMW forms of EPS I and EPS II are apparently essential for nodule invasion. Specificity for certain forms of EPS is not ubiquitous amongst rhizobia however, as both HMW and

LMW EPS of NGR234 and *R. trifolii* ANU794 are functional on *L. leucocephala* and *Trifolium pratense*, respectively (Djordjevic *et al.*, 1987). In addition, a strain of NGR234 containing the *exo* genes of *R. meliloti*, produces succinoglycans which can substitute for EPS of NGR234 during nodulation of *L. leucocephala* (Gray *et al.*, 1991). Increasing evidence suggests that EPS's mediate nodule invasion by preventing host-plant defense responses (Niehaus *et al.*, 1993; Pühler *et al.*, 1991), as in *M. sativa* where LMW EPS I of *R. meliloti* is able to suppress the host defense system (Niehaus *et al.*, 1996; Parniske *et al.*, 1994). These defense responses, which include abnormal thickness of the nodule apical cortex, and the increase in phenolic compounds bound to the nodule cell wall, are reminiscent of those induced by phyto-pathogens (Niehaus *et al.*, 1993).

Lipo-polysaccharides (LPS) which constitute the major structural component of the outer-membrane of gram-negative bacteria, also play a critical role in *Rhizobium*-legume symbioses (Figure 1). Type I LPS molecules (LPS I) are anchored into the outer membrane by their lipid A substituents (Carlson *et al.*, 1992), which are formed of an oligo-saccharide core (via 3-deoxy-D-*manno*-2-octulosonic-acid—Kdo), and the O-antigen, a strain-specific antigenic polysaccharide chain. Little is known about the biochemical function of gene products involved in their biosynthesis, although several *lps* genetic loci have been identified (García-de los Santos and Brom, 1997; Allaway *et al.*, 1996; Cava *et al.*, 1990).

In contrast to EPS's that are important for successful invasion of some indeterminate nodules, LPS's seem to be required in the early stages of symbiotic associations with determinate plants. For instance, LPS mutants of *B. japonicum* and *R. leguminosarum* bv. *phaseoli* are unable to nodulate *G. max* and *Phaseolus vulgaris* respectively (Stacey *et al.*, 1991; Noel *et al.*, 1986). In both cases, infection threads abort soon after initiation. Although complete LPS molecules are necessary for normal infection thread development in *P. vulgaris* (Cava *et al.*, 1989), they also seem to be important at later stages in the infection of hosts possessing indeterminate nodules. Inoculation of *P. sativum* with various classes of mutants of *R. leguminosarum* bv. *viciae* which produce LPS's with altered O-antigens, leads to abnormal nodule development in which few bacteria are released into the cytoplasm of the host cells. Abnormal, asynchronously-dividing bacteria, with altered peri-bacteroid membranes also form (Perotto *et al.*, 1994; de Maagd *et al.*, 1989*a*; Priefer *et al.*, 1989). Cytological evidence suggests that the formation of abnormal infection threads, fewer of which reach a restricted number of nodule cells, is a direct consequence of localized host-defense responses (Perotto *et al.*, 1994). Various alterations in the LPS structure probably affect the stability of the outer membrane, and thus result in abnormal bacteroid development. Since an intact O-antigen is required for sustained growth of the infection thread (Brink *et al.*, 1990), it is possible that it normally masks the oligo-saccharidic core. Then, when exposed in mutants incapable of synthesizing O-antigens, the LPS core becomes an elicitor of host plant-defense responses (Perotto *et al.*, 1994). This is consistent with the finding that normal nodule development occurs in the presence of EPS mutants capable of synthesizing long O-antigens (Kannenberg *et al.*, 1992).

In some cases, LPS-like structures can also substitute for missing EPS's (Putnoky *et al.*, 1990; Williams *et al.*, 1990). These alternative molecules, which are Kdo-rich polysaccharides similar to the capsular antigens of *E. coli* (Reuhs *et al.*, 1993), are known as K-antigens in rhizobia.

2.2. Acidic Capsular Polysaccharides (K Antigens, KPS)

The capsular polysaccharides (CPS) of bacteria form an extra-cellular layer at the surface of the cell wall (Figure 1). Although they also contain Kdo, the polysaccharide moiety of CPS is attached to phospholipids instead of lipid A. In animal pathogens, capsules protect bacteria from host-defense responses (Jann and Jann, 1990). Rhizobial production of KPS, a class of CPS structurally analogous to group II K antigens of *E. coli*, was first demonstrated in *R. fredii* USDA205 and *R. meliloti* AK631 (Reuhs *et al.*, 1993). In contrast to Exo$^-$ mutants of *R. meliloti* SU47, *R. meliloti* strain AK631 (an *exoB* mutant of Rm41) is still able to nodulate *M. sativa*. This phenotype is linked to the presence of *rkpZ* (formely known as *lpsZ*), a plasmid encoded gene which promotes the export of low-molecular-weight KPS (Reuhs *et al.*, 1995; Williams *et al.*, 1990).

Putnoky *et al.* (1990) showed that in *R. meliloti* strain Rm41, the locus *fix-23* is necessary for the synthesis of a strain-specific surface polysaccharide, later characterised as KPS (Petrovics *et al.*, 1993). KPS can substitute for EPS in *R. meliloti—M. sativa* symbioses. The *fix-23* locus contains *rkpABCDEF*, the products of which are probably involved in the synthesis of a lipid carrier or anchor for KPS (Petrovics *et al.*, 1993), and *rkpGHIJ* which probably encodes proteins involved in the modification and export of KPS (Kiss *et al.*, 1997). *R. fredii* strains USDA205, and USDA257 produce two distinct KPS's. One consists of Kdo and a mannose repeat (Kdo and galactose in strain USDA205), whereas the second KPS contains Kdo and 2-*O*-methylmannose (Forsberg and Reuhs, 1997). In contrast, the primary K antigen of *R. meliloti* AK631 consists of disaccharide repeats of 4-deoxy-4-aminuronic acid and a variant of Kdo (Reuhs *et al.*, 1993).

KPS's are present in various rhizobia, including *R. meliloti, R. fredii* and NGR234 (Reuhs, 1996). Unlike EPS's, KPS's are strain-specific antigens. They are highly variable in composition, linkage, and substitutions as well as in size even within the same species. Surprisingly, despite significant structural differences, KPS and EPS of *R. meliloti* AK631 apparently perform the same function during nodule invasion. As KPS's but not EPS's of AK631 induce the expression of genes related to flavonoid production in the leaves of *M. sativa* (Becquart-de Kozak *et al.*, 1997) it is possible that they promote infection via different pathways, however. Furthermore, modulation of KPS synthesis in *R. fredii* by plant compounds (Reuhs *et al.*, 1994), indicates that plant signal molecules released in response to KPS could also play a role in further modifications or even be responsible for the presence/absence of the bacteroid capsule (Olsen *et al.*, 1992).

2.3. Periplasmic Cyclic β-Glucans

Cyclic β-glucans are almost exclusively found in the *Rhizobiaceae*. Depending on the species, growth stage and culture conditions, these compounds, which normally accumulate in the periplasm of rhizobia, may also be exported into the extra-cellular medium (Geiger *et al.*, 1991; Breedveld *et al.*, 1990; Figure 1). *Rhizobium* and *Agrobacterium* species synthesize cyclic β-(1,2)-glucans with 17 to 25 glucose residues, some of which may be substituted with non-sugar moieties such as phosphoglycerol (Breedveld and Miller, 1994). In *Bradyrhizobium* species, the glucose residues (11 to 13) are linked by β-(1,3)- and β-(1,6)-glycosidic bonds and may also be substituted with phosphocholine (Rolin *et al.*, 1992). In *R. meliloti*, the nodule development genes *ndvA* and *ndvB*, are involved in the transport and synthesis of cyclic β-(1,2)-glucans (Ielpi *et al.*, 1990; Stanfield *et al.*, 1988). *ndvA* and *ndvB* mutants induce ineffective pseudo-nodules on *M. sativa* which lack infection threads, suggesting that periplasmic cyclic β-(1,2)-glucans are probably required at the same stage of nodule infection as EPS (Dylan *et al.*, 1990*b*; Dylan *et al.*, 1986). Spontaneous pseudo-revertants of *R. meliloti ndv* mutants which are symbiotically effective, fail to synthesize the periplasmic β-(1,2)-glucans however (Dylan *et al.*, 1990b). Further analysis showed that restoration of nodulation is linked to modifications of the surface of the cells, including protein composition and/or in the amount of highly polymerized EPS (Nagpal *et al.*, 1992; Soto *et al.*, 1992).

Despite the fact that cells that fail to produce cyclic β-glucans have altered cell surface properties which lead to reduced motility, as well as greater resistance to some bacteriophages or sensitivity to antibiotics, the exact role of β-glucans during plant invasion remains uncertain. It has been suggested that the failure of *ndv* mutants to form symbiotic associations results from their inability to adapt to changes in osmolarity (Quandt *et al.*, 1992; Dylan *et al.*, 1990*a*). A nodulation deficient *ndvC* mutant of *B. japonicum* altered in cyclic β-glucans lacking β-(1,6) glycosidic linkages, is still able to grow in hypo-osmotic medium however (Bhagwat *et al.*, 1996). It was suggested that cyclic β-glucans would primarily function as suppressors of the host-plant defenses. In fact, production of anti-microbial phytoalexins in *G. max* in response to β-glucans of fungal pathogens can be progressively inhibited by the addition of the structurally related bradyrhizobial β-(1,3)(1,6)-glucans (Mithöfer *et al.*, 1996). Furthermore, mutants of *R. leguminosarum* bv. *viciae* lacking cyclic β-(1,2)-glucans induce defense responses in host plants (Yang *et al.*, 1992*b*).

Another property of cyclic β-glucans is their capacity to form inclusion complexes with hydrophobic guest molecules, such as flavonoids (Breedveld and Miller, 1994). These glucan-guest molecule complexes are probably able to stimulate nodulation by increasing the solubility of the signal molecules released by host plants. Observations that exogenous addition of cyclic β-(1,2)-glucans to *R. leguminosarum* bv. *trifolii*—*T. repens* and *R. meliloti*—*M. sativa* systems leads to an increased number of nodules, supports this hypothesis (Dylan *et al.*, 1990*b*; Abe *et al.*, 1982). More over, the cyclic β-(1,2)-glucans of *R. fredii*, as well as the β-(1,3) and β-(1,6)-glucans of *B. japonicum*, are functionally equivalent to those of *R. meliloti* in promoting nodulation of *M. sativa* (Bhagwat *et al.*, 1992).

3. SECRETION OF BACTERIAL PROTEINS DURING NODULATION

3.1. NodO, a Signal Protein Secreted via a Type I Exporter

A number of protein secretion systems have been described in gram-negative bacteria (type I to type IV). Unlike the general secretory pathway (type II) and the type IV secretion machinery, type I and type III systems are *sec*-independent and direct proteins across inner and outer membranes without periplasmic intermediates. Bacterial toxins, such as the α-hemolysin of *E. coli*, are released in the extra-cellular space via type I exporters, whereas type III secreted proteins are mostly pathogenicity determinants, some of which are targeted to the eukaryotic cell cytosol (Hueck, 1998; Wandersman, 1996; Pugsley, 1993).

In various rhizobia, a number of secreted proteins play a role in nodulation. Amongst these, the flavonoid-inducible protein NodO of *R. leguminosarum* bv. *viciae* contains the characteristic nona-peptide tandem repeat RTX similar to that found in various bacterial toxins, and which is involved in Ca^{2+}-binding (Economou *et al.*, 1990; de Maagd *et al.*, 1989*b*). NodO is essential for nodulation of *P. sativum* and *Vicia sativa* in the absence of a functional *nodE* (Economou *et al.*, 1994; Downie and Surin,

1990). In addition, transfer of *nodO* to a *nodE* mutant of *R. leguminosarum* bv. *trifolii*, results in extension of the transconjugant's host-range to include *V. sativa* (Economou *et al.*, 1994). Similarly, a homologue of *nodO* in *Rhizobium* sp. BR816 (van Rhijn *et al.*, 1996) extends the host-range of a diversity of *Rhizobium* spp. and suppresses the nodulation deficiency of mutants lacking specific substituents at the non-reducing terminus of the Nod-factors (Vlassak *et al.*, 1998).

Although NodE (a β-keto-acyl-synthase) is required for the addition of unsaturated fatty acids to Nod-factors (Demont *et al.*, 1993; Spaink *et al.*, 1991), NodO is not directly involved in their biosynthesis and/or modification (Spaink *et al.*, 1991). Nonetheless, NodO is able to compensate for biologically ineffective Nod-factors produced by various mutant strains, probably by forming ion channels in the plasma membrane of plant cells (Sutton *et al.*, 1994). In fact, depolarization of the root-hair membranes and modifications in cytoplasmic Ca^{2+} levels are amongst the earliest changes measured in plant cells treated with Nod-factors (Gerhing *et al.*, 1997; Ehrhardt *et al.*, 1996; Fellé *et al.*, 1996; Ehrhardt *et al.*, 1992). In this respect, NodO could stimulate cation fluxes across the plant plasma membrane, and thus induce depolarisation and signal transduction even in the absence of biologically active Nod-factors (Sutton *et al.*, 1994).

The C-terminal domain of NodO is required for its secretion via the type I transport system (Figure 1) encoded by the *prsDE* genes of *R. leguminosarum* bv. *viciae* (Finnie *et al.*, 1997; Sutton *et al.*, 1996). At least three more proteins are exported via the Prs system of *R. leguminosarum*, two of which (PlyA and PlyB) are involved in processing bacterial EPS's (Finnie *et al.*, 1998). Although a *plyA plyB* double mutant is not symbiotically impaired, the Prs deficient mutant forms non-fixing nodules on *P. sativum* suggesting that at least one secreted protein other than NodO, PlyA and PlyB is necessary for effective symbiosis (Finnie *et al.*, 1998; 1997). Interestingly, a similar secretion system is present in a wide range of rhizobia, even though their genomes do not include *nodO* homologues (Scheu *et al.*, 1992). For instance, York and Walker (1998; 1997) found that a Prs type I secretion system of *R. meliloti* is involved in export of the ExsH endoglycanase which contributes to the production of the low-molecular-weight succinoglycan.

3.2. Type III Secretion Systems of Rhizobia

Delivery of proteins from bacteria to eukaryotic cells, via type III secretion systems (TTSS), is a key virulence mechanism in many plant and animal pathogens (Hueck, 1998). Recently, similar systems were also shown to play a role in the host-specific nodulation of legumes by rhizobia (Viprey *et al.*,

1998; Freiberg *et al.*, 1997; Meinhardt *et al.*, 1993). Genes of phyto-pathogenic bacteria responsible for eliciting hypersensitive responses and pathogenesis in plants are called *hrp* (Willis *et al.*, 1991; Lindgren *et al.*, 1986). To emphasize the similarities between plant and animal pathogens, apparently conserved genes are now known as *hrc* (*hrp* conserved; Bogdanove *et al.*, 1996), and encode components of a type III secretion machinery (Alfano and Collmer, 1997). Surprisingly, homologues of all *hrc* genes are present in pNGR234*a*, the symbiotic plasmid of NGR234 (Freiberg *et al.*, 1997). To clearly distinguish between TTSS components of phyto-pathogenic bacteria and those of symbiotic rhizobia, we have named the rhizobial "Hrc" homologues "Rhc" (for *rh*izobia *c*onserved) (Viprey *et al.*, 1998). As is the case for *hrp* genes, *rhc* loci of NGR234 are grouped in a single cluster which includes homologues of the *nolXWBTUV* genes of *R. fredii* USDA257 (Meinhardt *et al.*, 1993). Based on similarities to the Hrp secretion apparatus of plant pathogens, we proposed a model of the TTSS of NGR234 which includes two cytoplasmic and five inner-membrane proteins (RhcN, Q, as well as RhcR, S, T, U, and RhcV, respectively), two outer membrane proteins (RhcC1 and RhcC2), and one lipoprotein-associated outer membrane protein (RhcJ) (Viprey *et al.*, 1998; Figure 1).

TTSS's of bacterial pathogens are induced by contact with host cells and target virulence proteins directly into the eukaryotic cytosol, conditioning compatibility or incompatibility reactions (Lee, 1997). Similarly, export of *s*ignal-*r*esponsive proteins (SR) into the extra-cellular medium is flavonoid dependent and linked to a functional *nolXWBTUV* locus in strains of *R. fredii* (Bellato *et al.*, 1997; Krishnan *et al.*, 1995). In NGR234, two flavonoid inducible and pNGR234*a* encoded proteins, NolX and y4xL are secreted via the TTSS apparatus (Viprey *et al.*, 1998). As expected of flavonoid-inducible genes, expression of the TTSS loci is dependent on the symbiotic regulator NodD1 (Freiberg *et al.*, 1997; Krishnan *et al.*, 1995). Transcriptional activation of *rhc* loci by flavonoids is probably mediated by intermediary regulators, however (Hanin *et al.*, 1998*a*; Viprey *et al.*, 1998; Bellato *et al.*, 1996). For instance, the NodD1 controlled y4xI, which is homologous to the *hrp* transcriptional regulator HrpG of *Xanthomonas campestris* (Wengelnik *et al.*, 1996), is possibly a key intermediary in the regulatory cascade between flavonoids and activation of the TTSS machinery (Viprey *et al.*, 1998).

Although TTSS's are present in the genomes of NGR234, various strains of *R. fredii*, and *B. japonicum* (Palacios *et al.*, 1998), they are probably not ubiquitous amongst rhizobia. In fact, homologues of *rhc* and *nol* were not detected in the genomes of *R. meliloti* 2011 and *R. loti* NZP4010 (Viprey *et al.*, 1998). The precise function of proteins secreted via the TTSS, including the NolX and y4xL of NGR234 as well as the SR proteins of

USDA257, remains unknown. Obviously, they play an important role in the invasion of some host and non-host legumes. For instance, inactivation of the *nolXWBTUV* locus in USDA257 leads to the extension of the mutant host-range (Meinhardt *et al.*, 1993). Successful invasion of *Tephrosia vogelii* by NGR234 requires a functional TTSS, whereas nodulation of *Pachyrhizus tuberosus* is impaired when NolX and y4xL are actively secreted (Viprey *et al.*, 1998). Thus, proteins secreted via the TTSS seem to represent another determinant of host-specificity, possibly by directing compatible or incompatible reactions between symbionts and host-plants, and/or avoiding host-plant defense responses.

4. CONCLUSIONS AND PERSPECTIVES

Combined genetic and biochemical data show that successful *Rhizobium*-legume symbioses require the exchange of many molecular signals. Amongst these, Nod-factors synthesized by rhizobia largely control symbiotic specificity (Fellay *et al.*, 1995). A number of bacterial proteins such as NodO, y4xL and NolX which are secreted in response to plant signals, are also involved in host-range determination however. The need for the invading symbionts to display cell surface components "compatible" with the host cells also seems ubiquitous amongst rhizobia. Various phenotypes observed with *exo* and *lps* mutants on plants producing different nodule types, suggests that EPS's are probably essential during infection thread penetration whereas LPS's are needed for proper endocytosis as well as during bacteroid differentiation (Brewin *et al.*, 1991). Although the establishment of all known *Rhizobium*-legume symbiosis requires the presence of biologically active Nod-factors, invading rhizobia and legume-hosts have also evolved a variety of strategies that either improve nodulation efficiency or contribute to symbiotic specificity. Much remains to be done: (a) complete characterization of type I and type III secreted proteins and of their effects on their hosts; (b) identification of the plant symbiotic receptors, and; (c) elucidation of signal transduction pathways required for functional symbioses will explain recognition mechanisms between eukaryotic hosts and invading bacteria.

5. REFERENCES

Abe, M., Amemura, A., and Higashi, S., 1982, Studies on cyclic β-(1,2)-glucan obtained from periplasmic space of *Rhizobium trifolii* cells, *Plant Soil* **64**:315–324.

Alfano, J.R., and Collmer, A., 1997, The type III (Hrp) secretion pathway of plant pathogenic bacteria: trafficking harpins, Avr proteins, and death, *J. Bacteriol.* **179**(18):5655–5662.

Allaway, D., Jeyaretnam, B., Carlson, R.W., and Poole, P.S., 1996, Genetic and chemical characterization of a mutant that disrupts synthesis of the lipopolysaccharide core tetrasaccharide in *Rhizobium leguminosarum*, *J. Bacteriol.* **178**(21):6403–6406.
Aman, P., McNeil, M., Franzen, L.-E., Darvill, A.G., and Albersheim, P., 1981, Structural elucidation, using HPLC-MS and GLC-MS, of the acidic exopolysaccharide secreted by *Rhizobium meliloti* strain Rm1021. *Carbohydr. Res.* **95**:263–282.
Battisti, L., Lara, J.C., and Leigh, J.A., 1992, Specific oligosaccharide form of the *Rhizobium meliloti* exopolysaccharide promotes nodule invasion in alfalfa, *Proc. Natl. Acad. Sci. USA* **89**(12):5625–5629.
Becker, A., Kleickmann, A., Arnold, W., and Pühler, A., 1993, Analysis of the *Rhizobium meliloti exoH/exoK/exoL* fragment: ExoK shows homology to excreted endo-β-1,3–1,4 glucanases and ExoH resembles membrane proteins, *Mol. Gen. Genet.* **238**(1–2):145–154.
Becker, A., Ruberg, S., Kuster, H., Roxlau, A.A., Keller, M., Ivashina, T., Cheng, H.P., Walker, G.C., and Pühler, A., 1997, The 32-kilobase *exp* gene cluster of *Rhizobium meliloti* directing the biosynthesis of galactoglucan: genetic organization and properties of the encoded products, *J. Bacteriol.* **179**(4):1375–1384.
Becquart-de-Kozak, I., Reuhs, B.L., Buffard, D., Breda, C., Kim, J.S., Esnault, R., and Kondorosi A., 1997, Role of the K-antigen subgroup of capsular polysaccharides in the early recognition process between *Rhizobium meliloti* and alfalfa leaves, *Mol. Plant-Microbe Interact.* **10**(1):114–123.
Bellato, C.M., Balatti, P.A., Pueppke, S.G., and Krishnan, H.B., 1996, Proteins from cells of *Rhizobium fredii* bind to DNA sequences preceding *nolX*, a flavonoid-inducible *nod* gene that is not associated with a *nod* box, *Mol. Plant-Microbe Interact.* **9**(6):457–463.
Bellato, C., Krishnan, H.B., Cubo, T., Temprano, F., and Pueppke, S.G., 1997, The soybean cultivar specificity gene *nolX* is present, expressed in a *nodD*-dependent manner, and of symbiotic significance in cultivar-non specific strains of *Rhizobium* (*Sinorhizobium*) *fredii*, *Microbiol.* **143**:1381–1388.
Bhagwat, A.A., Gross, K.C., Tully, R.E., and Keister, D.L., 1996, β-glucan synthesis in *Bradyrhizobium japonicum*: characterization of a new locus (*ndvC*) influencing β-(1,6) linkages, *J. Bacteriol.* **178**(15):4635–4642.
Bhagwat, A.A., Tully, R.E., and Keister, D.L., 1992, Isolation and characterization of an *ndvB* locus from *Rhizobium fredii*, *Mol. Microbiol.* **6**(15):2159–2165.
Bogdanove, A.J., Beer, S.V., Bonas, U., Boucher, C.A., Collmer, A., Coplin, D.L., Cornelis, G.R., Huang, H.C., Hutcheson, S.W., Panopoulos, N.J., and Van Gijsegem, F., 1996, Unified nomenclature for broadly conserved *hrp* genes of phytopathogenic bacteria, *Mol. Microbiol.* **20**(3):681–683.
Borthakur, D., Barbur, C.E., Lamb, J.W., Daniels, M.J., Downie, J.A., and Johnston, A.W.B., 1986, A mutation that blocks exopolysaccharide synthesis prevents nodulation of peas by *Rhizobium leguminosarum* but not of beans by *R. phaseoli* and is corrected by cloned DNA from *Rhizobium* or the phytopathogen *Xanthomonas*, *Mol. Gen. Genet.* **203**(2):320–323.
Breedveld, M.W., and Miller, K.J., 1994, Cyclic β-glucans of members of the family *Rhizobiaceae*, *Microbiol. Rev.* **58**(2):145–161.
Breedveld, M.W., Zevenhuizen, L.P.T.M., and Zehnder, A.J.B., 1990, Excessive excretion of cyclic β-(1,2)-glucans by *Rhizobium trifolii* TA-1, *Appl. Environ. Microbiol.* **56**(7):2080–2086.
Brewin, N.J., 1991, Development of the legume root nodule, *Annu. Rev. Cell Biol.* **7**:191–226.
Brink, B.A., Miller, J., Carlson, R.W., and Noel, K.D., 1990, Expression of *Rhizobium leguminosarum* CFN42 genes for lipopolysaccharide in strains derived from different *R. leguminosarum* soil isolates, *J. Bacteriol.* **172**(2):548–555.

Carlson, R.W., Bhat, U.R., and Reuhs, B., 1992, *Rhizobium* lipopolysaccharides: Their structure and evidence for their importance in the nitrogen-fixing symbiotic infection of their host legumes, in: *Plant Biotechnology and Development*, (P.M. Gresshoff, ed.), CRC Press, Boca Raton, pp. 33–44.

Cava, J.R., Elias, P.M., Turowski, D.A., and Noel, K.D., 1989, *Rhizobium leguminosarum* CFN42 genetic regions encoding lipopolysaccharide structures essential for complete nodule development on bean plants, *J. Bacteriol.* **171**(1):8–15.

Cava, J.R., Tao, H., and Noel, K.D., 1990, Mapping of complementation groups within a *Rhizobium leguminosarum* CFN42 chromosomal region required for lipopolysaccharide synthesis, *Mol. Gen. Genet.* **221**(1):125–128.

Chakravorty, A.K., Zurowski, W., Shine, J., and Rolfe, B.G., 1982, Symbiotic nitrogen fixation: molecular cloning of *Rhizobium* genes involved in exopolysaccharide synthesis and effective nodulation, *J. Mol. Appl. Genet.* **1**(6):585–596.

Chen, H., Batley, M., Redmond, J., and Rolfe, B.G., 1985, Alteration of the effective nodulation properties of a fast-growing broad host range *Rhizobium* due to changes in exopolysaccharide synthesis, *J. Plant Physiol.* **120**:331–349.

Clover, R.H., Kieber, J., and Signer, E.R., 1989, Lipopolysaccharide mutants of *Rhizobium meliloti* are not defective in symbiosis, *J. Bacteriol.* **171**(7):3961–3967.

Cohn, J., Day, R.B., and Stacey, G., 1998, Legume nodule organogenesis, *Trends Plant Sci.* **3**(3):105–110.

de Maagd, R.A., Rao, A.S., Mulders, I.H.M., Goosen-de Roo, L., van Loosdrecht, M.C.M., Wijffelman, C.A., and Lugtenberg, B.J.J, 1989*a*, Isolation and characterization of mutants of *Rhizobium leguminosarum* bv. *viciae* 248 with altered lipopolysaccharides: Possible role of surface charge or hydrophobicity in bacterial release from the infection thread, *J. Bacteriol.* **171**(2):1143–1150.

de Maagd, R.A., Wijfjes, A.H., Spaink, H.P., Ruiz-Sainz, J.E., Wijffelman, C.A., Okker, R.J., and Lugtenberg, B.J., 1989*b*, *nodO*, a new *nod* gene of the *Rhizobium leguminosarum* biovar *viciae* sym plasmid pRL1J1, encodes a secreted protein, *J. Bacteriol.* **171**(12):6764–6770.

Demont, N., Debellé, R., Aurelle, H., Dénarié, J., and Promé, J.-C., 1993, Role of the *Rhizobium meliloti nodF* and *nodE* genes in the biosynthesis of lipo-oligosaccharidic nodulation factors, *J. Biol. Chem.* **268**(27):20134–20142.

Dénarié, J., Debellé, F., and Rosenberg, C., 1992, Signalling and host range variation in nodulation, *Annu. Rev. Microbiol.* **46**:497–531.

Dénarié, J., Debellé, F., and Promé, J.-C., 1996, *Rhizobium* lipo-chitooligosaccharide nodulation factors. signaling molecules mediating recognition and morphogenesis, *Annu. Rev. Biochem.* **65**:503–535.

Diebold, R., and Noel, K.D., 1989, *Rhizobium leguminosarum* exopolysaccharide mutants: biochemical and genetic analyses and symbiotic behavior on three hosts, *J. Bacteriol.* **171**(9): 4821–4830.

Djordjevic, S.P., Chen, H., Batley, M., Redmond, J.W., and Rolfe, B.G., 1987, Nitrogen fixation ability of exopolysaccharide synthesis mutants of *Rhizobium* sp. strain NGR234 and *Rhizobium trifolii* is restored by the addition of homologous exopolysaccharides, *J. Bacteriol.* **169**(1):53–60.

Downie, J.A., and Surin, B.P., 1990, Either of two *nod* gene loci can complement the nodulation defect of a *nod* deletion mutant of *Rhizobium leguminosarum* bv *viciae*, *Mol. Gen. Genet.* **222**(1):81–86.

Dylan, T., Ielpi, L., Stanfield, S., Kashyap, L., Douglas, C., Yanofsky, M., Nester, E., Helinski, D.R., and Ditta, G., 1986, *Rhizobium meliloti* genes required for nodule development are related to chromosomal virulence genes *in Agrobacterium tumefaciens*, *Proc. Natl. Acad. Sci. USA* **83**(12):4403–4407.

Dylan, T., Helinski, D.R, and Ditta, G.S., 1990*a*, Hypoosmotic adaptation in *Rhizobium meliloti* requires β-(1,2)-glucan, *J. Bacteriol.* **172**(3):1400–1408.

Dylan, T., Nagpal, P., Helinski, D.R., and Ditta, G.S., 1990*b*, Symbiotic pseudorevertants of *Rhizobium meliloti ndv* mutants, *J. Bacteriol.* **172**(3):1409–1417.

Economou, A., Hamilton, W.D.O., Johnston, A.W.B., and Downie, J.A., 1990, The *Rhizobium* nodulation gene *nodO* encodes a Ca^{2+}-binding protein that is exported without N-terminal cleavage and is homologous to haemolysin and related proteins, *EMBO J.* **9**(2):349–354.

Economou, A., Davies, A.E., Johnston, A.W.B., and Downie, J.A., 1994, The *Rhizobium leguminosarum* biovar *viciae nodO* gene can enable a *nodE* mutant of *Rhizobium leguminosarum* biovar *trifolii* to nodulate vetch, *Microbiol.* **140**:2341–2347.

Ehrhardt, D.W., Atkinson, E.M., and Long, S.R., 1992, Depolarization of alfalfa root hair membrane potential by *Rhizobium meliloti* Nod factors, *Science* **256**:998–1000.

Ehrhardt, D.W., Wais, R., and Long, S.R., 1996, Calcium spiking in plant root hairs responding to *Rhizobium* nodulation signals, *Cell* **85**(5):673–681.

Fellay, R., Rochepeau, P., Relic, B., and Broughton, W.J., 1995, Signals to and emanating from *Rhizobium* largely control symbiotic specificity, in: *Pathogenesis and Host Specificity in Plant Diseases. Histopathological, Biochemical, Genetic and Molecular Bases*, Volume 1 (U.S. Singh, R.P. Singh, and K. Kohmoto, eds.), Pergamon Elsevier Science Ltd., Oxford, pp. 199–220.

Fellé, H.H., Kondorosi, E., Kondorosi, A., and Schultze, M., 1996, Rapid alkalinization in alfalfa root hairs in response to rhizobial lipochitooligosaccharide signals, *Plant J.* **10**(2):295–301.

Finnie, C., Hartley, N.M., Findlay, K.C., and Downie J.A., 1997, The *Rhizobium leguminosarum prsDE* genes are required for secretion of several proteins, some of which influence nodulation, symbiotic nitrogen fixation and exopolysaccharide modification, *Mol. Microbiol.* **25**(1):135–146.

Finnie, C., Zorreguieta, A., Hartley, N.M., Downie, J.A., 1998, Characterization of *Rhizobium leguminosarum* exopolysaccharide glycanases that are secreted via a type I exporter and have a novel heptapeptide repeat motif, *J. Bacteriol.* **180**(7):1691–1699.

Fischer, R.F., and Long, S.R., 1992, *Rhizobium*-plant signal exchange, *Nature* **357**:655–660.

Forsberg, L.S., and Reuhs, B.L., 1997, Structural characterization of the K antigens from *Rhizobium fredii* USDA257: evidence for a common structural motif, with strain-specific variation, in the capsular polysaccharides of *Rhizobium* spp., *J. Bacteriol.* **179**(17):5366–5371.

Freiberg, C., Fellay, R., Bairoch, A., Broughton, W.J., Rosenthal, A., and Perret, X., 1997, Molecular basis of symbiosis between *Rhizobium* and legumes, *Nature* **387**:394–401.

García-de los Santos, A., and Brom, S., 1997, Characterization of two plasmid-borne *lps* β loci of *Rhizobium etli* required for lipopolysaccharide synthesis and for optimal interaction with plants, *Mol. Plant-Microbe Interact.* **10**(7):891–902.

Gehring, C.A., Irving, H.R., Kabbara, A.A., Parish, R.W., Boukli, N.M., and Broughton, W.J., 1997, Rapid, plateau-like increases in intracellular free calcium are associated with Nod-factor-induced root-hair deformation, *Mol. Plant-Microbe Interact.* **10**(7):791–802.

Geiger, O., Weissborn, A.C., Kennedy, E.P., 1991, Biosynthesis and excretion of cyclic glucans by *Rhizobium meliloti* 1021, *J. Bacteriol.* **173**(9):3021–3024.

Glazebrook, J., and Walker, G.C., 1989, A novel exopolysaccharide can function in place of the Calcofluor-binding exopolysaccharide in nodulation of alfalfa by *Rhizobium meliloti*, *Cell* **56**(4):661–672.

Glucksmann, M.A., Reuber, T.L., and Walker, G.C., 1993, Genes needed for the modification, polymerization, export, and processing of succinoglycan by *Rhizobium meliloti*: A model for succinoglycan biosynthesis, *J. Bacteriol.* **175**(21):7045–7055.

Gonzàles, J.E., Reuhs, B.L, and Walker, G.C., 1996, Low molecular weight EPS II of *Rhizobium meliloti* allows nodule invasion in *Medicago sativa. Proc. Natl. Acad. Sci. USA* **93**(16):8636–8641.

Gray, J.X., Zhan, H., Levery, S.B., Battisti, L., Rolfe, B.G., and Leigh, J.A., 1991, Heterologous exopolysaccharide production in *Rhizobium* sp. strain NGR234 and consequences for nodule development, *J. Bacteriol.* **173**(10):3066–3077.

Hanin, M., Jabbouri, S., Broughton, W.J., and Fellay, R., 1998*a*, SyrM1 of *Rhizobium* sp. NGR234 activates transcription of symbiotic loci and controls the level of sulfated Nod factors, *Mol. Plant-Microbe Interact.* **11**(5):343–350.

Hanin, M., Jabbouri, S., Broughton, W.J., Fellay, R., and Quesada-Vincens, D., 1998*b*, Molecular aspects of host-specific nodulation, in: *Plant-Microbe Interactions*, (G. Stacey and N.T. Keen, eds.), American Phytopathology Society, St Paul, MN, in press.

Her, G.-R., Glazebrook, J., Walker, G.C., and Reinhold, V.N., 1990, Structural studies of a novel exopolysaccharide produced by a mutant of *Rhizobium meliloti* strain Rm1021. *Carbohydr. Res.* **198**:305–312.

Hirsch, A.M., 1992, Developmental biology of legume nodulation, *New Phytol.* **122**:211–237.

Hotter, G.S., and Scott, D.B., 1991, Exopolysaccharide mutants of *Rhizobium loti* are fully effective on a determinate nodulating host but are ineffective on an indeterminate nodulating host, *J. Bacteriol.* **173**(2):851–859.

Hueck, C.J., 1998, Type III protein secretion systems in bacterial pathogens of animals and plants, *Microbiol. Mol. Biol. Rev.* **62**(2):379–433.

Ielpi, L., Dylan, T., Ditta, G.S., Helinski, D.R., and Stanfield, S.W., 1990, The *ndvB* locus of *Rhizobium meliloti* encodes a 319kDa protein involved in the production of β-(1,2)-glucan, *J. Biol. Chem.* **265**(5):2843–2851.

Jann, B., and Jann, K., 1990, Structure and biosynthesis of the capsular antigens of *Escherichia coli*, *Curr. Top. Microbiol. Immunol.* **150**:19–42.

Kannenberg, E.L., Rathbun, E.A., and Brewin, N.J., 1992, Molecular dissection of structure and function in the lipopolysaccharide of *Rhizobium leguminosarum* strain 3841 using monoclonal antibodies and genetic analysis, *Mol. Microbiol.* **6**(17):2477–2487.

Kijne, J.W., 1992, The *Rhizobium* infection process, in: *Biological Nitrogen Fixation*, (G. Stacey, R.H. Burris, and H.J. Evans, eds.), Chapman and Hall, New York, pp. 349–398.

Kim, C.H., Tully, R.E., and Keister, D.L., 1989, Exopolysaccharide-deficient mutants of *Rhizobium fredii* HH303 which are symbiotically effective, *Appl. Environ. Microbiol.* **55**(7):1852–1854.

Kiss, E., Reuhs, B.L., Kim, J.S., Kereszt, A., Petrovics, G., Putnoky, P., Dusha, I., Carlson, R.W., and Kondorosi, A., 1997, The *rkpGHI* and *-J* genes are involved in capsular polysaccharide production by *Rhizobium meliloti*, *J. Bacteriol.* **179**(7):2132–2140.

Krishnan, H.B., Kuo, C.-I., and Pueppke, S.G., 1995, Elaboration of flavonoid-induced proteins by the nitrogen-fixing soybean symbiont *Rhizobium fredii* is regulated by both *nodD1* and *nodD2*, and is dependent on the cultivar-specificity locus, *nolXWBTUV*, *Microbiol.* **141**:2245–2251.

Lee, C.A., 1997, Type III secretion systems: machines to deliver bacterial proteins into eukaryotic cells?, *Trends Microbiol.* **5**(4):149–156.

Leigh, J.A., and Lee, C.C., 1988, Characterization of polysaccharides of *Rhizobium meliloti exo* mutants that form ineffective nodules, *J. Bacteriol.* **170**(8):3327–3332.

Leigh, J.A., and Walker, G.C., 1994, Exopolysaccharides of *Rhizobium*: Synthesis, regulation and symbiotic function, *Trends Genet.* **10**(2):63–67.

Leigh, J.A., Signer, E.R., and Walker, G.C., 1985, Exopolysaccharide-deficient mutants of *Rhizobium meliloti* that form ineffective nodules, *Proc. Natl. Acad. Sci. USA* **82**(18):6231–6235.

Leigh, J.A., Reed, J.W., Hanks, J.F., Hirsch, A.M., and Walker, G.C., 1987, *Rhizobium meliloti* mutants that fail to succinylate their Calcofluor-bonding exopolysaccharide are deficient in nodule invasion, *Cell* **51**(4):579–587.

Lindgren, P.B., Peet, R.C., and Panopoulos, N.J., 1986, Gene cluster of *Pseudomonas syringae* pv. *"phaseolicola"* controls pathogenicity of bean plants and hypersensitivity of non host plants, *J. Bacteriol.* **168**(2):512–522.

Meinhardt, L.W., Krishnan, H.B., Balatti, P.A., and Pueppke, S.G., 1993, Molecular cloning and characterization of a sym plasmid locus that regulates cultivar-specific nodulation of soybean by *Rhizobium fredii* USDA257, *Mol. Microbiol.* **9**(1):17–29.

Mithöfer, A., Bhagwat, A.A., Feger, M., and Ebel, J., 1996, Suppression of fungal β-glucan-induced plant defence in soybean (*Glycine max* L.) by cyclic 1,3–1,6-β-glucans from the symbiont *Bradyrhizobium japonicum*, *Planta* **199**(2):270–275.

Mylona, P., Pawlowski, K., and Bisseling, T., 1995, Symbiotic nitrogen fixation, *Plant Cell* **7**(7):869–885.

Nagpal, P., Khanuja, S.P.S., and Stanfield, S.W., 1992, Suppression of the *ndv* mutant of *Rhizobium meliloti* by cloned *exo* genes, *Mol. Microbiol.* **6**(4):479–488.

Niehaus, K., Kapp, D., and Pühler, A., 1993, Plant defence and delayed infection of alfalfa pseudonodules induced by an exopolysaccharide (EPS I)-deficient *Rhizobium meliloti* mutant, *Planta* **190**(4): 415–425.

Niehaus, K., Baier, R., Becker, A., and Pühler, A., 1996, Symbiotic suppression of the *Medicago sativa* defense system- the key of *Rhizobium meliloti* to enter the host plant?, in: *Biology of Plant-Microbe Interactions*, (G. Stacey, B. Mullin, and M. Gresshoff, eds.), International Society for Molecular Plant-Microbe Interactions, St Paul, MN, pp. 349–352.

Noel, K.D., VandenBosch, K.A., and Kulpaca, B., 1986, Mutations in *Rhizobium phaseoli* that lead to arrested development of infection threads, *J. Bacteriol.* **168**(3):1392–1401.

Olsen, P., Collins, M., and Rice, W., 1992, Surface antigens present on vegetative *Rhizobium meliloti* cells may be diminished or absent when the cells are in the bacteroid form, *Can. J. Microbiol.* **38**:506–509.

Palacios, R., Boistard, P., Dávila, G., Fonstein., M., Göttfert, M., Perret, X., Ronson, C., and Sobral, B., 1998, Genome structure in nitrogen-fixing organisms, in: *Biological Nitrogen Fixation for the 21st Century*, (C. Elmerich, A. Kondorosi, and W.E. Newton, eds.), Kluwer Academic Pub., Dordrecht, pp. 541–547.

Parniske, M., Kosch, K., Werner, D., and Müller, P., 1993, *ExoB* mutants of *Bradyrhizobium japonicum* with reduced competitivity on *Glycine max*, *Mol. Plant-Microbe Interact.* **6**(1):99–106.

Parniske, M., Schmidt, P.E., Kosch, K., and Müller, P., 1994, Plant defense responses of host plants with determinate nodules induced by EPS-defective *exoB* mutants of *Bradyrhizobium japonicum*, *Mol. Plant-Microbe Interact.* **7**(5):631–638.

Parveen, N., Webb, D.T., and Borthakur, D., 1997, The symbiotic phenotypes of exopolysaccharide-defective mutants of *Rhizobium* sp. strain TAL1145 do not differ on determinate- and indeterminate-nodulating tree legumes, *Microbiol.* **143**:1959–1967.

Perotto, S., Brewin, N.J., and Kannenberg, E.L., 1994, Cytological evidence for a host defense response that reduces cell and tissue invasion in pea nodules by lipopolysaccharide-defective mutants of *Rhizobium leguminosarum* strain 3841, *Mol. Plant-Microbe Interact.* **7**(1):99–112.

Petrovics, G., Putnoky, P., Reuhs, B., Kim, J., Thorp, T.A., Noel, K.D., Carlson, R.W., and Kondorosi, A., 1993, The presence of a novel type of surface polysaccharide in *Rhizobium meliloti* requires a new fatty acid synthase-like gene cluster involved in symbiotic nodule development, *Mol. Microbiol.* **8**(6):1083–1094.

Priefer, U.B., 1989, Genes involved in lipopolysaccharide production and symbiosis are clustered on the chromosome of *Rhizobium leguminosarum* biovar *viciae* VF39, *J. Bacteriol.* **171**(11):6161–6168.

Pugsley, A.P., 1993, The complete general secretory pathway in gram-negative bacteria, *Microbiol. Rev.* **57**(1):50–108.

Pühler, A., Arnold, W., Buendia-Claveria, A., Kapp, D., Keller, M., Niehaus, K., Quandt, J., Roxlau, A., and Weng, W.M., 1991, The role of the *Rhizobium meliloti* exopolysaccharide EPS I and EPS II in the infection process of alfalfa nodules, in: *Advances in Molecular Genetics of Plant-Microbe Interactions*, Volume 1, (H. Hennecke and D.P.S. Verma, eds.), Kluwer Academic, Dordrecht, pp. 189–194.

Putnoky, P., Petrovics, G., Kereszt, A., Grosskopf, E., Ha, D.T.C., Banfalvi, Z., and Kondorosi, A., 1990, *Rhizobium meliloti* lipopolysaccharide and exopolysaccharide can have the same function in the plant-bacterium interaction, *J. Bacteriol.* **172**(9):5450–5458.

Quandt, J., Hillemann, K., Niehaus, W., Arnold, W., and Pühler, A., 1992, An osmorevertant of a *Rhizobium meliloti ndvB* deletion mutant forms infection threads but is defective in bacteroid development, *Mol. Plant-Microbe Interact.* **5**(5):420–427.

Relic, B., Perret, X., Estrada-García, M.T., Kopcinska, J., Golinowski, W., Krishnan, H.B., Pueppke, S.G., and Broughton, W.J., 1994, Nod factors are a key to the legume door, *Mol. Microbiol.* **13**(1):171–178.

Reuber, T.L., and Walker, G.C., 1993a, Biosynthesis of succinoglycan, a symbiotically important exopolysaccharide of *Rhizobium meliloti*, *Cell* **74**(2):269–280.

Reuber, T.L., and Walker, G.C., 1993b, The acetyl substituent of succinoglycan is not necessary for alfalfa nodule invasion by *Rhizobium meliloti* Rm 1021, *J. Bacteriol.* **175**(11):3653–3655.

Reuhs, B.L., 1996, Acidic capsular polysaccharides (K antigens) of *Rhizobium*, in: *Biology of Plant-Microbe Interactions*, (G. Stacey, B. Mullin, and P.M. Gresshoff, eds.), International Society for Molecular Plant-Microbe Interactions, St Paul, MN, pp. 349–352.

Reuhs, B.L., Carlson, R.W., and Kim, J.S., 1993, *Rhizobium fredii* and *Rhizobium meliloti* produce 3-deoxy-D-*manno*-2-octulosonic-acid containing polysaccharides that are structurally analogous to group II K antigens (capsular polysaccharides) found in *Escherichia coli*, *J. Bacteriol.* **175**(11):3570–3580.

Reuhs, B.L., Kim, J.S., Badgett, A., and Carlson, R.W., 1994, Production of cell-associated polysaccharides of *Rhizobium fredii* USDA205 is modulated by apigenin and host root extract, *Mol. Plant-Microbe Interact.* **7**(2):240–247.

Reuhs, B.L., Williams, M.N.V., Kim, J.S., Carlson, R.W., and Côté, F., 1995, Suppression of the Fix^- phenotype of *Rhizobium meliloti exo*B mutants by *lpsZ* is correlated to a modified expression of the K polysaccharide, *J. Bacteriol.* **177**(15):4289–4296.

Rolin, D.B., Pfeffer, P.E., Osman, S.F., Szwergold, B.S., Kappler, F., and Benesi, A.J., 1992, Structural studies of a phosphocholine substituted β-(1,3)(1,6) macrocyclic glucan from *Bradyrhizobium japonicum* USDA110, *Biochim. Biophysic. Acta.* **1116**(3):215–225.

Scheu, A.K., Economou, A., Hong, G.F., Ghelani, S., Johnston, A.W.B., and Downie, J.A., 1992, Secretion of the *Rhizobium leguminosarum* nodulation protein NodO by haemolysin-type systems, *Mol. Microbiol.* **6**(2):231–238.

Soto, M.J., Lepek, V., Lopez-Lara, I.M., Olivares, J., and Toro, N., 1992, Characterization of a *Rhizobium meliloti ndvB* mutant and a symbiotic revertant that regains wild-type properties, *Mol. Plant-Microbe Interact.* **5**(4):288–293.

Spaink, H.P., Sheeley, D.M., van Brussel, A.A.N., Glushka, J., York, W.S., Tak, T., Geiger, O., Kennedy, E.P., Reinhold, V.N., and Lugtenberg, B.J.J., 1991, A novel highly unsaturated fatty acid moiety of lipo-oligosaccharide signals determines host specificity of *Rhizobium*, *Nature*, **354**:125–131.

Stacey, G., So, J.-S., Roth, R.E., Lakshmi S.K., B., and Carlson, R.W., 1991, A lipopolysaccharide mutant of *Bradyrhizobium japonicum* that uncouples plant from bacterial differentiation, *Mol. Plant-Microbe Interact.* **4**(4):332–340.

Stanfield, S.W., Ielpi, L., O'Brocta, D., Helinski, D.R., and Ditta, G.S., 1988, The *ndvA* gene product of *Rhizobium meliloti* is required for β-(1,2)-glucan production and has homology to the ATP-binding export protein HlyB, *J. Bacteriol.* **170**(8):3523–3530.

Sutton, J.M., Lea, E.J.A., and Downie, J.A., 1994, The nodulation-signaling protein NodO from *Rhizobium leguminosarum* biovar *viciae* forms ion channels in membranes, *Proc. Natl. Acad. Sci. USA* **91**(21):9990–9994.

Sutton, J.M., Peart, J., Dean, G., and Downie, J.A., 1996, Analysis of the C-terminal secretion signal of the *Rhizobium leguminosarum* nodulation protein NodO; a potential system for the secretion of heterologous proteins during nodule invasion, *Mol. Plant-Microbe Interact.* **9**(8):671–680.

van Rhijn, P., Luyten, E., Vlassak, K., and Vanderleyden, J., 1996, Isolation and characterization of a pSym locus of *Rhizobium* sp. BR816 that extends nodulation ability of narrow host range *Phaseolus vulgaris* symbionts to *Leucaena leucocephala*, *Mol. Plant-Microbe Interact.* **9**(1):74–77.

Vlassak, K.M., Luyten, E., Verreth, C., van Rhijn, P., Bisseling, T., and Vanderleyden, J., 1998, The *Rhizobium* sp. BR816 *nodO* gene can function as a determinant for nodulation of *Leucaena leucocephala*, *Phaseolus vulgaris*, and *Trifolium repens* by a diversity of *Rhizobium* spp., *Mol. Plant-Microbe Interact.* **11**(5):383–392.

Viprey, V, Del Greco, A., Golinowski, W., Broughton, W.J., and Perret, X., 1998, Symbiotic implications of type III secretion machinery in *Rhizobium*, *Mol. Microbiol.* **28**(6):1381–1389.

Wandersman, C., 1996, Secretion across the bacterial outer membrane, in: *Escherichia coli* and *Salmonella*: cellular and molecular biology, (F.C. Neidhardt, R. Curtis III, J.L. Ingraham, E.C.C. Lin, K.B. Low, B. Magasanik, M. Riley, W.S. Reznikoff, M. Schaechter, and H.E. Umbarger, eds.), 2nd ed., ASM Press, Washington, D.C., pp. 955–966.

Wengelnik, K., Van den Ackerveken, G., and Bonas, U., 1996, HrpG, a key *hrp* regulatory protein of *Xanthomonas campestris* pv. *vesicatoria* is homologous to two-component response regulators, *Mol. Plant-Microbe Interact.* **9**(8):704–712.

Whitfield, C., and Valvano, M.A., 1993, Biosynthesis and expression of cell-surface polysaccharides in gram-negative bacteria, *Adv. Microb. Physiol.* **35**:135–246.

Williams, M.N.V., Hollingsworth, R.I., Klein, S., and Signer, E.R., 1990, The symbiotic defect of *Rhizobium meliloti* exopolysaccharide mutants is suppressed by *lpsZ*, a gene involved in lipopolysaccharide biosynthesis, *J. Bacteriol.* **172**(5): 2622–2632.

Willis, D.K., Rich, J.J., and Hraback, E.M., 1991, *hrp* genes of phytopathogenic bacteria. *Mol. Plant-Microbe Interact.* **4**(2):132–138.

Yang, C., Signer, E.R., and Hirsch, A.M., 1992*a*, Nodules initiated by *Rhizobium meliloti* exopolysaccharide mutants lack a discrete, persistent nodule meristem, *Plant Physiol.* **98**(1):143–151.

Yang, W.-C., Canter Cremers, H.C.J., Hogendijk, P., Katinakis, P., Wijffelman, C.A., Franssen, H., van Kammen, A., and Bisseling, T., 1992*b*, *In-situ* localization of chalcone synthase mRNA in pea root nodule development, *Plant J.* **2**(2):143–151.

York, G.M., and Walker, G.C., 1997, The *Rhizobium meliloti exoK* gene and *prsD/prsE/exsH* genes are components of independent degradative pathways which contribute to production of low-molecular-weight succinoglycan, *Mol. Microbiol.* **25**(1):117–134.

York, G.M., and Walker, G.C., 1998, The *Rhizobium meliloti* ExoK and ExsH glycanases specifically depolymerize nascent succinoglycan chains, *Proc. Natl. Acad. Sci. USA* **95**(9):4912–4917.

Zevenhuizen, L.P.T.M., and van Neerven, A.R.W., 1983, (1–2)-β-D-glucan and acidic oligosaccharides produced by *Rhizobium meliloti*, *Carbohydr. Res.* **118**:127–134.

Zhan, H., Levery, S.B., Lee, C.C., and Leigh, J.A., 1989, A second exopolysaccharide of *Rhizobium meliloti* strain SU47 that can function in root nodule invasion, *Proc. Natl. Acad. Sci. USA* **86**(9):3055–3059.

Part III

Obligate Intracellular Bacteria

Chapter 18

Chlamydia

Internalization and Intracellular Fate

M. Scidmore-Carlson and T. Hackstadt

1. INTRODUCTION

Chlamydiae are obligate intracellular bacteria that cause a wide spectrum of disease in both humans and non-human species. Within the family *Chlamydiaceae*, there is a single genus and four species: *Chlamydia trachomatis, Chlamydia psittaci, Chlamydia pneumoniae* and *Chlamydia pecorum.* Diseases of humans caused by *Chlamydia trachomatis* include sexually transmitted diseases and endemic blinding trachoma. *C. pneumoniae* is a widespread cause of community-acquired pneumonia (Grayston *et al.*, 1989) and is of intense current interest due to possible associations with atheroschlerosis (Kuo *et al.*, 1995). *C. psittaci* is primarily a zoonosis that occasionally infects humans and *C. percorum* is also an animal pathogen but infections of humans have not been reported. There are at least 15 serologically distinguished serovars of *C. trachomatis* with several sub-types now recognized. Different serovars are associated with distinct diseases. Infections caused by serovars A-C are associated primarily with endemic

---

M. SCIDMORE-CARLSON and T. HACKSTADT Host-Parasite Interactions Section, National Institute of Allergy and Infectious Diseases, National Institutes of Health, Rocky Mountain Laboratories, Hamilton, Montana 59840.

Subcellular Biochemistry, Volume 33: Bacterial Invasion into Eukaryotic Cells, edited by Oelschlaeger and Hacker. Kluwer Academic / Plenum Publishers, New York, 2000.

blinding trachoma, the leading cause of preventable blindness worldwide. Infections caused by serovars D-K are typically localized to the genital tract and are the most common cause of sexually transmitted disease. The lymphogranuloma venereum (LGV) serovars, L1, L2, and L3 also cause sexually transmitted infections, however, unlike diseases caused by the trachoma serovars which remain localized to mucosal epithelium, diseases caused by LGV are more systemic and invade the inguinal lymph nodes. Based upon the ability to cause systemic infections as well as differences in interactions with eukaryotic cells *in vitro*, the human *C. trachomatis* serovars are considered to comprise two biovars, trachoma and lymphogranuloma venereum (LGV) (Schachter and Caldwell, 1980).

Chlamydiae undergo a unique developmental cycle characterized by two functionally and morphologically distinct cell types. The infectious extracellular developmental form is termed an elementary body (EB). EBs are approximately 0.3 μm in diameter, display a condensed nucleoid, and exhibit no measurable metabolic activity. The intracellular replicative form is termed a reticulate body (RB). RBs are approximately 1 μm in diameter, display a dispersed chromatin structure, and are metabolically active. EBs are endocytosed into a tightly membrane bound vacuole, termed an inclusion, that expands in size as the organisms replicate. Shortly after internalization, the EB begins to differentiate into the larger, more pleomorphic RB. RBs divide by binary fission until 18 or more hours post-infection, at which time increasing proportions of the RBs begin to differentiate back into EBs while the remainder continue to divide. At the end of the developmental cycle, the organisms are released from the host cell to re-initiate infection of neighboring cells. The environmental signals that trigger the differentiation from EB to RB or from RB to EB are unknown. All chlamydial species undergo a similar developmental cycle differing only in the time that it takes to complete a cycle. The developmental cycle of the fast growing LGV serovars is approximately 40–44 hr while it is close to 72 hrs for the slower growing trachoma serovars (Moulder, 1991).

2. ENTRY

2.1. Parasite-Mediated Endocytosis

Because chlamydiae enter non-professional phagocytes 10–100 fold more efficiently than *E. coli* or polystyrene latex beads (Byrne and Moulder, 1978), the process has been termed "parasite-mediated phagocytosis" to emphasize the belief that surface structures of EBs enhance the internalization process. A number of biochemical studies involving chemical modification of EBs or the host cell have been performed in attempts to determine the nature of any chlamydial cell wall component that may function as a ligand and the cognate host cell receptor(s). Much of this data has been difficult to consolidate into a single mechanism of chlamydial entry due to the diversity of chlamydial species or biovars, host cell types, and experimental conditions employed. Thus the precise mechanisms of chlamydial uptake are not clearly defined. A large measure of the inconsistency in the literature is most likely due to the presence of multiple mechanisms for chlamydial entry.

EBs are considered to be metabolically inert and are not believed to contribute energetically to the entry process. Inhibition of chlamydial protein synthesis has no effect on entry. In addition, UV-irradiated EBs (Byrne and Moulder, 1978; Byrne, 1976) as well as purified EB cell walls (Eissenberg *et al.*, 1983; Levy and Moulder, 1982) attach to and enter host cells at rates equivalent to non-treated organisms implying that it is a structural component(s) of the cell wall that is required for attachment and internalization of chlamydiae. The identity of this component remains uncertain, however. Mild heat treatment of *C. psittaci* 6BC (Byrne and Moulder, 1978), *C. trachomatis* serovars B and E (Vretou *et al.*, 1989), and LGV strains (Byrne and Moulder, 1978) reduces infectivity. However, serovar A attachment to McCoy cells is only minimally reduced by similar mild heat treatment (Lee, 1981). Trypsinization of EBs has no effect on *C. psittaci* (Byrne, 1978) or LGV strains of *C. trachomatis* (Hackstadt and Caldwell, 1985; Bose and Paul, 1982) but decreases serovar B attachment for Hela cells (Su *et al.*, 1988).

Polycations are required for infectivity of chlamydiae. It has been proposed that polycations are required to neutralize or bridge the mutual repulsion of the parasite and the host cell, both of which are negatively charged (Hatch *et al.*, 1981). DEAE-dextran has also been shown to enhance attachment by neutralizing the negative charge of the host cells for non-LGV strains (Kuo *et al.*, 1973). However, DEAE-dextran has no effect on LGV strains. Hydrophobic interactions are also thought to be required (Su *et al.*, 1990). Trachoma biovars infect tissue culture cells less effectively than the LGV serovars and centrifugation of the inoculum onto the monolayer enhances infectivity of trachoma biovar organisms.

Because of the efficient attachment and internalization of EBs to a variety of host cells not encountered during natural infection, attachment is thought to occur via high affinity binding to a ubiquitous host cell ligand.

2.2. Chlamydial Adhesins

2.2.1. Heparan Sulfate

Recently, Stephens and colleagues have proposed a novel trimolecular mechanism of chlamydial attachment. The model suggests that a heparan sulfate like molecule present on the EB surface binds both a chlamydial ligand and a host cell ligand simultaneously to facilitate binding and uptake of chlamydiae (Zhang and Stephens, 1992). Data to support this hypothesis includes the inhibition of infectivity by heparin or heparan (Zhang and Stephens, 1992; Kuo and Grayston, 1976), by fibronectin or platelet factor (heparan-sulfate binding proteins) and by heparitinase treatment of EBs but not the host cell. Inhibition of infectivity by treatment of the organisms with heparitinase is restored by the exogenous addition of heparan sulfate (Zhang and Stephens, 1992). In addition, heat treatment causes release of bound heparan sulfate from the organism suggesting that the inhibition of infectivity observed when organisms are subjected to mild heat treatment might be due to loss of heparan sulfate that coats the organisms. The efficiency of heparin inhibition of attachment varies among chlamydial species and strains, however. Whereas both attachment and inclusion formation by an LGV strain of *C. trachomatis* are profoundly reduced by heparan sulfate (Zhang and Stephens, 1992; Kuo and Grayston, 1976), the effects of heparan sulfate on trachoma biovars of *C. trachomatis* (Chen and Stephens, 1997; Davis and Wyrick, 1997; Kuo and Grayston, 1976) or the GPIC strain of *C. psittaci* (Gutierrez-Martin, 1997) are different. For the trachoma biovars and the GPIC strain of *C. psittaci,* the ability to form a productive infection is inhibited to a much greater degree by infection in the presence of heparan sulfate than are the measured rates of attachment. These results suggest both glycosaminoglycan-dependent and -independent means of attachment to eukaryotic cells and further suggest that infection in the presence of heparan sulfate may have downstream effects from attachment by inhibiting either entry or development.

Although it appears that heparan sulfate plays a role in chlamydial infectivity, the biosynthetic source of the heparan sulfate remains somewhat controversial. Zhang and Stephens (1992) have shown that chlamydiae are able to infect and produce infectious progeny from a Chinese hamster ovary cell line deficient in synthesis of heparan sulfate suggesting that chlamydiae do not acquire heparan sulfate from the host cell. In addition, radiolabeled GAGs were isolated from this mutant cell line following metabolic labeling and infection with chlamydiae (Zhang and Stephens, 1992). In contrast, Su and colleagues (1996) demonstrated that heparitinase treatment of the host cells but not the organisms, caused loss of chlamydial infectivity.

2.2.2. MOMP

Given the abundance of MOMP in the chlamydial outer membrane as well as existence of monoclonal antibodies to MOMP that inhibit attachment, MOMP has long been a prime candidate for a chlamydial adhesin. Monoclonal antibodies specific to antigenic determinants located in the variable regions II and IV of serovar B MOMP as well as cleavage by trypsin at two sites within the same variable regions inhibit attachment of chlamydiae to HeLa cells (Su *et al.*, 1990; Su *et al.*, 1988). The authors suggest that MOMP functions as an adhesin by promoting both electrostatic and hydrophobic interactions with the host cell.

A mechanism for the chlamydial major outer membrane protein (MOMP) function in the attachment process and possible identification of the host ligand has been provided by studies of recombinant maltose binding protein (MBP)—MOMP fusions. *C. trachomatis*, mouse pneumonitis strain, MOMP expressed in *E. coli* as a C-terminal fusion with MBP, form high molecular weight colloidal particles visible by scanning electron microscopy. MBP-MOMP bound specifically to human epithelial cells at 4°C but was not internalized upon shifting the temperature to 37°C. An excess of heparin or heparan sulfate but not chondroitin sulfate inhibited binding of MBP-MOMP to eukaryotic cells. Binding to HeLa cells was inhibited by removal of cell surface proteoglycans by heparitinase treatment but not chondroitinase treatment and mutant CHO cell lines defective in glycosaminoglycan synthesis displayed reduced binding of both MBP-MOMP and native EBs. Furthermore, MBP-MOMP competitively inhibited EB binding to the cell surface as efficiently as did UV-treated mouse pneumonitis EBs. The data suggest that MOMP may be the chlamydial receptor for glycosaminoglycans. This model differs from that of Zhang and Stephens (1992) primarily in that synthesis of a heparan sulfate-like glycosaminoglycan by chlamydiae is not proposed and suggests that host cell surface glycosaminoglycans are necessary and sufficient to promote chlamydial attachment (Su *et al.*, 1996).

In addition to protein components of MOMP functioning in the attachment of chlamydiae to host cells, Kuo and colleagues suggest that the glycan of MOMP, consisting of an N-linked high mannose type oligosaccharide, also plays a role (Kuo *et al.*, 1996; Swanson and Kuo, 1994b). Purified glycan inhibits binding of viable EBs to HeLa cells and conversely, EBs prevent binding of the glycan to HeLa cells. More specifically, analysis with oligosaccharides of defined structures demonstrated that attachment and infectivity of *C. trachomatis* was mediated by the glycan of MOMP. Finally, maltose binding protein, a mammalian C-type lectin that binds mannose, inhibits infection of *C. trachomatis* by binding to the glycan of MOMP thereby preventing entry of the organism (Swanson *et al.*, 1998).

2.2.3. Other Putative Adhesins

Omp2 of *C. psittaci* strain guinea pig inclusion conjuctivitis has also been proposed to be a chlamydial adhesin. Although antibodies to Omp2 fail to inhibit chlamydial attachment to host cells, Omp2 present in detergent extracted outer membrane protein preparations, binds specifically to glutaraldehyde fixed Hela cells (Ting *et al.*, 1995). Other putative adhesins include 18 and 32kDa glycoproteins of *C. trachomatis* (Swanson and Kuo, 1994a; Swanson and Kuo, 1991), the 38kDa chlamydial cytoadhesin (Joseph and Bose, 1991a; Joseph and Bose, 1991b), the chlamydial Hsp70 (Schmiel *et al.*, 1991), the 28kDa GrpE-like protein (Schmiel et. al., 1995) and a glycolipid (Stuart *et al.*, 1991).

2.3. Host Factors

The identity of the host cell ligand mediating chlamydial attachment has similarly remained undetermined. The difficulty in identifying the host cell ligand suggests that chlamydiae have the ability to bind to more than one receptor. The receptor is trypsin sensitive and present on a wide range of cell types. Modification of carbohydrates by either periodate oxidation or trypsinization of L cells inhibits binding of chlamydiae implicating that a glycoprotein is involved either directly or indirectly in chlamydial attachment (Byrne, 1976; Hatch *et al.*, 1981). Several investigators have attempted to identify the specific sugars that mediate chlamydial binding to host cells with differing results. Although Levy observed inhibition of *C. psitttaci* 6BC and LGV serovar L1 attachment to L cells, HeLa, HEp2, Vero and McCoy cells in the presence of wheat germ agglutinin (WGA) (Levy, 1979; Soderlund and Kihlstrom, 1983a), this finding was not confirmed for 6BC binding to L cells (Hatch *et al.*, 1981). In addition, WGA had no effect on the attachment of LGV serovars L2 and L3 to Hela cells (Bose and Paul, 1982). Therefore the data regarding the requirement for N-acetylglucosamine remains inconclusive. Sialic acid containing glycoproteins have also been implicated as a ligand for chlamydial attachment for non-LGV trachoma biovars but the data is similarly inconclusive. Neuraminidase treatment has been shown to inhibit inclusion formation of trachoma serovars A, B, C, E (Kuo *et al.*, 1973) and K (Bose *et al.*, 1983) but not for LGV serovars (Kuo *et al.*, 1973). EBs can bind to insect cells that are devoid of N-acetylneuramic suggesting that sialic residues are not required for chlamydial attachment (Allan and Pearce, 1987). Finally heparan sulfate proteoglycans have also been postulated to be putative host cell receptors as treatment of host cells with heparinitinase reduces binding of EBs. In addition, reduced binding is observed for CHO cells deficient for heparan sulfate proteoglycans (Su *et al.*, 1996). Competitive inhibition studies using both trachoma and LGV serovars suggest that they bind to the same receptor on the host cell membrane (Vretou *et al.*, 1989).

The possibility of lipids being involved in the attachment process has also been addressed. Iodinated *C. trachomatis* and *C. pneumoniae* organisms bound to phosphatidylethanolamine that had been purified from Hela cells and separated by thin layer chromatography (Krivan *et al.*, 1991). In addition, chlamydiae bound to purified asialo-GM_1 and asialo-GM_2 suggesting that lipids might also be involved in the attachment of chlamydiae (Krivan *et al.*, 1991).

2.4. Mechanism of Uptake

A single uptake mechanism for chlamydiae has not been defined. As discussed below, chlamydiae are capable of entry by both phagocytosis and by receptor mediated endocytosis. For the highly invasive LGV strain serovar L2, evidence has been presented that this serovar enters cells via phagocytosis or a microfilament-dependent zipper mechanism. When chlamydiae are centrifuged onto host cells at 4°C and then warmed to 37°C to initiate endocytosis, the EBs are endocytosed in vesicles which are not clathrin coated. Endocytosis was not inhibited by inhibitors of receptor mediated endocytosis (monodansylcadaverine or amantadine) but was reduced in half by cytochalasin D, an inhibitor of host cell microfilament formation and by vincristine and vinblastine, inhibitors of microtubule function (Ward and Murray, 1984). Taken together, the evidence is suggestive of a microfilament dependent uptake mechanism. Uptake by receptor mediated endocytosis has also been postulated. Ultrastructural analysis of uptake by trachoma serovar E and by *C. psittaci* CAL-10 demonstrated that EBs preferentially bound to microvilli and were associated with clathrin coated pits (Hodinka *et al.*, 1988; Hodinka and Wyrick, 1986), both characteristics of receptor mediated endocytosis. Uptake of serovars L1 and E has also been shown to be partially inhibited by methylamine hydrochloride and by monodansylcadaverine (Soderlund and Kihlstrom, 1983b). Because inhibition by either cytochlasin D (Ward and Murray, 1984) or by amines (Soderlund and Kihlstrom, 1983b) is not complete, it suggests that in *in vitro* culture systems, chlamydiae may be internalized by several distinct uptake mechanisms (Reynolds and Pearce, 1991). Both the method of inoculation and the cell culture system have been shown to influence the mechanism of uptake. Uptake of *C. psittaci* strain GPIC without centrifugation was shown to be primarily by pinocytosis since it was affected by neither cytochalasin D nor temperature. However, when centrifugation was used to aid infection, uptake was sensitive to both cytochalasin D and temperature suggesting centrifugation changed the mode of uptake to a phagocytic uptake mechanism (Prain and Pearce, 1989). Interestingly, Wyrick and colleagues (1989) have shown that the substratum cells are grown on also influences the mechanism of uptake. When serovar E and LGV L2 were infected on human endometrial gland epithelial cells grown on collagen coated filters to create a polarized epithelia, EBs were observed in coated pits and coated vesicles. On the other hand, chlamydiae were not observed in coated pits when the cells were grown on plastic (Wyrick *et al.*, 1989).

2.5. Signal Transduction

The uptake of many pathogenic bacteria induces signal transduction mechanisms in host cells involving host protein phosphorylation (Finlay and Cossart, 1997). Chlamydiae are no exception. Immunoblot analysis using anti-phosphotyrosine specific antibodies identified the specific tyrosine phosphorylation of between 3–5 proteins upon infection of eukaryotic cells with either LGV serovar L2 or MoPn biovar of *C. trachomatis* (Fawaz *et al.*, 1997; Birkelund *et al.*, 1994). The phosphorylation of these proteins was inoculation dependent and was more pronounced when the multiplicities of infection were high. Phosphorylation was specific to uptake of chlamydiae since uptake of *E. coli* or latex beads did not induce phosphorylation of these same proteins nor were these proteins phosphorylated when chlamydiae were prevented from binding by the addition of heparin. Phosphoamino analysis showed that one protein was phosphorylated on both tyrosine and serine residues (Fawaz *et al.*, 1997). Indirect immunofluorescence analysis using the anti-phosphotyrosine antibody demonstrated the colocalization of the tyrosine phosphorylated proteins and the chlamydiae during the first 8 hr post infection. Later during the developmental cycle, when a mature inclusion was visible, the phosphotyrosine labeling was observed surrounding the intact inclusion. Cortactin, a tyrosine phosphorylated actin binding protein was also shown to colocalize with EBs. However, no increase in phosphorylation of cortactin was observed upon infection, therefore phosphorylation of cortactin is not induced upon chlamydial uptake. Tyrosine phosphorylation inhibitors (genistein, herbimycin, lavendustin A, staurosporine, dihydroxycinnamate and aminobenzoid acid) did not inhibit phosphorylation of these proteins. Therefore it is not known whether phosphorylation of these proteins is required for attachment, binding or trafficking of chlamydiae. In addition, the identity of these phosphorylated proteins has not been determined nor is it known whether these proteins are of chlamydial or host origin.

3. ESTABLISHMENT OF THE REPLICATION COMPETENT VACUOLE

3.1. Chlamydial Vacuoles are Non-Fusogenic with Endocytic Vesicles

Intracellular survival of chlamydiae is dependent on the ability to enter and replicate inside the host cell. Part of this survival mechanism involves inhibition of lysosomal fusion with the chlamydial inclusion. At no time during the developmental cycle is fusion with lysosomes observed; chlamydial inclusions do not contain acid phosphatase, LAMP1, cathepsin D, or vacuolar H^+-ATPase (Heinzen *et al.*, 1996; Taraska *et al.*, 1996) nor do secondary lysosomes containing ferritin fuse with chlamydial vacuoles (Wyrick and Brownridge, 1978). Inhibition of lysosomal fusion is limited to the chlamydial vacuole since co-infection of chlamydiae with yeast or *E. coli* does not inhibit fusion of yeast or *E. coli* containing vesicles with lysosomes (Eissenberg and Wyrick, 1981). In macrophages, the initial inhibition of lysosomal fusion is a property of the chlamydial cell wall. Purified cell walls are taken up by macrophages and are found within phagosomes that are devoid of ferritin for at least 3 hours post-infection. In contrast, cell walls subjected to mild heat treatment are found within ferritin containing lysosomes within 30 min. post-infection (Eissenberg and Wyrick, 1981). Inhibition of lysosomal fusion is also dependent on the metabolic activity of viable chlamydia. Prevention of chlamydial protein synthesis by treatment with chloramphenicol results in the eventual fusion of chlamydia-containing vacuoles with lysosomes in HeLa cells (Scidmore *et al.*, 1996b). However, fusion of lysosomes with vacuoles containing chloramphenicol inhibited EBs is very slow, taking up to 72 hrs for the majority of chlamydiae to be found in lysosomes. Similarly, *C. psittaci* containing vacuoles do not fuse with lysosomes for at least 24 hr in the presence of chloramphenicol (Friis, 1972). In addition, growth in the presence of chloramphenicol, rifampicin or nalidixic acid for as long as 30 hr was completely reversible (Tribby *et al.*, 1973). These results suggest that chlamydial avoidance of lysosomal fusion may be due to two distinct mechanisms; an initial delayed fusion with lysosomes due to some intrinsic property of the cell wall, and a longer term inhibition of lysosomal fusion brought about as a result of chlamydiae actively modifying the host's response through the actions of chlamydial protein(s).

The chlamydial inclusion is also non-fusogenic with endocytic vesicles.

By either indirect immunofluorescence or electron microscopy, markers of early endosomes (transferrin and transferrin receptor), late endosomes (mannose-6-phosphate receptor) or fluid phase markers (FITC-dextran and lucifer yellow) are not detected in the chlamydial inclusion (Hackstadt *et al.*, 1998; van Ooij *et al.*, 1997). Therefore, chlamydiae do not prevent lysosomal fusion by inhibiting maturation of their vesicle at an early endocytic stage as has been shown for *Mycobacterium* sp. (Clemens and Horwitz, 1995; Xu *et al.*, 1994) and *Ehrlichia chaffeensis* (Barnewall *et al.*, 1997). Instead, chlamydiae appear to be non-fusogenic with any host vesicles during the brief interval before the onset of chlamydial protein synthesis. This inhibition of endocytic fusion is likely related to the limited fusogenicity of EBs and their cell walls with lysosomes.

3.2. Cytoskeletal Requirements for Intracellular Development

Within 2 hrs post infection, endocytosed EBs are trafficked to a peri-Golgi location. Both host microtubules and microfilaments appear to have roles in the intracellular trafficking of chlamydiae although the role of the cytoskeleton appears to be different depending on the strain analyzed. Disruption of MT or MF by the use of specific inhibitors had no effect on the infectivity of LGV serovar L2, suggesting that an intact cytoskeleton is not important for uptake and replication of LGV serovars. Whereas cytochalasin D had no effect on serovar L2 infectivity, cytochalasin D inhibited the infectivity of serovar E. However, drugs that disrupt microtubules, colchicine and nocodozole, did not inhibit serovar E infection (Schramm and Wyrick, 1995). Between 2–4 hrs post-infection, colocalization of chlamydiae and microtubules is observed (Clausen *et al.*, 1997). Coincident with trafficking to a peri-Golgi region is a colocalization with the microtubule organizing center. Trafficking along the microtubules has been postulated to be dynein dependent (Clausen *et al.*, 1997). Dynein is a minus-ended microtubule motor that functions in the transport of molecules from the host cell periphery to the microtubule organizing center, consistent with the direction of EB movement. Drugs that inhibit dynein activity inhibit aggregation of EBs to the peri-Golgi region when the drugs are added during the first few hours post infection. In contrast to Schramm and collegues (1995), Clausen and collegues (1997) suggest that both serovars E and L2 are trafficked to a peri-Golgi region via a MT dependent mechanism.

Disruption of the host cytoskeleton inhibits both the aggregation of EBs to the peri-Golgi region and the homotypic fusion of chlamydial containing vacuoles (Schramm and Wyrick, 1995; Ridderhof and Barnes, 1989) as typically occurs when cells are multiply infected with

C. trachomatis. Disruption of the host cytoskeleton caused the appearance of multiple dispersed inclusions, rather than one large inclusion, when cells were infected with *C. trachomatis*, L2, at a high multiplicity of infection. No effect was observed on the trachoma serovar E. Aggregated chlamydiae also colocalized with F-actin and clathrin (Majeed and Kihlstrom, 1991). Calcium is thought to play a role in the redistribution of EBs and colocalization of F-actin (Majeed *et al.*, 1993). Consistent with the calcium dependent role in chlamydial trafficking, annexins III, IV and V translocate to the aggregated EBs (Majeed *et al.*, 1994). Annexins are a family of Ca^{2+} binding proteins that bind to phospholipids in the presence of Ca^{2+}.

3.3. Establishment of the Chlamydial Vacuole with an Exocytic Pathway

Trafficking studies using 6-[N-(7-nitrobenzo-2-oxa-1,3-diazol-4-yl)aminocaproyl sphingosine] (C_6-NBD-ceramide) have been instrumental in defining the interaction of the chlamydial inclusion with host vesicle trafficking systems. C_6-NBD-ceramide is a fluorescent analogue of ceramide that is a vital stain for the Golgi apparatus and has been used extensively to study sphingolipid trafficking in viable cells. At 4°C, NBD-ceramide intercalates non-specifically into virtually every cellular membrane except the inclusion membrane. Increasing the temperature to 37°C, causes trafficking of NBD-ceramide to the Golgi apparatus where it is enzymatically converted into NBD-sphingomyelin and NBD-glucosylcerobroside. NBD-gluocosylcerobroside and NBD-sphingomylein are then trafficked to the plasma membrane by a vesicular mediated mechanism (Lipsky and Pagano, 1985).

In chlamydia-infected cells, substantial amounts of NBD-sphingomyelin endogenously synthesized from exogenously added NBD-ceramide is trafficked to the chlamydial inclusion where it is incorporated into the cell walls of the bacteria. Quantitation of extracted lipids from infected cells by thin layer chromatography (Scidmore *et al.*, 1996a) and direct quantitation by microphotometry of viable cells (Hackstadt *et al.*, 1995) demonstrate that up to 40–50% of the NBD-sphingomyelin is trafficked to and retained by the intracellular chlamydiae. Once acquired by the chlamydiae, the NBD-sphingomyelin is no longer exchanged or transported intracellularly.

Trafficking of endogenous sphingomyelin as well as NBD-sphingomyelin to the plasma membrane is via a vesicle-mediated mechanism. Trafficking of the fluorescent probe to the chlamydial inclusion also displays characteristics of a vesicular mediated pathway. It is time, temperature and energy dependent (Hackstadt *et al.*, 1996). In addition, drugs that inhibit normal vesicular trafficking such as monensin and BFA also inhibit the trafficking of NBD-sphingomyelin to the chlamydial inclusion. C_6-NBD-sphingomyelin is exclusively present in the lumenal leaflet of exocytic vesicles and does not flip-flop through the bilayer (Pagano, 1989). Therefore, the NBD-sphingomyelin must be delivered to the intracellular chlamydiae as a result of vesicular fusion of exocytic vesicles. Photo-oxidation of cells labeled with N-(4,4-di-fluoro-5, 7-dimethyl-4-bora-3a,4a-diaza-s-indacene-3-pentanoyl) sphingosine demonstrated the actual fusion of vesicles containing the lipid probe with the chlamydial inclusion (Hackstadt *et al.*, 1996). By fluorescence microscopy little fluorescence is observed in the chlamydial inclusion membrane presumably due to the rapid acquisition of the fluorescent probe by the chlamydiae. NBD-sphingomyelin, incorporated directly into the plasma membrane is not trafficked to the chlamydial inclusion demonstrating that NBD-sphingomeylin is not trafficked to the inclusion through the fusion with endocytic vesicles but instead it is trafficked via vesicles originating from the trans-Golgi network (Hackstadt *et al.*, 1996).

The absence of endocytic markers and the presence of Golgi-derived lipids in the chlamydial inclusion suggest that the chlamydial inclusion may be best characterized as an aberrant exocytic vesicle in which fusion with the plasma membrane is inhibited or delayed. Interaction with this exocytic pathway has been proposed to represent a pathogenic mechanism whereby chlamydiae are effectively isolated from cellular pathways that involve maturation of endosomes to lysosomes, and therefore constitutes a protected site supporting replication.

Although chlamydiae interrupt the export of sphingomyelin from the Golgi apparatus, the export of secretory proteins does not appear to be disrupted. The post-translational processing and trafficking of vesicular stomatitis virus glycoprotein G, transferrin receptor and human histocompatibility class I antigen is not interrupted in chlamydia-infected cells nor are these proteins trafficked to the chlamydial inclusion (Scidmore *et al.*, 1996a). Therefore, chlamydiae do not cause a generalized disruption of exocytic trafficking. In polarized cells, sphingomyelin and glucocerebroside are differentially trafficked to the apical and basolateral membranes and trafficking of sphingolipids to the apical surface is thought to occur via cholesterol rafts. Perhaps chlamydiae interrupt a cognate apical pathway as the proteins that were analyzed are trafficked to the basolateral membrane in polarized cells. It should be remembered that the natural site of chlamydial infection is polarized epithelial surfaces.

Chlamydia-containing vesicles initiate interactions with this exocytic pathway within the first two hours post infection as determined by the ability to fuse with NBD-sphingomyelin-containing exocytic vesicles (Scidmore *et al.*, 1996a). Inhibition of chlamydial translation or transcription within the first 2 hr post-infection, inhibits not only the acquisition of NBD-sphingomyelin but the trafficking of EBs to a peri-Golgi region. These results suggest that early chlamydial gene expression is required for fusogenicity with sphingomyelin containing exocytic vesicles as well as for interactions with the host cytoskeleton.

3.4. Route of Entry

Controversy remains as to whether the route of entry is important for determining the intracellular fate of chlamydiae. Initial experiments in murine macrophages (Wyrick and Brownridge, 1978) or L cells (Friis, 1972) demonstrated that opsonized *C. psittaci* were trafficked to lysosomes. In contrast more recent experiments suggest that the intracellular trafficking of serovar L2 chlamydiae opsonized with monoclonal antibodies recognized by the FcγIII was identically to non-opsonized EBs in HeLa cells or CHO cells transfected with the FcγIII receptor (Scidmore *et al.*, 1996b; Su *et al.*, 1991). These data suggest that the route of entry is not important in the intracellular development since chlamydiae can override the normal trafficking of Fc receptors to lysosomes so long as they are able to synthesize protein to alter the host's response. This latter finding is also consistent with the fact that EBs can enter and replicate by a variety of different mechanisms in a variety of different cell types. Modification of the early chlamydial inclusion by expression of chlamydial proteins thus appears to be the key factor in establishing a replication competent vacuole (Scidmore *et al.*, 1996b).

4. CHARACTERISTICS OF THE CHLAMYDIAL VACUOLE

4.1. Permeability Properties

In contrast to the *T. gondii* parasitophorous vacuole membrane which permits passive diffusion of molecules up to 1300–1900 Da (Schwab *et al.*, 1994), the chlamydial inclusion does not contain pores that permit passive diffusion of tracer molecules as small as 520 Da (Heinzen and Hackstadt, 1997). Although the smallest molecule analyzed was 520 Da, it seems likely that the inclusion membrane may be a barrier to even smaller molecules. Because a limiting membrane binds chlamydiae during their entire intracellular developmental cycle, all metabolites required for growth and replication must cross this barrier. Unlike intracytoplasmic bacterial parasites which have direct access to the nutrient rich environment of the host cytoplasm (Falkow *et al.*, 1992; Moulder, 1985), parasites that are sequestered within a membrane-bound vacuole must have a specialized mechanism for acquiring nutrients from the host. Possible means of nutrient exchange are by i.) fusion of the vacuole with nutrient-laden vesicles of the endosomal/lysosomal pathway that are involved in fluid-phase uptake or turnover of endogenous components; ii.) open channels through the parasitophorous vacuole to the cytoplasm that allow the free exchange of low molecular weight molecules (Desai, 1993; Schwab *et al.*, 1994); or, iii.) the membrane could contain transport proteins of host or parasite origin that specifically bind and deliver metabolites from the cytoplasm to the lumen of the vacuole. Specific transporters located within the parasitophorous vacuole membrane as a mechanism for nutrient exchange by vacuole-bound parasites have not yet been documented.

4.2. Acidification

Chlamydiae replicate in vesicles that are not highly acidified as suggested by the neutral pH optimum for metabolism of purified EBs (Hatch *et al.*, 1985) and the resistance to inhibition of replication by lysosomotropic amines (Heinzen *et al.*, 1996; Schramm *et al.*, 1996) that raise the pH of acidic vesicles to inhibit growth of those organisms occupying a more acidic vesicle (Hackstadt and Williams, 1981). In addition, chlamydial inclusions do not acquire the pH fluorescent dye, acridine orange (Heinzen *et al.*, 1996). The pH of chlamydial containing vesicles was examined directly by examining the emissions ratios of EBs that had been conjugated to SNAFL (5-[and6-]- carboxyseminaphthofluorescein-1, succinimidyl ester) (Schramm *et al.*, 1996). SNAFL is a fluorescent, pH sensitive probe that does not affect the infectivity of EBs. Up to 12 hr post-infection, the pH of chlamydia-containing vesicles remained above 6, while the pH of vesicles containing heat-killed organisms was closer to 5. The lack of acidification of the chlamydial vacuole is consistent with the absence of the vacuolar proton ATPase (Heinzen *et al.*, 1996). The presence of a Na^+, K^+-ATPases in the nascent chlamydial inclusion is postulated to contribute to the slight acidification, but the presence of an ATPase in the inclusion membrane has not been confirmed (Schramm *et al.*, 1996).

4.3. Protein Constituents of the Inclusion Membrane

As described above, early chlamydial gene expression is required for the establishment of a replication competent vacuole. Insertion of chlamydial derived proteins into the inclusion membrane is a likely mechanism for controlling vesicular interactions within the host cell. To date, no host proteins have been identified in the inclusion membrane. Using convalescent sera from guinea pigs infected with *C. psittaci* strain GPIC, researchers identified three chlamydial derived inclusion membrane proteins, IncA, IncB and IncC (Bannantine *et al.*, 1998; Rockey *et al.*, 1995; Rockey and Rosquist, 1994). At least one of the proteins, IncA, is exposed on the cytoplasmic face of the chlamydial inclusion and is phosphorylated on serine or threonine residues by the host cell (Rockey *et al.*, 1997). Distantly related homologues have been identified in *C. trachomatis*. Recently, three additional *C. trachomatis* inclusion membrane proteins have been identified (Scidmore-Carlson *et al.*, 1998). None of the inclusion membrane proteins are similar to any other proteins in the databases. All vesicular fusion events in eukaryotic cells require SNARE/SNAP like factors (Rothman and Wieland, 1996). Perhaps, chlamydiae regulate vesicular interactions in the host cell by recruitment and binding of these fusion factors to the chlamydial inclusion membrane.

5. CONCLUSIONS

Whereas the majority of intracellular parasites are believed to occupy vesicles in which maturation of endosomes to lysosomes is arrested to provide an appropriate environment for parasite replication, chlamydiae appear to escape from the endocytic pathway altogether by establishing interactions with an exocytic pathway. Such an interaction has been proposed to represent a pathogenic mechanism by which chlamydiae create an intracellular niche supportive of chlamydial replication. Many fundamental questions remain in regard to the properties of the inclusion and the environment within. Clearly, chlamydiae direct the responses of the host cell to favor parasite survival. Identification of those signals that govern responses of the host cell to the chlamydial inclusion remains significant challenges in chlamydial biology.

6. REFERENCES

Allan, I., and Pearce, J.H., 1987, Association of *Chlamydia trachomatis* with mammalian and cultured insect cells lacking putative chlamydial receptors. *Microb. Pathog.* **2**:63–70.

Bannantine, J.P., Rockey, D.D., and Hackstadt, T., 1998, Tandem genes of *Chlamydia psittaci* that encode proteins localized to the inclusion membrane. *Mol. Microbiol.* **28**:1017–1026.

Barnewall, R.E., Rikihisa, Y., and Lee, E.H., 1997, *Ehrlichia chaffeensis* inclusions are early endosomes which selectively accumulate transferrin receptor. *Infect. Immun.* **65**:1455–1461.

Birkelund, S., Johnsen, H., and Christiansen, G., 1994, *Chlamydia trachomatis* serovar L2 induces protein tyrosine phosphorylation during uptake by HeLa cells. *Infect. Immun.* **62**:4900–4908.

Bose, S.K., and Paul, R.G., 1982, Purification of *Chlamydia trachomatis* lymphogranuloma venereum elementary bodies and their interaction with HeLa cells. *J. Gen. Microbiol.* **128**:1371–1379.

Bose, S.K., Smith, G.B., and Paul, R.G., 1983, Influence of lectins, hexoses, and neuraminidase on the association of purified elementary bodies of *Chlamydia trachomatis* UW-31 with HeLa cells. *Infect. Immun.* **40**:1060–1067.

Byrne, G.I., 1976, Requirements for ingestion of *Chlamydia psittaci* by mouse fibroblasts (L cells). *Infect. Immun.* **14**:645–651.

Byrne, G.I., 1978, Kinetics of phagocytosis of *Chlamydia psittaci* by mouse fibroblasts (L cells): separation of the attachment and ingestion stages. *Infect. Immun.* **19**:607–612.

Byrne, G.I., and Moulder, J.W., 1978, Parasite-specified phagocytosis of *Chlamydia psittaci* and *Chlamydia trachomatis* by L and HeLa cells. *Infect. Immun.* **19**:598–606.

Chen, J.C., and Stephens, R.S., 1997, *Chlamydia trachomatis* glycosaminoglycan-dependent and independent attachment to eukaryotic cells. *Microb. Pathog.* **22**:23–30.

Clausen, J.D., Christiansen, G., Holst, H.U., and Birkelund, S., 1997, *Chlamydia trachomatis* utilizes the host cell microtubule network during early events of infection. *Mol. Microbiol.* **25**:441–449.

Clemens, D.L., and Horwitz, M.A., 1995, Characterization of the *Mycobacterium tuberculosis* phagosome and evidence that phagosomal maturation is inhibited. *J. Exp. Med.* **181**:257–270.

Davis, C.H., and Wyrick, P.B., 1997, Differences in the association of *Chlamydia trachomatis* serovar E and serovar L2 with epithelial cells in vitro may reflect biological differences in vivo. *Infect. Immun.* **65**:2914–2924.

Desai, S.A., Krogstad, D.J., and McClesky, E.W., 1993, A nutrient-permeable channel on the intraerythrocytic malaria parasite. *Nature*. **362**:643–646.

Eissenberg, L.G., and Wyrick, P.B., 1981, Inhibition of phagolysosome fusion is localized to *Chlamydia psittaci*-laden vacuoles. *Infect. Immun.* **32**:889–896.

Eissenberg, L.G., Wyrick, P.B., Davis, C.H., and Rumpp, J.W., 1983, *Chlamydia psittaci* elementary body envelopes: ingestion and inhibition of phagolysosome fusion. *Infect. Immun.* **40**:741–751.

Falkow, S., Isberg, R.R., and Portnoy, D.A., 1992, The interaction of bacteria with mammalian cells. *Annu. Rev. Cell Biol.* **8**:333–363.

Fawaz, F.S., van Ooij, C., Homola, E., Mutka, S.C., and Engel, J.N., 1997, Infection with *Chlamydia trachomatis* alters the tyrosine phosphorylation and/or localization of several host cell proteins including cortactin. *Infect. Immun.* **65**:5301–5308.

Finlay, B.B., and Cossart, P., 1997, Exploitation of mammalian host cell functions by bacterial pathogens. *Science*. **276**:718–725.

Friis, R.R., 1972, Interaction of L cells and *Chlamydia psittaci*: entry of the parasite and host responses to its development. *J. Bacteriol.* **110**:706–721.

Grayston, J.T., Kuo, C.-C., Campbell, L.A., and Wang, S.-P., 1989, Chlamydia pneumoniae sp. nov. for Chlamydia sp. strain TWAR. *Int. J. Syst. Bacteriol.* **39**:88–90.

Gutierrez-Martin C.B., Ojcius, D.M., Hsia, R., Hellio, R., Bavoil, P.M., and Dautry-Varsat, A., 1997, Heparin-mediated inhibition of *Chlamydia psittaci* adherence to HeLa cells. *Microb. Pathog.* **22**:47–57.

Hackstadt, T., and Caldwell, H.D., 1985, Effect of proteolytic cleavage of surface-exposed proteins on infectivity of *Chlamydia trachomatis*. *Infect. Immun.* **48**:546–551.

Hackstadt, T., Rockey, D.D., Heinzen, R.A., and Scidmore, M.A., 1996, *Chlamydia trachomatis* interrupts an exocytic pathway to acquire endogenously synthesized sphingomyelin in transit from the Golgi apparatus to the plasma membrane. *EMBO J.* **15**:964–977.

Hackstadt, T., Scidmore, M.A., and Rockey, D.D., 1995, Lipid metabolism in *Chlamydia trachomatis* infected cells: directed trafficking of Golgi-derived sphingolipids to the chlamydial inclusion. *Proc. Natl. Acad. Sci. USA*. **92**:4877–4881.

Hackstadt, T., Scidmore-Carlson, M., and Fischer, E., 1998, Rapid dissociation of the *Chlamydia trachomatis* inclusion from endocytic compartments. *In* Proceedings of the ninth international symposium on Human Chlamydial infection. R. Stephens, G. Byrne, G. Christiansen, I. Clarke, J. Grayston, R. Rank, G. Ridgway, P. Saikku, J. Schachter, and W. Stamm, editors. International Chlamydia Symposium, San Francisco. 127–130.

Hackstadt, T., and Williams, J.C., 1981, Biochemical stratagem for obligate parasitism of eukaryotic cells by *Coxiella burnetii*. *Proc. Natl. Acad. Sci. USA*. **78**:3240–3244.

Hatch, T.P., Miceli, M., and Silverman, J.A., 1985, Synthesis of protein in host-free reticulate bodies of *Chlamydia psittaci* and *Chlamydia trachomatis*. *J. Bacteriol.* **162**:938–942.

Hatch, T.P., Vance, Jr., D.W., and Al-Hossainy, E., 1981, Attachment of *Chlamydia psittaci* to formaldehyde-fixed and unfixed L cells. *J. Gen. Microbiol.* **125**:273–283.

Heinzen, R.A., and Hackstadt, T., 1997, The *Chlamydia trachomatis* parasitophorous vacuolar membrane is not passively permeable to low molecular weight compounds. *Infect. Immun.* **65**:1088–1094.

Heinzen, R.A., Scidmore, M.A., Rockey, D.D., and Hackstadt, T., 1996, Differential interaction with endocytic and exocytic pathways distinguish parasitophorous vacuoles of *Coxiella burnetii* and *Chlamydia trachomatis*. *Infect. Immun.* **64**:796–809.

Hodinka, R.L., Davis, C.H., Choong, J., and Wyrick, P.B., 1988, Ultrastructural study of endocytosis of *Chlamydia trachomatis* by McCoy cells. *Infect. Immun.* **56**:1456–1463.

Hodinka, R.L., and Wyrick, P.B., 1986, Ultrastructural study of mode of entry of *Chlamydia psittaci* into L-929 cells. *Infect. Immun.* **54**:855–863.

Joseph, T.D., and Bose, S.K., 1991a, Further characterization of an outer membrane protein of *Chlamydia trachomatis* with cytadherence properties. *FEMS Microbiol. Lett.* **84**:167–172.

Joseph, T.D., and Bose, S.K., 1991b, A heat-labile protein of *Chlamydia trachomatis* binds to HeLa cells and inhibits the adherence of chlamydiae. *Proc. Natl. Acad. Sci. USA*. **88**:4054–4058.

Krivan, H.C., Nilsson, B., Lingwood, C.A., and Ryu, H., 1991, *Chlamydia trachomatis* and *Chlamydia pneumoniae* bind specifically to phosphatidylethanolamine in HeLa cells and to GalNacbeta1-4GalLbeta1-4Glc sequences found in asialo-GM1 and asialo-GM2. *Biochem. Biophys. Res. Commun.* **175**:1082–1089.

Kuo, C., Takahashi, N., Swanson, A.F., Ozeki, Y., and Hakomori, S., 1996, An N-linked high-mannose type oligosaccharide, expressed at the major outer membrane protein of *Chlamydia trachomatis*, mediates attachment and infectivity of the microorganism to HeLa cells. *J. Clin. Invest.* **98**:2813–2818.

Kuo, C.-C., and Grayston, J.T., 1976, Interaction of *Chlamydia trachomatis* organisms and HeLa 229 cells. *Infect. Immun.* **13**:1103–1109.

Kuo, C.-C., Wang, S.-P., and Grayston, J.T., 1973, Effect of polycations, polyanions, and neuraminidase on the infectivity of trachoma-inclusion conjunctivitis and lymphogranuloma venereum organisms in HeLa cells: sialic acid residues as possible receptors for trachoma-inclusion conjunctivitis. *Infect. Immun.* **8**:74–79.

Kuo, C.C., Grayston, J.T., Campbell, L.A., Goo, Y.A., Wissler, R.W., and Benditt, E.P., 1995, *Chlamydia pneumoniae*(TWAR) in coronary arteries of young adults (15–34 years old). *Proc. Natl. Acad. Sci., U.S.A.* **92**:6911–6914.

Lee, C.K., 1981, Interaction between a trachoma strain of *Chlamydia trachomatis* and mouse fibroblasts (McCoy cells) in the absence of centrifugation. *Infect. Immun.* **31**: 584–591.

Levy, N.J., 1979, Wheat germ agglutinin blockage of chlamydial attachment sites: antagonism by *N*-Acetyl-D-glucosamine. *Infect. Immun.* **25**:946–953.

Levy, N.J., and Moulder, J.W., 1982, Attachment of cell walls of *Chlamydia psittaci* to mouse fibroblasts (L cells). *Infect. Immun.* **37**:1059–1065.

Lipsky, N.G., and Pagano, R.E., 1985, Intracellular translocation of fluorescent sphingolipids in cultured fibroblasts: endogenously synthesized sphingomyelin and glucocerebroside analogues pass through the Golgi apparatus en route to the plasma membrane. *J. Cell Biol.* **100**:27–34.

Majeed, M., Ernst, J.D., Magnusson, K.E., Kihlstrom, E., and Stendahl, O., 1994, Selective translocation of annexins during intracellular redistribution of *Chlamydia trachomatis* in HeLa and McCoy cells. *Infect. Immun.* **62**:126–134.

Majeed, M., Gustafsson, M., Kihlstrom, E., and Stendahl, O., 1993, Roles of Ca^{2+} and F-actin in intracellular aggregation of *Chlamydia trachomatis* in eucaryotic cells. *Infect. Immun.* **61**:1406–1414.

Majeed, M., and Kihlstrom, E., 1991, Mobilization of F-Actin and clathrin during redistribution of *Chlamydia trachomatis* to an intracellular site in eucaryotic cells. *Infect. Immun.* **59**:4465–4472.

Moulder, J.W., 1985, Comparative biology of intracellular parasitism. *Microbiol. Rev.* **49**:298–337.

Moulder, J.W., 1991, Interaction of chlamydiae and host cells in vitro. *Microbiol. Rev.* **55**:143–190.

Pagano, R.E., 1989, A Fluorescent derivative of ceramide: Physical properties and use in studying the golgi apparatus of animal cells. *In* Fluorescence microscopy of living cells in culture. Vol. 29. Y.-l. Wang and T.L. D., editors. Academic Press, inc., San Diego. 75–85.

Prain, C.J., and Pearce, J.H., 1989, Ultrastructural studies on the intracellular fate of *Chlamydia psittaci* (strain guinea pig inclusion conjunctivitis) and *Chlamydia trachomatis* (strain lymphogranuloma venereum 434): modulation of intracellular events and relationship with endocytic mechanism. *J. Gen. Microbiol.* **135**:2107–2123.

Reynolds, D.J., and Pearce, J.H., 1991, Endocytic mechanisms utilized by chlamydiae and their influence on induction of productive infection. *Infect. Immun.* **59**:3033–3039.

Ridderhof, J.C., and Barnes, R.C., 1989, Fusion of inclusions following superinfection of HeLa cells by two serovars of *Chlamydia trachomatis*. *Infect. Immun.* **57**:3189–3193.

Rockey, D.D., Grosenbach, D., Hruby, D.E., Peacock, M.G., Heinzen, R.A., and Hackstadt, T., 1997, *Chlamydia psittaci* IncA is phosphorylated by the host cell and is exposed on the cytoplasmic face of the developing inclusion. *Mol. Microbiol.* **24**:217–228.

Rockey, D.D., Heinzen, R.A., and Hackstadt, T., 1995, Cloning and characterization of a *Chlamydia psittaci* gene coding for a protein localized to the inclusion membrane of infected cells. *Mol. Microbiol.* **15**:617–626.

Rockey, D.D., and Rosquist, J.L., 1994, Protein antigens of *Chlamydia psittaci* present in infected cells but not detected in the infectious elementary body. *Infect. Immun.* **62**:106–112.

Rothman, J.E., and Wieland, F.T., 1996, Protein sorting by transport vesicles. *Science.* **272**:227–234.

Schachter, J., and Caldwell, H.D., 1980, Chlamydiae. *Annu. Rev. Microbiol.* **34**:285–309.

Schmiel, D.H., Knight, S.T., Raulston, J.E., Choong, J., Davis, C.H., and Wyrick, P.B., 1991, Recombinant *Escherichia coli* clones expressing *Chlamydia trachomatis* gene products attach to human endometrial epithelial cells. *Infect. Immun.* **59**:4001–4012.

Schmiel, D.H., Raulston, J.E., Fox, E., and Wyrick, P.B., 1995, Characteristization, expression and envelope association of a *Chlamydia trachomatis* 28 kDa protein. *Microb. Pathogen.* **19**:227–236.

Schramm, N., Bagnell, C.R., and Wyrick, P.B., 1996, Vesicles containing *Chlamydia trachomatis* serovar L2 remain above pH 6 within HEC-1B cells. *Infect. and Immun.* **64**:1208–1214.

Schramm, N., and Wyrick, P.B., 1995, Cytoskeletal requirements in *Chlamydia trachomatis* infection of host cells. *Infect. Immun.* **63**:324–332.

Schwab, J.C., Beckers, C.J.M., and Joiner, K.A., 1994, The parasitophorous vacuole membrane surrounding intracellular *Toxoplasma gondii* functions as a molecular sieve. *Proc. Natl. Acad. Sci. USA*. **91**:509–513.

Scidmore, M.A., Fischer, E.R., and Hackstadt, T., 1996a, Sphingolipids and glycoproteins are differentially trafficked to the *Chlamydia trachomatis* inclusion. *J. Cell Biol.* **134**:363–374.

Scidmore, M.A., Rockey, D.D., Fischer, E.R., Heinzen, R.A., and Hackstadt, T., 1996b, Vesicular interactions of the *Chlamydia trachomatis* inclusion are determined by chlamydial early protein synthesis rather than route of entry. *Infect. Immun.* **64**:5366–5372.

Scidmore-Carlson, M., Shaw, E.I., Dooley, C.A., and Hackstadt, T., 1998, Identification and characterization of putative *Chlamydia trachomatis* inclusion membrane proteins. *In* Proceedings of the ninth international symposium on Human Chlamydial infection. R.S. Stephens, G.I. Byrne, G. Christiansen, I.N. Clarke, J.T. Grayston, R.G. Rank, G.L. Ridgway, P. Sikku, J. Schachter, and W.E. Stamm, editors. International Chlamydia Symposium, San Frnacisco. 103–106.

Soderlund, G., and Kihlstrom, E., 1983a, Attachment and internalization of a *Chlamydia trachomatis* lymphogranuloma venereum strain by McCoy cells: kinetics of infectivity and effect of lectins and carbohydrates. *Infect. Immun.* **42**:930–935.

Soderlund, G., and Kihlstrom, E., 1983b, Effect of methylamine and monodansylcadaverine on the susceptibility of McCoy cells to *Chlamydia trachomatis* infection. *Infect. Immun.* **40**:534–541.

Stuart, E.S., Wyrick, P.B., Choong, J., Stoler, S.B., and MacDonald, A.B., 1991, Examination of chlamydial glycolipid with monoclonal antibodies: cellular distribution and epitope binding. *Immunology*. **74**:740–747.

Su, H., Raymond, L., Rockey, D.D., Fischer, E., Hackstadt, T., and Caldwell, H.D., 1996, A recombinant *Chlamydia trachomatis* major outer membrane protein binds to heparan sulfate receptors on epithelial cells. *Proc. Natl. Acad. Sci. U.S.A.* **93**:11143–11148.

Su, H., Spangrude, G.J., and Caldwell, H.D., 1991, Expression of Fc gammaRIII on HeLa 229 cells: Possible effect on in vitro neutralization of Chlamydia trachomatis. *Infect. Immun.* **59**:3811–3814.

Su, H., Watkins, N.G., Zhang, Y.-X., and Caldwell, H.D., 1990, *Chlamydia trachomatis*-host cell interactions: role of the chlamydial major outer membrane protein as an adhesin. *Infect. Immun.* **58**:1017–1025.

Su, H., Zhang, Y.-X., Barrera, O., Watkins, N.G., and Caldwell, H.D., 1988, Differential effect of trypsin on infectivity of *Chlamydia trachomatis*: loss of infectivity requires cleavage of major outer membrane protein variable domains II and IV. *Infect. Immun.* **56**:2094–2100.

Swanson, A.F., Ezekowitz, R.A., Lee, A., and Kuo, C.-C., 1998, Human mannose-binding protein inhibits infection of HeLa cells by *Chlamydia trachomatis*. *Infect Immun.* **66**:1607–1612.

Swanson, A.F., and Kuo, C.-C., 1991, The characterization of lectin-binding proteins of Chlamydia trachomatis as glycoproteins. *Microb. Pathogen.* **10**:465–473.

Swanson, A.F., and Kuo, C.-C., 1994a, The 32-kDa glycoprotein of *Chlamydia trachomatis* is an acidic protein that may be involved in the attachment process. *FEMS Microbiol. Lett.* **123**:113–118.

Swanson, A.F., and Kuo, C.-C., 1994b, Binding of the glycan of the major outer membrane protein of *Chlamydia trachomatis* to HeLa cells. *Infect. Immun.* **62**:24–28.

Taraska, T., Ward, D.M., Ajioka, R.S., Wyrick, P.B., Davis-Kaplan, S.R., Davis, C.H., and Kaplan, J., 1996, The late chlamydial inclusion membrane is not derived from the endocytic pathway and is relatively deficient in host proteins. *Infect. Immun.* **64**:3713–3727.

Ting, L.-M., Hsia, R.-C., Haidaris, C.G., and Bavoil, P.M., 1995, Interaction of outer envelope proteins of *Chlamydia psittaci* GPIC with the HeLa cells surface. *Infect. Immun.* **63**:3600–3608.

Tribby, I.I.E., Friis, R.R., and Moulder, J.W., 1973, Effect of chloramphenicol, rifampicin, and nalidixic acid on Chlamydia psittaci growing in L cells. *J. Infect. Dis.* **127**:155–163.

van Ooij, C., Apodaca, G., and Engel, J., 1997, Characterization of the *Chlamydia trachomatis* vacuole and its interaction with the host endocytic pathway in HeLa cells. *Infect. Immun.* **65**:758–766.

Vretou, E., Goswami, P.C., and Bose, S.K., 1989, Adherence of multiple serovars of *Chlamydia trachomatis* to a common receptor on HeLa and McCoy cells is mediated by thermolabile protein(s). *J. Gen. Microbiol.* **135**:3229–3237.

Ward, M.E., and Murray, A., 1984, Control mechanisms governing the infectivity of *Chlamydia trachomatis* for HeLa cells: mechanisms of endocytosis. *J. Gen. Microbiol.* **130**:1765–1780.

Wyrick, P.B., and Brownridge, E.A., 1978, Growth of *Chlamydia psittaci* in macrophages. *Infect. Immun.* **19**:1054–1060.

Wyrick, P.B., Choong, J., Davis, C.H., Knight, S.T., Royal, M.O., Maslow, A.S., and Bagnell, C.R., 1989, Entry of genital *Chlamydial trachomatis* into polarized human epithelial cells. *Infect. Immun.* **57**:2378–2389.

Xu, S., Cooper, A., Sturgill-Koszycki, S., van Heyningen, T., Chatterjee, D., Orme, I., Allen, P., and Russell, D.G., 1994, Intracellular trafficking in *Mycobacterium tuberculosis* and Mycobacterium avium-infected macrophages. *J Immunol.* **153**:2568–2578.

Zhang, J.P., and Stephens, R.S., 1992, Mechanism of *C. trachomatis* attachment to eukaryotic host cells. *Cell.* **69**:861–869.

Chapter 19

Interaction of Rickettsiae with Eukaryotic Cells

Adhesion, Entry, Intracellular Growth, and Host Cell Responses

Marina E. Eremeeva, Gregory A. Dasch, and David J. Silverman

1. INTRODUCTION

The genus *Rickettsia* includes obligate intracellular gram-negative bacteria which are associated with a variety of arthropod vectors. Based on antigenic, genetic, growth, and epidemiological characteristics, the genus is divided into the typhus group, which includes two human pathogens, *Rickettsia prowazekii* and *R. typhi*, and *R. canada*, which is known only from ticks, and the spotted fever group (SFG), which encompasses more than twenty distinct isolates (Raoult and Roux, 1997; Weiss and Moulder, 1984). The former scrub typhus group of rickettsiae was recently separated into a new genus, *Orientia,* which contains only a single species, *O. tsutsugamushi* (Tamura *et al.*, 1995). Rickettsial species may cause diseases of different

MARINA E. EREMEEVA and DAVID J. SILVERMAN University of Maryland, School of Medicine, Baltimore, Maryland. **GREGORY A. DASCH** Naval Medical Research Institute, Bethesda, Maryland.

Subcellular Biochemistry, Volume 33: Bacterial Invasion into Eukaryotic Cells, edited by Oelschlaeger and Hacker. Kluwer Academic / Plenum Publishers, New York, 2000.

severity in humans which share the common symptoms of fever, rash, intoxication and vasculitis.

Endothelial cells of the middle and small vessels are the primary target for rickettsiae during *in vivo* infection. *R. rickettsii*, the etiological agent of Rocky Mountain spotted fever (RMSF), also invades underlying smooth muscle cells. Multiplication of rickettsiae results in spread of the infection to adjacent cells and to distant endothelium. The infection elicits strong humoral and cellular immune responses which are mediated by lymphocytes and macrophages.

The mechanism of rickettsia-caused cellular injury is not yet understood. Rickettsiae do not express any secreted toxic factor important in the course of disease, and their lipopolysacharides do not have potent toxic activity (Walker *et al.*, 1984*a*; Schramek *et al.*, 1977). It seems that rickettsiae directly cause the injury to the host tissues because they have the ability to invade and to colonize the endothelium, and eventually to destroy it (Winkler and Turco, 1988; Walker and Cain, 1980). The exact sequence of events is not known fully, although an oxidant-associated injury mechanism, in particular, has been suggested to explain the pathogenic effects exerted on eukaryotic cells by *R. rickettsii* (Silverman, 1984). A comparison of the biology of different rickettsiae and their interaction with eukaryotic cells is an important approach to understanding how rickettsiae may cause disease. Details of the biology, intracellular growth and interaction of rickettsia with phagocytic and non-phagocytic cells have been summarized in several exellent reviews, and different models of rickettsia-host cell interaction have been proposed (Turco and Winkler, 1994; Balayeva, 1990; Tamura, 1988; Winkler and Turco, 1988; Silverman, 1986*b*; Winkler, 1986; Wisseman, 1986; Weiss, 1982; Gudima, 1971).

We review here salient features of the cellular and molecular interactions which occur between rickettsiae and the non-phagocytic cells they are invading. The review is organized into two sections. The first describes the morphological and physiological processes which occur during entry and early stages of intracellular growth of rickettsiae. The second part is focused on host cell reactions to rickettsial infection, and particularly on proinflammatory, vascular and transcriptional responses. Unfortunately, the information available about these processes is still very fragmentary, and many aspects of the rickettsia-host cell interaction have been studied only for *R. prowazekii* and *R. rickettsii* in fibroblasts and endothelial cells. Since species of the genus *Rickettsia* display significant variability in their biological and molecular characteristics, the reliability of attempts to extrapolate from only two species to the other rickettsial species and to other host cell types will also be considered briefly.

2. HEMOLYSIS OF ERYTHROCYTES AS A MODEL SYSTEM FOR STUDY OF ADHESION AND ENTRY OF RICKETTSIAE

The ability of typhus group rickettsiae to lyse erythrocytes has been exploited extensively as a sensitive model for investigating the main principles of the rickettsia-host cell interaction (see Winkler and Turco, 1988 for more references). Although typhus group rickettsiae can hemolyze erythrocytes of many mammalian species, including those of humans *in vitro* (Winkler, 1985; Wojciechowski *et al.*, 1963), the significance of hemolysis in the pathogenesis of rickettsial diseases is not yet established. *In vitro* hemolytic activity may be an incidental abortive attempt by rickettsiae to parasitize the erythrocyte, because the rickettsiae fail to induce phagocytosis in a totally non-phagocytic cell (Winkler and Ramm, 1975; Ramm and Winkler, 1973*b*). Alternatively, bound rickettsiae may exploit erythrocytes as a free ride for disseminating within blood vessels and for reaching distant sites rapidly for invasion (Balayeva, 1990). Inoculation of white mice with a lethal dose of hemolytically active *R. prowazekii* does not result in intravenous erythrocyte hemolysis (Wisseman, 1968) even though mouse erythrocytes are hemolyzed by rickettsiae *in vitro* (Clarke and Fox, 1948). On the other hand, lethal injection of rabbits with *R. prowazekii* is primarily associated with intravenous hemolysis (Paterson *et al.*, 1954). Hemolysis has been documented in severe cases of RMSF, Mediterranean spotted fever (MSF), murine typhus, and scrub typhus in patients with glucose-6-phosphate dehydrogenase (G6PD) deficiency; it may also occur in severe cases of RMSF in individuals with normal levels of G6PD (Walker, 1988).

Rickettsial interaction with erythrocytes is a two step process: adsorption of rickettsiae to a cholesterol-containing receptor in the erythrocyte membrane and lysis (Ramm and Winkler, 1976). Adsorption occurs within 4 min at 34°C (Ramm and Winkler, 1973*a*), and requires maintenance of a proton-motive force in the rickettsiae, since it can be arrested by treatment with a number of metabolic inhibitors and by inhibition of oxidative phosphorylation (Winkler, 1974). The energetic status of the target erythrocyte seems to be less important, since rickettsiae can adhere even to metabolically inactivated red blood cells and erythrocyte ghosts (Winkler and Ramm, 1975; Ramm and Winkler, 1973*b*). Hemolytic activity only occurs with intact metabolically active rickettsiae and requires intimate rickettsia-erythrocyte contact (Ramm and Winkler, 1973*a*) so that purification and separation of a hemolysin from rickettsiae has been difficult. Rickettsia-induced lysis of the erythrocyte is a temperature and energy-dependent process, and requires active participation of both partners (Winkler and Miller, 1980; Ramm and Winkler, 1973*b*). After lysis occurs, rickettsiae are released from the membrane of the damaged blood cell, since phagocytosis is absent. The rickettsiae can then initiate a new hemolytic round if intact erythrocytes are still present (Winkler, 1977). Consequently, hemolysis *in vitro* consists of multiple hemolytic events, and is limited only by the number of available target cells.

Lysis of erythrocytes by *R. prowazekii* results in the release of hemoglobin and the accumulation of free fatty acids and lysophosphatides, the latter characteristic products of phospholipase A (PLA_2) activity (Winkler and Miller, 1980). Both the inner and outer leaflet of the erythrocyte membrane are accessible to rickettsia-associated PLA_2 activity since both phosphatidylcholine and phosphatidylethanolamine are substrates in human erythrocytes, and these fractions are preferentially localized in the outer leaflet and inner leaflet, respectively (Winkler and Miller, 1980). These findings suggest that *R. prowazekii* can also lyse other cellular membranes from the outside and inside using a similar PLA_2-associated mechanism. This hypothesis was experimentally confirmed during *R. prowazekii* infection of L929 mouse fibroblasts (Winkler and Daugherty, 1989; Winkler and Miller, 1982). Moreover, this phospholipase activity may overwhelm cell uptake mechanisms since L929 cells are lysed rather than invaded by *R. prowazekii*, when 50 or more rickettsiae per cell are used for infection. Cell lysis even with small numbers of *R. prowazekii* can be achieved when host cells are pretreated with cytochalasin before contact with the rickettsiae (Winkler and Turco, 1988; Winkler and Miller, 1984, 1982).

3. ADHESION, ENTRY, AND INTRACELLULAR GROWTH OF RICKETTSIAE IN EUKARYOTIC CELLS

Rickettsiae infect a very wide range of cells, from the epithelial cells found in the louse gut and tick salivary glands, through reptilian, avian and rodent cells, to human fibroblasts and endothelial cells. These observations suggest that rickettsial invasion may be a very nonspecific process that is mediated by a receptor with a relatively low and broad affinity (Winkler and Turco, 1988). If this is so, adhesion appears to trigger a very efficient general mechanism for uptake which results in rapid internalization of the bound rickettsiae.

3.1. Adhesion

Attachment of rickettsiae to the cell surface of eukaryotic cells is the first event occurring during rickettsial invasion. Association of the rickettsiae with eukaryotic cells is temperature dependent and is a linear function of time and quantity of rickettsiae in the inoculum. Rickettsial adhesion is a very rapid process which is immediately followed by uptake of the bound microorganisms (Walker and Winkler, 1978). Thus the study of adhesion requires special approaches to separate the sequential stages which occur during invasion by the rickettsiae. Two experimental approaches were developed to overcome these difficulties.

In the first approach, *R. prowazekii* was labeled metabolically with [α-^{32}P]-ATP using the unique rickettsial ATP/ADP translocase to measure uptake events occurring with viable and dead rickettsiae and to distinguish bound and internalized rickettsiae conveniently (Walker and Winkler, 1978). Adhesion of the [α-^{32}P]-ATP metabolically labeled *R. prowazekii* to L929 mouse fibroblasts, BSC-1 fibroblasts and human endothelial cells (HUVEC) was examined (Walker, 1984; Walker and Winkler, 1978). The respective roles of rickettsial and eukaryotic cell factors in *R. prowazekii* binding were evaluated separately using inhibitors specific for bacterium or host cell. Adhesion of viable *R. prowazekii* either to fibroblasts or HUVEC was not influenced by pretreatment of the eukaryotic cells with N-ethylmaleimide (NEM), NaF, cytochalasin B or *p*-chloromercuriphenylsulfonic acid, but it was greatly dependent on temperature, time of incubation and the number of rickettsiae per cell (Walker, 1984; Walker and Winkler, 1978). Treatment of rickettsiae with the calcium ionophore A23187 stimulated their binding to HUVEC (Walker, 1984), indicating the importance of calcium levels for invasion by *R. prowazekii*. In contrast, inactivation of rickettsiae with NEM, formalin or UV-light significantly reduced adhesion; however, this occurs without complete inhibition. Because uptake of inactivated rickettsiae is much less efficient than uptake of viable *R. prowazekii,* it is thought that inactivated rickettsiae probably bind to sites different from those occupied by native microorganisms (Walker and Winkler, 1978). Adhesion of *R. prowazekii* to the host cell is irreversible, and once it has occurred, it is no longer sensitive to the inhibitory effect of NEM (Winkler and Turco, 1988; Winkler and Miller, 1984). These results suggest that there is a specific ligand-receptor interaction involved in the attachment of *R. prowazekii* to non-phagocytic cells (Walker, 1984). However, because similar results were obtained in experiments with fibroblasts and endothelial cells, unique cell-specific surface receptors are not involved.

Binding of *R. conorii* to L929 mouse fibroblasts was also examined by a flow cytometric method (Li and Walker, 1992). In contrast to the radiolabeled rickettsial technique, which measures the mean of aggregate interactions, this method yields highly quantitative individual data on large numbers of cell association events as measured by their fluorescence intensity. Because immunofluorescent staining of *R. conorii* was done with paraformaldehyde-fixed and nonpermeabilized samples, the signals detected were assumed to be due exclusively to surface-bound rickettsiae (Li and Walker, 1992). Attachment of *R. conorii* to L929 cells occurs rapidly after rickettsiae and host cells are mixed together. Like *R. prowazekii*, attachment of *R. conorii* to fibroblasts exhibits a dose-dependent relationship during the first 30 min of incubation. Active roles for both rickettsiae and host cells are important in adhesion of *R. conorii*, since binding was susceptible to heat inactivation of the rickettsiae, treatment of either host cells or rickettsiae with 1% paraformaldehyde or digestion with 0.25% trypsin for 15 min.

3.2. Entry of Rickettsiae into the Host Cell

Bound viable rickettsiae appear to perturb the host cell plasma membrane in a way that results in their internalization at the site of adhesion (Walker and Winkler, 1978). Although the exact signal for entry is not known, the invasion of *R. prowazekii* in HUVEC is probably calcium dependent (Walker, 1984). Moreover, the uptake of *R. prowazekii* by fibroblasts and HUVEC is sensitive to treatment with cytochalasins, NEM, NaF, and requires metabolically active rickettsiae (Walker, 1984; Walker and Winkler, 1978). Accordingly, this process was termed induced phagocytosis (Walker and Winkler, 1978) in contrast with two other alternative models for rickettsial entry: direct penetration of a passive host cell by active rickettsiae or phagocytosis of passive rickettsiae by an active host cell (Winkler and Turco, 1988). The current hypothesis, developed primarily for *R. prowazekii*, suggests that induced phagocytosis is coupled with release from the phagosome by means of a rickettsial phospholipase (Winkler and Turco, 1988). The general applicability of this model to other species of *Rickettsia* is often assumed but had not been evaluated until recently for *R. rickettsii*, *R. conorii*, and *O. tsutsugamushi*.

3.2.1. Electron Microscopic Observations

Ultrastructural studies of the rickettsial entry process have been described for *R. conorii* and *R. rickettsii* invading cultured Vero cells and HUVEC, respectively (Eremeeva *et al.*, 1998; Teysseire *et al.*, 1995), and *O. tsutsugamushi* infecting mouse fibroblasts and HUVEC (Tamura 1988; Ng *et al.*, 1985; Urakami *et al.*, 1983). The entry of *O. tsutsugamushi* into phagocytic cells has been documented well also (Rikihisa and Ito, 1982; Ewing *et al.*, 1978). Several individual steps of rickettsial entry have been differentiated: (i) adhesion of the rickettsia to the host cell membrane; (ii) uptake by engulfment of the rickettsia by host cell membrane extensions, or due to invagination of the host cell membrane; (iii) inclusion of the rickettsia in a phagocytic vacuole; (iv) lysis of the vacuole and escape of the rickettsia into the cytoplasm; and (v) intracytosolic growth of the free-living rickettsia.

In our studies with *R. rickettsii* invasion of HUVEC (Eremeeva *et al.*, 1998), we employed the approach of Teysseire *et al.* (1995) where a high dose of rickettsiae was inoculated onto cultured HUVEC on ice followed by low speed centrifugation at 4°C. As was described previously, such conditions increase rickettsial contact with the host cell and hence adhesion of the rickettsiae, but the low temperature arrests their internalization. A subsequent temperature shift to 37°C allows initiation of the uptake of the bound rickettsiae in a somewhat synchronized fashion. As a result, monolayers, examined right before the temperature shift or during the first minutes after the temperature shift was initiated, contained many rickettsiae that were attached to the host cell membrane (Figure 1). *R. rickettsii* bound to the cell either apically or laterally, thus suggesting a random distribution of host cell receptor ligands on the rickettsial surface (Figure 1). The host cell membrane also exhibited remarkable changes at the site of rickettsial adhesion, which could be recognized by increases in its electron density. *R. conorii* sometimes binds to a pedestal-like structure in Vero cells (Teysseire *et al.*, 1995) which appears similar to that formed during invasion of enteropathogenic *E. coli* strains. In contrast, no special changes in the surface of the host membrane were observed during adhesion of *O. tsut-*

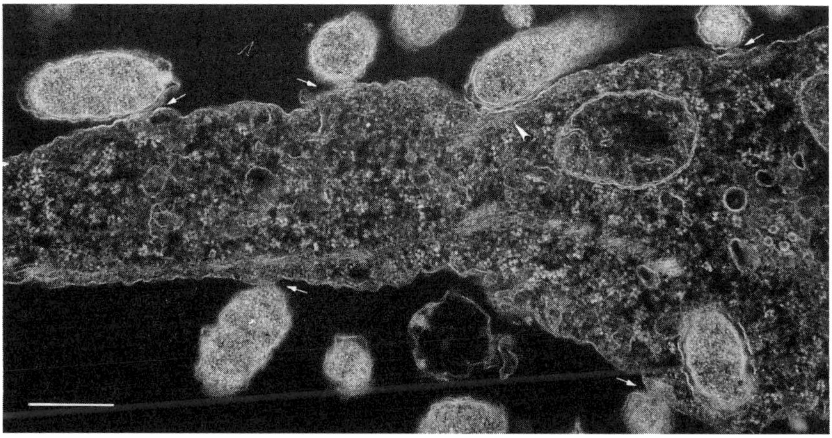

FIGURE 1. Adhesion of *R. rickettsii* Smith strain to HUVEC. A confluent monolayer of HUVEC held on ice was inoculated with *R. rickettsii* at a multiplicity of infection of 100 rickettsiae per cell, centrifuged for 20 min at 4°C and incubated in a 37°C water bath for 5 min before fixation. Attached rickettsiae are indicated by arrows. A high concentration of actin microfilaments at a site of rickettsial attachment is indicated by the arrowhead. Scale bar, 0.5 μm.

sugamushi to L929 mouse fibroblast cells (Tamura 1988; Urakami *et al.*, 1983). However, despite these structural differences, all bound rickettsiae likely must establish intimate contact with host cell receptors to initiate entry. Whether rickettsiae deliver an unidentified mediator(s) that initiates and/or facilitates the internalization of rickettsiae, as occurs with pathogenic bacteria exploiting a type III protein secretion system for their internalization (Hueck, 1998), has not been determined.

Perhaps in response to a putative rickettsial signal factor, the host cell membrane forms protrusions around both sides of the attached rickettsiae. Subsequently, the membrane protrusions grow and surround the rickettsia and engulf it within a phagocytic vacuole. Thus, formation of a complete phagocytic vacuole is a distinct step during *R. rickettsii* invasion of HUVEC. The detection of ruthenium red staining on extracellular rickettsiae and the outer surface of the host membrane and its absence in rickettsia-containing phagocytic vacuoles proves that the invasion vacuole closes completely prior to lysis and release of the rickettsiae into the cytoplasm, much as was described to occur during invasion of *R. conorii* into Vero cells (Teysseire *et al.*, 1995). A criticism of this conclusion is that the viability of rickettsiae enclosed in the vacuole cannot be determined so that the image depicted may illustrate only the trafficking of nonviable microorganisms (Hackstadt, 1996). However, in experiments with *O. tsutsugamushi* it was demonstrated that the formation of protrusions occurs only at sites of adhesion of morphologically and metabolically intact rickettsiae, while plasmolyzed or UV-killed organisms could attach to the cell membrane but did not induce phagocytosis (Tamura, 1988; Urakami *et al.*, 1983). Differences in the binding properties of viable and heat-inactivated rickettsiae were also demonstrated in ultrastructural studies on invasion of *R. conorii* into Vero cells (Walker *et al.*, 1998).

Both internalization and escape of *R. rickettsii* from the phagocytic vacuole in HUVEC are very rapid processes, since 10 to 30 min after internalization was allowed, a highly significant number of rickettsiae were observed in the cytoplasm. Rickettsiae were either completely surrounded by a phagocytic vacuole membrane, or in the process of escaping the phagocytic vacuole, or free in the cytoplasm without any surrounding membrane. During *R. conorii* infection of Vero cells between 40 and 90% of the bacteria were internalized by 3 and 12 min or later, respectively, and after 12 min, about 45% of the rickettsiae were already free in the cytoplasm (Teysseire *et al.*, 1995). After the rickettsiae are released from their phagocytic vacuoles, they start their intracellular growth cycle. Interestingly, the unipolar nucleation of actin tails by *R. rickettsii* starts even before the complete lysis of the phagocytic membrane occurs (Figure 2).

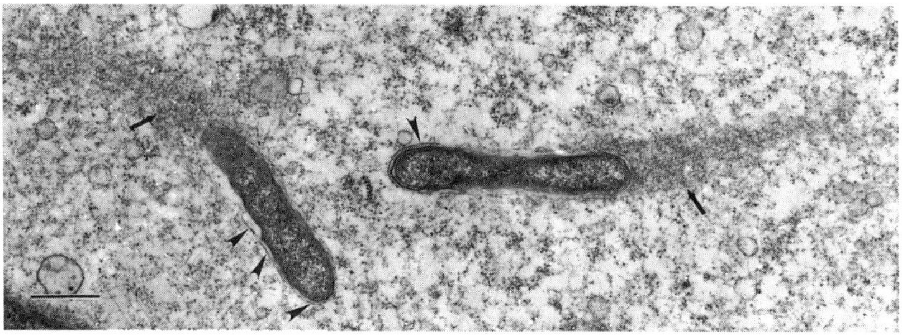

FIGURE 2. Formation of actin tails by *R. rickettsii* Smith strain after entry into the cytoplasm of HUVEC. A confluent monolayer was infected as described in the legend for Figure 1 and incubated at 37°C for 15 min before fixation. Actin tails are indicated by arrows. Fragments of phagosomal membrane in different stages of lysis are indicated by arrowheads. Scale bar, 0.5 μm.

3.2.2. Components of the Rickettsial Cell Involved in Invasion

The rickettsial structures that bind to the host cell and mediate its invasion have not been identified. The sensitivity to heat and protease treatment of the adhesion of *R. conorii* to cells suggests that surface-exposed protein(s) may play an important role in this interaction (Li and Walker, 1992). It is possible that different SFG species recognize the same host cell receptor, since *R. rickettsii* competitively inhibited the attachment of *R. conorii* to L929 cells by 51% in experiments when they were mixed in equal numbers. The species-specific rOmpA and rOmpB proteins have been suggested as the rickettsial ligands (adhesins) because they are surface-exposed and thermomodifiable. Since other surface array proteins have weak sequence homology with flagellae and other adhesins and have been proposed to have similar structural and functional properties (Kuen and Lubitz, 1996), this putative role for the rOmp proteins is very interesting to examine further. If this hypothesis is correct, it may be expected that *R. prowazekii* and *R. typhi* will not display the same type of adhesive characteristics as SFG rickettsiae. These typhus rickettsiae only have rOmpB proteins which form intermolecular disulfide bonds and which exhibit high resistance to trypsin treatment, while SFG rickettsiae possess non-disulfide linked rOmpA and rOmpB which appear to be more susceptible to proteases. Differences existing between the interactions of *R. prowazekii* and *R. typhi* and those of SFG rickettsiae toward erythrocytes are consistent with these molecular distinctions.

A role for rickettsia-associated PLA_2 in invasion is also likely since the release of free fatty acids and lysophosphatides occurs during both the hemolysis of erythrocytes and the entry of L929 cells by *R. prowazekii* (Winkler, 1985; Winkler and Miller, 1982, 1980). Suggestive evidence was obtained also for the existence of phospholipase activity during the infection of different types of eukaryotic cells by the non-hemolytic rickettsia, *R. rickettsii* (Silverman *et al.*, 1992; Walker *et al.*, 1983). Plaque formation by *R. rickettsii* in primary chicken fibroblasts and Vero cells was inhibited by addition of phentermine, a phospholipase inhibitor, to the agarose overlay (Walker *et al.*, 1983). Based on this observation, the authors suggested that phentermine inhibits a phospholipase-associated rickettsial cytopathic effect; however, no direct evidence was provided that phentermine inhibited the PLA_2 activity seen during rickettsial infection nor were potential direct deleterious effects on the rickettsiae evaluated (Winkler, 1985; Smith and Winkler, 1977). Pretreatment of *R. rickettsii* with 5–10 μM *p*-bromophenacyl bromide (*p*-BPB) or antiserum to *Ophiophagus hannah* venom (anti-PLA_2 serum) in cell free medium was found to have a greater effect on the entry of rickettsiae into Vero cells than did pretreatment of the host cells with these reagents alone (Silverman *et al.*, 1992). Inhibitory effects of the treatment with *p*-BPB were observed on spread of rickettsiae during the first 24 hr after infection and on plaque-forming activity by *R. rickettsii.* Although *p*-BPB is thought to be a specific PLA_2 inhibitor, rickettsial PLA_2 was not shown to be inhibited here and the pharmacological effects of *p*-BPB in this particular cell system were not evaluated. This is important since the inhibition of host cell glycolysis with NaF and protein synthesis with emetine changes the rate of development, size and morphology of plaques formed by *R. prowazekii* in Vero cell monolayers (Policastro *et al.*, 1996).

The 15 and 16 kDa proteins and 60 kDa protein of *R. rickettsii* could be detected by Western blotting using anti-eukaryotic PLA_2 and anti-phospholipase C antisera, respectively (Manor *et al.*, 1994). Structures on the *R. rickettsii* cells reacting with these antibodies were also detected by electron microscopy using immuno-gold labeling (Manor *et al.*, 1994). This is suggestive, if indirect, evidence for rickettsia-associated phospholipase(s). These antibodies have not been tested for their value in the affinity purification and isolation of putative *R. rickettsii* phospholipase(s). Interestingly, Winkler *et al.* (1994) demonstrated the predominant increase of phospholipase C (PLC) activity compared to PLA_2 and lysophospholipase activities at 30 to 48 hr growth of *R. prowazekii* in L929 cells based on analysis of hydrolytic products from choline-labeled phospholipids. Unfortunately, activation of host cell PLC cannot be ruled out completely in this experiment, since considerable PLC activity was also seen in uninfected controls.

In an important paper, Ojcius *et al.* (1995) compared the attributes of the PLA_2 and hemolytic activity of viable *R. prowazekii*. As expected for an extracellular phospholipase, the rickettsial PLA_2 activity was primarily active on negatively charged phospholipids. PLA_2 activity also showed the same pH dependence as hemolytic activity, being most active at pH 6–7 and declining at alkaline and acidic pH's, whether calcium was present or not. Both hemolytic activity and PLA_2 activity were sensitive to reduction with dithiothreitol and were magnesium dependent. However, in contrast to PLA_2 activity, hemolytic activity was shown to be enhanced significantly by calcium at either acidic or basic pH. Thus, it is likely that *R. prowazekii* produces more than one membranolytic factor. These observations may help to explain some puzzling observations. First, early phagosomes that are presumed to harbor the rickettsiae before their escape may be acidic. So the decrease in rickettsial hemolytic activity at acidic pH in the absence of calcium is difficult to understand unless the rickettsiae prevent acidification of the phagosome or escape before it can be acidified. On the other hand, calcium-dependent membranolytic factors and phospholipases are well known. Second, SFG rickettsiae lack hemolytic activity but they may very well have a calcium-dependent factor that accounts for their ability to escape the phagosome. More than one factor may also be present in typhus rickettsiae but the one associated with hemolytic activity may be absent in the SFG rickettsiae. For example, PLA_2 activity alone is usually insufficient to mediate hemolysis unless other factors such as lipases or proteases are present (Walker *et al.*, 1984*b*). In experiments with endothelial cells which did not affect host cell pinocytotic activity, treatment of the rickettsiae with the calcium ionophore A23187, which acts like a calcium channel, stimulated rickettsial adhesion and inhibited both rickettsial entry and rickettsial hemolytic activity (Walker, 1984). Consequently, the role of calcium in rickettsia-induced host cell membrane lysis and cell signaling and in rickettsial physiology appears complex and essential for the invasion process.

3.3. Intracellular Growth of Rickettsiae

3.3.1. Varied Patterns of Intracellular Growth of Different Species of Rickettsiae

After penetrating and lysing the phagocytic vacuole membrane, rickettsiae start their intracellular growth cycle, which is unique for different species and sometimes even for the strain of rickettsia (Wisseman, 1986; Gudima, 1971). Although different types of eukaryotic cells vary in their ability to maintain the growth and multiplication of rickettsiae, the course of infection observed depends greatly on the rickettsial passage history, and for the purpose of comparative studies, on the accuracy of the titration of the inoculum. Even if cloned cell lines are used as the experimental system, the population of target cells may be very heterogeneous, and this may be reflected as differentials in sensitivity toward rickettsial infection and replication.

The contrast observed in the growth patterns of typhus group and SFG rickettsiae is the most obvious (Silverman *et al.*, 1980; Wisseman *et al.*, 1976*a*, 1976*b*; Wisseman and Waddell, 1975). After infection *R. prowazekii* starts to multiply in the cytoplasm of its host cells and undergoes clearly distinguished and measurable lag, exponential, and stationary growth phases with a generation time of 8.8–8.9 hr at 34°C (Wisseman, 1986; Wisseman and Waddell, 1975). Immediately after inoculation the few intracytoplasmic *R. prowazekii* that are present do not cause visible damage to the host cell. Multiplying rickettsiae then slowly occupy the whole intracellular space and cause the destruction of intracellular structures, particularly degenerative changes in the endoplasmic reticulum and nucleus and a decline in the number and morphology of the mitochondria (Silverman *et al.*, 1980; Gudima, 1971). The infected cells become overfilled with rickettsiae (Figure 3), reaching >800 microorganisms per cell, and finally lyse on days 2 to 4 post infection and release rickettsiae into the extracellular medium, from which they can initiate a new cycle of infection in viable cells. Subsequent rounds of infection and proliferation result in asynchronous lysis of the host cells (Wisseman *et al.*, 1976*b*; Wisseman and Waddell, 1975).

This type of infection was observed for both virulent and nonvirulent strains of *R. prowazekii* in non-phagocytic cells (Wisseman, 1986). In contrast, macrophages and macrophage-like cells supported the growth of virulent strains, but inhibited the intracellular growth of the attenuated Madrid E strain (Penkina *et al.*, 1995; Winkler and Daugherty, 1983; Turco and Winkler, 1982; Gudima, 1979; Gambrill and Wisseman, 1973). The nature of this phenomenon is not completely understood. Both attenuated E and virulent Breinl strains exhibited serine and glycine auxotrophy in Vero cells, although dependence of strain E on these amino acids is much higher (Austin *et al.*, 1987). After addition of serine and glycine to the medium, the growth of the E strain in Vero cells was still repressed compared to the growth of strain Breinl, though their growth was similar in irradiated L929 cells. The nonvirulent E strain also has an alteration in lysine methylation patterns compared to virulent strains (Ching *et al.*, 1993; Rodionov *et al.*, 1991), which may influence its sensitivity to cytokines (Turco and Winkler, 1994*a*). The underlying metabolic basis for these differences in growth properties may be due in part to a defective *metL* gene in the E strain (Andersson and Andersson, 1997).

Different isolates of *R. typhi* exhibit very similar growth characteristics, but they differ from those of *R. prowazekii* (Wisseman, 1986). Growth of *R. typhi* was also confined to the cytoplasm, but shortly after invasion of the cytoplasm, the rickettsiae began to escape from the host cell and to

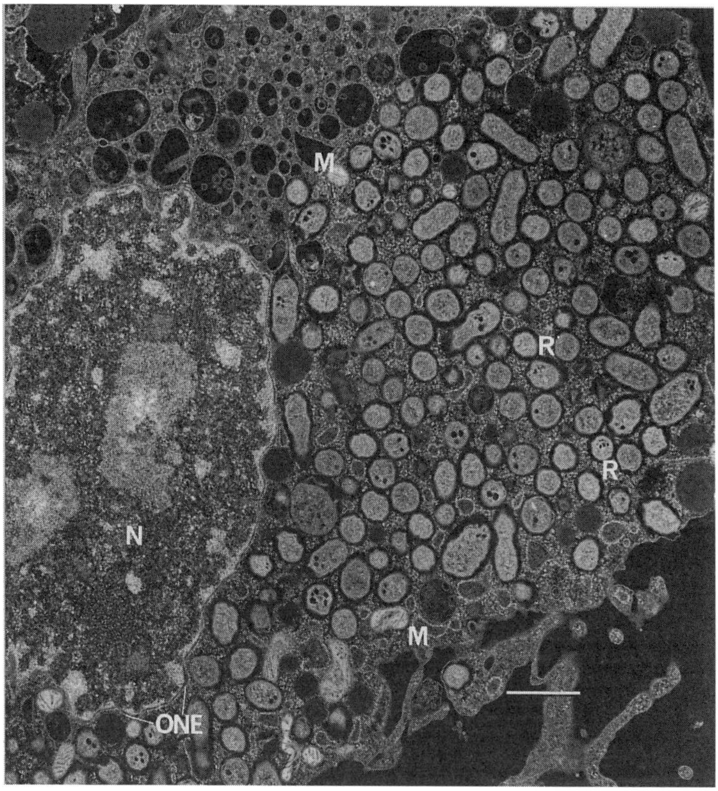

FIGURE 3. Electron microscopic observations of *R. prowazekii* Breinl strain grown in primary chicken fibroblasts, 72 hr after infection. Abbreviations: R, rickettsia; M, mitochondria; N, nucleus; ONE, outer nuclear envelope. Scale bar, 1.0 μm.

infect other cells in the culture, causing a linear increase in the number of infected cells. *R. canada* was originally assigned to the typhus group based on its conserved lipopolysacharide group antigen, but it has some phylogenetic affinities with SFG rickettsiae (Roux and Raoult, 1995; Stothard and Fuerst, 1995). *R. canada* grows in the cytoplasm, but it can also invade and replicate in the nucleus like SFG rickettsiae (Burgdorfer and Brinton, 1970). *O. tsutsugamushi* remains largely within the cytoplasm during its growth phase, although rarely it may also be seen infecting the nucleus (Tamura, 1988; Urakami *et al.*, 1982).

The growth of *R. rickettsii* has been studied in detail as a paradigm for SFG rickettsiae although different isolates from this group display a great variety of growth characteristics. *R. rickettsii* infection in chicken embryo and L929 cell cultures starts exponentially without any measurable lag phase (Wisseman *et al.*, 1976*a*). However, very soon after infection, *R. rickettsii* begins to escape from the host cell into the medium in very large numbers, resulting in: (i) failure of large numbers of rickettsiae to accumulate in the cytoplasm of infected cells; (ii) sustained rapid division of microorganisms in the cytoplasm; (iii) substantial accumulation of extracellular rickettsiae; and (iv) rapidly spreading infection in the culture, with most cells infected in 48 to 72 hr (Wisseman *et al.*, 1976*a*). In some cells, SFG rickettsiae invade the nucleus where they multiply to form compact masses (Figure 4) (Wisseman *et al.*, 1976*a*; Burgdorfer *et al.*, 1968). Thus, *R. rickettsii* is characterized by its capacity for bidirectional traffic across the cell cytoplasmic membrane and dominantly monodirectional passage across the nuclear envelope (Wisseman *et al.*, 1976*a*). Light and electron microscopic observations on cell cultures early in the infection detected few *R. rickettsii* in the cell cytoplasm or in the nucleus and no concomitant pathological changes in cell structures (Silverman 1984; Silverman and Bond, 1984; Wisseman *et al.*, 1976*a*). In contrast, later stages of infection are characterized by typical changes in cellular structure, notably the dilatation of the rough endoplasmic reticulum and outer nuclear membrane and accumulation of electron-dense material within the cisternae of intracellular membranes (Figure 4) (Silverman, 1984).

In initial studies, infections of cell cultures with *R. sibirica*, *R. conorii* and *R. akari* were characterized as benign compared with those of *R. rickettsii*, with maximum accumulation of rickettsiae at days 5–7 and cell death at days 7–10 after infection (see Gudima, 1971 for more references). However, ultrastructural changes similar to those observed in *R. rickettsii*-infected cells were also found in L929 cells infected with *R. japonica* Katayama strain (Iwasama *et al.*, 1992) and *R. sibirica* 246 strain (Popov *et al.*, 1986), indicating that common cytopathic mechanisms are likely to occur during SFG rickettsial growth in the cell cultures. The appearance and accumulation of pathological changes in the cultures infected with *R. japonica* or *R. sibirica* are merely delayed compared with those occurring in *R. rickettsii*-infected cultures, with the greatest effects observed at days 5–6 after infection. *R. conorii* infection also produces fewer necrotic effects in Vero cells when compared with *R. rickettsii*. These cytopathic effects were not due to a secreted rickettsial toxin since it failed to pass a membrane barrier (Walker *et al.*, 1984*a*).

The degree to which different rickettsiae exert their effects on eukaryotic cells can be quantified by comparison of their plaque forming activity (Wike *et al.*, 1972; Weinberg *et al.*, 1969; Kordova 1966), a technique which may be useful to some extent for distinguishing pathogenic and nonpathogenic rickettsiae (Hackstadt, 1996). Plaque forming activity is an important taxonomic characteristic (Weiss and Moulder, 1984). Among SFG *R. rick-*

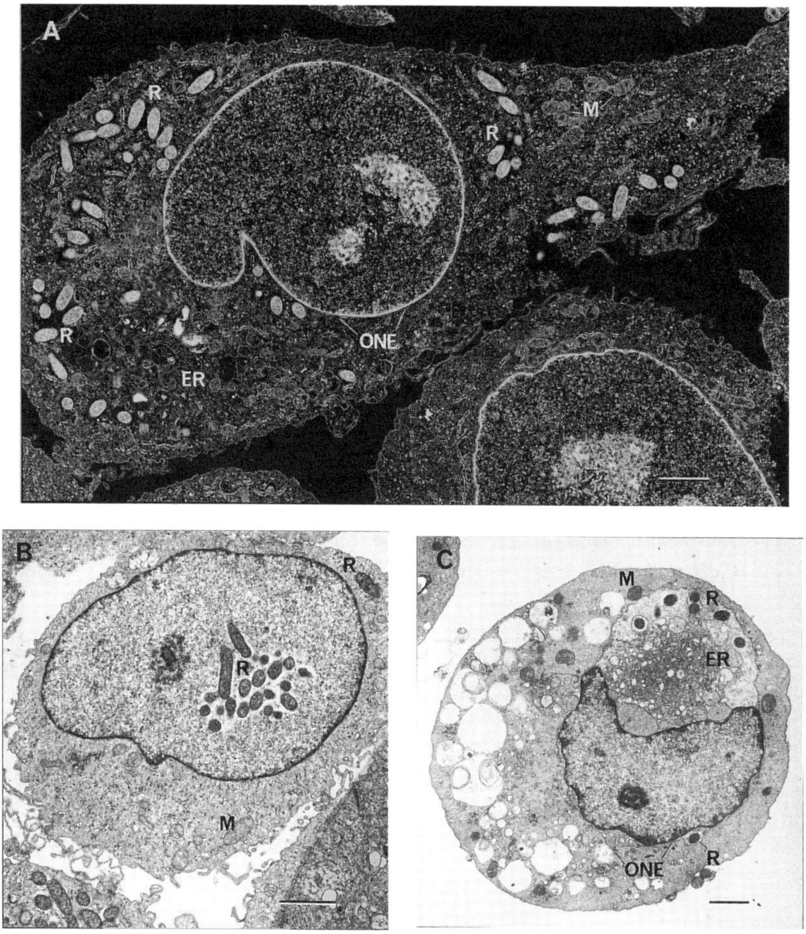

FIGURE 4. Electron microscopic observations on growth of SFG rickettsiae in the EA.hy 926 human endothelial cell line. (A). Intracytosolic multiplication of *R. sibirica* 246 strain, 72 hr after infection. (B). Intranuclear growth of *R. sibirica* 246 strain, 72 hr after infection. (C). Cytopathic effect of *R. rickettsii* Smith strain, 72 hr after infection. Abbreviations: R, rickettsiae; M, mitochondria; ER, endoplasmic reticulum; ONE, outer nuclear envelope. Scale bar, 1.5 μm.

ettsii formed clearly distinguished plaques of 2–3 mm in diameter at 6 to 7 days after inoculation, although some variability in size and plaque morphology exists depending on the cell used, diluent, incubation temperature, centrifugation of infected monolayers and the density of host cells (Walker *et al.*, 1982; Walker and Cain, 1980; Johnson and Pedersen, 1978; Wike *et al.*, 1972). Plaque-forming ability of other SFG rickettsiae also varies in different cell lines (Johnson and Pedersen, 1978; McDade *et al.*, 1969; Kordova,

1966) and they typically require a few days more than *R. rickettsii* to produce visible plaques. *R. prowazekii* forms plaques of significantly smaller size and less-defined morphology within 10 to 13 days (Wike *et al.*, 1972; McDade *et al.*, 1969). Interestingly, a wide selection of strains from three typhus group rickettsiae, *R. prowazekii*, *R. typhi*, and *R. canada*, exhibited very similar plaque-forming properties even though their growth characteristics in cell culture differ (Woodman *et al.*, 1978). *O. tsutsugamushi* may take as long as 20 days to produce extremely small lytic plaques in different cell cultures (Hanson, 1987; Oaks *et al.*, 1977; McDade and Gerone, 1970). Thus plaque formation is a sensitive indicator of subtle differences in the way different rickettsiae interact with their host cells. These differences may occur during attachment to the host cell, entry, intracellular growth, cell-to-cell spread, or membrane lysis and release. The molecular or genetic basis of these phenotypic differences among *Rickettsia* is not known yet.

3.3.2. Actin Polymerization and Rearrangement of the Cell Cytoskeleton During Rickettsial Infection

An active role for the cell cytoskeleton in the course of rickettsial infection is well documented since uptake from and release of rickettsiae into the extracellular space is inhibited by cytochalasins B and D (Heinzen *et al.*, 1993; Walker, 1984; Winkler and Miller, 1982). Microfilament-associated intracellular movement is another aspect of the rickettsia-cell interaction (Heinzen *et al.*, 1993; Teysseire *et al.*, 1992*b*; Todd *et al.*, 1983). Intracellular movement was first reported for *R. rickettsii* in cultured rat fibroblasts (Schaechter *et al.*, 1957), and later it was documented for *R. conorii* (Kokorin, 1968) and *O. tsutsugamushi* (Ewing *et al.*, 1978).

Filamentous actin tails of up to 70 μm length were detected at one pole of intracellular *R. rickettsii* and *R. conorii* in Vero cells and HUVEC, respectively, following staining with fluorescent-labeled phallotoxins (Figure 5) (Heinzen *et al.*, 1993, Teysseire *et al.*, 1992*b*). Nucleation of actin monomers occurs very rapidly after rickettsiae escape from the phagocytic vacuole into the cytoplasm. Microorganisms coated with F-actin are seen by 15 min after entry, and formation of fully-formed actin tails can be detected by 15 to 30 min post infection in both Vero cells and HUVEC (Eremeeva *et al.*, 1998; Heinzen *et al.*, 1993). Formation of actin tails is arrested following chloramphenicol treatment of infected cells, thus suggesting that *de novo* rickettsial protein synthesis is required.

Rickettsia is the third distinct genetic group of bacteria with actin-based motility, but the mechanisms of actin polymerization it uses are

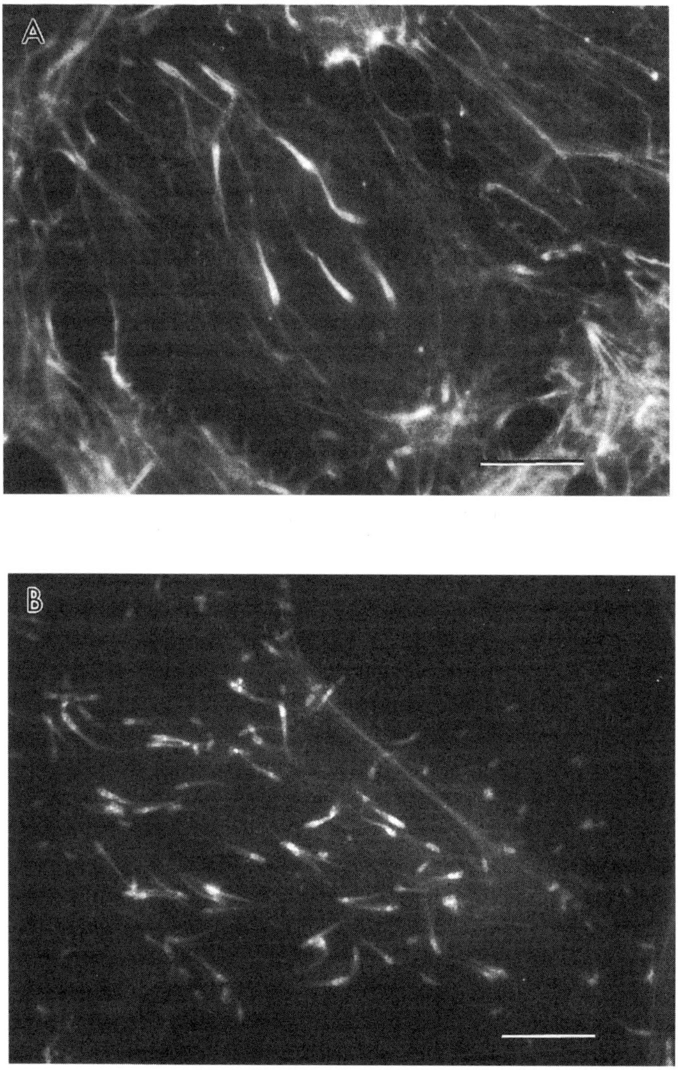

FIGURE 5. Microimmunofluorescence detection of formation of actin tails by *R. rickettsii* Smith strain in human endothelial cells. (A). Actin tail-associated mobility of *R. rickettsii* in the cytoplasm of EA.hy 926 cells, 72 hr after infection. Rickettsiae were detected with fluorescein-labeled antibody and actin was stained with rhodamin-labeled phalloidin. (B). Actin tail-associated release of *R. rickettsii* from HUVEC, 72 hr after infection. Rickettsiae were detected with Texas red-labeled antibody and actin was stained with fluorescein-labeled phalloidin. Scale bar, 10 μm.

poorly understood compared to those employed by *Shigella flexneri* and *Listeria monocytogenes* (see accompanying chapters in this volume). In those species the bacterial gene products essential for actin nucleation have been identified. Some unsuccessful attempts were made to define similarities between the *Listeria* and *Shigella* components and those of *Rickettsia* (Hackstadt, 1996). The difficulties encountered probably relate to the peculiarities of the strict intracellular parasitism of *Rickettsia*, limited information on rickettsial proteins, and the lack of efficient systems for manipulating the rickettsial genome. Recently, sequencing of the *R. prowazekii* genome was completed and a rudimentary transformation system for *R. prowazekii* was developed (Rachek *et al.*, 1998; Andersson and Andersson, 1997). These important advances will increase our progress in understanding rickettsial motility and the molecular biology of this microorganism in general.

The ability of rickettsiae to induce actin polymerization probably does not contribute greatly to their virulence, since formation of actin tails was detected for both virulent and nonvirulent strains of *R. rickettsii* and several other isolates of SFG rickettsiae including *R. montana*, *R. australis* and *R. parkeri* (Heinzen *et al.*, 1993). Moreover, F-actin tails were not detected in association with pathogenic *R. prowazekii*, or for *R. canada*, whose pathogenicity is uncertain (Heinzen *et al.*, 1993). Only a few *R. typhi* cells displayed actin tails which were generally much shorter than the long tails produced by SFG rickettsiae (Heinzen *et al.*, 1993, Teysseire *et al.*, 1992*b*). Consequently, the ability to nucleate long actin tails can be used as a taxonomic characteristic for distinguishing the SFG and typhus group of rickettsiae (Hackstadt, 1996; Heinzen *et al.*, 1993). The ability of SFG rickettsiae to penetrate the nucleus has been correlated with actin-based motility, since *R. prowazekii* is restricted to the cytoplasm (Heinzen *et al.*, 1993). However, this hypothesis seems inadequate since *R. canada* penetrates the nucleus but does not display actin tails, and *R. typhi* possesses short tails but does not invade the nucleus. Whether some differences exist in the capacity for actin polymerization among SFG isolates which may explain their variable cytotoxic effects toward eukaryotic cells *in vitro* awaits further examination. It seems more likely that some additional components of the rickettsial cell are involved.

The disturbances in host cell actin microfilaments induced during uptake of the rickettsiae and initiation of intensive actin tail polymerization by SFG rickettsiae suggest that a strong interplay of the host cell cytoskeleton and rickettsiae occurs during infection. However, the impact of rickettsia-associated actin polymerization on cellular integrity, and whether some changes in cellular actin pool accompany rickettsial multiplication and cell-to-cell spread is unknown. A decrease in the mRNA pool for actin was detected in human endothelial cells after *ex vivo* infection of umbilical cords with *R. rickettsii* for 4hr (Courtney *et al.*, 1996). The significance of these findings for the pathogenesis of rickettsial diseases also remains unstudied.

4. EUKARYOTIC CELL RESPONSES INDUCED BY RICKETTSIAL INFECTION

Vascular injury is the major initial lesion observed in rickettsia-caused diseases. The injury results from penetration and replication of the rickettsiae in endothelial cells. The damage inflicted on infected endothelial cells and resulting dissemination of rickettsiae into the surrounding and underlying tissues are detected in histological studies of severe cases. This damage results in enhanced permeability of tissue fluids into the infection foci, infiltration by monocytes and polynuclear neutrophils and inflammatory reactions, and the formation of microthrombi. Because multiple factors are involved in these vascular effects caused by the rickettsia-endothelium interaction *in vivo*, *in vitro* models have been employed to understand individual aspects of this complex process. When and how rickettsiae initiate changes in the infected cells, and what potential solutions for the control and prevention of these changes may be possible, are the major issues that have been addressed.

4.1. Changes in the Expression of Cellular Receptors

Inflammatory processes involving endothelial cells generally result in the expression of receptors belonging to the selectin family, E-selectin and P-selectin, and to the immunoglobulin superfamily, intercellular adhesion molecules [ICAM]-1 and -2 and vascular cell adhesion molecule [VCAM]-1 (Springer, 1995).

In vitro studies of human endothelial cells infected with *R. rickettsii* (Sporn *et al.*, 1993) and *R. conorii* (Dignat-George *et al.*, 1997) demonstrated increased surface expression of E-selectin which peaked between 4–8hr after infection and declined to near baseline levels within 24hr. Upregulation of ICAM-1 and VCAM-1 expression also occurred. Like E-selectin expression, their expression was time- and rickettsia inoculum size-dependent. However, peak levels of ICAM-1 and VCAM-1 were reached more slowly and showed greater increases than found for E-selectin (Dignat-George *et al.*, 1997). VCAM-1 levels significantly increased during the first 8hr and then stayed on a plateau between 8 and 25hr after infection, while ICAM-1 levels continued to increase throughout the 25hr of the experiment. The expression of each adhesion receptor correlated with the presence of viable intracellular rickettsiae and required *de novo* protein synthesis by the infected cells, since either inactivation of rickettsiae with formalin or pretreatment of HUVEC with cycloheximide inhibited the increase in expression (Dignat-George *et al.*, 1997).

The increased expression of E-selectin correlated with greater neutrophil and HL60 cell adhesion to HUVEC infected with *R. rickettsii* for 6–8 hr (Sporn *et al.*, 1993). Increased expression of VCAM-1 and ICAM-1 was associated with greater adhesion of mononuclear leukocytes to *R. conorii*-infected endothelial cells (Dignat-George *et al.*, 1997). Some of these changes occur *in vivo* since an increase in the circulatory concentration of ICAM-1 in the blood of a patient with a lethal case of RMSF has been reported (Sessler *et al.*, 1995). The significance of these changes in the pathogenesis of rickettsial diseases and how rickettsiae induce their expression remain unknown.

4.2. Changes in Proinflammatory Cytokines, the Procoagulant and Fibrinolytic Systems, and Prostaglandins

The development of pathological changes at the site of rickettsial invasion results in the production of multipotent soluble mediators, particularly those with proinflammatory and procoagulant activities, and factors which affect vascular permeability and tone.

4.2.1. Synthesis of Proinflammatory Cytokines

A cell-associated time-dependent increase in levels of IL-1α expression was demonstrated in cultured human endothelial cells infected with *R. conorii* and *R. rickettsii* during the first 18–24 hr of infection (Sporn and Marder, 1996; Kaplanski *et al.*, 1995). The changes in surface expression were associated with an increase in the steady state level of IL-1α mRNA in cells infected for 4 hr with *R. rickettsii* (Sporn and Marder, 1996). *R. conorii*-infected HUVEC also secreted high levels of IL-8 and IL-6, whose accumulation in the cell culture medium was time-dependent and became significant after 6–8 hr and 24 hr of infection, respectively (Kaplanski *et al.*, 1995). The production of either cytokine required the presence of intracellular viable rickettsiae. Furthermore, the secretion of IL-8 and IL-6 was shown to depend on cellular IL-1α induction, since the latter could be blocked by addition of IL-1 receptor antagonist or anti-IL-1α antibodies to the cell culture (Kaplanski *et al.*, 1995). These findings suggest IL-1α is an autocrine factor regulating HUVEC responses to *R. conorii* and *R. rickettsii* infection.

Rickettsia-associated induction of inflammatory cytokines was evaluated also in monocyte-derived macrophages (Manor and Sarov, 1990). Increased TNF-α levels were induced by *R. conorii* Casablanca strain and Israeli spotted fever rickettsiae and reached maximum levels at 24 hr postinfection. Later it was found that *R. conorii* Malish strain readily induced the synthesis of high levels of TNF-α by murine peritoneal macrophages and the macrophage-like cell line P388D1 (Jerrells and Geng, 1994). The stimulation was time and dosage dependent, but in contrast to the other cytokines, TNF-α was induced not only with viable, but with heat-inactivated rickettsiae. This last fact is consistent with a role for rickettsial LPS in TNF-α stimulation. In agreement with this hypothesis, *O. tsutsugamushi*, in which LPS has not been detected (Amano *et al.*, 1987), was unable to induce any TNF-α production (Jerrells and Geng, 1994). Synthesis of TNF-α by infected endothelial cells was not detected (Kaplanski *et al.*, 1995), thus suggesting that its effect at the site of infection could be due primarily to the secretory activity of infected infiltrating macrophages.

4.2.2. Changes in the Procoagulant System

Endogenously produced IL-1α affects more than the regulation of cytokine synthesis in infected cells since it is required for increased tissue factor (TF) expression in *R. rickettsii*-infected HUVEC (Sporn and Marder, 1996). TF is a membrane-bound glycoprotein that is involved in the activation of the extrinsic pathway of coagulation via a cascade of interactions involving factors VII and VIIa, IX and X which results in conversion of prothrombin to thrombin and of fibrinogen to fibrin (Bach, 1988).

The secretion of TF, measured by clotting assay and flow cytometric analysis, first increased between 4 and 8 hr after inoculation with *R. conorii*, remained almost constant for the next 4 hr, and then progressively decreased by 24 hr (Teysseire *et al.*, 1992*a*). Similar kinetics of TF expression were shown to occur during *R. rickettsii* infection in HUVEC, where TF levels increased by 8 hr and then declined by 24 hr after infection (Sporn *et al.*, 1994). As expected, the changes in TF mRNA levels peaked at 3 to 4 hr after initiation of infection, preceding the peak of TF expression (Shi *et al.*, 1998; Sporn *et al.*, 1994). The increase in TF expression did not require *de novo* cell protein synthesis and involved activation of transcriptional factor NF-κB by an unknown signal transduction pathway (Shi *et al.*, 1998, see section 4.3 for more details). Changes in thrombomodulin antigen expression on the surface of *R. conorii*-infected HUVEC also occur at about the same time (Teysseire *et al.*, 1992*a*). Its level, as measured by the thrombin-dependent activation of protein C or flow cytometric analysis, steadily decreased between 4 and 24 hr after inoculation with rickettsiae.

At 8 to 12 hr after contact with rickettsiae, infected human endothelial cells also release von Willebrand factor (vWF) from Weibel-Palade (WP) bodies as shown in experiments with *R. rickettsii* (Sporn *et al.*, 1991) and *R. conorii* (Teysseire *et al.*, 1992*a*). This process is characterized by an increase in the amount of the large 220 kDa mature multimers of vWF in the cell culture, with a concomitant decrease in these forms in the cells. On the other hand, *R. rickettsii* did not influence the distribution of intracellular immature dimers of vWF, indicating that infection specifically caused an increase in regulated secretion of vWF from the WP bodies (Sporn *et al.*, 1991).

It is possible that *in vivo* infection causes similar perturbances of the endothelium, since a 3.7 fold increase in tissue factor mRNA was demonstrated in the endothelium of an intact human umbilical vein 4 hr after it had been infected with *R. rickettsii* (Courtney *et al.*, 1996). Moreover, 12 patients with MSF exhibited elevated levels of plasma thrombomodulin and vWF, both of which progressively decreased during the treatment of the disease (George *et al.*, 1993). Thus, these changes in procoagulant factors, as first described in *in vitro* models, have relevance to the clinical description of SFG rickettsioses *in vivo*.

Unfortunately, similar studies of cultured endothelial cells following infection with typhus group rickettsiae have not been done, so little is known about any changes in the procoagulant cascade caused by *R. prowazekii* and *R. typhi*. However, levels of platelet-activating factor (PAF), whose synthesis typically involves PLA_2 activity, are significantly altered in cultured HUVEC from 24 to 72 hr post infection with *R. prowazekii* Madrid E (Walker and Mellott, 1993). When infected cells were treated with ATP or thrombin, they displayed even more dramatic increases in PAF levels 24–72 hr after infection when compared to uninfected but treated cells. However, the changes in PAF synthesis did not correlate with the number of intracellular rickettsiae. The newly synthesized PAF remains associated with the endothelial monolayer and is not released into the medium. PAF is known to promote adhesion of leukocytes to the endothelium and is a potent activator of a number of cell types including platelets, neutrophils, monocytes, and macrophages. Morphological studies documented a fourfold increase in adhesion of platelets to cultured endothelial cells infected with *R. rickettsii* for 48 to 96 hr (Silverman, 1986*a*). Thus, these results suggest that similar coagulative system changes are caused by distinct species of *Rickettsia*. In agreement with this hypothesis, thromboxane-dependent platelet activation and thrombin generation associated with increasing endothelial disfunction were detected in the blood of patients suffering from MSF, thus supporting the importance of the coagulation system in the pathogenesis of rickettsial disease *in vivo* (Davi *et al.*, 1995).

4.2.3. Changes in the Fibrinolytic System

In vivo evidence for the activation of the fibrinolytic system during MSF and RMSF includes the elevation of levels of fibrin degradation products, tissue-type plasminogen activator (tPA), and fibrinogen, together with activation of the fibrinolytic system within 24 hr after onset of fever (Rao *et al.*, 1988; Vicente *et al.*, 1986). *R. conorii* and *R. rickettsii* infected HUVEC were examined for changes in levels of tPA and tissue-type plasminogen inhibitor (tPI). However, different batches of cells either did not exhibit significant changes in tPA levels, or expressed lower and much slower increases of tPA in response to *R. conorii*-infection compared to the levels found in uninfected cells (Drancourt *et al.*, 1990, Raoult *et al.*, 1987). In contrast, the secretion of tPI into the cell culture medium increased following rickettsial inoculation and continued up to 48 hr after infection, thus causing the imbalance in the tPA and tPI pair (Drancourt *et al.*, 1990). Although full interpretation of these results requires more experimentation, the absence of correlation between the *in vitro* and *in vivo* observations suggests that rickettsia-induced changes in the fibrinolytic system may involve other cells or factors *in vivo* beside endothelial cells and that the *in vitro* model of infection lacks important factors that are operative *in vivo*.

4.2.4. Secretion of Prostaglandins

The dramatic changes which occur in the vascular system during acute typhus and SFG rickettsioses suggest that a significant role exists for vascular mediators in the rickettsia-endothelial cell interaction. Typhus group rickettsiae, particularly *R. prowazekii*, displayed significant PLA_2 activity during their growth cycle in mouse fibroblasts (Winkler and Daugherty, 1989). Cultured human endothelial cells were inoculated with *R. prowazekii* at a multiplicity of infection of 1000 hemolytically active rickettsiae per cell (Walker *et al.*, 1990). Under these conditions, approximately 19 rickettsiae were associated with each endothelial cell, with an average of 11.5 inside the cell after 1 hr incubation, and infected monolayers secreted increased levels of prostaglandins I_2 (PGI_2) compared to those found with control cells. PGI_2 secretion depended on the inoculum dosage of hemolytically active rickettsiae, but was abolished when rickettsiae were pretreated with formalin or NEM, a PLA_2 inhibitor. Hemolytically active *R. prowazekii* also induced the secretion of PGE_2. Its accumulation was linearly related to that of PGI_2; however, the latter was the major product secreted by *R. prowazekii*-infected endothelial cells with a ratio of PGI_2 to PGE_2 of 19:1. Intracellular growth of *R. prowazekii* in HUVEC for 6 days was accompanied by a decrease in surviving endothelial cells but with a concomitant increase in the levels of PGI_2 and PGE_2 secretion (Walker *et al.*, 1990). Thus, *R. prowazekii* stimulates prostaglandin secretion by infected endothelial cells at the moment of invasion and throughout its growth cycle. It has been hypothesized that intracellular rickettsiae use a PLA_2 to escape the cells, and that the arachidonate released due to PLA_2 activity is converted into the secreted prostaglandins.

In addition to vascular endothelial cells, dosage and time-dependent increases in secreted PGE_2 and another arachidonate-derived factor, leukotriene B_4 (LTB_4), were demonstrated in mouse polymorphonuclear leukocytes, and peritoneal and alveolar macrophages infected with *R. prowazekii* (Walker and Hoover, 1991; Walker *et al.*, 1991). Infection of monocyte-derived macrophages with *R. conorii* or Israeli spotted fever rickettsiae also was accompanied by dosage and time-dependent secretion of PGE_2 (Manor and Sarov, 1990).

In mice inoculated with a lethal dose of hemolytically active *R. prowazekii*, plasma levels of PGI_2 rose steadily, peaking 1 hr before death, while control animals exhibited no change in plasma PGI_2 levels (Walker *et al.*, 1990). Whether stimulation of the vascular endothelium with rickettsiae alone is sufficient for this reaction, or macrophage activation is also important for increasing the concentration of circulating prostaglandins is unknown.

Changes in prostaglandin secretion by cultured endothelial cells infected with SFG rickettsiae have not been measured directly yet. However, a significant portion of acute and convalescent sera from patients suffering with RMSF could depress the basal rate of PGI_2 secretion by cultured HUVEC or alter the ability of the cells to respond to thrombin by secreting PGI_2 (Walker and Triplett, 1991). The nature and function of a putative circulating anti-PG synthesis factor *in vivo* present in response to *R. rickettsii* infection remains problematic.

4.3. Expression of the Transcription Factor NF-κB

As described in section 4.2, upon entry of rickettsiae into the cytoplasm, vascular endothelial cells and macrophages rapidly increase the expression of procoagulant and proinflammatory molecules. These changes are probably due to a significantly altered transcriptional response by the host cells to infection.

R. rickettsii-infection of endothelial cells was shown to activate the transcriptional factor NF-κB (Sporn *et al.*, 1997), which regulates TF expression (Shi *et al.*, 1998). Activation of NF-κB required cellular uptake of *R. rickettsii* and could be inhibited by treatment with cytochalasin B. NF-κB activation by *R. rickettsii* exhibited sustained, biphasic kinetics: activation was first detected at 1.5 hr, it peaked at 3 to 7 hr, and it then declined by 14 hr. The highest level of activation observed was at 24 hr after infection, but NF-κB expression was not examined at later times. Two activated NF-κB species were present following *R. rickettsii* infection, the p50 homodimer and the p50–p65 heterodimer comprised of a 50 kDa and a 65 kDa subunit of NF-κB. Furthermore, *R. rickettsii* infection also resulted in an increase in the steady-state level of IκBα mRNA (Sporn *et al.*, 1997), which encodes the inhibitory subunit of NF-κB. In other systems both phosphorylation and degradation of IκBα are usually necessary for NF-κB activation and occur as a result of cell stimulation.

The nature of *R. rickettsii*-mediated activation of NF-κB was examined also with cytoplasmic extracts of HUVEC which had been separated from the cellular membrane fraction and proteasome (Sahni *et al.*, 1998). *R. rickettsii*-activation of NF-κB in the cell-free system resembled that occurring in cultured endothelial cells. After rickettsiae were added to the cell extract, NF-κB activation reached a maximum at 1 hr and then declined by 2 hr, and was dependent on the number of *R. rickettsii* organisms used for the reaction. Cell-free activated species of NF-κB also consisted of p50–p50 homodimers and p50–p65 heterodimers, but the mechanism of their assembly remains unknown. Surprisingly, none of the commonly known mechanisms of NF-κB activation could be identified when *R. rickettsii* was used as a stimulus (Sahni *et al.*, 1998). In the cell-free system, NF-κB was activated without participation of cellular membranes or the proteasome, it did not require an exogenous ATP supply, and phosphorylation of the inhibitory subunit IκBα did not occur. Activation may occur as a result of direct interaction between intracytoplasmic rickettsia and NF-κB. Whether this interaction is mediated by an unknown rickettsial protease or by other secreted factors is not known.

The early steps of *R. rickettsii*-induced NF-κB activation, particularly in the intact cellular system, are likely to involve a proteasome-mediated mechanism, since it could be inhibited upon treatment with the specific proteasome inhibitors, TPCK and MG 132 (Clifton *et al.*, 1998; Shi *et al.*, 1998). Surprisingly, such treatment resulted in extensive death of infected cells, which underwent apoptotic changes accompanied by evident nuclear DNA fragmentation (Clifton *et al.*, 1998). Apoptotic changes following *R. rickettsii*-infection also occurred in human embryonic fibroblasts transfected with a suppressor mutant inhibitory subunit IκBα, but non-transfected cells were not killed. Accordingly, invading *R. rickettsii* appears to inhibit host cell apoptosis via a mechanism which is dependent on NF-κB activation and thus modulates the host cell's response to its own advantage, allowing the targeted cell to continue as the site of infection (Clifton *et al.*, 1998). It is unknown whether inhibition of apoptosis is utilized by all species of rickettsiae, or how inhibition of apoptosis is effected and the infected cell begins the sequence of events which leads eventually to its necrosis.

4.4. Role of Nitric Oxide

Nitric oxide (NO) is important because of its toxicity for microorganisms and its role as a second messenger in signaling pathways (Nathan, 1992). NO production comprises the effector mechanism of antirickettsial activity occurring in mouse L929 fibroblasts (Feng and Walker, 1993; Turco and Winkler, 1993), mouse macrophage-like RAW264.7 cells (Turco and Winkler, 1994*b*) and mouse endothelial cells (Walker *et al.*, 1997). In these cellular systems NO synthesis occurs as a result of treatment of cultured cells with IFN-γ alone, or in combination with TNF-α prior to infection with *R. prowazekii* or *R. conorii.* Accumulation of NO in these cells is likely a result of activation of inducible nitric oxide synthase (*i*-NOS), since *i*-NOS mRNA was detected as early as 4 hr after cytokine stimulation, it increased at 8 hr and it slowly decreased by 72 hr (Walker *et al.*, 1997). The stimulatory effect of cytokines could be mimicked by addition of sodium nitroprusside, a source of NO. In contrast, treatment with aminoguanidine and N^G-monomethyl-L-arginine inhibited NO production and relieved its supression of the rickettsial infection (Walker *et al.*, 1997). Extracellularly released NO inhibits the ability of *R. prowazekii* to invade mouse cytokine-activated L929 fibroblasts and RAW264.7 cells by killing them directly (Turco *et al.*, 1998).

NO-mediated killing was implicated in clearance of *R. conorii* infection in immunocompetent C3H/HeN mice, since depletion of IFN-γ and TNF-α resulted in animal mortality due to an overwhelming rickettsial infection (Feng *et al.*, 1994). Consequently, one may hypothesize that IFN-γ secreted by T-lymphocytes and natural killer cells and TNF-α secreted by macrophages act in a synergistic fashion to stimulate synthesis of NO in rickettsia-infected cells, and this in turn kills the intracellular rickettsiae (Feng *et al.*, 1994).

Similar mechanisms may not function in cells of human origin. With the exception of hepatocytes (Nussler *et al.*, 1995) human phagocytic cells do not possess *i*-NOS (see Schoedon *et al.*, 1995 for more references). Substantial amounts of NO can be produced by epithelial cells and smooth muscle cells, while the data on human endothelial cells are conflicting (Schoedon *et al.*, 1995, Nathan, 1992). Furthermore, exposure of human endothelial cells to inflammatory stimuli, particularly to IFN-γ and/or LPS, decreases cellular levels of constitutive NOS (cNOS) and mRNA for cNOS without compensating *de novo* expression of *i*-NOS mRNA (MacNaul and Hutchinson, 1993). From this perspective, the absence of measurable NO synthesis in cytokine-stimulated human microvascular endothelial cells (HMEC-1) infected with *R. conorii* (Walker *et al.*, 1996) and in HUVEC and EA.hy 926 human endothelial cells infected with *R. rickettsii* (our unpublished observations) is not surprising. However, the role of NO-mediated mechanisms, both toxic and as a second messenger, in human cells infected with rickettsiae cannot be ruled out completely *in vitro* and particularly *in vivo*. It is possible that spread of rickettsial infection in underlying muscle cells or inflammatory cytokines secreted by infected endothelial cells may induce NO production by smooth muscle cells, and this exogenous NO may act on the endothelium *in vivo*.

4.5. Reactive Oxygen Species

Toxic reactive oxygen species (ROS) are typically associated with killing of bacteria by phagocytic cells. Details on rickettsia-macrophage interactions have been reviewed extensively (Winkler and Turco, 1994) and are not discussed here. We will summarize here the role of ROS during rickettsial infection in non-phagocytic cells, particularly vascular endothelial cells.

Hydrogen peroxide was proposed to be the rickettsicidal factor which is responsible for killing of *R. conorii* in cytokine-treated human microvascular endothelial cells HMEC-1 since rickettsial growth significantly increased in response to catalase treatment (Walker *et al.*, 1996). On the other hand, ROS, particularly superoxide radical and peroxides, appear to cause the cytopathic effect elicited by *R. rickettsii* in cultured HUVEC (Hong *et al.*, 1998; Santucci *et al.*, 1992; Silverman and Santucci, 1988). Four to 5 hr after exposure of HUVEC to *R. rickettsii*, elevated levels of peroxide were detected in the cell culture medium and infected monolayer (Hong *et al.*, 1998). Accumulation of superoxide was first detected after endothelial cells were exposed to 6 rickettsiae per cell for 1 hr (Santucci *et al.*, 1992) at a time when attachment and internalization of rickettsiae were occurring. At later stages of infection, intensive intracellular multiplication of rickettsiae was correlated with a dramatic accumulation of intracellular peroxides in the monolayer, and by irreversible necrotic changes in the structure of infected cells (Silverman and Santucci, 1988). The origin and timing of the appearance and accumulation of necrotic changes, and failure of the cellular antioxidant system to neutralize ROS (Devamanoharan *et al.*, 1994; Silverman and Santucci, 1990) are not fully understood. Whether this mechanism of cytotoxicity is typical for other *Rickettsia* species, and whether different species have similar capabilities for inducing and resisting the conditions associated with oxidative injury are unknown.

Upon treatment of *R. rickettsii*-infected human endothelial cells with 100 to 500 μM of the antioxidant α-lipoic acid, enhanced cellular antioxidant pools (reduced glutathione and glutathione peroxidase) and reduced intracellular peroxide levels were obtained (Eremeeva and Silverman, 1998*a*, 1998*b*). The effects of α-lipoic acid were dosage-dependent. α-Lipoic acid treatment of the endothelial cells resulted in an increase in their viability at 72 and 96 hr after infection with *R. rickettsii* compared to untreated infected cells, and prevented the occurrence of morphological changes in cells infected for 72 hr. Since α-lipoic acid did not alter directly the functional activity of rickettsiae, its protective effect probably involves complex changes in the metabolism of the infected host cell. Further studies are required to identify these mechanisms and to evaluate the potential of α-lipoic acid as an adjunct treatment for RMSF.

The detection of superoxide and hydrogen peroxide during attachment and invasion of endothelial cells by *R. rickettsii* suggests a potential role for ROS as signal transduction messengers (Crawford *et al.*, 1996). It is known that ROS-transmitted signals are sensed by cytoplasmic GTP-binding proteins from the Ras superfamily, which are key regulators of the structure and function of the actin cytoskeleton (Ridley, 1994). Invasion of *Salmonella* and *Shigella* into cells is mediated by different sets of GTP-binding proteins (Finlay and Cossart, 1997). The significance of ROS signaling for uptake of different species of *Rickettsia*, for modulating actin tail polymerization by the rickettsiae, and its role in altering endothelial cell function following rickettsial infection are unknown.

5. CONCLUSION AND PERSPECTIVES

This review has considered morphological, biochemical and cellular aspects of rickettsial interaction with their host cells. Although Rickettsia have been known for almost a century, the precise mechanisms they possess to invade and exploit eukaryotic cells to enable their multiplication are only partially elucidated. Similarly, how rickettsial infection and replication causes disease in humans is poorly understood. For many years this ignorance was blamed on the strict obligate intracellular lifestyle of rickettsiae, the difficulties posed by their cultivation, the high risk of laboratory infection by these agents, and the absence of sophisticated systems for their genetic manipulation.

Many of the powerful new techniques and approaches of molecular and cellular biology have been applied rapidly to the study of rickettsiae, but with some restrictions still, particularly in molecular genetics. However, greater efforts to utilize existing molecular and cellular tools are necessary to answer a number of significant questions about the pathogenesis of rickettsial diseases. Individual steps in the rickettsia-cell interactions need to be dissected, particularly the binding of the rickettsial ligand and its host cell receptor. Much is yet to be learned about the strategies that are used by the intracellular rickettsiae to survive, to multiply and to spread, and to avoid the host's defensive immune mechanisms, and about the structure and biochemistry of the rickettsial virulence factors that are involved in these interactions.

With the equivocal exception of *R. prowazekii*, humans are accidental hosts for rickettsiae, which typically circulate in zoonotic cycles or exclusively in arthropod hosts. Rickettsiae preferentially invade the endothelium and cause febrile diseases in humans, but colonize other types of cells in their arthropod vectors, usually without such dramatic effects on cell integrity. Increases in the virulence of *R. rickettsii* upon feeding or warming of their tick hosts are well known (Hayes and Burgdorfer, 1982; Spencer and Parker, 1924, 1923). The [illegible] of arthropod cell lines is a valuable tool for studying rickettsia-host cell interactions under conditions most resembling their natural environment. Although no differences were observed in the rickettsial antigens that were synthesized at 34°C in tick and mammalian cells, *R. rickettsii* grew to higher titers and survived longer in tick cells at high multiplicity of infection (Policastro *et al.*, 1997). Changes in the pattern of expression of several rickettsial antigens were also reported for *R. rickettsii* cultivated in tick cells following a temperature shift from 28° to 34°C (Policastro *et al.*, 1997). These phenomena and other aspects of rickettsial interaction with arthropod cells are quite significant since they may influence in unknown ways the pathogenicity of rickettsiae for vertebrates.

Until recently, different aspects of rickettsial infection of mammalian cell lines were examined primarily by using selected chemicals to inhibit given rickettsial or host cellular functions, or by arresting their respective metabolisms in different ways. Experiments with chemicals may be difficult, since it is often impossible to determine effective concentrations which do not influence rickettsial or host cell viability directly. Moreover, the potential for unexpected indirect side effects is always possible and is difficult to predict. The use of transgenic and knock-out cell lines with defined characteristics and/or known altered functions or pathways is an underused and powerful alternative for overcoming this problem.

Similarly, the use of isogenic virulence mutants of rickettsiae is another valuable genetic approach which has not been exploited fully. Two isogenic strains of *R. prowazekii* differing in virulence, the attenuated Madrid E strain and its spontaneous virulent revertant EVir strain, display significant differences in their biology and are the first and most famous rickettsial examples for such studies (Balayeva 1990; Balayeva and Nikolskaya, 1973).

The completion of the sequence of the *R. prowazekii* genome (Andersson and Andersson, 1997), coupled with recent work which significantly advances the possibility of an efficient transformation system for *R. prowazekii* (Rachek *et al.*, 1998), hopefully will facilitate and accelerate the identification of the molecular basis for the attenuation of the E strain and its reversion to virulence. Moreover, successful completion of such an analysis will greatly stimulate efforts to select or directly construct similar mutants for other species of *Rickettsia*. Analysis of rickettsial pathogenicity using the wide range of molecular techniques now available for dissecting bacterial virulence including site-specific mutagenesis, gene complementation, and differential gene expression will also be fruitful lines of investigation.

We believe this combination of new tools and approaches will elicit more insight into the molecular basis for rickettsial virulence and the underlying pathogenesis of diseases caused by the rickettsiae. This insight wil provide the means for improving the prevention of rickettsial diseases and facilitate the development of better methods for treating patients with rickettsial infections.

ACKNOWLEDGMENTS. This work was supported by Public Health Service grant AI 17416 from the National Institute of Allergy and Infectious Diseases (to D.J.S.) and by the Naval Medical Research and Development Command, Research Task 61102A.001.01.BJX.1293 (to G.A.D.). The opinions and statements contained herein are the private ones of the authors and are not to be construed as official or reflecting the views of the Naval Department or the Naval Service at large.

We are grateful to Vsevolod L. Popov for collaboration on structural studies of *R. rickettsii* invasion into HUVEC, for preparation of several electron photomicrographs and for constructive suggestions on the manuscript, and to Perry Comegys for excellent photographic work.

6. REFERENCES

Amano, K., Tamura, A., Ohashi, N., Urakami, H., Kaya, S., and Fukushi, K., 1987, Deficiency of peptidoglycan and lipopolysaccharide components in *Rickettsia tsutsugamushi*, *Infect. Immun.* **55**:2290–2292.

Andersson, J.O., and Andersson, S.G.E., 1997, Genomic rearrangements during evolution of the obligate intracellular parasite *Rickettsia prowazekii* as inferred from an analysis of 52,015 bp nucleotide sequence, *Microbiol.* **143**:2783–2795.

Austin, F.E., Turco, J., and Winkler, H.H., 1987, *Rickettsia prowazekii* requires host cell serine and glycine for growth, *Infect. Immun.* **55**:240–244.

Bach, R., 1988, Initiation of coagulation by tissue factor, *Crit. Rev. Biochem.* **23**:339–368.

Balayeva, N.M., 1990, Interaction of rickettsiae with the eukaryotic cells, *Zh. Mikrobiol. Epidemiol. Immunobiol.* (in Russian) **1990(2)**:80–86.

Balayeva, N.M., and Nikolskaya, V.N., 1973, Analysis of lung culture of *Rickettsia prowazeki* E strain with regard to its capacity of increasing virulence in passages on the lungs of white mice, *J. Hyg. Epidemiol. Microbiol. Immunol.* **17**:294–303.

Burgdorfer, W., and Brinton, L.P., 1970, Intranuclear growth of *Rickettsia canada*, a member of the typhus group, *Infect. Immun.* **2**:112–114.

Burgdorfer, W., Anacker, R.L., Bird, R.G., and Bertram, D.S., 1968, Intranuclear growth of *Rickettsia rickettsii*, *J. Bacteriol.* **96**:1415–1418.

Ching, W.-M., Wang, H., Davis, J., and Dasch, G.A., 1993, Amino acid analysis and multiple methylation of lysine residues in the surface protein antigen of *Rickettsia prowazekii*, in: *Techniques in Protein Chemistry* IV (R.H. Angeletti, ed.), Academic Press, New York, pp. 307–314.

Clarke, D.H., and Fox, J.P., 1948, The phenomenon of *in vitro* hemolysis produced by the rickettsiae of typhus fever, with a note on the mechanism of rickettsial toxicity in mice, *J. Exp. Med.* **88**:25–41.

Clifton, D.R., Gross, R.A., Sahni, S.K., van Antwerp, D., Baggs, R.B., Marder, V.J., Silverman, D.J., and Sporn, L.A., 1998, NF-κB-dependent inhibition of apoptosis is essential for host cell survival during *Rickettsia rickettsii* infection, *Proc. Natl. Acad. Sci. USA.* **95**:4646–4651.

Courtney, M-A., Haidaris, P.J., Marder, V.J., and Sporn L.A., 1996, Tissue factor mRNA expression in the endothelium of an intact umbilical vein, *Blood* **87**:174–179.

Crawford, L.E., Milliken, E.E., Irani, K., Zweier, J.L., Becker, L.C., Johnson, T.M., Eissa, N.T., Crystal, R.G., Finkel, T., and Goldschmidt-Clermont, P.J., 1996, Superoxide-mediated actin response in post-hypoxic endothelial cells, *J. Biol. Chem.* **271**:26863–26867.

Davi, G., Giammarresi, C., Vigneri, S., Ganci, A., Ferri, C., Di Francesco, L., Vitale, G., and Mansueto, S., 1995, Demonstration of *Rickettsia conorii*-induced coagulative and platelet activation *in vivo* in patients with Mediterranean spotted fever, *Thrombosis Haemostasis* **74**:631–634.

Devamanoharan, P.S., Santucci, L.A., Hong, J.E., Tian, X.-J., and Silverman, D.J., 1994, Infection of human endothelial cells by *Rickettsia rickettsii* causes a significant reduction in the levels of key enzymes involved in protection against oxidative injury, *Infect. Immun.* **62**:2619–2621.

Dignat-George, F., Teysseire, N., Mutin, M., Bardin, N., Lesaule, G., Raoult, D., and Sampol, J., 1997, *Rickettsia conorii* infection enhances vascular cell adhesion molecule-1 and intercellular adhesion molecule-1-dependent mononuclear cell adherence to endothelial cells, *J. Infect. Dis.* **175**:1142–1152.

Drancourt, M., Alessi, M-C., Levy, P-Y., Juhan-Vague, I., and Raoult, D., 1990, Secretion of tissue-type plasminogen activator and plasminogen activator inhibitor by *Rickettsia conorii-* and *Rickettsia rickettsii*-infected cultured endothelial cells, *Infect. Immun.* **58**:2459–2463.

Eremeeva, M.E., and Silverman, D.J., 1998*a*, Effects of the antioxidant α-lipoic acid on human umbilical vein endothelial cells infected with *Rickettsia rickettsii*, *Infect. Immun.* **66**:2290–2299.

Eremeeva, M.E., and Silverman, D.J., 1998*b*, *Rickettsia rickettsii* infection of the EA.hy 926 endothelial cell line: morphological response to infection and evidence for oxidative injury, *Microbiol.* **144**: in press.

Eremeeva, M.E., Popov, V.L., Walker, D.H., and Silverman, D.J., 1998, Ultrastructural studies of *Rickettsia rickettsii* entry into human endothelial cells, personal communication.

Ewing, E.P., Jr., Takeuchi, A., Shirai, A., and Osterman, J.V., 1978, Experimental infection of mouse peritoneal mesothelium with scrub typhus rickettsiae: an ultrastructural study, *Infect. Immun.* **19**:1068–1075.

Feng, H.-M., and Walker, D.H., 1993, Interferon-gamma and tumor necrosis factor-alpha exert their antirickettsial effect via induction of synthesis of nitric oxide, *Am. J. Pathol.* **143**:1016–1023.

Feng, H.-M., Popov, V.L., and Walker, D.H., 1994, Depletion of gamma interferon and tumor necrosis factor alpha in mice with *Rickettsia conorii*-infected endothelium: impairment of rickettsicidal nitric oxide production resulting in fatal overwhelming rickettsial disease, *Infect. Immun.* **62**:1952–1960.

Finlay, B.B., and Cossart, P., 1997, Exploitation of mammalian host cell functions by bacterial pathogens, *Science* **276**:718–725.

Gambrill, M.R., and Wisseman, C.L., Jr., 1973, Mechanisms of immunity in typhus infections. II. Multiplication of typhus rickettsiae in human macrophage cell cultures in the nonimmune system: influence of virulence of rickettsial strains and of chloramphenicol, *Infect. Immun.* **8**:519–527.

George, F., Brouqui, P., Boffa, M.C., Mutin, M., Drancourt, M., Brisson, C., Raoult, D., and Sampol, J., 1993, Demonstration of *Rickettsia conorii*-induced endothelial injury *in vivo* by measuring circulating endothelial cells, thrombomodulin, and von Willebrand factor in patients with Mediterranean spotted fever, *Blood* **82**:2109–2116.

Gudima, O.S., 1979, Quantitative study on the reproduction of virulent and vaccine *Rickettsia prowazeki* strains in cells of different origin, *Acta Virol.* **23**:421–427.

Gudima, O.S., 1971, Cytopathic action of *Rickettsia*, *Zh. Mikrobiol. Epidemiol. Immunobiol.* (in Russian) **1971(48)**:39–44.

Hackstadt, T., 1996, The biology of rickettsiae, *Infect. Agents Dis.* **5**:127–143.

Hanson, B.A., 1987, Improved plaque assay for *Rickettsia tsutsugamushi*, *Am. J. Trop. Med. Hyg.* **36**:631–638.

Hayes, S.F., and Burgdorfer, W., 1982, Reactivation of *Rickettsia rickettsii* in *Dermacentor andersoni* ticks: an ultrastructural analysis, *Infect. Immun.* **37**:779–785.

Heinzen, R.A., Hayes, S.F., Peacock, M.G., and Hackstadt, T., 1993, Directional actin polymerization associated with spotted fever group rickettsia infection of Vero cells, *Infect. Immun.* **61**:1926–1935.

Hong, J.E., Santucci, L.A., Tian, X., and Silverman, D.J., 1998, Superoxide dismutase-dependent, catalase-sensitive peroxides in human endothelial cells infected by *Rickettsia rickettsii*, *Infect. Immun.* **66**:1293–1298.

Hueck, C.J., 1998, Type III secretion systems in bacterial pathogens of animals and plants, *Microbiol. Mol. Biol. Rev.* **62**: 379–433.

Iwamasa, K., Okada, T., Tange, Y., and Kobayashi, Y., 1992, Ultrastructural study of the response of cells infected *in vitro* with causative agent of spotted fever group rickettsiosis in Japan, *APMIS* **100**:535–542.

Jerrells, T.R., and Geng, P., 1994, The role of tumor necrosis factor in host defense against scrub typhus rickettsiae. II. Differential induction of tumor necrosis factor-alpha production by *Rickettsia tsutsugamushi* and *Rickettsia conorii*, *Microbiol. Immunol.* **38**:713–719.

Johnson, J.W., and Pedersen, C.E., Jr., 1978, Plaque formation by strains of spotted fever rickettsiae in monolayer cultures of various cell types, *J. Clin. Microbiol.* **7**:389–391.

Kaplanski, G., Teysseire, N., Farnarier, C., Kaplanski, S., Lissitzky, J-C., Durand, J-M., Soubeyrand, J., Dinarello, C.A., and Bongrand, P., 1995, IL-6 and IL-8 production from cultured human endothelial cells stimulated by infection with *Rickettsia conorii* via a cell-associated IL-1α-dependent pathway, *J. Clin. Invest.* **96**:2839–2844.

Kokorin, I.N., 1968, Biological peculiarities of the development of rickettsiae, *Acta Virol.* **12**:31–35.

Kordova, N., 1966, Plaque assay of rickettsiae, *Acta Virol.* **10**:278.

Kuen, B., and Lubitz, W., 1996, Analysis of S-layer proteins and genes, in: *Crystalline Bacterial Cell Surface Proteins*, (U.B. Sleytr, P. Messner, D. Pum, and M. Sara, eds.), R.G. Landes Company, Austin, TX pp. 77–102.

Li, H., and Walker, D.H., 1992, Characterization of rickettsial attachment to host cells by flow cytometry, *Infect. Immun.* **60**:2030–2035.

MacNaul, K.L., and Hutchinson, N.I., 1993, Differential expression of iNOS and cNOS mRNA in human vascular smooth muscle cells and endothelial cells under normal and inflammatory conditions, *Biochem. Biophys. Res. Commun.* **196**:1330–1334.

Manor, E., and Sarov, I., 1990, Tumor necrosis factor alpha and prostaglandin E_2 production by human monocyte-derived macrophages infected with spotted fever group rickettsiae, *Ann. N.Y. Acad. Sci.* **590**:157–167.

Manor, E., Carbonetti, N.H., and Silverman, D.J., 1994, *Rickettsia rickettsii* has proteins with cross-reacting epitopes to eukaryotic phospholipase A2 and phospholipase C, *Microb. Pathog.* **17**:99–109.

McDade, J.E., and Gerone, P.J., 1970, Plaque assay for Q fever and scrub typhus rickettsiae, *Appl. Microbiol.* **19**:963–965.

McDade, J.E., Stakebake, J.R., and Gerone, P.J., 1969, Plaque assay system for several species of *Rickettsia*, *J. Bacteriol.* **99**:910–912.

Nathan, C., 1992, Nitric oxide as a secretory product of mammalian cells, *FASEB J.* **6**:3051–3064.

Ng, F.K.P., Oaks, S.C. , Jr., Lee, M., Groves, M.G., and Lewis, G.E., Jr., 1985, A scanning and transmission electron microscopic examination of *Rickettsia tsutsugamushi*-infected human endothelial, MRC-5, and L-929 cells, *Japan. J. Med. Sci. Biol.* **38**:125–139.

Nussler, A.K., DiSilvio, M., Liu, Z.Z., Geller, D.A., Freeswick, P., Dorko, K., Bartoli, F., and Billiar, T.R., 1995, Further characterization and comparison of inducible nitric oxide synthase in mouse, rat, and human hepatocytes, *Hepatology* **21**:1552–1560.

Oaks, S.C., Jr., Osterman, J.V., and Hetrick, F.M., 1977, Plaque assay and cloning of scrub typhus rickettsiae in irradiated L-929 cells, *J. Clin. Microbiol.* **6**:76–80.

Ojcius, D.M., Thibon, M., Mounier, C., and Dautry-Varsat, A., 1995, pH and calcium dependence of hemolysis due to *Rickettsia prowazekii*: comparison with phospholipase activity, *Infect. Immun.* **63**:3069–3072.

Paterson, P.Y., Wisseman, C.L., Jr., and Smadel, J.E., 1954, Studies of rickettsial toxins. I. Role of hemolysis in fatal toxemia of rabbits and rats, *J. Immunol.* **72**:12–23.

Penkina, G.A., Ignatovich, V.F., and Balaeva, N.M., 1995, Interaction of *Rickettsia prowazekii* strains of different virulence with white rat macrophages, *Acta Virol.* **39**:205–209.

Policastro, P.F., Munderloh, U.G., Fischer, E.R., and Hackstadt, T., 1997, *Rickettsia rickettsii* growth and temperature-inducible protein expression in embryonic tick cell lines, *J. Med. Microbiol.* **46**:839–845.

Policastro, P.F., Peacock, M.G., and Hackstadt, T., 1996, Improved plaque assays for *Rickettsia prowazekii* in Vero76 cells, *J. Clin. Microbiol.* **34**:1944–1948.

Popov, V.L., Dyuisalieva, R.G., Smirnova, N.S., Tarasevich, I.V., and Rybkina, N.N., 1986, Ultrastructure of *Rickettsia sibirica* during interaction with the host cell, *Acta Virol.* **30**:494–498.

Rachek, L.O., Tucker, A.M., Winkler, H.H., and Wood, D.O., 1998, Transformation of *Rickettsia prowazekii* to rifampin resistance, *J. Bacteriol.* **180**:2118–2124.

Ramm, L.E., and Winkler, H.H., 1976, Identification of cholesterol in the receptor site for rickettsiae on sheep erythrocyte membranes, *Infect. Immun.* **13**:120–126.

Ramm, L.E., and Winkler, H.H., 1973*a*, Rickettsial hemolysis: adsorption of rickettsiae to erythrocytes, *Infect. Immun.* **7**:93–99.

Ramm, L.E., and Winkler, H.H., 1973*b*, Rickettsial hemolysis: effect of metabolic inhibitors upon hemolysis and adsorption, *Infect. Immun.* **7**:550–555.

Rao, A.K., Schapira, M., Clements, M.L., Niewiarowski, S., Budzynski, A.Z., Schmaier, A.H., Harpel, P.C., Blackwelder, W.C., Scherrer, J-R., Sobel, E., and Colman, R.W., 1988, A prospective study of platelets and plasma proteolytic systems during the early stages of Rocky Mountain spotted fever, *N. Engl. J. Med.* **318**:1021–1028.

Raoult, D., and Roux, V., 1997, Rickettsioses as paradigms of new or emerging infectious diseases, *Clin. Microbiol. Rev.* **10**:694–719.

Raoult, R., Arnoux, D., Drancourt, M., and Ardissone, J.P., 1987, Enzyme secretion by human endothelial cells infected with *Rickettsia conorii*, *Acta Virol.* **31**:352–356.

Ridley, A.J., 1994, Signal transduction through the GTP-binding proteins Rac and Rho, *J. Cell. Sci. Suppl.* **18**:127–131.

Rikihisa, Y., and Ito, S., 1982, Entry of *Rickettsia tsutsugamushi* into polymorphonuclear leukocytes, *Infect. Immun.* **38**:343–350.

Rodionov, A.V., Eremeeva, M.E., and Balayeva, N.M., 1991, Isolation and partial characterization of the M(r) 100 kDa protein from *Rickettsia prowazekii* strains of different virulence, *Acta Virol.* **35**:557–565.

Roux, V., and Raoult, D., 1995, Phylogenetic analysis of the genus *Rickettsia* by 16S rDNA sequencing, *Res. Microbiol.* **146**:385–396.

Sahni, S.K., van Antwerp, D.J., Eremeeva, M.E., Silverman, D.J., Marder, V.J., and Sporn, L.A., 1998, Proteasome-independent activation of nuclear factor κB in cytoplasmic extracts from human endothelial cells by *Rickettsia rickettsii*, *Infect. Immun.* **66**:1827–1833.

Santucci, L.A., Gutierrez, P.L., and Silverman, D.J., 1992, *Rickettsia rickettsii* induces superoxide radical and superoxide dismutase in human endothelial cells, *Infect. Immun.* **60**:5113–5118.

Schaechter, M., Bozeman, F.M., and Smadel, J.E., 1957, Study on the growth of rickettsiae. II. Morphologic observations of living rickettsiae in tissue culture cells, *Virology* **3**:160–172.

Schoedon, G., Schneemann, M., Walter, R., Blau, N., Hofer, S., and Schaffner, A., 1995, Nitric oxide and infection: another view, *Clin. Infect. Dis.* **21(Suppl 2)**:S152–S157.

Schramek, S., Brezina, R., and Kazar, J., 1977, Some biological properties of an endodotoxic lipopolysaccharide from the typhus group rickettsiae, *Acta Virol.* **21**:439–441.

Sessler, C.N., Schwartz, M., Windsor, A.C., and Fowler, A.A., 3rd, 1995, Increased serum cytokines and intercellular adhesion molecule-1 in fulminant Rocky Mountain spotted fever, *Crit. Care Med.* **23**:973–976.

Shi, R-J., Simpson-Haidaris, P.J., Lerner, N.B., Marder, V.J., Silverman, D.J., and Sporn, L.A., 1998, Transcriptional regulation of endothelial cell tissue factor expression during *Rickettsia rickettsii* infection: involvement of the transcription factor NF-κB, *Infect. Immun.* **66**:1070–1075.

Silverman, D.J., 1986*a*, Adherence of platelets to human endothelial cells infected by *Rickettsia rickettsii*, *J. Infect. Dis.* **153**:694–700.

Silverman, D.J., 1986*b*, Infection and injury of human endothelial cells by *Rickettsia rickettsii*, *Ann. Inst. Pasteur/Microbiol.* **137A**:336–341.

Silverman, D.J., 1984, *Rickettsia rickettsii*-induced cellular injury of human vascular endothelium *in vitro*, *Infect. Immun.* **44**:545–553.

Silverman, D.J., and Bond, S.B., 1984, Infection of human vascular endothelial cells by *Rickettsia rickettsii*, *J. Infect. Dis.* **149**:201–206.

Silverman, D.J., and Santucci, L.A., 1990, A potential protective role for thiols against injury caused by *Rickettsia rickettsii*, *Ann. N. Y. Acad. Sci.* **590**:11–117.

Silverman, D.J., and Santucci, L.A., 1988, Potential for free radical-induced lipid peroxidation

as a cause of endothelial cell injury in Rocky Mountain spotted fever, *Infect. Immun.* **56**:3110–3115.
Silverman, D.J., Santucci, L.A., Meyers, N., and Sekeyova, Z., 1992, Penetration of host cells by *Rickettsia rickettsii* appears to be mediated by a phospholipase of rickettsial origin, *Infect. Immun.* **60**:2733–2740.
Silverman, D.J., Wisseman, C.L., Jr., and Waddell, A., 1980, *In vitro* studies of rickettsia-host cell interactions: ultrastructural study of *Rickettsia prowazekii*-infected chicken embryo fibroblasts, *Infect. Immun.* **29**:778–790.
Smith, D.K., and Winkler, H.H., 1977, Characterization of a lysine-specific active transport system in *Rickettsia prowazekii*, *J. Bacteriol.* **129**:1349–1355.
Spencer, R.R., and Parker, R.R., 1924, Rocky Mountain spotted fever. Experimental studies on tick virus, *Public Health Rep.* **39**:3027–3040.
Spencer, R.R., and Parker, R.R., 1923, Rocky Mountain spotted fever: infectivity of fasting and recently fed ticks, *Public Health Rep.* **38**:333–339.
Sporn, L.A., and Marder, V.J., 1996, Interleukin-1α production during *Rickettsia rickettsii* infection of cultured endothelial cells: potential role in autocrine cell stimulation, *Infect. Immun.* **64**:1609–1613.
Sporn, L.A., Sahni, S.K., Lerner, N.B., Marder, V.J., Silverman, D.J., Turpin, L.C., and Schwab, A.L., 1997, *Rickettsia rickettsii* infection of cultured human endothelial cells induces NF-kappaB activation, *Infect. Immun.* **65**:2786–2791.
Sporn, L.A., Haidaris, P.J., Shi, R-J., Nemerson, Y., Silverman, D.J., and Marder, V.J., 1994, *Rickettsia rickettsii* infection of cultured human endothelial cells induces tissue factor expression, *Blood* **83**:1527–1534.
Sporn, L.A., Lawrence, S.O., Silverman, D.J., and Marder, V.J., 1993, E-selectin-dependent neutrophil adhesion to *Rickettsia rickettsii*-infected endothelial cells, *Blood* **81**:2406–2412.
Sporn, L.A., Shi, R-J., Lawrence, S.O., Silverman, D.J., and Marder, V.J., 1991, *Rickettsia rickettsii* infection of cultured endothelial cells induces release of large von Willebrand factor multimers from Weibel-Palade bodies, *Blood* **78**:2595–2602.
Springer, T.A., 1995, Traffic signals on endothelium for lymphocyte recirculation and leukocyte emigration, *Annu. Rev. Physiol.* **57**:827–872.
Stothard, D.R., and Fuerst, P.A., 1995, Evolutionary analysis of the spotted fever and typhus groups of *Rickettsia* using 16S rRNA gene sequences, *System. Appl. Microbiol.* **18**:52–61.
Tamura, A., 1988, Invasion and intracellular growth of *Rickettsia tsutsugamushi*, *Microbiol. Sci.* **5**:228–232.
Tamura, A., Ohashi, N., Urakami, H., and Miyamura, S., 1995, Classification of *Rickettsia tsutsugamushi* in a new genus, *Orientia* gen. nov., as *Orientia tsutsugamushi* comb. nov., *Int. J. Syst. Bacteriol.* **45**:589–591.
Teysseire, N., Boudier, J-A., and Raoult, D., 1995, *Rickettsia conorii* entry into Vero cells, *Infect. Immun.* **63**:366–374.
Teysseire, N., Arnoux, D., George, F., Sampol, J., and Raoult, D., 1992*a*, von Willebrand factor release and thrombomodulin and tissue factor expression in *Rickettsia conorii*-infected endothelial cells, *Infect. Immun.* **60**:4388–4393.
Teysseire, N., Chiche-Portiche, C., and Raoult, D., 1992*b*, Intracellular movements of *Rickettsia conorii* and *R. typhi* based on actin polymerization, *Res. Microbiol.* **143**:821–829.
Todd, W.J., Burgdorfer, W., and Wray, G.P., 1983, Detection of fibrils associated with *Rickettsia rickettsii*, *Infect. Immun.* **41**:1252–1260.
Turco, J., and Winkler, H.H., 1994*a*, Cytokine sensitivity and methylation of lysine in *Rickettsia prowazekii* EVir and interferon-resistant *R. prowazekii* strains, *Infect. Immun.* **62**:3172–3177.
Turco, J., and Winkler, H.H., 1994*b*, Relationship of tumor necrosis factor alpha, the nitric oxide

synthase pathway, and lipopolysaccharide to the killing of gamma interferon-treated macrophagelike RAW264.7 cells by *Rickettsia prowazekii*, *Infect. Immun.* **62**:2568–2574.

Turco, J., and Winkler, H.H., 1993, Role of the nitric oxide synthase pathway in inhibition of growth of interferon-sensitive and interferon-resistant *Rickettsia prowazekii* strains in L929 cells treated with tumor necrosis factor alpha and gamma interferon, *Infect. Immun.* **61**:4317–4325.

Turco, J., and Winkler, H.H., 1982, Differentiation between virulent and avirulent strains of *Rickettsia prowazekii* by macrophage-like cell lines, *Infect. Immun.* **35**:783–791.

Turco, J., Liu, H., Gottlieb, S.F., and Winkler, H.H., 1998, Nitric oxide-mediated inhibition of the ability of *Rickettsia prowazekii* to infect mouse fibroblasts and mouse macrophage-like cells, *Infect. Immun.* **66**:558–566.

Urakami, H., Tsuruhara, T., and Tamura, A., 1983, Penetration of *Rickettsia tsutsugamushi* into cultured mouse fibroblasts (L cells): an electron microscopic observation, *Microbiol. Immunol.* **27**:251–263.

Urakami, H., Tsuruhara, T., and Tamura, A., 1982, Intranuclear *Rickettsia tsutsugamushi* in cultured mouse fibroblasts (L cells), *Microbiol. Immunol.* **26**:445–447.

Vicente, V., Alberca, I., Ruiz, R., Herrero, I., Gonzalez, R., and Portugal, J., 1986, Coagulation abnormalities in patients with Mediterranean spotted fever, *J. Infect. Dis.* **153**:128–131.

Walker, D.H., 1988, Pathology and pathogenesis of the vasculotropic rickettsioses, in: *Biology of Rickettsial Diseases*, Volume I (D.H. Walker, ed.), CRC Press, Boca Raton, FL, pp. 115–138.

Walker, D.H., and Cain, B.G., 1980, The rickettsial plaque. Evidence for direct cytopathic effect of *Rickettsia rickettsii*, *Lab. Invest.* **43**:388–396.

Walker, D.H., Popov, V.L., and Feng, H.-M., 1998, personal communication.

Walker, D.H., Popov, V.L., Crocquet-Valdes, P.A., Welsh, C.J.R., and Feng, H.M., 1997, Cytokine-induced, nitric oxide-dependent, intracellular antirickettsial activity of mouse endothelial cells, *Lab. Invest.* **76**:129–138.

Walker, D.H., Popov, V.L., Welsh, C.J. R., and Feng, H.-M., 1996, Mechanisms of rickettsial killing within cytokine-stimulated endothelial cells, in: *Rickettsiae and Rickettsial Diseases*, Proceedings of the Vth International Symposium, Stara Lesna, September 1–6, 1996 (J. Kazar and R. Toman, eds.), VEDA, Bratislava, pp. 51–56.

Walker, D.H., Firth, W.T., and Hegarty, B.C., 1984*a*, Injury restricted to cells infected with spotted fever group rickettsiae in parabiotic chambers, *Acta Tropica Basel* **41**:307–312.

Walker, D.H., Tidwell, R.R., Rector, T.M., and Geratz, J.D., 1984*b*, Effect of synthetic protease inhibitors of the amidine type on cell injury by *Rickettsia rickettsii*, *Antimicrob. Agents Chemother.* **25**:582–585.

Walker, D.H., Firth, W.T., Ballard, J.G., and Hegarty, B.C., 1983, Role of phospholipase-associated penetration mechanism in cell injury by *Rickettsia rickettsii*, *Infect. Immun.* **40**:840–842.

Walker, D.H., Firth, W.T., and Edgell, C.-J.S., 1982, Human endothelial cell culture plaques induced by *Rickettsia rickettsii*, *Infect. Immun.* **37**:301–306.

Walker, T.S., 1984, Rickettsial interactions with human endothelial cells *in vitro*: adherence and entry, *Infect. Immune.* **44**:205–210.

Walker, T.S., and Hoover, C.S., 1991, Rickettsial effects on leukotriene and prostaglandin secretion by mouse polymorphonuclear leukocytes, *Infect. Immun.* **59**:351–356.

Walker, T.S., and Mellott, G.E., 1993, Rickettsial stimulation of endothelial platelet-activating factor synthesis, *Infect. Immun.* **61**:2024–2029.

Walker, T.S., and Triplett, D.A., 1991, Serologic characterization of Rocky Mountain spotted fever. Appearance of antibodies reactive with endothelial cells and phospholipids, and factors that alter protein C activation and prostacyclin secretion, *Am. J. Clin. Pathol.* **95**:725–732.

Walker, T.S., and Winkler, H.H., 1978, Penetration of cultured mouse fibroblasts (L cells) by *Rickettsia prowazeki*, *Infect. Immun.* **22**:200–208.

Walker, T.S., Dersch, M.W., and White, W.E., 1991, Effects of typhus rickettsiae on peritoneal and alveolar macrophages: rickettsiae stimulate leukotriene and prostaglandin secretion, *J. Infect. Dis.* **163**:568–573.

Walker, T.S., Brown, J.S., Hoover, C.S., and Morgan, D.A., 1990, Endothelial prostaglandin secretion: effects of typhus rickettsiae, *J. Infect. Dis.* **162**:1136–1144.

Weinberg, E.H., Stakebake, J.R., and Gerone, P.J., 1969, Plaque assay for *Rickettsia rickettsii*, *J. Bacteriol.* **98**:398–402.

Weiss, E., 1982, The biology of rickettsiae, *Annu. Rev. Microbiol.* **36**:345–370.

Weiss, E., and Moulder, J.W., 1984, The rickettsias and chlamydias, in: *Bergey's Manual of Systematic Bacteriology*, Volume 1 (N.R. Krieg, and J.G. Holt, eds.), The Williams and Wilkins Co., Baltimore, pp. 687–739.

Wike, D.A., Tallent, G., Peacock, M.G., and Ormsbee, R.A., 1972, Studies of the rickettsial plaque assay technique, *Infect. Immun.* **5**:715–722.

Winkler, H.H., 1986, Early events in the interaction of the obligate intra-cytoplasmic parasite, *Rickettsia prowazekii*, with eurcaryotic cells: entry and lysis, *Ann. Inst. Pasteur/Microbiol.* **137A**:333–336.

Winkler, H.H., 1985, Rickettsial phospholipase A activity, in: *Rickettsiae and Rickettsial Diseases* (J. Kazar, ed.), Publishing House of the Slovak Academy of Sciences, Bratislava, pp. 185–194.

Winkler, H.H., 1977, Rickettsial hemolysis: adsorption, desorption, readsorption, and hemagglutination, *Infect. Immun.* **17**:607–612.

Winkler, H.H., 1974, Inhibitory and restorative effects of adenine nucleotides on rickettsial adsorption and hemolysis, *Infect. Immun.* **9**:119–126.

Winkler, H.H., and Daugherty, R.M., 1989, Phospholipase A activity associated with the growth of *Rickettsia prowazekii* in L929 cells, *Infect. Immun.* **57**:36–40.

Winkler, H.H., and Daugherty, R.M., 1983, Cytoplasmic distinction of avirulent and virulent *Rickettsia prowazekii*: fusion of infected fibroblasts with macrophage-like cells, *Infect. Immun.* **40**:1245–1247.

Winkler, H.H., and Miller, E.T., 1984, Activated complex of L-cells and *Rickettsia prowazekii* with N-ethylmaleimide-insensitive phospholipase A, *Infect. Immun.* **45**:577–581.

Winkler, H.H., and Miller, E.T., 1982, Phospholipase A and the interaction of *Rickettsia prowazekii* and mouse fibroblasts (L-929 cells), *Infect. Immun.* **38**:109–113.

Winkler, H.H., and Miller, E.T., 1980, Phospholipase A activity in the hemolysis of sheep and human erythrocytes by *Rickettsia prowazekii*, *Infect. Immun.* **29**:316–321.

Winkler, H.H., and Ramm, L.E., 1975, Adsorption of typhus rickettsiae to ghosts of sheep erythrocytes, *Infect. Immun.* **11**:1244–1251.

Winkler, H.H., and Turco, J., 1994, Rickettsiae and macrophages, *Immunology Series* **60**:401–414.

Winkler, H.H., and Turco, J., 1988, *Rickettsia prowazekii* and the host cell: entry, growth and control of the parasite, *Curr. Top. Microbiol. Immunol.* **138**:81–107.

Winker, H.H., Day, L., and Daugherty, R., 1994, Analysis of hydrolytic products from choline-labeled host cell phospholipids during growth of *Rickettsia prowazekii*, *Infect. Immun.* **62**:1457–1459.

Wisseman, C.L., Jr., 1986, Selected observations on rickettsiae and their host cells, *Acta Virol.* **30**:81–95.

Wisseman, C.L., Jr., 1968, Some biological properties of rickettsiae pathogenic for man, *Zbl. Bakt. Paras. Infekt. Hyg. I. Orig. Abt.* **206**:299–313.

Wisseman, C.L., Jr., and Waddell, A.D., 1975, *In vitro* studies on rickettsia-host cell interactions: intracellular growth cycle of virulent and attenuated *Rickettsia prowazeki* in chicken embryo cells in slide chamber cultures, *Infect. Immun.* **11**:1391–1401.

Wisseman, C.L., Jr., Edlinger, E.A., Waddell, A.D., and Jones, M.R., 1976*a*, Infection cycle of *Rickettsia rickettsii* in chicken embryo and L-929 cells in culture, *Infect. Immun.* **14**:1052–1064.

Wisseman, C.L., Jr., Waddell, A.D., and Silverman, D.J., 1976*b*, *In vitro* studies on rickettsia-host cell interactions: lag phase in intracellular growth cycle as a function of stage of growth of infecting *Rickettsia prowazeki,* with preliminary observations on inhibition of rickettsial uptake by host cell fragments, *Infect. Immun.* **13**:1749–1760.

Wojciechowski, E., Mikolajczyk, E., and Frygin, C., 1963, Hemolytic activity of typhus rickettsiae on erythrocytes of different species, *Exp. Med. Microbiol.* **15**:238–247.

Woodman, D.R., Weiss, E., Dasch, G.A., and Bozeman, F.M., 1977, Biological properties of *Rickettsia prowazekii* strains isolated from flying squirrels, *Infect. Immun.* **16**:853–860.

Chapter 20

Ehrlichial Strategy for Survival and Proliferation in Leukocytes

Yasuko Rikihisa

ABSTRACT

Ehrlichial organisms classified in the family *Rickettsiaceae*, are uniquely adapted to survive and proliferate in primary host defensive cells such as monocytes and granulocytes. Recent success in culturing ehrlichial organisms in leukemia cell lines enabled us to apply molecular and cell biologic techniques for their characterization. Ehrlichial major outer membrane protein antigens are encoded by a large number of polymorphic multigenes conserved among related species. Both ligand and receptor for monocytic ehrlichial binding are proteins. For internalization and proliferation of ehrlichial organisms Ca^{2+}-calmodulin and protein tyrosine phosphorylation are required. Monocytic and granulocytic ehrlichial organisms enter and maintain distinct cytoplasmic compartments which do not fuse with each other or lysosomes. Binding of monocytic ehrlichiae to macrophage surface induces rapid protein kinase A activation, leading to interference of signal

YASUKO RIKIHISA Department of Veterinary Biosciences, College of Veterinary Medicine, Ohio State University, Columbus, Ohio 43210, E-mail: rikihisa.1@osu.edu

Subcellular Biochemistry, Volume 33: Bacterial Invasion into Eukaryotic Cells, edited by Oelschlaeger and Hacker. Kluwer Academic / Plenum Publishers, New York, 2000.

transduction in the host cells such as Jak-Stat pathway or down regulation of class II antigen in response to interferon-γ. Altogether ehrlichial organisms upon contact with phagocytic leukocytes maim multiple innate microbicidal activity directed to them.

1. INTRODUCTION

Ehrlichiae are small gram-negative pleomorphic cocci that are obligatory intracellular bacteria. Ehrlichiae replicate in the membrane-bound vacuoles (parasitophorous vacuoles) in the cytoplasm of chiefly monocytes or granulocytes. Over the last decade, several new *Ehrlichia* spp. or strains have been isolated and/or characterized. *E. chaffeensis* which is the human monocytic ehrlichiosis agent and human granulocytic ehrlichiosis (HGE) agent discovered recently in the U.S. are two such examples. Ehrlichial organisms are vector-borne, transmitted either by the tick or the trematode. Ehrlichiae infect various domestic and wild animals, and humans, and cause a disease called ehrlichiosis, a febrile systemic disease often accompanied with hematological abnormalities, lymphadenopathy, and elevation of liver enzyme activity. Depending on the ehrlichial species involved, serologic tests, direct microscopic observation of stained blood smears, and/or PCR of blood specimens is currently used to diagnose ehrlichial diseases. Doxycycline (intravenous or oral) is generally utilized to treat human and canine ehrlichiosis. Successful cultivation of monocytic (monocyte-tropic) and granulocytic (granulocyte-tropic) ehrlichiae in myelocytic or promyelocytic leukemia cell lines (starting in 1984 and in 1995, respectively) has been significantly advancing our understanding of these organisms.

Small (0.2 to 0.4 μm), dense forms resembling chlamydial elementary bodies and relatively large, light forms (0.8 to 1.5 μm) resembling reticular bodies have been noted in several species of *Ehrlichia*; however, a chlamydia-like life cycle has not been demonstrated (Rikihisa, 1991, 1990a). Ehrlichiae are surrounded by thin outer and inner membranes. Unlike members of the genus *Rickettsia*, members of the genus *Ehrlichia* show no thickening of either leaflet of the outer membrane (Rikihisa, 1991). Outer membranes are more ruffled in the HGE agent than in *E. sennetsu* which is the human Sennetsu fever agent or *E. chaffeensis* (Rikihisa *et al.*, 1997). By transmission electron microscopy, members of the genus *Ehrlichia* do not appear to contain significant amounts of peptidoglycan (Rikihisa, 1991). Ehrlichiae have distinct ribosomes and a fine meshwork of DNA strands (Rikihisa, 1991, 1990a). Clumps of ribosomes are found in the cytoplasm as well as beneath the cytoplasmic membrane. Genome sizes of *E. chaffeensis*, *E. sennetsu*, and *E. risticii* which is the Potomac horse fever agent are small, approximately one fourth that of *Escherichia coli*, at 1163, 867, and 881 kilobases, respectively, as determined by pulsed-field gel electrophoresis (Rydkina *et al.*, 1996). Host cell-free *E. ristcii* and *E. sennetsu* can metabolize L-glutamine, but not glucose, or glucose-6-phosphate, suggesting that they lack a conventional glycolytic pathway (Weiss *et al.*, 1988). These host cell-free ehrlichiae metabolize L-glutamine well at pH 7.2–8.0 but poorly at pH below 6.5 (Weiss *et al.*, 1988).

In *Bergey's Manual of Systematic Bacteriology* ehrlichiae are classified in the family *Rickettsiaceae* based on morphology and biological characteristics (Ristic and Huxoll, 1984). Since then the 16S ribosomal RNA (rRNA) gene of almost every known *Ehrlichia* species has been sequenced and antigenic compositions of several ehrlichial organisms have been analyzed. Based on the 16S rRNA gene sequence homology, *Ehrlichia* species' closest relatives are, indeed, *Rickettsia* spp. The relatedness of 16S rRNA gene sequences of the genus *Rickettsia* to those of *Ehrlichia, Cowdria*, and *Anaplasma* spp. is 83 to 84%. *Ehrlichia* spp. are as divergent among themselves as between themselves and *Rickettsia* spp. 16S rRNA gene sequences revealed that *Ehrlichia* spp. along with several related genera can be divided into three distinct genetic groups, with about 7 to 15% sequence divergence among the groups (Figure 1). Group 1 consists of *E. canis, E. chaffeensis, E. muris* which infects wild mice, *E. ewingii*, which is the agent

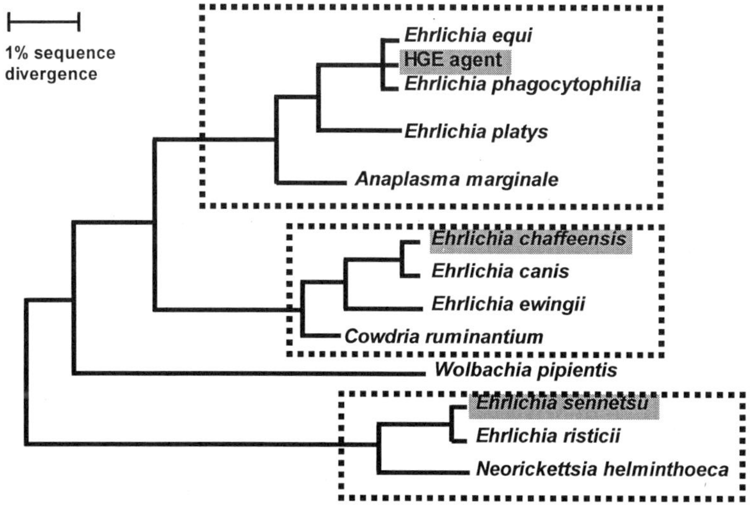

FIGURE 1. Dendrogram based on 16S rDNA sequence similarity. HGE agent, *E. chaffeensis*, and *E. sennetsu* belong to 3 separate clusters within the tribe *Ehrlichieae* which includes the tribe *Wolbachieae* and genus *Anaplasma*.

of canine granulocytic ehrlichiosis and *Cowdria ruminantium* which is the agent of Heartwater of ruminants. Group 2 consists of *E. equi* which is the equine granulocytic ehrlichiosis agent, *E. phagocytophila* which infects granulocytes of ruminants and causes Tick-borne fever, HGE agent, and *E. platys* which infects canine platelets. *Anaplasma marginale*, which has a tropism for bovine erythrocytes, is closest to group 2 in 16S rRNA gene sequence comparison (97% similarity). Group 3 consists of *E. sennetsu*, *E. risticii*, *Stellantchasmus falcatus* agent (SF agent) isolated from the metacercaria stage of a trematode *S. falcatus* encysting in grey mullet fish in Japan (Wen *et al.*, 1996), and *Neorickettsia helminthoeca* which causes Salmon poisoning disease in Canidae. Members of the genus *Wolbachia* which are symbionts of varieties of insects and a microfilaria, and are not known to cause a disease in vertebrates, fall among *Ehrlichia* spp. Incidentally, based on 16S rRNA gene sequence comparison, genera *Haemobartonella* and *Eperythrozoon* that are classified in the family *Anaplasmataceae* in *Bergey's Manual of Systematic Bacteriology*, are not related to the order *Rickettsiales* at all (Rikihisa *et al.*, 1997). Full nucleotide sequences of HSP60 gene have been determined for several ehrlichial species including *E. chaffeensis*, HGE agent, *C. ruminantium*, *E. risticii*, and *E. sennetsu*, and these sequences show a relatedness similar to that observed among the 16S rRNA gene sequences, though the sequence variation is generally greater (Zhang *et al.*, 1997; Somner *et al.*, 1997, 1993).

The *E. canis* group, except for *E. ewingii* and *C. ruminantium*, infects monocytes and macrophages. *E. ewingii* infects granulocytes and *C. ruminantium* infects vascular endothelial cells and neutrophils. The *E. phagocytophila* group, except for *E. platys* and *A. marginale*, infects granulocytes. The *N. helminthoeca* group infects monocytes and macrophages. *E. risticii* in addition infects intestinal epithelial cells and mast cells (Rikihisa *et al.*, 1985). Within each group, ehrlichial organisms are antigenically highly cross-reactive and share several homologous surface antigens. For example, *E. canis, E. chaffeensis, E. muris*, *C. ruminantium*, and Venezuelan human ehrlichia (Perez *et al.*, 1996) have a 23 to 34 kDa major surface antigen complex. *E. sennetsu*, *E. risticii*, and the SF agent group share a 51 to 55 kDa major surface antigen. *E. equi, E. phagocytophila*, and HGE agent share 43 to 49 kDa major antigens. Among different groups of ehrlichiae, however, there is little antigenic cross-reactivity. The major common antigen among all *Ehrlichia* spp. is HSP60, although the molecular size of HSP60 varies slightly among different species (Zhang *et al.*, 1997). The HSP60 antigen of *Ehrlichia* spp. is also cross-reactive with that of *Rickettsia* spp. but not with GroEL of *Escherichia coli* (Zhang *et al.*, 1997). Vectors which may have co-evolved with parasitic or symbiotic ehrlichiae, are also different among the three genetic and antigenic groups. The *E. canis* group is transmitted by a tick other than *Ixodes* spp., *E. phagocytophila*, *E. equi*, and HGE agent are transmitted by *Ixodes* ticks, and the *N. helminthoeca* group is transmitted by the fluke. For *E. sennetsu* and *E. risticii*, the mode of transmission is unknown, but it has been suspected to be the trematodes (Barlough *et al.*, 1998; Reubel *et al.*, 1998; Wen *et al.*, 1996; Fukuda *et al.*, 1973; Misao and Katsuta, 1956). The various ehrlichiae, the diseases they cause, and fundamental findings regarding ehrlichiae have been described in review articles (Rikihisa, 1999a & b; 1997, 1996, 1991; Dumler and Bakken, 1998; Walker and Dumler, 1996) and to *Bergey's Manual of Systematic Bacteriology* (Ristic and Huxoll, 1984). In this review, I have focused on ehrlichial major surface protein antigens and ehrlichial interaction with host cells in an effort to develop the understanding of how these organisms colonize leukocytes.

2. OUTER MEMBRANE PROTEINS OF *Ehrlichia* SPP.

We obtained the outer membrane fraction for analysis by the use of N-lauroylsarcosine sodium salt (sarkosyl) on Percoll-density gradient-purified ehrlichial organisms (Ohashi *et al.*, 1998a). In 0.15 % sarkosyl, ehrlichial outer membrane is insoluble but the cytoplasmic membrane and cytoplasm are solubilized (Ohashi *et al.*, 1998a). By SDS-PAGE of the outer membrane fraction, several overlapping proteins of MW 23–31 kDa were shown to be predominant in the outer membranes of *E. chaffeensis* and *E. canis*. *E. sennetsu* major outer membrane proteins (OMPs) are 51–55 kDa. HGE agent major OMPs are 42 to 49 kDa (Zhi *et al.*, 1997). These major OMPs are also major antigens recognized by the infected host by western immunoblot analysis (Zhi *et al.*, 1997). We began cloning the 28 kDa range of major outer membrane proteins of *E. chaffeensis* by N-terminal amino acid sequence analysis, PCR cloning, sequencing, and expressing the gene product (Ohashi *et al.*, 1998a). These outer membrane proteins are found to be a polymorphic multi-gene family of proteins coded by homologous but nonidentical genes. Three hypervariable regions are identified in the amino acid sequences. There are more than 10 copies in *E. chaffeensis* Arkansas strain which are tandemly arranged 1 kb apart with an intergenic spacer containing a putative promotor and a ribosome binding site. Among 7 genes the deduced and completely sequenced amino acid sequences show a divergence between 85.7–58.8 %, and by N-terminal amino acid sequence analysis, 5 of them are found not being predominantly expressed at the protein level in a human monocytic leukemia cell line THP-1 cell culture system. In the case of *Neisseria gonorrhoeae*, although there are multiple copies of the pilin gene scattered around the chromosome, there is usually only one copy that has a promoter and is expressed (Zhang *et al.*, 1992). These genes have internally repeated DNA sequences, allowing homologous recombination events to occur between the minicasettes of the gene copies, between the exogenote and the chromosome during transformation, and between 2 copies of chromosomes. This contributes to the pilin variation of *N. gonorrhoeae.* Previously an unusual phenomenon that monoclonal antibodies reacted with 2–3 bands of 30 kDa range of whole *E. chaffeensis* organisms was reported (Chen *et al.*, 1996). Although either protein degradation, the monoclonal being a polyclonal antibody, or other reasons are possible, this phenomenon is probably (not yet proven) related to the multi-gene family nature of the outer membrane proteins of *E. chaffeensis*. *E. canis* OMPs are also encoded by the multi-gene family of proteins (Ohashi *et al.*, 1998b) homologous to *E. chaffeensis* OMP-1. *E. canis omp-1s* are also arranged in tandem similar to *E. chaffeensis omp-1s* arrangement. These two sets of *omp-1s* are homologous (62.7–77.5%) to an immunodominant 32 kDa protein (MAP1 or Cr32) gene of *C. ruminantium*, which was reported as a single copy gene by Southern blotting of *Hind* III-digested genomic DNA of 3 strains of *C. ruminantium* (van Viet *et al.*, 1994). MAP-1 was shown to have strain variation (Reddy *et al.*, 1996). *E. canis* and *E. chaffeensis omp-1s*, and *C. ruminantium map-1* are related to *Wolbachia pipientis* surface protein gene *wsps* (46.9–41.5% amino acid sequence homology). *E. canis* and *E. chaffeensis omp-1s, C. ruminantium map-1*, and *W. pipientis wsp* are all homologous to the gene of the major surface protein (MSP4), a 31 kDa protein of *Anaplasma maginale* (42.1–48.8% amino acid sequence homology). Both *wsp* and *msp-4* are reported to be a single copy gene. *A. maginale msp-2* and HGE agent *p44* both of which are multi-gene families, are related (45.2–45.5 % amino acid sequence homology) to each other, but far from the major outer membrane protein genes of *E. canis*, *E. chaffeensis*, *C. ruminantium*, and *W. pipientis* (13.6–29.3 % amino acid sequence homology). If *Wolbachia* spp. really have only a single copy outer membrane protein gene, the presence of multigene copies in *Ehrlichia* spp. may have provided the evolutionary advantage for survival in the face of an advanced host immune system in vertebrate hosts. Phylogenetic relationships of these homologous outer membrane protein genes so far sequenced are shown in Figure 2.

OMPs (approximately 44 kDa) of HGE agent are also a multigene family of proteins homologous to MSP-2 of *A. marginale.* OMP-1 (P44) gene copies of HGE agents are widely distributed in the genome and do not make a cluster like that of *E. chaffeensis*. The arrangement of *p44* gene copies of HGE agent is more complex than that of *E. chaffeensis* OMP-1. MSP4 and MAP2 are to be considered the most dominant and conserved antigens in all isolates and the major protective antigens of *A. maginale*

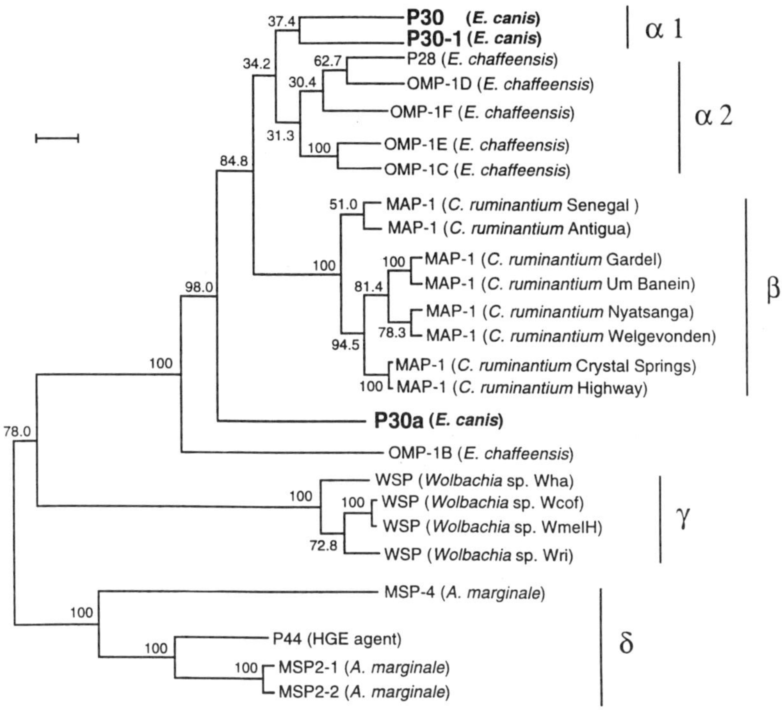

FIGURE 2. Phylogenetic classification among P30, P30–1, and P30a of *E. canis*, and the major outer membrane proteins of the closely related rickettsiae based on the amino acid sequence similarities. Scale bar shows 10% divergence in the amino acid sequences. Bootstrap values from 100 analyses are shown at the branch points of the tree. Bars with symbols show representative clusters (Ohashi *et al.*, 1998b).

(French *et al.*, 1998; Palmer *et al.*, 1994), although the mechanism has not been elucidated. MSP2 was reported as a polymorphic multi-gene family of proteins, but the complete genomic DNA sequences are available for only 2 genes. In our laboratory, by RT-PCR using conserved sequences of OMP-1 (P44) genes of HGE agent as primer sites, at least 6 nonidentical copies of OMP-1 mRNA are found to be coexpressed in HL-60 cells (Zhi *et al.*, 1999). Therefore, although by SDS-PAGE, only a few major outer membrane proteins appear to be expressed, actually homologous proteins of similar molecular sizes are overlapping in the bands. Notably, it was reported that during persistent *A. marginale* infection of cattle, multiple pleomorphic *msp-2* gene copies are expressed and a new variant gene expression takes place during subsequent parasitemia (French *et al.*, 1998). Despite the small genome size of *Ehrlichia* spp., presence of these multiple copies of *omp-1* suggests an important role of these genes in ehrlichial survival and possibly immunoavoidance. By SDS-PAGE, several OMP-1 homologous proteins were found to be expressed in *E. chaffeensis* cultivated in DH82 cells or THP-1 cells. It remains to be seen which copies are expressed in humans, deer (a reservoir of *E. chaffeensis*), and ticks. The recombinant P28 and native outer membrane proteins were surface preferentially labeled with ^{125}I and solubilized with octyl glycoside and incubated with THP-1 or DH82 monocytes briefly prefixed with formalin. By autoradiography of host cell-bound proteins separated by SDS-PAGE, we showed that this recombinant P28 and native outer membrane 29, 28, and 25 kDa proteins bind to the host macrophages (Zhang *et al.*, 1998b). It will be of interest to determine the kind of signal that is transmitted by the binding of ehrlichial outer membrane proteins to the macrophage surface. Yu *et al.* (1997) recently cloned a 120 kDa protein of *E. chaffeensis.* We did not detect this protein as the major protein in the outer membrane fraction.

3. EHRLICHIAL BINDING AND HOST CELL SURFACE RECEPTORS

Ehrlichia spp. are gram-negative small cocci which lack pili or capsule, suggesting that they bind to the host cell surface via their outer membrane. Although very few ehrlichial organisms are seen bound to the host cell surface, among pleomorphic organisms smaller forms appear to bind. *E. risticii* binding to P388D_1 macrophages at 4°C is dose dependent, and treatment of either ehrlichiae or host cells with 1% paraformaldehyde for 30 min or 0.25% trypsin for 15 min prevents ehrlichial binding, indicating that both the ehrlichial ligand and host cell receptor are likely surface proteins (Messick and Rikihisa, 1993). Fab fragments of polyclonal anti-*E.risticii* IgG prevents ehrlichial binding to the macrophage. Anti-*E.risticii* IgG did not prevent binding or internalization of ehrlichiae but all internalized ehrlichiae are destroyed (Messick and Rikihisa, 1994). Thus, it is most likely that an ehrlichial surface protein antigen is the ligand and unless ehrlichiae enter through the receptor of this ligand, they are destroyed. Since HGE agent infects only granulocytes, and *E. chaffeensis, E. canis, E. risticii*, and *E. sennetsu* infect monocytes and macrophages, the receptor for these 2 groups of *Ehrlichia* spp. may be different. Alternatively, they may use the homologous surface protein as the receptor for internalization, but subsequent cytoplasmic events such as fusion with transferrin receptor containing endosomes or lysosomal fusion may be different, allowing only one group of ehrlichiae to survive in monocytes or granulocytes after internalization. *E. risticii* can replicate in monocytes/macrophages but not in horse granulocytes. We found that *E. risticii* cannot bind to horse granulocytes but can bind to monocytes at 4°C. Although *E. risticii* can enter into granulocytes at 37°C, they are subsequently destroyed (Messick and Rikihisa, 1993). Therefore, binding and internalization through ehrlichia-specific ligand receptor interaction appears to be essential for subsequent ehrlichial survival.

For intracellular bacteria, a few host surface receptors have been elucidated. An evolving paradigm is that many bacteria take advantage of the host cell surface adhesin molecules for their binding and internalization. E-cadherin was identified as the receptor for internalin, a surface protein required for entry of a gram-positive bacterium, *Listeria monocytogenes* (Mengaud *et al.*, 1996). Gram-negative bacteria, *Yersinia pseudotuberculosis, Y. enterocolitica*, enteropathogenic *E. coli, Shigella flexneri*, *Salmonella typhimurium*, and *S. typhi* bind β1 integrin (Isberg, 1991). Integrins are a family of cell surface receptors for extracellular matrix proteins. *Legionella pneumophila* binds CR1 and CR3 (integrin) on the surface of macrophages (Payne *et al.*, 1987). The cytoplasmic tail of these receptors performs an active function in internalization by interacting with the cytoskeleton and inducing its rearrangement. Binding and internalization through this route does not activate the host cells, since this is a normal physiologic receptor. *E. risticii* internalization is not blocked with anti-CR3 (Rikihisa, unpublished observation) but is inhibited by monodansylcadaverine similar to virus entry. This mechanism is unlike all the above-mentioned bacteria which are resistant to monodansylcadaverine (Moulder, 1985). Thus, the ehrlichial receptor is unlikely to be β-integrin but is rather a kind of receptor which directs "receptor-mediated endocytosis". Phosphotyrosine colocalized with *E. risticii* inclusions (Zhang and Rikihisa, 1997) which may be related to the ehrlichial receptor. Incubation of human monocytes with *E. chaffeensis* raises protein kinase A activity 25-fold over the basal level within 30 min of incubation (Lee and Rikihisa, 1998) (Figure 3), suggesting that ehrlichial receptor may be coupled with G protein-adenylate cyclase activation system. Increase in cytoplasmic cyclic AMP levels were previously seen in *E. risticii*-infected P388D_1 macrophages (van Heeckeren *et al.*, 1993) and horse intestinal tissue (Rikihisa *et al.*, 1992). This activation requires only the binding of a protein component of *E. chaffeensis*, ehrlichial internalization or a carbohydrate residue is not required (Lee and Rikihisa, 1998). This rapid activation of protein kinase A then leads to interference of Jak-Stat pathway induced by interferon-γ (Lee and Rikihisa, 1998). Inhibition of the upregulation of class II antigen expression in response to interferon-γ by *E. risticii* infection on mouse macrophages (Messick and Rikihisa, 1992) may be also mediated by protein kinase A

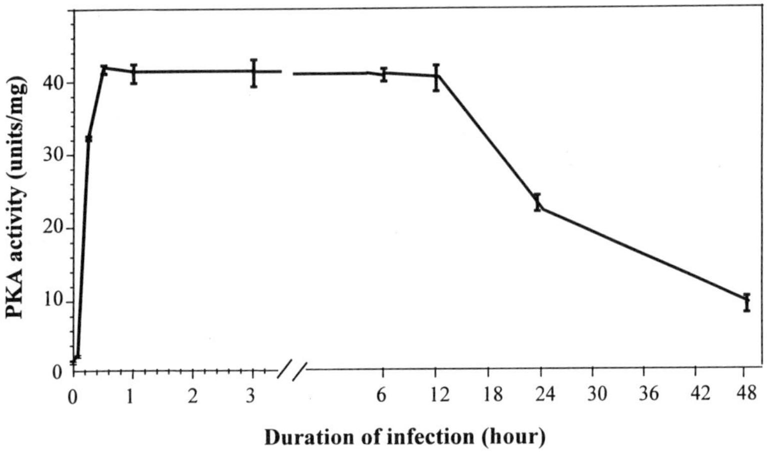

FIGURE 3. THP-1 cells pretreated with PMA were infected with *E. chaffeensis* for the indicated periods of time. Cell lysates were prepared and assayed for protein kinase A activity. Results are expressed as the means ± standard deviations of three independent experiments. One unit is defined as the amount of enzyme that transfers 1 nmol phosphate in 1 min from ATP to the synthetic PKA-substrate Kemptide at 30°C, pH 7.5 (Lee, E. and Rikihisa, Y. 1998).

activation. On the other hand, HGE agent not only does not induce reactive oxygen intermediate (ROI) production in neutrophils, but also blocks ROI generation by neutrophils in response to phorbol myristic acetate (Mott and Rikihisa, 1999). These studies indicate that ehrlichial organisms interfere with host signal transduction to prevent the activation of the bactericidal machinery in leukocytes. Inhibition occurs rapidly at the step of ehrlichial binding to the host cell, suggesting that the interaction of ehrlichial outer membrane proteins and the host cell receptor may have a critical role in this process

4. EHRLICHIAL INTERNALIZATION AND REPLICATION

Most *E. risticii* internalization takes place within 3 h of incubation with $P388D_1$ host cells (Rikihisa *et al.*, 1994). *E. risticii* internalization is resistant to cytochalasin D but highly sensitive to monodansylcadaverine indicating that ehrlichia enters macrophages by receptor-mediated endocytosis, rather than by phagocytosis (Messick and Rikihisa, 1993). Ehrlichial internalization is also sensitive to taxol and colchicine which are inhibitors of microtubule functions (Rikihisa *et al.*, 1994). Ehrlichial internalization and proliferation but not binding are blocked by protein tyrosine kinase inhibitors, but not by serine-threonine kinase inhibitors, indicating requirement of protein tyrosine phosphorylation for ehrlichial internalization and proliferation (Zhang and Rikihisa, 1997). Phosphotyrosine colocalizes with *E. risticii* inclusions, and 52 and 54 kDa proteins are tyrosine phosphorylated in *E. risticii*-infected host cells (Zhang and Rikihisa, 1997). Ehrlichial internalization and proliferation, but not binding are also blocked by Ca^{2+} channel blockers, and calmodulin antagonists, but not by the inhibitors or activators of protein kinase C. Verapamil and W-7 are reversible inhibitors of Ca^{2+} channel and calmodulin, respectively. These compounds showed inhibitory effects only when they coexist with both ehrlichiae and macrophages (Rikihisa *et al.*, 1995). Pretreatment of macrophages with these compounds and then washing before interaction with ehrlichiae or pretreatment of ehrlichiae and washing before interaction with macrophages failed to inhibit (Rikihisa *et al.*, 1995). A calmodulin antagonist, trifluoperazine, and verapamil are reported to inhibit chlamydial infection also (Shainklin-Kestenbaum *et al.*, 1989; Kihlstrom and Solderlund, 1983), but the mechanism has not been investigated. Inhibitors of the Ca^{2+}-calmodulin system, thus may be blocking the site common to both chlamydia and ehrlichia. Several cytoplasmic proteins (clathrin, adaptins, annexins) reversibly associate with membranes in Ca^{2+}-dependent manner and regulate vesicular traffic. Calmodulin antagonists and Ca^{2+} channel blockers block clathrin network formation, thus cytoplasmic membrane traffic. Ehrlichiae may be killed due to the inability to maintain inclusion compartments in the presence of these inhibitors. Alternatively, the primary site of inhibition may be a host enzyme, transglutaminase (Ca^{2+}-dependent enzyme that cross-links proteins), since this enzyme activity is essential for *E. risticii* uptake and survival (Rikihisa, *et al.*, 1994; Messick and Rikihisa, 1993) and for chlamydial uptake (Solderlund and Kihlstrom, 1983). *E. risticii* is killed by Ca^{2+} ionophore or an inhibitor of cytoplasmic Ca^{2+} mobilization, TMB-8. Regardless of time of treatment, pretreatment of murine macrophages with Ca^{2+} ionophore, A23187 irreversively makes them refractory to infection without generation of nitric oxide (NO·) or ROI (Rikihisa, *et al.*, 1995), which suggests that this treatment activates an ehrlichiacidal mechanism in macrophages. A23187 at the concentrations which do not kill the host cells, directly inhibits glutamine metabolism in Percoll-density gradient-purified *E. risticii* in SPK buffer (Ca^{2+}-free). Host cell-free *E. risticii* and *E. sennetsu* can metabolize L-glutamine to CO_2 and generate ATP through oxidative phosphorylation (Weiss *et al.*, 1990, 1988). Thus, Ca^{2+} and H^+ concentrations in ehrlichiae as well as in ehrlichial inclusions probably influence ehrlichial ATP synthesis. A23187 equilibrates Ca^{2+} concentration in *Ehrlichia* with its surroundings, which probably makes them incapable of generating ATP.

5. THE EHRLICHIAL INCLUSION COMPARTMENT

E. chaffeensis and *E. sennetsu* are genetically divergent (14.4% 16S rRNA gene sequence divergence) obligatory intracellular bacteria of human monocytes/macrophages. HGE agent is an obligate intracellular bacterium of human granulocytes (16S rRNA gene sequence divergence from *E. chaffeensis* and *E. sennetsu* are 7.5 % and 14.7 % respectively). The outer membrane protein profiles of these 3 organisms are quite distinct and there is very little antigenic cross-reactivity among them except the HSPs. The association between the outer membrane and the inclusion membrane is very different among different *Ehrlichia* spp. (Rikihisa, *et al.*, 1991). *E. sennetsu* and *E. risticii* are very tightly and individually surrounded on its entire circumference with the inclusion membrane. Usually, a part of the outer membrane of each HGE agent, which is loosely packed, attaches to the inclusion membrane (Rikihisa, *et al.*, 1997). The outer membrane of only *E. chaffeensis* and *E. canis* which reside in the peripheral area of the inclusion, attaches to the inclusion membrane. *E chaffeensis* and *E. canis* which occupies the central area of the inclusion does not appear to be in contact with the inclusion membrane. When THP-1 cells are coinfected with either *E. chaffeensis* and HGE agent or *E. chaffeensis* and *E. sennetsu*, they enter into and retain separate compartments, suggesting the host cell receptor of these 3 organisms are distinct and once they enter they do not share the compartment (Mott *et al.*, 1999) (Figure 4). We recently demonstrated that inclusions of *E. chaffeensis* are early endosomes which still retain some Class I and Class II surface antigens and β_2 microglobulin, but clathrin is absent (Barnewall *et al.*, 1997). The inclusions were positive for a cytoplasmic small GTPase, rab5 and EEA1 (early endosomal antigen 1) (Mott *et al.*, 1999). Although weak presence of a subunit of vacuolar type H^+-ATPase and accumulation of a basic lysosomotropic amine, DAMP are detected, neither lysosomal membrane glycoprotein (CD63) nor lysosomal membrane associated protein (LAMP-1) is detected in ehrlichial inclusions (Barnewall *et al.*, 1997). This result is in agreement with our previous EM study using acid phosphatase cytochemistry and secondary lysosomes preloaded with electron dense tracers, which showed the lack of lysosomal fusion selectively with inclusions containing *E. risticii* (Well and Rikihisa, 1988). Notably, *E. chaffeensis* inclusions are strongly labeled with anti-transferrin receptor (TfR) antibody. TfR is not present at the time of internalization but progressively accumulates in the ehrlichial inclusions until eventually all cytoplasmic TfR is localized there. Thus, *E. chaffeensis* appears to reside in early endosomes and take advantage of a natural Tf-TfR recycling pathway which does not fuse with lysosomes (Barnewall *et al.*, 1997), thereby avoiding lysosomal fusion. *E. sennetsu* also accumulates TfR in its inclusion (Figure 5). This phenomenon seems to

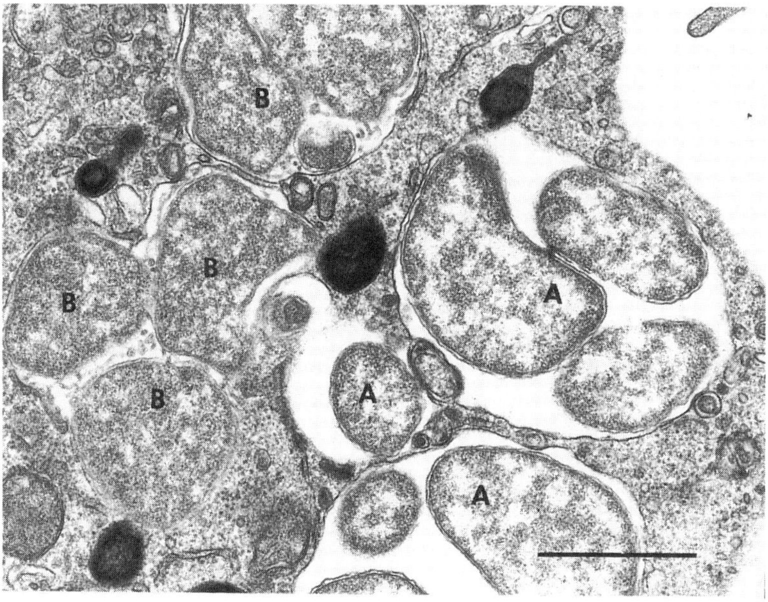

FIGURE 4. Transmission electron micrograph of HL-60 cells simultaneously infected with HGE agent (A) and *E. chaffeensis* (B). Note *E. chaffeensis* has tubular extracellular appendages (arrow) and HGE agent resides loosely in the clear matrix of the inclusion. None of the inclusions contain both organisms. Bar = 1 μm (Rikihisa, Y., 1998).

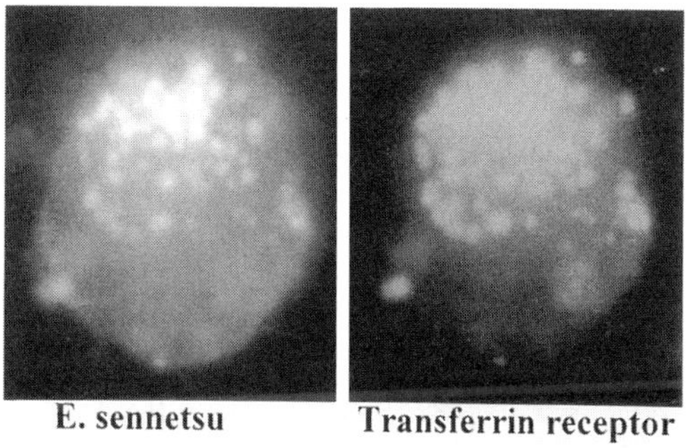

FIGURE 5. Double immunofluorescence labeling of *E. sennetsu*-infected $P388D_1$ cells with rabbit anti-*E. sennetsu* antibody and FITC-anti-rabbit IgG, and mouse anti-transferrin receptor and Rhodamine anti-mouse IgG. Note almost complete colocalization of two labels. Magnification. ×2,700 (Zhang, Y. and Y. Rikihisa, unpublished data).

be also very important in considering the extreme sensitivity of *E. chaffeensis, E. risticii*, and *E. sennetsu* to cytoplasmic iron depletion by a cytoplasmic iron chelator, deferoxamine or interferon-γ (Barnewall *et al.*, 1994; Park and Rikihisa, 1992). Monocytic ehrlichiae may acquire iron in the host cells directly from iron-transferrin in the inclusion. Furthermore, we found that *E. sennetsu* and *E. risticii* infection upregulates the expression of TfR mRNA in the host cell, thus host cells infected with monocytic ehrlichiae have more TfR protein and TfR-endosomes (Barnewall *et al.*, 1997).

In contrast, our result suggests that HGE agent resides in a compartment which neither accumulates TfR nor fuses with lysosomes (Mott *et al.*, 1999). TfR, Rab5, EEA1, and vacuolar type ATPase were absent in HGE agent inclusions. More vesicle associated membrane protein 2 (VAMP-2, synaptobrevin) was found on HGE agent inclusions than on *E. chaffeensis* inclusions. The HGE agent inclusion was negative for C_6-NBD ceramide, cation-independent mannose-6-phosphate receptor. These results indicate that the HGE agent resides in an unique cytoplasmic compartment, which is distinct from the cytoplasmic compartment occupied by *E. chaffeensis* or *E. sennetsu*, within the same cell, and from those occupied by any other intracellular organisms studied so far. Thus, the inclusion of HGE agent may be able to avoid lysosomal fusion by lacking a significant identity of any host cell cytoplasmic compartment. Unlike *E. chaffeensis* or *E. sennetsu*, HGE agent in HL-60 cells did not increase in TfR mRNA or iron responsive protein (IRP) levels throughout infection. The HGE agent in HL-60 cells was only partially inhibited by deferoxamine treatment (Barnewall *et al.*, 1999). These results suggest that the HGE agent utilizes a mechanism of iron acquisition independent from host transferrin-TfR iron acquisition.

6. EHRLICHIACIDAL MECHANISM BY INTERFERON-γ

Gamma interferon (IFN-γ) is an important pleiotropic cytokine produced by T-cells and natural killer cells. IFN-γ levels in the blood were significantly increased in *E. muris (*which is closely related to *E. chaffeensis*)-infected mice at day 5 post-infection (Kawahara *et al.*, 1996). This rapid production of IFN-γ and an increase of NK cell population in *E. muris*-infected mice suggest that NK cells may be the source of this IFN-γ. IFN-γ is known to induce multiple microbicidal activities in macrophages and neutrophils. ROI are among them. ROI, however, has no role in ehrlichiacidal mechanism induced by IFN-γ, since none of the ROI scavengers inhibited ehrlichial killing. IFN-γ activates indoleamine 2′, 3′-dioxygenase which causes tryptophan depletion (Carlin, 1989). This is the major inhibitory mechanism of *Chlamydia* which lacks cytochromes, thus the capability of oxidative phosphorylation (Moulder, 1985). Tryptophan depletion is not the anti-ehrlichial mechanism, since tryptophan supplementation does not inhibit the anti-*E. risticii* activity (Park and Rikihisa, 1992). In rodent macrophages, IFN-γ induces transcription and translation of the gene of cytosolic NO· synthase enzyme which generates lipophilic NO· radical from L-arginine (Iyengar *et al.*, 1987). NO· binds to catalytically active Fe-S centers of various enzymes and inactivates these enzymes (Sneyder and Bredt, 1992). This in turn kills the microorganism which is highly dependent on these enzymes unless it can acquire iron competitively from the host. We found that NO· generation is the primary ehrlichiacidal mechanism in murine macrophages induced by IFN-γ (Park and Rikihisa, 1992), but not in human monocytes (Barnewall and Rikihisa, 1994). Ehrlichiae are obligate aerobic bacteria which lack the conventional metabolic pathway for glycolysis, but can generate ATP from L-glutamine via the electron transport chain coupled with oxidative phosphorylation (Weiss *et al.*, 1990). Thus, iron is essential for cytochromes and probably other iron containing enzymes of ehrlichiae. *E. risticii* and *E. chaffeensis* are inhibited by deferoxamine (Barnewall and Rikihisa, 1994; Park and Rikihisa, 1992), a cell permeable iron chelator, an actinomycete siderophore, which suggests that intracellularly, monocytic ehrlichiae have access to the labile iron pool, but do not produce siderophores of high iron binding affinity. *E. sennetsu* and *E. chaffeensis* are readily killed in human monocytes by recombinant human IFN-γ treatment given prior to or at the early stage of infection. TfR-Tf-mediated iron uptake is the major iron acquisition mechanism by the host cells (Baynes *et al.*, 1987). Iron-saturated (holo-), but not iron-free (apo-) Tf reverses the inhibition (Barnewall and Rikihisa, 1994). Although IFN-γ does not induce NO· synthase in human monocytes, IFN-γ treatment reduces the intracellular labile iron pool drastically which is accessible to ehrlichiae in the inclusion, by reducing the number of the surface TfR up to 80% (Byrd and Horowitz, 1989). We found IFN-γ also down regulates TfR mRNA, thus monocytic ehrlichiae are killed in human cells by depletion of cytoplasmically available iron. How monocytic ehrlichiae acquire iron under these conditions is unknown. Since monocytic ehrlichial inclusions accumulate Tf and TfR (Barnewall *et al.*, 1997), monocytic ehrlichiae may acquire iron directly from Tf present in their inclusion where iron is released at slightly acidic pH (Nunez *et al.*, 1990). This iron sequestration by ehrlichiae may result in reduction in host cytosolic iron levels, which increases the binding activity of IRP to iron responsive element (IRE) present in nontranslated 3′ loops of TfR mRNA, leading to inhibition of TfR mRNA degradation by RNase, therefore, upregulation of TfR mRNA (Barnewall *et al.*, 1999). This in turn further helps ehrlichial iron uptake via

Tf. It is significant that, however, when IFN-γ is added after establishment of infection, it is no longer effective (Barnewall and Rikihisa, 1994). This change may be related to the ability of cytoplasmic ehrlichiae to upregulate TfR mRNA in macrophages so that they can counteract the down regulation of TfR induced with IFN-γ. Alternatively, this may be due to inhibition of the Jak-Stat pathway. We found that *E. chaffeensis*, more specifically the binding of its protein component to monocytes, blocks Jak-Stat signal transduction (Lee and Rikihisa, 1998). Tyrosine phosphorylation of Jak1 and Stat1α were blocked when IFN-γ was added as early as 30 min after infection. At least a part of this inhibition may be attributed to the increased host cell protein kinase A activity. *E. chaffeensis* infection stimulates protein kinase A activity of the host cells by 25-fold within 30 min of infection. Addition of a protein kinase A inhibitor abrogated this inhibition (Lee and Rikihisa, 1998).

7. INDUCTION OF HOST CELL CYTOKINE GENE EXPRESSION BY EHRLICHIAE

The presence of very few ehrlichial organisms in the blood and tissue suggests that clinical signs of ehrlichiosis are mediated by the host, probably amplified by some cytokine production. Unlike various extracellular bacteria, however, various facultative intracellular bacteria, rickettsia, or chlamydia interaction with mouse peritoneal macrophages with *E. risticii* does not evoke generation of ROI, prostaglandins, or major proinflammatory cytokines especially TNF-α or IL-6 (van Heeckeren *et al.*, 1993). With *E. chaffeensis* infection of human peripheral blood monocytes or human myelocytic leukemia cell line, THP-1 cells, IL-1β and IL-8, a neutrophil chemokine are the only proinflammatory cytokines generated (Lee and Rikihisa, 1996). In addition, the immunosuppressive cytokine IL-10 is induced. Thus, in this manner ehrlichiae can delay the induction of the protective host immune response. Live or intact *E. chaffeensis* organism is not required for mRNA expression of these cytokines. When we compared *E. chaffeensis* components which induce the cytokine gene expression, a periodate-sensitive carbohydrate residue of organisms is responsible for inducing this limited cytokine gene expression (Lee and Rikihisa, 1996). Ehrlichial proteins did not induce any of the cytokines examined (Lee and Rikihisa, 1996). Thus, it appears that the interaction between ehrlichial protein ligand and the host receptor does not induce signal transduction toward host monocyte and macrophage activation. Once antibody against *E. chaffeensis* is added, however, TNF-α, IL-6, and IL-1 production as intense as that with *E. coli* lipopolysaccharide (LPS) stimulation takes place

(Lee and Rikihisa, 1997), indicating Fc receptor-mediated proinflammatory cytokine gene activation is not blocked by *E. chaffeensis* infection. These genes are activated by sustained degradation of IkB-α and activation and nuclear translocation of NF-κB nuclear transcription factor (Lee and Rikihisa, 1997). This result suggests that when the patients develop antibodies against *E. chaffeensis*, these pro-inflammatory cytokines may be generated and cause damages. This pro-inflammatory cytokine generation may also help get rid of ehrlichiae and activate protective immune responses against ehrlichiae. Thus, cytokines and modulation of cytokine production by antibodies against ehrlichiae may have significant roles in immune protection and pathology depending on the timing of antibody development with respect to the progression of ehrlichial infection and on which ehrlichial antigens are targeted.

8. HSP60 AND HSP70 OF EHRLICHIAE

We cloned, sequenced, and expressed the *E. sennetsu* GroEL operon. A gene encoding heat shock protein 70 homolog (HSP70) was isolated and sequenced from a gene library of *E. sennetsu* (Zhang *et al.*, 1998). The ehrlichial HSP70 gene encoded a 636-amino acid protein with an approximate molecular weight of 68,805 Da and was homologous to DnaK of *Escherichia coli*. DNA sequence resembling –35 and –10 promoter sequences of *E. coli dnaK* was observed in the upstream region of the ehrlichial HSP70 gene. By using RT-PCR analysis, mRNA levels of ehrlichial HSP70 and HSP60 were examined under temperature shift from 28 to 37°C and from 37 to 40°C. HSP70 mRNA induction level was greater than that of HSP60 mRNA following a 37 to 40°C temperature shift, while reverse was true following a 28°C to 37°C temperature shift. The result suggests that HSP60 and HSP70 play different roles during transfer from vector temperature to human body temperature and during a febrile condition in ehrlichial disease.

The HSP60 homolog 55 kDa protein of the HGE agent was recognized by monospecific polyclonal antiserum against the recombinant *E. sennetsu* 55 kDa antigen (Zhang *et al.*, 1997). Most patient sera, however, either did not recognize or weakly recognized the 55 kDa protein from different HGE agent strains, with the exception of the convalescent-phase serum from one patient, which strongly reacted to a 55 kDa protein in all HGE agent strains. This may explain our previous finding that both acute- and convalescent-phase sera from this patient displayed a high cross-reactivity between the HGE agent and *E. chaffeensis* antigen (Zhi *et al.*, 1997). Previous studies suggested that HSP60 and its homologous antigens might function as a common protective antigen in the immune response against bacteria infection because of their high immunoreactivity and antigenic cross reactivity. The failure to recognize HSP60 homologous antigen in most HGE patients may be related to the pathogenesis of human granulocytic ehrlichiosis.

9. CONCLUDING REMARKS

Presence of a polymorphic multi-gene family of major surface proteins is a unique characteristic of this group of bacteria. This characteristic may provide phenotypic diversity within a limited genome size of this group of bacteria and result in an efficient strategy for adapting to rapidly responding immune system defenses. Important questions about functions of these proteins and the mechanism of this gene expression remain to be answered. The study of binding, internalization, and proliferation of ehrlichiae has provided us with new insights toward the unique requirements of this organism for intracellular survival. Bidirectional signal transduction and alteration of signal transduction at ehrlichiae and host cell interface may be particularly relevant in understanding a dynamic mechanism by which these organisms counter host defense and generate a wide range of consequences, from clearance of infection, persistent subclinical infection, to fatal severe disease. In summary, the analysis of the infection mechanism of ehrlichial organisms facilitates the understanding the molecular mechanisms and evolution of intracellular parasitism and may yield important practical applications.

ACKNOWLEDGMENT. Most of studies in the author's laboratory reported in this review were supported by grants AI30010 and AI40934 from the National Institutes of Health.

10. REFERENCES

Barlough, J.E., Reubel, G.H., Madigan, J.E., Vredevoe, L.K., Miller, P.E., Miller, and Rikihisa, Y., 1998, Detection of *Ehrlichia risticii*, the agent of Potomac horse fever, in freshwater stream snails (*Pleuroceridae*: *Juga* spp.) of Northern California, *J. Appl. Environ. Microbiol.* **64**:2888–2893.

Barnewall, R.E., and Rikihisa, Y., 1994, Abrogation of interferon-γ induced inhibition of *Ehrlichia chaffeensis* infection in human monocytes with iron transferrin, *Infect. Immun.* **62**:4804–4810.

Barnewall, R.E., Rikihisa, Y., and Lee, E.H., 1997, *Ehrlichia chaffeensis* inclusions are early endosomes which selectively accumulate transferrin receptor. *Infect. Immun.* **65**:1455–1461.

Barnewall, R.E., Ohashi, N., and Rikihisa, Y., 1999, *Ehrlichia chaffeensis* and *E. sennetsu*, but not the Human granulocyte ehrlichiosis agent colocalize with transferrin receptor and up-regulate transferrin receptor mRNA by activating iron-responsive protein 1. *Infect Immun.* **67**:2258–2265.

Baynes, R., Bukofzer, G., Bothwell, T., Bezwoda, W., and Macfarlane, B., 1987, Transferrin receptors and transferrin iron uptake by cultured human blood monocytes, *Eur. J. Cell. Biol.* **43**:372–376.

Byrd, T.F., and Horowitz, M.A., 1989, Interferon gamma-activated human monocytes down regulate transferrin receptors and inhibit the intracellular multiplication of *Legionella pneumonphila* by limiting the availability of iron, *J. Clin. Invest.* **83**:1457–1465.

Carlin, J.M., Borden, E.C., and Byrne, G.I., 1989, Interferon-induced indoleamine 2,3-dioxygenase activity inhibits *Chlamydia psittaci* replication in human macrophages, *J. Interferon Res.* **9**:329–337.

Chen, S-M., Popov, V.L,. Feng, H-M., and Walker, D.H., 1996, Antigenic diversity among strains of *Ehrlichia chaffeensis*, in: *Proceedings of the 5th International Symposium on Rickettsiae and Rickettsial Diseases* (Bratislava, Slovak Republic), Slovak Academy of Sciences, Bratislava, Slovak Republic, International Society of Rickettsiae and Rickettsial Diseases, pp. 329–334.

Dumler, J.S., and Bakken, J.S. 1998, Human ehrlichiosis: Newly recognized infections transmitted by ticks, *Annu. Rev. Med.* **49**:201–213.

French, D.M., McElwain, T.F., McGuire, T.C., and Palmer, G., 1998, Expression of *Anaplasma marginale* major surface protein 2 variants during persistent cyclic rickettsemia, *Infect. Immun.* **66**:1200–1207.

Fukuda, T., Sasahara, T., and Kitao, T., 1973, Studies on the causative agent of Hyuganetsu disease, XI. Characteristics of rickettsia-like organism isolated from metacercaria of *Stellantchasmus falcatus* parasitic in grey mullet, *J. Jap. Assoc. Infect. Dis.*, **47**:474–482 (text in Japanese).

Iyengar, R., Stuehr, D.J., and Marletta, M.A., 1987, Macrophage synthesis of nitrite, nitrate, and N-nitrosamines: Precursors and role of the respiratory burst, *Proc. Natl. Acad. Sci. USA* **84**:6369–6373.

Isberg; R.R., 1991, Discrimination between intracellular uptake and surface adhesion of bacterial pathogens, *Science* **252**:934–938.

Kawahara, M., Suto, C., Shibata, S., Futohashi, M., and Rikihisa, Y., 1996, Impaired antigen specific responses and enhanced polyclonal stimulation in mice infected with *Ehrlichia muris*, *Microbiol. Immunol.* **40**:575–587.

Kihlstrom, E., and Soderlund, G., 1983, Trifluoperazine inhibits the infectivity of *Chlamydia trachomatis* for McCoy cells, *FEMS Microbiology letters*. **20**:119–123.

Lee, E., and Rikihisa, Y., 1996, Lack of TNF-α, IL-6, and GM-CSF but presence of IL-1β, IL-8, and IL-10 expression in human monocytes exposed to viable or killed *Ehrlichia chaffeensis*, *Infect. Immun.* **64**:4211–4219.

Lee, E., and Rikihisa, Y., 1997, Anti-*Erhlichia chaffeensis* antibody induces potent proinflammatory cytokine mRNA expression in human monocytes exposed to *E. chaffeensis* through sustained reduction of IkB-α and activation of NF-kB, *Infect. Immun.* **65**:2890–2897.

Lee, E., and Rikihisa, Y., 1998, Protein kinase A-mediated inhibition of interferon-gamma-induced tyrosine phosphorylation of Janus kinases and latent cytoplasmic transcription factors in human monocytes by *Ehrlichia chaffeensis*, *Infect. Immun.* **66**:2514–2520.

Mengaud, J., Ohayon, H., Gounon, P., Mege, R-M., and Cossart, P., 1996, E-cadherin is the receptor for internalin, a surface protein required for entry of *L.monocytogenes* into epithelial cells, *Cell* **84**:923–932.

Messick, J.B., and Rikihisa, Y., 1992, Suppression of I-A^d on P388D_1 cells by *Ehrlichia risticii* infection in response to gamma interferon, *Vet. Immunol. Immunopathol.* **32**:225–241.

Messick, J.B., and Rikihisa, Y., 1993, Characterization of *Ehrlichia risticii* binding, internalization, and growth in host cells by flow cytometry, *Infect. Immun.* **61**:3803–3810.

Messick, J.B., and Rikihisa, Y., 1994, Inhibition of binding, entry or survival of *Ehrlichia risticii* in P388D_1 cells by anti-*E. risticii* IgG and Fab fragments, *Infect. Immun.* **62**:3156–3161.

Misao, T., and Katsuta, K. 1956, Epidemiology of infectious mononucleosis, *Jap. J. Clin. Exp. Med.* **33**:73–82.

Mott, J., Barnewall R., and Rikihisa, Y., 1999, Human granulocytic ehrlichiosis agent and *Ehrlichia chaffeensis* reside in different cytoplasmic compartments in HL-60 cells, *Infect. Immun.* **67**:1368–1378.

Mott, J., and Rikihisa, Y., 1999, Human granulocytic ehrlichiosis agent inhibition of superoxide anion generation in human neutrophils require contact of viable organisms with the host cell. American Society for Microbiology, Chicago, IL, June, 1999, Abstract in press.

Moulder, J.W., 1985, Comparative biology of intracellular parasitism, *Microbiol. Rev.* **49**:298–337.

Nunez, M-T, Gaete, V., Watkins, J.A., and Glass, J., 1990, Mobilization of iron from endocytic vesicles, *J. Biol. Chem.* **265**:6688–6692.

Ohashi, N., Unver, A., and Rikihisa, Y., 1998a, Cloning and expression of immunodominant 30 kDa major outer membrane protein of *Ehrlichia canis* and application of the recombinant protein for serodiagnosis, *J. Clin. Microbiol.* **36**:2671–2680.

Ohashi, N., Zhi, N., Zhang, Y., and Rikihisa, Y., 1998b, Immunodominant major outer membrane proteins of *Ehrlichia chaffeensis* are encoded by a polymorphic multigene family, *Infect. Immun.* **66**:132–139.

Palmer, G.H., Aid, G., Barbet, A.F., McGuire, T.C., and McElwain, T.F., 1994, The immunoprotective *Anaplasma marginale* major surface protein 2 is encoded by a polymorphic multigene family, *Infect. Immun.* **62**:3808–3816.

Park, J., and Rikihisa, Y., 1991, Inhibition of *Ehrlichia risticii* growth in murine peritoneal macrophage by gamma interferon, calcium ionophore, and concanavalin A, *Infect. Immun.* **59**:3418–3423.

Park, J., and Rikihisa, Y., 1992, L-arginine-dependent killing of intracellular *Ehrlichia risticii* by macrophages treated with interferon-gamma, *Infect. Immun.* **60**:3504–3508.

Payne, N.R., and Horwitz, M.A., 1987, Phagocytosis of *Legionella pneumophila* is mediated by human monocyte complement receptors, *J. Exp. Med.* **166**:1377–1389.

Perez, M., Rikihisa, Y., and Wen, B., 1996, Antigenic and genetic characterization of an*Ehrlichia canis*-like agent isolated from a human in Venezuela, *J. Clin. Microbiol.* **34**:2133–2139.

Reddy, R.G., Sulsona, C.R., Harrison, R.H., Mahan, S.M., Burridge, M.J., and Barbet A.F., 1996, Sequence heterogeneity of the major antigenic protein 1 genes from *Cowdria ruminantium* isolates from different geographical areas, *Clin. Diagn. Lab. Immunol.* **3**:417–422.

Reubel, G.H., Barlough, J.E., and Madigan, J.E., 1998, Production and characterization of *E. risticii*, the agent of Potomac horse fever, from snails (Pleuroceridae: *Juga* spp.) in aquarium culture and genetic comparison to equine strains, *J. Clin. Microbiol.* **36**:1501–1511.

Rikihisa, Y., 1990a, Ultrastructure of rickettsiae with special emphasis on ehrlichia, in: *Ehrlichiosis: A Vector-Borne Disease of Animals and Humans*, (J.C. Williams and I. Kakoma eds.), Kluwer Academic Pubishers, Boston, Mass., p. 22–31

Rikihisa, Y. 1990b, Growth of *Ehrlichia risticii* in human colonic epithelial cells, in: *Rickettsiology: Current issues and perspectives*, 590, (K. Hechemy, ed.), New York Acad. Sci., NY., pp. 104–110.

Rikihisa, Y., 1991, The tribe *Ehrlichieae* and ehrlichial diseases, *Clin. Microbiol. Rev.* **4**:286–308.

Rikihisa, Y., 1995, Ehrlichial Diseases, in: *Zoonotic Diseases*, *Clin. Microbiol.* **22**, 413–419 (In Japanese).

Rikihisa, Y., 1996, Ehrlichiae, in: *Proceedings of the 5th International Symposium on Rickettsiae and Rickettsial Diseases* (Bratislava, Slovak Republic), Slovak Academy of Sciences, Bratislava, Slovak Republic, International Society of Rickettsiae and Rickettsial Diseases, pp. 272–286.

Rikihisa, Y., 1997, Ehrlichiae, emerging human pathogens, in: *Proceedings of the 2nd International Symposium on Lyme Disease*, Shizuoka, Japan, pp. 332–345.

Rikihisa, Y., 1999a, Ehrlichiae of veterinary importance, in: *Proceedings of the 6th International Symposium on Rickettsiae and Rickettsial Diseases* (Marseille, France), Elsevior, Paris, International Society of Rickettsiae and Rickettsial Diseases, In press.

Rikihisa, Y., 1999b, Clinical and biological aspects of infections caused by *Ehrlichia chaffeensis*, *Microb. Infect.* **3**:1–10.

Rikihisa, Y., Ewing, S.A., and Fox, J.C., 1994, Western immunoblot analysis of *E. chaffeensis, E. canis* or *E. ewingii* infection of dogs and humans, *J. Clin. Microbiol.* **32**:2107–2112.

Rikihisa, Y., Kawahara, M., Wen, B., Kociba, G., Fuerst, P., Kawamori, F., Suto, C., Shiba, S., and Futohashi, M., 1997, Western immunoblot analysis of *Haemobartonella muris* and comparison of 16S rDNA sequences of *H. muris, H. felis*, and *Eperythrozoon suis*, *J. Clin. Microbiol.* **35**:823–829.

Rikihisa, Y., Johnson, G.J, Wang, Y.-Z., Reed, S.M., Fertel, R., and Cooke, H.J., 1992, Loss of adsorptive capacity for sodium chloride as a cause of diarrhea in Potomac horse fever, *Res. Vet. Sci.* **52**:353–362.

Rikihisa, Y., Pretzman, C.I., Johnson, G.C., Reed, S.M., Yamamoto, S., and Andrews, F. 1988, Clinical and immunological responses of ponies to *Ehrlichia sennetsu* and subsequent *Ehrlichia risticii* challenge, *Infect. Immun.* **56**:2960–2966.

Rikihisa, Y., Zhang, Y., and Park, J., 1994, Role of clathrin, microfilament, and microtubule on infection of mouse peritoneal macrophages with *Ehrlichia risticii*, *Infect. Immun.* **62**:5126–5132.

Rikihisa, Y., Zhang, Y., and Park, J., 1995, Role of Ca^{2+} and calmodulin on ehrlichial survival in macrophages, *Infect. Immun.* **63**:2310–2316.

Rikihisa, Y., Zhi, N., Wormser, G. Wen, B., Horowitz, H.W., and Hechemy, K.E., 1997, Direct isolation and cultivation of human granulocytic ehrlichia from a human patien, *J. Infect. Dis.* **175**:210–213.

Ristic, M., and Huxsoll, D., 1984, Tribe II, *Ehrlichiae.*, in: *Bergey's Manual of Systematic Bacteriology*, Vol. 1., (N.R. Krieg and J.G. Holt, eds.), The Williams and Wilkins Co., Baltimore, MD., p. 704–711.

Rydkina, E., Roux, V., Balayeva, N., and Raoult, D., 1996, Determination of the genome size of ehrlichiae by pulsed-field gel electrophoresis, in: *Proceedings of the 5th International Symposium on Rickettsiae and Rickettsial Diseases* (Bratislava, Slovak Republic), Slovak Academy of Sciences, Bratislava, Slovak Republic, International Society of Rickettsiae and Rickettsial Diseases, pp. 318–323.

Shainklin-Kestenbaum, R., Winikoff, Y., Kol, R., Chaimovitz, C., and Sayou, I., 1989, Inhibition of growth of *Chlamydia trachomatis* by the calcium antagonist verapamil, *J. Gen. Microbiol.* **135**:1619–1623.

Snyder, S.H., and Bredt, D.S., 1992, Biological roles of nitric oxide, *Sci. Am.* 68–77.

Solderland, G., and Kihlstrom, E., 1983, Effect of methylamine and monodansylcadaverine on the susceptibility of McCoy cells to *Chlammydia trachomatis* infection, *Infect. Immun.* **40**:531–541.

Sumner, J.W., Nicholson, W.L., and Massung, R.E., 1997, PCR amplification and comparison of nucleotide sequences from the *groESL* heat shock operon of *Ehrlichia* spp., *J. Clin. Microbiol.* **35**:2087–2092.

Sumner, J.W., Sims, K.G., Jones, D.C., and Anderson, B.E., 1993, *Ehrlichia chaffeensis* expresses an immunoreactive protein homologous to the *Escherichia coli* GroEL protein, *Infect. Immun.* **61**:3536–3539.

van Heeckeren, A., Rikihisa, Y., Park, J., and Fertel, R., 1993, Tumor necrosis factor-α, Interleukin-1α, Interleukin-6 and prostaglandin E_2 production in murine peritoneal macrophages infected with *Ehrlichia ristici, Infect. Immun.* **61**:4333–4337.

van Viet, A.M., Jongejan, F., von Kleef, M., and van der Zeijst, B.A.M., 1994, Molecular cloning, sequence analysis, and expression of the gene encoding the immunodominant 32-kilodalton protein of *Cowdria ruminantium, Infect. Immun.* **62**:1451–1456.

Walker, D.H., and Dumler, J.S., 1996, Emergence of ehrlichiosis as a human health problems, *Emerging Infect. Dis.* **2**:18–29.

Weiss, E., Dasch, G.A., Kang, Y-H., and Westfall, H.N., 1988, Substrate utilization by *Ehrlichia sennetsu* and *Ehrlichia risticii* separated from host constituents by renografin gradient centrifugation, *J. Bacteriol.* **170**:5012–5017.

Weiss, E., Williams, J.C., Dasch, G.A., and Kang, Y-H., 1989, Energy metabolism of monocytic *Ehrlichia, Proc. Natl. Acad. Sci. USA* **866**:1674–1678.

Wells, M.Y., and Rikihisa, Y., 1988, Lack of lysosomal fusion with phagosomes containing *Ehrlichia risticii* in $P388D_1$ cells: Abrogation of inhibition with oxytetracycline, *Infect. Immun.* **56**:3209–3215.

Wen, B., Rikihisa, Y., Yamamoto, S., and Fuerst, P., 1996, Characterization of SF agent, an *Ehrlichia* sp. isolated from *Stellantchasmus falcatus* fluke, by 16S rRNA base sequence, serological, and morphological analysis, *Int. J. Syst. Bacteriol.* **46**:149–154.

Yu, X-J., Crocquet-Valdes, P., and Walker, D.H., 1997, Cloning and sequencing of the gene for a 120 kDa immunodominant surface protein of *Ehrlichia chaffeensis*, *Gene* **184**:149–154.

Zhang, Q-Y., Deryckere, D., Lauer, P., and Koomey, M., 1992, Gene conversion in *Neisseria gonorrhoeae*: evidence for its role in pilus antigenic variation, *Proc. Nat. Acad. Sci. USA.* **89**:5366–5370.

Zhang, Y., Ohashi, N., Lee, E., and Rikhisa, Y., 1997, *Ehrlichia sennetsu groE* operon and antigenic properties of the GroEL homolog, *FEMS Immunol. Med. Microbiol.* **18**:39–46.

Zhang, Y., and Rikihisa, Y., 1997, Tyrosine phosphorylation is required for ehrlichial internalization and replication in $P388D_1$ cells, *Infect. Immun.* **65**:2959–2964.

Zhang, Y., Ohashi, N., and Rikihisa, Y., 1998a, Cloning of the heat shock protein 70 (HSP70) of *Ehrlichia sennetsu* and differential expression of HSP70 and HSP60 mRNA after temperature upshift, *Infect. Immun.* **66**:3106–3112.

Zhang, Y., Rikihisa, Y., Ohashi, N., and Zhi, N., 1998b, Binding of major outer membrane proteins of *Ehrlichia chaffeensis* to DH82 cell surface, submitted.

Zhi, N., Rikihisa, Y.H., Kim, Y., Wormser, G.P., and Horowitz, H.W., 1997, Comparison of major antigenic proteins of six strains of human granulocytic ehrlichiosis agents by Western immnoblot analysis, *J. Clin. Microbiol.* **35**:2606–2611.

Zhi, N., Ohashi, N., Rikihisa, Y., Horowitz, H.W., Womser, G.P., Hechemy, K., 1998, Cloning and expression of 44 kDa major outer membrane protein antigen gene of human granulocytic ehrlichiosis agent and application of the recombinant protein to serodiagnosis, *J. Clin. Microbiol.* **36**:1666–1673.

Zhi, N., Ohashi N., and Rikihisa, Y., 1999, Characterization of the expressed genes in p44 multigene family encoding major antigenic outer membrane proteins of the human granulocytic ehrlichiosis agent in HL-60 cells, American Society for Microbiology, Chicago, IL, June, 1999, Abstract in press.

Part IV

New Approaches and Applications

Chapter 21

DNA Vaccine Delivery by Attenuated Intracellular Bacteria

Guido Dietrich and Werner Goebel

1. SUMMARY

Vaccination by intramuscular or intradermal injection of antigen-encoding DNA is a promising new approach leading to strong cellular and humoral immune responses. Since bone-marrow derived antigen presenting cells (APC) seem to induce these immune responses after migration to the spleen, it is desirable to deliver DNA vaccines directly to splenic APC. Recently, attenuated intracellular bacteria have been exploited for the introduction of DNA vaccine vectors into different cell types *in vitro* as well as *in vivo* and offer an attractive alternative to the direct inoculation of naked plasmid DNA.

GUIDO DIETRICH Lehrstuhl für Mikrobiologie, Theodor-Boveri-Institut für Biowissenschaften, Universitat Würzburg, 97074 Würzburg, Germany and Chiron Behring GmbH & Co, Preclinical Vaccine Research, 35006 Marburg, Germany. **WERNER GOEBEL** Lehrstuhl für Mikrobiologie, Theodor-Boveri-Institut für Biowissenschaften, Universität Würzburg, 97074 Würzburg, Germany.

Subcellular Biochemistry, Volume 33: Bacterial Invasion into Eukaryotic Cells, edited by Oelschlaeger and Hacker. Kluwer Academic / Plenum Publishers, New York, 2000.

2. DNA VACCINATION

The discovery that the injection of plasmid DNA can lead to the direct transfection of muscle cells *in vivo* (Wolff *et al.*, 1990) led to the development of DNA vaccination. DNA vaccination, in short, is the introduction of antigen-encoding genes into living organisms. DNA vaccines consist of a plasmid vector containing a promoter which is active in eukaryotic cells, the antigen-encoding gene and a polyadenylation/transcription termination sequence. The plasmid DNA can either be dissolved in saline and injected into muscle tissue with a syringe and needle or alternatively be bound to gold particles and propelled into the dermis with a ballistic apparatus like the gene gun. Both mechanisms result in the uptake of the DNA and in the subsequent expression of plasmid-encoded antigens by body cells (for review: Donnelly *et al.*, 1997). Compared to conventional vaccines, DNA vaccines may have several distinct advantages: (1) Plasmid DNA can be produced and isolated easily and at relatively low costs. (2) Plasmid DNA is stable, its transport and storage are possible even at elevated temperatures. (3) The injection of pure plasmid DNA into living organisms seems to be safe and not to have any side effects. Even during pregnancy and immune suppression no risks should be expected. (4) DNA vaccination leads to the expression of antigens in their native form, which allows very efficient processing and presentation together with MHC (major histocompatibility) class I and MHC class II, inducing cellular as well as humoral immune responses (Conry *et al.*, 1994; Sedegah *et al.*, 1994; Ulmer *et al.*, 1993; Wang *et al.*, 1993). (5) The antigen-expression is lasting quite long which might allow the elicitation of life-long immunity with just a single dose (Wolff *et al.*, 1992). (6) DNA vaccines might be applied directly after birth and induce protection even in the presence of maternal antibodies (Bot *et al.*, 1997; Hassett *et al.*, 1997; Martinez *et al.*, 1997). (7) DNA vaccines are ideally suited for the construction of polyvalent vaccines (Hanke *et al.*, 1998; Wild *et al.*, 1998; Boyer *et al.*, 1997). (8) Coexpression of cytokines may allow the enhancement as well as the modulation of the immune response into a desired direction (Kim *et al.*, 1997; Xiang and Ertl, 1995). (9) With the DNA vaccination based expression-library-immunisation, the identification of protective T-cell- as well as B-cell-antigens of pathogenic microorganisms is possible (Barry *et al.*, 1995).

Since the first report of a protective immune response by DNA vaccination (Tang *et al.*, 1992), DNA vaccines have been used to elicit immune responses against bacteria (Tascon *et al.*, 1996; Huygen *et al.*, 1996, Barry *et al.*, 1995) and their exotoxins (Anderson *et al.*, 1996), viruses (Kuhöber *et al.*, 1996; Xiang *et al.*, 1994; Ulmer *et al.*, 1993), parasites (Xu *et al.*, 1995; Sedegah *et al.*, 1994), tumor antigens (Conry *et al.*, 1994) and allergy antigens (Hsu *et al.*, 1996). However, it still is not fully understood, how these immune responses are accomplished and which cell types are involved. Surprisingly, myocytes and keratinocytes seem to play a minor or even no role for the immune response after the intramuscular injection of plasmid DNA or the introduction into the skin via the gene gun (Torres *et al.*, 1997). Bone-marrow derived professional antigen presenting cells instead seem to induce the immune responses after migration to the spleen (Torres *et al.*, 1997; Iwasaki *et al.*, 1997, Corr *et al.*, 1996; Doe *et al.*, 1996).

Despite its many advantages, DNA vaccination still faces several problems: the efficiency of the DNA-uptake seems to be quite low and dose-dependant, so that high amounts of DNA have to be injected to elicit protective immune responses (Deck, 1997). Furthermore, the intramuscular injection of plasmid DNA does not seem to induce immune responses at distal mucosal surfaces (Deck *et al.*, 1997). Reports on the successful mucosal application of DNA vaccines are scarce, Fynan *et al.* (1993) described the application as nose drops of plasmid DNA encoding a very strong antigen which led to protective immunity. In muscle tissue there is a very limited number of antigen presenting cells (Hohlfeld and Engel, 1994), therefore protection against pathogenic microorganisms might only be possible with highly immunogenic antigens (Corr *et al.*, 1996). This might also be the reason for the problems associated with DNA vaccination of primates. Only in a single study the protection of chimpanzees against infection with HIV could be shown after multiple injections of several DNA vaccine vectors encoding four different antigens of HIV (Boyer *et al.*, 1997).

3. ALTERNATIVE METHODS FOR DNA VACCINE DELIVERY

The central role of professional antigen presenting cells in the spleen for the immune response after intramuscular or intradermal injection (Torres *et al.*, 1997; Iwasaki *et al.*, 1997, Corr *et al.*, 1996; Doe *et al.*, 1996) makes the direct introduction of DNA vaccines into these cells desirable. Encapsulation of plasmid DNA in inert biodegradable carrier particles like poly(DL-lactide-coglycolide) microspheres (Jones *et al.*, 1997) or complexation with liposomes (Klavinskis *et al.*, 1997; Smyth Templeton *et al.*, 1997) could offer possible strategies to enhance the delivery of DNA vaccines into these cells. These approaches also have the advantage that they allow the oral or intranasal application of the DNA vaccines, leading to transfection of mucosal as well as splenic cells and to strong immune responses at both sites (Jones *et al.*, 1997; Klavinskis *et al.*, 1997). Another approach for the introduction of plasmid DNA into splenic cells is to exploit the capacity of intracellular bacterial pathogens to enter such cells. Instead of purifying the plasmid DNA and applying it as naked DNA, intracellular bacteria which are carrying the plasmid and unable to grow inside the host cell are used as vehicles for plasmid delivery.

3.1. *In Vitro* DNA Vaccine Delivery by Attenuated Intracellular Bacteria

Recently, attenuated mutants of *Shigella flexneri* (Sizemore *et al.*, 1997; Powell *et al.*, 1996; Sizemore *et al.*, 1995), invasive *Escherichia coli* (Courvalin *et al.*, 1995), *Salmonella typhimurium* (Darji *et al.*, 1997) and *Listeria monocytogenes* (Dietrich *et al.*, 1998) were utilized for plasmid delivery into cultured mammalian cells.

In the first such report, Sizemore *et al.* (1995) described the use of an attenuated strain of *S. flexneri* as DNA vehicle. This auxotrophic mutant, *S. flexneri* 15D, is a derivative of *S. flexneri* 2a carrying a deletion in the *asd* gene (encoding aspartate β-semialdehyde dehydrogenase). This enzyme is essential for the synthesis of the bacterial cell wall constituent diaminopimelinic acid (DAP). DAP auxotrophs undergo lysis unless the amino acid is present in the culture medium (Nakayama *et al.*, 1988). Since DAP is not present in mammalian cells, *S. flexneri* 15D can infect BHK cells and escape from the phagocytic vacuole, but is unable to multiply in the host cell cytosol, resulting in a rapid decrease of viable intracellular bacteria (Sizemore *et al.*, 1995). This strain was transformed with the eukaryotic expression vector pCMVβ, coding for β-galactosidase under the control of the immediate early promoter/enhancer from the human cytomegalovirus (CMV). Strain 15D(pCMVβ) was able to infect BHK cells at similar ratios as 15D alone and could be shown to be suited for the cytosolic delivery of the plasmid, leading to β-galactosidase expression in 1–2% of the cells (Sizemore *et al.*, 1995). β-galactosidase activity throughout the cytoplasm of the transfected BHK cells clearly indicated that the protein had been synthesized by the mammalian cell after the plasmid DNA had been released by the bacteria (Sizemore *et al.*, 1995). Infection of BHK cells with an isolate of 15D(pCMVβ) unable to synthesize functional IpaB and IpaC did not lead to β-galactosidase expression, demonstrating that the invasion of the bacteria is a key step for successful plasmid delivery (Sizemore *et al.*, 1995). Sizemore *et al.* (1995) could also apply *S. flexneri* 15D(pCMVβ) successfully for the delivery of the plasmid DNA into P815 cells, ruling out the possibility that the plasmid transfer is restricted to a single cell type. The amount of DNA that was transferred to each individual cell, leading to strong expression of β-galactosidase, was calculated to be not more than 20×10^{-9} μg (Sizemore *et al.*, 1995). As an alternative to the *asd* mutant, Powell *et al.* (1996) used the auxotrophic and simultaneuosly spreading-negative mutant strain *S. flexneri* CVD1203 (obtained by chromosomal deletions of the *aroA* and *virG* genes) for the delivery of plasmid DNA into eukaryotic cells. Transformed with the expression vector pSV-βGal, encoding the *lacZ* gene under the control of the virus-derived eukaryotic SV40 promoter, these bacteria were shown to carry the plasmid DNA into HeLa, Henle-407 and CaCo2 cells. In these cell lines, plasmid delivery by the *S. flexneri* strain CVD1203 yielded β galactosidase-expressing cells with an efficiency ranging from 0.1 to 10% of the infected cells (Powell *et al.*, 1996).

In a similar approach Courvalin *et al.* (1995) exploited an invasive, but auxotrophic strain of *E. coli* for the transfer of eukaryotic β-galactosidase expression vectors. The auxotrophic mutant strain *E. coli* BM2710 constructed by Courvalin *et al.* (1995) was also blocked in the synthesis of DAP, due to a deletion of the *dapA* gene. The transfer of the 200 kb virulence plasmid pWR110 of *S. flexneri* serotype 5 strain M90T (Sansonetti *et al.*, 1982) conferred an invasive phenotype to this auxotrophic *E. coli* mutant strain, leading to the cytosolic localisation of the bacteria in the host cell (Courvalin *et al.*, 1995; Sansonetti *et al.*, 1983). Transformed with either replicative or integrative eukaryotic β-galactosidase expression vectors, this strain was used for the delivery of the plasmids into HeLa, CHO, A549 or Cos cells. Similarly to the plasmid delivery by the attenuated Shigellae, transfection rates were about 1–3% after the mammalian cells had been infected at a multiplicity of infection (moi) of 500 bacteria per cell, which ensures infection of most of the host cells (Courvalin *et al.*, 1995). Again, β-galactosidase activity was observed in the entire host cell cytosol, demonstrating that β-galactosidase synthesis was performed by the host cell after plasmid delivery by the bacteria (Courvalin *et al.*, 1995).

S. flexneri and *E. coli* carrying the Shigella virulence plasmid can enter the host cell cytosol and therefore introduce the plasmid DNA directly into this compartment, from where it can subsequently be transported into the nucleus where transcription of plasmid-encoded genes can occur. In contrast to these bacteria, *S. typhimurium* is normally retained in a specialized phago(lyso)some (Carroll *et al.*, 1979). Despite this phago(lyso)somal localization of Salmonella, Darji *et al.* (1997) showed that the attenuated *S. typhimurium aroA* SL7207 strain can also deliver plasmid vectors into primary macrophages. The *S. typhimurium aroA* SL7207 strain is blocked in the synthesis of aromatic amino acids and therefore unable to survive inside mammalian cells. SL7207 was used for the delivery of vector pCMVβ into primary peritoneal macrophages of BALB/c mice, leading to expression of β-galactosidase in about 30% of the mammalian cells (Darji *et al.*, 1997). The β-galactosidase synthesis was again of host cell origin since Darji *et al.* (1997) could demonstrate the splicing of *lacZ*-RNA by the eukaryotic host cells. Furthermore, these authors showed that β-galactosidase was synthesized in the presence of tetracycline and at a point in time when most of the intracellular bacteria were already lysed. Although the data suggest that delivery of plasmid DNA by Salmonella results in the expression of plasmid-encoded genes by the mammalian host cells it remains unclear how the plasmid DNA is transported from the phagolysosome into the nucleus of the host cell. One possibilty is that the infection by Salmonella leads to leakage of the phagosome. Alternatively, a specific, but yet unknown transport process may be responsible for the translocation of the plasmid DNA across the phagosomal membrane. The observation that plasmid delivery by Salmonella is only functional in primary macrophages but not in cell lines (Darji *et al.*, 1997; Sizemore *et al.*, 1995), suggests that this property may be specific for mature macrophages (Lowrie, 1998).

A major disadvantage of *S. flexneri*, *E. coli* and *S. typhimurium* as DNA vaccine carriers could be the gram-negative nature of these bacteria, all of which produce lipopolysaccharide (LPS); which besides being toxic might lead to a reduced synthesis of plasmid-encoded genes by the host cells. Therefore, we used the gram-positive bacterium *Listeria monocytogenes* as an alternative carrier for eukaryotic expression vectors (Dietrich *et al.*, 1998). *L. monocytogenes* is—like *S. flexneri*—able to enter the host cell cytosol (Portnoy *et al.*, 1992). The wild type *L. monocytogenes* strain EGD was attenuated by a deletion of the genes *mpl*, *actA* and *plcB* resulting in strain *L. monocytogenes* Δ2 (Dietrich *et al.*, 1998). These genes are necessary for intra- and intercellular movement after the bacteria escape from the host cell phagosome into the cytosol (Portnoy *et al.*, 1992). Delivery of the plasmid DNA into the cytosol was achieved by specific autolysis of the bacteria in this compartment. To this end, we cloned the *ply118* gene, encoding the lysis protein of the *Listeria*-specific bacteriophage A118 (Loessner *et al.*, 1995) under the control of the listerial *actA*-promoter, which is active only in the host cell cytosol (Dietrich *et al.*, 1998). This attenuated suicide strain of *L. monocytogenes* was used as vehicle for the delivery of plasmid vectors carrying either the *gfp* (coding for the green fluorescent protein, GFP) or the *cat*-gene (encoding chloramphenicol acetyl tranferase, CAT) under the control of the CMV-promoter. Efficient plasmid delivery as well as GFP- and CAT-expression were obtained in the murine macrophage-like cell line P388D$_1$ (Dietrich *et al.*, 1998).

Fluorescence in the GFP-expressing P388D$_1$-macrophages was distributed over the entire cytosol after delivery of the GFP-expression vector, indicating that the protein was indeed expressed by the host cells (Dietrich *et al.*, 1998). Furthermore, expression of GFP and CAT was highest at 48 h post infection, *i.e.* at a point in time when most of the intracellular bacteria had undergone autolysis (Dietrich *et al.*, 1998). Lysin-mediated plasmid delivery was even more efficient than the killing of the intracellular bacteria by antibiotic treatment (Dietrich *et al.*, 1998). Bacterial lysis and plasmid DNA delivery do not seem to inhibit host cell proliferation since GFP-expressing cells were observed to divide at a normal rate (Dietrich *et al.*, 1998). When we used the attenuated suicide *L. monocytogenes* strain for the delivery of eukaryotic GFP- or CAT-expression vectors into the (mouse)-fibroblast cell line L929, we obtained somewhat lower transfection efficiencies, with 0.01% of the fibroblast cells expressing GFP. Again, the efficiency of the delivery was much lower when the intracellular bacteria where killed by treatment with antibiotics. In this case only 0.005% of the fibroblast cells exhibited expression of GFP (G. Dietrich, S. Gfroerer and I. Gentschev, unpublished data).

The delivery of plasmid DNA by attenuated suicide *L. monocytogenes* is also working in primary cells. The infection of bone-marrow derived macrophages (BMM) from BALB/c mice with bacteria carrying a GFP-expression vector led to strong GFP-expression by the macrophage cells (G. Dietrich and I. Gentschev, unpublished data). The expression of heterologous antigens obtained by this procedure results in efficient processing and presentation together with MHC class I. This was demonstrated with the plasmid construct p3LOVA118 which codes for the major H-2K^b T cell epitope from chicken ovalbumin ($OVA_{257-264}$) (Kovacsovics-Bankowski and Rock, 1995), expressed under the control of the CMV-promoter. When this plasmid was delivered to BMM by the attenuated suicide *L. monocytogenes*, the T cell epitope was presented efficiently together with MHC class I molecules, as shown by the activation of a $OVA_{257-264}$-specific T cell hybridoma (Dietrich *et al.*, 1998). Thus, delivery of DNA by attenuated suicide *L. monocytogenes* leads to efficient expression and presentation of plasmid-encoded antigens and should be well suited for the elicitation of immune responses *in vivo*.

3.2. *In Vivo* DNA Delivery by Attenuated Intracellular Bacteria

Shigella and Salmonella have also been used for plasmid delivery in animal models. *S. flexneri* 15D carrying pCMVβ (10^6–10^7 bacteria) were applied to mice intranasally twice in four week intervals (Sizemore *et al.*, 1997). Four weeks after the last application, the mice had developed β-galactosidase-specific T cells and antibodies (Sizemore *et al.*, 1997), in a dose-dependent manner, *i.e.* higher numbers of bacteria resulted in better immune responses. These responses could be further enhanced by the co-inoculation of DAP (Sizemore *et al.*, 1997) which increases the invasivity of *S. flexneri* 15D (Sizemore *et al.*, 1997). The same authors could also demonstrate the delivery of plasmid pCMVβ into the surface of guinea pig eyes by *S. flexneri* 15D, by the β-galactosidase positive staining of corneal cells.

Vector pCMVβ was also delivered after intraperitoneal injection of *S. typhimurium aroA* (pCMVβ) into BALB/c mice, leading to up to 30% of β-galactosidase expressing peritoneal macrophages (Darji *et al.*, 1997). The β-galactosidase activity was again distributed over the entire cytosol and not restricted to phagosomal vacuoles as would be expected if the protein had been synthesized by the bacteria and not by the mammalian host cells. Oral vaccination of BALB/c mice with 10^8 bacteria resulted in strong cytotoxic T cell and helper T cell as well as antibody responses. Interestingly, immunization with β-galactosidase expressing *S. typhimurium* led to much weaker $CD8^+$ T cell and no detectable $CD4^+$ T cell responses or antibody titers (Darji *et al.*, 1997), again suggesting that after the immunization with *S. typhimurium* carrying pCMVβ, the vector had indeed been transferred from the bacteria to the host cells and β-galactosidase was expressed by the mammalian cells. Even 35 days post infection, at a point in time when no surviving bacteria could be isolated from spleens of infected mice, β-galactosidase expressing adherent spleen cells could be detected, further supporting the assumption that the protein had been expressed by the host cells, showing a surprisingly long persistence of the plasmid under the *in vivo* conditions.

S. typhimurium aroA was also used for the delivery of plasmid vectors encoding two antigens of *L. monocytogenes*, ActA and LLO (listeriolysin) (Darji *et al.*, 1997). After a single oral immunization of BALB/c mice with 10^8 bacteria, the mice showed specific $CD8^+$ and $CD4^+$ T cell and antibody responses to LLO or ActA, respectively. The cytotoxic T cell responses and the antibody titers declined though, after a peak on day 35, while the titer of helper T cells remained high at least until day 77. The helper T cells were of the Th_1 type since high levels of IFN-γ, but no IL-4, were produced. By three oral boosts with 10^8 bacteria the decline of cytotoxic T cell and antibody responses was converted into a constant increase, showing that boosting with DNA vaccine delivering bacteria is interestingly possible and effective. Mice immunized orally once with *S. typhimurium* carrying the LLO expression vector were protected only partially against a lethal *L. monocytogenes* infection, while boosting resulted in full protection. In contrast, mice that had been immunized with bacteria delivering the ActA expression vector were surprisingly not protected against *L. monocytogenes*, although a strong cytotxic $CD8^+$ T cell response—the main cause of protection against *L. monocytogenes* (Ladel *et al.*, 1994; Kaufmann, 1993; Berche *et al.*, 1987)—had been observed. Possibly, the ActA protein is not a protective antigen (Darji *et al.*, 1997). So far, only the extracellular proteins LLO (Pamer *et al.*, 1991), p60 (Pamer *et al.*, 1994) and a metalloprotease (Busch *et al.*, 1997) have been clearly identified as potent T cell antigens of *L. monocytogenes*.

3.3. Potential Risks and Side Effects of DNA Vaccine Delivery by Attenuated Intracellular Bacteria

One of the crucial issues concerning the use of attenuated bacteria as carriers for DNA vaccine is the safety of the attenuated strains (Higgins and Portnoy, 1998). Salmonella is a very attractive organism since mutant strains have been in use as live vaccines for humans and animals (Hackett, 1993) and shown to be safe. Auxotrophic mutants of invasive *E. coli* K-12 or *S. flexneri* are an attractive alternative because of their ability to introduce the plasmid DNA directly into the host cell cytosol, but their biosafety has to be determined. Furthermore, all these gram-negative bacteria express LPS and might therefore have endotoxic effects.

L. monocytogenes Δ2 used in our previous study (Dietrich *et al.*, 1998) as carrier strain is highly attenuated in mice, showing an intravenous LD_{50} that is three orders of magnitude higher than that of wild-type *L. monocytogenes* EGD (i.e., 1×10^7 bacteria for Δ2 compared to 1×10^4 for EGD; Dietrich *et al.*, 1998). Intracellular expression of the phage lysin PLY118 under the control of the *actA*-promoter considerably increases this attenuation (The intravenous LD_{50} of lysin-expressing *L. monocytogenes* Δ2 is more than 10^8 in BALB/c-mice; Schlereth *et al.*, unpublished data).

Compared to vaccination with pure DNA, delivery of vaccines by attenuated intracellular bacteria should enhance the specificity of the introduction into certain cell types, *e.g.* macrophages.

Several biosafety aspects concerning the introduction of plasmid DNA into mammalian systems have also been discussed. Bacterial DNA sequences may provoke undesired side effects, *e.g.* enhanced inflammation or the development of anti-DNA-antibodies (Pisetsky, 1997; Roman *et al.*, 1997). Since plasmid DNA can persist for more than 19 months in mice after intramuscular injection (Wolff *et al.*, 1992), prolonged expression of antigens could eventually lead to inflammation, autoimmunity or immune tolerance (Donnelly *et al.*, 1997; Hasset and Whitton, 1996; Donnelly *et al.*, 1995). Another concern is the possible integration of plasmid DNA into the host cell genome (Donnelly *et al.*, 1997; Ertl and Xiang 1996; Hasset and Whitton, 1996; Tascon *et al.*, 1996; Donnelly *et al.*, 1995; Vogel and Sarver, 1995) which might enhance the risk of oncogenesis (Donnelly *et al.*, 1997).

We have analysed possible plasmid integration after its delivery by attenuated suicide *L. monocytogenes*, taking advantage of the expression of the neomycine phosphotransferase II gene (*neo*) carried also on the vectors used in our studies. Infection of $P388D_1$-macrophages with the vector-carrying bacteria and subsequent selection on the antibiotic G418 yielded resistant clones with a rate of one in 10^7 infected cells (Dietrich *et al.*, 1998). The clones contained multiple tandemly arranged plasmid copies integrated into the genomic DNA of the macrophage cells (Dietrich *et al.*, 1998). Taking into account that in this experiment only about 0.01% of the cells, (as judged by the expression of GFP) had been transfected, this would mean that 0.1% of the transfected macrophages carried plasmid DNA integrated into their genome. Similarly plasmid delivery into L929 fibroblasts by attenuated *L. monocytogenes* resulted in resistant clones with a rate of 1 per 10^6 infected cells (Dietrich *et al.*, unpublished data). Again, Southern Blot analysis revealed the integration of multiple plasmid copies in concatameric arrangement. The calculated rate of integration of plasmid DNA related to the transfected cells was about 2% suggesting that the integration into the genome occurs in different host cells at a relatively high frequency.

Similar results were obtained when invasive auxotrophic *E. coli* were used for the delivery of different vectors into CHO cells and A549 cells (a human pulmonary carcinoma cell line) (Courvalin *et al.*, 1995). In these experiments the insertion of autonomously replicating and integrating vectors into the genome of the A549 host cells was demonstrated. Furthermore, the methylation pattern of the integrated plasmid DNA resembled that of eukaryotic DNA.

These data clearly demonstrate that plasmid DNA of the type commonly used in DNA vaccination may well integrate into the genome of transfected host cells *in vitro*.

4. DISCUSSION

The delivery of plasmid DNA encoding microbial antigens (generally termined "DNA vaccine") by attenuated intracellular bacteria is a novel approach which may add to the numerous advantages of DNA vaccines by facilitating the application of the DNA (oral vaccination) and by increasing the efficiency of the immune response. In addition, production of such DNA vaccine strains may be less expensive than highly purified vaccine DNA. Auxotrophic strains of *S. flexneri* and invasive *E. coli* as well as attenuated suicide *L. monocytogenes* are able to deliver plasmid DNA directly into the cytosol of infected host cells due to their ability to enter this host cell compartment (Dietrich *et al.*, 1998; Sizemore *et al.*, 1995; Courvalin *et al.*, 1995; Figure 1). Subsequently, the delivered plasmid DNA can enter the host cell nucleus, which is a prerequisite for the transcription of plasmid-encoded antigen genes (Lowrie, 1998). The release of the plasmid DNA from these carrier bacteria is accomplished by desintegration of the bacteria due to the lack of DAP (Sizemore *et al.*, 1995; Courvalin *et al.*, 1995) or

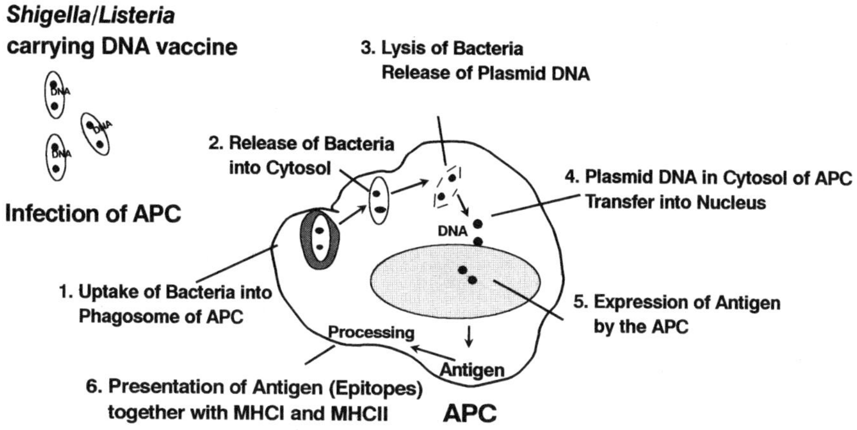

FIGURE 1. Delivery of DNA vaccines by attenuated Shigella and Listeria. Bacteria carrying DNA vaccine vectors are taken up by phagocytosis and subsequently enter the host cell cytosol. In this host cell compartment the bacteria lyse, thereby releasing the plasmid DNA, which can be transported into the nucleus, leading to expression of plasmid-encoded antigens by the host cell. Antigens can finally be processed and presented together with major histocompatibility class I and II (MHC I and MHC II). With kind permission from Dr. Ivaylo Gentschev.

by the specific expression of a phage lysin (Dietrich *et al.*, 1998). Surprisingly, such plasmid DNA can also be delivered by attenuated *S. typhimurium*. The process of how the vaccine DNA is transferred into the nucleus in this case remains to be elucidated since Salmonella is confined to the host cell phagosome and the plasmid DNA has to be transported across the phagosomal membrane (Figure 2; Lowrie, 1998; Darji *et al.*, 1997; Carroll *et al.*, 1979).

Clearly, the major advantage of DNA vaccine delivery by attenuated intracellular bacteria is the possible oral application which avoids injection of the DNA by needle or gene gun. These bacteria can colonize the lumen of the intestine, cross the epithelium and have direct access to the antigen-presenting cells in the gut-associated lymphoid tissue (GALT) (Pascual *et al.*, 1997; Sizemore *et al.*, 1995) thus resulting in mucosal immunity. With suitable vaccine strains delivery of DNA-encoded antigens should also be feasible to other mucosal sites, *e.g.* nose or eyes (Pascual *et al.*, 1997; Sizemore *et al.*, 1995; Courvalin *et al.*, 1995). Delivery of vaccine DNA to the genitourinary tract, with subsequent stimulation of immune responses against certain pathogens that enter through these surfaces—such as sexually transmitted pathogens—might also be feasible (Pascual *et al.*, 1997).

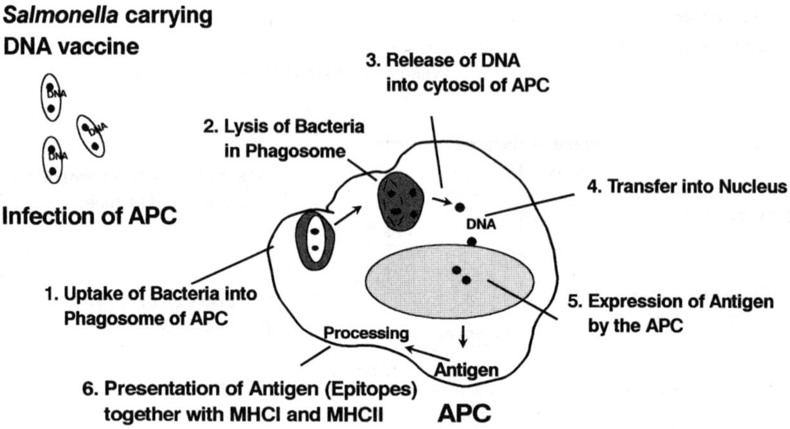

FIGURE 2. Delivery of DNA vaccines by attenuated Salmonella. After phagocytosis of the *Salmonellae* the bacteria presumably remain in the phagosomal compartment of the host cell. Here the bacteria lyse and release the plasmid DNA, which subsequently has to be transported across the phagosomal membrane in order to enter the host cell nucleus. With kind permission from Dr. Ivaylo Gentschev.

Bacterial carriers may introduce plasmid DNA into specific tissues or cell types (Higgins and Portnoy, 1998). The preferential delivery of DNA vaccine vectors to antigen presenting cells like macrophages in immunologically relevant organs by attenuated Salmonella might account for the strong immune response observed even after a single vaccination with an attenuated *S. typhimurium* strain, delivering small amounts of DNA (Darji *et al.*, 1997). The surprisingly strong immune response might in this case be potentiated by the adjuvant effect of bacterial components including LPS or unmethylated sequence motifs in the prokaryotic DNA (Darji *et al.*, 1997).

There are concerns regarding the safety of bacterial carriers for the immunised host (Higgins and Portnoy, 1998). Attenuation of the intracellular bacteria used as carrier organisms for DNA vaccine has to be balanced to cause minimal damage to the host by toxic bacterial components and the efficiency of plasmid delivery and antigen expression which requires a certain persistance of the bacteria within the host (Sizemore *et al.*, 1995). The potential adverse effects caused by LPS of the attenuated gram-negative Shigella, Escherichia and Salmonella might be circumvented by the use of gram-positive bacteria like *L. monocytogenes* (Dietrich *et al.*, 1998).

The data of Dietrich *et al.* (1998) and those of Courvalin *et al.* (1995) clearly indicate that vaccine DNA, at least when delivered by attenuated bacteria, can integrate into the host cell genome. No such integration was previously detected when plasmid DNA was injected into mouse muscle tissue, although this DNA persisted in the mice for more than 19 months (Nichols *et al.*, 1995; Wolff *et al.*, 1992). Integration of heterologous M13mp18 DNA into the mouse genome occured, however, when this DNA was ingested orally (Schubbert *et al.*, 1997). Using attenuated *L. monocytogenes* as plasmid carrier, rather high frequencies of plasmid integration were found in different cell types (up to 2% of the infected host cells) (Dietrich *et al.*, unpublished data). This may be due, at least in part, to the high plasmid copies delivered to the host cells by the attenuated *L. monocytogenes* and *E. coli* and may not generally hold true for all DNA vaccination strategies. Although such integration of vaccine DNA may also occur *in vivo*, the delivery of this DNA by the attenuated suicide *L. monocytogenes* probably evokes a strong CTL response against listerial antigens, especially p60 and listeriolysin, which will also be produced in the infected host cells (Davis *et al.*, 1997; Pamer, 1994; Harty and Bevan, 1992). Host cells infected by *L. monocytogenes*, including those with integrated plasmid DNA, should therefore finally be eradicated by CTLs which will minimize the risk associated with genomic plasmid integration. This does not seem to be the case, when Salmonella is used for the delivery, since Darji *et al.* (1997) observed β-galactosidase expressing cells in the spleen of infected mice for at least 35 days post infection.

Although the transfection efficiency *in vivo* is relatively high, it may not be sufficient for the induction of long-lasting protection (Darji *et al.*, 1997) and boostering may be necessary. There are possibilities to enhance DNA vaccine delivery by attenuated intracellular bacteria, *e. g.* plasmid delivery into the host cell cytosol by attenuated *S. typhimurium* can be substantially improved by the simultaneous secretion of a suitable cytolysin by the carrier strain (Catic *et al.*, submitted). This will allow a better release of plasmid DNA from the phagosomal vacuole into the host cell cytosol.

The delivery of DNA vaccines by attenuated intracellular bacteria combines the numerous advantages of live vaccine carriers with those of naked DNA vaccines. Therefore this novel approach offers an attractive alternative to vaccination with pure plasmid DNA and should be further explored *in vitro* as well as *in vivo*.

5. REFERENCES

Anderson, R., Gao, X.-M., Papakonstantinopoulou, A., Roberts, M., and Dougan, G., 1996, Immune response in mice following immunization with DNA encoding fragment C of tetanus toxin, *Infect. Immun.* **64**:3168–3173.

Barry, M.A., Lai, W.C., and Johnston, S.A., 1995, Protection against mycoplasma infection using expression-library immunization, *Nature* **377**:632–635.

Berche, P., Gaillard, J.-L., and Sansonetti, P.J., 1987, Intracellular growth of *Listeria monocytogenes* as a prerequisite for *in vivo* induction of T cell-mediated immunity, *J. Immunol.* **138**:2266–2271.

Bot, A., Antohi, S., Bot, S., Garcia-Sastre, A., and Bona, C., 1997, Induction of humoral and cellular immunity against influenza virus by immunization of newborn mice with a plasmid bearing a hemagglutinin gene, *Intl. Immunol.* **9**:1641–1650.

Boyer, J.D., Ugen, K.E., Wang, B., Agadjanyan, M., Gilbert, L., Bagarazzi, M.L., Chattergoon, M., Frost, P., Javadian, A., Williams, W.V., Refaeli, Y., Ciccarelli, R.B., McCallus, D., Coney, L., and Weiner, D.B., 1997, Protection of chimpanzees from high-dose heterologous HIV-1 challenge by DNA vaccination, *Nature Med.* **3**:526–532.

Busch, D.H., Bouwer, H.G., Hinrichs, D., and Pamer, E.G., 1997, A nonamer peptide derived from *Listeria monocytogenes* metalloprotease is presented to cytolytic T lymphocytes, *Infect. Immun.* **65**:5326–5329.

Cardenas, L., and Clements, J.D., 1992, Oral immunization using live attenuated *Salmonella spp.* as carriers of foreign antigens, *Clin. Microbiol. Rev.* **5**:328–342.

Carroll, M.E.W., Jackett, P.S., Aber, V.R., and Lowrie, D.B., 1979, Phagolysosome formation, cyclic adenosine 3′–5′ monophosphate and the fate of *Salmonella typhimurium* in mouse peritoneal macrophages, *J. Gen. Microbiol.* **110**:421–429.

Catic, A., Dietrich, G., Gentschev, I., Goebel, W., Kaufmann, S.H.E., and Hess, J., Enhanced major histocompatibility complex class I antigen presentation of macrophages mediated via DNA delivery or protein expression by attenuated *Salmonella typhimurium* in presence of phagosomal escape function, Submitted.

Conry, R.M., LoBuglio, A.F., Kantor, J., Schlom, J., Loechel, F., Moore, S.E., Sumerel, L.A., Barlow, D.L., Abrams, S., and Curiel, D.T., 1994, Immune responses to carcinoembryonic antigen polynucleotide vaccine, *Cancer Res.* **54**:1164–1168.

Corr, M., Lee, D.J., Carson, D.A., and Tighe, H., 1996, Gene vaccination with naked plasmid DNA: mechanism of CTL priming, *J. Exp. Med.* **184**:1555–1560.

Courvalin, P., Goussard, S., and Grillot-Courvalin, C., 1995, Gene transfer from bacteria to mammalian cells, *CR Acad. Sci.* **318**:1207–1202.

Darji, A., Guzman, C.A., Gerstel, B., Wachholz, P., Timmis, K.N., Wehland, J., Chakraborty, T., and Weiss, S., 1997, Oral somatic transgene vaccination using attenuated *S. typhimurium*, *Cell* **91**:765–775.

Davis, H.L., Brazolot Millan, C.L., and Watkins, S.C., 1997, Immune-mediated destruction of transfected muscle fibers after direct gene transfer with antigen-expressing plasmid DNA, *Gene Ther.* **4**:181–188.

Deck, R.R., DeWitt, C.M., Donnelly, J.J., Liu, M.A., and Ulmer, J.B., 1997, Characterization of humoral immune responses induced by an influenza hemagglutinin DNA vaccine, *Vaccine* **15**:71–78.

Dietrich, G., Bubert, A., Gentschev, I., Sokolovic, Z., Simm, A., Catic, A., Kaufmann, S.H.E., Hess, J., Szalay, A.A., and Goebel, W., 1998, Delivery of antigen-encoding plasmid DNA into the cytosol of macrophages by attenuated suicide *Listeria monocytogenes*, *Nature Biotechnol.* **16**:181–185.

Doe, B., Selby, M., Barnett, S., Baenziger, J., and Walker, C.M., 1996, Induction of cytotxic T lymphocytes by intramuscular immunization with plasmid DNA is facilitated by bone marrow-derived cells, *Proc. Natl. Acad. Sci. USA* **93**:8578–8583.

Donnelly, J.J., Friedman, A., Martinez, D., Montgomery, D.L., Shiver, J.W., Motzel, S.L., Ulmer, J.B., and Liu, M.A., 1995, Preclinical efficacy of a prototype DNA vaccine: enhanced protection against antigenic drift in influenza virus, *Nature Med.* **1**:583–587.

Donnelly, J.J., Ulmer, J.B., Shiver, J.W., and Liu, M.A., 1997, DNA Vaccines, *Ann. Rev. Immunol.* **15**:617–648.

Ertl, H.C.J., and Xiang, Z., 1996, Novel Vaccine Approaches, *J. Immunol.* **156**:3579–3582.

Fynan, E.F., Webster, R.G., Fuller, D.H., Haynes, J.R., Santoro, J.C., and Robinson, H.L., 1993, DNA vaccines: protective immunizations by parenteral, mucosal and gene-gun inoculations, *Proc. Natl. Acad. Sci. USA* **90**:1478–1482.

Hackett, J., 1993, Use of *Salmonella* for heterologous gene expression and vaccine delivery systems, *Curr. Opin. Biotechnol.* **4**:611–615.

Hanke, T., Schneider, J., Gilbert, S.C., Hill, A.V.S., and McMichael, A., 1998, DNA multi-CTL epitope vaccines for HIV and *Plasmodium falciparum*: immunogenicity in mice, *Vaccine* **16**:426–435.

Harty, J.T., and Bevan, M.J., 1992, CD8 T cells specific for a single nonamer epitope of *Listeria monocytogenes* are protective *in vivo*, *J. Exp. Med.* **175**:1531–1538.

Hassett, D.E., and Whitton, J.L., 1996, DNA immunization, *Trends Microbiol.* **4**:307–312.

Hassett, D.E., Zhang, J., and Whitton, J.L., 1997, Neonatal DNA immunization with a plasmid encoding an internal viral protein is effective in the presence of maternal antibodies and protects against subsequent viral challenge, *J. Virol.* **71**:7881–7888.

Higgins, D.E., and Portnoy, D.A. 1998. Bacterial delivery of DNA evolves, *Nature Biotechnol.* **16**:138–139.

Hohlfeld, R., and Engel, A.G., 1994, The immunobiology of muscle, *Immunol. Today* **15**:269–273.

Hsu, C.H., Chua, K.Y., Tao, M.H., Lai, Y.L., Wu, H.D., Huang, S.K., and Hsieh, K.H., 1996, Immunoprophylaxis of allergen-induced immunoglobulin E synthesis and airway hyperresponsiveness *in vivo* by genetic immunization, *Nature Med.* **2**:540–544.

Huygen, K., Content, J., Denis, O., Montgomery, D.L., Yawman, A.M., Deck, R.R., DeWitt, C.M., Orme, I.M., Baldwin, S., D'Souza, C., Drowart, A., Lozes, E., Vandenbussche, P., Van Vooren, J.-P., Liu, M.A., and Ulmer, J.B., 1996, Immunogenicity and protective efficacy of a tuberculosis DNA vaccine, *Nature Med.* **2**:893–898.

Iwasaki, A., Torres, C.A.T., Ohashi, P.S., Robinson, H.L., and Barber, B.H., 1997, The dominant role of bone-marrow derived cells in CTL induction following plasmid DNA immunization at different sites, *J. Immunol.* **159**:11–14.

Jones, D.H., Corris, S., McDonald, S., Clegg, J.C.S., and Farrar, G.H., 1997, Poly(DL-lactide-co-glycolide)-encapsulated Plasmid DNA elicits systemic and mucosal antibody responses to encoded protein after oral administration, *Vaccine* **15**:814–817.

Kaufmann, S.H.E., 1993, Immunity against intracellular bacteria, *Annu. Rev. Immunol.* **11**:129–163.

Kim, J.J., Bagarazzi, M.L., Trivedi, N., Hu, Y., Kazahaya, K., Wilson, D.M., Cicarelli, R., Chattergoon, M.A., Dang, K., Mahalingam, S., Chalian, A.A., Agadjanyan, M.G., Boyer, J.D., Wang, B., and Weiner, D.B., 1997, Engineering of *in vivo* immune responses to DNA immunization via codelivery of costimulatory molecules, *Nature Biotechnol.* **15**:641–646.

Klavinskis, L.S., Gao, L., Barnfield, C., Lehner, T., and Parker, S., 1997, Mucosal immunization with DNA-liposome complexes, *Vaccine* **15**:818–820.

Kuhöber, A., Pudollek, H.-P., Reifenberg, K., Chisari, F.V., Schlicht, H.-J., Reimann, J., and Schirmbeck, R., 1996, DNA immunization induces antibody and cytotxic T cell responses to Hepatitis B core antigen in H-2^b mice, *J. Immunol.* **156**:3687–3695.

Ladel, C.H., Flesch, I.E.A., Arnoldi, J., and Kaufmann, S.H.E., 1994, Studies with deficient knock-out mice reveal impact of both MHC I and MHC II dependent T cell responses on *Listeria monocytogenes* infection, *J. Immunol.* **153**:3116–3122.

Loessner, M.J., Wendlinger, G., and Scherer, S., 1995, Heterogenous endolysins in *Listeria monocytogenes* bacteriophages: a new class of enzymes and evidence for conserved holin genes within siphoviral lysis casettes, *Mol. Microbiol.* **16**:1231–1241.

Lowrie, D.B., 1998, DNA vaccination exploits normal biology, *Nature Med.* **4**:147–148.

Martinez, X., Brandt, C., Sadallah, F., Tougne, C., Barrios, C., Wild, F., Dougan, G., Lambert, P.H., and Siegrist, C.A., 1997, DNA immunization circumvents deficient induction of T helper type 1 and cytotoxic T lymphocyte responses in neonates and during early life, *Proc. Natl. Acad. Sci. USA* **94**:8726–8731.

Nakayama, K., Kelly, S.M., and Curtiss III, R., 1988, Construction of an Asd+ expression-cloning vector: stable maintenance and high level expression of cloned genes in a *Salmonella* vaccine strain, *Biotechnology* **6**:693–697.

Nichols, W.W., Ledwith, B.J., Manam, S.V., and Troilo, P.J., 1995, Potential DNA vaccine integration into the host cell genome, *Ann. N.Y. Acad. Sci.* **772**:30–39.

Pamer, E.G., Harty, T.J., and Bevan, M.J., 1991, Precise prediction of a dominant class I MHC-restricted epitope of *Listeria monocytogenes*, *Nature* **353**:852–855.

Pamer, E.G., 1994, Direct sequence identification and kinetic analysis of an MHC class I-restricted *Listeria monocytogenes* CTL epitope, *J. Immunol.* **152**:686–694.

Pascual, D.W., Powell, R.J., Lewis, G.K., and Hone, D.M., 1997, Oral bacterial vaccine vectors for the delivery of subunit and nucleic acid vaccines to the organized lymphoid tissue of the intestine, *Behr. Inst. Mitt.* **98**:143–152.

Pisetski, D.S., 1997, Immunostimulatory DNA: a clear and present danger? *Nature Med.* **3**:829–831.

Powell; R.J., Lewis, G.K., and Hone, D.M., 1996, Introduction of eukaryotic expression cassettes into animal cells using a bacterial vector delivery system, in: *Vaccines 96: Molecular approaches to the control of infectious disease*, Cold Spring Harbor Laboratory, Cold Spring Harbor, NY, pp. 183–187.

Roman, M., Martin-Orozco, E., Goodman, J.S., Nguyen, M.-D., Sato, Y., Ronaghy, A., Kornbluth, R.S., Richman, D.D., Carson, D.A., and Raz, E., 1997, Immunostimulatory DNA sequences function as T helper-1-promoting adjuvants, *Nature Med.* **3**:849–854.

Sansonetti, P.J., Kopecko, D.J., and Formal, S.B., 1982, Involvement of a large plasmid in the invasive ability of *Shigella flexneri*, *Infect. Immun.* **35**:852–860.

Sansonetti, P.J., Hale, T.L., Dammin, G.J., Kapper, C., Collins, H.H., and Formal, S.B., 1983, Alteration in the pathogenicity of *Escherichia coli* K-12 following the transfer of plasmid and chromosomal genes from *Shigella flexneri*, *Infect. Immun.* **39**:1392–1402.

Schubbert, R., Renz, D., Schmitz, B., and Doerfler, W., 1997, Foreign (M13) DNA ingested by mice reaches peripheral leukocytes, spleen, and liver via intestinal wall mucosa and can be covalently linked to mouse DNA, *Proc. Natl. Acad. Sci. USA* **94**:961–966.

Sedegah, M., Hedstrom, R., Hobart, P., and Hoffmann, S.L., 1994, Protection against malaria by immunization with circumsporozoite protein plasmid DNA, *Proc. Natl. Acad. Sci. USA* **90**:9866–9870.

Sher, A., and Coffman, R.L., 1992, Regulation of immunity to parasites by T cells and T cell-derived cytokines, *Annu. Rev. Immunol.* **10**:385–409.

Sizemore, D.R., Branstrom, A.A., and Sadoff, J.C., 1995, Attenuated *Shigella* as a DNA delivery vehicle for DNA-mediated immunization, *Science* **270**:299–302.

Sizemore, D.R., Brandstrom, A.A., and Sadoff, J.C., 1997, Attenuated bacteria as a DNA delivery vehicle for DNA-mediated immunization, *Vaccine* **15**:804–807.

Smyth Templeton, N., Lasic, D.D., Frederik, P.M., Strey, H.H., Roberts, D.D., and Pavlakis, G.N., 1997, Improved DNA: liposome complexes for increased systemic delivery and gene expression, *Nature Biotechnol.* **15**:647–652.

Tang, D.C., DeVit, M., and Johnston, S.A., 1992, Genetic immunization is a simple method for eliciting an immune response, *Nature* **356**:152–154.

Tascon, R.E., Colston, M.J., Ragno, S., Stavropoulos, E., Gregory, D., and Lowrie, D.B., 1996, Vaccination against tuberculosis by DNA injection, *Nature Med.* **2**:888–892.

Tian, J., Clare-Salzler, M., Herschenfeld, A., Middleton, B., Newman, D., Mueller, R., Anta, S., Evans, C., Atkinson, M.A., Mullen, Y., Sarvetnick, N., Tobin, A.J., Lehmann, P.V., and Kaufmann, D.L., 1996, Modulating autoimmune responses to GAD inhibits disease progression and prolongs islet graft survival in diabetes-prone mice, *Nature Med.* **2**:1348–1353.

Torres, C.A.T., Iwasaki, A., Barber, B.H., and Robinson, H.L., 1997, Differential dependence on target site tissue for gene gun and intramuscular DNA immunizations, *J. Immunol.* **158**:4529–4532.

Ulmer, J.B., Donnelly, J.J., Parker, S.E., Rhodes, G.H., Felgner, P.L., Dwarki, V.J., Gromkowski, S.H., Deck, R.R., DeWitt, C.M., Friedman, A., Hawe, L.A., Leander, K.R., Martinez, D., Perry, H.C., Shiver, J.W., Montgomery, D.L., and Liu, M.L., 1993, Heterologous protection against influenza by injection of DNA encoding a viral protein, *Science* **259**:1745–1749.

Wang, B., Ugen, K.E., Srikantan, V., Agadjanvan, M.G., Dang, K., Rafaeli, Y., Sato, A.I., Boyer, J.D., Williams, W.V., and Weiner, D.B., 1993, Gene inoculation generates immune responses against human immunodefiency virus type 1, *Proc. Natl. Acad. Sci. USA* **90**:4156–4160.

Wild, J., Grüner, B., Metzger, K., Kuhröber, A., Pudollek, H.-P., Hauser, H.-J., Schirmbeck, R., and Reimann, J., 1998, Polyvalent vaccination against hepatitis B surface and core antigen using a dicistronic expression plasmid, *Vaccine* **16**:353–360.

Wolff, J.A., Malone, R.W., Williams, P., Chong, W., Acsadi, G., Jani, A., and Felgner, P., 1990, Direct gene transfer into mouse muscle *in vivo*, *Science* **247**:1465–1468.

Wolff, J.A., Ludtke, J.J., Acsadi, G., Williams, P., and Jani, A., 1992, Long-term persistence of plasmid DNA and foreign gene expression in mouse muscle, *Hum. Mol. Gen.* **1**:363–369.

Xiang, Z.Q., Spitalnik, S., Tran, M., Wunner, W.H., Cheng, J., and Ertl, H.C.J., 1994, Vaccination with a plasmid vector carrying the rabies virus glycoprotein gene induces protective immunity against rabies virus, *Virology* **199**:132–140.

Xiang, Z.Q., and Ertl, H.J.C., 1995, Manipulation of the immune response to a plasmid-encoded viral antigen by coinoculation with plasmids expressing cytokines, *Immunity* **2**:129–135.

Xu, D., and Liew, F.Y., 1995, Protection against leishmaniasis by injection of DNA encoding a major surface glycoprotein, gp63, of *L. major*, *Immunology* **84**:173–176.

Chapter 22

Vaccines against Intracellular Pathogens

Raúl G. Barletta, Ruben O. Donis, Ofelia Chacón, Homayoun Shams, and Jeffrey D. Cirillo

1. ABSTRACT

Vaccination against intracellular pathogens presents unique problems that are specific to the growth environment used by these organisms. For all vaccines it is important to determine the best antigen(s) and inoculation method that will induce the proper strength and type of immune response as well as protect against subsequent challenge. With intracellular pathogens, however, the need for a cell-mediated immune response, limited direct access of the immune system to the infectious agent and potential for control of antigen processing and presentation in the host cell by the pathogen make vaccine design even more complex. The majority of the vaccines in use today, including those used for intracellular pathogens, were developed using traditional methods and the efficacies and inoculation methods determined empirically. The advent of molecular biology and the

RAÚL G. BARLETTA Department of Veterinary and Biomedical Sciences, University of Nebraska, Lincoln, Nebraska 68583 and Center of Biotechnology, University of Nebraska, Lincoln, Nebraska 68583. **RUBEN O. DONIS, HOMAYOUN SHAMS, AND JEFFREY D. CIRILLO** Department of Veterinary and Biomedical Sciences, University of Nebraska, Lincoln, Nebraska 68583. **OFELIYA CHACÓN** Department of Veterinary and Biomedical Sciences, University of Nebraska, Lincoln, Nebraska 68583 and Deparment of Veterinary Pathobiology, Texas A&M University, College Station, Texas 77843. *Corresponding author, E-mail: braul@crcvms.unl.edu

Subcellular Biochemistry, Volume 33: Bacterial Invasion into Eukaryotic Cells, edited by Oelschlaeger and Hacker. Kluwer Academic / Plenum Publishers, New York, 2000.

development of a better understanding of the mechanisms of immune protection should allow a more directed approach to vaccine design. Using *Salmonella* and mycobacteria as model intracellular pathogens, we review recent advances in our understanding of potential mechanisms of immune protection and methods of vaccine design and delivery. We propose directions for further study and strategies for the design and delivery of vaccines against intracellular pathogens based on current technology.

2. INTRODUCTION

Intracellular pathogens such as viruses, certain bacteria and parasitic protozoa are etiologic agents of endemic, epidemic or emerging infectious diseases worldwide. Development of effective vaccines against these diseases, including HIV and other viral pathogens, tuberculosis, malaria, and typhoid would have a major impact on public health throughout the world. Worldwide smallpox eradication and polio eradication in the Americas, are successes attesting to the impact of effective vaccines in the context of regional vaccination plans, providing a paradigm for future endeavors (Ada, 1997). The optimal design and methods for delivery of vaccines against intracellular pathogens requires an understanding of the biology of the diseases that they cause and mechanisms of immunological protection against them. During the course of infection with an intracellular pathogen, there are several stages at which interference by the immune system may prevent the induction of disease. These stages of infection would include adherence to host cells, entry and intracellular growth. Interference with the adherence and entry of intracellular pathogens may be the most effective method of disease prevention. This is due to the fact that during these stages of infection the immune system has access to the pathogen; whereas, after entry the pathogen may limit antigen presentation by the host cell (Rescigno *et al.*, 1998; VanHeyningen *et al.*, 1997; Clemens and Horwitz, 1995) and remain relatively invisible to the immune system (Polyak *et al.*, 1997). In addition, arresting infection prior to the intracellular replication of the pathogen may reduce the severity of disease symptoms due to a reduced pathogen load. Prevention of adherence may be accomplished by induction of a strong humoral immune response. Opsonization with specific antibodies against bacterial ligands for the host cell receptors may reduce their ability to adhere (Fluckiger *et al.*, 1998). This observation suggests that, despite the primary localization of intracellular pathogens within host cells during disease, a strong humoral response prior to infection may be protective. Thus, the ability to induce a humoral response, particularly involving IgA which is present at the mucosal sites where initial infections occur, may prevent establishment of productive infections.

Entry into host cells may also be prevented or interfered with by a strong humoral response. This has been shown quite well for entry of intracellular pathogens into cells that do not carry Fc receptors (Fluckiger *et al.*, 1998). With monocytic cells, however, that carry Fc receptors, effects upon entry by the presence of a strong humoral response may not be as easily determined. It is conceivable that opsonization with antibodies may actually increase the adherence and entry of a particular pathogen into monocytes. However, this mechanism of entry may not be the one that is most preferred by the pathogen, since uptake via Fc-mediated mechanisms usually results in degradation of the phagocytosed particle (Hostoffer *et al.*, 1994; Mosser, 1994). Thus, a protective response may result if a strong humoral response interferes with the intracellular pathogen preferred mechanism of entry. This mechanism may be Fc-mediated and it is detrimental to the survival of the pathogen.

Protection against a pathogen after entry requires a proper cell-mediated immune response. This fact makes the determination of appropriate antigens for use in vaccination critical. Factors that are important to consider for antigens to be used in vaccines include chemical and structural characteristics of the antigen itself, location of the antigen in the pathogen and the site of antigen delivery to the host. Intracellular parasites reside either in a membrane bound compartment (vesicle) or in the cytoplasm of host cells as free entities. Salmonella and mycobacteria are pathogens that normally reside within vacuoles in the cell; whereas, *Shigella dysenteriae* and *Listeria monocytogenes* replicate within the cytoplasm (see the Shigella and Listeria chapter). Major histocompatability (MHC) class I molecules present immunodominant peptides derived from cytoplasmic parasites. These in turn will engage the $CD8^+$ T cells. In contrast, intracellular vesicular parasites will supply peptides to the Class II molecules, via the endocytic pathway of antigen processing (Verjans *et al.*, 1995; Verma *et al.*, 1995; Brett *et al.*, 1993). Clearance of intracellular parasites residing in vacuoles within the cell poses a major problem for the immune system of the host, for two reasons: 1) the induced Class II-dependent effector immune responses, primarily antibodies, are not effective in clearance of the infection, and 2) since Class II expression is in general restricted to professional antigen-presenting cells, such as macrophages, B lymphocytes and dendritic cells, only these cell types will engage $CD4^+$ T cells. Consequently, the parasite may be able to persist in the host for prolonged periods of time.

In addition, some cytoplasmic pathogens have evolved strategies to evade antigen presentation via the Class I pathway and persist in the host.

Listeria monocytogenes has been shown to down-regulate expression of MHC Class I in infected macrophages (Schuller *et al.*, 1998). Other pathogens utilize similar strategies, including blocking peptide transport by TAP (Transport Associated Protein) and interference with Class I expression and function (Schuller *et al.*, 1998; York *et al.*, 1994). Thus, specific characteristics of the pathogen itself may determine how well different antigens are presented to the immune system. The characteristics of the antigens themselves that result in MHC class I and II presentation remain somewhat unclear despite the general rule that cytoplasmic antigens present via class I MHC and vacuolar antigens present via class II MHC. This is due to the observation that antigens expressed extracellularly or in vacuoles may also be presented via class I MHC (Harding, 1996; Jondal *et al.*, 1996). However, it may be possible to target antigens to presentation by a particular MHC class using fusion to different peptides (Kim *et al.*, 1997). Antigens that are presented by CD1 molecules rather than the MHC, such as lipoglycans and other non-peptide antigens, may induce a greater variety of T cell responses (Sugita *et al.*, 1998). Unlike antigens that would induce a protective humoral response, those that induce a cell-mediated response are not necessarily cell-surface antigens. Targeting of specific antigens to the cytosol may allow the induction of a class I MHC expression resulting in $CD4^+$ and/or $CD8^+$ T cell activation against the intracellular pathogen. However, unknown characteristics of the antigen may still affect the efficiency of presentation.

A strong $CD8^+$ T cell response does not necessarily indicate that the host is protected against future infection (White *et al.*, 1996). The combination of $CD4^+$ and $CD8^+$ T cell responses is likely to be required for complete protection. In the case of $CD4^+$ T helper cells, a decision is made early in the immune response to determine whether a $CD4^+$ Th1 (cellular activating) or Th2 (humoral activating) subset response will occur (Mosmann and Coffman, 1989). A Th1 response is thought to be required to eliminate intracellular pathogens after entry. The presence of IL-12 results in the production of a Th1 response and IL-4 causes a Th2 response to occur (Emoto *et al.*, 1997). These data indicate that the presence or induction of different cytokines during vaccination against intracellular pathogens may determine whether the vaccine will be protective. Thus, the method of vaccination is dependent upon the vulnerability of the particular pathogen at each stage of infection. The complexity of inducing an immune response that will eliminate intracellular pathogens after entry makes the prevention of earlier stages of infection, adherence and entry, attractive. However, the ability to regulate the type of cell-mediated immune response against intracellular pathogens through targeting of the antigen to specific sites in antigen presenting cells or through secretion of lymphokines is desirable for modern vaccines. A summary of many of the imporant immunological considerations involved in the production of vaccines against intracellular pathogens is shown in Figure 1.

In addition to the immunological aspects of protection there are many factors that should be considered to make the vaccine practical for general use. The vaccine should be extremely safe with a low frequency of complications. Schedules should be planned to allow the greatest numbers of individuals to be vaccinated. In order to accomplish this goal the costs of the vaccine should be as low as possible. It is preferable that the vaccine can be given at birth when most children are accessible to health care personnel. The vaccine should be protective from as soon after vaccination as possible for as long as possible. The route of delivery of the vaccine should take into consideration the many different environments and cultures in which it must be administered. This makes administration without needles preferable either by oral or aerosol routes. The physical presentation of the vaccine and its thermal stability (requirement for a cold chain) are also important considerations. In addition, it would be useful to have a method that allows differentiation between naturally infected and vaccinated individuals. There are a number of additional considerations that are specific to the development of live recombinant vaccines. The vaccine vehicle must be amenable to transfer and stable maintenance of recombinant DNA. Expression of recombinant antigens should be in the proper conformation and in the proper location within the vaccine to allow induction of the preferred immune response. It is necessary that the final vaccine does not carry antibiotic resistance markers in order to maintain recombinant antigens. This prevents the possibility of transfer of these resistance markers to other pathogenic organisms in the environment. The traditional approach to vaccine development usually includes the search for avirulent organisms that will give cross-reactive immunity, ideally with no side effects, isolation of a subunit or excreted antigen that confers immunity to the whole organism, isolation of an avirulent organism from the virulent strain by changes in growth conditions or mutagenesis, or by attenuation of the pathogen through heat, chemical or lyophilization treatment. Since we now have a better understanding of the methods of antigen presentation and the immunological basis of protection against different pathogens the opportunity exists to develop improved vaccines for a number of different infections. These developments provide the basis for the production of improved live recombinant vaccine strains.

We have chosen to focus our discussion of the development of vaccines against intracellular pathogens on the two model organisms Salmonella and mycobacteria. These pathogens were chosen because of their medical importance and distant relationship (see the Salmonella and Mycobacteria

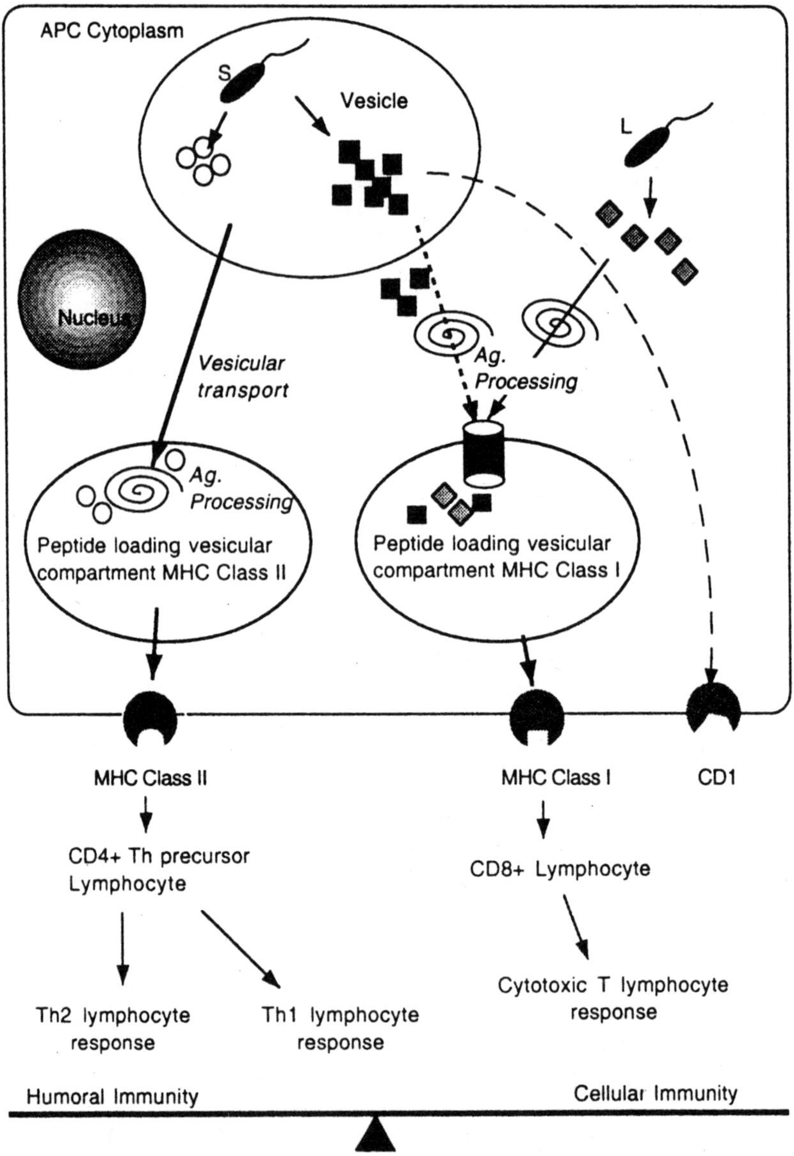

FIGURE 1. Vaccination to induce cellular and humoral immune responses. The intracellular bacteria either remain within a vesicular compartment (S) or enter the cytosolic compartment (L). Antigen from free cytosolic bacteria (filled diamonds) will be processed and translocated to the Class I peptide loading compartment. Antigens from bacteria within vesicles can remain in the vesicle (open circles) or be transported into the cytosol across the membrane (closed squares). Antigens that remain in the vacuole will be presented primarily through Class II MHC pathway, for recognition by $CD4^+$ T cells that play an essential role in antibody responses (Th2 pathway) and cellular immunity (Th1 pathway) (Mosmann and Coffman, 1989).

chapter). The differences in the mechanisms of pathogenesis used by these intracellular pathogens should allow consideration of a broad range of factors relevant to vaccination against organisms that have the ability to enter host cells. The pathogenic mycobacteria are gram-positive organisms that primarily infect macrophages; whereas, Salmonella is a gram-negative bacterium that infects primarily epithelial cells and also has the ability to survive in macrophages. Thus, these organisms differ in their genetic relatedness and in their preferred niche within the host. The fact that both Salmonella (Germanier and Furer, 1975) and mycobacteria (Stover *et al.*, 1991) are being developed as live recombinant vaccine vehicles makes understanding protection against them and the properties of the immune response that may be induced by them potentially important for vaccination against a number of different pathogens.

3. PROTECTION AGAINST SALMONELLA

Salmonella are gram-negative facultative intracellular enterobacteria whose medical importance stems from the severe and fatal diseases they cause in humans and animals. All isolates of Salmonella are considered as belonging to the same species, named *Salmonella enterica* (Christensen *et al.*, 1998). *S. typhi* is a host-restricted human pathogen which causes enteric fever (typhoid fever) in humans. Typhoid fever remains a significant public health problem; there are an estimated 16.6 million cases annually and approximately 600,000 deaths worldwide (Bernard Ivanoff, WHO, Geneva, Switzerland). A spectrum of virulence is found among the more than 30,000 different Salmonella serotypes. Severe cases require hospitalization, yet Salmonella infections can be inapparent (subclinical) and infections often lead to a long term carrier state, with continued or intermittent fecal shedding. Mortality following human Salmonella infections is uncommon in healthy adults, but is significant in infants and children, as well as the elderly (Slutsker *et al.*, 1998). Salmonella are second only to *Campylobacter jejuni*

Bacterial antigens that enter the cytoplasm (closed squares) will be processed to be presented by MHC Class I molecules leading to stimulation of $CD8^+$ T cells which differentiate into cytotoxic lymphocytes. In the case of atypical MHC molecules, such as CD1, a wide variety of T cell responses may be induced and antigens from both vacuoles and the cytosol may be presented (Sugita *et al.*, 1998). Antigen presentation by CD1 is thought to primarily result in induction of a cell-mediated immune response. In summary, antigen routing at the subcellular level has a profound impact on the balance between the two arms of the immune response: humoral and cellular.

as a cause of food poisoning. Infection with virulent Salmonella induces immunological memory which results in significant protection from disease symptoms upon re-exposure to the agent (Jones and Falkow, 1996). Although adaptive immune responses act in concert with non-specific or innate immunity, the latter will not be discussed here. Salmonella infection usually takes place orally and proceeds through an obligatory stage of growth in the lumen of the distal segments of the small intestine. Thus effector immune mechanisms active in this milieu, e.g. secretory IgA (sIgA) can limit bacterial proliferation and damage to the intestinal mucosa, preventing symptoms of enteritis (Michetti *et al.*, 1994; Michetti *et al.*, 1992). In addition, sIgA may be able to limit bacterial invasion of the epithelial enterocyte layer and M cells. Mice bearing subcutaneous Sal4 hybridoma tumors and secreting monoclonal sIgA against a carbohydrate epitope exposed on the bacterial surface into their gastrointestinal tracts were protected against oral challenge with *S. typhimurium* (Michetti *et al.*, 1992). Thus sIgA alone can prevent infection by an invasive enteric pathogen, presumably by immune exclusion at the mucosal surface, because these mice were fully susceptible to intraperitoneal injection with *S. typhimurium*.

The relevant immune effectors after Salmonella have translocated across the gut epithelium, whether bacteria are colonizing the Peyer's patches or disseminating to the spleen and other organs, are the so-called "serum" immunoglobulins (IgM and IgG) and T cells. The role of serum IgG in protection has been more controversial (Eisenstein, 1998). Serum antibodies against O and Vi antigens used in passive protection studies with mouse strains inherently susceptible to Salmonella have prevented typhoid fever. Others have shown a correlation between serum anti-flagellum (H antigen) antibodies and protection against typhoid fever (Gilman *et al.*, 1977). However, several reports described failure to induce protective immunity using non-replicating *S. typhimurium* vaccines, which induce essentially a Th2 $CD4^+$ T helper cell responses leading to IgG antibody production in vaccinated mice (Collins, 1969). These results and others utilizing passive transfer of serum and T lymphocytes indicate that antibody alone is insufficient to protect Salmonella-hypersusceptible mice; cellular immunity is necessary in this model to achieve protection (Mastroeni *et al.*, 1993; Blanden *et al.*, 1966; Collins *et al.*, 1966). Available data suggest that effector lymphocytes involved in mediating protection have the capacity to lyse Salmonella-infected target cells (Sztein *et al.*, 1995). The importance of cellular elements in protection and elimination of intracellular bacteria, including *S. typhimurium*, was demonstrated in the classical experiments of Mackaness *et al.* (1967). However, the relative contribution of $CD8^+$ cytotoxic T lymphocytes (CTLs) vs. $CD4^+$ helper cells to an early restriction of septicemia or in the subsequent clearance of bacteria entrenched in cytoplasmic vesicles within infected cells remains to be elucidated (Mastroeni *et al.*, 1993).

4. VACCINATION AGAINST *Salmonella*

Although much of the current interest in Salmonella vaccines stems from their potential as heterologous antigen delivery vectors, development of avirulent strains of Salmonella was initially aimed at preventing Salmonella infections causing typhoid and food poisoning. These early efforts, were based on the observation, that convalescent hosts are protected for extended periods of time. As a result, both killed and attenuated strains of *S. typhi* and later *S. typhimurium* have been developed, to protect human and animal health. The first typhoid fever vaccine was reported as early as 1896 using heat-killed *S. typhi* (Siler, 1941). Large scale parenteral immunization against typhoid fever was carried out in Indian army soldiers, and during World War I, in British and American soldiers. Low typhoid fever morbidity and mortality in these troops during Word War I was attributed to the efficacy of the killed typhoid vaccine (Siler, 1941). In 1960s, World Health Organization (WHO) sponsored several properly controlled clinical trails to ascertain the efficacy of two subcutaneous doses of heat-phenol inactivated vaccine and acetone-inactivated vaccine in the areas where typhoid fever was endemic. The results were very encouraging, up to 85% protection was observed in some cohorts (Joo, 1971). In a long term special trial in schoolchildren, the results were less favorable, and the discrepancy in results was clear and attributed to the food transmission route (Tapa and Cvjetanovic, 1975).

Typhoid fever vaccines are still not widely used in *S. typhi* endemic areas or by travelers visiting such areas. The major reason for this situation is the limited efficacy of licensed vaccines; the Vi polysaccharide vaccine and the live attenuated Ty21a. The Ty21a mutant is the only licensed live attenuated *S. typhi* vaccine. Clinical trials have yielded mixed results: protection levels are disappointing. The reasons for the failures are not immediately apparent from analysis of the data of standard immunological parameters in vaccinees. Intestinal IgA and serum IgG antibodies to *S. typhi* lipopolysaccharide are produced in human subjects in response to the *S. typhi* Ty21a attenuated typhoid oral vaccine (Forrest *et al.*, 1991). Vaccination with a higher number of viable *S. typhi* organisms than currently contained in the licensed vaccine were required for the induction of intestinal IgA responses (Kantele *et al.*, 1991). A number of experimental vaccines have been studied, aimed at identifying factors controlling safety and protective efficacy. Much of the work was carried out in mice, often using virulence-attenuated *S. typhimurium* as a vaccine and oral or parenteral lethal infections with wild type (WT) *S. typhimurium* to assess protection. Many of the studies measured immune response parameters to determine humoral and cellular effector levels at different anatomical sites following vaccination. We will review here only some of the most recent observations in these models, referring the reader to excellent previous reviews (Jones and Falkow, 1996; Levine *et al.*, 1996).

4.1. Subunit Vaccines against Salmonella

A vaccine composed of the Vi capsular polysaccharide of *S. typhi*, was licensed in the United States and other countries since the early 1990's. Clinical studies done in the United States, Europe, and the developing world, in which children and adults unexposed to typhoid or those living in endemic areas were given single dose of 25 micrograms of the purified polysaccharide by intramuscular injection. Only minor reactions to the vaccine were reported by <10% of subjects. The vaccines were shown to be immunogenic, as 90% of subjects mounted an antibody response which lasted about 3 years. Protective efficacy was evaluated in two human clinical trials carried out in areas in which typhoid is endemic; the efficacy was 55% and 75%, respectively, in adults and in children older than 5 years. The Vi vaccine seems an improvement over the other typhoid vaccines in regard to safety, immunogenicity, and efficacy. The Vi capsular polysaccharide is a well-standardized antigen, it is effective in a single parenteral dose, and may be used in children 2 years of age or older (Plotkin and Bouveret-Le Cam, 1995).

4.2. Attenuated Salmonella Vaccines

The limitations of the killed typhoid vaccine and the observations made during the early 1960's with the *Listeria monocytogenes* mouse model regarding the importance of T lymphocytes in immunity against listeriosis, favored the development of live Salmonella vaccines (Mackaness, 1967; Mackaness and Blanden, 1967). Studies showed the importance of cellular immunity for host protection in the murine typhoid model (Blanden *et al.*, 1966; Collins *et al.*, 1966). Several attenuated Salmonella strains were introduced for use as live oral Salmonella vaccines to induce cellular and humoral immunity to protect against gastroenteritis and typhoid fever (Table 1). The properties of many, but not all, of these live oral vaccines compared very favorably against the traditional whole-cell killed vaccines, as reviewed elsewhere (Eisenstein and Sultzer, 1983; Collins, 1974). For example, a streptomycin-dependent *S. typhi* was tested as a live vaccine can-

Table I
Features of *Salmonella* Vaccine Candidates

Definition	Nature	Strain	Genotype	Route	Efficacy	References
Heat killed	Wcell	*S. typhi*	WT	Parent.	Clin. trial	(Germanier, 1984)
Acetone-inactivated	Wcell	*S. typhi*	WT Ty 2	Parent. Oral	Clin. trial	(Cvejetanovich and Uemura, 1965)
Heat-phenol inactivated	Wcell	*S. typhi*	WT Ty 2	Parent.	Clin. trial	(Cvejetanovich and Uemura, 1965)
Cell free Vi antigen	Sub.	*S. typhi*	N/A	Parent.	Clin. trial	(Hornick *et al.*, 1970)
Streptomycin-dependent	Atten.	*S. typhi* (Ty 2)	N/D	Oral	Clin. trial	(Mel *et al.*, 1974; Reitman, 1967)
Sensitive to galactose & defective LPS in absence of galactose	Atten.	*S. typhi* (Ty21a)	Δ*galE* *rpoS*	Oral	Clin. trial	(Germanier and Furer, 1971)
Mutations in genes necessary for intracellular survival	Atten.	*S. typhi*	Δ*phoP* Δ*phoQ*	Oral	Pilot Clin. trial	(Hohmann *et al.*, 1996a; Hohmann *et al.*, 1996b)
Requires p-mainobenzoic acid	Atten.	*S. typhi*	Δ*aro*	Oral	Clin. trial	(Hoiseth and Stocker, 1981)
Mutation in adenylate cyclase & cAMP receptor proteins genes	Atten.	*S. tm*	Δ*cva* Δ*crp*	Oral	Mice	(Curtiss and Kelly, 1987)
Mutations in genes necessary for intracellular survival	Atten.	*S. tm*	Δ*phoP* Δ*phoQ*	Oral	Mice	(Galan and Curtiss, 1989; Miller *et al.*, 1989)
Mutation in the genes for colonization in deep tissues	Atten.	*S. tm*	Δ*cdt*	Oral	Mice	(Zhange *et al.*, 1997)

Abbreviations: WT: wild type; Parent.: Parenteral; Wcell: Whole cell; Sub.: Subunit; N/A: not applicable; Atten.: Attenuated; N/D: not determined; clin.: clinical; *S. tm*: *S. typhimurium*.

didate. Although the attenuating mutations in this vaccine strain did not alter the relevant antigens of the bacterial cell, no protective efficacy was demonstrated, probably due to the insufficient replication capacity *in vivo* (Levine *et al.*, 1976). An extensive body of experimental data indicates that oral vaccination with live attenuated Salmonella vectors induces a serum IgG response accompanied by a secretory IgA (sIgA) response at mucosal surfaces, especially the intestine. These responses have been measured by isotype-specific coproantibody assays or similar assays using mucus from other mucosal surfaces. *S. typhi galE* mutants were introduced by Germanier and Furer in 1970's as avirulent and immunogenic vaccine candidates (Germanier and Furer, 1975). This mutant, designated Ty21a, lacked UDP-galactose 4-epimerase activity, and was defective in LPS synthesis in the absence of galactose. Following safety and efficacy clinical trials the mutant was licensed as an oral typhoid fever vaccine in most countries of the world. Recent studies have shown that the *galE* mutation is not responsible for the attenuated phenotype of Ty21a (Robbe-Saule *et al.*, 1995). Indeed a gene which encodes a RNA polymerase sigma factor (*rpoS*) was found to be mutated in *S. typhi* live oral vaccine Ty21a (Robbe-Saule *et al.*, 1995).

An aromatic amino acid-dependent *S. typhimurium aroA* mutant strain (3-phosphoshikimate-1-carboxyvinyltransferase deficient) was found to be avirulent and was introduced as a vaccine candidate by Hoiseth and Stocker in 1981. Aromatic-dependent *Salmonella* (Δ*aroA*) have mutations in chorismate pathway genes and cannot synthesize aromatic compounds, including amino acids. Deletions in any of several genes in the aromatic amino acid synthesis pathway have similar phenotypes, although deletion of more than one gene increases the safety of the organism by reducing the chance of reversion to WT by horizontal gene transfer. Interestingly, a complete definition of the molecular basis of virulence attenuation in *aroA S. typhimurium* strain SL3261 is still unavailable. Although the *aroA* mutation is attenuating, complementation with WT *aroA* does not restore virulence in SL3261 (Lockman and Curtiss, 1990). The aromatic compound biosynthetic pathway was targeted for mutagenesis in many serotypes besides *S. typhimurium*; Δ*aro* strains of *S. typhi* (Tacket *et al.*, 1992b; Hone *et al.*, 1991), *S. dublin* (Stocker *et al.*, 1983), and *S. enteritidis* (Cooper *et al.*, 1990), are avirulent but retain significant capacity to survive in the host to be immunogenic. Orally vaccinated individuals were resistant to disease after exposure to the WT strains. Protection levels varied widely, however, ranging from resistance to inoculation with 300 to 10,000 median lethal doses (LD50). *S. typhi aro* mutants passed safety and efficacy clinical trials in human populations and might ultimately be adopted for use in human populations (Tacket *et al.*, 1992b). Attenuated derivatives of *S. typhimurium* have been constructed by introducing stable mutations into genes *aroA* and *aroD* rendering the strains dependent on aromatic compounds for growth. An *aroA* and *aroD* mutant strain derived from the fully calf virulent *S. typhimurium* strain ST4/74 was used to vaccinate orally neonatal calves (Jones *et al.*, 1991). The animals tolerated well the vaccine and resisted challenge infection with WT *S. typhimurium*, suggesting that vaccination with attenuated strains of *S. typhimurium* is a safe and effective means of protecting young calves against *S. typhimurium* infection (Jones *et al.*, 1991). O'Callaghan *et al.* (1990) compared the immune responses to *purA* mutants with that of the better characterized *aroA* mutants. The mutant strains were administered intravenously as live attenuated vaccines to BALB/c mice. *S. typhimurium* strain HWSH *aroA*-immunized mice were well protected against intravenous challenge with the wild-type virulent homologous strain for at least 10 weeks. In contrast, *purA*-immunized mice succumbed to challenge with WT *S. typhimurium*. Mice immunized with either vaccine were able to mount *S. typhimurium*-specific T-cell proliferative responses and produced anti-*S. typhimurium* humoral antibodies in their serum. The antibody titer was greater in those mice immunized with the *aroA* mutant (O'Callaghan *et al.*, 1990). This study suggests that the magnitude of the humoral antibody response may be a relevant predictor of protection.

Deletion of genes encoding global regulatory proteins such as adenylate cyclase (D*cya*), cyclic AMP receptor proteins genes (*Dcrp*), the two-component transcriptional regulatory system PhoQ/PhoP (*phoP* and *phoQ* genes) renders *Salmonella* avirulent and immunogenic in *S. typhi* (Tacket *et al.*, 1992a) and *S. typhimurium* (Galan and Curtiss, 1989). These mutants have been tested as oral vaccines in volunteers or animal models and their efficacy in induction of protective immunity against *S. typhi* (typhoid fever) and *S. typhimurium*-induced disease (typhoid-like disease) was demonstrated. Measurement of immunological parameters which correlate with protection such as humoral mucosal and cellular immunity has demonstrated that *cya* and *crp* mutants are comparable to that of aromatic (*aroA*, *aroD*, etc.) mutants (Nardelli-Haefliger *et al.*, 1996; Tacket *et al.*, 1992a; Curtiss and Kelly, 1987). The OmpR protein is a positive regulator required for the expression of *ompC* and *ompF*, and its mutation was found to attenuate virulence in *S. typhimurium*. BALB/c mice inoculated with *ompR S. typhimurium* by the oral or intravenous route showed no signs of disease and were well protected against challenge with virulent SL1344 (Dorman *et al.*, 1989). *S. typhimurium ompR*-dependent genes are required for full expression of the virulent phenotype, since mutants deficient in both *ompC* and *ompF* are attenuated *in vivo* (Chatfield *et al.*, 1991). OmpR and EnvZ proteins make up a two-component regulatory system controlling the expression of outer membrane protein genes, including *ompC*, *ompF*, and *tppB*. Mutants in *ompR* and *envZ* have altered intracellular trafficking and show a dramatic reduction in *Salmonella*-induced filaments, the peculiar lysosomal glycoprotein-containing tubules which are induced specifically by intracellular Salmonella. The *ompR* mutants were able to invade and replicate in HeLa cells to levels comparable to those in WT strain SL1344 (Mills *et al.*, 1998).

Another interesting candidate vaccine is the *S. typhimurium htrB* mutant, which has a defect in lipid A acylation, the bioactivity of the lipopolysaccharide (LPS) is reduced and consequently is more sensitive to the intracellular antibacterial environment of murine macrophages. *In vivo*, *htrB* mutants are severely limited in their ability to colonize lymphoid organs and to cause systemic disease in mice. Immunogenicity is reported to be satisfactory (Jones *et al.*, 1991). Besides the mutants described above, which targeted defined regulatory genes or metabolic pathways, multiple mutants have been constructed. Chatfield *et al.* (1992) reported the construction of *aroA-htrA* double mutants, which were shown to be safe and effective in mice. These mutants have increased margins of safety by reducing the probability of mutation or horizontal gene transfer events which could lead to reversion to virulence. In summary, various attenuated oral vaccine strains have demonstrated to be genetically stable, possess limited capacity to replicate *in vivo*, are without side effects in the host, and do not establish long term persistence. Most importantly, these strains retain the capacity to reach lymphoid tissues where they interact with professional antigen presenting cells, such as dendritic cells (Hopkins and Kraehenbuhl, 1997).

4.3. Vaccination Using Recombinant Salmonella

Since the late 18th. century, numerous approaches have been developed to deliver immunizing moieties to the host with the intent of reaching the relevant antigen presenting cell and lymphocyte. Non-replicating parenteral delivery systems, which are usually formulated with adjuvant, induce serum antibody responses but have a limited capacity to induce secretory mucosal antibodies and memory $CD8^+$ lymphocytes (Ada, 1997). This category includes the classical inactivated vaccines, toxoids and bacterins, as well as the new subunit vaccines derived from recombinant DNA based expression systems; e.g. the yeast-derived hepatitis B vaccine (McAleer *et al.*, 1992). DNA based delivery systems are very promising candidates for memory $CD8^+$ induction (Manickan *et al.*, 1997; Felgner and Rhodes, 1991). These non-spreading molecules generate transcripts from which antigenic proteins are synthesized within the host, either directly in antigen presenting cells (APC) or taken up by APCs following expression in somatic cells. The DNA vaccination approach has shown excellent protective efficacy in animals under experimental conditions (Manickan *et al.*, 1997). Live replicating microorganisms, especially viral agents, but also some bacteria, such as *Listeria monocytogenes*, are among the most potent inducers of antigen specific memory-phenotype lymphocytes bearing CD8 and high level of CD44 markers (Murali-Krishna *et al.*, 1998; McGregor *et al.*, 1970). Viral delivery systems such as recombinant poxviruses and togaviruses have been extensively characterized, and some have been licensed as animal vaccines (Hariharan *et al.*, 1998). More recently, bacterial delivery systems have been the subject of intense exploration, and the results suggest that they may possess unique advantages. Existing attenuated bacteria, such as *S. typhi* and *Mycobacterium bovis* (Bacillus Calmette-Guerin) were joined by others such as *Listeria monocytogenes* (Weiskirch and Paterson, 1997; see also the chapter by Dietrich and Goebel), *Bacillus anthracis* (Sirard *et al.*, 1997), and *Streptococcus gordonii* (Di Fabio *et al.*, 1998) as delivery vectors for heterologous antigens.

Comparison of heterologous antigen-presenting abilities of different isogenic *S. typhimurium* vaccine strains bearing a variety of different virulence-attenuating mutations (Dunstan *et al.*, 1998) demonstrated that equivalent levels of the C fragment of tetanus toxin, the heterologous antigen analyzed, were expressed. All attenuated *S. typhimurium* strains were capable of penetrating the intestinal epithelium, colonizing the Peyer's patches and spleens of orally vaccinated BALB/c mice. The *purA* mutant showed the lowest level of colonization and persistence in the Peyer's patches while the Δ*htrA* showed lowest level of colonization and persistence in the spleen. The level of specific antibody elicited and protection against toxin challenge was directly correlated with the mutants' ability to colonize the Peyer's patches. Tetanus fragment C-specific IgG1 and IgG2a was significantly lower in animals immunized with *S. typhimurium DpurA*, the mutant with limited Peyer's patch colonization capabilities. In a study using an attenuated strain of *S. typhi Dcya, Dcrp-cdt* expressing a gene encoding a hepatitis B virus (HBV) core-pre-S protein, the induction of specific secretory immunoglobulin A (sIgA) at genital and rectal surfaces in orally vaccinated volunteers was evaluated. Anti-LPS sIgA was found in vaginal and cervical secretions as well as saliva of volunteers (Nardelli-Haefliger *et al.*, 1996).

The detailed knowledge of Salmonella genetics and molecular biology has allowed the manipulation of the bacterial genome and the controlled expression of foreign antigens. Thus, avirulent mutants of *S. typhi* and *S. typhimurium*, originally developed as candidate vaccines for typhoid fever and gastroenteritis, are being used as a vehicle for heterologous antigen delivery. Several genetically engineered or recombinant (r-) Salmonella vaccines have been tested for their capability to induce immune responses against antigens expressed by a transgene inserted in the Salmonella genome. For example, r-*S. typhi* has been used successfully in animal models to induce protective immunity against tetanus toxin, and r-*S. typhimurium* has been exploited to induce antibodies against HBV proteins. As shown in Table 2, there are a number of systems in which model Salmonella vaccines expressing heterologous proteins have been used to demonstrate protective immunity (or immune correlates of protection) against diseases caused by intracellular parasites.

4.4. The Ideal Salmonella Vaccine

Vaccination of large numbers of individuals with the licensed Ty21a strain of *S. typhi* has shown that live oral Salmonella vaccines can be used in the human population without significant untoward side effects. Although the Ty21a strain vaccine has shown some limitations in its ability to protect against typhoid fever, it may be possible to use this strain to elicit immunity against other intracellular pathogens (Germanier, 1984). Advances in the understanding of the interactions of Salmonella with enterocytes and the molecular mediators of diarrheal disease pathogenesis will also contribute to specifically focus the immune response to the relevant molecular targets. It would be desirable to identify the pathogenic determinants of host specificity of *S. typhimurium* diarrhea. This information could lead to the development of transgenic murine models of *S. gastroenteritis* which would boost development of immunization approaches to prevent food poisoning.

A number of pathogenic bacteria of plants and animals have evolved a specialized protein secretion system (type III) to deliver a battery of effector proteins into the host cell to interfere with or stimulate cellular responses for their own benefit (Galan and Bliska, 1996). *S. typhimurium* encodes a type III system in a pathogenicity island located at centisome 63 of the chromosome. This type III secretion apparatus is required for signaling leading to a variety of cellular responses, such as rearrangement of the actin cytoskeleton, gene transcription, and in some cells, triggering of apoptosis (Galan, 1996). These cellular alterations are essential for virulence, because they allow Salmonella to translocate into the host cells. Several substrates of the centisome 63 type III secretion system of Salmonella have been identified. Some of these proteins are involved in the secretion process itself or in the translocation of effector molecules into the

Table II
Recombinant Salmonella Vaccines

Strain	Genotype	Heterologous antigen	Immunity	References
S. tm	Δ*aroA*	Influenza nucleoprotein	$CD4^+$ T cell	(Tite *et al.*, 1990)
S. tm	Δ*aroA*	Fragment C of tetanus toxin	humoral, CMI	(Dunstan *et al.*, 1998)
S. tm	Δ*aroA*Δ*aroD*	Fragment C of tetanus toxin	humoral, mucosal, CMI	(Roberts *et al.*, 1998)
S. tm	Δ*cya*Δ*crp*	Fragment C of tetanus toxin	humoral, CMI	(Dunstan *et al.*, 1998)
S. tm	Δ*htrA*	Fragment C of tetanus toxi	humoral, CMI	(Dunstan *et al.*, 1998)
S. tm	Δ*ompR*	Fragment C of tetanus toxin	humoral, CMI	(Dunstan *et al.*, 1998)
S. tm	Δ*purA*	Fragment C of tetanus toxin	humoral, CMI	(Dunstan *et al.*, 1998)
S. tm	Δ*aroA*Δ*aroD*	Hepatitis B core antigen	humoral, mucosal	(Londono *et al.*, 1996)
S. tm	Δ*aroA*	*P. gingivalis* hemagglutinin	humoral, mucosal	(Dusek *et al.*, 1994)
S. tm	Δ*aroA*	Transgene vaccination	CTL	(Darji *et al.*, 1997)
S. tm	Δ*aroA*Δ*sptP*	Flu NP, and LCMV NP	CTL	(Russmann *et al.*, 1998)
S. typhi	Δ*aroC*Δ*aroD*	Fragment C of tetanus toxin	humoral, CMI	(Galan *et al.*, 1997)
S. typhi	Δ*cya*Δ*crp*Δ*cdt*	Hepatitis B core antigen	humoral, mucosal	(Nardelli-Haefliger *et al.*, 1996; Tacket *et al.*, 1997)

Abbreviations: *S. tm*: *S. typhimurium*; CMI: Cell-mediated immunity; CTL: Cytotoxic T lymphocyte.

target animal cell. Other secreted proteins are effector molecules, e.g.; Salmonella protein tyrosine phosphatase (SptP), a tyrosine phosphatase that is translocated into host cells (Fu and Galan, 1998). SptP delivered into the cytosol of host cells results in disruption of the actin cytoskeleton (Fu and Galan, 1998).

Recent studies have exploited the ability of Salmonella to translocate SptP to the cytosol of infected animal cells for the purposes of vaccination (Russmann *et al.*, 1998). If the heterologous protein is fused to SptP, it may be delivered to the cytosol of the cell for processing and presentation by MHC Class I. SptP consists of two functional domains; an N-terminal domain of unknown function and a C-terminal domain with tyrosine phosphatase activity. To avoid disruption of the SptP structure, a minigene encoding Class I restricted CTL epitopes from the lymphocytic choriomeningitis (LCM) virus was inserted into the interdomain region of SptP, yielding a chimeric protein, SptP-LCM. *S. typhimurium* expressing the SptP-LCM translocated the chimeric protein to the cytosol of the host cell. BALB/c mice vaccinated with attenuated *aroA S. typhimurium* expressing SptP-LCM chimeric proteins developed vigorous CTL responses to the epitope encoded by the minigene. Moreover, these vaccinated mice resisted challenge with a lethal dose of LCMV. Vaccination of C57BL/6 mice, which fail to recognize the LCMV epitope encoded in the SptP chimera, succumbed to challenge, in agreement with the hypothesis that protection was haplotype-restricted. These results suggest that SptP and possibly other targets of the *S. typhimurium* type III secretion apparatus can be exploited as molecular couriers for the delivery of heterologous proteins to the cytosol of antigen presenting cells (Russmann *et al.*, 1998).

Intracellular pathogens have evolved very sophisticated mechanisms of immune evasion, therefore, successful vaccination presents a magnificent challenge (Schuller *et al.*, 1998). Bacterial vaccines such as Salmonella offer some unique features in terms of interaction with the lymphoid tissues of the gut. In the light of its ability to stimulate sIgA responses and Th1 responses which preferentially home to the gut, it is anticipated that pathogens such as *Helicobacter pylori* may be prime candidates for success. Indeed preliminary data supports this view (Corthesy-Theulaz *et al.*, 1998). Substantial progress has been made to introduce mutations which render other Salmonella serotypes safe and effective as vaccine delivery vectors for use in animals (Russmann *et al.*, 1998; Villarreal-Ramos *et al.*, 1998). Indeed, the available safety and efficacy data for certain Salmonella vaccine prototypes suggest that field trials could be carried out in the near future. However, real challenges posed by regulatory issues and public acceptance remain to be overcome before Salmonella becomes a licensed vaccine delivery system for use in veterinary and human health care.

5. PROTECTION AGAINST MYCOBACTERIAL DISEASES

Mycobacteria cause a number of diseases in humans including tuberculosis, which is the leading cause of death from an infectious disease in the world (Bloom and Murray, 1992). Though, leprosy has low mortality, its morbidity is quite high in affected areas, estimated to be 10–12 million in the 1980s (Noordeen, 1991). Hence, the significance of mycobacterial diseases make them prime candidates for the development of effective vaccines. A critical feature of a good vaccine against tuberculosis and leprosy is the elicitation of protective immunity. However, protective immunity in mycobacterial infections is a complex process. The protective cell-mediated response to mycobacterial infection is thought to involve primarily $CD4^+$ T cells (Orme *et al.*, 1993a). These cells induce a Th1 response through secretion of IL-2 and IFNγ (Barnes *et al.*, 1993; Orme *et al.*, 1993b), resulting in macrophage activation and elimination of mycobacteria. Despite these observations there appears to be a role for other T cell subpopulations, natural killer (NK) cells (Bloom, 1994), and humoral immunity (Glatman-Freedman and Casadevall, 1998). For example, antigen-specific cytotoxic $CD8^+$ T cells apparently lyse infected macrophages by a Fas-independent, granule-dependent mechanism that results in death of intracellular bacteria (Stenger *et al.*, 1997). IFNγ mediated induction of $CD8^+$ T cells conferring protective immunity against tuberculosis has also been documented (Tascon *et al.*, 1998). Non-peptide mycobacterial antigens can also be recognized by specific T-cells. CD1b-restricted presentation of mycolic acids and lipoarabinomanans to $CD4^-$ $CD8^-$ γδT-cell clones occurs as well as the activation by nonpeptidic phosphorylated mycobacterial ligands of the major human γδT cell subset expressing Vγ9 and Vδ2 T cell receptor regions (Lang *et al.*, 1995; Sieling *et al.*, 1995; Beckman *et al.*, 1994). The γδT cell subset may influence local cellular traffic through cytokine secretion and promote the accumulation of lymphocytes and monocytes as well as limit the access of inflammatory cells to the site of infection (Munk and Emoto, 1995). There is also evidence that prevention of phagosome-lysosome fusion may be reduced during antibody-mediated phagocytosis (Armstrong and Hart, 1975).

Evidence for the role of macrophages and T lymphocytes in protection against mycobacteria is derived from clinical and pathological observations. Upon infection, fewer than 10% of inhaled mycobacteria reach the bronchioles and alveoli. In the lung these bacteria are phagocytosed by alveolar macrophages. Although these cells are responsible for destruction or inhibition of over 90% of the inhaled bacilli, the surviving microorganisms are sufficient to initiate an intracellular replication. Intracellularly, these microorganisms reside in non-fused phagocytic vacuoles (Clemens *et al.*,

1995; Sturgill-Koszycki *et al.*, 1994). Intracellular mycobacteria multiply and eventually kill their host cell. Released bacteria act as chemoatractants for nonactivated monocytes and macrophages, either directly or through induction of diverse cytokines (IL-1β, TNF-α, MCP-1, MIP-1 α, and osteopontin) by host cells (Nau *et al.*, 1997; Fenton and Vermeulen, 1996). Migrating macrophages become infected and continue secreting cytokines. Thus, a primary granuloma forms through accumulation of mycobacterial laden macrophages. After 2–3 weeks, killing occurs through delayed type hypersensitivity (DTH) resulting in the formation of a caseous center in the granuloma. In the presence of a CMI response, bacilli released from this region are ingested by activated macrophages. However, *M. tuberculosis* can persist for long periods of time in a dormant state (latency) inside the granuloma potentially leading to reactivation when the host's immune system is compromised (Fenton and Vermeulen, 1996; Dannenberg, 1993).

When bacterial replication is not arrested and the DTH response is exaggerated, liquefaction of the caseous center occurs. Bacilli will find a favorable extracellular environment in the accumulated caseous material and continue to multiply. This process will inevitably result in a high concentration of mycobacterial antigens, potentiating a DTH response. Nearby bronchi may erode and cavities start to form in the lung tissue. Cavitated lesions may allow microorganisms into the bronchus, resulting in continuous discharge of infectious bacilli in the sputum. Occasionally, cavities may erode the wall of a vein, resulting in milliary disease (Fenton and Vermeulen, 1996; Dannenberg, 1993). Without effective antimicrobial treatment it is likely that bacilli will spread to other regions of the lung. Uncontrolled, this process will lead to caseous bronchopneumonia and death (Dannenberg, 1993). Since the discovery of the tuberculin test to measure DTH, the relationship between DTH and protective immunity has been debated. Although studies suggest that these processes may be mediated by T cells exhibiting a Th 1-cytokine profile, it is not clear if the relevant role of specific cytokines, such as IFN-γ, is the same for both of them (Tsukada *et al.*, 1991). Whether DTH is absolutely necessary for protective immunity in tuberculosis has not been determined. It appears that low-level hypersensitivity is associated with protection, while a persistent vaccine-associated hypersensitivity does not necessarily correlate with vaccine-derived protection (Fine *et al.*, 1994).

6. VACCINES AGAINST MYCOBACTERIA

A subunit vaccine composed of defined mycobacterial proteins may be protective against mycobacteria, particularly if proteins are found that induce both a humoral and CMI responses (Elhay and Andersen, 1997). It has been suggested that a synthetic peptide vaccine composed of multiple immunogenic peptides may be protective (Patarroyo *et al.*, 1986). In addition, proteins and lipoproteins secreted by mycobacteria may be recognized by the immune system early in infection (Andersen, 1994). If these antigens can be used to modulate the early immune response they may serve as useful vaccine candidates. The most abundant of these proteins are a group of antigens identified in both *M. bovis* (MPB) and *M. tuberculosis* (MPT): MPB59 (MPT59), MPB64 (MPT64), MPB70 (MPT70) and MPB83 (MPT83). Different mycobacterial species elicit DTH responses of varying degrees to these antigens (Roche *et al.*, 1994; Harboe *et al.*, 1986). The most promising study with secreted antigens has been with the *M. tuberculosis* 30 kDa protein (antigen 85B). This antigen, alone or in combination with other extracellular proteins, induces a strong CMI responses and protection against aerosol challenge in guinea pigs (Horwitz *et al.*, 1995). Other secreted proteins with potential protective B- and T-cell epitopes include the 19 kDa lipoprotein (Young and Garbe, 1991), the 38 kDa lipoglycoprotein (da Fonseca *et al.*, 1998; Vordemeier *et al.*, 1991), and the 40 kDa antigen (Andersen *et al.*, 1992). A group of low molecular weight proteins secreted early during *M. tuberculosis* growth in culture present additional vaccine candidates. One of these proteins, 6 kDa ESAT-6, has been implicated in memory and shown to elicit a DTH response (Elhay *et al.*, 1998). Other secreted or cell-surface antigens of immunological interest are a group of proline-rich proteins found in *M. bovis*, *M. tuberculosis* and *M. leprae*, which are recognized by antisera from infected individuals or animals (Berthet *et al.*, 1998; Bigi *et al.*, 1995). Somatic mycobacterial antigens are also processed in macrophages and hence elicit a CMI response. Nevertheless, the issue of whether somatic antigens could be protective is controversial, while substantial evidence exists in favor of a protective role for secreted antigens (Baldwin *et al.*, 1998). Other potentially important antigens, whose subcellular localization has not been established, are the glycine-rich PE and PPE families discoverd in the genome project since they may be a source of antigenic variation or inhibition of antigen processing (Cole *et al.*, 1998). Thus, there are many candidate antigens for development of a subunit vaccine against tuberculosis, yet it is still unclear which of them will be the most protective in humans.

Another approach for a subunit vaccine is the use of DNA vaccination. In one of these approaches, a retroviral vector expressing the *Mycobacterium leprae* 65 kDa heat shock protein was transfected into the macrophage cell line J774. The resulting transfectants presented the antigen to both MHC class I and class II restricted T cells. These cells protected BALB/c mice against challenge with virulent *M. tuberculosis* H37Rv (Lowrie *et al.*, 1997). A requirement for protection is that the vaccine elicits a large population of antigen specific cytotoxic $CD8^+$ T cells prior to challenge (Baldwin *et al.*, 1998). DNA vaccination against tuberculosis may be used for the identification of protective antigens and determination of antigens that are relevant to skin test reactions. In this context, mixtures of plasmids carrying various *M. tuberculosis* genes could be tested for protective and/or diagnostic efficacy (Ulmer, 1996). Aspects for future research on DNA vaccination against tuberculosis, include selection of the best adjuvants and administration routes, and evaluation of the potential side effects of long term persistence of plasmid DNA in the host (e.g., chromosomal integration leading to mutagenesis, or an auto-immune response), problems common to DNA vaccination approaches (Elanschezhiyan *et al.*, 1997). Promising results have been obtained in challenge experiments using mice and guinea pigs with a DNA vaccine encoding the *M. tuberculosis* antigen 85A (Baldwin *et al.*, 1998). Since peptide epitope recognition by T cells is MHC restricted (Launois *et al.*, 1994; Daugelat *et al.*, 1992), it is likely that more than a single peptide will be necessary for protection to be achieved using a subunit vaccine. In conclusion, with the correct antigens inducing a good CMI and possibly humoral response, a protective subunit vaccine against tuberculosis is feasible, as documented by studies with protein and DNA vaccines (Baldwin *et al.*, 1998; Horwitz *et al.*, 1995).

Environmental non-tuberculous mycobacteria play a role in providing certain degree of natural immunity against tuberculosis and leprosy. Several live or, preferentially, heat-killed atypical mycobacteria, alone or in combination with live BCG have been tested as vaccine candidates (Stanford and Stanford, 1994). Trials with a heat-killed *Mycobacterium vaccae* vaccine were carried out with human volunteers vaccinated in a three-dose intradermal schedule. This vaccine elicited lymphocyte proliferative responses to *Mycobacterium avium* antigens without tuberculin skin test conversions, thus showing immunological competence and compatibility with current diagnostics (von Reyn *et al.*, 1997). However, in another trial, this vaccine was unable to protect cattle from an experimental challenge with virulent *M. bovis* (Buddle *et al.*, 1995). Recombinant *M. smegmatis* strains expressing BCG antigens and *Mycobacterium w* have also been tested (Falcone *et al.*, 1995; Mustafa, 1988).

6.1. Vaccination against Mycobacteria with BCG

Bacillus Calmette-Guérin (BCG) was created by attenuation of *M. bovis* through 231 passages on medium containing ox bile. Extensive research from its development in 1921 indicates that BCG is avirulent in normal immune-competent individuals with no apparent reversion frequency to a virulent form. Hence, BCG has been rated as one of the safest vaccines ever developed. This vaccine is inexpensive to produce and can be safely administered to young children (Baldwin *et al.*, 1998). BCG was the first vaccine ever given orally, but due to the development of lymphadenitis in children, oral use was discontinued globally in 1976 and replaced by systemic inoculation. BCG has been widely used for several decades as an anti-tuberculosis vaccine in humans. Regarding efficacy, BCG vaccination has had a significant effect on the prevention of tuberculosis, reducing the risk of infection by 50% and deaths by 70% (Colditz *et al.*, 1994). Furthermore, BCG is also effective in eliciting protection against leprosy (Fine, 1988). Up to now, attempts to improve BCG by combining it with killed *M. leprae* have not resulted in a significant improvement in protection (Convit *et al.*, 1992). Recently, concerns have been raised regarding the use of BCG in immunocompromised patients. However, only very few cases of untoward complications have been observed in such high-risk populations. There are variations among the repertoire of BCG substrains used for vaccination that may affect vaccination efficiency. Significant differences were observed in the growth of strains Glaxo, Japanese, Pasteur, Prague and Russian in mice. Infection of mice with Prague and Japanese BCG substrains resulted in the lowest recovery of microorganisms. The Japanese strains ranked the poorest in terms of inducing a cytotoxic activity and production of anti-PPD antibody responses. Similarly, the Japanese BCG substrain failed to protect mice against a second challenge with either Pasteur or Japanese recombinant strains expressing beta-galactosidase (Lagranderie *et al.*, 1996).

The molecular bases for the attenuation and efficacy of BCG remain unknown. There are many antigens shared by *M. tuberculosis*, *M. bovis* and *M. bovis* BCG. Earlier studies established that the genes for the immunodominant antigens of 71, 65, 19 and 14 kDa are highly homologous in the *M. tuberculosis* complex including BCG (Lu *et al.*, 1987). These findings provide a basis for the immunity against tuberculosis provided by BCG vaccination. Recent studies focused on potential differences between BCG and virulent *M. tuberculosis* or *M. bovis*. Every BCG substrain carries a deletion of the gene encoding the secreted protein ESAT-6 (Harboe *et al.*, 1996). Physical mapping of BCG Pasteur and comparison to the map of *M. tuberculosis* H37Rv confirmed the deletion for the gene encoding ESAT-6 from BCG and identified other deletions (Philipp *et al.*, 1996). Three distinct genomic regions were found to be deleted from BCG by subtractive genomic hybridization techniques (Mahairas *et al.*, 1996). The critical region for BCG attenuation was attributed to the 9.5 kb region denominated RD1 which was found conserved in all virulent strains, but deleted from all BCG substrains. Furthermore, re-introduction of RD1 into BCG repressed the expression of 10 proteins and resulted in a two dimensional gel electrophoretic pattern of total proteins almost identical to the virulent strains. According to this study, attenuation of BCG may be due to the deletion of a DNA region involved in negative regulation of gene expression. In conclusion, BCG, although it is still widely administered in several countries around the world, is not an ideal vaccine against tuberculosis. BCG will convert vaccinees to a positive skin test, which has deterred its routine use in the United States and highlights the need to develop more effective vaccines compatible with diagnostics.

6.2. Vaccination Using Recombinant BCG

The development of genetic systems for the manipulation of mycobacteria and testing of the first recombinant BCG vaccine candidates (Aldovini and Young, 1991; Jacobs *et al.*, 1991; Stover *et al.*, 1991) triggered several studies on the use of BCG as a vaccine vehicle to express foreign antigens. Many antigens have been expressed in r-BCG using a series of autonomously replicating or integrating mycobacterial vectors (Table 3). Overall, these studies indicate that r-BCG elicits strong humoral and CMI response to a wide variety of antigens upon both oral and systemic immunization.

6.3. The Ideal Mycobacterial Vaccine

An ideal vaccine against tuberculosis and leprosy should be able to protect against the widest possible range of pathogenic bacterial isolates eliciting a strong CMI requiring presentation of both peptide and non-peptide antigens, long-lasting immunological memory, and possibly humoral immunity. Given this complexity, we believe that live attenuated vaccines would offer the best possibilities for an ideal mycobacterial vaccine. Furthermore, this type of vaccine could potentially be administered orally as enteric-coated lyophilized tablets, thus avoiding the use of needles (Barletta *et al.*, 1990). Additionally, a live attenuated vaccine could be even made compatible with current or experimental new diagnostics (Cirillo *et al.*, 1995). Although recent studies with subunit vaccines have been promising, so far these vaccines conferred shorter survival than BCG when tested in *M. tuberculosis* aerogenically challenged guinea pigs, underlying the problems to be solved with this type of vaccines (Baldwin *et al.*, 1998). However, subunit vaccines also may be important for disease control in immunodeficient individuals or in areas where the standard skin test has diagnostic significance. Furthermore, a subunit vaccine may be useful to boost individuals previously vaccinated with BCG or those at risk of reactivation disease (Baldwin *et al.*, 1998).

Table III
Representative Recombinant BCG Vaccines Expressing Foreign Antigens

Target	Main Heterologous Antigens Tested	Model	Route	Immunity	References
TB	Lysteriolysin fusion protein of *Listeria monocytogenes*	Cell lines	*In vitro*	$CD8^+$ T cells	(Hess *et al.*, 1998)
SIV	SIV Nef peptide	Mouse	Oral	CMI	(Lagranderie *et al.*, 1997)
Measles	Measles virus nucleoprotein	monkey	ID, IN	CMI	(Zhu *et al.*, 1997)
Schisto.	*Schistosoma mansoni* 28 GST (Glutathione-S-Transferase)	Mouse	IV, IP, SubQ, IN	Humoral	(Kremer *et al.*, 1996)
Malaria	Circumsporozoite protein oligopeptide as fusion protein with *M. kansasii* alpha antigen	Mouse	SubQ	Humoral	(Matsumoto *et al.*, 1996)
Leish.	Gp63 of *Leishmania major*, as fusion protein with N-terminal region of beta-lactamase	Mouse	IV, SubQ	CMI	(Abdelhak *et al.*, 1995)
HIV	V3 principal neutralizing epitope of human immunodeficiency virus (HIV)	Mouse G. pig	SubQ	CMI Humoral	(Honda *et al.*, 1995)
Pneum.	Pneumococal surface protein A. Secreted protein or chimeric exported lipoprotein	Mouse	IP	Humoral	[Langermann, 1994 #452]
Lyme	Outer surface protein A of *Borrelia burgdorferi*	Mouse	IN	Humoral Mucosal	(Langermann *et al.*, 1994)
HIV	HIV type I Env and Gag polypeptides, expressed using mycobacterial *hsp70* promoter.	Mouse	ID, IV	CMI Humoral	(Aldovini and Young, 1991)
β-gal Tetanus HIV	β-Gal, tetanus toxin fragment C and HIV type I gp120 antigen expressed under the control of the *hsp60* promoter.	Mouse	ID, IP, IV	CMI Humoral	(Stover *et al.*, 1991)

Abbreviations: IV: Intravenous; IP: Intraperitoneal; SubQ: Subcutaneous; IN: Intranasal; ID: Intradermal; IC: Intracerebral; G. pig: Guinea pig; CMI: Cell-mediated immunity.

The development of a new live-attenuated vaccine also poses many research challenges. The development of genetic systems to manipulate *M. bovis* BCG, and *M. tuberculosis* offers the greatest promise for the development of new and effective vaccines against mycobacterial infections. Methods for allelic replacement and transposon mutagenesis in slow-growing mycobacteria have been reviewed (Pelicic *et al.*, 1998). Recently, using conditionally replicating genetic elements, two groups have reported major breakthroughs in the genetic manipulation of *M. tuberculosis* (Bardarov *et al.*, 1997; Pelicic *et al.*, 1997). This technology coupled with the knowledge of the *M. tuberculosis* genetic blueprint (Cole *et al.*, 1998) should provide novel strategies for the rational development of live attenuated vaccines. The most promising approach for vaccine development would be to start with virulent *M. tuberculosis* microorganisms and generate attenuated mutants. These mutants should carry at least two attenuating deletion mutations and no-antibiotic resistant markers, so as to avoid unwanted reversions or transfers of drug-resistance. Other important issues to consider are a) the degree of attenuation: mutants that are too attenuated may not induce protective immunity as it was the case for *S. typhimurium phoP* mutants (Fields *et al.*, 1989); and b) whether the vaccine strain could be enhanced by co-administration of cytokines or by endowing the engineered strain with the capability to produce cytokines or phagosome membrane disrupters as lysteriolysin. In this case, temporal expression of these genes may be critical and may have to be regulated by promoters solely active inside phagocytic cells, as for example the promoter of the *mig* gene, first identified in *M. avium* (Plum and Clark-Curtiss, 1994) with a corresponding homologue (*fadD19*) found in *M. tuberculosis* genome sequence (Cole *et al.*, 1998).

Candidate genes for attenuating mutations may be those directly involved in the pathogenic process or those encoding housekeeping enzymes whose suppression may alter the ability of *M. tuberculosis* to survive and/or replicate within phagocytes. We summarize in Table 4 those genes from which more extensive work with mycobacterial systems has been reported. The best documented examples of mutations leading to attenuated phenotypes *in vivo* are the inactivation of the *erp* gene (Berthet *et al.*, 1998) and previous work on the inactivation of genes conferring auxotrophic requirements for leucine (*leuD*) or methionine (*met*) (Bange *et al.*, 1996; Guleria *et al.*, 1996; McAdam *et al.*, 1995). Additional candidate genes and previous work has been summarized elsewhere (Cole *et al.*, 1998). Other candidates are unidentified mycobacterial genes such as those involved in the inhibition of phagosome acidification (Via *et al.*, 1998; Sturgill-Koszycki *et al.*, 1994). It may be also possible to identify and inactivate genes responsible for subtle changes in the structure of

Table IV
Potential Attenuating Mutations in *M. tuberculosis*

Gene	Product of Function	Current Status on the Generation of Attenuated Phenotypes	References
ahpC	Alkyl hydroperoxidase (alkyl hydroperoxide reductase).	Overexpression of AhpC shown to be a compensatory mechanism in *KatG* mutant strains (resistant to isoniazid) that protects *M. tuberculosis* against toxic peroxides.	(Sherman *et al.*, 1996) (Heym *et al.*, 1997) (Chen *et al.*, 1998)
asd *lysA*	Aspartate semi-aldehyde dehydrogenase: auxotrophy for diaminopimelate, lysine, methionine, and threonine.	Tested as a lethal-suppresser combination in *M. smegmatis*.	(Pavelka and Jacobs, 1996)
erp	Exported Repetitive Protein	Inactivation of the *M. tuberculosis* or *M. bovis* BCG *erp* gene results in impaired multiplication in cultured macrophages and mice.	(Berthet *et al.*, 1998)
fadD19 *(mig)*	Acyl-CoA Synthase. Biosynthesis of fatty acids.	Demonstrated expression of *M. avium* gene upon phagocytosis. Induction upon acidic conditions. Identified as *fadD19* in the tuberculosis sequencing project.	(Plum and Clark-Curtiss, 1994) (Plum *et al.*, 1997). (Cole *et al.*, 1998)

(continued)

Table IV
Continued

Gene	Product of Function	Current Status on the Generation of Attenuated Phenotypes	References
ilv	Biosynthesis of isoleucine, leucine, and valine.	BCG mutant conferred protection against challenge with *M. tuberculosis* in mice and was avirulent for SCID mice.	(Guleria *et al.*, 1996)
katG	catalase-peroxidase	Decreased virulence of isoniazid resistant strains observed. *katG* deleted mutant displayed reduced virulence in mice.	(Zhang *et al.*, 1992) (Wilson *et al.*, 1995) (Rouse *et al.*, 1996) (Li *et al.*, 1998)
leuD	Biosynthesis of leucine.	BCG mutants constructed and shown to be attenuated in mice and macrophages. BCG mutant conferred protection.	(Bange *et al.*, 1996; McAdam *et al.*, 1995) (Guleria *et al.*, 1996)
mas	Mycocerosic acid synthase. Biosynthesis of mycocerosic acid and mycosides.	Generation and biochemical characterization of BCG knock-out mutant.	(Azad *et al.*, 1996)
met	Locus in the biosynthesis of methionine	BCG mutant conferred protection against challenge with *M. tuberculosis* in mice and was avirulent for SCID mice.	(Guleria *et al.*, 1996)
sigA (rpoV)	Essential sigma factor	Point mutation associated with loss of virulence in *M. bovis*	(Collins *et al.*, 1995)
pps	Synthesis of cell wall lipids.	Generation and characterization of BCG knock-out mutant.	(Azad *et al.*, 1997)
ureC	Urease.	Associated with modulation of the pH within phagosomes. Gene inactivation in BCG determined minor attenuating effects.	(Clemens *et al.*, 1995) (Reyrat *et al.*, 1996)

lipoarabinomannan since these changes influence receptor-mediated uptake and ultimately intracellular fate and overall virulence (Kang and Schlesinger, 1998).

Development of new vaccines will require rigorous testing in animal models before being widely used in human populations. There are several animal models of tuberculosis reviewed elsewhere (Bloom, 1994). The mouse model is cost effective, disease is dose dependent, animals produce a strong immune response, and there are a series of recombinant and immunodeficient strains available. However, the guinea pig has the advantage that animals can be reproducibly infected with the development of lesions similar to human tubercles followed by caseating necrosis (Baldwin *et al.*, 1998). The rabbit model is more suitable for the study of the spread of tuberculosis to tissues and organs. Rabbits are more susceptible to tuberculosis and disease progression follows a similar pattern to disease in infants and immunocompromised individuals. Other models being developed that closely resemble human disease include swine and primates. The swine model seems to be appropriate for the study of tuberculous meningitis (Bolin *et al.*, 1997). Primates may provide the best model for latency and chronic infection (Walsh *et al.*, 1996). Suitable animals models for leprosy are more elusive and may not replicate human disease well. Recently, rhesus and sooty mangabey monkeys (Gormus *et al.*, 1998) have all been used, but do not replicate human disease well.

7. CONCLUDING REMARKS

Recent advances in molecular biology and in our understanding of the immune response have provided us with the necessary tools to develop improved vaccines against intracellular pathogens. The technology now exists to allow determination of the proper antigens for vaccination, location of expression of the antigen and the method of delivery to allow protection. Through the use of recombinant vaccine delivery vehicles such as mycobacteria and Salmonella we can introduce nearly any antigen in any form to allow presentation to the immune system through a pathway of our choosing. This technology will surely allow the production of improved vaccines well into the future. The difficulty now lies in improving methods for testing to determine vaccine safety and efficacy; achieving public acceptance and developing guidelines for obtaining governmental approval. In order for the vaccines that are developed to truly have an impact on public health there must also be widespread vaccination programs throughout all areas of the world. Thus, it may be necessary for developed countries to facilitate vaccination programs elsewhere in order to improve conditions in their own countries. International collaborations and continued financial support for vaccine technology are necessary to eradicate important infectious diseases.

Acknowledgments. We thank Zhengu Feng for helpful discussions and assistance with computer analysis of sequences from the *M. tuberculosis* genome project, and Stephanie Williams for help with library materials. O.C. was supported by a loan fellowship from Colciencias and Colfuturo (Santafé de Bogotá, Colombia). Research in the laboratory of J.D.C. is supported by the Center for Indoor Air Research; and R.G.B. by NRICGP-USDA #95-37204-2148 and BARD-USDA #IS-2564-95C.

8. REFERENCES

Abdelhak, S., Louzir, H., Timm, J., Blel, L., Benlasfar, Z., Lagranderie, M., Gheorghiu, M., Dellagi, K., and Gicquel, B., 1995, Recombinant BCG expressing the leishmania surface antigen Gp63 induces protective immunity against *Leishmania major* infection in BALB/c mice, *Microbiology* **141**:1585–1592.

Ada, G., 1997, Overview of vaccines, *Mol. Biotechnol.* **8**:123–134.

Aldovini, A., and Young, R.A., 1991, Humoral and cell-mediated immune responses to live recombinant BCG-HIV vaccines, *Nature* **351**:479–482.

Andersen, A.B., andersen, P., and Ljungqvist, L., 1992, Structure and function of a 40,000-molecular-weight protein antigen of *Mycobacterium tuberculosis*, *Infect. Immun.* **60**:2317–2323.

Andersen, P., 1994, Effective vaccination of mice against *Mycobacterium tuberculosis* infection with a soluble mixture of secreted mycobacterial proteins, *Infect. Immun.* **62**:2536–2544.

Armstrong, J.A., and Hart, P.D., 1975, Phagosome-lysosome interactions in cultured macrophages infected with virulent tubercle bacilli, *J. Exp. Med.* **142**:1–16.

Azad, A.K., Sirakova, T.D., Fernandes, N.D., and Kolattukudy, P.E., 1997, Gene knockout reveals a novel gene cluster for the synthesis of a class of cell wall lipids unique to pathogenic mycobacteria, *J. Biol. Chem.* **272**:16741–16745.

Azad, A.K., Sirakova, T.D., Rogers, L.M., and Kolattukudy, P.E., 1996, Targeted replacement of the mycocerosic acid synthase gene in *Mycobacterium bovis* BCG produces a mutant that lacks mycosides, *Proc. Natl. Acad. Sci. USA* **93**:4787–4792.

Baldwin, S.L., D'Souza, C., Roberts, A.D., Kelly, B.P., Frank, A.A., Lui, M.A., Ulmer, J.B., Huygen, K., McMurray, D.M., and Orme, I.M., 1998, Evaluation of new vaccines in the mouse and guinea pig model of tuberculosis, *Infect. Immun.* **66**:2951–2059.

Bange, F.C., Brown, A.M., and Jacobs, W.R., Jr., 1996, Leucine auxotrophy restricts growth of *Mycobacterium bovis* BCG in macrophages, *Infect. Immun.* **64**:1794–1799.

Bardarov, S., Kriakov, J., Carriere, C., Yu, S., Vaamonde, C., McAdam, R.A., Bloom, B.R., Hatfull, G.F., and Jacobs, W.R., Jr., 1997, Conditionally replicating mycobacteriophages: a system for transposon delivery to *Mycobacterium tuberculosis*, *Proc. Natl. Acad. Sci. USA* **94**:10961–10966.

Barletta, R.G., Snapper, S.B., Cirillo, J.D., Connell, N.D., Kim, D.D., Jacobs, W.R., Jr., and

Bloom, B.R., 1990, Recombinant BCG as a candidate oral vaccine, *Res. Microbiol.* **141**:931–939.

Barnes, P.F., Abrams, J.S., Lu, S., Sieling, P.A., Rea, T.H., and Modlin, R.L., 1993, Patterns of cytokine production by mycobacterium-reactive human T-cell clones, *Infect. Immun.* **61**:197–203.

Beckman, E.M., Porcelli, S.A., Morita, C.T., Behar, S.M., Furlong, S.T., and Brenner, M.B., 1994, Recognition of a lipid antigen by CD1-restricted alpha beta+ Tcells, *Nature* **372**:691–694.

Berthet, F.X., Lagranderie, M., Gounon, P., C., L.-W., Ensergueix, D., Chavarot, P., Thouron, F., Maranghi, E., Pelicic, V., Portnoi, D., Marchal, G., and Gicquel, B., 1998, Attenuation of virulence by disruption of the *Mycobacterium tuberculosis erp* gene., *Science* **282**:759–762.

Bigi, F., Alito, A., Fisanotti, J.C., Romano, M.I., and Cataldi, A., 1995, Characterization of a novel *Mycobacterium bovis* secreted antigen containing PGLTS repeats, *Infect. Immun.* **63**:2581–2586.

Blanden, R.V., Mackaness, G.B., and Collins, F.M., 1966, Mechanisms of acquired resistance in mouse typhoid, *J. Exp. Med.* **124**:585–600.

Bloom, B.R., 1994, *Tuberculosis: pathogenesis, protection, and control*, ASM Press, Washington, D. C.

Bloom, B.R., and Murray, C.J.L., 1992, Tuberculosis: Commentary on a reemergent killer, *Science* **257**:1055–1064.

Bolin, C.A., Whipple, D.L., Khanna, K.V., Risdahl, J.M., Peterson, P.K., and Molitor, T.W., 1997, Infection of swine with *Mycobacterium bovis* as a model of human tuberculosis, *J. Infect. Dis.* **176**:1559–1566.

Brett, S.J., Rhodes, J., Liew, F.Y., and Tite, J.P., 1993, Comparison of antigen presentation of influenza A nucleoprotein expressed in attenuated AroA-*Salmonella typhimurium* with that of live virus, *J. Immunol.* **150**:2869–2884.

Buddle, B.M., Keen, D., Thomson, A., Jowett, G., McCarthy, A.R., Heslop, J., De Lisle, G.W., Stanford, J.L., and Aldwell, F.E., 1995, Protection of cattle from bovine tuberculosis by vaccination with BCG by the respiratory or subcutaneous route, but not by vaccination with killed *Mycobacterium vaccae*, *Res. Vet. Sci.* **59**:10–16.

Chatfield, S.N., Dorman, C.J., Hayward, C., and Dougan, G., 1991, Role of *ompR*-dependent genes in *Salmonella typhimurium* virulence: mutants deficient in both *ompC* and *ompF* are attenuated *in vivo*, *Infect. Immun.* **59**:449–452.

Chatfield, S.N., Strahan, K., Pickard, D., Charles, I.G., Hormaeche, C.E., and Dougan, G., 1992, Evaluation of *Salmonella typhimurium* strains harbouring defined mutations in *htrA* and *aroA* in the murine salmonellosis model, *Microb. Pathog.* **12**:145–151.

Chen, L., Xie, Q.W., and Nathan, C., 1998, Alkyl hydroperoxide reductase subunit C (AhpC) protects bacterial and human cells against reactive nitrogen intermediates, *Mol. Cell* **1**:795–805.

Christensen, H., Nordentoft, S., and Olsen, J.E., 1998, Phylogenetic relationships of *Salmonella* based on rRNA sequences, *Int. J. Syst. Bacteriol.* **48**:605–610.

Cirillo, J.D., Stover, C.K., Bloom, B.R., Jacobs, W.R., Jr., and Barletta, R.G., 1995, Bacterial vaccine vectors and bacillus Calmette-Guérin, *Clin. Infect. Dis.* **20**:1001–1009.

Clemens, D.L., and Horwitz, M.A., 1995, Characterization of the *Mycobacterium tuberculosis* phagosome and evidence that phagosomal maturation is inhibited, *J. Exp. Med.* **181**: 257–270.

Clemens, D.L., Lee, B.Y., and Horwitz, M.A., 1995, Purification, characterization, and genetic analysis of *Mycobacterium tuberculosis* urease, a potentially critical determinant of host-pathogen interaction, *J. Bacteriol.* **177**:5644–5652.

Colditz, G.A., Brewer, T.F., Berkey, C.S., Wilson, M.E., Burdick, E., Fineberg, H.V., and Mosteller, F., 1994, Efficacy of BCG vaccine in the prevention of tuberculosis, *JAMA* **271**:698–702.

Cole, S.T., Brosch, R., Parkhill, J., Garnier, T., Churcher, C., Harris, D., Gordon, S.V., Eiglmeier, K., Gas, S., Barry, C.E., 3rd, Tekaia, F., Badcock, K., Basham, D., Brown, D., Chillingworth, T., Connor, R., Davies, R., Devlin, K., Feltwell, T., Gentles, S., Hamlin, N., Holroyd, S., Hornsby, T., Jagels, K., Krogh, A., McLean, J., Moule, S., Murphy, L., Oliver, K., Osborne, J., Quail, M.A, Rajandream, R.-A., Rogers, J, Rutter, S., Seeger, K., Skelton, J., Squares, R., Squares, S., Sulston, J.E., Taylor, K., Whitehead, S., and Barrell, B.G., 1998, Deciphering the biology of *Mycobacterium tuberculosis* from the complete genome sequence, *Nature* **393**:537–544.

Collins, D.M., Kawakami, R.P., de Lisle, G.W., Pascopella, L., Bloom, B.R., and Jacobs, W.R.J., 1995, Mutation of the principal sigma factor causes loss of virulence in a strain of the *Mycobacterium tuberculosis* complex, *Proc. Natl. Acad. Sci. USA* **92**:8036–8040.

Collins, F.M., 1969, Effect of specific immune mouse serum on the growth of *Salmonella enteritidis* in mice preimmunized with living or ethyl alcohol-killed vaccines, *J. Bacteriol.* **97**:676–683.

Collins, F.M., 1974, Vaccines and cell-mediated immunity, *Bacteriol. Rev.* **38**:371–402.

Collins, F.M., Mackaness, G.B., and Blanden, R.V., 1966, Infection-immunity in experimental salmonellosis, *J Exp Med* **124**:601–619.

Convit, J., Sampson, C., Zuniga, M., Smith, P.G., Plata, J., Silva, J., Molina, J., Pinardi, M.E., Bloom, B.R., and Salgado, A., 1992, Immunoprophylactic trial with combined *Mycobacterium leprae*/BCG vaccine against leprosy: preliminary results, *Lancet* **339**:446–450.

Cooper, G.L., Nicholas, R.A., Cullen, G.A., and Hormaeche, C.E., 1990, Vaccination of chickens with a *Salmonella enteritidis aroA* live oral Salmonella vaccine, *Microb. Pathog.* **9**:255–265.

Corthesy-Theulaz, I.E., Hopkins, S., Bachmann, D., Saldinger, P.F., Porta, N., Haas, R., Zheng-Xin, Y., Meyer, T., Bouzourene, H., Blum, A.L., and Kraehenbuhl, J.P., 1998, Mice are protected from *Helicobacter pylori* infection by nasal immunization with attenuated *Salmonella typhimurium* $phoP^c$ expressing urease A and B subunits, *Infect. Immun.* **66**:581–586.

Curtiss, R., III., and Kelly, S.M., 1987, *Salmonella typhimurium* deletion mutants lacking adenylate cyclase and cyclic AMP receptor protein are avirulent and immunogenic, *Infect. Immun.* **55**:3035–3043.

Cvejetanovich, B., and Uemura, K., 1965, The present status of field and laboratory studies of typhoid and paratyphoid vaccines with special reference to studies sponsored by W.H.O., *Bull. W.H.O* **32**:29–36.

da Fonseca, D.P., Joosten, D., van der Zee, R., Jue, D.L., Singh, M., Vordermeier, H.M., Snippe, H., and Verheul, A.F., 1998, Identification of new cytotoxic T-cell epitopes on the 38 kilodalton lipoglycoprotein of *Mycobacterium tuberculosis* by using lipopeptides, *Infect. Immun.* **66**:3190–3197.

Dannenberg, A.M., Jr., 1993, Immunopathogenesis of pulmonary tuberculosis, *Hosp. Pract.* **28**:51–58.

Darji, A., Guzman, C.A., Gerstel, B., Wachholz, P., Timmis, K.N., Wehland, J., Chakraborty, T., and Weiss, S., 1997, Oral somatic transgene vaccination using attenuated *S. typhimurium*, *Cell* **91**:765–775.

Daugelat, S., Gulle, H., Schoel, B., and Kaufmann, S.H., 1992, Secreted antigens of *Mycobacterium tuberculosis*: characterization with T lymphocytes from patients and contacts after two-dimensional separation, *J. Infect. Dis.* **166**:186–190.

Di Fabio, S., Medaglini, D., Rush, C.M., Corrias, F., Panzini, G.L., Pace, M., Verani, P., Pozzi, G., and Titti, F., 1998, Vaginal immunization of Cynomolgus monkeys with *Streptococcus gordonii* expressing HIV-1 and HPV 16 antigens, *Vaccine* **16**:485–492.

Dorman, C.J., Chatfield, S., Higgins, C.F., Hayward, C., and Dougan, G., 1989, Characterization of porin and *ompR* mutants of a virulent strain of *Salmonella typhimurium*: *ompR* mutants are *attenuated in vivo*, *Infect. Immun.* **57**:2136–2140.

Dunstan, S.J., Simmons, C.P., and Strugnell, R.A., 1998, Comparison of the abilities of different attenuated *Salmonella typhimurium* strains to elicit humoral immune responses against a heterologous antigen, *Infect Immun* **66**:732–740.

Dusek, D.M., Progulske-Fox, A., and Brown, T.A., 1994, Systemic and mucosal immune responses in mice orally immunized with avirulent *Salmonella typhimurium* expressing a cloned *Porphyromonas gingivalis* hemagglutinin, *Infect Immun* **62**:1652–1657.

Eisenstein, T.K., 1998, Intracellular pathogens: the role of antibody-mediated protection in Salmonella infection [letter], *Trends. Microbiol.* **6**:135–136.

Eisenstein, T.K., and Sultzer, B.M., 1983, Immunity to Salmonella infection, *Adv. Exp. Med. Biol.* **162**:261–296.

Elanschezhiyan, M., Karem, K.L., and Rouse, B.T., 1997, DNA vaccines-a modern gimmick or a boon to vaccinology?, *Crit. Rev. Immunol.* **17**:139–154.

Elhay, M.J., and andersen, P., 1997, Immunological requirements for a subunit vaccine against tuberculosis, *Immunol. Cell. Biol.* **75**:595–603.

Elhay, M.J., Oettinger, T., and andersen, P., 1998, Delayed-type hypersensitivity responses to ESAT-6 and MPT64 from *Mycobacterium tuberculosis* in the guinea pig, *Infect. Immun.* **66**:3454–3456.

Emoto, M., Emoto, Y., and Kaufmann, S.H., 1997, Bacille Calmette Guerin and interleukin-12 down-modulate interleukin-4-producing CD4+ NK1+ T lymphocytes, *Eur. J. Immunol.* **27**:183–188.

Falcone, V., Bassey, E., Jacobs, W., Jr., and Collins, F., 1995, The immunogenicity of recombinant *Mycobacterium smegmatis* bearing BCG genes, *Microbiology* **141**:1239–1245.

Felgner, P.L., and Rhodes, G., 1991, Gene therapeutics, *Nature* **349**:351–352.

Fenton, M.J., and Vermeulen, M.W., 1996, Immunopathology of tuberculosis: roles of macrophages and monocytes, *Infect. Immun.* **54**:683–690.

Fields, P.I., Groisman, E.A., and Heffron, F., 1989, A Salmonella locus that controls resistance to microbicidal proteins from phagocytic cells, *Science* **243**:1059–1061.

Fine, P.E.M., 1988, BCG vaccination against tuberculosis and leprosy, *Brit. Med. Bull.* **44**:691–703.

Fine, P.E.M., Sterne, J.A.C., Pönnighaus, J.M., and Rees, R.J.W., 1994, Delayed-type hypersensitivity, mycobacterial vaccines and protective immunity, *Lancet* **344**:1245–1249.

Fluckiger, U., Jones, K.F., and Fishetti, V.A., 1998, Immunoglobulins to group A streptococcal surface molecules decrease adherence to and invasion of human pharyngeal cells, *Infect. Immun.* **66**:974–979.

Forrest, B.D., LaBrooy, J.T., Beyer, L., Dearlove, C.E., and Shearman, D.J., 1991, The human humoral immune response to *Salmonella typhi* Ty21a, *J. Infect. Dis.* **163**:336–345.

Fu, Y., and Galan, J.E., 1998, The *Salmonella typhimurium* tyrosine phosphatase SptP is translocated into host cells and disrupts the actin cytoskeleton, *Mol. Microbiol.* **27**:359–368.

Galan, J.E., 1996, Molecular genetic bases of Salmonella entry into host cells, *Mol. Microbiol.* **20**:263–271.

Galan, J.E., and Bliska, J.B., 1996, Cross-talk between bacterial pathogens and their host cells, *Annu. Rev. Cell. Dev. Biol.* **12**:221–255.

Galan, J.E., and Curtiss, R.D., 1989, Virulence and vaccine potential of *phoP* mutants of *Salmonella typhimurium*, *Microb. Pathog.* **6**:433–443.

Galan, J.E., Gomez-Duarte, O.G., Losonsky, G.A., Halpern, J.L., Lauderbaugh, C.S., Kaintuck, S., Reymann, M.K., and Levine, M.M., 1997, A murine model of intranasal immunization to assess the immunogenicity of attenuated *Salmonella typhi* live vector vaccines in stimulating serum antibody responses to expressed foreign antigens, *Vaccine* **15**:700–708.

Germanier, R., 1984, in *Bacterial Vaccines*, (R. Germanier, ed.), Academic Press, Orlando, FL, pp. 137–166.

Germanier, R., and Furer, E., 1971, Immunity in experimental salmonellosis. II. Basis for the avirulence and protective capacity of *galE* mutants of *Salmonella typhimurium*, *Infect. Immun.* **4**:663–673.

Germanier, R., and Furer, E., 1975, Isolation and characterization of *galE* mutant Ty21a of *Salmonella typhi*: a candidate strain for a live oral typhoid vaccine, *J. Infect. Dis.* **141**:553–558.

Gilman, R.H., Hornick, R.B., Woodard, W.E., DuPont, H.L., Snyder, M.J., Levine, M.M., and Libonati, J.P., 1977, Evaluation of a UDP-glucose-4-epimeraseless mutant of *Salmonella typhi* as a live oral vaccine, *J. Infect. Dis.* **136**:717–723.

Glatman-Freedman, A., and Casadevall, A., 1998, Serum therapy for tuberculosis revisited: Reappraisal of the role of antibody-mediated immunity against *Mycobacterium tuberculosis*, *Clin. Microbiol. Rev.* **11**:514–532.

Gormus, B.J., Baskin, G.B., Xu, K., Bohm, R.P., Mack, P.A., Ratterree, M.S., Cho, S.N., Meyers, W.M., and Walsh, G.P., 1998, Protective immunization of monkeys with BCG or BCG plus heat-killed *Mycobacterium leprae*: clinical results, *Lepr. Rev.* **69**:6–23.

Guleria, I., Teitelbaum, R., McAdam, R.A., Kalpana, G., Jacobs, W.R., Jr., and Bloom, B.R., 1996, Auxotrophic vaccines for tuberculosis, *Nat. Med.* **2**:334–337.

Harboe, M., Nagai, S., Patarroyo, M.E., Torres, M.L., Ramirez, C., and Cruz, N., 1986, Properties of proteins MPB64, MPB70, and MPB80 of *Mycobacterium bovis* BCG, *Infect. Immun.* **52**:293–302.

Harboe, M., Oettinger, T., Wiker, H.G., Rosenkrands, I., and andersen, P., 1996, Evidence for occurrence of the ESAT-6 protein in *Mycobacterium tuberculosis* and virulent *Mycobacterium bovis* and for its absence in *Mycobacterium bovis* BCG, *Infect. Immun.* **64**:16–22.

Harding, C.V., 1996, Class I MHC presentation of exogenous antigens, *J. Clin. Immunol.* **16**:90–96.

Hariharan, M.J., Driver, D.A., Townsend, K., Brumm, D., Polo, J.M., Belli, B.A., Catton, D.J., Hsu, D., Mittelstaedt, D., McCormack, J.E., Karavodin, L., Dubensky, T.W., Jr., Chang, S.M., and Banks, T.A., 1998, DNA immunization against herpes simplex virus: enhanced efficacy using a Sindbis virus-based vector, *J. Virol.* **72**:950–958.

Hess, J., Miko, D., Catic, A., Lehmensiek, V., Russell, D.G., and Kaufmann, S.H., 1998, *Mycobacterium bovis* Bacille Calmette-Guerin strains secreting listeriolysin of *Listeria monocytogenes*, *Proc. Natl. Acad. Sci. USA* **95**:5299–5304.

Heym, B., Stavropoulos, E., Honore, N., Domenech, P., Saint-Joanis, B., Wilson, T.M., Collins, D.M., Colston, M.J., and Cole, S.T., 1997, Effects of overexpression of the alkyl hydroperoxide reductase AhpC on the virulence and isoniazid resistance of *Mycobacterium tuberculosis*, *Infect. Immun.* **65**:1395–1401.

Hohmann, E.L., Oletta, C.A., Killeen, K.P., and Miller, S.I., 1996a, *phoP*/*phoQ*-deleted *Salmonella typhi* (Ty800) is a safe and immunogenic single-dose typhoid fever vaccine in volunteers, *J. Infect. Dis.* **173**:1408–1414.

Hohmann, E.L., Oletta, C.A., and Miller, S.I., 1996b, Evaluation of a *phoP*/*phoQ*-deleted, *aroA*-deleted live oral *Salmonella typhi* vaccine strain in human volunteers, *Vaccine* **14**:19–24.

Hoiseth, S.K., and Stocker, B.A., 1981, Aromatic-dependent *Salmonella typhimurium* are nonvirulent and effective as live vaccines, *Nature* **291**:238–239.

Honda, M., Matsuo, K., Nakasone, T., Okamoto, Y., Yoshizaki, H., Kitamura, K., Sugiura, W., Watanabe, K., Fukushima, Y., Haga, S., Katsura, Y., Tasaka, H., Komuro, K., Yamada, T., Asano, T., Yamazaki, A., and Yamazaki, S., 1995, Protective immune responses induced by secretion of a chimeric soluble protein from a recombinant *Mycobacterium bovis* bacillus Calmette-Guerin vector candidate vaccine for human immunodeficiency virus type 1 in small animals, *Proc. Natl. Acad. Sci. USA* **92**:10693–10697.

Hone, D.M., Harris, A.M., Chatfield, S., Dougan, G., and Levine, M.M., 1991, Construction of genetically defined double *aro* mutants of *Salmonella typhi*, *Vaccine* **9**:810–816.

Hopkins, S.A., and Kraehenbuhl, J.P., 1997, Dendritic cells of the murine Peyer's patches colocalize with *Salmonella typhimurium* avirulent mutants in the subepithelial dome, *Adv. Exp. Med. Biol.* **417**:105–109.

Hornick, R.B., Greisman, S.E., Woodward, T.E., DuPont, H.L., Dawkins, A.T., and Snyder, M.J., 1970, Typhoid fever: pathogenesis and immunologic control. 2, *N. Engl. J. Med.* **283**:739–746.

Horwitz, M.A., Lee, B.W., Dillon, B.J., and Harth, G., 1995, Protective immunity against tuberculosis induced by vaccination with major extracellular proteins of *Mycobacterium tuberculosis*, *Proc. Natl. Acad. Sci. USA* **92**:1530–1534.

Hostoffer, R.W., Krukovets, I., and Berger, M., 1994, Enhancement by tumor necrosis factor-alpha of Fc alpha receptor expression and IgA-mediated superoxide generation and killing of *Pseudomonas aeruginosa* by polymorphonuclear leukocytes, *J. Infect. Dis.* **170**:82–87.

Jacobs, W.R., Jr., Kalpana, G.V., Cirillo, J.D., Pascopella, L., Udani, R.A., Jones, W.D., Jr., Barletta, R.G., and Bloom, B.R., 1991, Genetic systems for the mycobacteria, *Meth. Enzymol.* **204**:537–555.

Jondal, M., Schirmbeck, R., and Reimann, J., 1996, MHC class I-restricted CTL responses to exogenous antigens, *Immunity* **5**:295–302.

Jones, B.D., and Falkow, S., 1996, Salmonellosis: host immune responses and bacterial virulence determinants, *Annu. Rev. Immunol.* **14**:533–561.

Jones, P.W., Dougan, G., Hayward, C., Mackensie, N., Collins, P., and Chatfield, S.N., 1991, Oral vaccination of calves against experimental salmonellosis using a double *aro* mutant of *Salmonella typhimurium*, *Vaccine* **9**:29–34.

Joo, I., 1971, *Scientific Publication # 226*, World Health Organization.

Kang, B.K., and Schlesinger, L.S., 1998, Characterization of mannose receptor-dependent phagocytosis mediated by *Mycobacterium tuberculosis* lipoarabinomannan, *Infect. Immun.* **66**:2769–2777.

Kantele, A., Arvilommi, H., Kantele, J.M., Rintala, L., and Makela, P.H., 1991, Comparison of the human immune response to live oral, killed oral or killed parenteral *Salmonella typhi* Ty21a vaccines, *Microb. Pathog.* **10**:117–126.

Kim, D.T., Mitchell, D.J., Brockstedt, D.G., Fong, L., Nolan, G.P., Fathman, C.G., Engleman, E.G., and Rothbard, J.B., 1997, Introduction of soluble proteins into the MHC class I pathway by conjugation to an HIV *tat* peptide, *J. Immunol.* **159**:1666–1668.

Kremer, L., Riveau, G., Baulard, A., Capron, A., and Locht, C., 1996, Neutralizing antibody responses elicited in mice immunized with recombinant bacillus Calmette-Guerin producing the *Schistosoma mansoni* glutathione S-transferase, *J. Immunol.* **156**:4309–4317.

Lagranderie, M., Balazue, A.M., Gicquel, B., and Gheorghiu, M., 1997, Oral immunization with recombinant *Mycobacterium bovis* BCG simian immunodeficiency virus *nef* induces local and systemic cytotoxic T-lymphocytes responses in mice, *J. Virol.* **71**:2303–2309.

Lagranderie, M.R., Balazue, A.M., Deriaud, E., Leclerc, C.D., and Gheorghiu, M., 1996, Comparison of immune responses of mice immunized with five different *Mycobacterium bovis* BCG vaccine strains, *Infect. Immun.* **64**:1–9.

Lang, F., Peyrat, M.A., Constant, P., Davodeau, F., David-Ameline, J., Poquet, Y., Vie, H., Fournie, J.J., and Bonneville, M., 1995, Early activation of human Vγ9Vδ2 T cell broad cytotoxicity and TNF production by nonpeptidic mycobacterial ligands, *J. Immunol.* **154**:5986–5994.

Langermann, S., Palaszynski, S., Sadziene, A., Stover, C.K., and Koenig, S., 1994, Systemic and mucosal immunity induced by BCG vector expressing outer-surface protein A of *Borrelia burgdorferi*, *Nature* **372**:552–555.

Launois, P., DeLeys, R., N'Diaye Niang, M., Drowart, A., Adrian, M., Dierckx, P., Cartel, J.L., Sarthou, J.L., Van Vooren, J.P., and Huygen, K., 1994, T-cell-epitope mapping of the major secreted mycobacterial antigen Ag85A in tuberculosis and leprosy, *Infect. Immun.* **62**:3679–3687.

Levine, M.M., DuPont, H.L., Hornick, R.B., Snyder, M.J., Woodward, W., Gilman, R.H., and Libonati, J.P., 1976, Attenuated, streptomycin-dependent *Salmonella typhi* oral vaccine: potential deleterious effects of lyophilization, *J. Infect. Dis.* **133**:424–429.

Levine, M.M., Galen, J., Barry, E., Noriega, F., Chatfield, S., Sztein, M., Dougan, G., and Tacket, C., 1996, Attenuated *Salmonella* as live oral vaccines against typhoid fever and as live vectors, *J. Biotechnol.* **44**:193–196.

Li, Z., Kelley, C., Collins, F., Rouse, D., and Morris, S., 1998, Expression of *katG* in *Mycobacterium tuberculosis* is associated with its growth and persistence in mice and guinea pigs, *J. Infect. Dis.* **177**:1030–1035.

Lockman, H.A., and Curtiss, R.D., 1990, Occurrence of secondary attenuating mutations in avirulent *Salmonella typhimurium* vaccine strains, *J. Infect. Dis.* **162**:1397–1400.

Londono, L.P., Chatfield, S., Tindle, R.W., Herd, K., Gao, X.M., Frazer, I., and Dougan, G., 1996, Immunisation of mice using *Salmonella typhimurium* expressing human papillomavirus type 16 E7 epitopes inserted into hepatitis B virus core antigen, *Vaccine* **14**:545–552.

Lowrie, D.B., Silva, C.L., Colston, M.J., Ragno, S., and Tascon, R.E., 1997, Protection against tuberculosis by a plasmid DNA vaccine, *Vaccine* **15**:834–838.

Lu, M.C., Lien, M.H., Becker, R.E., Heine, H.C., Buggs, A.M., Lipovsek, D., Gupta, R., Robbins, P.W., Grosskinsky, C.M., Hubbard, S.C., and Young, R.A., 1987, Genes for immunodominant protein antigens are highly homologous in *Mycobacterium tuberculosis*, *Mycobacterium africanum*, and the vaccine strain *Mycobacterium bovis* BCG, *Infect. Immun.* **55**:2378–2382.

Mackaness, G.B., 1967, The relationship of delayed hypersensitivity to acquired cellular resistance, *Br. Med. Bull.* **23**:52–54.

Mackaness, G.B., and Blanden, R.V., 1967, Cellular immunity, *Prog. Allergy* **11**:89–140.

Mahairas, G.G., Sabo, P.J., Hickey, M.J., Singh, D.C., and Stover, C.K., 1996, Molecular analysis of genetic differences between *Mycobacterium bovis* BCG and virulent *M. bovis*, *J. Bacteriol.* **178**:1274–1282.

Manickan, E., Karem, K.L., and Rouse, B.T., 1997, DNA vaccines—a modern gimmick or a boon to vaccinology?, *Crit. Rev. Immunol.* **17**:139–154.

Mastroeni, P., Villarreal-Ramos, B., and Hormaeche, C.E., 1993, Adoptive transfer of immunity to oral challenge with virulent salmonellae in innately susceptible BALB/c mice requires both immune serum and T cells, *Infect. Immun.* **61**:3981–3984.

Matsumoto, S., Yanagi, T., Ohara, N., Wada, N., Kanbara, H., and Yamada, T., 1996, Stable expression and secretion of the B-cell epitope of rodent malaria from *Mycobacterium bovis BCG* and induction of long-lasting humoral response in mouse, *Vaccine* **14**:54–60.

McAdam, R.A., Weisbrod, T.R., Martin, J., Scuderi, J.D., Brown, A.M., Cirillo, J.D., Bloom, B.R., and Jacobs, W.R., Jr., 1995, *In vivo* growth characteristics of leucine and methionine auxotrophic mutants of *Mycobacterium bovis* BCG generated by transposon mutagenesis, *Infect. Immun.* **63**:1004–1012.

McAleer, W.J., Buynak, E.B., Maigetter, R.Z., Wampler, D.E., Miller, W.J., and Hilleman, M.R., 1992, Human hepatitis B vaccine from recombinant yeast. 1984 [classical article], *Biotechnology* **24**:500–502.

McGregor, D.D., Koster, F.T., and Mackaness, G.B., 1970, The short lived small lymphocyte as a mediator of cellular immunity, *Nature* **228**:855–856.

Mel, D.M., Arsic, B.L., Radovanovic, M.L., Kaljalovic, R., and Litvinjenko, S., 1974, Safety tests in adults and children with live oral typhoid vaccine, *Acta. Microbiol. Acad. Sci. Hung.* **21**:161–166.

Michetti, P., Mahan, M.J., Slauch, J.M., Mekalanos, J.J., and Neutra, M.R., 1992, Monoclonal secretory immunoglobulin A protects mice against oral challenge with the invasive pathogen *Salmonella typhimurium*, *Infect. Immun.* **60**:1786–1792.

Michetti, P., Porta, N., Mahan, M.J., Slauch, J.M., Mekalanos, J.J., Blum, A.L., Kraehenbuhl, J.P., and Neutra, M.R., 1994, Monoclonal immunoglobulin A prevents adherence and invasion of polarized epithelial cell monolayers by *Salmonella typhimurium* [see comments], *Gastroenterology* **107**:915–923.

Miller, S.I., Kukral, A.M., and Mekalanos, J.J., 1989, A two-component regulatory system (PhoP PhoQ) controls *Salmonella typhimurium* virulence, *Proc. Natl. Acad. Sci. USA* **86**:5054–5058.

Mills, S.D., Ruschkowski, S.R., Stein, M.A., and Finlay, B.B., 1998, Trafficking of porin-deficient *Salmonella typhimurium* mutants inside HeLa cells: *ompR* and *envZ* mutants are defective for the formation of *Salmonella*-induced filaments, *Infect. Immun.* **66**:1806–1811.

Mosmann, T.R., and Coffman, R.L., 1989, TH1 and TH2 cells: different patterns of lymphokine secretion lead to different functional properties, *Annu. Rev. Immunol.* **7**:145–173.

Mosser, D.M., 1994, Receptors on phagocytic cells involved in microbial recognition, *Immunol. Ser.* **60**:99–114.

Munk, M.E., and Emoto, M., 1995, Functions of T-cell subsets and cytokines in mycobacterial infections, *Eur. Respir. J.* **Suppl. 20**:668S-675S.

Murali-Krishna, K., Altman, J.D., Suresh, M., Sourdive, D.J., Zajac, A.J., Miller, J.D., Slansky, J., and Ahmed, R., 1998, Counting antigen-specific CD8 T cells: a reevaluation of bystander activation during viral infection, *Immunity* **8**:177–187.

Mustafa, A.S., 1988, Identification of T-cell-activating recombinant antigens shared among three candidate antileprosy vaccines, killed *M. leprae*, *M. bovis* BCG, and *Mycobacterium w*, *Int. J. Lepr. Other Mycobact. Dis.* **56**:265–273.

Nardelli-Haefliger, D., Kraehenbuhl, J.P., Curtiss, R., 3rd, Schodel, F., Potts, A., Kelly, S., and De Grandi, P., 1996, Oral and rectal immunization of adult female volunteers with a recombinant attenuated *Salmonella typhi* vaccine strain, *Infect. Immun.* **64**:5219–5224.

Nau, G.J., Guilfoile, P., Chupp, G.L., Berman, J.S., Kim, S.J., Kornfeld, H., and Young, R.A., 1997, A chemoattractant cytokine associated with granulomas in tuberculosis and silicosis, *Proc. Natl. Acad. Sci. USA* **94**:6414–6419.

Noordeen, S.K., 1991, A look at world leprosy, *Lep. Rev.* **62**:72–86.

O'Callaghan, D., Maskell, D., Tite, J., and Dougan, G., 1990, Immune responses in BALB/c mice following immunization with aromatic compound or purine-dependent *Salmonella typhimurium* strains, *Immunology* **69**:184–189.

Orme, I.M., anderson, P., and Boom, W.H., 1993a, T cell response to *Mycobacterium tuberculosis*, *J. Infect. Dis.* **167**:1481–1497.

Orme, I.M., Roberts, A.D., Griffin, J.P., and Abrams, J.S., 1993b, Cytokine secretion by CD4 T lymphocytes acquired in response to *Mycobacterium tuberculosis* infection, *J. Immunol.* **151**:518–525.

Patarroyo, M.E., Parra, C.A., Pinilla, C., del Portillo, P., Torres, M.L., Clavijo, P., Salazar, L.M., and Jimenez, C., 1986, Immunogenic synthetic peptides against mycobacteria of potential immunodiagnostic and immunoprophylactic value, *Lepr. Rev.* **57 Suppl 2**:163–168.

Pavelka, M.S., Jr., and Jacobs, W.R., Jr., 1996, Biosynthesis of diaminopimelate, the precursor of lysine and a component of peptidoglycan, is an essential function of *Mycobacterium smegmatis*, *J. Bacteriol.* **178**:6496–6507.

Pelicic, V., Jackson, M., Reyrat, J.M., Jacobs, W.R., Jr., Gicquel, B., and Guilhot, C., 1997, Efficient allelic exchange and transposon mutagenesis in *Mycobacterium tuberculosis*, *Proc. Natl. Acad. Sci. USA* **94**:10955–10960.

.Pelicic, V., Reyrat, J.M., and Gicquel, B., 1998, Genetic advances for studying *Mycobacterium tuberculosis* pathogenicity, *Mol. Microbiol.* **28**:413–420.

Philipp, W.J., Nair, S., Guglielmi, G., Lagranderie, M., Gicquel, B., and Cole, S.T., 1996, Physical mapping of *Mycobacterium bovis* BCG pasteur reveals differences from the genome map of *Mycobacterium tuberculosis* H37Rv and from *M. bovis*, *Microbiol.* **142**:3135–3145.

Plotkin, S.A., and Bouveret-Le Cam, N., 1995, A new typhoid vaccine composed of the Vi capsular polysaccharide, *Arch. Intern. Med.* **155**:2293–2299.

Plum, G., Brenden, M., Clark-Curtiss, J.E., and Pulverer, G., 1997, Cloning, sequencing, and expression of the *mig* gene of *Mycobacterium avium*, which codes for a secreted macrophage-induced protein, *Infect. Immun.* **65**:4548–4557.

Plum, G., and Clark-Curtiss, J.E., 1994, Induction of *Mycobacterium avium* gene expression following phagocytosis by human macrophages, *Infect. Immun.* **62**:476–483.

Polyak, S., Chen, H., Hirsch, D., George, I., Hershberg, R., and Sperber, K., 1997, Impaired class II expression and antigen uptake in monocytic cells after HIV-1 infection, *J. Immunol.* **159**:2177–2188.

Reitman, M., 1967, Infectivity and antigenicity of streptomycin-dependent *Salmonella typhosa*, *J. Infect. Dis.* **117**:101–107.

Rescigno, M., Citterio, S., Thery, C., Rittig, M., Medaglini, D., Pozzi, G., Amigorena, S., and Ricciardi-Castagnoli, P., 1998, Bacteria-induced neo-biosynthesis, stabilization, and surface expression of functional class I molecules in mouse dendritic cells, *Proc. Natl. Acad. Sci. USA* **95**:5229–5234.

Reyrat, J.M., Lopez-Ramirez, G., Ofredo, C., Gicquel, B., and Winter, N., 1996, Urease activity does not contribute dramatically to persistence of *Mycobacterium bovis* bacillus Calmette-Guerin, *Infect. Immun.* **64**:3934–3936.

Robbe-Saule, V., Coynault, C., and Norel, F., 1995, The live oral typhoid vaccine Ty21a is a *rpoS* mutant and is susceptible to various environmental stresses, *FEMS Microbiol. Lett.* **126**:171–1716.

Roberts, M., Li, J., Bacon, A., and Chatfield, S., 1998, Oral vaccination against tetanus: comparison of the immunogenicities of *Salmonella* strains expressing fragment C from the *nirB* and *htrA* promoters, *Infect. Immun.* **66**:3080–3087.

Roche, P.W., Peake, P.W., Billman-Jacobe, H., Doran, T., and Britton, W.J., 1994, T-cell determinants and antibody binding sites on the major mycobacterial secretory protein MPB59 of *Mycobacterium bovis*, *Infect. Immun.* **62**:5319–5326.

Rouse, D.A., DeVito, J.A., Li, Z., Byer, H., and Morris, S.L., 1996, Site-directed mutagenesis of the *katG* gene of *Mycobacterium tuberculosis*: effects on catalase-peroxidase activities and isoniazid resistance, *Mol. Microbiol.* **22**:583–592.

Russmann, H., Shams, H., Poblete, F., Fu, Y., Galan, J.E., and Donis, R.O., 1998, Delivery of epitopes by the salmonella type III secretion system for vaccine development, *Science* **281**:565–568.

Schuller, S., Kugler, S., and Goebel, W., 1998, Suppression of major histocompatibility complex class I and class II gene expression in *Listeria monocytogenes*-infected murine macrophages, *FEMS Immunol. Med. Microbiol.* **20**:289–299.

Sherman, D.R., Mdluli, K., Hickey, M.J., Arain, T.M., Morris, S.L., Barry, C.E., 3rd, and Stover, C.K., 1996, Compensatory *ahpC* gene expression in isoniazid-resistant *Mycobacterium tuberculosis*, *Science* **272**:1641–1643.

Sieling, P.A., Chatterjee, D., Porcelli, S.A., Prigozy, T.I., Mazzaccaro, R.J., Soriano, T., Bloom, B.R., Brenner, M.B., Kronenberg, M., Brennan, P.J., and Modlin, R.L., 1995, CD1-restricted T cell recognition of microbial lipoglycan antigens, *Science* **269**:227–230.

Siler, J.F., 1941, *Immunization to typhoid fever*, Johns Hopkins University Press.

Sirard, J.C., Fayolle, C., de Chastellier, C., Mock, M., Leclerc, C., and Berche, P., 1997, Intra-cytoplasmic delivery of listeriolysin O by a vaccinal strain of *Bacillus anthracis* induces CD8-mediated protection against *Listeria monocytogenes*, *J. Immunol.* **159**:4435–4443.

Slutsker, L., Altekruse, S.F., and Swerdlow, D.L., 1998, Foodborne diseases. Emerging pathogens and trends, *Infect. Dis. Clin. North. Am.* **12**:199–216.

Stanford, J.L., and Stanford, C.A., 1994, Immunotherapy of tuberculosis with *Mycobacterium vaccae* NCTC 11659, *Immunobiology* **191**:555–563.

Stenger, S., Mazzaccaro, R.J., Uyemura, K., Cho, S., Barnes, P.F., Rosat, J.P., Sette, A., Brenner, M.B., Porcelli, S.A., Bloom, B.R., and Modlin, R.L., 1997, Differential effects of cytolytic T cell subsets on intracellular infection, *Science* **276**:1684–1687.

Stocker, B.A., Hoiseth, S.K., and Smith, B.P., 1983, Aromatic-dependent "*Salmonella* sp." as live vaccine in mice and calves, *Dev. Biol. Stand.* **53**:47–54.

Stover, C.K., de la Cruz, V.F., Fuerst, T.R., Burlein, J.E., Benson, L.A., Bennet, L.T., Bansal, G.P., Young, J.F., Lee, M.H., Hatfull, G.F., Snapper, S.B., Barletta, R.G., Jacobs, W.R., Jr., and Bloom, B.R., 1991, New use of BCG for recombinant vaccines, *Nature* **351**:456–460.

Sturgill-Koszycki, S., Schlesinger, P.H., Chakraborty, P., Haddix, P.L., Collins, H.L., Fok, A.K., Allen, R.D., Gluck, S.L., Heuser, J., and Russell, D.G., 1994, Lack of acidification in *Mycobacterium* phagosomes produced by exclusion of the vesicular proton-ATPase, *Science* **263**:678–681.

Sugita, M., Moody, D.B., Jackman, R.M., Grant, E.P., Rosat, J.P., Behar, S.M., Peters, P.J., Porcelli, S.A., and Brenner, M.B., 1998, CD1–a new paradigm for antigen presentation and T cell activation, *Clin. Immunol. Immunopathol.* **87**:8–14.

Sztein, M.B., Tanner, M.K., Polotsky, Y., Orenstein, J.M., and Levine, M.M., 1995, Cytotoxic T lymphocytes after oral immunization with attenuated vaccine strains of *Salmonella typhi* in humans, *J. Immunol.* **155**:3987–3993.

Tacket, C.O., Hone, D.M., Curtiss, R.d., Kelly, S.M., Losonsky, G., Guers, L., Harris, A.M., Edelman, R., and Levine, M.M., 1992a, Comparison of the safety and immunogenicity of delta *aroC* delta *aroD* and delta *cya* delta *crp Salmonella typhi* strains in adult volunteers, *Infect. Immun.* **60**:536–541.

Tacket, C.O., Hone, D.M., Losonsky, G.A., Guers, L., Edelman, R., and Levine, M.M., 1992b, Clinical acceptability and immunogenicity of CVD 908 *Salmonella typhi* vaccine strain, *Vaccine* **10**:443–446.

Tacket, C.O., Kelly, S.M., Schodel, F., Losonsky, G., Nataro, J.P., Edelman, R., Levine, M.M., and Curtiss, R., 3rd, 1997, Safety and immunogenicity in humans of an attenuated *Salmonella typhi* vaccine vector strain expressing plasmid-encoded hepatitis B antigens stabilized by the Asd-balanced lethal vector system [published erratum appears in *Infect Immun* 1997 Sep;65(9):3981], *Infect. Immun.* **65**:3381–3385.

Tapa, S., and Cvjetanovic, B., 1975, Controlled field trial on the effectiveness of one and two doses of acetone-inactivated and dried typhoid vaccine, *Bull. World Health Organ.* **52**:75–80.

Tascon, R.E., Stavropoulos, E., Lukacs, K.V., and Colston, M.J., 1998, Protection against *Mycobacterium tuberculosis* infection by CD8+ T cells requires the production of gamma interferon, *Infect. Immun.* **66**:830–834.

Tite, J.P., Gao, X.M., Hughes-Jenkins, C.M., Lipscombe, M., O'Callaghan, D., Dougan, G., and Liew, F.Y., 1990, Anti-viral immunity induced by recombinant nucleoprotein of influenza A virus. III. Delivery of recombinant nucleoprotein to the immune system using attenuated *Salmonella typhimurium* as a live carrier, *Immunology* **70**:540–546.

Tsukada, H., Kawamura, I., Arakawa, M., Nomoto, K., and Mitsuyama, M., 1991, Dissociated development of T cells mediating delayed-type hypersensitivity and protective T cells against *Listeria monocytogenes and* their functional difference in lymphokine production, *Infect Immun* **59**:3589–3595.

Ulmer, J.B., 1996, DNA vaccines, *Curr. Opin. Immunol.* **8**:531–536.

VanHeyningen, T.K., Collins, H.L., and Russell, D.G., 1997, IL-6 produced by macrophages infected with *Mycobacterium* species suppresses T cell responses, *J. Immunol.* **158**: 330–337.

Verjans, G.M., Janssen, R., UytdeHaag, F.G., van Doornik, C.E., and Tommassen, J., 1995, Intracellular processing and presentation of T cell epitopes, expressed by recombinant *Escherichia coli* and *Salmonella typhimurium*, to human T cells, *Eur. J. Immunol.* **25**:405–410.

Verma, N.K., Ziegler, H.K., Wilson, M., Khan, M., Safley, S., Stocker, B.A., and Schoolnik, G.K., 1995, Delivery of class I and class II MHC-restricted T-cell epitopes of listeriolysin of *Listeria monocytogenes* by attenuated Salmonella, *Vaccine* **13**:142–150.

Via, L.E., Fratti, R.A., McFalone, M., Pagan-Ramos, E., Deretic, D., and Deretic, V., 1998, Effects of cytokines on mycobacterial phagosome maturation, *J. Cell Sci.* **111**:897–905.

Villarreal-Ramos, B., Manser, J., Collins, R.A., Dougan, G., Chatfield, S.N., and Howard, C.J., 1998, Immune responses in calves immunised orally or subcutaneously with a live *Salmonella typhimurium aro* vaccine, *Vaccine* **16**:45–54.

von Reyn, C.F., Arbeit, R.D., Yeaman, G., Waddell, R.D., Marsh, B.J., Morin, P., Modlin, J.F., and Remold, H.G., 1997, Immunization of healthy adult subjects in the United States with inactivated *Mycobacterium vaccae* administered in a three-dose series, *Clin. Infect. Dis.* **24**:843–848.

Vordemeier, H.M., Harris, D.P., Roman, E., Lathigra, R., Moreno, C., and Ivanyi, J., 1991, Identification of T cell stimulatory peptides from the 38 kDa protein of *Mycobacterium tuberculosis*, *J. Immunol.* **147**:1023–1029.

Walsh, G.P., Tan, E.V., dela Cruz, E.C., Abalos, R.M., Villahermosa, L.G., Young, L.J., Cellona, R.V., Nazareno, J.B., and Horwitz, M.A., 1996, The Philippine cynomolgus monkey (*Macaca fasicularis*) provides a new nonhuman primate model of tuberculosis that resembles human disease, *Nat. Med.* **2**:430–436.

Weiskirch, L.M., and Paterson, Y., 1997, *Listeria monocytogenes*: a potent vaccine vector for neoplastic and infectious disease, *Immunol. Rev.* **158**:159–169.

White, D.W., Wilson, R.L., and Harty, J.T., 1996, $CD8^+$ T cells in intracellular bacterial infections of mice, *Res. Immunol.* **147**:519–524.

Wilson, T.M., de Lisle, G.W., and Collins, D.M., 1995, Effect of *inhA* and *katG* on isoniazid resistance and virulence of *Mycobacterium bovis*, *Mol. Microbiol.* **15**:1009–1015.

York, I.A., Roop, C., andrews, D.W., Riddell, S.R., Graham, F.L., and Johnson, D.C., 1994, A cytosolic herpes simplex virus protein inhibits antigen presentation to CD8+ T lymphocytes, *Cell* **77**:525–535.

Young, D.B., and Garbe, T.R., 1991, Lipoprotein antigens of *Mycobacterium tuberculosis*, *Res. Microbiol.* **142**:55–65.

Zhang, X., Kelly, S.M., Bollen, W.S., and Curtiss, R., 3rd, 1997, Characterization and immunogenicity of *Salmonella typhimurium* SL1344 and UK-1 delta *crp* and delta *cdt* deletion mutants, *Infect. Immun.* **65**:5381–5387.

Zhang, Y., Heym, B., Allen, B., Young, D., and Cole, S., 1992, The catalase-peroxidase gene and isoniazid resistance of *Mycobacterium tuberculosis*, *Nature* **358**:591–593.

Zhu, Y.D., Fennelly, G., Miller, C., Tarara, R., Saxe, I., Bloom, B., and McChesney, M., 1997, Recombinant bacille Calmette-Guerin expressing the measles virus nucleoprotein protects infant rhesus macaques from measles virus pneumonia, *J. Infect. Dis.* **176**:1445–1453.

Chapter 23

Identification and *in Situ* Detection of Intracellular Bacteria in the Environment

Bettina C. Brand, Rudolf I. Amann, Michael Steinert, Dorothee Grimm, and Jörg Hacker

1. INTRACELLULAR MICROORGANISMS IN THE ENVIRONMENT

Today it is generally accepted that our knowledge of bacterial diversity in the environment has been severely limited by the need to obtain pure cultures prior to characterization by testing for multiple physiological and biochemical properties. In addition, the morphology of microorganisms is in general too simple to serve as a basis for a reliable and proper classification; only in rare cases does it allow the *in situ* identification of individual population members by microscopy (Woese, 1987). Viable plate count or most probable-number techniques have been used for quantification of active cells in different environments but are always selective and can

---

BETTINA C. BRAND, MICHAEL STEINERT, DOROTHEE GRIMM, and JÖRG HACKER Institute for Molecular Biology of Infectious Diseases, University of Würzburg, D-97070 Würzburg, Germany. **RUDOLF I. AMANN** Max-Planck-Institute for Marine Microbiology, Dept. Molecular Ecology, D-28359 Bremen, Germany.

Subcellular Biochemistry, Volume 33: Bacterial Invasion into Eukaryotic Cells, edited by Oelschlaeger and Hacker. Kluwer Academic / Plenum Publishers, New York, 2000.

therefore not yield sufficient documentation of the true community structure (Table 1). For aquatic habitats as well as soils and sediments it has been frequently reported that direct microscopic counts exceed viable-cell counts by several orders of magnitude (Torsvik *et al.*, 1990; Ferguson *et al.*, 1984; Jones, 1977). This phenomenon is known as the "great plate count anomaly" described by Staley and Konopka (1985). Any estimation of the numbers of bacteria in the environment, whether they are pathogens, indicator organisms or genetically modified microorganisms, must allow for the fact that a proportion of the target organisms have entered the non-culturable but viable fraction of the microbial population. This accounts especially for bacterial endosymbionts colonizing free-living and parasitic protozoa although the roles such endosymbionts play in host survial, infectivity, and invasiveness are unclear (Fritsche *et al.*, 1993).

Consequently, in the last decade, several attempts have been made to establish direct methods for the analysis of the bacterial community. Immunofluorescence microscopy has been widely applied to the detection and enumeration of particular microorganisms when conventional techniques have proved difficult (Chantler and McIllmurray, 1988). Fluorochromes can be coupled to an antibody that binds directly with a target antigen on the cell or to a second antibody that recognizes an antibody produced against the microorganisms. Successful application of fluorescent antibodies can be affected by a range of factors, including specificity, cross reactivity, non-specific staining, accessibility (particularly when microorganisms are located within host cells), expression of the antigen encoding genes, stability of the antigen under environmental conditions and the inability of this technique to distinguish between viable and non-viable cells.

Methodologies involving the detection of target nucleic acid sequences have been used extensively ranging from oligonucleotides to functional recombinant genes. DNA hybridization techniques (colony hybridization, solution hybridization) require either the cultivation of microorganisms or total DNA extraction. Subsequent probing for a specific trait can be used to monitor that characteristic on a presence or absence basis (Somerville *et al.*, 1989; Holben *et al.*, 1988). Another approach for identification is to use the polymerase chain reaction (PCR) to increase the relative concentration of target following extraction of total DNA including the target nucleic sequences from an environmental sample. PCR can be used to amplify DNA molecules present at essentially undetectable levels to quantities that permit detection of an identifying sequence from which the presence of the target organism may be inferred. This method was clearly improved by the development of a magnetic bead-based system for DNA isolation utilizing monodisperse beads with the aim of producing a general approach for PCR-ready DNA (Rudi *et al.*, 1997). One of the limitations of PCR is that it can be used to detect specific genetic traits but not to determine whether the trait was in its original host or had been transferred to other members of the microbial population. The appearance of PCR artifacts is a potential risk in the PCR-mediated analysis of complex microorganisms as it may suggest the existence of organisms that do not actually exist in the sample investigated: (i) depending on the quality of the DNA *in vitro* recombinations of two or more wild-type rRNA genes so-called chimeric sequences can be formed at frequencies of several percent (Wang and Wang, 1996; Kopczinsky *et al.*, 1994; Liesack *et al.*, 1991). (ii) Deletion mutants due to stable secondary structures as they are present in ribosomal RNA's (Gutell *et al.*, 1994). (iii) Point mutants due to misincorporation by DNA polymerases (Eckert and Kunkel, 1991).

One example of various powerful molecular techniques described in this chapter, fluorescently labeled rRNA-targeted nucleic acid probes, today allow an *in situ* identification of individual microbial cells in their natural habitat. The technique relies on the specific hybridization of the nucleic acid probes to the naturally amplified intracellular rRNA. The following chapters will discuss methodical aspects of this technique and its application especially the *in situ* detection of pathogenic bacteria. While extremely useful for identification, rRNA-based techniques usually do not provide the researcher with much information about *in situ* functions of bacteria within their respective ecosystems. Molecular techniques for the *in situ* visualization of mRNA (polyDIG-transcript-HRP-Fab-TSA) of key functional proteins in single microbial cells provide the opportunity to narrow this gap (Wagner *et al.*, 1998).

Table I
Percentage of Culturable Bacteria in Different Environments in Comparison with Total Cell Counts (Amann *et al.*, 1995)

Habitat	Culturable cells (%)[a]
Seawater	0.001–0.1
Freshwater	0.25
Mesotrophic lake	0.1–1
Unpolluted estuarine waters	0.1–3
Activated sludge	1–15
Sediments	0.25
Soil	0.3

[a] Culturable bacteria are measured as CFU.

2. rRNA MOLECULES: SCOPE FOR THE DETECTION OF INTRACELLULAR BACTERIA

2.1. Probes and Their Design

2.1.1. The Background

The sequencing of 16S and 23S ribosomal RNA (rRNA) molecules is currently the gold standard for the classification of microbial isolates. Comparative analyses of rRNA primary structures are for the first time in the history of microbiology facilitating the reconstruction of universal phylogenetic trees and yield a stable framework for bacterial taxonomy (Woese, 1987). Today, the 16S rRNA sequence of more than 50% of the validly described species of bacteria has been determined and can, together with many more rRNA sequences directly retrieved from the environment, be publicly accessed via two large databases (Maidak *et al.*, 1997; Van de Peer *et al.*, 1997). More than ten years ago, it has been proposed by David Stahl, Norman Pace and coworkers to use an rRNA approach for studies in microbial ecology. They suggested to study microbial diversity in a culture-independent way by direct rRNA sequence retrieval and to use nucleic acid probes complementary to rRNA or rRNA genes as tools for monitoring microbial community structure and dynamics in the environmental samples (Olsen *et al.*, 1986). It should be realized that the *in situ* identification of intracellular bacteria in the environment by rRNA based techniques is just one application of the original rRNA approach.

2.1.2. Specific Traits of rRNA

It might, therefore, be useful to outline why rRNA molecules are such unique targets for nucleic acid probes (Stahl and Amann, 1991): (i) They are functionally conserved and present in all organisms. (ii) The primary structures of 16S and 23S rRNA molecules are composed of sequence regions of higher and lower evolutionary conservation. (iii) Their natural amplification with high copy numbers of usually more than 10,000 per cell greatly increases the sensitivity of rRNA-targeted probing. This last point is also the reason for the potential of one particular assay format which will be outlined in detail later, namely the fluorescence-*in-situ*-hybridization. Only when targeted to the naturally amplified rRNA molecules the relatively insensitive fluorescently monolabeled oligonucleotides can be used for a reliable detection of individual microbial cells. The same probes fail to detect nucleic acids present at lower copy numbers, e.g., mRNA or chromosomal genes.

2.1.3. Probe Design

By referring to the rRNA databases or by comparative analysis of a newly retrieved rRNA sequence oligonucleotide probes can be designed in a directed way. Due to the patchy evolutionary conservation of rRNA primary structures the specificity of rRNA-targeted oligonucleotide probes can be tailored to the needs of the investigator reaching usually from the species to the domain (Bacteria, Archaea, Eukarya) level. It is also possible to design so-called group-specific oligonucleotides complementary to sequence regions characteristic for phylogenetic entities like genera, families or subclasses. These have been successfully used for rapid classification of bacteria (Amann *et al.*, 1995).

For the design of rRNA-targeted probes one should consider several aspects which have been reviewed in detail elsewhere (Stahl and Amann, 1991). Important is, for example, that any probe will only be as good as the sequence data base and the alignment used for its construction. Most important is of course the correctness of the target sequence. The number, position and quality of the mismatches determine the effectiveness of a probe in discrimination of target and non-target cells. One centrally located mismatch in an 18mer oligonucleotide can be sufficient. Furthermore, the location of the probe target site on the 16S or 23S rRNA secondary structure models has to be exactly determined. Both paired and unpaired regions can serve as targets. However, care should be taken that a probe is not complementary to both halves of a long helix (>4 nucleotides). This would automatically result in self-complementarity of parts of the probe sequence which could severely influence its performance in the hybridization.

Today, the probe selection should be performed in a computer-assisted way using appropriate software. Wolfgang Ludwig and coworkers from the Department of Microbiology of the Technical University München have, for example, developed ARB, a program package that includes all tools necessary for probe design (Strunck *et al.*, 1998). In ARB, aligned rRNA sequences together with higher order structure information, documentation and a phylogenetic tree are stored in a central database. The target sequence(s) are selected directly in the tree assuring phylogenetic meaningfulness. Parameters like probe length (usually 18 nucleotides) and G + C content (usually >50%) can be predefined and then the PROBE_DESIGN tool searches for potential target sites.

An ordered list is generated from which selected sites can be evaluated with the PROBE_CHECK tool against the complete database.

2.1.4. Optimizing the Hybridization Conditions

Even though theoretical probe target groups and dissociation temperatures can be calculated all newly designed probes must, nevertheless, be further tested by hybridization against nucleic acids of multiple reference organisms (e.g. "phylogrid analysis"; Devereux *et al.*, 1992; Manz *et al.*, 1992). Firstly, the dissociation temperature, the "melting point", at which 50% of the maximally bound probe is still on the target has to be determined. An optimal hybridization stringency is usually close but below this dissociation temperature. Probe specificity and sensitivity are strongly dependent on the exact hybridization conditions (Stahl and Amann, 1991). Parameters like hybridization and washing temperatures, concentrations of monovalent cations and denaturing agents (e.g., formamide) have to be carefully optimized. Secondly, the laboratory testing has to analyze additional target strains and, e.g., closely related nontarget species for which rRNA sequences are not yet available. Only if a probe works nicely with known reference organisms can it be applied with good confidence to complex environments.

2.1.5. Hybridization Assays

Whereas the temperature of dissociation from a defined DNA or RNA target site is a stable characteristic of a given probe other aspects of the detection of microorganisms in the environment are strongly dependent on the hybridization assay and the marker used to label the probe. We will here limit our discussion of various formats to the readily and inexpensively available oligonucleotides and two widely used techniques that have been successfully used to quantify defined population in complex microbial communities by hybridization with rRNA-targeted oligonucleotide probes. This is on the one hand quantitative dot blot hybridization (Stahl *et al.*, 1988), and on the other hand fluorescence-*in-situ*-hybridization (Amann *et al.*, 1990*a,b*; DeLong *et al.*, 1989). The former is based on the extraction of total nucleic acid directly from an environmental sample, its immobilization on nylon membranes (numerous samples can be treated simultaneously), followed by the quantification of a specific type of 16S rRNA relative to the total 16S rRNA detected with a universal probe. In this assay the probe is usually labeled with ^{32}P and hybridization signals are today often quantified by a Phosphorimager. The recorded signal is a composite of the abundance of a particular population and its average cellular rRNA content. It is consequently difficult to convert the relative abundance of a specific rRNA to absolute cell numbers even though estimates can be obtained.

In contrast, fluorescence-*in-situ*-hybridization has the potential to yield exact cell numbers in an environmental sample and, e.g., the exact localization of an endosymbiotic bacterium in its host. Fluorescence-*in-situ*-hybridization is based on the diffusion and specific binding of oligonucleotide probes that carry in the standard assay a single fluorochrome at the 5′ end to their intracellular targets, the 16S or 23S rRNA molecules. After an incubation time of one to several hours fluorescent-dye labeled oligonucleotide probes have hybridized specifically to the ribosomes. Fluorochromes are advantageous compared to other labels because of their superior spatial resolution and the instantaneous detectability in epifluorescence microscopes or confocal laser scanning microscopes that also allow for three-dimensional reconstruction (e.g., Amann *et al.*, 1996; Schramm *et al.*, 1996; Aβmus *et al.*, 1995; Manz *et al.*, 1995). Furthermore, the reliability of *in situ* identification can be significantly increased by the simultaneous hybridization with two or three oligonucleotide probes targeting the same population (Amann *et al.*, 1996; Amann, 1995*a*). Consequently, over the last years most whole cell hybridization studies have been performed with fluorescent oligonucleotides. Labeling of oligonucleotides with fluorescent dyes and purification has been described in detail (Amann, 1995*b*). This is today a routine exercise that is marketed in a very satisfactory manner by various biotechnology companies.

2.1.6. Methodological Aspects of Fluorescence-*in-situ*-Hybridization

Unlike regular PCR-based molecular detection fluorescence-*in-situ* hybridization identifies bacterial cells directly in their natural environment and yields important, non averaging information on localization and special associations. The combination of PCR-assisted 16S rRNA sequence retrieval and fluorescent oligonucleotide probing has been used for phylogenetic analysis and *in situ* identification of individual cells of hitherto uncultured bacterial endosymbionts of protozoa (e.g., Amann *et al.*, 1997, 1991; Springer *et al.*, 1996, 1993, 1992; Embley *et al.*, 1992*a,b*). Culturable as well as hitherto unculturable endosymbiotic bacteria could be specifically detected and enumerated in their host cells. Other recent environmental applications encompassed enumeration and analysis of spatial distribution of bacterial populations in activated sludge (e.g., Springer *et al.*, 1996, 1993, 1992; Amann *et al.*, 1996; Wagner *et al.*, 1993) and biofilms (e.g., Neef *et al.*, 1996; Schramm *et al.*, 1996; Szewzyk *et al.*, 1994; Manz *et al.*, 1993; Ramsing *et al.*, 1993; Amann *et al.*, 1992*a*).

The full potential of fluorescence-*in-situ*-hybridization can only be exploited by those who realize potential pitfalls. Important is, for example, that during hybridization the cells are exposed to elevated temperature, detergents and osmotic gradients. Thus, fixation is essential for maintaining the morphological integrity of the cells. Most but not all microorganisms are stabilized in their morphology and permeabilized for oligonucleotide probes by standard fixations like aldehydes (formalin, paraformaldehyde, glutaraldehyde) and/or alcohols (methanol, ethanol). A variety of fixatives have been evaluated. Glutaraldehyde results in considerable autofluorescence of the specimen. Autofluorescence is minimized by fixation in fresh formaldehyde solutions, e.g., a solution of 4% paraformaldehyde (Sigma, St. Louis, MO) in phosphate buffered saline (PBS). This fixation protocol was developed and optimized for gram-negative bacteria. Probe permeability of paraformaldehyde fixed cells of certain gram-positive bacteria is often limited. Here, probe penetration can be enhanced by lysozyme/EDTA treatment of fixed cells prior to hybridization (Hahn *et al.*, 1992), fixation by ethanol only (Roller *et al.*, 1994), or by short time fixation in a alcohol/formaldehyde mixture (DeLong *et al.*, 1989).

Secondly, even though the dissociation temperature should be the same independent whether an rRNA-targeted oligonucleotide probe is used *in situ* or on membranes it is important to realize that during *in situ* hybridization certain target sites might be inaccessible for probing (Amann *et al.*, 1995). This is likely caused by protein-rRNA or rRNA-rRNA interactions in the fixed ribosomes. Consequently, a probe working nicely in quantitative slot blot hybridization does not necessarily work for *in situ* hybridization.

Additionally, probe-conferred fluorescence is determined by the rRNA content of fixed cells which is again influenced by the growth rate (Wallner *et al.*, 1993; DeLong *et al.*, 1989). It has even been shown that the direct correlation between the growth rates of bacterial cells, the average ribosome contents, and the probe-conferred fluorescence can be used to estimate growth rates of individual cells *in situ* (Poulsen *et al.*, 1993). When a new probe is developed initial *in situ* hybridization experiments should therefore be performed with rapidly growing cells. When examining environmental samples one has usually not the option to activate the cells before fixation even though exactly this has recently been suggested by Ouverney *et al.* (1998). It is therefore a problem commonly encountered in oligotrophic environmental samples (especially in environments with background fluorescence; e.g. Hahn *et al.*, 1992) that hybridization signals are at or below the detection limit of a regular epifluorescence microscope or flow cytometer (Wallner *et al.*, 1993). This has prompted studies to develop more sensitive whole cell hybridization techniques.

Signal amplification by hybridization with multiple monolabeled oligonucleotides is possible (Amann *et al.*, 1990*a*), but often restricted to two- or threefold increase by the limited availability of target sites with identical specificity within the 16S and 23S rRNA molecules. By substituting the standard fluorochromes fluorescein and tetramethylrhodamine by the brighter and more photostable carbocyanine dye CY3 Glöckner and coworkers (1996) could detect significantly more bacteria in lake water samples. The authors furthermore argue that detection yields can be increased by the use of specific high quality filter sets that are more sensitive than standard rhodamine filter sets and allow for color discrimination of specific fluorochrome and background fluorescence.

Alternatively, *in situ* identification of whole fixed bacterial cells can also be achieved with digoxigenin- and enzyme-labeled oligonucleotides (Amann *et al.*, 1992*b*; Zarda *et al.*, 1991). If their detection is based on the enzymatic transformation of a suitable substrate, e.g. the nonfluorescent colored diaminobenzidine, both labels can be detected very sensitively. Recently, Schönhuber *et al.* (1997) described the application of the fluorescein-labeled tyramide as substrate for horseradish peroxidase for the ultrasensitive staining of cells via the tyramide system amplification. Applications of enzymatic assays may be hindered by limited permeability of fixed cells (especially of gram-positives) for the relatively large anti-digoxigenin-antibody or oligonucleotide-enzyme conjugate. Nevertheless, nonfluorescent enzyme-linked probe assays may be the only way to detect specific populations in environments with strong autofluorescence like plant tissues.

3. IDENTIFICATION AND *IN SITU* DETECTION OF DIFFERENT INTRACELLULAR ORGANISMS

3.1. Legionella

The environmental pathogen *Legionella pneumophila*, the etiologic agent of Legionnaires' disease, normally inhabits aquatic environments or wet soil, surviving most of the time as intracellular parasites of amoebae and ciliates (Brand and Hacker, 1996). Intracellular growth of *L. pneumophila* in trophozoites of a variety of amoebae has been shown under laboratory conditions (Kurtz *et al.*, 1982). Consequently, legionellae contained in amoebae, and especially in amoebal cysts could survive environmental temperature extremes, chlorination, and other adverse conditions. Overall, infected amoebae containing legionellae would be present in the drift from contaminated aquatic environments (Barbaree *et al.*, 1986) and provide an excellent vehicle whereby concentrated infectious particles could be delivered to humans. Development of legionellosis has been attributed to the inhalation of viable organisms in fine aerosols into the lung, where they invade and multiply within alveolar macrophages and other phagocytic cells (Jepras *et al.*, 1985; Horwitz and Silverstein, 1980). Cellbiological and molecular aspects of the life cycle of Legionella are described in detail in the chapter by Y. Abu Kwaik.

Isolation and reliable culturing of Legionella on selective media is fastidious especially because the bacterium is able to form viable but not culturable cells which cannot be cultured without previous passage through hosts cells e.g. amoebae (Steinert *et al.*, 1997; Paszko-Kolva *et al.*, 1992; Hussong *et al.*, 1987). Due to slow growth and lack of suitable phenotypic tests, identification of *Legionella* spp. remains difficult.

To examine the behaviour of *Legionella* spp. in complex microbial communities and diverse ecosystems two oligonucleotides were constructed which are complementary to the regions of 16S rRNA characteristic for *Legionellaceae*. In an evaluation the results of classical techniques employing selective media, immunofluorescence and *in situ* hybridization were in good accordance for routine environmental and clinical isolates (Manz *et al.*, 1995). *L. pneumophila* suspended in drinking water at approximately 10^4 colony forming units per ml could be rapidly detected by a combination of membrane filtration and whole-cell-hybridization. Even after an incubation for one year one third of the cells detected by phase contrast microscopy showed detectable levels of probe-conferred fluorescence. *In situ* hybridization also facilitated visualization of *Legionella* spp. in model biofilms and within infected cells of the ciliated protozoan *Tetrahymena pyriformis* (Manz *et al.*, 1995).

Among members of the family *Legionellaceae*, *L. pneumophila* is involved in more than 95% of cases of severe atypical pneumonia (Ruf *et al.*, 1990). The construction of a new 16S rRNA-targeted oligonucleotide specific for *L. pneumophila* proved useful for the differentiation from other *Legionella* and non-legionella species and for the detection of *L. pneumophila* within amoebae (Figure 1) (Grimm *et al.*, 1998). Moreover, *in situ* hybridization was shown to be an excellent tool to specifically detect *L. pneumophila* in environmental samples of diverse nature exhibiting a wide range of biological, physical and chemical conditions. An ecological study comprised 20 different habitats including a total of 31 independent samples: 93.5% of all samples contained bacteria; 77% and 22% were contaminated with *Legionella* spp. and *L. pneumophila*, respectively (Grimm, unpublished). This study demonstrated the importance of the *Legionella*-specific oligonucleotides for the recognition of natural reservoirs for disease and the monitoring of disinfection procedures in the near future.

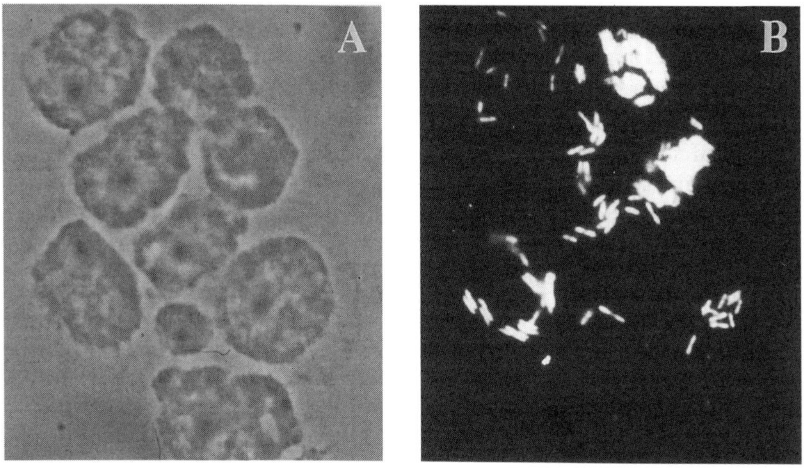

FIGURE 1. Phase contrast (A) and epifluorescence (B) micrographs of *L. pneumophila* Corby-infected *Acanthamoeba castellanii* cells 16 h postinfection. The micrograph (B) represents exposure of the sample after hybridization with the CY3-labeled probe LEGPNE1. Magnifications, ×955. Either amoebae are filled with bacteria localized in particular areas of the cells which may be the phagosomes, or bacteria are present as single cells.

3.2. Mycobacterium

Mycobacteria are nonmotile, rod-shaped, acid-alcohol-fast staining, facultative intracellular bacteria with a gram-positive type of cell wall (Wayne and Kubica, 1986). Infections like tuberculosis (TB), leprosy but also infections with MOTT bacilli (mycobacteria other than tubercle bacilli) are important sources of morbidity and mortality. After a steady decrease in the last decades, mycobacterial disease has reemerged as a serious public health problem. Increasing case rates and outbreaks of multidrug-resistant tuberculosis have been reported (CDC, 1993). In addition, the picture of nontuberculous mycobacterial disease has radically changed with the AIDS epidemic (Falkinham, 1996).

The pathogenesis of the nontuberculous mycobacteria is different from that of *M. tuberculosis*. *M. tuberculosis* bacilli are spread primarily by airborne particles (droplet nuclei) expelled by persons who have infectious TB (Wayne, 1994) while many MOTT bacilli are mostly free living saprophytes (Falkinham, 1996; Grange, 1991). Opportunistic infections with *M. avium*, *M. mariunum*, *M. kansasii*, *M. intracellulare*, *M. scrofulaceum*, *M. chelonae*, and *M. fortuitum* have been associated with various water environments including drinking water. In addition, *M. chelonae* and *M. fortuitum* are often inoculated by contaminated needles while, *M. ulcerans* most likely enters the skin by intimate contact with vegetation like spiky grasses (Falkinham, 1996; Grange, 1991). Immundeficiency is apparently the main risk factor for bacteremia and disseminated multiorgan bacterial disease with these agents. Immunocompetent individuals suffer primarily from pulmonary infections or skin infections associated with trauma (Falkinham, 1996). Recent studies suggest that the interaction of *M. avium* with environmental amoebae (*Acanthamoeba*) enhances its virulence as seen in the beige mouse model (Cirillo *et al.*, 1997). The *M. avium* bacteria replicate in association with *Acanthamoeba*, inhibit phagolysosome fusion of the host cell and may survive disinfection trials within the cyst wall of the amoebae (Steinert *et al.*, 1998*a*; Cirillo *et al.*, 1997).

The interaction of mycobacteria with host phagocytic cells is one key aspect to their pathogenesis. The interaction between *M. avium* and macrophages results in the production of immune-supressive cytokines that inhibit the effector function of T-cells, natural killer cells and macrophages (Bermudez, 1994). For *M. avium* it has been demonstrated that virulence mechanisms against macrophages include prevention of the acidification of phagocytic vesicles and the limited fusion of the phagosome with endosomal and lysosomal compartments. High catalase activity, resistance to inhibitory serum constituents and the exudation of lipids also have been suggested as virulence factors (Falkinham, 1996). Since plasmids were more common in *M. avium* complex isolates recovered from AIDS patients than isolates from immunocompetent induviduals or the environment, plasmids may contain virulence genes (Crawford and Bates, 1986).

Much progress has been accomplished in the direct detection of acid-fast bacilli (especially with *M. tuberculosis)* from clinical specimens. Gas-Liquid Chromatography/Mass Spectrometry for the detection of tuberculostearic acid, PCR with specific repeated DNA elements IS6110, the 65 kDa gene or the a 583-bp amplicon within the 16S rRNA as target sequence appear to be superior to time consuming culture procedures and an enzyme immunosorbent assay (EIA) (Salfinger and Pfyffer, 1994). Alternative amplification procedures like the strand-displacement amplification, nucleic-acid-sequence based amplification, ligase chain reaction, and Q replicase amplification are in the process of development (Salfinger and Pfyffer, 1994). Environmental analyses of water, biofilms, soil, dust and aerosols still rely on selective culture procedures. Combinations of different culture media and radiometric systems have improved the culture methodology (Peters *et al.*, 1995; Salfinger and Pfyffer, 1994). Recently, a method based on 16S rRNA gene-targeted PCR and oligonucleotide probing was developed for detecting *M. chlorophenolicum* in soil (Briglia *et al.*, 1996). More sophisticated technologies are needed for more detailed future ecologic studies of mycobacteria.

3.3. Chlamydia

Recently, tiny coccoid bacteria (diameter approximately 0.5 μm) observed as obligate intracellular parasites of acanthamoebae (Horn *et al.*, 1998) which had been isolated from human nasal mucosa (Michel *et al.*, 1994) were submitted to rRNA sequence analysis in order to enable a reliable identification and a sound classification of these organisms not cultivable on cell free media (Amann *et al.*, 1997). The parasites were released from infected host cells by freezethawing and filter-purified. The almost full length rRNA operon could be amplified by polymerase chain reaction. After sequencing the 16S and 23S rDNA sequences were comparatively analyzed. A specific probe was developed, labeled with peroxidase and successfully used for *in situ* detection of the tiny coccoid cells. This proved that the sequences were indeed retrieved from the bacterial parasite. Interestingly, the 16S and 23S rRNA of strain Bn9 showed the highest degree of relationship with the *Chlamydiaceae*-family. With 16S rRNA sequence similarities to different members of the genus *Chlamydia* between 86 to 87% this relationship is not very close. The three hitherto sequenced species of the genus *Chlamydia*, *C. pneumoniae*, *C. psittaci*, and *C. trachomatis*, have 16S rRNA similarity values between 94 and 96%. This indicates that the Bn9 parasites are likely not another species within the genus *Chlamydia*, but representatives of another genus within the family *Chlamydiaceae*. The provisional name "*Parachlamydia acanthamoebae*" has been suggested for this new species. Further *in situ* hybridization also demonstrated that a second strain, Berg17, binds the specific probe. This was additional evidence for a close relationship between the two intracellular parasites. Recently, the 16S rRNA of another chlamydia-like obligate intracellular strain of free-living amoebae, Hall's coccus, was shown to be highly similar to that of "*Parachlamydia acanthamoebae*" (Birtles *et al.*, 1997)

3.4. Rickettsia

The rRNA approach also allowed to phylogenetically analyze and identify two interesting species of Rickettsia-related obligate endosymbionts of ciliates, *Holospora obtusa* (Amann *et al.*, 1991) and *Caedibacter caryophila* (Springer *et al.*, 1993). Members of the genus Holospora live in the nuclei of their specific host cells. *H. elegans* and *H. undulata* infect the micronuclei of *Paramecium caudatum*, whereas *H. obtusa* infects the macronucleus in other strains of the same host species. Even though these bacterial endosymbionts have been microscopically detected more than a century ago they refused all attempts of cultivation. In 1991, *Holospora obtusa* was the first bacterial endosymbiont to be phylogenetically analyzed and identified *in situ* by the rRNA approach (Amann *et al.*, 1991). The 16S rRNA sequence was retrieved by PCR and cloning and shown to have the highest similarity of approximately 85% with the 16S rRNA sequences of members of the genus *Rickettsia*. Fluorescent-*in-situ*-hybridization with an oligonucleotide probe assigned the retrieved sequence to the macronucleolar endosymbionts and outruled the possibility that the sequences originated from bacteria ingested in the food vacuoles. The 16S rRNA-targeted probe constructed from the *H. obtusa* sequence also hybridized to *H. undulata* and *H. elegans* and was thereby shown to be genus-specific. Therefore, a highly variable region of the 23S rRNA of *H. obtusa* was sequenced and used for construction of a more specific probe that hybridized solely to *H. obtusa*.

In a similar way, the phylogenetic position of *Caedibacter caryophila*, a so far uncultured killer symbiont occurring in the macronuclei of certain *Paramecium caudatum* strains, was elucidated. This bacterium is toxic for susceptible strains of paramecia and may contain unusual refractile inclusions, the so-called R-body. Based on its 16S rRNA gene *C. caryophila* is a member of the alpha-subclass of Proteobacteria moderately related to *Holospora obtusa* (86% 16S rRNA similarity) and to the genus *Rickettsia*. *In situ* detection of these gram-negative cells is straightforward and easy and consequently hybridization with fluorescently labeled oligonucleotide probes would be a convenient way to screen for the environmental distribution of these and other *Rickettsia*-related endosymbionts of protozoa.

3.5. Listeria

Listeria spp. are gram-positive facultative anaerobic rods (Seeliger and Jones, 1986). These bacteria exhibit a temperature-dependent peritrichous flagellation that confers a characteristic tumbling motility when the bacteria are cultured at 20 to 25°C (Peel *et al.*, 1988). Out of 8 known species, only *L. monocytogenes* and *L. ivanovii* are pathogenic (Seeliger and Jones, 1986). Originally, listeriosis was considered a veterinary disease. Epidemiologic investigations however, demonstrated that human listeriosis is a food borne disease, most commonly associated with dairy products (Schuchat *et al.*, 1991). Infections in healthy adults show mostly mild influenza-like symptoms or are even asymptomatic and high carriage rates have been observed. In contrast, the clinical syndromes of immunocompromised individuals include central nervous system infections and bacteremia. In pregnant women, listeriosis can result in stillbirth (Farber and Peterkin, 1991).

L. monocytogenes has been isolated from soil, water, sewage sludge, and decaying vegetation, as well as from domestic animals (Schuchat *et al.*, 1991). The ability of this organism to grow in refrigerated food is a substantial problem for the food industry (Frank *et al.*, 1990; Gilbert *et al.*, 1989) and it is complicated by the fact that these bacteria have a decreased susceptibility to biocides when associated with biofilms (Ren and Frank, 1993). Trophozoites and cysts of amoebe may also hamper the control and eradication, since *in vitro* studies showed that *L. monocytogenes* is able to grow within Acathamoeba and Tetrahymena (Ly and Muller, 1990*a*, 1990*b*).

To monitor the incidence of *L. monocytogenes* in food rapid methods need to be designed. The low initial inoculum, which is difficult to detect but easily grows to a substantial life threatening dose, is the main obstacle. Therefore, the analysis is typically done after enrichment procedures (Farber and Peterkin, 1991). Monoclonal antibody-based enzyme-linked immunosorbent assays and the detection of *L. monocytogenes* by means of polymerase chain reaction amplification, followed by agarose gel electrophoresis or dot blot analysis have been widely reported (Farber and Peterkin, 1991). A specific probe based on unique regions of the 16S rRNA was used in the development of a commercial hybridization assay (Gene-Trac)(King *et al.*, 1989). In this assay a fluorescein labeled detector probe and a polydeoxyadenosine-tailed capture probe hybridize to adjacent regions of the target 16S rRNA. These complexes are then captured on dipsticks where a horseradish peroxidase-labeled anti-fluorescein antibody visualizes the hybridization. Another method detects *L. monocytogenes* bacilli after direct colony hybridization on hydrophobic grid-membrane filters by using a digoxigenin-labelled listeriolysin O DNA probe (Yan *et al.*, 1996). A completely different approach employes a recombinant derivative of a broad-host-range bacteriophage, specific for the genus *Listeria*. The luciferase reporter bacteriophage A511::*luxAB* transduces bacterial bioluminescence into infected cells (Loessner *et al.*, 1997). This system is applicabable for food and environmental samples and detects very low initial contamination rates within 24 hours. This brief overview demonstrates that the methodology for the detection of Listeria is still developing. Molecular methods that can be automated and done on a large scale by food processors seem to be the future direction.

3.6. Shigella

Shigellae, the causative agents of bacillary dysentery, are usually acquired by drinking water contaminated with feces or by person-to-person fecal transmission. The species involved in gastroenteritis are *S. dysenteriae*, *S. flexneri*, *S. boydii*, and *S. sonnei*. The culture of the organisms from stool specimens, biochemical tests and confirmatory latex agglutination are the standard methods for detection (Pillay *et al.*, 1997; Rowe and Gross, 1984). PCR inhibitors in feces lower the sensitivity of detection by means of DNA amplification. However, by combining immunomagnetic separation (IMS) with PCR sensitivity in fecal specimens can be increased. IMS captures and separates Shigella bacteria by O-antigen specific monoclonal antibodies coated on magnetic beads (Achi *et al.*, 1996).

Although the bloody diarrhea caused by Shigella is normally self-limiting, neurological symptoms or kidney failure (hemolytic uremic syndrome, HUS) sometimes follow. Malnourished infants are at highest risk for such complications. Only 200 of the gram-negative rods are required to cause infection (Salyers and Whitt, 1994). All virulent Shigella bacilli harbor a 140 MDa (220 kb) or 120 MDa plasmid which carries most of the structural genes necessary for adherence to integrins, invasion (invasion plasmid antigens, *ipa*), and cell to cell spread (intracellular spread, *icsAB*) (Menard, 1996; see also chapter 11). Various other virulence associated loci and genes that regulate plasmid encoded virulence determinants are located on the bacterial chromosome (*vacB, vacC, ompR envz, kcpA, virR, fur, stxAB*) (Saylers and Whitt, 1994; Sansonetti, 1992).

Epidemiological studies have shown that surface waters can act as a source of infection (Khan *et al.*, 1979) and that the association of Shigella and free living protozoa may provide bacteria with increased resistance to free chlorine residuals (King *et al.*, 1988). Parachamber experiments have shown that Shigella bacilli are able to grow on secreted products of *Acanthamoeba polyphaga*. Furthermore, high temperatures (37°C) in direct amoeba cocultures seem to shift the equlibrium of digestion of the bacteria versus growth of the bacteria towards bacterial multiplication (Steinert *et al.*, 1998*b*). Until recently, only routine bacteriological methods were used for detection in the environment. This is now believed to be insufficient. Monitoring in the environment requires techniques which take into account that the numbers of shigellae are usually very low. Furthermore, bacilli from several *Shigella* species have been shown to enter into a viable but nonculturable (VBNC) state under specific conditions (Rahman *et al.*, 1996). Such VBNC shigellae may still pose health problems, since they retain virulence factors and are likely to be resuscitated to a culturable state under appropriate conditions. Studies of laboratory microcosms have shown that nonculturable shigellae can be detected by PCR and fluorescent-antibody techniques (Islam *et al.*, 1993).

3.7. General Aspects

The importance of the environment as a reservoir for bacterial pathogens has been recognized for several decades. However, until recently the isolation and analysis of pure cultures has been the primary experimental approach. Important issues like the localization in the microenvironment, the physiologic activity in the environment, and interactions with other species remained unknown. Due to the progress of modern molecular technology these issues can now be examined as has been exemplified with *L. pneumophila* (Grimm *et al.*, 1998). By applying immunolabeling and episcopic differential interference contrast microscopy microcolonies of legionellae were found in aquatic biofilms (Rogers and Keevil, 1992). The introduction of reporter genes can reveal the transcriptional activity of genes under defined conditions (Abu Kwaik *et al.*, 1997) and the development of coculture systems and recombinant DNA technology has enabled the investigation of host-parasite relationships on a molecular level (Wintermeyer *et al.*, 1995). Several of these studies showed that the multiplication of Legionella within amoebae enhances infection rates in macrophages (Cirillo *et al.*, 1994). Furthermore, growth within amoebae increases the resistance of this pathogen to antimicrobial agents (Barker *et al.*, 1995). The importance of protozoa in the developement of pathogenicity and intracellular survival strategies seems not to be restricted to Legionella. Intracellular growth within protozoa may also prime other environmental bacteria for virulence (Barker and Brown, 1994). Studies on *M. avium*, *Chlamydia* ssp., and *Listeria monocytogenes* suggest this evolutionary role of protozoa (Cirillo *et al.*, 1997; Essig *et al.*, 1997; Ly and Muller, 1990*b*). Since protozoa also play an integral part in cycling of nutrients in aquatic food chains (including man-made water systems), their role in the maintenance of infectious agents may even exceed the restricted number of environmental intracellular pathogens (Steinert *et al.*, 1998*b*).

4. PERSPECTIVES

The rRNA approach together with other molecular techniques displays great potential for an analysis of microbial diversity which is not affected by the limits of pure-culture techniques. The PCR-mediated analysis of 16S rRNA is a powerful tool even there is no general guideline for a "good PCR-mediated analysis of 16S rRNA from environmental samples". Therefore, it is necessary to compare the results from different nucleic acid extractions, PCR amplification and cloning experiments. Specific oligonucleotides derived from environmental 16S rRNA sequences should be used as dye-labeled probes for 16S rRNA targeted *in situ* hybridization of fixed sample material for the quantitative analysis as single cells can be specifically detected and counted in the microscope.

With the growing 16S rRNA sequence data set, which enables more and more accurate phylogenetic affiliations and possible information of metabolic pathways from the phylogenetically nearest neighbor, improvements in *ex situ* culture techniques should become easier. In addition, revolutionary insights into ecology will be obtained by studying microorganisms in their niches. Correlations between the community composition, the spatial relationships of the different members and the function of the populations can now be determined. Applying these techniques to human diseases of suspected infectious etiology may rapidly elucidate novel candidate pathogens.

With the increasing sensitivity of *in situ* hybridization techniques, detection of low-copy-number nucleic acids will become possible. *In situ* monitoring of the lateral transfer of plasmids in natural environments and the detection of specific mRNA are just two of many possible applications.

5. REFERENCES

Abu Kwaik, Y., Gao, L.Y., Harb, O.S., and Stone, B.J., 1997, Transcriptional regulation of the macrophage-induced gene (*gspA*) of *Legionella pneumophila* and phenotypic characterization of a null mutant, *Mol. Microbiol.* **24**:629–642.

Achi, R., Mata, L., Siles, X., and Lindberg, A.A., 1996, Immunomagnetic separation and detection show shigellae to be common faecal agents in children from urban marginal communities of Costa Rica, *J. Infect.* **32**:211–218.

Amann, R.I., 1995*a*, Fluorescently labeled, rRNA-targeted oligonucleotide probes in the study of microbial ecology, *Mol. Ecol.* **4**:543–554.

Amann, R.I., 1995*b*, *In situ* identification of microorganisms by whole cell hybridization with rRNA-targeted nucleic acid probes, in: *Molecular Microbial Ecology Manual*, (A.D.L. Akkerman, J.D. van Elsas, and F.J. de Bruijn, eds.), Kluwer Academic Publishers, Dordrecht, Netherlands, 3.3.6., p. 1–15.

Amann, R.I., Binder, B.J., Olson, R.J., Chisholm, S.W., Devereux, R., and Stahl, D.A., 1990*a*. Combination of 16S rRNA-targeted oligonucleotide probes with flow cytometry for analyzing mixed microbial populations, *Appl. Environ. Microbiol.* **56**:1919–1925.

Amann, R.I., Krumholz, L., and Stahl, D.A., 1990*b*, Fluorescent-oligonucleotide probing of whole cells for determinative, phylogenetic, and environmental studies in microbiology, *J. Bacteriol.* **172**:762–770.

Amann, R., Springer, N., Ludwig, W., Görtz, H.-D., and Schleifer, K.-H., 1991, Identification *in situ* and phylogeny of uncultured bacterial endosymbionts, *Nature* (London) **351**:161–164.

Amann, R.I., Stromley, J., Devereux, R., Key, R., and Stahl, D.A., 1992*a*, Molecular and microscopic identification of sulfate-reducing bacteria in multispecies biofilms, *Appl. Environ. Microbiol.* **58**:614–623.

Amann, R., Zarda, B., Stahl, D.A., and Schleifer, K.-H., 1992*b*, Identification of individual prokaryotic cells by using enzyme-labeled, rRNA-targeted oligonucleotide probes, *Appl. Environ. Microbiol.* **58**:3007–3011.

Amann, R., Ludwig, W., and Schleifer, K.-H., 1995, Phylogenetic identification and *in situ* detection of individual microbial cells without cultivation, *Microbiol. Rev.* **59**:143–169.

Amann, R., Snaidr, J., Wagner, M., Ludwig, W., and Schleifer, K.-H., 1996, *In situ* visualization of high genetic diversity in a natural microbial community, *J. Bacteriol.* **178**:3496–3500.

Amann, R., Springer, N., Schönhuber, W., Ludwig, W., Schmidt, E.N., Müller, K.-D., and Michel, R., 1997, Obligate intracellular bacterial parasites of acanthamoebae related to *Chlamydia* spp., *Appl. Environ. Microbiol.* **63**:115–121.

Arnheim, N., and Ehrlich, H.A., 1992, Polymerase chain reaction strategy, *Annu. Rev. Biochem.* **61**:131–169.

Aβmus, B., Hutzler, P., Kirchhof, G., Amann,R., Lawrence, J.R., and Hartmann, A., 1995, *In situ* localization of *Azospirillum brasilense* in the rhizosphere of wheat using fluorescently labeled, rRNA-targeted oligonucleotide probes and scanning confocal laser microscopy, *Appl. Environ. Microbiol.* **61**:1013–1019.

Barbaree, J.M., Fields, B.S., Feeley, J.C., Gorman, G.W, and Martin, W.T., 1986, Isolation of protozoa from water associated with a legionellosis outbreak and demonstration of intracellular multiplication of *Legionella pneumophila*, *Appl. Environ. Microbiol.* **51**:422–424.

Barker, J., and Brown, M.R.W., 1994, Trojan horses of the microbial world: protozoa and the survival of bacterial pathogens in the environment, *Microbiol.* **140**:1253–1259.

Barker, J., Scaife, H., and Brown, M.R.W., 1995, Intraphagocytic growth induces an antibiotic-resistant phenotype of *Legionella pneumophila*, *Antimicrob. Agents Chemother.* **39**:2684–2688.

Bermudez, L.E., 1994, Immunobiology of *Mycobacterium avium* infection, *Eur. J. Clin. Microbiol. Infect. Dis.* **13**:1000–1006.

Birtles, R.J., Rowbotham, T.J., Storey, C., Marrie, T.J., and Raoult, D., 1997, Chlamydia-like obligate parasite of free-living amoebae, *Lancet* **349**:925–926.

Brand, B.C., and Hacker, J., 1996, The biology of *Legionella* infection., in: *Host response to intracellular pathogens*, (S.H.E. Kaufmann, ed.), R.G. Landes Company, Austin, pp. 291–312.

Briglia, M., Eggen, R.I.L., DeVos, W.M., and Van Elsas, J.D., 1996, Rapid and sensitive method for the detection of *Mycobacterium chlorophenolicum* PCP-1 in soil based on 16S rRNA gene-targeted PCR, *Appl. Environ. Microbiol.* **62**:1478–1480.

Centers for Disease Control and Prevention, 1993, Initial therapy for tuberculosis in the era of multidrug resistance, *MMWR.* **42** (RR-7): 1–8.

Chantler, S., and McIllmurray, M.B., 1988, Labeled antibody methods for detection and identification of microorganisms, *Methods in Microbiology* **19**:273–332.

Cirillo, J.D., Falkow, S., and Thomkins, L.S., 1994, Growth of *Legionella pneumophila* in *Acanthamoeba castellanii* enhances invasion, *Infect. Immun.* **62**:3254–3261.

Cirillo, J.D., Falkow, S., Tompkins, L.S., and Bermudez, L.E., 1997, Interaction of *Mycobacterium avium* with environmental amoebae enhances virulence, *Infect. Immun.* **65**:3759–3767.

Crawford, J.T., and Bates, J.H., 1986, Analysis of plasmids in *Mycobacterium avium-intracellulare* isolates from persons with acquired immunodeficiency syndrome, *Am. Resp. Dis.* **134**:659–661.

DeLong, E.F., Wickham, G.S., and Pace, N.R., 1989, Phylogenetic stains: ribosomal RNA-based probes for the identification of single microbial cells, *Science* **243**:1360–1363.

Devereux, R., Kane, M.D., Winfrey, J., and Stahl, D.A., 1992, Genus- and group-specific hybridization probes for determinative and environmental studies of sulfate-reducing bacteria, *System. Appl. Microbiol.* **15**:601–610.

Eckert, K.A., and Kunkel, T.A., 1991, DNA polymerase fidelity and the polymerase chain reaction, *PCR Methods Appl.* **1**:17–24.

Embley, T.M., Finlay, B.J., and Brown, S., 1992*a*, RNA sequence analysis shows that the symbionts in the ciliate *Metopus contortus* are polymorphs of a single methanogen species, *FEMS Microbiol. Lett.* **97**:57–62.

Embley, T.M., Finlay, B.J., Thomas, R.H., and Dyal, P.L., 1992*b*, The use of rRNA sequences and fluorescent probes to investigate the phylogenetic positions of the anaerobic ciliate *Metopus palaeformis* and its archaeobacterial endosymbiont, *J. Gen. Microbiol.* **138**:1479–1487.

Essig, A., Heinemann, M., Simnacher, U., and Marre, R., 1997, Infection of *Acanthamoeba castellanii* by *Chlamydia pneumoniae*, *Appl. Environ. Microbiol.* **63**:1396–1399.

Falkinham, J.O., 1996, Epidemiology of infection by nontuberculous Mycobacteria, *Clin. Microbiol. Rev.* **9**:177–215.

Farber, J.M., and Peterkin, P.I., 1991. *Listeria monocytogenes*, a food-borne pathogen, *Microbiol. Rev.* **55**:476–511.

Ferguson, R.L., Buckley, E.N., and Palumbo, A.V., 1984, Response of marine bacterioplankton to differential filtration an confinement, *Appl. Environ. Microbiol.* **47**:49–55.

Frank, J.F., Gillett, R.A.N., and Ware, G.O., 1990, Association of *Listeria* spp. contamination in the dairy processing plant environment with the presence of staphylococci, *J. Food Prot.* **53**:928–932.

Fritsche, T.R., Gautom, R.K., Seyedirashti, S., Bergeron, D.L., and Lindquist, T.D. 1993, Occurence of bacterial endosymbionts in *Acanthamoeba* spp. isolated from corneal and environmental specimens and contact lenses, *J. Clin. Microbiol.* **31**:1122–1126.

Gilbert, R.J., Miller, K.L., and Roberts, D., 1989, *L. monocytogenes* and chilled foods, *Lancet* **i**:383–384.

Glöckner, F.O., Amann, R., Alfreider, A., Pernthaler, J., Psenner, R., Trebesius, K., and Schleifer, K.-H., 1996, An optimized *in situ* hybridization protocol for planktonic bacteria, *Syst. Appl. Microbiol.* **19**:403–406.

Grange, J.M., 1991, Environmental mycobacteria and human disease, *Lepr. Rev.* **62**:353–361.

Grimm, D., Merkert, H., Ludwig, W., Schleifer, K.-H., Hacker, J., and Brand, B.C., 1998, Specific detection of *Legionella pneumophila*: construction of a new 16S rRNA-targeted oligonuclotide probe, *Appl. Environ. Microbiol.* **64**:2686–2690.

Gutell, R.R., Larsen, N., and Woese, C.R., 1994, Lessons from an evolving rRNA: 16S and 23S rRNA structures from a comparative perspective, *Microbiol. Rev.* **58**:10–26.

Hahn, D., Amann, R.I., Ludwig, W., Akkermans, A.D.L., and Schleifer, K.-H., 1992, Detection of micro-organisms in soil after *in situ* hybridization with rRNA-targeted, fluorescently labelled oligonucleotides, *J. Gen. Microbiol.* **138**:879–887.

Holben, W.E., Jansson, J.K., Chelm, B.K., and Tiedje, J.M., 1988, DNA probe method for the detection of specific microorgnisms in the soil community, *Appl. Environ. Microbiol.* **54**:703–711.

Horn, M., Wagner, M., Fritsche, T., and Schleifer, K.-H., 1998, Phylogenetic studies on *Acanthamoeba* and nonculturable bacterial endosymbionts using 18S and 16S rDNA sequence analysis. VAAM, General Meeting, Frankfurt, Germany.

Horwitz, M.A., and Silverstein, S.C., 1980, Legionnaires' disease bacterium (*Legionella pneumophila*) multiplies intracellularly in human monocytes, *J. Clin. Invest.* **66**:441–450.

Hussong, D., Colwell, R.R., O'Brien, M., Weiss, E., Pearson, A.D., Weiner, R.M., and Burge, W.D., 1987, Viable *Legionella pneumophila* not detectable by culture on agar media, *Bio/Technology* **5**:947–950.

Islam, M.S., Hasan, M.K., Miah, M.A., Sur, G.C., Felsenstein, A., Venkatesan, M., Sack, R.B., and Albert, M.J., 1993, Use of the polymerase chain reaction and fluorescent-antibody methods for detecting viable but nonculturable *Shigella dysenteriae* type 1 in laboratory microcosms, *Appl. Environ. Microbiol.* **59**:536–540.

Jepras, R.I., Fitzgeorge, R.B., and Baskerville, A., 1985, A comparison of virulence of two strains of *Legionella pneumophila* based on experimental aerosol infection of guinea pigs, *J. Hyg.* **95**:29–38.

Jones, J.G., 1977, The effects of environmental factors on estimated viable and total populations of planktonic bacteria in lakes and experimental enclosures, *Freshwater Biol.* **7**:67–91.

Khan, M.U., Curlin, G.T., and Huq, M.I., 1979, Epidemiology of *Shigella dysenteriae* type 1 infections in Dacca [sic] urban area, *Trop. Geogr. Med.* **31**:213–223.

King, C.H., Shotts, E.B., Wooley, R.E., and Porter, K.G., 1988, Survival of coliforms and bacterial pathogens within protozoa during chlorination, *Appl. Environ. Microbiol.* **54**:3023–3033.

King, W., Raposa, S., Warshaw, J., Johnson, A., Halbert, D., and Klinger, J.D., 1989, A new colorimetric nucleic acid hybridization assay for *Listeria* in foods, *Int. J. Food Microbiol.* **8**:225–232.

Kopczinsky, E.D., Bateson, M.M., and Ward, D.M., 1994, Recognition of chimeric small-subunit ribosomal DNAs composed of genes from uncultivated microorganisms, *Appl. Environ. Microbiol.* **60**:746–748.

Kurtz, J.B., Bartlett, C.L.R., Newton, U.A., White, R.A., and Jones, N.L., 1982, *Legionella pneumophila* in cooling towers in London and a pilot trial of selected biocides, *J. Hyg.* **88**:369–381.

Liesack, W., Weyland, H., and Stackebrandt, E., 1991, Potential risks of gene amplification by PCR as determined by 16S rDNA analysis of a mixed-culture of strict barophilic bacteria, *Microb. Ecol.* **21**:191–198.

Loessner, M.J., Rudolf, M., and Scherer, S., 1997, Evaluation of luciferase reporter bacteriophage A511::*luxAB* for detection of *Listeria monocytogenes* in contaminated foods, *Appl. Environ. Microbiol.* **63**:2961–2965.

Ly, T.M.C., and Muller, H.E., 1990a, Interactions of *Listeria monocytogenes, Listeria seeligeri* and *Listeria innocua* with protozoans, *J. Gen. Appl. Microbiol.* **36**:143–150.

Ly, T.M.C., and Muller, H.E., 1990b. Ingested *Listeria monocytogenes* survive and multiply in protozoa, *J. Med. Microbiol.* **33**:51–54.

Maidak, B.L., Olsen, G.J., Larsen, N., Overbeek,R., McCaughey, M.J., and Woese, C.R., 1997, The RDP (Ribosomal Database Project), *Nucleic Acids Res.* **25**:109–110.

Manz, W., Amann, R., Ludwig, W., Wagner, M., and Schleifer, K.-H., 1992, Phylogenetic oligodeoxynucleotide probes for the major subclasses of proteobacteria: problems and solutions, *System. Appl. Microbiol.* **15**:593–600.

Manz, W., Szewzyk, U., Eriksson, P., Amann, R., Schleifer, K.H., and Stenström, T.-A., 1993, *In situ* identification of bacteria in drinking water and adjoining biofilms by hybridization with 16S and 23S rRNA-directed fluorescent oligonucleotide probes, *Appl. Environ. Microbiol.* **59**:2293–2298.

Manz, W., Szewzyk, R., Szewzyk, U., Hutzler, P., Amann, R., and Schleifer, K.H., 1995, *In situ* identification of *Legionellaceae* using specific rRNA-targeted oligonucleotide probes and confocal laser scanning microscopy, *Microbiol.* **141**:29–39.

Menard, R., Dehio, C., and Sansonetti, P., 1996, Bacterial entry into epithelial cells: the paradigm of *Shigella*, *Trends Microbiol.* **4**:220–226.

Michel, R., Hauröder-Philippczyk, B., Müller, K.-D., and Weishaar, I., 1994, *Acanthamoeba* from human nasal mucosa infected with an obligate intracellular parasite, *Eur. J. Parasitol.* **30**:104–110.

Neef, A., Zaglauer, A., Meier, H., Amann, R., Lemmer, H., and Schleifer, K.-H., 1996, Population analysis in a denitrifying sand filter: conventional and *in situ* identification of *Paracoccus* sp. in methanol-fed biofilms, *Appl. Environ. Microbiol.* **62**:4329–4339.

Olsen, G.J., Lane, D.J., Giovannoni, S.J., Pace, N.R., and Stahl, D.A., 1986, Microbial ecology and evolution: a ribosomal RNA approach, *Annu. Rev. Microbiol.* **40**:337–365.

Ouverney, C.C., and Fuhrman, J.A., 1997, Increase in fluorescence intensity of 16S *rRNA in situ* hybridization in natural samples treated with chloramphenicol, *Appl. Environ. Microbiol.* **63**:2735–2740.

Paszko-Kolva, C., Shahamat, M., and Colwell, R.R., 1992, Longterm survival of *Legionella pneumophila* serogroup 1 under low-nutrient conditions and associated morphological changes, *FEMS Microbiol. Ecol.* **102**:45–55.

Peel, M., Donachle, W., and Shaw, A., 1988, Temperature-dependent expression of flagella of *Listeria monocytogenes* studied by electron microscopy, SDS-PAGE and western blotting, *J. Gen. Microbiol.* **143**:2171–2178.

Peters, M., Muller, C., Rush-Gerdes, S., Seidel, C., Gobel, U., Pohle, H.D., and Ruf, B., 1995, Isolation of atypical mycobacteria from tap water in hospitals and homes: is this a possible source of disseminated MAC infection in AIDS patients?, *J. Inf.* **31**:39–40.

Pillay, D.G., Karas, A.J., and Sturm, A.W., 1997, An outbreak of Shiga bacillus dysentery in KwaZulu/Natal, South Africa, *J. Infect.* **34**:107–111.

Poulsen, L.K., Ballard, G., and Stahl, D.A., 1993, Use of rRNA fluorescence *in situ* hybridization for measuring the activity of single cells in young and established biofilms, *Appl. Environ. Microbiol.* **59**:1354–1360.

Rahman, I., Shahamat, M., Chowdhury, M.A.R., and Colwell, R.R., 1996, Potential virulence of viable but nonculturable *Shigella dysenteriae* type 1, *Appl. Environ. Microbiol.* **62**:115–120.

Ramsing, N.B., Kühl, M., and Jörgensen, B.B., 1993, Distribution of sulfate-reducing bacteria, O_2 and H_2S in photosynthetic biofilms determined by oligonucleotide probes and microelectrodes, *Appl. Environ. Microbiol.* **59**:3820–3849.

Ren, T., and Frank, J.F., 1993, Susceptibility of starved planktonic and biofilm *Listeria monocytogenes* to quaternary ammonium sanitizer as determined by direct viable and agar plate counts, *J. Food Prot.* **56**:573–576.

Rogers, J., and Keevil, C.W., 1992, Immunogold and fluorescein immunolabelling of *Legionella pneumophila* within an aquatic biofilm visualized by using episcopic differential interference contrast microscopy, *Appl. Environ. Microbiol.* **58**:2326–2330.

Roller, C., Wagner, M., Amann, R., Ludwig, W., and Schleifer, K.-H., 1994. *In situ* probing of gram-positive bacteria with high DNA G+C content using 23S rRNA-targeted oligonucleotides, *Microbiol.* **140**:2849–2858.

Rowe, B., and Gross, R.J., 1984, Shigella, in: *Bergey's manual of systematic bacteriology*, (N.R. Krieg, G. Holt, eds.) Williams and Wilkins, Baltimore, pp. 423–427.

Rudi, K., Kroken, M., Dahlberg, O.J., Deggerdal, A., Jakobsen, K.S., and Larsen, F., 1997, Rapid, universal method to isolate PCR-ready DNA using magnetic beads, *BioTechniques* **22**:506–511.

Ruf, B., Schürmann, D., Horbach, I., Fehrenbach, F.J., and Pohle, H.D., 1990, Prevalence and diagnosis of *Legionella* pneumonia: a 3-year prospective study with emphasis on application of urinary antigen detection, *J. Infect. Dis.* **162**:1341–1348.

Salfinger, M., and Pfyffer, G.E., 1994, The new diagnostic Mycobacteriology Laboratory, *Eur. J. Clin. Microbiol. Infect. Dis.* **13**:961–979.

Sansonetti, P., 1992, Molecular and cellular biology of *Shigella flexneri* invasiveness: from cell assay systems to shigellosis, *Curr. Top. Microbiol. Immunol.* **180**:1–19.

Saylers, A.A., and Whitt, D.D., 1994, Bacterial pathogenesis, ASM, Washington, D.C., pp. 169–181.

Schönhuber, W., Fuchs, B., Juretschko, S., and Amann, R., 1997, Improved sensitivity of whole cell hybridization by the combination of horseradish peroxidase-labeled oligonucleotides and tyramide signal amplification, *Appl. Environ. Microbiol.* **63**:3268–3273.

Schramm, A., Larsen, L.H., Revsbech, N.P., Ramsing, N.B., Amann, R., and Schleifer, K.-H., 1996, Structure and function of a nitrifying biofilm as determined by *in situ* hybridization and microelectrodes, *Appl. Environ. Microbiol.* **62**:4641–4647.

Schuchat, A., Swaminathan, B., and Broome, C.V., 1991, Epidemiology of human listeriosis, *Clin. Microbiol. Rev.* **4**:169–183.

Seeliger, H.P.R., and Jones, D., 1986, *Listeria,* in: *Bergey's Manual of Systematic Bacteriology*, **2**:1235–1245.

Somerville, C., Knight, I.T., Straube, W.L., and Colwell, R.R., 1989, Simple rapid method for the direct isolation of nucleic acids from aquatic environments, *Appl. Environ. Microbiol.* **55**:548–554.

Springer, N., Ludwig, W., Drozanski, V., Amann, R., and Schleifer, K.-H., 1992, The phylogenetic status of *Sarcobium lyticum*, an obligate intracellular bacterial parasite of small amoebae, *FEMS Microbiol. Lett.* **96**:199–202.

Springer, N., Ludwig, W., Amann, R., Schmidt, H.J., Görtz, H.-D., and Schleifer, K.-H., 1993, Occurrence of fragmented 16S rRNA in an obligate bacterial endosymbiont of *Paramecium caudatum*, *Proc. Natl. Acad. Sci. USA* **90**:9892–9895.

Springer, N., Amann, R., Ludwig, W., Schleifer, K.-H., and Schmidt, H., 1996, *Polynucleobacter necessarius*, an obligate bacterial endosymbiont of the hypotrichous ciliate *Euplotes aediculatus*, is a member of the β-subclass of Proteobacteria, *FEMS Microb. Lett.* **135**:333–336.

Stahl, D.A., and Amann, R.I., 1991, Development and application of nucleic acid probes in bacterial systematics, in: *Sequencing and Hybridization Techniques in Bacterial Systematics*, (E. Stackebrandt and M. Goodfellow, eds.), John Wiley and Sons, Chichester, England, pp. 205–248.

Stahl, D.A., Flesher, B., Mansfield, H.R., Montgomery, L., 1988, The use of phylogenetically based hybridization probes for studies of ruminal microbial ecology, *Appl. Environ. Microbiol.* **54**:1079–1084.

Staley, J.T., and Konopka, A., 1985, Measurement of *in situ* activities of nonphotosynthetic microorganisms in aquatic and terrestrial habitats, *Annu. Rev. Microbiol.* **39**:32–346.

Steinert, M., Emödy, L., Amann, R., and Hacker, J., 1997, Resuscitation of viable but nonculturable *Legionella pneumophila* Philadelphia JR32 by *Acanthamoeba castellanii*, *Appl. Environ. Microbiol.* **63**:2047–2053.

Steinert, M., Birkness, K., White, E., Fields, B., and Quinn, F., 1998*a*, *Mycobacterium avium* bacilli grow saprozoically in coculture with *Acanthamoeba polyphaga* and survive within the cyst wall, *Appl. Environ.Microbiol.* **64**:2256–2261.

Steinert, M., Birkness, K., White, E., Quinn, F., and Fields, B., 1998*b*, Survival of bacterial pathogens within *Acanthamoeba polyphaga*, 98th ASM General Meeting, Atlanta, Abstr. N49, p. 374.

Strunk, O., Gross, O., Reichel, B., May, M., Hermann, S., Stuckman, N., Nonhoff, B., Lenke, M., Ginhart, A., Vilbig, A., Ludwig, T., Bode, A., Schleifer, K.-H., and Ludwig, W., 1998, ARB: a software environment for sequence data, http://www.mikro.biologie.tu-muenchen.de.

Szewzyk, U., Manz, W., Amann, R., Schleifer, K.-H., and Stenström, T.-A., 1994, Growth and *in situ* detection of a pathogenic *Escherichia coli* in biofilms of a heterotrophic water bacterium by use of 16S- and 23S-rRNA-directed fluorescent oligonucleotide probes, *FEMS Microbiol. Ecol.* **13**:169–175.

Torsvik V., Goksoyr, J., and Daae, F.L., 1990, High diversity of DNA of soil bacteria, *Appl. Environ. Microbiol.* **56**:782–787.

Van de Peer, Y., Jansen, J., De Rijk, P., and De Wachter, R., 1997, Database on the structure of small ribosomal subunit RNA, *Nucleic Acids Res.* **25**:111–116.

Wagner, M., Amann, R., Lemmer, H., and Schleifer, K.-H., 1993, Probing activated sludge with proteobacteria-specific oligonucleotides: inadequacy of culture-dependent methods for describing microbial community structure, *Appl. Environ. Microbiol.* **59**:1520–1525.

Wagner, M., Schmid, M., Juretschko, S., Trebesius, K.-H., Bubert, A., Goebel, W., and Schleifer,

K.-H., 1998, *In situ* detection of a virulence factor mRNA and 16S rRNA *in Listeria monocytogenes*, *FEMS Microbiol. Lett.* **160**:159–168.

Wallner, G., Amann, R., and Beisker, W., 1993, Optimizing fluorescent *in situ* hybridization of suspended cells with rRNA-targeted oligonucleotide probes for the flow cytometric identification of microorganisms, *Cytometry* **14**:136–143.

Wang, G.C.Y., and Wang, Y., 1996, The frequency of chimeric molecules as a consequence of PCR co-amplification of 16S rRNA genes from different bacterial species, *Microbiology* **142**:1107–1114.

Wayne, L.G., 1994, Dormancy of *Mycobacterium tuberculosis* and latency of disease, *Eur. J. Clin. Microbiol. Infect. Dis.* **13**:908–914.

Wayne, L.G., and Kubica, G.P., 1986, Mycobacteriae, in: *Bergey's Manual of Systematic Bacteriology*, **2**:1436–1457.

Wintermeyer, E., Ludwig, B., Steinert, M., Schmitt, B., Fischer,G., and Hacker, J., 1995, Influence of site-specific altered Mip proteins on intracellular survival of *Legionella pneumophila* in eukaryotic cells, *Infect. Immun.* **63**:4576–4583.

Woese, C.R., 1987, Bacterial evolution, *Microbiol. Rev.* **51**:221–271.

Yan, W., Malik, M.N., Peterkin, P.I., and Sharpe, A.N., 1996, Comparison of the hydrophobic grid-membrane filter DNA probe method and the Health Protection Branch standard method for the detection of *Listeria monocytogenes* in foods, *Int. J. Food Microbiol.* **30**:379–84.

Zarda, B., Amann, R., Wallner,G., and Schleifer, K.-H., 1991, Identification of single bacterial cells using digoxigenin-labeled, rRNA-targeted oligonucleotides, *J. Gen. Microbiol.* **137**:2823–2830.

Chapter 24

New Approaches for Diagnosis of Infections by Intracellular Bacteria

Reinhard Marre

1. INTRODUCTION

Laboratory diagnosis of infectious diseases usually is achieved by microscopy, culture, antigen and antibody detection or, in order to overcome diagnostic gaps of single tests, a combination of these methods. During the last years, however, microbiologists increasingly became aware of bacterial species which cannot satisfactorily be detected in a clinical microbiology laboratory by standard methods either due to their low sensitivity and specificity or due to the need to apply expensive and time consuming methods. The introduction of molecular-genetic methods in clinical microbiology seemed to solve many problems: Since speed of detection was no longer dependent on the generation time and on growth conditions of the pathogen and since specific detection of the *in vitro* amplified DNA or RNA was possible, a new world of non or barely *in vitro* culturable pathogens appeared. With increasing experience the limits of this technology became obvious, requiring a meticulous quality control program.

REINHARD MARRE Department of Medical Microbiology and Hygiene, University of Ulm, D-89070 Ulm, Germany.

Subcellular Biochemistry, Volume 33: Bacterial Invasion into Eukaryotic Cells, edited by Oelschlaeger and Hacker. Kluwer Academic / Plenum Publishers, New York, 2000.

Before introducing the molecular-genetic methods into clinical microbiology these techniques have been used successfully in research laboratories for various scientific purposes. In a clinical microbiology laboratory the requirements are different. The diagnostic kits have to be evaluated with respect to sensitivity, specificity, positive and negative predictive value, handling problems, time of analysis, hands-on-time and total costs. Additional problems to be coped with are that clinical specimen vary in their composition, may contain inhibitors or may be contaminated. Furthermore, the detection of the nucleic acid of a pathogen is not always indicative of an infection with this pathogen. The nucleic acid might persist in spite of successful therapy or could only indicate harmless colonisation. In terms of evaluation it became obvious that so-called "gold standards" faded and new definitions of the laboratory diagnosis of infectious diseases had to be worked out.

In the field of clinical microbiology, molecular genetic methods are used for identification and detection of infectious agents, toxin genes or resistance markers. It can be divided into hybridisation and amplification methods or a combination of both. Hybridisation techniques primarily serve for culture confirmation and direct detection of pathogens in clinical specimens. A special application is the *in-situ*-hybridisation which allows to analyse the distribution of micro-organisms in tissue specimens. In order to increase sensitivity, the hybridisation reaction often is coupled with detection systems mainly based on enzymatic reaction, chemiluminescence and fluorescence (Table 1). The advantage of hybridisation tests for direct detection of pathogens is their convenience, but, due to a low sensitivity, they largely have been replaced by DNA or RNA amplification methods. Excep-

Table I
Comparison of Detection Sensitivities of Nucleic Acid Based Techniques for Microorganisms (Podzorski and Pershing, 1995)

Method	Approx. copy no. detectable
Ethidium bromide	10^8
Radiolabeled oligonucleotide probes	10^6
Radiolabeled full length probes	10^4
Enzyme-coupled probes	10^4
Chemiluminescent probes	10^4
Compound or branched probes	10^4
Nucleic acid amplification	<10

tions are *in-situ*-hybridisation tests and culture confirmation tests (which do not require high sensitivities) of micro-organisms, the identification of which is time-consuming and require a number of special identification kits, e.g. *Mycobacterium tuberculosis* and Mycobacteria Other Than Tuberculosis (MOTT). In addition, hybridisation tests are used for identification of *in vitro* amplified DNA or RNA.

Nucleic acid amplification tests principally have the advantage of high sensitivity and, when coupled with an amplificate identification step (probe or sequencing), high specificity as well. Because of the versatility, the automatisation power, the possibility to further characterize the amplificates, *in vitro* nucleic acid amplification tests (NAT) will significantly contribute to the laboratory diagnosis of infectious diseases. However, besides an impressible number of different in-house amplification tests which are more or less well characterized and evaluated, only a few tests have been so far commercialized in the field of bacteriology. These mainly concern the laboratory diagnosis of infectious diseases such as tuberculosis, gonorrhoea and infections caused by *Chlamydia trachomatis*, infectious diseases which are difficult to diagnose clinically or which cannot sufficiently be detected by commonly available standard methods other than NAT. However, amplification tests are also of help for the detection of *Borrelia burgdorferi*, enterohemorrhagic *Escherichia* coli, *Bartonella* species, *Ehrlichia* species, *Legionella* species, *Mycoplasma pneumoniae*, *Chlamydia pneumoniae* and *Chlamydia psittaci* in clinical specimens and they support the diagnosis of methicillin resistance in *Staphylococcus aureus* and of rifampicin resistance in *Mycobacterium tuberculosis* (Podzorski and Pershing, 1995). In addition, nucleic acid amplificatin tests were also developed to detect pathogens in food (e.g., *Listeria monocytogenes*) and water (e.g., *Legionella pneumophila*).

2. AMPLIFICATION METHODS

In order to speed-up the detection time, to increase sensitivity and specificity, to optimize the test performance and last but not least out of commercial reasons (patent protection, obligatory combination of test kit and technical equipment) different amplification and detection methods were developed. They are based on either target amplification or probe amplification and are described and critically evaluated by Podzorski and Pershing (1995).

The polymerase chain reaction (PCR) is an exponential amplification of a target sequence (Mullis and Falcona, 1987). The specificity of the reaction results from the selection of a specific target, appropriate primers and identification of the amplificate. Concerning the PCR kits marketed by Roche, the amplified sequence is detected by hybridization accomplished with an enzyme-linked immunosorbent assay-based detection system.

The strand displacement amplification (SDA) procedure, developed and patented by Becton Dickinson, is an isothermal DNA amplification test which can be divided into a linear amplification step resulting in the synthesis of fragments with a hemiphosphoisothioat *hinc* II site and an exponential amplification step with repeated cycles with polymerization and strand displacement (Walker *et al.* 1992). Detection of amplification is achieved by hybridization with a fluorescence-labeled probe. In spite of the complicated chemical process most of the reactions can be run in one tube only. According to the information of the manufacturers the SDA is quick (duration of the test 2.5 hr), gives quantitative data and, since, the reaction takes place in sealed microplates, is protected against contamination.

The transcription based amplification system (TAS) or transcription mediated amplification technique (TMA) has been developed by Gen Probe (Kwoh *et al.*, 1989). In contrast to the SDA and commonly used PCR the target sequence is a RNA. TMA can be divided into two steps: (1) transcription of the target RNA into cDNA, (2) transcription of the cDNA into multiple copies of RNA. This reaction requires a reverse transcriptase

Table II
Target Sequences for Detection of Selected Intracellular Pathogens by Amplification Tests

Species	Test	Manufacturer	target
Chlamydia trachomatis	LCR	Abbott	cryptic plasmid
	PCR	Roche	cryptic plasmid
	TMA	Gen-Probe	rRNA
	Q beta replicase	Gene-Trak	rRNA
	SDA	Becton Dickinson	cryptic plasmid
Mycobacterium tuberculosis	PCR	Roche	16S rRNA gene
	PCR	Digene	*mtp 40* gene
	LCR	Abbott	chromosomal gene of the protein antigen b
	Q beta replicase	Gene-Trak	rRNA
	TMA	Gen-Probe	rRNA
	SDA	Becton Dickinson	IS6110
Neisseria gonorrhoeae	PCR	Roche	cytosin DNA methyl transferase gene
	LCR	Abbott	*opaI*

to transcribe the RNA into DNA, a RNaseH degrading the RNA template and the T7 polymerase binding to the T7 promoter sequence of the oligonucleotide primer. This technique allows rapid amplification (10 million-fold within 1–2 hr) under isothermal conditions. The RNA copies can be detected by any of the procedures described above. The selection of RNA as target sequence increased the sensitivity since multiple copies of rRNA are in one cell. In addition, RNA molecules are less stable, thus decreasing the risk of contamination.

The ligase chain reaction and the Qβ-replicase test are probe amplification tests. In the ligase chain reaction, developed and marketed by Abbott, at first two oligonucleotide primers anneal to the target DNA (Wu and Wallace, 1989). When complementary at the ligation junction, the ligase joins the pair, forming a longer product that is bound to the template. The ligation products are subsequently used as templates for primers which are again ligated to give new templates, which, in the test kit marketed by Abbott, is detected by a microparticle enzyme immunoassay.

The Qβ replicase system (Gene-Trak) is even more rapid requiring theoretically only 13 minutes to yield 10^{12} copies (Lizardi *et al.*, 1988) of the target molecule. It is a probe amplification system based on the use of the midivariant-1, a template for the Qβ replicase, as a reporter into which the probe oligonucleotide had been inserted. A crucial step is the removal of the unbound reporter probes, since they could act as an amplification template. Therefore, a poly(dG) or a poly(dA) target-specific capture probe is hybridizied with target RNA or DNA. The poly(dG) or (dA) tail is used for capture by a oligo(dC) or oligo (dT) derivatized paramagnetic bead complex. This complex consisting of the target DNA or RNA hybridizes with the target-specific substrate so that the capture probe is bound to a magnet, unbound target-specific substrate probes can thus be removed. The advantages of the Qβ replicase dual capture assay are the speed of amplification, the high sensitivity and its convenience. In addition, the technique could be used for a simultaneous multiple target detection system. The main disadvantage is that contaminating, not hybridized reporter probes could serve as templates for amplification.

3. SPECIFIC APPLICATIONS OF NUCLEIC ACID AMPLIFICATION TESTS

3.1. *Chlamydia trachomatis*

Detection of *Chlamydia trachomatis*, a pathogen causing urethritis, salpingitis, epididymitis, conjunctivitis, and, in newborns, pneumonia has been optimised significantly during the last ten years. Since cultural detection of *Chlamydia trachomatis* requires optimal transportation conditions of the specimens to avoid loss of viability and since it requires the availability of cell cultures in specialized laboratories, culture techniques have been replaced by non-culture methods, such as antigen detection by immunofluorescence (DFA) or enzyme-linked immunosorbent assays (EIA), in the first generation of non-culture tests. Sensitivities of these tests were 88 (DFA) and 98% (EIA) when using urethral or cervical specimens from woman and 70% with urethral specimens from men. The specificities ranged from 84 to 98% (Chernesky *et al.*, 1986). A slight increase of sensitivity and specificity was attained by a DNA/RNA hybridization test, marketed by Gen-Probe, which is based on the detection of the chlamydial rRNA by a probe labeled with acridinium. This test does not detect *Chlamydia psittaci* and *Chlamydia pneumoniae.*

With the introduction of nucleic acid amplification tests it was possible to use urine specimens instead of urethral swabs and cervical swabs which had to be obtained by invasive, occasionally painful procedures. The ligase chain reaction (LCx, Abbott Laboratories) the transcription mediated amplification test (AMP CT, Gen Probe, San Diego, CA, USA) PCR (Cobas AMPLICOR PCR, Roche Diagnostic Systems, Basel, Switzerland) and the Qβ replicase amplified assay (Gene Trak, Framingham, MA, USA) have been tested for detection of *C. trachomatis* in clinical specimens. The ligase chain reaction and the PCR use a sequence of the cryptic plasmid of *C. trachomatis*, which is found in most isolates of *C. trachomatis;* however exceptions may occur (An *et al.*, 1992). Compared to direct hybridisation tests and antigen detection tests these assays were able to reliably detect a urethral *C. trachomatis* infection by analysing a first-catch urine specimen (Table 3). The disadvantage of both tests is the need of thermocycling and, because they are based on DNA amplification, the need of a stringent control of inhibition, which might be a serious problem and has been reported in up to 14% of the specimens (Liu *et al.*, 1997; An *et al.*, 1995; Bass *et al.*, 1993).

Detection systems based on rRNA theoretically and practically have a higher sensitivity since the breakdown of one cell only would give 1.000 to 10.000 copies of the target molecule (Mouton *et al.*, 1997). The AMP CT and the Qβ replicase test took this into acount. Their lower limit of detectibility is about 1.000 molecules. The amplification is performed under isothermal conditions. Sensitivities and specificities are given in Table 3. The Qβ replicase assay was not evaluated in larger clinical studies. First publications allowed the conclusion that the sensitivity and specificity of the Qβ replicase is in the range of the AMP CT test (An *et al.*, 1995). In addition, it offers the advantage of a quantification of the number of bacterial cells.

Table III
Molecular Genetic Methods for Detection of *Chlamydia trachomatis*

Test (Manufacturer)	Specimens characteristics	Sensitivity	Specificity	Definition of true positives	Reference
PCR (Roche)	urethral swabs of male patients	79,5	100	culture positive specimens	(Stary *et al.*, 1996)
PCR (Roche)	first catch urine of male patients	75,0	100	culture positive specimens	
Hybridisation (Gen-probe)	urethral swabs of male patients	61,5	100	culture positive specimens	
TMA (Gen-Probe)	First catch urine of female patients	91,4	99,6	culture positive specimens or specimens positive by two amplification methods	(Pasternack *et al.*, 1997)
PCR (Roche)	First catch urine of female patients	97,1	99,8	culture positive specimens or specimens positive by two amplification methods	
Culture	cervical specimens	85,7	100	culture positive specimens or specimens positive by two amplification methods	
LCR (Abbott)	First catch urine of male or female patients	83,7	99,9	specimens positive by at least two different tests	(Goessens *et al.*, 1997)

(continued)

Table III
Continued

Test (Manufacturer)	Specimens characteristics	Sensitivity	Specificity	Definition of true positives	Reference
TMA (Gen-Probe)	First catch urine of male or female patients	85,4	99,1	specimens positive by at least two different tests	
PCR (Roche)	First catch urine of male or female patients	92,7	99,4	specimens positive by at least two different tests	
Culture	Urethral or cervical specimens of male or female patients	57,7	99,3	specimens positive by at least two different tests	
TMA (Gen-Probe)	First catch urine of females	84,3	98,8	Combination of test results	(Mouton *et al.*, 1997)
Cell culture	Urethral and cervical specimens of females	72,5	99,2	Combination of test results	
TMA (Gen-Probe)	First catch urine of males	100	99,2	Combination of test results	
Cell culture	urethral swabs of males	57,4	99	Combination of test results	
PCR (Roche)	conjunctival swabs	88	100	culture positive specimens	(Kowalski *et al.*, 1995)

Because RNA is less stable and might be degraded quickly by naturally occurring RNases, RNA based detection methods might better reflect an active infection. On the other hand, Chlamydia may persist in a metabolically nearly inactive form in phagocytes (Beatty *et al.* 1992) which could be better detected by a PCR technique.

3.2. *Chlamydia pneumoniae and Chlamydia psittaci*

Chlamydia pneumoniae has been identified as a common respiratory pathogen causing tracheitis, bronchitis, pneumonia, sinusitis, otitis and conjunctivitis (Kuo *et al.*, 1995). In addition, *Chlamydia pneumoniae* has been associated with coronary heart disease on the basis of its detection in atherosclerotic specimen by PCR, immunocytology and culture (Ramirez *et al.*, 1996; Campbell *et al.*, 1995; Kuo *et al.*, 1995; Kuo *et al.*, 1993). *Chlamydia psittaci* is a pathogen of birds rarely causing pneumonia in man. Since cell culture techniques often fail to detect *C. pneumoniae* (Bomann *et al.* 1997), and since specimens are rarely cultured for detection of *C. psittaci* due to special safety regulations when handling viable *C. psittaci*, a number of DNA amplification techniques have been developed in recent years, however, these techniques usually are in-house tests and may thus have a considerable degree of variation between different laboratories. Three different target sequences have been used for detection of *C. pneumoniae* by PCR. These are a 474 bp *Pst*I fragment of unknown function (Campbell *et al.*, 1992), a region of the 16S rRNA (Gaydos *et al.*, 1992) and a base sequence within a 53 kDa protein gene (Kubota *et al.*, 1996). None of the target sequences could be found in *Chlamydia* species other than *C. pneumoniae* or in species frequently encountered in respiratory specimens. In order to increase sensitivity primers were labeled with fluorescein (Wilson *et al.*, 1996) or amplificates detected by an enzyme immunosorbent assay (Gaydos *et al.*, 1994). Related to copies of DNA/RNA all methods appeared to have a similar sensitivity which can be increased ca. 100 fold using a nested PCR, so that only two inclusion forming bodies can be detected (Bomann *et al.*, 1997; Khan *et al.*, 1996). An interesting development is the nested multiplex PCR based on the amplification of a genus specific target sequence and a species (*C. pneumoniae, C. psittaci, C. trachomatis*) specific target sequence of the 16S rRNA gene (Messmer *et al.*, 1997; Rasmussen *et al.*, 1992). As with *C. trachomatis* DNA/RNA amplification techniques are superior to culture and antigen detection with respect to sensitivity (Bomann *et al.*, 1997). At the moment there is no consensus about the target sequence to be used in a routine clinical microbiology laboratory. Methodological differences might account for the high variability in detection of *C. pneumoniae* in arteriosclerotic lesions ranging from 0 to 40% positivity rates (Weiss *et al.*, 1996; Campbell *et al.*, 1992). These contradicting, at the moment not explainable results are of specific concern regarding the excited discussion on the contribution of *C. pneumoniae* to inflammation and arteriosclerosis.

3.3. *Neisseria gonorrhoeae*

In the case of *Neisseria gonorrhoeae*, a fastidious microorganism, is the causative micro-organism of gonorrhoe, a sexually transmitted disease. It is able to grow on artificial microbiological media but cultural detection of Neisseria may be difficult under unfavourable specimen transport conditions. Therefore, in addition to microscopy and culture, non-cultural methods such as antigen detection, hybridization techniques and NAT assays are in use in clinical microbiological laboratories in order to overcome the problem of the viability of *Neisseria gonorrhoeae.* Target sequences are the chromosomal cystosin DNA methyl transferase gene used by the PCR marketed by Roche or the *opaI* gene used by the LCR marketed by Abbott. The *opaI* gene is repeated up to 11 times in a *N. gonorrhoeae* genome (Burczak *et al.*, 1995). Sensitivity of the LCR ranged from 96.2 to 100% in cervical specimens and 98.1 to 99% in male urethral specimens (Ching *et al.*, 1995). The corresponding data for culture were 73.1 to 92.1% and 96.1 to 97.1%. Similar to *Chlamydia trachomatis*, detection of *Neisseria gonorrhoeae* was also possible without significant loss of the test performance when using first catch urine as specimen, thus offering a practical alternative for STD screening and surveillance (Smith *et al.*, 1995). In contrast to *Chlamydia trachomatis*, however, an antibiotic resistance of *Neisseria gonorrhoeae* cannot be reliable predicted. Therefore, cultural detection of *Neisseria gonorrhoeae* cannot be totally replaced by NAT methods and should mainly be regarded as supplementary to culture.

3.4. Legionella

Legionella species are pathogens of protozoa of the aquatic environment. Only very few of the more than 40 *Legionella* species can cause respiratory tract infections if inhaled by susceptible individuals. Legionella can be detected in clinical specimens by culture on supplemented media but it takes 3 to 4 days in average before colonies become visible. Sensitivity of cultural detection ranges between 30 to 70% depending of the pre-treatment procedure and the kind of specimen used. In order to speed up and to facilitate detection of Legionella a radio-labeled rRNA probe was developed. The probe was complementary to a genus specific rRNA sequence and labeled with ^{125}I. The probe was shown to have a sensitivity of 56% and a specificity of more than 99% (Edelstein *et al.*, 1987; Edelstein, 1986). The probe detected most of the clinically relevant species other than *L. pneumophila* also, but the sensitivity was less. Since the sensitivity of 56% was regarded as unsatisfactory, amplification tests were developed and evaluated. Target sequences were fragments of the 16S rRNA gene (Stone *et al.*, 1996; Lisby and Dessau, 1994), the *mip* gene (Koide and Saito, 1995; Lindsay *et al.*, 1994) and a 5S rRNA gene (Murdock *et al.*, 1996; Maiwald *et al.*, 1995; Koide and Saito, 1995). The Qβ replicase assay used the 16S rRNA as a target sequence (Stone *et al.*, 1996). Respiratory tract secretions but also serum and urine were used as specimens (Murdoch *et al.*, 1996; Lindsay *et al.*, 1994;), the latter two showed acceptable sensitivities with 73% if the specimen was obtained in the first 4 days of the disease. All published reports suffered from low numbers of patients. With an effective nosocomial infection control the number of patients suspected to have legionnaire`s disease might also be low in most hospitals so that, in contrast to *Chlamydia trachomatis*, the necessity to establish an amplification test for detection of *Legionella* in clinical specimens is restricted to laboratories which receive high numbers of specimens of hospitals with a high incidence of Legionella infections. Amplification tests, however, which allow a quantification of Legionella might be helpful for controlling tap water.

3.5. *Mycobacterium tuberculosis*

Laboratory diagnosis of tuberculosis including identification and susceptibility testing of the isolates requires several weeks before definitive diagnosis can be made. Therefore, clinical management of a patient with presumed tuberculosis is mainly based on the clinical presentation, x-ray and comperatively little sensitive and specific microbiological tests such as a Ziehl-Neelsen stain. Since tuberculosis is a potentially life-threatening disease with high prevalence especially in underdeveloped countries and since early diagnosis and treatment would help to combat the spread of tuberculosis, there is an urgent need to use molecular genetic tools to speed up laboratory diagnosis. Different tests are meanwhile on the market or in a pre-market stage. These are principally the same test as for detection of *Chlamydia trachomatis*. They are either based on amplification of a target sequence or on amplification of primers. These tests are the AMPLICON MTB (Roche) and the SHARP Signal System (Digene Diagnostic. Inc.) based on PCR, the MTD (Gen Probe) based on a transcription mediated amplification assay, the *Mycobacterium tuberculosis* LCx (Abbott Laboratories Diagnostic Division, Chicago, Ill.), a ligase chain reaction, a Strand Displacement Amplification (SDA) Probe Tec System by Becton Dickinson and a Q beta replicase amplification assay (Gene Trak). The target sequences are given in Table 2, the sensitivities and specificties are given in Table 4. Since alternative, well established methods for the detec-

Table IV
Molecular Genetic Methods for Detection of *Mycobacterium tuberculosis*

Test (Manufacturer)	Specimens characteristics	Sensitivity	Specificity	Definition of true positives	Literature
PCR (Roche)	respiratory and nonrespiratory specimens	83,5	98,8	combination of tests and clinical presentation	(Reischl *et al.*, 1998)
PCR (Roche)	respiratory specimens	73,8	97,2	combination of tests and clinical presentation	(Huang *et al.*, 1996)
PCR (Sharp Digene)	respiratory specimens	69,0	91,6	combination of tests and clinical presentation	
Microscopy	respiratory specimens	67	98	culture	(Rajalahti *et al.*, 1998)
PCR (Roche)	respiratory specimens	83	99	culture	
PCR (Roche)	respiratory specimens	76,4	99,8	clinical diagnosis	(Shah *et al.*, 1998)
Q beta replicase (Gene-Trak)	respiratory specimens	79	98	combination of tests and clinical presentation	(Smith *et al.*, 1997b)
Q beta replicase (Gene-Trak)	respiratory specimens	84	97	combination of tests and clinical presentation	(Smith *et al.*, 1997a)

TMA (Gen-Probe)	smear positive respiratory specimens	100	100	culture and clinical presentation	(Piersimoni *et al.*, 1997)
PCR (Roche)	smear positive respiratory specimens	97	100	culture and clinical presentation	
TMA (Gen-Probe)	smear negative respiratory specimens	87,5	98,8	culture and clinical presentation	
PCR (Roche)	smear negative respiratory specimens	66,7	99,6	culture and clinical presentation	
Microscopy	respiratory specimens	77	100	culture and clinical presentation	
Culture	respiratory specimens	87	100	culture and clinical presentation	
LCR (Abbott)	respiratory specimens	90,8	100	combination of tests and clinical presentation	(Ausina *et al.*, 1997)
Culture	respiratory specimens	88,2	100	combination of tests and clinical presentation	
Microscopy	respiratory specimens	82,6	92	combination of tests and clinical presentation	

tion of *Mycobacterium tuberculosis* exist, the diagnostic gain of a nucleic acid amplification based detection system can be evaluated. Similar to other diagnostic tests false positive and false negative results can be obtained by nucleic acid amplification tests. False positives may result from the presence of non-viable *Mycobacterium tuberculosis* after the patient had been on antituberculosis medication (Kennedy *et al.*, 1994), false negatives are possible due to inhibitors in the specimens. Therefore, a NAT result has to be interpreted with the same restrictions as the result of negative cultures and negative or positive Ziehl-Neelsen stains. As is true for other amplification tests also, the amplification tests for *M. tuberculosis* are no stand-alone test and should be interpreted together with clinical and further laboratory data. The considered approval by FDA for the Gen Probe MTD and the Roche Amplicor was therefore limited to smear positive respiratory tract specimens from patients who have not been on antituberculosis medication for seven or more days and who have not been treated within the last 12 months, since the test results (sensitivity and positive predictive value) decreased considerably when smear negative specimens were studied and since amplification techniques may remain positive for many months after completion of therapy (Anonymous, 1997)

3.6. Miscellaneous Microorganisms

Bartonella henselae, the etiological agent of cat scratch disease, is a pleomorphic gram-negative bacillus which can be cultured from affected tissue. However, cultural detection and identification is difficult and antibody response variable (Bergmans *et al.*, 1997). Therefore, amplification of *Bartonella henselae* DNA has been proposed as an additional diagnostic test. Principally three different PCR protocols are in use. The target sequences are a part of the 16S rRNA gene, a portion of the *gltA* and *htrA* genes coding for either a citrate synthetase (PCR/CS) or a 60 kDA heat shock-like protein (PCR/HSP). The PCR/rRNA was followed by a hybridisation step to identify the amplificate since universal broad host-range PCR primers were used. Both, the PCR/rRNA and the PCR/CS were able to correctly identify most of the positive specimens (100% and 99%) (Avidor *et al.*, 1997). It has been recommended by the authors to use a two-step protocol, the first step being a PCR/rRNA in order to detect bacterial, and the second step being a PCR/CS to specifically detect *Bartonella henselae*.

Ehrlichia with the species *E. chaffeensis* causing the human monocytic Ehrlichiosis and *E. phagocytophilia* causing the human granulocytic Ehrlichiosis are intracellular pathogens transmitted by ticks. Both species belong to the family of *Rickettsiaceae*. The clinical picture of Ehrlichiosis is highly variable and may be mistaken for a Lyme disease, making an accurate clinical diagnosis difficult. Similar to the Bartonella infection, PCR would offer an alternative diagnostic approach. Most PCR assays use primers specific for the 16S rRNA, which however might be unable to differentiate between species of the *E. phagocytophilia* or the *E. canis* genogroup (Engvall *et al.*, 1996).

4. PERSPECTIVES

Nucleic acid amplification tests have contributed significantly to the diagnosis of infectious diseases. As with most microbiological methods the value of these methods highly depends on the expertise of the personnel working with these methods. Automation, control of contamination and control of inhibition will help to increase the acceptability of these techniques in clinical microbiological laboratories and to standardize the direct detection of pathogens in clinical specimens. For the future, it would be helpful if commercialized tests would be able to detect a broader spectrum of pathogens in the same specimens by amplification methods during the same test protocol such as the combination of a detection of both, *Neisseria gonorrhoeae* and *Chlamydia trachomatis*, or *Mycoplasma pneumoniae* and *Chlamydia pneumoniae*.

5. REFERENCES

An, Q., Liu, J., O'Brien, W., Radcliffe, G., Buxton, D., Popoff, S., King, W., Vera, G.M., Lu, L., Shah, J., and others, 1995, Comparison of characteristics of Q beta replicase-amplified assay with competitive PCR assay for *Chlamydia trachomatis*, *J. Clin. Microbiol.* **33**:58–63.

An, Q., Radcliffe, G., Vassallo, R., Buxton, D., O'Brien, W.J., Pelletier, D.A., Weisburg, W.G., Klinger, J.D., and Olive, D.M., 1992, Infection with a plasmid-free variant Chlamydia related to *Chlamydia trachomatis* identified by using multiple assays for nucleic acid detection. *J. Clin. Microbiol.* **30**:2814–2821.

Andersen, A.B., and Hansen, E.B., 1989, Stucture and mapping of antigenic domains of protein b, a 38000 molecular weight protein of *Mycobacterium tuberculosis*. *Infect. Immun.* **57**:2481–2488.

Anonymous, 1997, Rapid diagnostic tests for tuberculosis: what is the appropriate use? American Thoracic Society Workshop, *Am. J. Respir. Crit. Care Med.* **155**:1804–1814.

Ausina, V., Gamboa, F., Gazapo, E., Manterola, J.M., Lonca, J., Matas, L., Manzano, J.R., Rodrigo, C., Cardona, P.J., and Padilla, E., 1997, Evaluation of the semiautomated Abbott LCx *Mycobacterium tuberculosis* assay for direct detection of *Mycobacterium tuberculosis* in respiratory specimens, *J. Clin. Microbiol.* **35**:1996–2002.

Avidor, B., Kletter, Y., Abulafia, S., Golan, Y., Ephros, M., and Giladi, M., 1997, Molecular diagnosis of cat scratch disease: a two-step approach. *J. Clin. Microbiol.* **35**:1924–1930.

Bass, C.A., Jungkind, D.L., Silverman, N.S., and Bondi, J.M., 1993, Clinical evaluation of a new polymerase chain reaction assay for detection of *Chlamydia trachomatis* in endocervical specimens, *J. Clin. Microbiol.* **31**:2648–2653.

Beatty, W.L., 1993, Morphologic and antigenic characterization of interferon mediated persistent *Chlamydia trachomatis* infection *in vitro*, *Proc. Natl. Acad. Sci.* USA **90**:3998–4002.

Boman, J., Allard, A., Persson, K., Lundborg, M., Juto, P., and Wadell, G., 1997, Rapid diagnosis of respiratory *Chlamydia pneumoniae* infection by nested touchdown polymerase chain reaction compared with culture and antigen detection by EIA, *J. Infect. Dis.* **175**:1523–1526.

Burczak, J.D., Ching, S.F., Hu, H.Y., and Lee, H., 1995, Ligase chain reaction for the detection of infectious agents. In: *Molecular methods for virus detection*, (D. Wiebrauk and D.H. Farkas, eds.), New York, Academic Press Inc., pp. 315–327.

Campbell, L.A., O'Brien, E.R., Cappuccio, A.L., Kuo, C.C., Wang, S.P., Stewart, D., Patton, D.L., Cummings, P.K., and Grayston, J.T., 1995, Detection of *Chlamydia pneumoniae* TWAR in human coronary atherectomy tissues, *J. Infect. Dis.* **172**:585–590.

Campbell, L.A., Perez-Melgosa, M., Hamilton, D.J., Kuo, C.C., and Grayston, J.T., 1992, Detection of *Chlamydia pneumoniae* by polymerase chain reaction, *J. Clin. Microbiol.* **30**:434–439.

Chernesky, M.A., Mahony, J.B., Castriciano, S., Mores, M., Stewart, I.O., Landis, S.J., Seidelman, W., Sargeant, E.J., and Leman, C., 1986, Detection of *Chlamydia trachomatis* antigens by enzyme immunoassay and immunofluorescence in genital specimens from symptomatic and asymptomatic men and women, *J. Infect. Dis.* **154**:141–148.

Ching, S., Lee, H., Hook, E.W., Jacobs, M.R., and Zenilman, J., 1995, Ligase chain reaction for detection of *Neisseria gonorrhoeae* in urogenital swabs. *J. Clin. Microbiol.* **33**:3111–3114.

Edelstein, P.H., 1986, Evaluation of the Gen-Probe DNA probe for the detection of legionellae in culture, *J. Clin. Microbiol.* **23**:481–484.

Edelstein, P.H., Bryan, R.N., Enns, R.K., Kohne, D.E., and Kacian, D.L., 1987, Retrospective study of Gen-Probe rapid diagnostic system for detection of legionellae in frozen clinical respiratory tract samples, *J. Clin. Microbiol.* **25**:1022–1026.

Engvall, E.O., Pettersson, B., Persson, M., Artursson, K., and Johansson, K.E., 1996, A 16S rRNA-based PCR assay for detection and identification of granulocytic *Ehrlichia* species in dogs, horses, and cattle. *J. Clin. Microbiol.* **34**:2170–2174.

Essig, A., Zucs, P., Susa, M., Wasenauer, G., Mamat, U., Hetzel, M., Wieshammer, S., Brade, H., and Marre, R., 1995, Diagnosis of ornithosis by cell culture and polymerase chain reaction in a patient with chronic pneumonia, *Clin. Infect. Dis.* **21**:1495–1497.

Gaydos, C.A., Quinn, T.C., and Eiden, J.J., 1992, Identification of *Chlamydia pneumoniae* by DNA amplification of the 16S rRNA gene, *J. Clin. Microbiol.* **30**:796–800.

Gaydos, C.A., Roblin, P.M., Hammerschlag, M.R., Hyman, C.L., Eiden, J.J., Schachter, J., and Quinn, T.C., 1994, Diagnostic utility of PCR-enzyme immunoassay, culture, and serology for detection of *Chlamydia pneumoniae* in symptomatic and asymptomatic patients. *J. Clin. Microbiol.* **32**:903–905.

Goessens, W.H., Mouton, J.W., van-der-Meijden, W.I., Deelen, S., van Rijsoort, T.H., Lemmens-den, T.N., Verbrugh, H.A., and Verkooyen, R.P., 1997, Comparison of three commercially available amplification assays, AMP CT, LCx, and COBAS AMPLICOR, for detection of *Chlamydia trachomatis* in first-void urine. *J. Clin. Microbiol.* **35**:2628–2633.

Huang, T.S., Liu, Y.C., Lin, H.H., Huang, W.K., and Cheng, D.L., 1996, Comparison of the Roche AMPLICOR MYCOBACTERIUM assay and Digene SHARP Signal System with in-house PCR and culture for detection of *Mycobacterium tuberculosis* in respiratory specimens. *J. Clin. Microbiol.* **34**:3092–3096.

Kennedy, N., Gillespie, S.H., Saruni, A.O., Kisyombe, G., McNerney, R., Ngowi, F.I., and Wilson, S., 1994, Polymerase chain reaction for assessing treatment response in patients with pulmonary tuberculosis. *J. Infect. Dis.* **170**:713–716.

Khan, M.A., and Potter, C.W., 1996, The nPCR detection of *Chlamydia pneumoniae* and *Chlamydia trachomatis* in children hospitalized for bronchiolitis, *J. Infect.* **33**:173–175.

Koide, M., and Saito, A., 1995, Diagnosis of *Legionella pneumophila* infection by polymerase chain reaction, *Clin. Infect. Dis.* **21**:199–201.

Kowalski, R.P., Uhrin, M., Karenchak, L.M., Sweet, R.L., and Gordon Y.J., 1995, Evaluation of the polymerase chain reaction test for detecting chlamydial DNA in adult chlamydial conjunctivitis, *Ophthalmology* **102**:1016–9.

Kubota, Y., 1996, A new primer pair for detection of *Chlamydia pneumoniae* by polymerase chain reaction, *Microbiol. Immunol.* **40**:27–32.

Kuo, C.C., Grayston, J.T., Campbell, L.A., Goo, Y.A., Wissler, R.W., and Benditt E.P., 1995, *Chlamydia pneumoniae* (TWAR) in coronary arteries of young adults (15–34 years old). *Proc. Natl. Acad. Sci. USA.* **92**:6911–6914.

Kuo, C.C., Jackson, L.A., Campbell, L.A., and Grayston, J.T., 1995, *Chlamydia pneumoniae* (TWAR). *Clin. Microbiol. Rev.* **8**:451–461.

Kuo, C.C., Shor, A., Campbel, L.A., Fukushi, H., Patton, D.L., and Grayston, J.T., 1993, Demonstration of *Chlamydia pneumoniae* in atherosclerotic lesions of coronary arteries. *J. Infect. Dis.* **167**:841–849.

Kwoh, D.Y., Davis, G.R., Whitfield, K.M., Chappelle, H.L., DiMichele, J., and Gingeras, T.R., 1989, Transcription-based amplification system and detection of amplified human immunodeficiency virus type 1 with a bead-based sandwich hybridization format. *Proc. Natl. Acad. Sci. USA* **86**:1173–1177.

Lindsay, D.S., Abraham, W.H., and Fallon, R.J., 1994, Detection of *mip* gene by PCR for diagnosis of Legionnaires' disease. *J. Clin. Microbiol.* **32**:3068–3069.

Lisby, G, and Dessau, R., 1994, Construction of a DNA amplification assay for detection of *Legionella* species in clinical samples. *Eur J. Clin. Microbiol. Infect. Dis.* **13**:225–231.

Lizardi, P., Guerra, C., Lomeli, H., Tussie-Luna, I., and Kramer, F., 1988, Exponential amplification of recombinant RNA hybridization probes, *Biotechnology* **6**:1197–1202.

Maiwald, M., 1991, The Polymerase chain reaction in the bacteriological diagnostic laboratory—an enlarging spectrum of applications, *Klin. Lab.* **37**:194–200.

Messmer, T.O., Skelton, S.K., Moroney, J.F., Daugharty, H., and Fields, B.S., 1997, Application of a nested, multiplex PCR to psittacosis outbreaks. *J. Clin. Microbiol.* **35**:2043–2046.

Mouton, J.W., Verkooyen, R., van-der-Meijden, W.I., van-Rijsoort, T.H., Goessens, W.H., Kluytmans, J.A., Deelen, S.D., Luijendijk, A., and Verbrugh, H.A., 1997, Detection of *Chlamydia trachomatis* in male and female urine specimens by using the amplified *Chlamydia trachomatis* test. *J. Clin. Microbiol.* **35**:1369–1372.

Mullis, K.B., and Faloona, F.A., 1987, Specific synthesis of DNA *in vitro* via a polymerase-catalyzed reaction. *Methods Enzymol.* **155**:335–350.

Murdoch, D.R., Walford, E.J., Jennings, L.C., Light, G.J., Schousboe, M.I., Chereshsky, A.Y., Chambers, S.T., and Town, G.I., 1996, Use of the polymerase chain reaction to detect Legionella DNA in urine and serum samples from patients with pneumonia. *Clin. Infect. Dis.* **23**:475–480.

Pasternack, R., Vuorinen, P., and Miettinen, A., 1997, Evaluation of the gen-probe *Chlamydia trachomatis* transcription-mediated amplification assay with urine specimens from women. *J. Clin. Microbiol.* **35**:676–678.

Piersimoni, C., Callegaro, A., Nista, D., Bornigia, S., De-Conti, F., Santini, G., and De-Sio, G., 1997, Comparative evaluation of two commercial amplification assays for direct detection of *Mycobacterium tuberculosis* complex in respiratory specimens. *J. Clin. Microbiol.* **35**:193–196.

Podzorski, R.P., and Persing, D.H., 1995, Molecular detection and identification of microorganisms, In: *Manual of Clinical Microbiology* (P.R. Murray, E.J. Baron, M.A. Pfaller, F.C. Tenover, and R.H. Yolken, eds.), ASM Press, Washington, DC., pp. 130–157.

Rajalahti, I., Vuorinen, P., Nieminen, M.M., Miettinen, A., 1998, Detection of *Mycobacterium tuberculosis* complex in sputum specimens by the automated Roche Cobas Amplicor Mycobacterium Tuberculosis Test. *J. Clin. Microbiol.* **36**:975–978.

Ramirez, J.A., 1996, Isolation of *Chlamydia pneumoniae* from the coronary artery of a patient with coronary atherosclerosis. The *Chlamydia pneumoniae*/Atherosclerosis Study Group. *Ann. Intern. Med.* **125**:979–982.

Rasmussen, S.J., Douglas, F.P., and Timms, P., 1992, PCR detection and differentiation of *Chlamydia pneumoniae, Chlamydia psittaci* and *Chlamydia trachomatis. Mol. Cell. Probes* **6**:389–394.

Reischl, U., Lehn, N., Wolf, H., and Naumann, L., 1998, Clinical evaluation of the automated COBAS AMPICOR MTB assay for testing of respiratory and nonrespiratory specimens. *J. Clin. Microbiol.* **36**:2853–2860.

Rekrut, K., Howell, C., and Schuh, J., 1995, Comparative evaluation of acid fast smear, culture, and PCR, using Digene and Roche kits for the detection of *Mycobacterium tuberculosis*. 95th ASM General Meeting, Abstr. C-149, p. 26.

Shah, S., Miller, A., Mastellone, A., Kim, K., Colaninno, P., Hochstein, L., and D'Amato, R., 1998, Rapid diagnosis of tuberculosis in various biopsy and body fluid specimens by the AMPLICOR *Mycobacterium tuberculosis* polymerase chain reaction test. *Chest* **113**:1190–1194.

Smith, J.H., Radcliffe, G., Rigby, S., Mahan, D., Lane, D.J., and Klinger, J.D., 1997a, Performance of an automated Q-beta replicase amplification assay for *Mycobacterium tuberculosis* in a clinical trial. *J. Clin. Microbiol.* **35**:1484–1491.

Smith, J.H., Buxton, D., Cahill, P., Fiandaca, M., Goldston, L., Marselle, L., Rigby, S., Olive, D.M., Hendricks, A., Shimei, T., Klingler, J.D., Lane, D.J., and Mahan, D.E., 1997b, Detection of *Mycobacterium tuberculosis* directly from sputum by using a prototype automated Q-beta replicase assay. *J. Clin. Microbiol.* **35**:1477–1483.

Smith, K.R., Ching, S., Lee, H., Ohhashi, Y., Hu, H.Y., Fisher, H.C., and Hook, E.W., 1995, Evaluation of ligase chain reaction for use with urine for identification of *Neisseria gonorrhoeae* in females attending a sexually transmitted disease clinic. *J. Clin. Microbiol.* **33**:455–457.

Stary, A., Choueiri, B., Horting, M., Halisch, P., and Teodorowicz, L., 1996, Detection of *Chlamydia trachomatis* in urethral and urine samples from symptomatic and asymptomatic male patients by the polymerase chain reaction. *Eur J. Clin. Microbiol. Infect. Dis.* **15**:465–471.

Stone, B.B., Cohen, S.P., Breton, G.L., Nietupski, R.M., Pelletier, D.A., Fiandaca, M.J., Moe, J.G., Smith, J.H., Shah, J.S., and Weisburg, W.G., 1996, Detection of rRNA from four respiratory pathogens using an automated Q beta replicase assay. *Mol. Cell. Probes* **10**:359–370.

Walker, G.T., Fraiser, M.S., Schram, J.L., Little, M.C., Nadeau, J.G., and Malinowski, D.P., 1992, Strand displacement amplification-an isothermal, *in vitro* DNA amplification technique. *Nucleic Acids Res.* **20**:1691–196.

Weiss, S.M, Roblin, P.M., Gaydos, C.A., Cummings, P., Patton, D.L., Schulhoff, N., Shani, J., Frankel, R., Penney, K., Quinn, T.C., Hammerschlag, M.R., and Schachter, J., 1996, Failure to detect *Chlamydia pneumoniae* in coronary atheromas of patients undergoing atherectomy. *J. Infect. Dis.* **173**:957–962.

Wilson, P.A., Phipps, J., Samuel, D., and Saunders, N.A., 1996, Development of a simplified polymerase chain reaction-enzyme immunoassay for the detection of *Chlamydia pneumoniae*. *J. Appl. Bacteriol.* **80**:431–438.

Wu, D.Y., and Wallace, R.B., 1989, The ligation amplification reaction (LAR)—amplification of specific DNA sequences using sequential rounds of template dependent ligation. *Genomics* **4**:560–569.

Index